JET PROPULSION ENGINES

I. Thermodynamics and Physics of Matter. Editor: F. D. Rossini
II. Combustion Processes. Editors: B. Lewis, R. N. Pease, H. S. Taylor
III. Fundamentals of Gas Dynamics. Editor: H. W. Emmons
IV. Theory of Laminar Flows. Editor: F. K. Moore
V. Turbulent Flows and Heat Transfer. Editor: C. C. Lin
VI. General Theory of High Speed Aerodynamics. Editor: W. R. Sears
VII. Aerodynamic Components of Aircraft at High Speeds. Editors: A. F. Donovan, H. R. Lawrence
VIII. High Speed Problems of Aircraft and Experimental Methods. Editors: A. F. Donovan, H. R. Lawrence, F. Goddard, R. R. Gilruth
IX. Physical Measurements in Gas Dynamics and Combustion. Editors: R. W. Ladenburg, B. Lewis, R. N. Pease, H. S. Taylor
X. Aerodynamics of Turbines and Compressors. Editor: W. R. Hawthorne
XI. Design and Performance of Gas Turbine Power Plants. Editors: W. R. Hawthorne, W. T. Olson
XII. Jet Propulsion Engines. Editor: O. E. Lancaster

VOLUME XII
HIGH SPEED AERODYNAMICS
AND JET PROPULSION

JET PROPULSION ENGINES

EDITOR: O. E. LANCASTER

PRINCETON, NEW JERSEY
PRINCETON UNIVERSITY PRESS
1959

London: OXFORD UNIVERSITY PRESS

L. C. CARD 58-5030

PRINTED IN THE UNITED STATES OF AMERICA BY
THE MAPLE PRESS COMPANY, INC., YORK, PENNA.

FOREWORD

On behalf of the Editorial Board, I would like to make an acknowledgement to those branches of our military establishment whose interest and whose financial support were instrumental in the initiation of this publication program. It is noteworthy that this assistance has included all three branches of our Services. The Department of the Air Force through the Air Research and Development Command, the Department of the Army through the Office of the Chief of Ordnance, and the Department of the Navy through the Bureau of Aeronautics, Bureau of Ships, Bureau of Ordnance, and the Office of Naval Research made significant contributions. In particular, the Power Branch of the Office of Naval Research has carried the burden of responsibilities of the contractual administration and processing of all manuscripts from a security standpoint. The administration, operation, and editorial functions of the program have been centered at Princeton University. In addition, the University has contributed financially to the support of the undertaking. It is appropriate that special appreciation be expressed to Princeton University for its important over-all role in this effort.

The Editorial Board is confident that the present series which this support has made possible will have far-reaching beneficial effects on the further development of the aeronautical sciences.

Theodore von Kármán

PREFACE

Rapid advances made during the past decade on problems associated with speed flight have brought into ever sharper focus the need for a comprehensive and competent treatment of the fundamental aspects of the aerodynamic and propulsion problems of high speed flight, together with a survey of those aspects of the underlying basic sciences cognate to such problems. The need for a treatment of this type has been long felt in research institutions, universities, and private industry and its potential reflected importance in the advanced training of nascent aeronautical scientists has also been an important motivation in this undertaking.

The entire program is the cumulative work of over one hundred scientists and engineers, representing many different branches of engineering and fields of science both in this country and abroad.

The work consists of twelve volumes treating in sequence elements of the properties of gases, liquids, and solids; combustion processes and chemical kinetics; fundamentals of gas dynamics; viscous phenomena; turbulence; heat transfer; theoretical methods in high speed aerodynamics; applications to wings, bodies and complete aircraft; nonsteady aerodynamics; principles of physical measurements; experimental methods in high speed aerodynamics and combustion; aerodynamic problems of turbo machines; the combination of aerodynamic and combustion principles in combustor design; and finally, problems of complete power plants. The intent has been to emphasize the fundamental aspects of jet propulsion and high speed aerodynamics, to develop the theoretical tools for attack on these problems, and to seek to highlight the directions in which research may be potentially most fruitful.

Preliminary discussions, which ultimately led to the foundation of the present program, were held in 1947 and 1948 and, in large measure, by virtue of the enthusiasm, inspiration, and encouragement of Dr. Theodore von Kármán and later the invaluable assistance of Dr. Hugh L. Dryden and Dean Hugh Taylor as members of the Editorial Board, these discussions ultimately saw their fruition in the formal establishment of the Aeronautics Publication Program at Princeton University in the fall of 1949.

The contributing authors and, in particular, the volume editors, have sacrificed generously of their spare time under present-day emergency conditions where continuing demands on their energies have been great. The program is also indebted to the work of Dr. Martin Summerfield who guided the planning work as General Editor from 1949–1952. The cooperation and assistance of the personnel of Princeton University Press and of the staff of this office has been noteworthy. In particular, Mr. H. S. Bailey, Jr., the Director of the Press, and Mr. R. S. Snedeker, who has supervised the project at the Press and drawn all the figures, have been of great help. Special mention is also due Mrs. E. W. Wetterau of this office who has handled the bulk of the detailed editorial work for the program.

Coleman duP. Donaldson
General Editor

PREFACE TO VOLUME XII

This volume considers those principles and problems encountered in combining components to form a complete engine. It relies heavily upon the other volumes which deal with basic principles or principles and problems related to components of an engine.

Section A gives a concise history of the development of rockets and air flow jet engines. Section B gives definitions of thrust and various efficiencies and derives relationships for the performance of the different jet propulsion systems. Section C gives the performance analysis of turbojets based on the internal solution of matching the compressor, combustor, turbine, and nozzle. It includes a discussion of off-design performance and describes the problems of control and testing which are unique to a complete unit. Section D treats the turboprop in a somewhat similar manner. It gives the logic for interest in a turboprop and discusses the additional complications. Section E is devoted to the ramjet, its performance, controls, and methods of testing. Section F discusses the wave engines in general, and in particular the pulse jet and the comprex. Section G treats the liquid rocket engine, from the consideration of appropriate fuels (both monopropellent and bipropellent) to the designing and testing of the motor. Section H gives a similar treatment for solid propelled rockets, with special stress on the stability and characteristics of burning. The possibility of a variety of hybrid engines, part rocket, part turbine, or more generally, part jet and part rotating machinery, is introduced in Sections I and J which treat two such cases—the ram-rocket and the jet rotor. Each section derives the possible performance and outlines the possible use of these engines. Section K deals with the problems in making a nuclear jet power plant suitable for aircraft. It gives the theory related to the shielding, heat transfer, and the production and control of a small lightweight reactor. The final section does not quite give a peek into the future, but it gives a systematic procedure for exploring the many possibilities of the types of jet engines.

At this point, I want to express my appreciation for the kind cooperation of the many authors who contributed to make the volume possible, and especially I want to give my heartfelt thanks to Dr. Coleman duP. Donaldson and his staff, whose spark and tireless efforts have brought our works to fruition.

O. E. Lancaster
Volume Editor

CONTENTS

DeMarquis D. Wyatt and Bruce T. Lundin, Lewis Flight Propulsion Laboratory, National Advisory Committee for Aeronautics, Cleveland, Ohio

JET PROPULSION ENGINES

SECTION A

HISTORICAL DEVELOPMENT OF JET PROPULSION

CHAPTER 1. A SHORT HISTORY OF ROCKET PROPULSION UP TO 1945

F. J. MALINA

A,1. Introduction. A thorough history of the development of jet propulsion engines in its major aspects cannot be written today,[1] especially for the period starting about 1935. During this period the industrially advanced nations of the world, including France, Germany, Italy, Japan, the United Kingdom, the United States, and the Soviet Union, initiated broad developments, the results of which have not been completely released to the scientific world. It was during this period that the practical developments of jet propulsion flowered to an unprecedented degree. A broad structure of knowledge was built giving jet propulsion a scientific discipline comparable to that which had been developed earlier in the twentieth century for other types of heat engines, such as steam and internal combustion engines.

A characteristic of this period was the restricted dissemination of information because of world political conditions, with the result that jet propulsion developments in many cases took place in national isolation. Therefore, priority of discovery of principles and the first application of methods from related fields cannot be easily determined, and loses much of its significance. Work on jet propulsion engines within each country advanced along parallel lines.

A second characteristic of the period was the use of teams of research workers on a scale seldom found previously for other technical developments. This further complicates the determination of individual priority of discovery, so that it frequently appears more equitable to credit a community of workers for many of the achievements of this period.

In this short chapter we will attempt to summarize technical advances up to the end of World War II. Because of the situation prevailing during the most vigorous period of the development of jet propulsion, this history

[1] This chapter was written in 1950.

will suffer in being partial both in the sense of incompleteness and of emphasis on developments in those countries where information has been more widely disseminated.

A,2. Classification of Jet Propulsion Engines. Before discussing the history of jet propulsion, it is necessary to have clear definitions of the terms to be used.[2] It is customarily said that jet propulsion is the method of propulsion based on Newton's third law of motion, i.e. the principle of reaction. However, this is not sufficiently specific, since all forms of propulsion in a fluid medium depend on a force of reaction caused by the momentum imparted to a portion of the fluid. For example, the action of the conventional propeller consists essentially in increasing the momentum of the air or water passing through the propeller disk; the thrust of the propeller can be considered as the reaction of the increase of momentum. The propulsion of a rowboat, a swimming person, or a flying bird is also based on the reaction principle, the propulsive force being furnished by the reaction of the increased momentum of the surrounding fluid.

In order to discriminate between jet and other forms of propulsion it can be said that, in the case of jet propulsion, matter is ejected from the propelled body in order to create momentum. This matter may either be wholly carried within the body or taken from the surrounding medium through which the propelled body moves. A jet propulsion device of the first kind is called a "rocket," a device of the second kind is called a fluid flow engine. In the latter the fluid refers to that taken in from the surrounding medium.

From the above definitions it is seen that a ducted propeller belongs to a class of jet propulsion devices, whereas a free propeller does not. The Stipa airplane built by Caproni and flown in 1932 should be classified as a jet-propelled aircraft. The Stipa airplane can be considered as a predecessor of the jet-propelled aircraft of today. However, it is more satisfactory to restrict the term "jet propulsion" to thermal jet propulsion systems. It is evident that mixed cases are possible where part of the energy is created by mechanical means, and part by thermal means.

An interesting example in nature of locomotion by jet propulsion is found in the squid (*Loligo*), a mollusk of the class of the Cephalopoda. The squid has a kind of siphon under its neck and is able to eject water to propel itself, especially when a short burst of speed is desired. The squid is thus equipped with a mechanical jet propulsion device.

The various systems of jet propulsion are classified below according to their method of operation [*1*,*2*,*3*]. As pointed out above, the two main

[2] This article is based upon a lecture by Th. von Kármán in 1943, given as the initial lecture of a course in jet propulsion for Army and Navy personnel at the California Institute of Technology.

classes are (1) rockets, which do not use the surrounding medium for propulsion, and (2) devices which do, whether the medium be atmospheric air or water. There are several types in each class as illustrated in Tables A,2a and A,2b.

Table A,2a. Rockets—jet propulsion systems not utilizing the surrounding medium.

1. Solid propellant rocket motor
   a. Constant volume combustion
   b. Constant pressure combustion
      (i) Restricted burning
      (ii) Unrestricted burning
2. Liquid propellant rocket engine
   a. Gas pressure feed system
   b. Turbopump feed system
3. Gas or vapor propellant rocket engine

Table A,2b. Jet propulsion systems utilizing the surrounding medium.

1. Jet reaction of engine exhaust
2. Campini system
3. Turbojet
4. Turboprop
5. Ducted turboprop
6. By-pass engine
7. Compressorless system
8. Jet-driven rotor
9. Ducted radiation

A more complex system of classification for jet engines has been proposed by Zwicky ([*4*] and Sec. L). He arranges the members of the class of jet engines in a so-called *morphological box* or manifold. The various dimensions of the box represent significant known chemical, mechanical, and aerodynamic or hydrodynamic characteristics of jet engines. Engines with different combinations of these characteristics can then be described in a systematic manner. On the basis of this scheme of classification he has proposed a new system of nomenclature.

A,3. From Antiquity to the Beginning of the Twentieth Century. Historical research has not so far been able to determine the first application of the jet propulsion principle. The earliest known account in this field tells of the construction of a flying dove of wood by Archytas, the founder of theoretical mechanics. Archytas was a Greek who lived in Tarentum in southern Italy around the fourth century B.C. Aulus Gellius in his *Noctes Atticae* gave origin to the idea that the dove flew by means of expanding vapor contained within it—perhaps the dove was jet propelled [*5,6*].

There seems no doubt that Heron of Alexandria invented a device which operated on the reaction principle. The exact date of his work is not definitely known, but it is probable that he lived in the beginning of the first century B.C. [*6*]. Among numerous inventions described by Heron in his treatises is the *aelopile*, which consisted of a hollow sphere that was rotated about an axis by steam issuing from two jets, which were arranged in opposite directions in a plane perpendicular to the axis of rotation. Centuries passed before the reaction principle was applied again, after which a continuous line of development can be traced up to the present time.

The reaction principle was applied in the solid propellant rocket, commonly known as the black powder rocket. The earliest reference to what appears to have been a black powder rocket was recorded in the *Wu Ching Tsung Yao*, an official publication dated 1040 A.D. which describes various kinds of weapons used by the army during the Sung Dynasty [*7,8*]. A fire arrow, the Huo Yao Pien Chien, is mentioned, and the description and name clearly indicate the use of black powder. It is stated that five ounces of powder were applied to the end of the arrow. Fire arrows of this type appear to have been put to use between 950 and 1000 A.D. Another type of fire arrow is mentioned, the so-called San Kung Ch'uang Tzu Nu, which was projected by a crossbow. It is stated that the arrow may be projected by the force of black powder, if the elasticity of the crossbow permits. The construction of black powder rockets spread rapidly to other parts of Asia, and it is believed that they were first introduced to Europe during the Mongol invasions.

The propellant of the black powder rocket was a mixture of potassium nitrate (saltpeter), charcoal, and sulfur of varying proportions. This mixture was used until the development of new types of explosives and propellants for guns at the end of the nineteenth century.

The discovery of black powder is still a lively subject of historical research. Recently, Needham [*8*] and Tsao Thien-Chin appear to have found the earliest mention of black powder in the seventh-century book of the Thang dynasty, *Chen Yuan Miao Tao Yao Lueh* (*Important Information about the True Origin of the Mysterious Tao*). A mixture of saltpeter, sulfur, arsenic, and honey, as the source of carbon, is described, with the warning that the mixture had caused serious accidents and should not be made.

A few formulas for black powder which have been used at various times have been tabulated on a percentage basis by Davis [*9*] as shown in Table A,3.

Little change in the formula arrived at in 1560 for guns and cannon has been made by the modern manufacturers of black powder. This mixture is "too fast" for rockets. A typical mixture used at present in black powder rockets contains 60 per cent saltpeter, 30 per cent charcoal, and 10 per cent sulfur.[3]

It seems that John Bate, early in the seventeenth century, understood the individual functions of the three components of black powder when he wrote: "The Saltpeter is the Soule, the Sulphur the Life, and the Coales the Body of it." Saltpeter supplies the oxygen for the combustion of the charcoal, and the sulfur is called the *life* of the mixture because this element first catches fire and spreads it throughout the mass [*10*].

Table A,3. Black powder formulas.

	Saltpeter	*Charcoal*	*Sulfur*
8th century,[4] Marcus Graecus	66.66	22.22	11.11
	69.22	23.07	7.69
c.1252, Roger Bacon	37.50	31.25	31.25
1350, Arderne (laboratory receipt)	66.6	22.2	11.1
1360, Whitehorne	50.0	33.3	16.6
1560, Brusselle studies	75.0	15.62	9.38
1635, British Government contract	75.0	12.5	12.5
1781, Bishop Watson	75.0	15.0	10.0

In the sixteenth century, descriptions were printed of the method of fabricating powder rockets. Before that time, and even afterward, as far as special details were concerned, the manner of fabrication was a closely guarded trade secret, which was passed on from artisan to artisan. The rocket consisted of a paper tube constricted at one end to about two thirds of the tube diameter. The propellant was loaded into the tube by pounding (later pressing) successive layers around a tapered spindle. The conical cavity, extending from the constriction almost to the opposite end of the tube, provided a sufficiently large burning area for the charge to deliver a high enough thrust for propelling the rocket. This method of fabrication is still used for "sky rockets," with the exception that the exit hole for the gases is no longer provided by crimping the tube, but by inserting a perforated plug of clay.

The black powder rocket motor was brought to its highest state of development by Sir William Congreve [*11*], who was spurred into carrying out his work by the success achieved by Indian troops with rockets against the British during their campaigns toward the end of the eighteenth century. In 1817 Congreve listed seven types of rockets designed for the British army, one of which, for example, weighed 42 pounds and had a

[3] Lorrain Hanzelet (Jean Appier) has the following comments in his *La Pyrotechnie*, published in 1630: "It is necessary to have compositions according to the greatness or the littleness of the rockets, for that which is proper for the little ones is too violent for the large ones—because the fire, being lighted in a large tube, lights a composition of great amplitude and burns a great quantity of material, and no geometric proportionality applies."

[4] Date not certain. Modern scholars consider the thirteenth century to be closer to the truth.

range of 3000 yards. The principal difference between the Congreve rockets and the common fireworks rocket was the replacement of the paper tube by a cast iron tube. Congreve died in 1828 believing that rockets would replace artillery in warfare. After his death the use of rockets declined in spite of improvements made in their accuracy, principally by William Hale who developed at Woolwich around 1846 a rocket spun by tangential jets. At about this time the loading of rockets was greatly speeded up as manual loading by mallet and wedge was replaced by the hydraulic press and power drill [*12,13*].

Although empirical improvements were made in the construction of the powder rocket both in Asia and in Europe, the mode of operation of the device was not correctly understood until the seventeenth century. Vanoccio Biringuccio wrote the following in his *De la Pirotechnia,* which was published in Venice in 1540:

> "Now powder is earth, consisting of the four elementary principles, and when the sulfur conducts the fire into the dryest part of the powder, fire and air increase . . . the other elements also gird themselves for battle with each other and the rage of battle is changed by heat and moisture into a strong wind . . . "

The description of the reaction principle awaited the work of Galileo Galilei, and finally the formulation by Sir Isaac Newton of the laws of motion in his *Philosophiae naturalis principia mathematica,* which was published in 1687.

William's Gravesande, at the time of Newton, published a work entitled *Natural Philosophy* with the subtitle *An Introduction to Sir Isaac Newton's Philosophy.* In this book he illustrated Newton's third law of motion by means of a steam carriage consisting of a boiler topped by a converging nozzle, which erroneously came to be known as "Newton's steam carriage." It thus appears that s'Gravesande can be credited with being the first to describe a jet-propelled device with an understanding of the basic principle involved.

Jet propulsion engines from the time of the first application of the reaction principle, according to our system of classification, were all rocket devices using solid or gaseous propellants. Thermal jet propulsion engines did not make their appearance until the twentieth century. Jet propulsion for boats was suggested by John Fitch of the United States in 1790, and Rumsey in 1793 succeeded in devising a working system which propelled a boat on the Potomac at 4 miles per hour [*14*].

During the nineteenth century, gaseous propellant rockets were frequently described in patent literature; however, no important use of the proposals was made. Steam and compressed air were chosen as gases suitable, first, for ground locomotion and, later, also for propelling aircraft. Solid propellant rockets, beginning with their invention in the

10th century, were developed primarily for firework displays and as military ordnance. The development of military rockets was pushed into the background during the second half of the nineteenth century by the improvements achieved in gunnery. The spinning shell fired from guns was so much superior both in range and accuracy that most armies abandoned the use of rockets. Their use on a large scale as a military weapon did not return until World War II.

A,4. The Rocket from 1900 to 1945. Although the basic principle of operation of the rocket engine was known after Newton published the laws of motion in 1687, a further development in fundamental theory did not take place until this century.

As pointed out in the previous chapter, the solid propellant rocket was built in accordance with the "tricks of the trade." No systematic study was made to improve rocket performance either by applying fundamental advances made in chemistry or by improving mechanical design by applying new knowledge of thermodynamics and gas dynamics. Furthermore, the powder-rocket motor did not contribute to the steady growth of these sciences as, for example, the steam engine did after James Watt patented the principle of the condenser in 1769. After the development of the rifled gun, there was neither the pressing need to design better rockets nor the curiosity to probe its potentialities.

Around the beginning of this century, the great progress achieved in the construction of a heavier-than-air flying machine stimulated a more serious consideration of the possibility of travel outside the earth's atmosphere by means of the rocket. Furthermore, basic research into the structure and behavior of the atmosphere and of extraterrestrial phenomena began to require a vehicle which could carry instruments to heights greater than could be reached by the balloon. By the 1920's it became evident that the power characteristics of the piston engine-propeller combination would limit the speed of the aircraft as well as the load that could be lifted from the ground or water within a reasonable take-off distance. The rocket engine promised to be of use for both of these problems. In addition, the demand for a tremendous increase in fire power for ground armies and the need for recoilless artillery for aircraft during World War II brought the relatively inaccurate rocket back into service as a military weapon.

Before reviewing the progress made in the design of rocket engines, it is worthwhile to mention some of the fundamental inventions and scientific advances of the latter part of the nineteenth century upon which this progress depended.

In 1884 Paul Vieille invented the first smokeless powder, colloided *poudre blanche* made by treating a mixture of soluble and insoluble nitro-cotton with ether-alcohol. In 1888 Alfred Nobel invented *ballistite,* a colloidal mixture of nitroglycerin and soluble nitrocellulose. In 1895 Linde operated a plant producing liquid air in fairly large quantities, and in the same year Dewar successfully liquefied hydrogen.

The end of the nineteenth century saw the development of large scale steel-making processes and of steel alloys of high strength. In 1886 the process of electric welding was developed, and other welding techniques by the end of the century. Aluminum became a commercial metal with its production in the electric furnace from 1886. The plastics industry had its beginning in the latter part of the century [*14*]. By the beginning of the twentieth century the work of Carnot, Joule, Thomson (Lord Kelvin), and others had provided the essential basic theory required for understanding the thermodynamic aspects of heat engines, including the rocket motor.

Experience had been accumulated on the combustion of fuel with air in the internal combustion engine, and preliminary steps had been taken in the design of the gas turbine. Important contributions to the theory of the flow of compressible fluids had been made by Saint Venant, Wantzel, Reynolds, Riemann, Hugoniot, Mach, and others. The method of obtaining a high velocity jet by means of the convergent-divergent nozzle had been discovered by Carl de Laval.

In a British patent granted in 1889 [*15*] de Laval stated:

> "My invention relates to an improvement in turbines which are set in motion by means of a current of steam; and the object of the improvement is to increase, by complete expansion, the velocity of the steam current, thus producing the relatively largest quantity of *vis viva* of the steam.
>
> "I attain this object by the construction of the steam supply pipe in such a manner that the cross sections of the same are slowly increased near the turbine wheel and in the direction of the latter. The rate of increasing the cross sections is due to the proportion and distance between the smallest section and the largest one, in such a manner that in the steam passage between these two sections a permanent current of steam is produced under isoëntropical expansion."

A,5. Development of Rocket Engines. (From Literature Published up to 1940.) K. E. Ziolkowsky is generally considered the first to have proposed space travel on scientific grounds by means of the rocket. A paper, which he completed in 1898, entitled "Rockets in Cosmic Space," was published in the Russian journal, *Science Survey*, in 1903 [*16*]. He realized that if a rocket were to attain velocities necessary for space travel the rocket engine would have to be greatly superior to the black powder rocket, in both specific impulse and the absolute magnitude of thrust. A summary of his work was recently published by Kosmodemjanski [*17*].

Ziolkowsky proposed the possible use of liquid propellants and also suggested the use of the step-rocket principle, which was demonstrated by the launching of the American-developed WAC Corporal sounding rocket [*18*] from the German-developed V-2 [*19*] on February 24, 1949. The WAC Corporal, in attaining a height of about 250 miles, entered air-free space and thus practically demonstrated Ziolkowsky's proposals almost exactly 50 years later. (The step-rocket principle was actually used in firework rockets as early as 1700, not to improve range but to evoke a better display, and in 1855 Colonel Boxer at the Royal Laboratory, Woolwich, had begun the development of a two-step life-line rocket [*20*].)[5]

An American contemporary of Ziolkowsky, R. H. Goddard, was speculating at the same time on the conditions of the upper atmosphere, and in a notebook dating back to 1899 he mentioned the possibility of employing rockets for this work [*21*]. In 1907 he prepared a paper suggesting that heat from radioactive materials could be used to expel substances at high velocity from an orifice, thus furnishing jet propulsion sufficient to navigate in interplanetary space. Publication of the paper was refused by several popular scientific journals.

The idea of using hydrogen and oxygen as a propellant for an interplanetary rocket, and the construction of such a rocket according to the step-rocket principle occurred to Goddard in 1907. Beginning in 1914 he took out several patents on the "rocket apparatus." Then in 1919 the Smithsonian Institution published his paper entitled "A Method of Reaching Extreme Altitudes" [*22*], which reported the pioneering experiments he had made to determine accurately the mechanical energy of a solid propellant rocket. The rocket motor was of the constant volume type, in that a charge of nitrocellulose propellant or smokeless powder was burned very rapidly with the nozzle of the rocket plugged at the time of ignition.

The following paragraphs are taken from Goddard's paper:

> . "It was obviously desirable to perform certain experiments: First, with the object of finding just how inefficient an ordinary rocket is, secondly, to determine to what extent the efficiency could be increased in a rocket of new design. The term "efficiency" here means the ratio of the kinetic energy of the expelled gases to the heat energy of the powder, the kinetic energy being calculated from

[5] J. J. Driscoll has brought to my attention a book entitled "Artis Magnae Artilleriae" by Casimiro Siemienowics published in Amsterdam in 1650. It contains detailed information on the construction of black powder rockets in which the charges could be fired in parallel, and also arrangements for firing in series or a step-wise manner are shown.

the average velocity of ejection, which was obtained indirectly by observations on the *recoil* of the rocket.

"It was found that not only does the powder in an ordinary rocket constitute but a small fraction of the total mass ($\frac{1}{4}$ or $\frac{1}{5}$), but that, furthermore, the efficiency is only 2 per cent, the average velocity of ejection being about 1000 ft/sec. This was true even in the case of the Coston ship rocket, which was found to have a range of a quarter of a mile.

"Experiments were next performed with the object of increasing the average velocity of ejection of the gases. Charges of dense smokeless powder were fired in strong steel chambers, these chambers being provided with smooth tapered nozzles, the object of which was to obtain the work of expansion of the gases, much as is done in the de Laval steam turbine. The efficiencies and velocities obtained in this way were remarkably high, the highest efficiency, or rather "duty," being over 64 per cent, and the highest average velocity of ejection being slightly under 8000 ft/sec., which exceeds any velocity hitherto attained by matter in appreciable amounts.

"These velocities were proved to be real velocities, and not merely effects due to reaction against the air, by firing the same steel chambers *in vacuo*, and observing the recoil. The velocities obtained in this way were not much different from those obtained in air."

During World War I, experiments with this type of motor were made by Goddard and Hickman; however, at the end of the war in 1918 the development was dropped. Their work was resumed about twenty years later, and led to the development of the "bazooka" rocket weapon which was used in large numbers in World War II. The propulsion of a rocket by means of successive impulses was considered by Goddard in his paper of 1919 and further analyzed by Tsien and Malina in 1938 [*23*]. The use of the constant volume motor for long duration propulsion has not been found practical, mainly because of the difficulties in designing a mechanism for reloading the motor.

Interest in liquid propellant rocket motors received a great stimulus with the publication, by Oberth in 1923, of a book primarily concerned with theoretical aspects of interplanetary travel [*24*]. He suggested alcohol and liquefied hydrogen as likely fuels, with liquid oxygen as the oxidizer.

In 1928 Esnault-Pelterie published the lectures, which he had delivered the preceding year to the Astronomical Society of France, on the exploration of the upper atmosphere by means of rockets and on the possibilities of interplanetary flight [*25*]; and in 1930 he published an expanded version of his lectures, including a discussion of the thrust characteristics and the problem of combustion of the liquid propellant rocket motor [*26*]. Beginning in 1928, Rinin began the publication of a series of nine volumes reviewing various aspects of rocket propulsion with special reference to interplanetary flight. His fifth volume entitled *Theory of Reaction Propulsion* was published in 1929 [*27*]. In 1929, Oberth published a second book dealing with interplanetary flight and general aspects of rocket motor design [*28*].

In Austria, France, Germany, the United Kingdom, the United States, and the Soviet Union [*29,30,31,32*] the publications of Oberth, Esnault-Pelterie, and Rinin led to the establishment of societies of enthusiasts for conquering space. Some of the societies carried out experimental programs on rocket engines, but the work was handicapped by adverse financial conditions, and the tendency to tackle problems of design in a highly empirical fashion without calling upon knowledge available in related technical fields.

Between 1930 and 1933, the Verein für Raumschiffahrt (Society for Space Travel) in Germany made experiments with liquid propellant engines using a propellant combination of liquid oxygen and gasoline or water-diluted alcohol. Cooling of the motors was attempted by surrounding the whole motor with a cooling jacket filled with water or fuel. The coolant was not circulated at velocities high enough to absorb the heat transferred to the walls in high specific impulse motors [*30*]. This society was the precursor of the German Army group which, starting in 1932 under the command of Colonel Walter Dornberger, developed the V-2 long range rocket missile.

The American Rocket Society began experimentation on liquid oxygen-gasoline motors in 1932 and continued its activities for nearly a decade; a summary of the work of the society during this period was recently given by Wyld [*33*]. Experimental motor designs were tested, including a motor constructed of thick aluminum walls to absorb the heat transferred during short periods of operation, and a motor using regenerative cooling was tested in 1933. Several leading members of the American Rocket Society formed Reaction Motors, Inc., in 1941, which produced the liquid oxygen-alcohol engine for the Bell X-1 supersonic aircraft.

The possibility of augmenting the thrust of a rocket motor by allowing the jet to mix with the surrounding air in the presence of bodies which deflect the air set in motion by the jet was first tried experimentally by Mélot in 1924 [*34*]. The results of further studies of thrust augmentors were published in 1932 by Jacobs and Shoemaker [*35*], and in 1933 by Schubauer [*36*].

In 1933 Sänger published a review of the state of design theory of rocket motors in his book *Raketen-Flugtechnik* [*37*], and followed this with a paper in 1934 which gave some results of tests on a motor burning gaseous oxygen and diesel fuel oil. He reported obtaining an exhaust velocity above 10,000 ft/sec at a chamber pressure of 1500 lb/in^2. The motor was cooled by circulating the cooling fluid through copper tubes which were wound to form the walls of the combustion chamber and exhaust nozzle [*38*]. In a paper published in 1936 he put forward the following design rules [*39*]:

(a) The volume of necessary combustion space (in cm^3) must bear to the area of the smallest cross section of the nozzle (in cm^2) the relation from 50 to 5000 cm.
(b) Expansion nozzles for the exhaust gases of rocket motors should have an average angle of flare over 25° and under 270°.
(c) The circulation of the coolant along the walls in contact with the flame must take place in canals and must be so positive that it maintains unfailingly at every section of the wall in contact with the flame a velocity of circulation of prescribed amount.

In 1935 Rinin read a paper at the Volta Congress in Rome, briefly reviewing some thermodynamic aspects of the rocket motor and of thrust augmentors [*40*]. In the same year Damblanc published the results of his experiments with black powder motors used in France in signal, life-saving, and "anti-hail" rockets. The largest motor tested was constructed in two steps; each charge gave a peak thrust of about 200 lb, and the duration of each charge, provided with a central conical hole, was about 1.5 seconds. A discussion of the mode of burning of the powder charge was also presented [*41*].

Goddard published his second Smithsonian report in 1936 [*42*], which contains a general presentation of his work in developing a liquid propellant rocket. He states the following in regard to static tests carried out on a liquid oxygen-gasoline engine during the period 1930 to 1932:

> "Although combustion chambers had been constructed at Clark University (from 1920 to 1922) which operated satisfactorily, it appeared desirable to conduct a series of thorough tests in which the operating conditions were varied, the lift (thrust) being recorded as a function of time. Various modifications in the manner of feeding the liquids under pressure to the combustion chamber were tested, as well as variations in the proportions of the liquids, and in the size and shape of the chambers. The chief conclusions reached were that satisfactory operation of the combustion chambers could be obtained with considerable variation of conditions, and that larger combustion chambers afforded better operation than those of smaller size."

Furthermore, he wrote,

> "The combustion chamber finally decided upon for use in flights was $5\frac{3}{4}$ inches in diameter and weighed 5 pounds. The maximum lift

(thrust) was 289 pounds, and the period of combustion usually exceeded 20 seconds.

"The masses of liquids used during the lifting period were the quantities most difficult to determine. Using the largest likely value of the total mass of liquids ejected and the integral of the lift-time curve obtained mechanically, the velocity of the ejected gases was estimated to be over 5000 feet per second."

In 1936 a rocket research group was formed at the Guggenheim Aeronautical Laboratory of the California Institute of Technology (GALCIT)[6] with the purpose of studying the design fundamentals of solid and liquid propellant motors. This group consisted of F. J. Malina, H. S. Tsien, A. M. O. Smith, J. W. Parsons, W. Arnold, and E. S. Forman. The *Journal of the American Rocket Society* carried a report by Malina in 1938 describing tests made with a motor burning gaseous oxygen and ethylene [*43*], and a report by Parsons and Forman in 1939 giving the results of experiments on solid propellant motors burning smokeless powder under conditions of constant volume combustion [*44*]. In 1940 Malina published a paper in which the characteristics of the liquid propellant rocket motor based upon the theory of perfect gases were summarized, and a universal ideal thrust diagram was presented, from which have been derived the charts showing the variation of thrust coefficient with pressure ratio and nozzle area ratio. Tsien's derivation of the λ factor which corrects the calculated value of *velocity thrust* for the divergence of gas flow in an expanding nozzle was also included in the paper [*45*].

Vogelpohl in 1937 published a study of the thrust characteristics of the rocket motor in which he considered the effect on the thrust of the ratio of specific heats of the exhaust gases [*46*].

Experimental studies of rocket motors were initiated in 1939 by Truax at the U.S. Naval Engineering Experiment Station, Annapolis, Maryland, with the compressed air and gasoline combination [*47,48*].

In 1938 Bartocci published the results of extensive experiments made with gaseous oxygen at room temperature [*49*]. He studied in particular the thrust delivered by the de Laval nozzle as a function of pressure ratio for various values of the nozzle area ratio. His tests showed clearly the loss in thrust caused by under- and overexpansion of the gas in the nozzle.

A,6. Development of Solid Propellant Rocket Motors. (From Literature Published after 1940.) The revival of interest in rocket propulsion during the first part of this century was strongly influenced by ideas of the conquest of space. Investigation was concentrated on liquid

[6] The Jet Propulsion Laboratory is an outgrowth of the work of this group.

propellant engines since they promised to give the highest possible performance, and therefore work on solid propellant rocket motors was not given sustained attention. This position was radically changed by developments in aeronautics and ordnance in the 1930's.

From a thermodynamic point of view two kinds of solid propellant motors have been developed. In the first kind, combustion takes place at essentially constant volume, and a propulsive impulse is provided as already indicated in the previous article. In the second kind, combustion takes place at essentially constant pressure. The constant pressure rocket motor can further be classified into restricted- and unrestricted-burning types, the former having been developed to deliver a thrust force for a period from about 2 seconds to more than 30 seconds, and the latter for a period of less than 2 seconds.

A partial history can be pieced together from publications released after World War II on the development of various types of solid propellant rocket motors during the period preceding the outbreak of war in 1939 and the end of hostilities in 1945.

Developments in the United Kingdom and the United States. In 1935 the Research Department at Woolwich Arsenal, England, initiated investigations on motors using cordite smokeless powder as a propellant. By 1936, encouraging advances in technique had been made, and Sir Alwyn D. Crow was entrusted with the direction of an enlarged program of research and development [*50*]. An unrestricted-burning cordite motor of a new type was developed under this program which went into service in 1940 as the propulsive unit for an antiaircraft rocket. The motor consisted of a steel tube sprayed inside with a suspension of finely ground alumina in a solution of sodium silicate which was baked to provide a temperature-insulating layer. The cordite propellant charge, in the form of a hollow cylinder, was supported in the tube so that both the inner and the outer surfaces could burn. Motors of this type using 2-in.- and 3-in.-diameter cordite charges were developed. A larger motor was also constructed with a propellant charge composed of 11 tubular sticks of cordite disposed in the space between the rocket body tube and an internal axial tube, perforated to allow free flow of gas but which prevented a loss of cordite [*20*]. A description of the propellant used and other details of the cordite motors has been published recently by Wheeler, Whittaker, and Pike [*51*].

In the United States the military services did not support research on armament rockets between the two World Wars; however, L. A. Skinner of the Army Ordnance Department began tests in 1932 on his own initiative, with limited funds and facilities at Aberdeen, Maryland, He was primarily interested in the possibility of accelerating a shell by means of a solid or liquid propellant rocket motor [*52*].

The National Defense Research Committee (NDRC) upon the urging of C. N. Hickman, who had worked on armament rockets with Goddard in 1918, set up Section H in Division A of the NDRC in 1940. This section concentrated on the design of unrestricted-burning solid propellant motors for various armament purposes. Much work was done on the preparation of ballistite propellant charges formed by (1) solvent extrusion, (2) solventless extrusion, (3) casting, and (4) pressure molding. This work was done between 1940 and 1943 at Indian Head, Maryland, and then, under the direction of R. E. Gibson, at the Alleghany Ballistics Laboratory which was operated for Section H by George Washington University.

In 1941 Section H and the Army Ordnance Department began their cooperation on the project which developed the bazooka. The motor designed for the projectile used five tubular charges of solvent-extruded ballistite, each with a $\frac{3}{8}$-in. outside, and a $\frac{1}{16}$-in. inside diameter. The motor operated at a peak chamber pressure of about 500 atmospheres, and the burning time was about thirty thousandths of a second.

A great impetus to the NDRC rocket program was given by a memorandum submitted in 1941 to Vannevar Bush by C. C. Lauritsen upon his return from a visit to England, where he studied the work done by Sir Alwyn Crow's group. In the same year a development program was initiated at the California Institute of Technology with Lauritsen as Director of Research. The design of rocket motors was supervised by W. A. Fowler, and work on propellants and interior ballistics was carried out under B. H. Sage, D. S. Clark, and W. N. Lacey.

By the end of World War II this group had designed and produced several types of unrestricted-burning solid propellant motors using dry extruded ballistite. The basic features of the motors were similar to those developed in the United Kingdom. The largest motor was designed to propel the aircraft rocket called the "Tiny Tim" and the propellant charge consisted of four cruciform ballistite grains weighing a total of 146 pounds. The exhaust gases flowed through 24 nozzles spaced symmetrically around a central nozzle with a blowout disk which ruptured if the chamber pressure reached 2250 lb/in.^{2} [*52*].

The striking advance made during this period in the theory underlying the design of solid propellant motors of the unrestricted-burning type is clearly brought out in a book recently published by Wimpress [*53*]. Although the book lacks references to the builders of this body of knowledge, it must be recognized as a product of numerous research workers in the United Kingdom and the United States, who studied various types of solid propellant motors.

The development and large scale production of solventless double-base propellant used in American armament rockets has been described by Marsh in an article published in 1945 [*54*].

Active work on solid and liquid propellant rocket engines for assisting the take-off of aircraft was initiated in the United States by General H. H. Arnold in 1938 when he asked the National Academy of Sciences to arrange for an investigation of these devices. Investigations were begun in 1939 at the California Institute of Technology (GALCIT) under the direction of von Kármán by Malina, Parsons, and Forman. After preliminary studies, it was decided to concentrate on the restricted-burning type of solid propellant unit since thrusts for periods up to 30 seconds were required. In 1940, in a fundamental study, von Kármán and Malina proved theoretically that stable burning in this type of unit was possible, and gave propellant and design criteria for stability [*55*].

In August, 1941, assisted take-off tests were made on the Ercoupe airplane piloted by H. A. Boushey, Jr. of the U.S. Army Air Corps. The units used had an amine-black powder propellant formulated by Parsons. The charge was pressed into the combustion chamber in 22 increments under a pressure of 18 tons, each unit delivering a thrust of about 25 lb for a period of 12 seconds. Further development of this type of unit was discontinued when it was found to have unsatisfactory temperature limits under storage conditions [*56*].

However a unit which met minimum temperature limits was developed in 1942, and arrangements for its manufacture by the Aerojet Engineering Corporation were made in 1943. The propellant charge, formulated by Parsons, consisted of 75 per cent pulverulent potassium perchlorate mixed with 25 per cent asphalt; the oxidizer was added to the heated fuel and poured while hot into the combustion chamber, and then allowed to cool into a tough mass. The rocket unit operated at a chamber pressure of 1800 lb/in.^{2} at a temperature between 3000 and 4000°F and developed an exhaust velocity of 5500 ft/sec. Copious quantities of smoke in the jet were produced by the oxidizer used in the propellant [*55*,*56*].

The theory and design of the restricted-burning unit was summarized in a recent paper by Seifert, Mills, and Summerfield [*55*].

Developments in other countries. In Italy, between 1927 and 1929, research was carried out on the constant pressure type of solid propellant motor under the direction of G. A. Crocco [*57*]. A motor using a tabular charge of colloidal solventless powder was developed to operate at chamber pressures between 50 and 100 atmospheres. The specific impulse obtained ranged from 150 to 170 lb-sec/lb. This motor was a precursor of the type developed to a high degree in the United Kingdom and the United States, as described above.

In Germany, work was carried out on solid propellant motors of various types mainly by the following concerns: Rheinmetall-Borsig, Wilhelm Schmidding, and Dynamit A.G. Before the end of World War II, Japan had put into service a piloted antiship missile propelled by three solid propellant rockets [*58*]. In the Soviet Union, development and production of solid propellant rockets was undertaken on a large scale as evidenced by a variety of rocket weapons used during the war. Published information on developments in Germany, Japan, and the Soviet Union is not sufficient for a summary to be written with any degree of reliability at present.

A,7. Development of Liquid Propellant Rocket Engines. (From Literature Published after 1940.) The literature published up to 1940 shows that the basic problems of designing successful liquid propellant engines were correctly stated, at least in major outline. The properties of a number of propellants were fairly well known, especially those combinations using liquid oxygen as the oxidizer. In connection with the rocket motor the thermodynamics of the combustion process, and of the flow of the gaseous products through the exhaust nozzle to produce thrust, were understood; and a beginning had been made in solving the cooling problem, and the preliminary stages of development of suitable systems of propellant supply had also been reached. It may be noted that the greatest advance to be recorded in this article was the mechanical design of the complete engine, although a number of basic discoveries were made, especially in the domain of propellants, and the theory for guiding design was greatly expanded.

Developments in Germany. From about 1932 onward, work on liquid propellant engines was pursued intensively in Germany under a cloak of secrecy. Engines were developed for various applications around three types of oxidizer: liquid oxygen, hydrogen peroxide, and nitric acid. Work on liquid oxygen engines was concentrated at the army establishment in Peenemunde; on the hydrogen peroxide system at the H.W.K. (Hellmuth, Walter, Kommanditgesellschaft) firm in Kiel; and on the nitric acid system at the BMW (Bayerische Motoren Werke) firm in Munich and the Rheinmetall-Borsig firm in Berlin.

The principal center for the development of rocket engines based on liquid oxygen was established by the Army, under the leadership of Dornberger, at Peenemunde in 1937–1938, although work had been initiated in Berlin on the design of rocket missiles using this oxidizer in 1932. The research and technical development program at Peenemunde was directed by von Braun, Thiel, and Steinhoff. The work of the center culminated in the design and construction of the engine which propelled the A-4 rocket missile, better known as the V-2. Work on the engine started in about 1940 and incorporated the experience gained in testing smaller engines, the first designed in 1933 for the A-1 missile, which weighed 330 lb; the second designed in 1934 for the A-2 and delivering 660-lb thrust for 16 seconds; and the third designed in 1938 for the A-3, which weighed 1650 lb and had an engine delivering 3300-lb thrust for 45 seconds. The V-2 was first successfully launched in October, 1942, when it traversed a distance of about 120 miles.

The V-2 engine had a turbocentrifugal pump feed system which supplied the liquid oxygen and a fuel mixture of 75 per cent ethyl alcohol and 25 per cent water. The pumps were driven by an impulse turbine powered by steam from a hydrogen-peroxide generator developed by the Walter firm. The rocket motor, operating at a chamber pressure of 294 lb/in.^{2}, delivered 55,000-lb thrust at sea level for a period of 65 seconds. The average specific impulse of the motor was 223 lb-sec/lb. The motor was cooled regeneratively by the fuel, and in addition film cooling was resorted to by injecting a portion of the fuel through rows of small holes around the circumference of the chamber and exhaust nozzle. Expansion joints were provided in the outer wall of the motor to allow for the greater thermal expansion of the inner wall. The fuel and oxidizer were fed into the combustion chamber through eighteen separate injectors arranged at the end of the chamber. Ignition of the propellant mixture was accomplished either by a pyrotechnic or a liquid-operated torch that was inserted through the nozzle into the chamber [*19,33*].

In addition to the development of engines for missiles, the Peenemunde establishment constructed a liquid oxygen-alcohol assisted take-off engine delivering 2200-lb thrust for 20 seconds which went into service in 1942 [*59*].

The German development of rocket engines based on hydrogen peroxide has been carefully reviewed by Baxter [*60*]. The hydrogen peroxide system was developed almost exclusively by the firm of Walter at Kiel, which began work sometime after 1934 [*61,62*]. In 1936 the Deutsche Versuchsanstalt für Luftfahrt carried out flight tests on a Walter "cold" engine using a compressed air feed system, and a simple catalytic plate for producing decomposition of the peroxide. The following year, flight tests were made with an improved engine using liquid calcium permanganate as a catalyst and giving 650-lb thrust for 30 seconds.

In 1938 flight tests were made on the Heinkel 112 with an engine equipped with a turbopump, driven by decomposed peroxide, to replace the compressed air propellant feed system used previously. The turbopump system was adopted later for the engine used in the V-2. The thrust of the engine could be controlled by the pilot up to a maximum of 2200 lb.

The specific impulse obtained with 80 per cent hydrogen peroxide alone ("cold" engine) was about 155 lb-sec/lb. To obtain higher values, fuels were burned with the oxygen released by the decomposition of the peroxide. It was discovered that fuel mixtures based on hydrazine hydrate were self-igniting with liquid peroxide. A fuel mixture of 30 per cent hydrazine hydrate, 57 per cent methyl alcohol, and 13 per cent water when burned with the peroxide gave a specific impulse of about 210. To reduce hazards due to ignition delay, small amounts of potassium cuprocyanide were dissolved in the water making up the fuel mixture.

This fuel was used in the engine for the Me 163A interceptor aircraft. The rocket motor used in this engine, known as the Walter 109-509-A2 was regeneratively cooled by the fuel. The thrust delivered by the motor at sea level could be varied from about 500 to 3750 lb.

Work on engines utilizing nitric acid as an oxidizer was carried out by the Bayerische Motoren Werke (BMW). Hemesath, chief chemist at BMW, claims to be the inventor of fuels that are self-igniting (hypergol fuels) with nitric acid, and further states that at the BMW rocket station near Karlsfeld some 6000 fuel-nitric acid combinations were tested by 1945. An engine using a propellant combination with nitric acid as the oxidizer was developed for the Henschel 8-117 aircraft [*63*].

In 1936, Sänger was called from Vienna to set up a research institute at Trauen, Germany, to study the fundamentals of liquid propellant rocket engines leading to the construction of an engine delivering a thrust of 100 tons. In 1939, experimental work was undertaken on a number of problems, including studies of fuels containing powdered light metals, of liquid ozone as a possible oxidizer, and of motor cooling. Research on rocket engines was terminated in 1942 and efforts were concentrated on ramjet research [*64,65*].

Developments in the United States. The principal groups that worked on liquid propellant engines were the following: the Jet Propulsion Laboratory at the California Institute of Technology; the Aerojet Engineering Corporation; Reaction Motors, Inc; and Project TED ESS 3401 of the Navy Bureau of Aeronautics.

At the California Institute of Technology, under von Kármán and Malina, among various activities that were initiated in 1939 for the Army Air Forces (cf. Art. 6) and in 1944 for the Army Ordnance Department, was the development of the nitric acid-aniline engine in which the propellant was fed to the combustion chamber by gas pressure. Assisted take-off flight tests were carried out for the Army Air Forces on the A-20A airplane piloted by Dane in April of 1942 with two engines, each delivering 1000-lb thrust for 25 seconds [*56*]. The rocket motor was uncooled, and the spontaneously igniting propellant combination was fed into the combustion chamber through an impinging-jet injector by gas pressure from compressed nitrogen cylinders. By the end of 1942 a service-type engine based on this prototype was designed under M. Summerfield at the Aerojet Engineering Corporation, which was founded in that year by von Karman, Malina, Haley, Parsons, Forman, and Summerfield [*66,67*]. Haley was president of the corporation during its first formative years. By the end of the war the Aerojet Engineering Corporation had also delivered a service type assisted take-off engine of the acid-aniline type which delivered 1500-lb thrust for 38 seconds [*68*].

From the 1942 prototype, an engine for the WAC Corporal sounding rocket was developed in 1945 under Malina for the Army Ordnance

Department at the Jet Propulsion Laboratory. This engine incorporated an aniline-cooled motor built by the Aerojet Engineering Corporation to deliver 1500-lb thrust for 45 seconds. A compressed air propellant feed system was used [*18,56*].

Work on a monopropellant engine was initiated at the Jet Propulsion Laboratory and the Aerojet Engineering Corporation in 1944, and considerable progress had been achieved at the latter organization by the end of 1945 under the supervision of Zwicky [*69*].

Reaction Motors, Inc., with L. Lawrence, Jr. as first president, carried out developments primarily on engines using the liquid oxygen-alcohol propellant under contracts from the Navy Department. It constructed the 1500 N4C engine which powered the Air Force experimental supersonic airplane X-1. The engine consisted of four regeneratively cooled motors, each delivering a maximum thrust of 1500 lb. The total thrust of the engine could be varied between 1000 and 6000 lb in four steps. The propellant was supplied to the motors by means of centrifugal pumps driven by a gas turbine [*70,33*].

The Navy Department in 1941, under the leadership of Bolster of the Bureau of Aeronautics, initiated project TED EES 3401 at the U.S. Naval Engineering Experiment Station to carry out research on liquid propellant rocket engines for assisting the take-off of large flying boats [*71*]. Truax was assigned as officer in charge of the project [*72*]. During studies of chemical gas generation methods for pressurizing propellants, Stiff found that aniline and nitric acid ignited spontaneously. On the basis of this information Malina at the California Institute of Technology initiated development of the nitric acid-aniline engine, and later the Truax group also adopted this combination of the development of a prototype engine delivering 1500-lb thrust for 35 seconds. This engine, in which a compressed gas feed system was used, was incorporated in droppable assisted take-off units which were flight tested on the Navy flying boat PBY-2 by Gore in the spring of 1943.

The project also developed a prototype engine for the Gorgon, a radio-controlled winged missile. The propellant combination used was mixed nitric-sulfuric acids and monoethylaniline. The engine delivered a thrust of 350 lb for 2 minutes.

Goddard, beginning in 1942, worked at the U.S. Naval Engineering Experimental Station on a prototype liquid oxygen-gasoline engine.

The progress made during this period in the theory and design of liquid propellant engines, arising primarily from work in the United States, has been presented in books by Zucrow [*73*] and by Sutton [*74*].

Development in other countries. In 1929, in Italy, work was initiated on liquid propellant rocket motors by G. A. Crocco. Studies were made with the gasoline-nitrogen dioxide combination in a motor that was regeneratively cooled—the chamber by the oxidizer and the nozzle by the fuel. The motor was operated at a chamber pressure of 140 lb/in.^{2} for durations up to 10 minutes. Further development on the motor was discontinued in 1930 because of lack of funds [*57*].

In 1932, work was resumed on propellants of the monopropellant type for possible use in torpedo and rocket engines. Studies by R. M. Corelli and L. Crocco were made of the decomposition of stabilized liquid explosives such as dinitroglycerin. In 1933 Crocco suggested experiments on nitromethane. In addition to studies of decomposition, tests were also made of its stability with respect to shock and temperature [*57,75*].

There is at present insufficient published information on work undertaken in other countries for a summary to be made with any degree of reliability.

CHAPTER 2. ROCKET DEVELOPMENT SINCE 1945

R. C. TRUAX

A,8. Liquid Propellant Rockets. The end of World War II brought about an abrupt change in the course of rocket development in all major nations of the world. In Germany, where technical progress in rocketry overshadowed that of any other country, the development of rocket-propelled vehicles came to a dead stop. The United States, taking a cue from the German progress in rocket-propelled missiles and piloted aircraft, abandoned its preoccupation with assisted take-off and began many ambitious programs in the missile and aircraft fields. In other western countries, rocket development followed a somewhat similar course, but on a considerably lower scale of effort.

Available information on rocket development trends in the Soviet Union is very meagre. During World War II, the Soviets pioneered in the large scale use of solid propellant rocket artillery. There is some evidence to indicate that they have built on this base and expanded it by information acquired from Germany at the close of the war. It appears that the Russians picked up the development of many German guided missiles in a very direct fashion, utilizing prior developments to a maximum extent, and continuing at a high and more constant level of support.

Technical progress since 1945 in the field of liquid propellant rockets has been characterized by a host of minor improvements, rather than by major advances.

The turbopump system for feeding propellants to the combustion chamber has gained much wider acceptance. Numerous new power plants embodying pump-feed have been developed for applications where the weight-saving justifies the additional complexity. To avoid the necessity of supplying a third propellant to drive the pump turbine, gas generators which utilize the primary propellants for this purpose have been brought to a high level of perfection. The use of hydrogen peroxide for turbine drive is still preferred, however, in some cases. Single-stage pumps are now universally used for all discharge pressures. Simpler and more reliable shaft seals have been developed. Perhaps the most spectacular improvement has been the increase in allowable suction specific speed. This improvement permits, alternatively, lower suction pressures or smaller pump size and weight for a given flow rate and discharge head.

Improvement in combustion chamber design has been considerable. Very important reductions in size and weight over, say, the V-2 propulsion system have been brought about through better mechanical design and the use of higher combustion pressures.

In the direction of increased specific impulse, gains made have resulted primarily from the use of increased combustion pressures and from the use of improved propellants. The combination liquid-oxygen-gasoline, used since the earliest experiments of R. H. Goddard because of its cheapness and potentially high specific impulse has, in a sense, come into its own through cooling developments which permit it to be used at maximum performance. No marked improvement in specific impulse has been effected through better understanding of the process of mixing and combustion, although numerous high performance injection systems have been developed by empirical methods.

The years since the end of World War II have been a progressive increase in the maximum thrust of rocket power plants. From the fifty-five thousand pounds of the V-2 engine, thrust has increased to the order of hundreds of thousands of pounds. These giant engines consume more than a thousand pounds of propellant every second and in some applications generate millions of horsepower.

A great many propellant combinations have been investigated, both experimentally and theoretically. The three oxidizers in use prior to 1945, namely, liquid oxygen, nitric acid, and hydrogen peroxide, are still the only ones used for other than experimental purposes. Hydrocarbons have become very nearly standard for use with liquid oxygen. Aniline and various amines, much used with nitric acid during the war, are being replaced by other higher performance fuels such as hydrazine and its derivatives which, like aniline, ignite upon contact with nitric acid.

A moderate amount of work has been done with high performance combinations such as liquid hydrogen, liquid oxygen, fluorine-ammonia, fluorine-hydrogen, etc. Undesirable physical or logistic qualities have thus far prevented acceptance of these combinations for any practical applications.

The most spectacular advances in the field of liquid propellant rockets that have taken place since 1945 have been not in improvements in propulsion engines themselves, but in the application of rocket propulsion to various types of vehicles. The most important of these vehicles are piloted aircraft and guided missiles. In the United States only research aircraft have been rocket-propelled. A series of these craft have been produced. The first of these was the X-1, designed by the Bell Aircraft Corporation. The X-1 used a Reaction Motors liquid oxygen-alcohol engine developing 6000 lb of thrust. This airplane was the first to exceed the speed of sound in level flight. A second airplane, the Douglas D558-2 "Skyrocket," exceeded 1200 mph and ascended to an altitude of nearly 80,000 ft. An improved version of the X-1, designated X-1A, attained a speed of some 1600 miles per hour and an altitude substantially in excess of 80,000 ft. The X-1A and the D558-2 both use an improved version of the Reaction Motors engine used in the X-1. The X-2, which began its flight tests in 1955, uses a Curtiss-Wright rocket engine of about 15,000-lb thrust.

A fourth rocket airplane of the research type has been reported to be under development in the United States. This airplane is designated the X-15. Altitudes in excess of 100 miles and speeds of well over four thousand miles per hour are expected.

Rocket-propelled, man-carrying airplanes have also appeared in countries other than the United States. One of the most notable of these is the French "Trident." This airplane uses a large rocket engine in the fuselage as a primary source of propulsion and two small, wingtip-mounted turbojet engines for landing.

Numerous auxiliary rocket power plants for short-time performance improvement of manned aircraft have appeared, including both liquid and solid propellant engines for take-off assistance.

At least three research rockets have been developed since 1945 which should be noted. A rather small rocket, known as the Aerobee, was developed as an enlarged version of the WAC Corporal, the first rocket developed solely for atmospheric sounding. The Aerobee has been fired in considerable numbers. An improved version known as the Aerobee-Hi has also been produced and fired successfully to a height of 163 miles. The Aerobee is essentially a two-stage rocket having a low impulse solid propellant first stage, and a high impulse liquid propellant second stage.

A somewhat similar altitude research rocket, known as "Veronique" has been produced by the French. This rocket is notable for its unique launching system. Although the Aerobee and the WAC Corporal used a high acceleration launch coupled with a high launching tower to attain stability, the Veronique uses a system of cables which maintain the rocket in a vertical altitude until aerodynamic stability is attained.

The largest altitude rocket designed primarily for research is the U.S. Navy's "Viking." This is a single-stage rocket of ten to fourteen thousand pounds gross weight. Twelve of these rockets were fired in the period 1945 to 1955. Altitudes up to 158 miles were attained with payloads in the range of 500 to 1000 pounds.

The Viking and Aerobee programs led to the most ambitious scientific project in the rocket field—the creation of a small artificial satellite of the earth, a program which was announced in mid-1955.

The satellite itself, weighing some 20–30 pounds, is to be placed in an orbit by a three-stage rocket. The first stage has evolved from the Viking, the second stage from the Aerobee, and the third stage is expected to be a new solid propellant design. The all-up weight of the vehicle is expected to be in the neighborhood of 22,000 pounds. This rocket, known as the Vanguard, exemplifies modern rocket practice. The first stage uses a turbopump liquid oxygen-hydrocarbon engine, the second stage a pressure-fed nitric acid-hypergolic fuel system, and the third stage an internal-burning solid propellant power plant of high performance.

The period from 1945 to 1955 has seen many military applications of the liquid propellant rocket engine. Development of the Corporal, a late wartime project, has been completed and the missile placed in operational use. This is a short range surface-to-surface missile developed for the U.S. Army. The Nike, a solid-liquid combination rocket for antiaircraft use, is also in service.

Several much more ambitious projects have recently been initiated. Two intermediate range ballistic rockets, Thor and Jupiter, are under development for the Air Force and Navy, respectively. The latter is being developed by the Army primarily for Naval employment.

The most spectacular application of liquid propellant rocket techniques is, in many respects, the intercontinental ballistic missile. Two separate missiles in this range category are at present under active development for the U.S. Air Force. In these applications, the liquid propellant rocket reaches the ultimate in size, performance, and complexity.

The future of the liquid propellant rocket appears directed towards higher performance missiles, satellites and, eventually, interplanetary space ships. From a technical standpoint, only minor improvements in weight, reliability, and performance appear possible until means are found for applying effectively the energy latent in fission or fusion of atomic nuclei.

A,9. Solid Propellant Rockets. The end of World War II saw solid propellant rockets in large scale use. Hundreds of thousands were used by all participants for barrage purposes. Although their accuracy was not equal to that of rifled artillery, they could be fired very rapidly in great numbers from lightweight equipment. In the air, rockets were mounted on smaller airplanes for ground bombardment. Solid propellant rockets were used to assist take-off of heavily loaded aircraft. In Germany, some solid propellant rockets were used in experimental guided missiles.

The solid rockets of 1945 were largely short burning time devices. For unguided projectiles, this short burning time was required to give accurate trajectories. Propellant burning rates were quite high. In order to get long durations, end-burning types of charges were necessary. The necessity for inhibiting burning on the outer surfaces caused considerable difficulty, for any increase in burning area led to greater pressures and possible case rupture. Longer durations also led to increased case and nozzle weights, because all metal parts exposed to the flame had to have sufficient thermal capacity to avoid overheating.

Two basic types of solid propellants were in use by the end of the war, double-base and composite. The former was a composition of nitrocellulose and nitroglycerin. The charges were usually formed by solventless extrusion in large presses. The press size requirements put an upper limit on economical charge size.

Composite propellants were mechanical mixtures of fuel and oxidizer. These propellants were frequently castable, and thus amenable to production in large size grains. The most common propellant of this type was composed of a mixture of potassium perchlorate in asphalt. The consistency was controlled by the addition of petroleum oil. This particular propellant was widely used for JATO (jet-assisted take-off) units. It suffered from severe temperature limitations because of the change in viscosity of the asphalt with temperature. It also burned to give potassium chloride as one of the products of combustion. This material appeared as a dense white smoke which caused many operational difficulties.

Many improvements have been made in solid propellant rockets during the last ten years. These improvements have widened the field of application, reduced the cost, and improved the reliability. One of the most important improvements has been the development of the internal burning grain. This configuration causes the case to be insulated from the propellant gases by the unburned propellant for all but the last fraction of a second. This insulation permits a very considerable reduction in case weight. Nozzle weights have been reduced through careful design and the use of ceramics.

New propellants have been developed which give considerably higher performance. Ammonium perchlorate and ammonium nitrate have largely replaced potassium perchlorate in mixed propellants. This substitution not only has improved the specific impulse, but has eliminated the smoke problem. Propellant formulations are now available to cover a wide range of burning rates. Many propellants burn stably at much lower pressures than have been possible in the past. Variation of burning rate with charge temperature has been greatly reduced.

One of the drawbacks of solid propellant rockets stems from the fact that fuel and oxidizer are premixed in the rocket case. The propellant therefore is basically an explosive. The problem of safety is one of formulating the propellant so as to have a low sensitivity to those sources of ignition or detonation that might reasonably be encountered. All propellants are of course potentially reactive, so that the problem of stability in storage is an important one. Considerable progress was made in these areas in the years between 1945 and 1955. One company demonstrates the safety of a particular propellant by using it for ash trays!

The many improvements in solid propellant rockets have led to a marked widening of their sphere of application. With size and burning time restrictions largely eliminated, the solid rocket is competing with the liquid in the guided missile field. The solid rocket is still heavier, in general, than the turbopump type of liquid propellant rocket. It is also inferior in specific impulse to the liquid oxygen-gasoline rocket. In many cases, however, the greater simplicity of the solid propellant rocket permits the use of multiple stages to overcome the weight and specific impulse disadvantages.

Unguided solid propellant rockets are being provided as fighter armament, replacing machine guns and cannons. The most popular of these, known as "Mighty Mouse" has a diameter of only 2.3 inches. Many of these rockets are fired in salvo, permitting a very high rate of fire in the short periods of contact encountered in sonic and supersonic combat.

The unguided barrage rocket is being replaced by large rockets carrying nuclear warheads. The currently operational variety is known as "Honest John." It carries a 1500-lb warhead a distance of some ten miles.

The solid propellant rocket has a monopoly in the field of air-to-air guided missiles. Three missiles in this category have been disclosed by the United States Government. These are known as Sparrow, Sidewinder, and Falcon. The first two are Navy weapons, the last an Air Force development.

In the air-to-surface category, a missile known as "Bullpup" is under development. This missile also uses a solid propellant rocket motor.

All surface-to-air missiles in production use either a solid propellant booster or sustainer, or both. Terrier, a Navy Missile, is a two-stage rocket; it uses a short duration, high thrust booster and a lower thrust, longer duration sustainer. Both rockets use solid propellants. The Navy's Talos missile uses a solid booster and a ramjet sustainer. The Army Nike uses a solid booster and a liquid sustainer.

Surface-to-surface missiles less frequently have solid propellant rocket power plants, but even here the solid rocket is extremely useful. The turbojet-powered Regulus, Matador, and Snark all use solid rockets to obtain essentially zero-length launching. A comparatively short range ballistic missile, known as Sergeant, is also under development. This missile uses one of the largest solid propellant grains ever manufactured.

For experimental use, many applications of solid propellant rockets have been made. A very high performance rocket called "Deacon" has been used for atmospheric sounding. Since this rocket is small, and hence loses much of its impulse in overcoming aerodynamic drag, many have been launched from balloons, from altitudes of 80,000 ft or more. The Aerobee research rocket uses a solid propellant booster giving 18,000 pounds of thrust for 21 seconds. In the United Kingdom, an all-solid propellant atmospheric research rocket comparable to the Aerobee has been produced. A solid propellant rocket is scheduled to provide the final stage of propulsion for the satellite to be put in orbit during the International Geophysical Year.

Many other experimental devices, such as high speed sleds, torpedoes, aerial targets, etc., have used solid rockets.

In the field of long range ballistic missiles, the use of solid propellants appears to offer certain advantages in terms of launching manpower, readiness time, and general simplicity. If the difficulties of transporting, handling, and storing such large containers of explosive can be adequately solved, and if the inherent performance disadvantage and lack of precise controllability of the solid rocket can be overcome, at least some application in this field seems sure to follow.

CHAPTER 3. AIR FLOW JET ENGINES

A. D. BAXTER

A,10. Introduction. At the beginning of the twentieth century heavier-than-air machines for flight were making their first feeble hops a few feet from the ground. The principles of aerodynamics were sufficiently well understood for this to have occurred much earlier if a suitable propulsion plant had been available. The primary requisite for such a unit was a high power/weight ratio. However, the steam engine which had held undisputed sway in the power plant field for nearly 150 years was not easily adaptable, with its indirect heat supply, to this role. Consequently it was only the advent of a new form of motor which brought the aeroplane into the realm of practical achievement. The new plant was the internal combustion engine which converted fuel energy directly into heat energy of the working fluid.

This engine, in its early days, still far from satisfied the demands of aircraft designers and, even then, air flow jet propulsion in many of its forms was visualized [*76*] as a more advanced stage in power plant development. As with many inventions, the time was not ripe for its development and for some forty years the piston engine was improved and flourished, until it brought aircraft performance to a point where jet propulsion became quite attractive. At this stage the efficient conversion of heat energy directly into kinetic energy produced as great an advance as the original development of the internal combustion engine. The gas turbine was the medium for introducing this revolution which changed a barely considered idea into a practical form of propulsion almost overnight. Much of this very rapid change can be attributed to the comparative ease with which the main engine components can be developed and tested independently of one another, together with their amenability to analytical treatment.

This transformation in power plants came only toward the end of World War II, so that the real repercussions on aircraft design and high speed flight were not apparent until the postwar period. What is clear, however, is that a field of very high speed flight is opening up which invites the application of jet propulsion motors of even more simple design than the gas turbine. Ramjets are in this class, but their development, within the period covered by this survey, is almost negligible compared with what the turbojet has produced already and what the ramjet may be expected to produce in the future.

The history of this tremendous development in propulsion is very much confused in its early years by the interweaving of new ideas, so that attempts to produce a logical sequence of progress in one basic type may become disjointed if it is not realized that intermediate steps in an alternative type may have influenced development. As an example, the turbojet engine was not completely segregated from the turboprop in its development nor the ramjet from the ducted fan with afterburning. In the following articles some account of this will be taken while following the broad divisions of the subject as indicated in Table A,2b.

A,11. Piston Engine Jet Development. There are three elements common to all air flow jet propulsion systems, namely, air compression, combustion, and expansion. Two of the main distinguishing features are the method of compression and, where this is achieved mechanically, the form of the compressor and its drive. In modern developments the compressor is generally of the radial or axial continuous flow type driven almost exclusively by a turbine, from which we derive the classification of turbojet. It has not always been so and, perhaps naturally, the piston engine played a large part in early inventions.

Lorin, the French engineer, is regarded as the pioneer of jet propulsion and he wrote many papers advocating "propulsion by direct reaction" from 1908 onward. In that year he proposed a system using a reciprocating engine and exhaust jet [77]. Each cylinder, after performing its normal operations of suction, compression, and ignition, allowed the combustion gases to exhaust down a funnel to the atmosphere. All the gas energy, except that necessary for providing compression of the new charge, was directed into the jet. No power was taken from the crankshaft and the scheme represents the first suggestion for using part of the combustion gas energy of the reaction system for the purpose of compression. Its chief disadvantage was the hopelessly inadequate mass flow that the engine could deal with.

In 1909 another Frenchman, Marconnet, gave examples of jet reaction [*78*] in which compression was effected by a Rootes blower. Both of these pioneers and others concentrated much of their attention thereafter on the simple ram compression engine (Art. 13), but were forced to give some thought to mechanical compression in order to overcome the problem of static operation. In 1917 a British inventor, Harris, reverted to the mechanically driven compressor. His invention [*79*] used a centrifugal blower with forward-facing intake, driven by a small twin cylinder piston engine. Among its novel features was a proposal to burn powdered coal in the compressed air and to augment the thrust by ejector action of the resultant high temperature jet.

Later, Mélot propounded a double-acting free piston engine somewhat akin to the earlier arrangement of Lorin, but without the crank drive [*76*]. Combustion pressure on one side of the piston forced it to compress mixture on the other side. As the piston moved, it opened ports through which the burned gas was ejected and the major portion of its energy utilized in propulsion. In 1928, in Budapest, Fono made various jet propulsion claims [*80*] similar to those of Mélot and Lorin, but included one to interpose a single-stage axial compressor in what was essentially a Lorin duct.[7] The compressor was to be driven by wind vanes exposed to the slipstream, and for operation without forward motion a special driving motor was provided.

G. G. Smith has described various schemes of this period [*81*, pp. 45–47], including those of the Swiss, Schurter, and the German Junkers Company. The latter was probably the most advanced form of piston-jet combination proposed. It consisted of a multistage axial compressor driven by a multirow radial two-stroke engine. Some of the compressed air was used for scavenging and charging the cylinders, and the remainder was passed round them to provide cooling before entering a duct with the engine exhaust and flowing to a propulsive nozzle. Fuel was to be burned in the duct to increase the propulsive effort. Campini, an Italian jet propulsion inventor, thought along much the same lines, using reciprocating engines to drive centrifugal blowers or ducted fans [*82*]. His scheme was actually built and installed in a Caproni monoplane, but the airplane was not regarded as a success, having a maximum speed of only 230 mph even with afterburning [*83*].

A,12. Turbojet Development. Piston engine compressor drives were never extremely successful, being relatively heavy and complicated, and it remained for the purely rotary drive of the turbine to justify jet propulsion. The first appearance of a turbine was in 1921

[7] See Art. 13.

when Guillaume combined an axial compressor and a multistage turbine [*84*]. The patent drawings show obvious influences of piston engine techniques, but nevertheless include ideas which have been used in more modern designs. The next turbine proposal came from Whittle in a British patent in 1930 [*85*]. This contained all the main features of his later designs, but was allowed to lapse through lack of financial support.

From the early thirties there were small but growing bodies of engineers in various countries, who were beginning to appreciate the potentialities of the gas turbine as an aircraft power plant. In a short while, national interests were involved and most of the work became shrouded in the veil of secrecy which was lifted only at the end of World War II. Much parallel but completely isolated work was proceeding simultaneously in the belligerent countries and has since been well described by Schlaifer [*86*]. In the present article the sequence of events will be considered separately for each country, but first it is worth recording the milestones of the earliest jet propulsion flights as some measure of each nation's progress in the field.

The first flight in the world of a turbojet-powered aircraft was made by the Heinkel He178 on August 27, 1939 [*87*], four days before the outbreak of war between Germany and Poland. On the other hand, Japan was so far behind that she purchased German licenses and copied the Messerschmitt Me262 twin-jet fighter. The result was neither a successful aircraft nor engine, the combination making its first flight a few days before the end of the war in Asia in August, 1945. The Italian Caproni-Campini C.C.2 was reported to have flown in 1940 [*83*], but there was no reliable information concerning it until a widely publicized flight to the Italian Experimental Establishment at Guidonia in November, 1941. The aircraft did not use a gas turbine but was propelled by the Campini system already mentioned (Art. 11).

The first British flight was on May 14, 1941, when the Gloster aircraft E28/39, powered by the Whittle W.1 engine brought Whittle's long and arduous efforts to fruition [*88*]. A little under four and one-half years later, the direct descendants of this initial combination, the Gloster Meteor IV with two Rolls-Royce Derwent V engines, raised the world speed record from 469 mph to 606 mph [*89*]. In America the first jet flight came on October 1, 1942, when the Bell XP-59A, fitted with two General Electric Company type I.A engines, took the air [*90*, p. 108].

United Kingdom. Developments in Britain followed two distinct lines, one originating in the work of Air Commodore Sir Frank Whittle, utilizing the centrifugal air compressor, the other emanating from A. A. Griffith and associated with the axial compressor.

As long ago as 1926, Griffith, working at the Royal Aircraft Establishment, propounded an aerodynamic theory of turbine design based upon flow past airfoils as distinct from flow through passages. The subsequent experiments and reports have been described by Cox in a Wright Brothers lecture [*91*]. They include, in November, 1929, an official paper by Griffith on the prospects of the gas turbine driving an airscrew. He concluded that it would be lighter, less bulky, and more efficient than contemporary piston engines. His scheme was for a turbocompressor unit consisting of a number of mechanically independent wheels, each carrying a set of compressor blades surrounding a set of turbine blades. Such a contraflow contrarotating unit was built and tested in 1939, almost ten years later [*92*], but it is difficult to know why work was allowed to lapse for that period. In fact, no interest was taken in the gas turbine and axial compressor fields from the time of these early Griffith proposals until 1936.

In that year, the Royal Aircraft Establishment obtained authority to build an axial flow compressor and in the following year another paper on the same lines as the earlier one received more favorable treatment. Work began on a turbine-airscrew engine design and arrangements were made to obtain the collaboration of the Metropolitan Vickers Electrical Co., Ltd. The early history of this association has been described by Constant [*93*, p. 411] and the later developments by D. M. Smith [*94*]. At first, effort was concentrated on the turboprop engine and it was not until July, 1940, when the work of Whittle had clearly demonstrated the importance of first achieving an axial compressor turbojet engine, that design of such an engine commenced. The engine, designated F.2 and later developed into the Beryl [*95*], had distinctive features. They included a nine-stage compressor, a single annular combustion chamber with upstream injection, and a two-stage turbine, all on the same axis.

Until this period all other available effort was concentrated on the relatively advanced centrifugal type of engine, but a little more weight was added to the axial type when Armstrong Siddeley was given a contract in November, 1942. The result was a jet engine, the ASX, which ran five months later [*91*]. This was the first move toward a change in balance of effort, which gathered momentum only when the pressure of war demands eased and the advantages of the axial engine became more apparent. The RAE, however, had continued to foster this type and in addition to its association with Metropolitan Vickers, it cooperated with other firms on the axial compressor problems of the period. Relatively little was known of the requisites for high performance, but the series of compressors which were built and tested set future design on a much firmer basis [*93*, p. 452].

While ideas on axial engines were slowly crystallizing in the prewar days, the same struggle to win support was being waged for the centrifugal engine. Whittle has told the story in his Clayton lecture [*88*]. A company, known as Power Jets, Ltd., was formed, which in June, 1936, placed an order with the British Thomson-Houston Company for the manufacture of an engine to Whittle's requirements. The engine was a simple jet propulsion gas turbine having a single-stage, double-sided centrifugal compressor driven by a single-stage turbine. In spite of the simplicity it was a remarkable venture, going well beyond all previous engineering experience in each of its main components. A compression ratio of 4:1 in a single-stage centrifugal blower was specified, compared with the best contemporary superchargers of $2\frac{1}{2}$:1, the single-stage turbine wheel of $16\frac{1}{2}$ inches outside diameter had to give more than 3000 shaft horsepower with high efficiency, and the combustion intensity was many times greater than existing practice. The first tests were begun on April 12, 1937, and the engine continued to operate, with several intervals for reconstruction, until 1941, when a turbine failure ended the career of this forerunner of a long line of turbojet engines.

It had proved its worth long before this and by the summer of 1939 the Air Ministry had recognized that Whittle had the basis of a practicable aircraft engine. It commissioned Power Jets to supply an engine for flight test and the Gloster Aircraft Company to build an experimental aircraft to house it. This first flight engine, known as the W.1, was cleared for flight in April, 1941. By 1940, however, the Air Ministry had anticipated the possibilities of using jet engines in the war and Gloster was authorized to design a jet-propelled, twin-engined interceptor fighter.[8] As originally planned, the engines were to be a more advanced Whittle design—the W.2. Unfortunately the design proved too optimistic and the required component efficiencies were not achieved. Although it was unsuccessful, knowledge was accumulating rapidly and a whole series of engine designs and modifications were being investigated. Out of all this appeared the W.2.B, which must rank as a classic. It had the reverse flow combustion system which was typical of the Whittle designs and it was eventually developed to give nearly three times the thrust of the W.1 without occupying any more space.

At this time other firms were drawn into the engine production. The first to be so engaged was the Rover Company in the spring of 1940. Their original contract was for production, development, and manufacture of the W.2 engine so that Power Jets might be free to concentrate on research and technical development. This latter role absorbed all Power Jets' efforts as the war progressed and in 1944 it was reinforced by amalgamation with the RAE team in a government-controlled company, Power Jets (Research and Development) Ltd.[9] [*96*]. Rover finally passed their work over to Rolls-Royce in 1943, but before this they pioneered a change in design from the reverse flow combustion chamber

[8] The F9/40—prototype of the Meteor.

[9] In 1946 this company's functions were reduced to those of consultants, holders of patents on behalf of the government, and a training school in gas turbine technology. The majority of the personnel was transferred to a new National Gas Turbine Establishment under the direct control of the Ministry of Supply.

to the straight-through type which has since been universally adopted. Their prototype engine, W.2B.26, ran in March 1942, and served as a basis for the Derwent I which was later developed by Rolls-Royce [*97*].

Rolls-Royce's interest had been stimulated by the flight of the E28/39 aircraft and in 1942 their first venture, the WR.1,[10] was started as a subcontract from Power Jets. It served mainly as an introduction to the new field and it was not long before the skill and experience of the firm were fully deployed. It assumed responsibility for Meteor power plant production and development and before the end of 1945 had put up jet thrusts from 1600 lb in the Welland to 5000 lb in the Nene, had increased thrust-to-weight ratios from 1.86 to 2.8, and had improved engine reliability so that 180 hour overhaul periods were approved [*91*].

Another company early in the field was DeHavilland [*98*]. In January, 1941, with the aid of Halford, designer of many successful piston engines, work commenced on the H.1, later known as the "Goblin." It represented a departure from Whittle's practice in having only a single-sided impeller and also a straight-through combustion system.

In 1941, official control of jet propulsion work was placed under a Director of Special Projects, R. Cox, within the Ministry of Aircraft Production. He inaugurated a Gas Turbine Collaboration Committee with representatives from all the main parties concerned in this important development, in order to coordinate new ideas and experience. Its formation was a unique example of direct competitors sharing information with complete frankness for the national benefit. It was in marked contrast to the German attitude and undoubtedly helped to place Great Britain much further ahead technically than any other nation.

Before completing the record of work in Great Britain, mention must be made of the engineers' awareness of potential developments of the basic design. Thrust augmentation by ducted fan was demonstrated in 1943 when Metropolitan Vickers modified the F.2 engine by the addition of a second turbine driving a two-stage, contrarotating fan. This unit, the F.3, increased the F.2 thrust by 66 per cent without any increase in fuel consumption and was the most economical jet engine in the world at that time, with a specific consumption of 0.65 lb/hr/lb of thrust [*99*]. Other thrust-boosting methods were tried on the bench, including water and alcohol injection into the intakes and afterburning.[11] Development of afterburning was in hand in 1943 at the RAE and engines were flight tested in a Meteor [*100*] the following year. An estimated increase in thrust of 25 per cent was obtained in flight and this was limited only by lack of a flightworthy variable area propelling nozzle.

United States. American interest was first recorded in 1922 when

[10] The initials were those of Whittle and Rolls-Royce and symbolized the close technical liaison.

[11] Originally referred to as "reheat" in Great Britain.

the Bureau of Standards was asked to investigate the practicability of jets. A study prepared by Buckingham and later published by the National Advisory Committee for Aeronautics [*101*] concluded that "propulsion by the reaction of a simple jet cannot compete in any respect with airscrew propulsion at such flying speeds as are now in prospect." The speeds envisaged were of the order of 250 mph and computations indicated that at this speed the jet would take about four times as much fuel per thrust horsepower as the airscrew. This report discouraged any further interest for many years and it was not until 1938 that a modest program was initiated at Wright Field to develop the gas turbine. In the interim S. Moss, of the General Electric Company, through his pioneering work on exhaust gas turbosuperchargers [*102*] had been a lone figure in maintaining some interest in turbine problems, particularly of high temperature materials. His persistence was fortunate, for the experience gained in that period proved of great value when the urgent requirement arose.

In 1939 the NACA was charged with basic research into turbojet problems and during the next two years a number of gas turbine research contracts were placed by different government agencies, but as late as June, 1940, a very discouraging report [*103*] was submitted to the Secretary of the Navy by a committee including such authorities as von Kármán and Millikan. This report suggested that the gas turbine could not be considered a serious aspirant in the field of aeronautics until much better materials were available. Thus it was not until the full extent of the British developments was revealed in the summer of 1941 that real pressure was put on the work. Once liaison had been established, full advantage of British experience was taken. By September, 1941, drawings of the Whittle engine had been studied, contracts placed with the General Electric Company for engines and the Bell Aircraft Corporation for airframes [*104*]. The following month a Whittle engine with a Power Jets test team was flown across the Atlantic, thus beginning a two-way collaboration which continued throughout the war. The G.E. version of the Whittle engine (type I) had its first successful run just over six months after work commenced and six months later the Bell aircraft became the second type of Allied jet-propelled aircraft to fly. Following this, General Electric produced a series of type I engines with greater thrusts, ending in 1944 with the I-40 having a thrust of 3750 lb [*105*].

One reason for the early pessimistic outlook in America was the failure to appreciate that new flight techniques would arise with jet propulsion. In consequence, emphasis was placed on obtaining fuel consumptions at least comparable to those of piston engines and the first projects involved turbopropellers, ducted fans, and compound piston-turbine arrangements. Northrop put forward a scheme for a propeller turbine engine, the Turbodyne, in 1940, but its development proved slow and difficult and it was not until the end of 1944 that it actually ran on the test bed. Also in 1940 the NACA began construction of an experimental engine of the Campini ducted fan type (Art. 11). This operated successfully with afterburning in the fan exhaust during 1942, but it was abandoned the following year with another ducted fan designed by Allis Chalmers when the merits of pure turbojets had been appreciated. Two other groups were working at this time, one a private venture of Pratt & Whitney, whose PT-1 design consisting of a free piston reciprocating Diesel compressor with a gas turbine driving the propeller, was definitely regarded as a long term research project, the other a General Electric group at the firm's Schenectady plant, which was quite independent of the group working on the British turbojet designs. It was sponsored by the NACA gas turbine committee and considered only axial compressor engines. Its first unit was a turboprop, but a jet was started in 1943. This was designated TB-180 [*106*] and was comparable to the British F.2 engine, except that it had multiple straight-through combustion chambers. Its thrust was 4000 lb and it first ran in April, 1944.

During this period, the aircraft engine firms had only slight interest in the new forms of propulsion and it was not until they were convinced of their practicability by outside agencies that they became active in the field. This reluctance to depart from conventional power plants was also apparent in other countries. The first jet propulsion scheme came from the Lockheed Aircraft Corporation in 1940 but, like the Northrop turbopropeller, it suffered from ambitious attempts to give it a high efficiency and eventually it did not materialize. The only other turbojet scheme in this period came from the Turbo Engineering Corporation which received support from the Navy for a theoretical investigation in 1941. This resulted in a contract for a small booster turbojet to give 1100-lb thrust at sea level, but again it was very slow in development. A little later, Westinghouse was given a similar contract. It had been studying axial compressor engines for some time under the direction of the Navy Bureau of Aeronautics, but had no knowledge of any other projects. It produced the first American axial jet engine, the 19A, in March, 1943 [*107*]. This gave 1200-lb thrust in the remarkably small outside diameter of 19 inches, and completed a 100-hour endurance test by the following July.

From this time onward, firms engaged on the turbojet program increased rapidly. By the end of 1945 more than a dozen major engine firms were active in the field and at least the same number of aircraft firms. The latter had about twenty-two experimental jet fighter and bomber projects in various stages of design and five that had already reached the flight testing stage.

Germany. No comprehensive survey of German work has been prepared in Germany and consequently most of our information comes from the interrogation of the leading personnel carried out at the end of the war by various technical investigation teams. The results of their work have been reviewed both in America [*90*, pp. 6–54] and in England [*108*], but some gaps are inevitable. As in England, the first impulses to practical aero-gas turbines came from individuals. Among these, von Ohain is worthy of mention. His first patents in 1935 were filed in the name of Max Hahn and assigned to the Heinkel Company [*87*]. He later designed the He-S3 turbojet which made the first jet flight in 1939. Heinkel, however, in spite of this early experience, was slow in exploiting it and was afterward overtaken by other firms, so that von Ohain never received due credit for his efforts.

In 1936 the German Air Ministry (RLM) became interested through one of its young engineers, H. Schelp. He was later placed in charge of gas turbine work but did not, at first, receive great support from the engine industry. By the end of 1938, however, he had succeeded in getting Junkers and the Bayerische Motoren Werke (BMW) to start project work which eventually led to the operational engines Ju-004 [*109*] and BMW-003 [*110*]. These engines were outwardly very similar in many respects to the English and American axial compressor types, but were crude in both design and performance. Normal engine life did not exceed 50 hours, thrust-to-weight ratio was only 1.3:1, and specific consumption was high at 1.4 lb/hr/lb. Restrictions on the use of high temperature materials were, in part, responsible for this poor performance, but it also appeared that some sacrifice was made in the interests of production. These points resulted in some interesting design features, particularly noteworthy being the elaborate cooling air system which absorbed some 7 per cent of the compressor output. In this field and especially in turbine blade cooling the Germans were far ahead of other countries.

A remarkable aspect of all designs was the complete preoccupation with axial compressor types, dictated by a desire for small frontal area. Almost all compressor design was based upon work on aerodynamic blade theory [*111*] at Göttingen and elsewhere by Betz and Encke. This differed from British ideas in having all the pressure rise occur in the moving blades. While this may have resulted in a more simple construction, it had few aerodynamic attractions and compressor efficiencies were low.

The complete exclusion of centrifugal compressor engines was the result of a rigid policy laid down by the RLM. It failed because it was not backed by coordinated research and development as in Britain. Consequently design stagnated and new engines did not show any real improvement, except in size, over their predecessors.

Other countries. Interest in the gas turbine was displayed before the war, particularly in Sweden and Switzerland, and in both these countries the idea of aircraft turbojet engines was eventually conceived.

Brown Boveri were the chief exponents in Switzerland and they produced designs with axial compressors and multistage turbines for aircraft propulsion. Their schemes were patented in Germany in 1939 [*81*, p. 51] and may have had some influence on German designs. In Sweden, Lysholm proposed a number of gas turbine aircraft applications, both by airscrew and jet propulsion, in a comprehensive American patent in 1933 [*112*]. His early ideas were based on multistage centrifugal compressors coupled to double-rotation Ljungström radial flow turbines. A Bofors experimental unit of this type was built and tested in 1934, but some trouble was experienced with compressor surging. Lysholm then had the idea of a rotary screw compressor [*113*] and by 1936 he had incorporated it in an engine using a multistage axial flow turbine. Descriptions of many of his schemes were published in a paper presented by him in London in 1947 [*114*].

A,13. Ramjet Development. The continuous air flow thermal duct jet has been variously described as propulsive duct, Lorin tube, ramjet, athodyd,[12] and even "flying stove pipe." The American term, ramjet, is now almost universally adopted.

Lorin, whose advocacy of thermal jet propulsion has already been noted (Art. 11) gave much thought to the compressorless motor and from him it derived one of its early names. Its possibilities were noted by both him and Marconnet as early as 1909, but it was not until 1913 that he propounded the more or less classical streamlined form of modern days [*115*]. The idea was premature, of course, since flying was in its infancy and attainable speeds were so low that little compression could be obtained. Consequently, little or no practical investigation was carried out. In spite of this, or perhaps on account of it, ideas continued to abound. Another French development, that of Hayot [*116*], introduced an interesting feature which attempted to meet, at least in part, the low ram criticism. The whole motor was to be housed in the leading edge of a wing, the intake being on the under surface where pressure was high and the exhaust on the upper surface where it was low. Additionally, a proportion of the hot gas was to be led back to the intake where it acted as an injector pump.

Also belonging to this early period are the suggestions of an American inventor, Lake, who obtained patents in 1909 [*117*]. In his system, mixing of air with the combustion gas to augment the thrust was proposed for the first time. An unrecognized advantage of this scheme was the increase in propulsive efficiency which resulted from the lower jet velocity. This system anticipated by some ten years that of Mélot, whose "propulseur à trompe" [*118*] became much better known. This augmentation method

[12] A convenient contraction of *A*ero-*th*ermo-*dy*namic *d*uct, first used by the Germans during the war.

was described by various other inventors of the same period, but in their cases it was applied to mechanical compressor motors. Mélot himself later applied it to rockets (Art. 5).

Up to this time, inventors were content with motors suitable for subsonic flight, but in the later part of the 1920–1930 decade the work of Mach, Prandtl, and ballisticians generally inspired consideration of supersonic flight motors, even though aircraft speeds had not risen high enough to make subsonic ramjets a reasonable proposition. In 1928 Fono described a thermal jet motor for high altitude flight for aircraft with high supersonic speeds. One of the characteristics of the motor was that "the cross section of the compression nozzle tapers at the front in the direction of flow." In 1933, René Leduc produced designs [*119*] which were very similar to present day schemes. They included a suggestion for surrounding a projectile body by a tube which "suppresses the impact waves at supersonic speed." He also considered boundary layer control for both inner and outer surfaces of his duct entry by suitable connection of high and low pressure regions. The combination of jet propulsion and boundary layer suction has been proposed frequently in more recent times [*120,121,122*]. In 1938, Leduc exhibited a model ramjet-powered aircraft at the Paris Salon de l'Aviation. This "machine of the future" was estimated to have an output of 14,000 hp and a speed of 625 mph.

Wartime developments were mainly in the hands of the Germans, who had been quick to appreciate the possibilities of the ramjet as a power plant for fighter and guided missiles. The Walter company at Kiel had explored the problems as long ago as 1936, but its work was mainly on a research basis and construction of a full scale model was only started shortly before the end of the war in Europe [*62*]. Sänger, of the Air Research Institute at Brunswick, was the next to take up the work and his earliest studies were reported in 1938 [*123*]. These were not very favorable as he concluded that the ramjet was a very inflexible power plant. On further consideration he modified his opinions of the value of the motor and quickly became a leader in the field. By 1943 he had built a full scale ramjet which was flight-tested on the back of a Do17Z [*124*]. This successful demonstration was followed by the design and test of very large units and by the end of the war he had various schemes for pure ramjet propulsion [*125*].

Sänger's work inspired much interest among the aircraft designers and both Focke-Wulf and Heinkel produced designs powered by ramjets. Pabst worked with the former company, studying particularly the combustion problem of the ramjet [*126*]. From his work he evolved a vaporizing burner which gave a short flame; this led to a short motor with a length-to-diameter ratio of only $2\frac{1}{2}:1$ compared with Sänger's 6:1.

Lippisch, who was a consultant to the Messerschmitt company, became interested in this propulsion in 1943. He considered the use of solid fuel and carried out experiments with various shaped baskets packed with broken coal, charcoal, and briquettes [*127*]. His schemes were incorporated in several advanced aircraft designs including delta wing fighters [*128*]. Another German, Oswatitsch, studied the aerodynamic aspects and produced a theory on supersonic intake design using oblique shocks to give high efficiency. In tests he recovered 70 per cent of the isentropic pressure available in air initially flowing at a Mach number of 3 compared with a normal shock pressure recovery of 33 per cent [*129*].

In Britain there is little published information on wartime ramjet investigations and it can be assumed that very little effort was diverted from the main turbojet development for this purpose. G. G. Smith claims to have proposed a ramjet-propelled missile in 1943 for the bombardment of Germany from England [*81*, p. 288] and a little work on combustion systems has been reported [*93*, p. 470].

In America, interest in its possibilities was roused as soon as jet propulsion was demonstrated as a practical proposition. Theoretical studies were started on behalf of the Navy Bureau of Aeronautics and experimental work on subsonic ramjet combustion commenced at the Massachusetts Institute of Technology. In 1944 the work was extended to supersonic ramjets in the Applied Physics Laboratory, Johns Hopkins University, and since that date the effort has been actively pursued by groups of associated industrial and university contractors. The first successful flights of a ramjet at supersonic speeds were made in June, 1945 giving a demonstration of thrust exceeding drag in a six-inch diameter model [*130*]. The earliest information on aircraft ramjets relates to a subsonic unit constructed by Marquardt Aircraft Company. A pair of these were fitted on the wing tips of a Mustang aircraft and later to a Lockheed F-80 Shooting Star. This machine was the first piloted aircraft to fly with ramjet power only [*81*, p. 291]. Since that time there has been intense activity and considerable information has been released, but this applies to a period later than that covered by the present survey.

A,14. Development of Intermittent Jets. The intermittent jet engine has been variously described as a pulse jet, resojet, aero-resonator and buzz engine. Its principle is closely related to the well-known Kadenacy system for exhaust scavenging of two-stroke engines [*131*], but its origin goes back as far as that of the continuous flow ramjet. Two forms of compressorless intermittent explosion motors may be distinguished—nonresonant, where the frequency of firing is controlled mechanically, and resonant, where the operation is set by the natural frequency of the motor tube.

Only a few examples of the former class have been proposed, but these date from the time of Holzwarth's constant volume gas turbine combustion chamber. Typical of them is the proposal in 1931 of Goddard

[*132*] with inlet flap valves and plug exhaust valve coupled and linked with the spark ignition. A later idea, patented by The Fairey Aviation Company in England in 1944 [*133*], used only an automatic inlet valve and so avoided any difficulty with hot exhaust valves. It closely resembled the resonating type of motor, except for the heavy valve design which controlled the frequency, rather than the tube length.

Resonating types are first found in the patents of Marconnet in 1909 [*78*]. These give examples of both mechanical compression (Art. 11) and ram compression. The descriptions make it clear that he was fully aware of the possibilities of automatic operation. Independently, at the same period, Karavodine built and operated a resonating duct to provide energy for driving a small gas turbine. This combination had an efficiency of only 2.5 per cent [*134*] and may partially account for some of the subsequent neglect of the type, which continued for 20 years until the patent of Paul Schmidt [*135*]. Schmidt reached the typical tubular construction of the resonator jet duct by reasoning that pistonlike transfer of energy to an air column was more efficient than the ejector type of augmentation, proposed by Mélot [*118*]. From 1930 to 1940 he worked on the principle and during the latter part of the period received some German Government backing. This was not very enthusiastic until, early in the 1939–1945 war, he demonstrated a model giving nearly 1000-lb static thrust. Arrangements were then made for production by the Argus aero-engine firm and some intensive development and research was undertaken. The development culminated in the "flying bomb" motors used against England in 1944; the research was carried out at various institutes including the DFS, Ainring, and the LFA Brunswick, where Schultz-Grunow [*136*] examined the laws of wave propagation in resonator ducts.

No other country appears to have taken any serious interest in this engine until the impact of gas turbine jet propulsion turned thoughts to the possibilities of other thermal jet cycles. Late in 1943, Aerojet Engineering Corporation propounded an "aeropulse," and work was started under Zwicky with support from the Navy Bureau of Aeronautics. The motor was similar to Schmidt's tube except that it was to be completely filled with explosive mixture. By the spring of 1944, a motor having a small entry tube which acted as an acoustic valve had been built and was under test at the U.S. Naval Engineering Experimental Station, Annapolis. Later in the year copies of the German model were built and studies of the theoretical principle undertaken, but up to the end of 1945 no further major advances were made and the prospects for future applications were not very bright [*137*].

A,15. Other Forms of Air Flow Jet. The early use of piston engines as an intermediary in thermal jet propulsion has been noted (Art. 11), but it was not until 1932 that the use of exhaust gases as a means of providing jet reaction to supplement the shaft power was suggested [*138*]. It was developed practically in the 1935–1940 era when aircraft speeds had reached a reasonable figure and was applied to many high speed aircraft during the war [*139*]. In the same period it was realized that some jet reaction, or at least reduction in drag, could be obtained by ducting the air flow through radiators of piston engines [*140*]. This, too, was extensively used during the war period.

The principle of utilizing the exhaust has since been applied to gas turbine propeller engines where the exhaust gas will supply quite an appreciable thrust [*141*]. The first such engine to be tested was the T-31-GE, a General Electric Company axial compressor turboprop, in the spring of 1943. By 1945, similar engines had been developed in England by Armstrong Siddeley [*142*], Bristol [*143*], and Rolls-Royce [*144*]. The last of these was a modified Derwent jet engine. It was installed in a Meteor and flew in September, 1945, the first turboprop to power an aircraft in flight. Since that time there has been a considerable broadening of interest in such mixed thrust engines because of their improved efficiency in the medium flight speed range.

A similar case has been made out for the combination of piston engine and turbocompressor unit in a manner which permits of considerable exhaust thrust being developed [*145*]. Such systems are generally complicated, but are justified by their increased thermal efficiency. Many possible combinations of compound engines were described by Bachle [*146*]. Two early examples started in 1944 were the Allison V-1710-127 and the British Napier Nomad [*81*, pp. 280, 284].

Proposals for thermal jets to drive rotors have been made since very early days. In 1904, Nernst suggested a scheme [*147*, p. 1219] for conducting air through hollow rotating arms where centrifugal force would compress it. Fuel was then injected and burned before the hot gases expanded through nozzles, so producing a torque on the central shaft. A similar scheme was developed in 1925 at the RAE by Carter [*148*], in which a hollow airscrew was driven by thermal jets at its tips.

An early helicopter scheme embodying jet-driven rotors was Avery's "flying machine," dating from 1903 and illustrated by Smith [*81*, p. 36]. The next helicopter proposal of interest was a single blade rotor conceived by two Frenchmen, Papin and Rouilly, in 1915 [*149*, pp. 169–172]. A similar scheme was proposed in America by M. Knight about 1936 and tests of a wind tunnel model were made in the following year. The same principle was adopted by the German Doblhoff, who used an engine in the fuselage to drive a compressor. This compressor supplied air to the blade tips where fuel was added and burned to produce the reaction jets. Several models were successfully flown during World War II [*150*]. In England, Pullin proposed the application of a turbojet engine to the same end [*151*], but in his case the hot gases leaving the turbine were conducted to the blade tips without the addition of further heat. One other method of deriving rotor tip thrust envisaged in Germany was by the use of ramjets fitted to the blade extremities. This idea was put forward by Focke-Wulf in a remarkable rotating wing fighter design [*152*]. The wings were to have variable pitch so that the ram speed could be maintained at a high value for all aircraft speeds. Schemes similar to this and all the foregoing have been actively investigated since the close of World War II by several countries.

A,16. Postwar Turbojet Development. By the end of 1945 it was clear that jet propulsion would largely supersede the piston engine-propeller combination for high speed flight and consequently the development tempo, which might have been expected to slacken without the stimulus of war, actually increased. Unfortunately the secrecy of wartime conditions, after a temporary easement, was reimposed and it became very difficult to find what progress was being made. Careful consideration was given to all the theoretical combinations of engine components which might promise advantages [*153*], but no immediate action toward completely new types was apparent. In fact, the period became one of advance in detail more than in major features.

This steady progress is illustrated [*154*] by the increase in maximum engine static thrust from the 5000 lb of the Nene in 1945 (Art. 12) to the "over 15,000 lb" of the De Havilland Gyron in 1953. Similarly frontal area was reduced and the thrust area ratio increased from 500 lb/ft^2 in the Metro-Vickers Beryl to 1465 lb/ft^2 in the Armstrong Siddeley Sapphire 7. Static specific fuel consumptions were also greatly reduced and instead of figures around 1.0 lb/hr/lb the most economical engines gave consumptions around 0.76 lb/hr/lb, which was very close to the limit predicted only a few years earlier [*155*]. Reduction in engine weight was not so marked although several spectacular reductions pointed the way to lighter construction. The Bristol Orpheus [*156*] was a good example, having a thrust-to-weight ratio of 4.5 lb/lb compared with average values in 1945 of less than 3.0 lb/lb.

These examples of British progress were repeated in American designs and were the fruits of the intensive research efforts undertaken [*157*] in this period. The main work was on four lines, aimed at increasing compression ratio, cycle temperature, combustion efficiency and intensity, and the operating temperature of heat-resisting materials.

The first of these encouraged a most important postwar development—the diversion of effort from the centrifugal to the axial compressor turbojet. The necessity for this was quickly recognized in the United States, but it provoked much argument in Great Britain [*158*] and produced many detailed problems [*159*]. That it was a sound move has since been clearly demonstrated by the almost universal adoption of the axial layout in new designs. It also ended the era of simplicity in which engines were designed and manufactured within six months.

Limitations on the maximum pressure ratio of a single rotor axial compressor caused by poor off-design characteristics led to proposals for "two spool" engines in which independent turbines drove separate high and low pressure compressors [*160*]. This form has been largely responsible for the improved specific fuel consumption noted above. Such arrangements were first considered in 1938 by the Royal Air Establishment [*161*] but were abandoned in favor of a more simple engine layout. In 1944 a combination of axial and centrifugal compressors on one shaft was used on the Bristol Theseus turboprop [*143*] and in 1947 Rolls-Royce used separately driven axial and centrifugal compressors in the Clyde [*162*]. Efforts to reduce the size and weight of compressors turned attention to supersonic compressors [*163*] in which one or two stages would do the work of six or eight subsonic stages. The Germans were the first to consider the advantages of such compressors and Weise of the DVL, Berlin, took out patents in 1936 [*164*], but his experimental compressor was a failure. Westinghouse carried out some work in 1941–1942, but the modern work was initiated by Kantrowitz at the NACA Langley Laboratory [*165*]. So far, however, it does not appear to have been applied to an actual engine.

Higher maximum temperature cycles offered advantages in specific thrust output and could be achieved by better materials or cooling. At the end of the war the use of the high nickel-chromium, heat-resisting alloys [*166*] was firmly established in Great Britain for temperatures up to 1073°K. These were further developed so that operating temperatures up to 1200°K became possible [*167*]. The main advances were, however, in cooling techniques. The Germans had again been pioneers in this field, with both air cooling [*108*] and liquid cooling [*168*]. Extensive experimental work was carried out by NACA after 1945 and a series of reports issued [*169*]. Similar work in England at the National Gas Turbine Establishment resulted in experimental operation at turbine gas temperatures up to 1400°K [*170*]. Most of the work involved internal cooling flow, but one interesting proposal was for external turbine blade cooling in conjunction with afterburning [*171*]. In this system, liquid fuel was sprayed onto the blades and was burned further downstream.

The third line of development was in combustion systems. The position in 1945 [*172*] was far from satisfactory, and a great deal of both practical and theoretical study effort was applied to increasing combustion intensity, stability at altitude, ignition, and efficiency [*173*]. This was accompanied by a trend toward annular chambers, first using a number of individual flame tubes, but in later designs containing a single annular primary zone.

Apart from the progress in these main components, the exhaust system underwent considerable development. The advent of supersonic flight increased its importance both with reference to thrust augmentation by afterburning and to efficient propelling nozzle jet expansion. The first afterburning systems have been mentioned [*100*], but their full value could not be realized without the use of variable area propelling nozzles. Originally it was considered that two-position nozzles would meet the needs [*174*] and various geometrical forms with such descriptive titles as letter box, eyelid, caliper, and clamshell were tried. By 1950 it was found desirable to have fully variable nozzles and an iris construction was developed [*175*]. This had a high efficiency, the opening was always circular and concentric, and it could be conveniently cooled. Mechanically operated nozzles have, however, the disadvantages of considerable additional weight and complication and this was claimed to be overcome by the French [*176*], who produced an aerodynamically controlled nozzle in which a radial flow of compressed air injected at the jet exit section caused the effective area to vary.

All these nozzles have been of the simple convergent type, but at supersonic flight speeds convergent-divergent types are necessary for good efficiency. An annular supersonic nozzle with good off-design operation was proposed by Griffith in 1952 [*177*]. This made use of a Prandtl-Meyer type of expansion around the external lip of the annular throat. Its chief disadvantages were the difficulty of cooling the central bullet and the fact that the maximum diameter of the nozzle was necessarily greater than the jet diameter.

Further development efforts on turbojet exhaust systems were directed toward thrust deflection, thrust spoiling, and thrust reversal. The first has been tried in connection with both shortened take-off and landing [*178*], and the others with landing only. An early thrust spoiler was made in 1945 and described by Cox [*91*]. A more sophisticated system using "eyelid" shutters over the normal jet exhaust orifice and deflecting the gas through airfoil vanes on the side of the pipe was demonstrated by Rolls-Royce in 1956 [*179*]. In the meantime the French had applied their aerodynamic nozzle control system to thrust spoiling; by 1951 they had completed wind tunnel tests with a De Havilland Goblin [*180*].

The propelling nozzle was also the subject of considerable study in connection with the problem of jet noise. The first success in dealing with this was by the use of deflector "teeth" on the nozzle, proposed by Westley and others working at Cranfield, England [*181*]. This was developed by Rolls-Royce into the "petal" nozzle [*182*] which gave a marked reduction in noise level without loss in thrust.

A,17. Other Recent Forms and Applications. Postwar turboprop engines developed in parallel with the turbojets, taking advantage of all the improvements in compressor design, combustion, and materials to provide power/frontal area ratios increasing from 295 ehp/ft^2 in 1946 to 600 ehp/ft^2 in 1955, specific consumptions falling from 0.60 lb/ehp-hr to 0.40 lb/ehp-hr, and specific weights dropping from 0.80 lb/ehp to 0.53 lb/ehp [*183*]. Maximum single-shaft engine powers remained in the neighborhood of 4000 ehp, largely because of the limitations on power absorption by a single airscrew. An effective doubling of the power was provided by the use of twinned engines. Two separate engines drove either a single contrarotating propeller, as in the Bristol "coupled Proteus" [*184*], or two separate concentric propellers, as in the Armstrong Siddeley "Double Mamba" [*185*]. A later development for ensuring large power at altitude, described as the "supercharged turboprop," was put forward in 1954 by Bristols [*186*]. This concept required that the potential sea level power output of the engine should be very large, but that the mechanical design and stressing should limit the actual power to some lower figure. This figure, however, could be maintained to quite high altitudes because the theoretical performance could become more and more real with increasing altitude without exceeding the limiting stresses. The great advantage of this was the much lighter engine at height.

The ducted fan engine and its near relation, the bypass turbojet, were the subject of much design thought because of their good specific consumption and other virtues, but the first practical example was the French Turboméca Aspin ducted fan type which produced a thrust of 440 lb with a specific consumption of 0.628 lb/hr/lb [*187*]. This engine was flown in a Fouga aircraft in 1951. About the same time a much larger bypass engine, the Rolls-Royce Conway, was being developed and run at over 10,000-lb thrust [*188*].

While the main turbojet applications were to high speed aircraft, the foregoing were all intended for normal subsonic operation. At the same time new thoughts were forming on possible uses at very low speeds. The jet-driven rotor for helicopters was under active development (Art. 15) and an interesting variant of this was the proposal to mount actual turbojet engines at the rotor tips to provide the thrust [*189*]. A less ambitious but more promising scheme was the use of a single engine to provide both vertical lift and horizontal thrust as proposed by Fairey [*190*] and incorporated in their Rotodyne. This machine had gas turbines which provided shaft power to air compressors for driving rotor tip jets during hovering flight. These were declutched and the power applied to driving propellers in forward flight. This was one form of the so-called "convertiplane" which aroused serious thought about 1949 [*191*]. Another, and more direct, form achieved vertical take-off by starting with the aircraft pointing vertically and relying on the thrust generated by the propeller being greater than the all-up weight. When adequate height and speed had been obtained the aircraft could be turned on to a horizontal course. Preliminary designs of this type were made in the Research

Division of the Bureau of Aerodynamics and experiments on such an arrangement were carried out by Convair in 1954 [*192*] and were made possible by the low specific weight of gas turbines. Similar experiments with direct lift jets were made by Rolls-Royce with their "flying bedstead" about the same time [*193*]. The ultimate objective was to provide an aircraft with a large number of lifting jets which could be diverted to give horizontal thrust in flight, or alternatively could be switched off while other engines gave forward thrust.

A radically different approach to low speed landing and take-off was proposed by Davidson [*194*] with the "jet flap." In this the engine jet was blown out along the trailing edge of the wing through a narrow slot which deflected it downwards. This stabilized the air flow over the wing at very low forward speeds, and very high lift was achieved in combination with useful propulsive thrust. The use of such a jet for boundary layer control was not new and had been demonstrated as early as 1931 by Bamber [*195*], but neither he nor the others prior to Davidson had appreciated the vital principle of the combined lift and thrust effect. It was not until Attinello advanced the idea of supersonic blowing that the concept became practical. His ideas were like a breath of spring after a long winter. The Attinello system was soon tested and installed in both United States and British airplanes.

At the high speed end of the flight spectrum, ramjets developed rapidly, especially for propulsion of guided missiles [*196*], but no completely new ideas were put forward. Leduc (Art. 13) continued his work in France which was interrupted by the war and by 1949 had flown his 010 machine successfully. During the following years his integrated ramjet-aircraft flew on a large number of occasions at high subsonic speeds, with the ramjet as the sole power plant [*197*]. Proposals for hybrid power plants in which ramjets were combined with rockets or turbojets were mentioned from time to time [*197*], but the most interesting was the air-turborocket which has been actually constructed by Aerojet [*198*] and Experiment Incorporated. It consisted essentially of a turbojet in which the compressor was driven by a turbine running on rocket propellants and the compressed air was burned in a tailpipe with auxiliary fuel. The engine was much less dependent on altitude and forward speed than the conventional turbojet and had better consumption characteristics than the rocket.

The advent of nuclear fission as a practical source of energy led to many attractive but ill-considered proposals for its application to aircraft. Serious studies indicated that, at least in early designs, atomic energy would only be used indirectly as a substitute for the combustion system to provide heat energy which could be transferred to a more conventional gas turbine system. Nevertheless, in spite of the many peculiar problems it presented, several American firms undertook practical design studies and it was claimed that an atomic engine would be airborne in 1955 [*199*].

In conclusion, it may be observed that in the ten years following the 1939–1945 war, air flow jet engines supplanted piston engines in almost all high performance military aircraft, they made large inroads on the civil aviation field including the first regular turbojet passenger service in 1951 [*200*], they made supersonic flight a commonplace, and finally they practically doubled the world air speed record [*201*].

A,18. Cited References.

1. Roy, M. *Recherches théoriques sur le rendement et les conditions de réalisation des systèmes motopropulseurs à réaction.* Gauthier-Villars, Paris, 1930.
2. Malina, F. J. *J. Eng. Educ. 37,* 179 (1946).
3. Smith, G. G. *Gas Turbines and Jet Propulsion for Aircraft,* 5th ed. Flight Publ. Co., 1950.
4. Zwicky, F. *Aeronaut. Eng. Rev. 6,* 20 (1947).
5. Laufer, B. *Prehistory of Aviation.* Field Museum of Nat. History, Chicago, 1928.
6. Sarton, G. *Introduction to the History of Science,* Vol. I. Carnegie Inst. Washington, 1927.
7. Wang, L. *Isis 37,* 160 (1947).
8. Needham, J. *Science and Civilization in China,* Vol. V. Cambridge Univ. Press (to be published).
9. Davis, T. L. *The Chemistry of Powder and Explosives.* Wiley, 1943.
10. Bate, J. *The Mysteries of Nature and Art,* 2nd ed. London, 1635.
11. Congreve, W. *The Congreve Rocket System.* London, 1827.
12. Hale, W. *A Treatise on the Comparative Merits of a Rifled Gun and a Rotary Rocket.* London, 1863.
13. Gleason, F. W. F. *Ordnance 32,* 327, 397 (1948).
14. Lilley, S. *Men, Machines and History.* Cobbett Press, London, 1948.
15. de Laval, C. E. F. *Brit. Patent 7143,* 1889.
16. Ziolkowsky, K. E. *Na-ootchnoye Obozreniye (Science Survey),* 1903.
17. Kosmodenjanski, A. A. *The Famous Scientist K. E. Ziolkowsky.* Wajennoje Izdatelstwo, Ministerstwa Wooruzennych Sil SSR, Moscow, 1949.
18. Malina, F. J. *Army Ordnance 31,* 45 (1946).
19. Perring, W. G. A. *J. Roy. Aeronaut. Sci. 50,* 483 (1946).
20. Wheeler, W. H. *Encyclopedia Britannica (Rockets) Yearbook,* 1948.
21. Goddard, R. H. *Rocket Development.* Prentice-Hall, 1948. (Cf. *Introduction* by E. C. Goddard and G. E. Pendray.)
22. Goddard, R. H. *Smithsonian Inst. Misc. Collections 71,* (*2*), 1919.
23. Tsien, H. S., and Malina, F. J. *J. Aeronaut. Sci. 6,* 50 (1938).
24. Oberth, H. *Die Rakete in den Planetenraumen.* Oldenbourg, Munich, 1923.
25. Esnault-Pelterie, R. *L'exploration par fusée de la très haute atmosphère et la possibilité des voyages interplanetaires.* Société Astronomique de France, Paris, 1928.
26. Esnault-Pelterie, R. *L'Astronautique.* Lahure, Paris, 1950.
27. Rinin, N. A. *Theory of Reaction Propulsion.* Izdatelstwo Akademii Nauk USSR, Leningrad, 1932.
28. Oberth, H. *Wege zur Raumschiffahrt.* Oldenbourg, Munich, 1929.
29. Cleator, P. E. *Rockets Through Space.* Simon and Schuster, 1936.
30. Ley, W. *Rockets and Space Travel.* Viking Press, 1947.
31. Pendray, G. E. *The Coming Age of Rocket Power,* 2nd ed. Harper, 1948.

32. Ananoff, A. *L'Astronautique.* Fayard, Paris, 1950.
33. Wyld, J. H. *J. Am. Rocket Soc. 70,* 2 (1949).
34. Mélot, H. F. *L'Excelsior.* Mar. 1924.
35. Jacobs, E. N., and Shoemaker, J. *NACA Tech. Note 431,* 1932.
36. Schubauer, G. B. *NACA Tech. Note 442,* 1933.
37. Sänger, E. *Raketen-Flugtechnik.* Oldenbourg, Munich, 1933.
38. Sänger, E. *NACA Tech. Mem. 1012,* 1942.
39. Sänger, E. *J. Am. Rocket Soc. 35,* 2–12 (1936).
40. Rinin, N. A. *Acuinto Convengo Volta.* Rome, 1935.
41. Damblanc, L. *L'Aerophile 43,* 205, 241 (1935).
42. Goddard, R. H. *Smithsonian Inst. Misc. Collections 95,* (*3*), 1936.
43. Malina, F. J. *J. Am. Rocket Soc. 41,* 3 (1938).
44. Parsons, J. W., and Forman, E. S. *J. Am. Rocket Soc. 43,* 4 (1939).
45. Malina, F. J. *J. Franklin Inst. 230,* 433 (1940).
46. Vogelpohl, G. *Forsch. Gebiete Ingenieurw. 8,* 35 (1937).
47. Truax, R. C. *J. Am. Rocket Soc. 40,* 9 (1938).
48. Truax, R. C. *J. Am. Rocket Soc. 42,* 6 (1939).
49. Bartocci, A. *Aerotecnica 18,* 235 (1938).
50. Crow, Sir A. D. *Proc. Inst. Mech. Engrs. London 158,* 1 (1948).
51. Wheeler, W. H., Whittaker, H., and Pike, H. H. M. *J. Inst. Fuel London 20,* 137 (1947).
52. Burchard, J. E., ed. *Rockets, Guns and Targets.* Little, Brown and Co., 1948.
53. Wimpress, R. N. *Internal Ballistics of Solid Fuel Rockets.* McGraw-Hill, 1950.
54. Marsh, H. N. *Chem. Ind. 57,* 65 (1945).
55. Seifert, H. S., Mills, M. M., and Summerfield, M. *Am. J. Phys. 15,* 1 (1947).
56. *Engineering and Science.* Calif. Inst. Technol., July 1946.
57. Crocco, L. *J. Am. Rocket Soc. 80,* 32 (1950).
58. Osborne, N. T. *J. Am. Rocket Soc. 65,* (*3*), 1946.
59. Ehricke, K. A. *Rocketscience 4,* (*3*), 57 (1950).
60. Baxter, A. D. *Aircraft Eng. 19,* 249 (1947).
61. Hiscocks, R. D., Orr, J. L., and Green, J. J. *Brit. Int. Obj. Sub-Comm. Final Rept. 123, Item 25, 26, 5,* 1946.
62. Barham, C. L., and Simard, R. *Comb. Int. Obj. Sub-Comm. Rept. Item 5, File XXVIII-53,* 1946.
63. Liebhafsky, H. A., Norris, R. H., and Hull, E. H. *Comb. Int. Obj. Sub-Comm. Rept. Item 5, File XXVI-83,* 1946.
64. Sanger, E. *Interavia 3,* 557 (1948).
65. O'Mara, J. A. *Comb. Int. Obj. Sub-Comm. Rept. Item 6, File XXXII-66.*
66. Seifert, H. S., Mills, M. M., and Summerfield, M. *Am. J. Phys. 15,* 121 (1947).
67. Berggren, W. P., Ross, C. C., Young, R. B., and Hawk, C. E. *J. Am. Rocket Soc. 73,* 17 (1948).
68. Summerfield, M. *Soc. Automotive Engrs. MP Southern California,* May 1945.
69. Zwicky, F., and Ross, C. C. *Soc. Automotive Engrs. MP,* Apr. 1949.
70. Nesbitt, M. W. *J. Am. Rocket Soc. 68,* 1 (1946).
71. Bolster, C. M. *J. J. Cabot Fund, Publ. 9,* Norwich Univ., 1950.
72. Truax, R. C. *J. Am. Rocket Soc. 74,* 62 (1948).
73. Zucrow, M. J. *Principles of Jet Propulsion and Gas Turbines.* Wiley, 1948.
74. Sutton, G. P. *Rocket Propulsion Elements.* Wiley, 1949.
75. Corelli, R. M. *Chimica e industria (Milan) 31,* 436 (1949).
76. Gohlke. *Flugsport 31,* 1939. Transl. *Aircraft Eng. 14,* 32 (1942).
77. Lorin, R. *French Patent Specification 390,256,* 1908.
78. Marconnet. *French Patent Specification 412,478,* 1909.
79. Harris, H. S. *Brit. Patent Specification 118,123,* 1917.
80. Fono, A. *Austrian Patent Specification 560,075,* 1928.
81. Smith, G. G. *Gas Turbines and Jet Propulsion,* 5th ed. Illife and Sons, London, 1950.
82. Campini, C. C. *French Patent Specification 741,858,* 1932.
83. *Comb. Int. Obj. Sub-Comm., Item 5, File XII-24,* 1945.
84. Guillaume. *French Patent Specification 534,801,* 1921.

85. Whittle, F. *Brit. Patent Specification 347,206*, 1930.
86. Schlaifer, R. *Development of Aircraft Engines*. Harvard Univ. Press, 1950.
87. Bamford, L. P., and Robinson, S. T. *Comb. Int. Obj. Sub-Comm. Rept., Item 5, File XXIII-14*, 1945.
88. Whittle, F. *Proc. Inst. Mech. Eng. 152*, 419 (1945).
89. *Flight 48*, 520 (1945).
90. Neville, L. E., and Silsbee, N. F. *Jet Propulsion Progress*. McGraw-Hill, 1948.
91. Cox, H. R. *J. Aeronaut. Sci. 13*, 53 (1946).
92. Baxter, A. D., and Smith, C. W. R. *Brit. Aeronaut. Research Council Repts. and Mem. 2607*, 1951.
93. Constant, H. *Proc. Inst. Mech. Eng. 153*, 409 (1945).
94. Smith, D. M., et al. *Proc. Inst. Mech. Eng. 157*, 471 (1947).
95. *Flight 49*, 420A, 420B (1946).
96. *The Aeroplane 66*, 489 (1944).
97. *Flight 48*, 447 (1945).
98. *Engineering 161*, 175 (1946).
99. *Flight 49*, 420 (1946).
100. Baxter, A. D., and Kell, C. *Brit. Aeronaut. Research Council Repts. and Mem. 2422*, 1950.
101. Buckingham, E. Jet propulsion for airplanes. *NACA Rept. 159*, 1923.
102. Moss, S. A. *Trans. Am. Soc. Mech. Engrs. 66*, 351 (1944).
103. *Bur. of Ships, Tech. Bull. 2*, 1941.
104. *Jane's All the World's Aircraft 55(d)*. MacMillan, 1947.
105. Streid, D. *Aviation 45, 1*, 51–59 (1946).
106. *Aviation Week 47*, 36 (1947).
107. *Aircraft Eng. 18*, 254 (1946).
108. Adderley, J. W. *Brit. Int. Obj. Sub-Comm. Overall Rept. 12*, His Majesty's Stationery Office, 1949.
109. *Aircraft Eng. 17*, 342 (1945).
110. *Comb. Int. Obj. Sub-Comm. Rept., Item 5, File XXVI-30*, 1945.
111. Betz, A. *LFF Jahrbuch*, Part II, 1938.
112. Lysholm, A. J. R. *U.S. Patent Specification 2,085,761*, 1933.
113. Lysholm, A. J. R., Smith, R. B., and Wilson, W. A. *J. Soc. Automotive Engrs. 51*, 193 (1943).
114. Lysholm, A. J. R. *Proc. Inst. Mech. Eng. 157*, 498 (1947).
115. Lorin, R. *L'Aerophile 21*, 229 (1913).
116. Hayot. *German Patent Specification 330,014*, 1913.
117. Lake. *Austrian Patent Specification 63,081*, 1909.
118. Mélot, H. F. *French Patent Specification 523,427*, 1920.
119. Leduc, R. *French Patent Specification 770,326*, 1933.
120. Maingnet. *German Patent Specification 607,894/62b*, 1933.
121. Akimoff. *German Patent Specification 626,326*, 1934.
122. Fedden, A. H. R., and Owner, F. M. *Brit. Patent Specification 484,405*, 1936.
123. Sänger, E. Efficiencies and size ratios of Lorin engines. *Zent. Wissensch. Berichtsw. über Luftfahrtforsch., Berlin, Rept. FB 996*, 1938. Transl. *Brit. Int. Obj. Sub-Comm. Gr. 2/HEC 578*, 1946.
124. Sänger, E., and Bredit, I. A Lorin propulsion unit for jet fighters. *Zent. Wissensch. Berichtsw. uber Luftfahrtforsch., Berlin, Rept. UM 3509*, 1943. *NRCC Rept. Tech. Transl. TT-30*, 1947.
125. Sänger, E. The place of Lorin propulsion in jet fighters. *Zent. Wissensch. Berichtsw. über Luftfahrtforsch., Berlin, Rept. UM 3536*, 1944.
126. Pabst, O. E. *Focke-Wulf Repts. 09041/2/3. ADI (Tech.) Microfilm R2008 F328*, 1944.
127. Schwabl, H. Investigations on suitability of solid propellants for ramjets. *Zent. Wissensch. Berichtsw. über Luftfahrtforsch., Berlin, Rept. UM 2852. Transl. Army Air Force F-TS-1021-RE*, 1943.
128. Lippisch, A. P.13 high speed fighter with Lorin duct propulsion. *ADI (Tech.) Microfilm R3563 F529.*
129. Oswatitsch, K., and Bohm, H. Aerodynamic forces and flow with jet pro-

jectiles. *Zent. Wissensch. Berichtsw. über Luftfahrtforsch., Berlin, Rept. 1010,* 1944. *ADI (Tech.) Microfilm R2128 F526.*
130. Goss, W. H., and Cook, E. *SAE Quart. Trans. 2*, 642 (1948).
131. Davies, S. J. *Engineering 143*, 685 (1937).
132. Goddard, R. H. *U.S. Patent Specification 1,980,266*, 1931.
133. Tips, M. *Brit. Patent Specification 583,341*, 1944.
134. Davy, N. *The Gas Turbine.* Constable, 1914.
135. Schmidt, P. *German Patent Specification 523,655*, 1931.
136. Schultz-Grunow, F. Gas dynamic investigations of the pulse jet tube. *NACA Tech. Mem. 1131*, 1947.
137. Edelman, L. B. *SAE Quart. Trans. 1*, 204 (1947).
138. Oestrich, H. Prospects for jet propulsion of airplanes with special reference to exhaust gases. *NACA Misc. Paper 34*, 1932.
139. Turner, L. R., and White, M. D. Flight tests of NACA jet propulsion exhaust stacks on Supermarine Spitfire airplane. *NACA Rept. Mr (WRL-680)*, 1942.
140. Rauscher, M., and Phillips, W. H. *J. Aeronaut. Sci. 8*, 167 (1941).
141. Hooker, S. G. *J. Roy. Aeronaut. Soc. 50*, 298 (1946).
142. *The Aeroplane 70*, 393 (1946).
143. *Engineering 162*, 307 (1946).
144. *The Engineer 184*, 88 (1947).
145. Ricardo, H. R. *J. Roy. Aeronaut. Soc. 50*, 323 (1946).
146. Bachle, C. F. *Trans. SAE 53*, 345 (1945).
147. Stodola, A. *Steam and Gas Turbines*, Vol. II. Transl. by Loewenstein. Peter Smith, 1945.
148. Carter, B. C., and Coales, D. *Brit. Patent Specification 227,151*, 1925.
149. Lamé, M. L. V. *Le Vol Vertical.* Blondel La Rougery, 1934.
150. Kelley, B., and Mulvey, H. J. *Cent. Int. Obj. Sub-Comm. Rept., Item No. 5, File XXXI-5*, 1945.
151. Pullin, C. G. *Brit. Patent Specification 557,011*, 1942.
152. Liptrot, R. N. *Brit. Int. Obj. Sub-Comm. Overall Rept. 8*, His Majesty's Stationery Office, 1948.
153. Zwicky, F. *Aviation 46*, 49 (1947).
154. Baxter, A. D. *Aircraft Eng. 25*, 50 (1953).
155. Constant, H. *Proc. Inst. Mech. Eng. 163*, 163 (1950).
156. *Bristol Quart. 1*, (7), 17 (1955).
157. Silverstein, A. *J. Aeronaut. Sci. 16*, 197 (1949).
158. Moult, E. S., et al. *J. Roy. Aeronaut. Soc. 55*, 129 (1951).
159. Baxter, A. D. *Aircraft Eng. 24*, 186 (1952).
160. Bristol Aircraft Co. *Brit. Patent Specification 700,564*, 1952.
161. Constant, H. *Proc. Inst. Mech. Eng. 153*, 411 (1945).
162. *The Aeroplane 76*, 321 (1949).
163. Davidson, I. M. *Brit. Patent Specification 615,219*, 1945.
164. Weise, A. Unconventional compressors. *Navy Dept. BuShips 338, Part A*, 9 (1946).
165. Erwin, J. R., Wright, C. C., and Kantrowitz, A. Investigation of an experimental supersonic axial flow compressor. *NACA Research Mem. L6J01b*, 1947.
166. *Engineering 166*, 595 (1948).
167. *The Engineer 200*, 305 (1955).
168. Smith, A. G., and Pearson, R. D. *Proc. Inst. Mech. Eng. 163*, 221 (1950).
169. Ellerbrock, H. General discussion on heat transfer. *Proc. Inst. Mech. Eng.*, 1951.
170. Ainley, D. G. *J. Roy. Aeronaut. Soc. 60*, 563 (1956).
171. Power Jets, Ltd. *Brit. Patent Specification 731,343*, 1951.
172. Watson, E. A., and Clarke, J. S. *J. Inst. Fuel 21*, 1 (1947).
173. Watson, E. A., and Clarke, J. S. Combustion in gas turbines. *Inst. Mech. Eng. and Am. Soc. Mech. Engrs.*, 1955.
174. Lundin, B. T. Investigation of several clamshell variable area exhaust nozzles for turbojet engines. *NACA Research Mem. E9B02*, 1949.

175. Kilrain, W. A. *Am. Aviation 17*, 38 (1953).
176. *The Aeroplane 84*, 22 (1955).
177. Griffith, A. A. *Brit. Patent Specification 668,344*, 1952.
178. *The Aeroplane 88*, 194 (1955).
179. *Engineering 182*, 344 (1956).
180. *The Aeroplane 86*, 466 (1954).
181. Westley, R. An investigation of the noise field from a small jet and methods for its reduction. *College of Aeronautics, Cranfield, Rept. 53*, 1952.
182. Greatrex, F. B. *Fifth Intern. Aeronaut. Conference*, Los Angeles, 1955.
183. Baxter, A. D. *Proc. Inst. Mech. Eng. 171*, 1 (1957).
184. *The Aeroplane 75*, 162 (1948).
185. *The Aeroplane 76*, 362 (1949).
186. *Flight 68*, 393 (1955).
187. *The Aeroplane 80*, 170 (1951).
188. *Aviation Week 57*, (*20*), 21 (1952).
189. Fitzwilliams, O. L. *J. Helicopter Assoc. Great Brit. 5*, 4 (1952).
190. Fairey Aviation Co. *Brit. Patent Specification 735,106*, 1953.
191. *Proc. First Convertible Aircraft Congress*, Philadelphia, 1949.
192. *Aviation Week 60*, (*25*), 26 (1954).
193. *The Aeroplane 87*, 503 (1954).
194. Davidson, I. M. *J. Roy. Aeronaut. Soc. 60*, 25 (1956).
195. Bamber, M. J. Wind tunnel tests on airfoil boundary layer control using a backward opening slot. *NACA Rept. 385*, 1931.
196. *Interavia 8*, 401 (1953).
197. Cleaver, A. V. *J. Roy. Aeronaut. Soc. 60*, 705 (1956).
198. *The Aeroplane 84*, 659 (1955).
199. *Aviation Week 64*, (*18*), 48 (1956).
200. *Aviation Week 57*, (*18*), 83 (1952).
201. *Flight 69*, 313 (1956).

SECTION B

BASIC PRINCIPLES OF JET PROPULSION[1]

MAURICE ROY

CHAPTER 1. DEFINITIONS AND SIMPLIFICATIONS

B,1. Classification of Jet Propulsion Engines. In the absence of all contact with an external solid, a prime mover placed in a fluid of finite or even zero density can propel itself by ejecting a fluid or solid mass toward the rear. By convention we say that this mover is propelled "by jet propulsion," although the propulsive thrust really results from the effects of pressure and friction exerted on the wall of the hollow interior of the mover by the solids or fluids moving in the interior toward the exhaust nozzle.

The mechanical power plant that involves a propeller acts in a similar manner but externally, in the sense that it imparts to the outer medium an additional momentum directed toward the rear. Because the ejection of the exhaust gases of a turboprop (that is, a gas turbine driving a propeller) can contribute more or less, to the total thrust of the system, we include this engine, which constitutes a mixed form of jet propulsion, in the present exposition of basic principles.

In all jet propulsion engines, just as for all propulsion of thermal origin, the source of available energy is an exothermic chemical transformation of solids, liquids, or gases, *carried on board* and called propellants. For brevity, we include all these propellants under the single term "fuel."

For propulsion in the atmosphere, where ambient air is available, this air may be inducted by the jet engine in order to participate essentially in the chemical transformation of the fuel: the engine is then properly called an air flow jet engine in contrast to the rocket, which does not use any air and is the only jet engine which can be used for propulsion in vacuum.

A flow of ambient air may also be inducted exclusively for augmenting the thrust by increasing the mass flow of the discharged gases: thus we can conceive of an air flow jet engine with two flows, an example of

[1] This section was written in 1951.

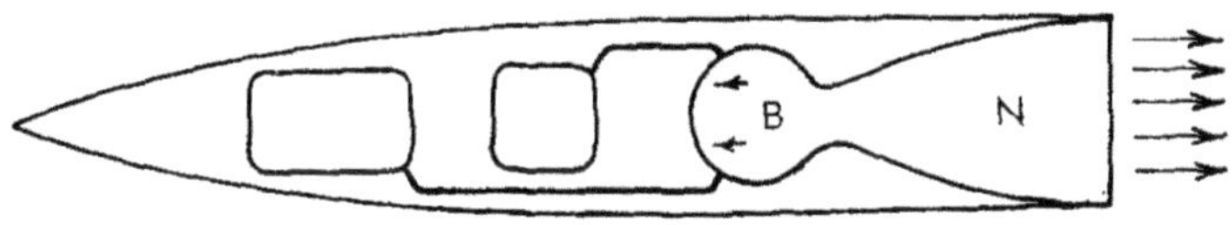

Rocket
(liquid propellants)

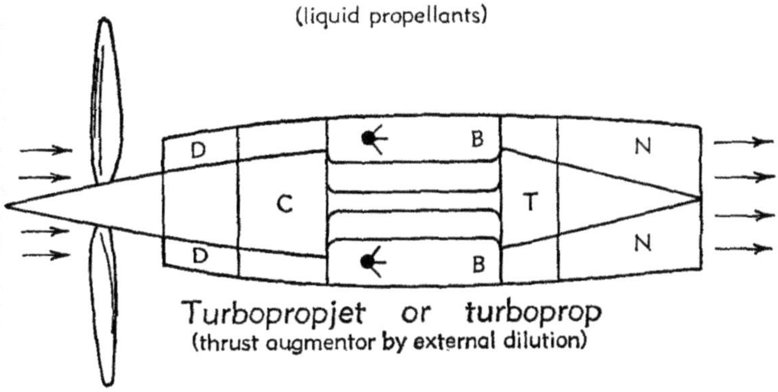

Turbopropjet or turboprop
(thrust augmentor by external dilution)

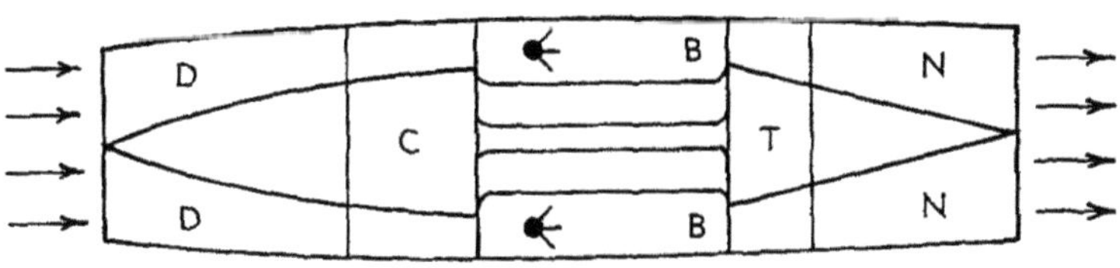

Turbojet

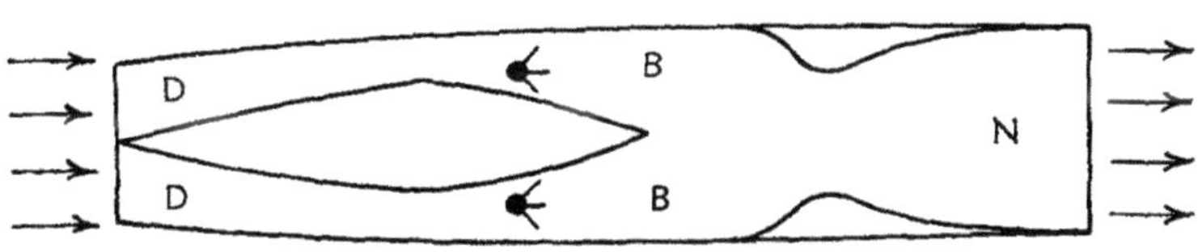

Ramjet

Pulse jet

Fig. B,1.

which is given by a turbojet with a ducted fan. In fact, we can even conceive of a rocket whose thrust is augmented by inducting ambient air—this air may be induced by an ejector effect or by the ram pressure due to flight velocity. The first system is called an ejector rocket and the latter a ramrocket. At low speeds, the latter operates as an ejector rocket.

Whereas the internal flow of a rocket is usually continuous, and even quasi-steady, that of an air flow jet engine can be continuous or discontinuous, i.e. intermittent. The machines which work on the internal flow of a jet engine and have a continuous flow require purely rotating machines, i.e. turbomachines; the engine is called a turbojet.

If any of these machines are reciprocating, the internal flow is discontinuous and the engine is called a motojet. Up to now this kind of jet engine has not been used, although it was proposed as early as 1908 by R. Lorin [*1*].

If compression and expansion of the internal flow is not produced by such rotating or reciprocating machinery but by oscillatory motion of the flow, i.e. wave motion, the jet engine is called a pulse jet.

If, in a continuous flow jet, only the compression resulting from ram effect in the inlet diffuser is utilized, the compression and expansion machinery is eliminated, and the turbojet becomes a "ramjet," the principle of which was proposed by Lorin in 1913 [*2*]. This very simple jet engine, which is also called an "athodyd" by abbreviation of "aero-thermodynamic duct," is necessarily inefficient at low speeds.

The principal types of engines as classified are represented schematically in Fig. B,1.

We can further subdivide this classification by distinguishing among the many types of air inlets, internal flows, exhaust nozzles, and ways of introducing fuel or active substances such as the "coolants" in the various flows, etc. Combinations then appear possible between the preceding fundamental types: for example, the well-known turbojet with "reheat" can be considered as a ramjet supercharged by a turbojet.

B,2. Thermodynamic Evolution and States of the Internal Flow. All air flow jet propulsion engines are composed of several constituent parts, in which the changes in all or part of the active substances (fuel, coolant, air) take place. These constituent parts include *stationary* members such as the inlet diffusers, burners, and tailpipes, and *moving* members such as the distributors, compressors, pumps, and turbines. Practically, it is necessary to study the change of state, i.e. thermodynamic evolution, of the internal flow across each such constituent member by fixing the state of the flow in a conventional entrance or exit section of the specified member, the state at the exit of each member being considered as the state at the inlet of the member following.

In any reference section chosen arbitrarily the area is designated by the symbol A_x; the corresponding state of the flow is symbolized by the subscript x. This state is characterized by physical parameters (pressure p, density ρ, absolute temperature T, etc.), by chemical parameters (hereafter, a single parameter denoted by κ, which is zero before the reaction and equal to 1 when the transformation is complete), and finally by a parameter of motion, composed of the specific relative kinetic energy $E_k = \frac{1}{2}V^2$ (V, velocity referred to the casing or stator of the engine), each one of these quantities being marked by the subscript x.

The state of flow is thus defined arbitrarily as p_x, T_x, κ_x, E_{k_x}. Actually, this state is variable from one point to another across the section A_x, as well as during the duration of the period τ of the "fixed set of running conditions," granting the existence of such a fixed set of conditions.

The quantities p_x, T_x, κ_x, and E_{k_x} then represent *mean values in the section A_x during the period τ*, these averages being weighted in relation to the elementary masses of fluid $\delta m \delta \tau$ which are flowing across the element of cross section δA_x during the element of period $\delta \tau$. With the exception of motojets and pulse jets, the conditions of the internal flows are practically steady and the preceding averages are then weighted only in relation to the steady mass flow δm across the element of cross section δA_x.

The calculation of the *thrust* of a propulsive engine is mainly based on the axial component of the relative flow velocity in the inlet and exit sections. The exit sections A_e are practically orthogonal to the x axis, and are in the end opposite to the direction of the velocity V_∞ of the engine in translation; therefore the relative exhaust velocity V_e is practically parallel to the x axis. Hence it is particularly the average (over the period) of V_e, mass-averaged with respect to δm, which comes into play in the calculation of the thrust. This *simple average*, denoted by the symbol $\overline{V}_e$, differs from the *quadratic* average $\overline{\overline{V}}_e$, which is defined by $E_{k_e} = \frac{1}{2}\overline{\overline{V}}_e^2$. It is important to remember that we always have

$$\overline{V}_e < \overline{\overline{V}}_e = \sqrt{2E_{k_e}} \tag{2-1}$$

With regard to the air inlet section A_i they are defined in such a way that the air flow velocity V_i is in each section practically parallel to the x axis, so that in the calculation of the thrust we have again $\overline{V}_i < \overline{\overline{V}}_i \equiv \sqrt{2E_{k_i}}$.

In order that the normal force of the flow, over the sections A_i and A_e which are practically perpendicular to the x axis, be identified with the reversible pressure p, we assume in these sections that the effect of viscosity is small and, in particular, that the thickness of the boundary layer in the exit flow is sufficiently thin in the sections A_e.

The preceding conditions and reservations are necessary in order to justify the general expressions to be discussed below.

If it is assumed that the operation is accomplished under a fixed set of conditions, then here it can only be a question of *horizontal* flight. In this case, gravity does not affect the internal operation and the state (with subscript ∞) of the undisturbed atmosphere, that is, the atmosphere far upstream of the engine and at the same altitude is defined by p_∞, T_∞ and the specific relative kinetic energy $E_{k\infty} = \frac{1}{2}V_\infty^2$.

B,3. Thrust and Drag. The jet propulsion engine is housed in a streamlined container which is a hollow body connected to the outer surface of the engine by the exhaust nozzle and possibly by the air inlets. In order to define thrust or drag, it is necessary to separate by convention the jet propulsion engine from the other bodies to which it can be connected. Thus such a convention is quite natural for a totally isolated engine, like a rocket without wings or tail surfaces, or even for a streamlined engine plainly separated from the airplane to which it is attached, by pods for example. The convention is more arbitrary for an engine more or less completely enclosed in a wing.

Whatever the conventional definition may be, here we make the following designations (Fig. B,3a where the forces are positive):

1. External drag F_{ext} is the axial resultant of the normal and tangential forces of the ambient atmosphere on the *external* surface Σ_{ext} of the engine
2. Internal thrust F_{int} is the axial resultant of the internal flow forces on the *interior* surface of the engine
3. Effective thrust F_{eff}, is the difference between the internal thrust and external drag $(F_{int} - F_{ext})$ possibly augmented by the thrust F_p of the propeller, i.e. the axial resultant of the forces of the ambient atmosphere on the propeller blades which protrude from the engine cowling.

We then have, by definition:

$$F_{eff} = F_{int} - F_{ext} + F_p \tag{3-1}$$

The solid external surface Σ_{ext} is bounded on one hand by its conventional frontiers with one or more bodies (airplane, wings, and tail surfaces) connected to the engine, and on the other hand by the fictitious surfaces of inlet A_i and of exit A_e.

In the case of an air inlet of the "scoop" type, the fictitious surface of inlet (and of area A_i) is drawn across the entering flow and practically follows a plane perpendicular to the x axis, while the portion of the streamline surface located downstream of the cross section A_i, which establishes the boundary of the inducted flow, must be considered a part of the external surface Σ_{ext} of the propulsive engine. Of course, the exact shape of this portion of the streamline surface varies with the "fixed set of running conditions" and accordingly with the flight speed V_∞. At zero speed, in particular, the drawing of a surface A_i practically perpendicular to the x axis may be impossible, because the flow of inducted air enters the inlet following various directions more or less oblique to the x axis. In such a case we would have to integrate the axial components of pressure and velocity V_i of the flow for each element of a better appropriate surface A_i, but for simplification we do not consider this particular case hereafter.

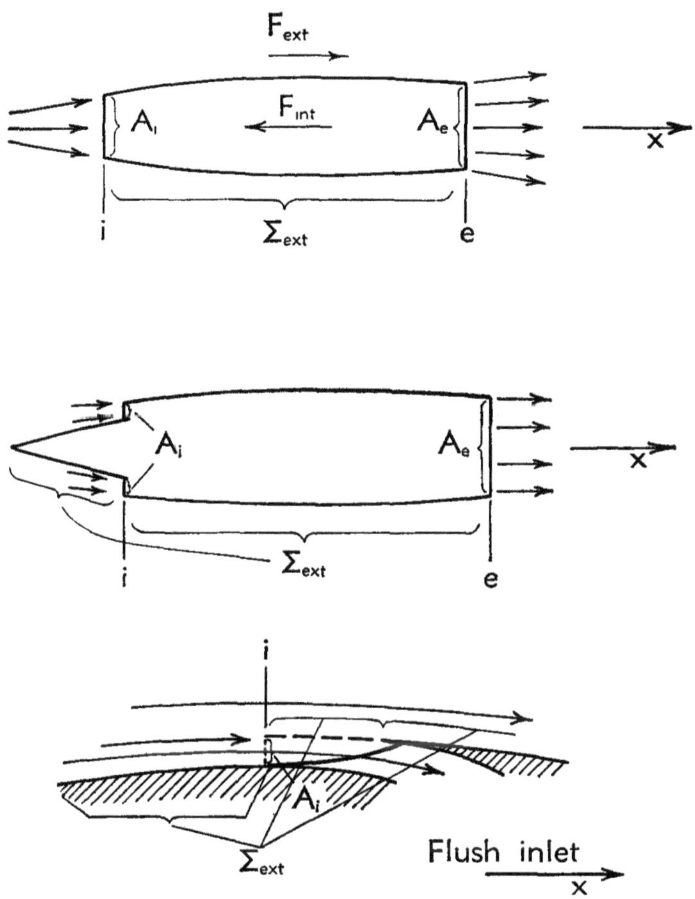

Fig. B,3a.

It is important to note, particularly as a result of the presence of shock waves in the neighborhood of the sections A_i and A_e, that the pressure of the internal flow in these sections is not always uniform or equal to the pressure p_∞ of the uniform atmosphere infinitely far upstream.

Therefore let us draw (Fig. B,3b), across the internal flow extended upstream of A_i and downstream of A_e, two sections A_∞ and A_f perpendicular to the x axis and where the pressure p_∞ exists, these sections being located as near as possible to A_i and A_e.

At A_∞, the air at p_∞ and T_∞ is practically at the axial and uniform relative speed V_∞. Let us designate by Σ'_{ext} and Σ''_{ext} the portions of the streamline surface, extended to A_∞ and A_f, which bound the internal flow, and which are to be added to Σ_{ext}; and by ΔF_{ext} and ΔF_{int} the corresponding additions to F_{ext} and F_{int} which correspond respectively to the axial resultant of the forces of the ambient and external flow, or the internal flow, on the additional lateral surfaces Σ'_{ext} and Σ''_{ext}.

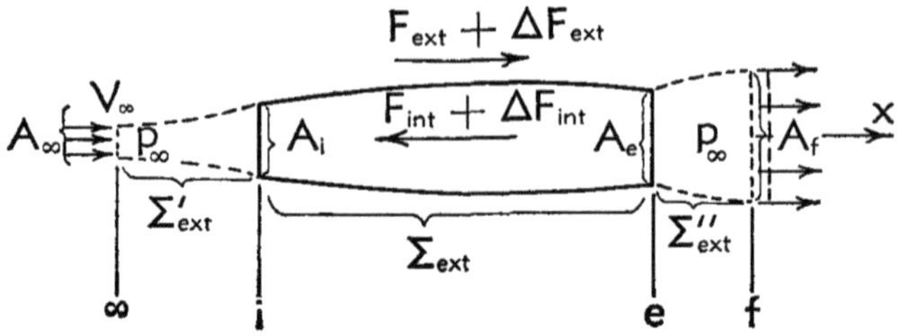

Fig. B,3b.

Since by definition $\Delta F_{ext} = \Delta F_{int}$, Eq. 3-1 is written as

$$F_{eff} = (F_{int} + \Delta F_{int}) - (F_{ext} + \Delta F_{ext}) + F_p \tag{3-2}$$

Let us suppose $F_{ext} + \Delta F_{ext}$ is known. The internal thrust F_{int}, or its corrected value $F_{int} + \Delta F_{int}$ can then be obtained from the momentum theorem, applied during the period τ to the portion of the flow included respectively between A_i and A_e, or between A_∞ and A_f.

Let us designate by m the average mass flow across a fixed section during the period τ, affixing the subscripts a, f, and w respectively, to the air, to the propellant or fuel, and finally to the coolants such as water or water plus methanol, which one can inject into the flow in order to reduce its temperature.

Noting that the fuel and the coolant can be considered in their reservoirs[2] to be at the pressure p_∞, and withdrawn at negligible speed, we have

$$F_{int} = (m_a + m_f + m_w)V_e - m_a V_i + (p_e A_e - p_i A_i) \tag{3-3}$$

$$F_{int} + \Delta F_{int} = (m_a + m_f + m_w)V_f - m_a V_\infty - p_\infty(A_\infty - A_f) \tag{3-4}$$

Either one of these equations can be used to calculate F_{int}.

If we call d_{ext} and Δd_{ext} the drags F_{ext} and ΔF_{ext} that are calculated by subtracting p_∞ from the normal pressures on the surfaces Σ_{ext} and $\Sigma'_{ext} + \Sigma''_{ext}$, and if we observe that the resultant of the uniform pressure p_∞ on the surfaces completed by A_∞ and A_f is zero, we can write

$$F_{ext} + \Delta F_{ext} = d_{ext} + \Delta d_{ext} - p_\infty(A_\infty - A_f)$$

[2] If these substances are stored in the liquid or solid state, the pressure in the reservoir is of negligible influence.

Then from Eq. 3-2 and 3-4, we have

$$F_{eff} = (m_a + m_f + m_w)V_f - m_a V_\infty - (d_{ext} + \Delta d_{ext}) + F_p$$

The "kinetic thrust of the jet" is designated by the term[3] F_k, where:

$$F_k = (m_a + m_f + m_w)V_f - m_a V_\infty \tag{3-5}$$

This differs from the effective thrust of the jet only by the sum $d_{ext} + \Delta d_{ext}$, which is generally small and which can be calculated as specified above.

In that which follows, F_k is taken as a measure of the thrust of the jet, and $F_k + F_p$ as a measure of the *total thrust*, propeller included. If the section A_f practically coincides with the actual exit section A_e, then V_f becomes the simple average V_e of the axial exhaust velocity.

It should be noted that in the case of supersonic propulsion, one might prefer to take $F_k - F_n$ as a measure of the principal part instead of F_k, with

$$F_n = n_e A_e - n_\infty A_\infty$$

where n designates the "dynalpy" $P + \rho V^2$, and the downstream section of flow is A_e instead of A_f with $P_e \neq P_\infty$.

B,4. Powers and Efficiencies. The engine being considered consumes, on an average and during unit time, the mass of fuel or propellant m_f (in one or more distinct bodies). All the powers are hereafter referred to this mass, which is evaluated as "specific power." Provided that we assume air to take part in the chemical transformation of the fuel, the relative air-fuel supply ratio m_a/m_f is *at least sufficient* to render possible a complete transformation ($\kappa = 1$).

In addition to the states ∞, i, e, and f considered above for air (subscript $_a$) and the burned gases (subscript $_g$) we distinguish the following states: 0, the state of the fuel, coolant, and air at rest ($E_{k0} = 0$) and at p_∞, T_∞, the initial state for the definition of the heating value Q_0 of the fuel at constant pressure and per unit mass of the fuel, in the presence of the respective proportions m_w/m_f and m_a/m_f of coolant and air; fi, the final state of the gases of complete combustion at p_∞ ($\kappa_{fi} = 1$), these gases having been brought back to rest ($E_{k_{fi}} = 0$) and to T_∞; fr, the final state of the actual combustion gases ($\kappa_{fr} = \kappa_g$), taken at $p_g = p_{fr} = p_\infty$, and brought back to T_∞ and to rest ($E_{k_{fr}} = 0$), under the constant presspre p_∞.

Using these notations and designating the specific enthalpy[4] and

[3] This term is called "universal thrust" by Zwicky [*3*], and "gross thrust" by the author [*4*]. The designation "kinetic" is used here in preference, in order to emphasize the momentum change.

[4] In the above equations, the changes in h_f and h_g include what is usually called "chemical energy" as well as the enthalpy of the working medium.

entropy by h and s, the (specific) *heating value* Q_0 is defined by

$$m_f Q_0 = m_a(h_a)_0 + m_f(h_f)_0 + m_w(h_w)_0 - (m_a + m_f + m_w)(h_g)_{fi} \quad (4\text{-}1)$$

Then the (specific) *energetic value* Φ_0 of the fuel, that is, the diminution of its (specific) free energy when undergoing the "reversible" transformation from 0 to fi at the stationary condition in one atmosphere at T_∞, is defined by

$$m_f\Phi_0 = m_f Q_0 + T_0[(m_a + m_f + m_w)(s_g)_{fi} - m_a(s_a)_0 - m_f(s_f)_0 - m_w(s_w)_0]$$
$$\equiv m_f(Q_0 + q_0) \quad (4\text{-}2)$$

Concerning the engine in translation at a speed V_∞, the masses m_f and m_w taken aboard and discharged in unit time possess the specific and absolute kinetic energy $E_{k_\infty} = \frac{1}{2}V_\infty^2$, taken away from the source of energy of fuel previously consumed, with the result that the total *available power* $m_f P_{av}$, related to the absolute coordinate system, is:

$$m_f P_{av} = m_f \Phi_0 + (m_f + m_w)E_{k_\infty} \quad (4\text{-}3)$$

On the other hand, the "combustion efficiency" η_b of the actual and incomplete combustion is defined by

$$\eta_b m_f Q_0 = m_a(h_a)_0 + m_f(h_f)_0 + m_w(h_w) - (m_a + m_f + m_w)(h_g)_{fr} \quad (4\text{-}4)$$

This being granted, then we denote: $m_f P_p$ as the *mechanical power* furnished to the eventual propeller on its shaft; and the quantity $m_f P_{pr}$ as the *effective power of propulsion*, or propulsive power, for which we identify the effective thrust to be the sum $F_k + F_p$. Then it follows that

$$m_f P_{pr} = (F_k + F_p)V = (F_k + F_p)\sqrt{2E_{k_\infty}} \quad (4\text{-}5)$$

The quantity $m_f P_{th}$ is the *thermal power* of the transformation of the active substances, starting with the states i and 0, to the state f, for the engine in translation at the speed V_∞ (to be defined later).

By convention, we define the *over-all efficiency* η_0 by the ratio of the effective power to the available power, or

$$\eta_0 = \frac{P_{pr}}{P_{av}} = \frac{(F_k + F_p)\sqrt{2E_{k_\infty}}}{m_f\Phi_0 + (m_f + m_w)E_{k_\infty}} \quad (4\text{-}6)$$

the *thermal efficiency* η_{th} by the ratio of the thermal power to the available power, or

$$\eta_{th} = \frac{P_{th}}{P_{av}} \quad (4\text{-}7)$$

and finally, the *propulsive efficiency* η_{pr} by the ratio of the effective power of propulsion to the thermal power, or

$$\eta_{pr} = \frac{P_{pr}}{P_{th}} \quad (4\text{-}8)$$

so that the first efficiency is the product of the other two:

$$\eta_0 = \eta_{th}\eta_{pr} \tag{4-9}$$

From Eq. 4-6 and by definition of the propeller efficiency η_p, the thrust F_p of the propeller is related to the mechanical power $m_f P_p$ that it receives on its shaft by the relation

$$F_p V_\infty = \eta_p m_f P_p \tag{4-10}$$

As for F_p, η_p is here relative to the propeller in operation on the engine under the fixed set of conditions considered, that is, the aerodynamic interaction of the engine on the propeller is taken into account by η_p, just as the external drag of the engine F_{ext} includes the interaction of the propeller on the engine.

Concerning the mechanical power $m_f P_p$ let us note that the expression for it may be derived from the principle of conservation of energy by reason of the assumed existence of a fixed set of running conditions

$$m_f P_p = m_a (h_a)_i + m_f (h_f)_0 + m_w (h_w)_0 - (m_a + m_f + m_w)(h_g)_f + m_a E_{k_i} - (m_a + m_f + m_w) E_{k_f} - m_f q^e_{if}$$

where the term $m_f q^e_{if}$ represents the total heat transferred in unit time by radiation and conduction to the external atmosphere by the engine and the internal flow from the section A_i to the fictitious section A (Fig. B,3b).

Then, from state ∞ to state i upstream of the engine, the flow of inducted air undergoes a steady state and adiabatic transformation, that is, at constant stagnation enthalpy, even if the flow crosses one or more stationary shock waves upstream of the inlet section A_i, so that:

$$(h_a)_i + E_{k_i} = (h_a)_\infty + E_{k_\infty}; \quad (h_a)_\infty = (h_a)_0$$

Taking into account these relations and Eq. 4-4, the above expression for $m_f P_p$ becomes:

$$m_f P_p = m_f(\eta_b Q_0 - q^e_{if} - q_f) - m_a E_{k_\infty} - (m_a + m_f + m_w) E_{k_f} \tag{4-11}$$

where, for simplification, we place

$$m_f q_f = (m_a + m_f + m_w)[(h_g)_f - (h_g)_{fr}] = (m_a + m_f + m_w) \int_{T_\infty}^{T_f} (c_p)_g dT \tag{4-12}$$

The *thermal power* $m_f P_{th}$ cited above is defined as the sum of two terms: the first is the total mechanical power that would be received through the same transformation $i \rightarrow f$ of the active substances for the engine in flight at the speed $V_\infty = \sqrt{2E_{k_\infty}}$, but by bringing the gases back to absolute rest by means of a perfect impulse turbine. This would add to $m_f P_p$ the additional mechanical power

$$(m_a + m_f + m_w)(E_{k_f} - E_{k_\infty})$$

The second is the supplement of the effective power of propulsion, which corresponds to the thrust due to the discharge of the supply $m_f + m_w$ at the speed V_∞, and would then be $(m_f + m_w)2E_{k_\infty}$.

From this definition, we therefore have

$$m_f P_{\text{th}} = m_f(\eta_b Q_0 - q_{if}^e - q_t) + (m_f + m_w)E_{k_\infty} \tag{4-13}$$

One will note that this expression, which is necessary, depends only on the transformation $i \rightarrow f$ of the internal flow, the consumed supply of the active substances taken aboard, and the flight speed $V_\infty = \sqrt{2E_{k_\infty}}$.

If the engine were to operate at the stationary condition, that is at zero speed ($V_\infty = 0$), with the same transformation $i \rightarrow f$ of the internal flow and with the same active substances, but taken and restored to absolute rest, the *thermal power at the stationary condition,* denoted by $m_f P_{\text{th}}^0$, would consist only of the mechanical power obtained by (1) adding to $m_f P_p$ the additional power $(m_a + m_f + m_w)E_{k_f}$ furnished by a perfect impulse turbine which would bring the discharged gases back to absolute rest, and (2) subtracting the power $m_a E_{k_\infty}$ absorbed by a perfect impulse fan which would pass the air, at p_∞, T_∞, from absolute rest to the state ∞.

By analogy with the definition of η_{th} (Eq. 4-7), we will call *thermal efficiency at the stationary condition,* denoted by η_{th}^0, the ratio of $m_f P_{\text{th}}^0$ to $m_f \Phi_0$, without the additional kinetic term. Then

$$\eta_{\text{th}}^0 = \frac{\eta_b Q_0 - q_{if}^e - q_f}{\Phi_0} \tag{4-14}$$

As we will see further on, it will be especially convenient to consider this term, which depends only on the transformation $i \rightarrow f$.

Let us note that the *maximum mechanical power* which can be furnished by this transformation completed by the return to absolute rest and at p_∞ of the discharged gases in the state f is:

$$m_f P_{\text{th}} - (m_f + m_w)2E_{k_\infty}$$

It is therefore convenient to place

$$P_p = \zeta \left[P_{\text{th}} - \left(1 + \frac{m_w}{m_f}\right) 2E_{k_\infty} \right] \tag{4-15}$$

where ζ represents the fraction of the preceding maximum power transferred to the propeller. The case $\zeta = 0$ corresponds to the case of a pure jet or rocket, whereas $\zeta = 1$ corresponds to a pure turboprop or a propeller driven by a rocket motor, the burned gases of either engine being exhausted with the relative velocity V_∞.

As ζ ranges between 0 and 1 (we evidently exclude ζ negative, which would correspond to a propeller operating like a windmill), an actual propeller necessarily corresponds to

$$P_{\mathrm{th}} - \left(1 + \frac{m_{\mathrm{w}}}{m_{\mathrm{f}}}\right) 2E_{\mathrm{k}_\infty} > 0 \tag{4-16}$$

Finally let us recall that, in Eq. 3-5, V_f represents a *simple average* of the axial exhaust velocity in one or several sections A_f, which very frequently coincides with the actual exit section or sections A_e, where this simple average is not necessarily equal to $\sqrt{2E_{\mathrm{k}_f}}$.

Also, if the change of state of the internal flow between the sections A_e and A_f is accomplished under quasi-steady conditions, almost adiabatically, with constant chemical composition, and with a very small change $(p_e - p_f)$ of pressure, we have

$$(h_{\mathrm{g}})_e - (h_{\mathrm{g}})_f \cong E_{\mathrm{k}_f} - E_{\mathrm{k}_e} \cong \left(\frac{\gamma_{\mathrm{g}} - 1}{\gamma_{\mathrm{g}}}\right) \frac{p_e - p_f}{(\rho_{\mathrm{g}})_e} \tag{4-17}$$

where γ_{g} represents the ratio of specific heats of the exhaust gases. We see that if $p_e - p_f$ is sufficiently small, it is admissible, to a first approximation, to let the states f and e and the corresponding section A_f and A_e coincide, even more so in Eq. 4-11 and 4-12 than in Eq. 3-5. For greater rigor, we forego this simplification here.

CHAPTER 2. GENERAL FORMULAS

B,5. Jet Engines with a Single Discharged Flow. We first suppose (Art. 5 to 11) that there is only a single discharged flow, i.e. that the exhaust nozzle, even if it is composed of distinct parts, provides for the discharge of a flow in which the various portions are characterized by the same state f.

Furthermore, for simplification, we will agree hereafter to identify the exhaust axial velocity (simple average) V_f with the quadratic average $\sqrt{2E_{\mathrm{k}_f}}$. Then:

$$V_f \cong \sqrt{2E_{\mathrm{k}_f}}$$

Let us place, with respect to the various supplies of active substances,

$$\delta = \frac{m_{\mathrm{a}}}{m_{\mathrm{f}}}, \quad \nu = \frac{m_{\mathrm{w}}}{m_{\mathrm{f}}}$$

δ representing the inverse of the usual fuel-air ratio.

Let us also place

$$\lambda_\infty^2 = \frac{E_{\mathrm{k}_\infty}}{\Phi_0} = \frac{V_\infty^2}{2\Phi_0}, \quad \lambda_f^2 = \frac{E_{\mathrm{k}_f}}{\Phi_0} = \frac{V_f^2}{2\Phi_0}$$

with $\Phi_0 = Q_0 + q_0$ while noting that q_0 is generally neglected for mixtures of air and gasoline or air and kerosene so that the specific heating value Q_0 is then substituted for the (specific) energetic value Φ_0. Consider also that our "thermal efficiency" η_{th} is then identical with the "combustion efficiency" in the usual theory of internal combustion engines, with the addition of the term $(1 + \nu)E_{k_\infty}$ which corresponds to the absolute kinetic energy of the active substances transported aboard at the velocity $V_\infty = \sqrt{2E_{k_\infty}}$.

Let us refer back, on the other hand, to the definition (Eq. 4-14) of the thermal efficiency η^0_{th} at the stationary condition and to the definition (Eq. 4-15) of the coefficient ζ of transmission of mechanical power to the propeller. The thermal efficiency η_{th} *in flight* is expressible then, from Eq. 4-7, 4-3, and 4-14, by:

$$\eta_{th} = \frac{\eta^0_{th} + (1 + \nu)\lambda^2_\infty}{1 + (1 + \nu)\lambda^2_\infty} \tag{5-1}$$

Then, from Eq. 4-11 and 4-10, *the square of the specific exhaust velocity* λ^2_f is expressible as

$$\lambda^2_f = \lambda^2_\infty + (1 + \zeta)\frac{\eta^0_{th} - (1 + \nu)\lambda^2_\infty}{1 + \nu + \delta} \tag{5-2}$$

Hence from Eq. 4-6, 3-5, 4-10, and 5-2, the *over-all efficiency* η_0 is

$$\eta_0 = \zeta\eta_p \frac{\eta^0_{th} - (1 + \nu)\lambda^2_\infty}{1 + (1 + \nu)\lambda^2_\infty} + \frac{2\lambda^2_\infty}{1 + (1 + \nu)\lambda^2_\infty} \left[(1 + \nu + \delta)\sqrt{1 + (1 - \zeta)\frac{\eta^0_{th} - (1 + \nu)\lambda^2_\infty}{(1 + \nu + \delta)\lambda^2_\infty}} - \delta\right] \tag{5-3}$$

Finally, from the definition of Eq. 4-8, the *propulsive efficiency* is:

$$\eta_{pr} = \zeta\eta_p \frac{\eta^0_{th} - (1 + \nu)\lambda^2_\infty}{\eta^0_{th} + (1 + \nu)\lambda^2_\infty} + \frac{2\lambda^2_\infty}{\eta^0_{th} + (1 + \nu)\lambda^2_\infty} \left[(1 + \nu + \delta)\sqrt{1 + (1 - \zeta)\frac{\eta^0_{th} - (1 + \nu)\lambda^2_\infty}{(1 + \nu + \delta)\lambda^2_\infty}} - \delta\right] \tag{5-4}$$

The obvious advantage that is offered in the use of η^0_{th}, in preference to η_{th} in the above equations, results from the fact that η^0_{th} varies much less than η_{th} with λ^2_∞ for a given set of conditions of *internal operation* $i \rightarrow f$ of the engine. If we regard η^0_{th} as fixed, for example, then Eq. 5-1 immediately expresses the variation of η_{th} with λ^2_∞; in particular, whatever η^0_{th} may be, η_{th} always tends toward 1 when λ_∞, and therefore the flight speed V_∞, increases indefinitely.

B,6. Hypervelocities. The necessary condition (Eq. 4-16) for the propeller to be able to effectively receive mechanical power, requires that, for a jet engine with mechanically driven propeller (for example, a turboprop), the specific velocity λ_∞ remain below the limit

$$\lambda^* = \sqrt{\frac{\eta_{th}^0}{1 + \nu}} \tag{6-1}$$

For $\lambda_\infty > \lambda^*$, the corresponding velocities can be called "hypervelocities," in the sense that they exclude any possibility of effective utilization of a propeller mechanically driven by an engine which is characterized by η_{th}^0.[5]

For example, consider an engine burning kerosene [$\Phi_0 = (6.6 \text{ km/sec})^2$] by means of air inducted from the outside, and without the addition of coolant ($\nu = 0$). For $\eta_{th}^0 = 0.4$, the hypervelocity threshold corresponds to

$$V^* = \sqrt{2 \times 0.4(6.6)^2} = 5.9 \text{ km/sec}$$

For a propellant mixture of the $H_2 + O_2$ or alcohol + liquid oxygen type, propellants for which Φ_0 is 11 to 4 times less, the threshold V^* would be reduced about 50 to 70 per cent, but would nevertheless remain extremely high.

B,7. Optimum Combination of Propeller and Jet. For convenience in the discussion of the general formulas (Eq. 5-2, 5-3 and 5-4) let us consider the case of a turboprop and let us assume that δ, η_p, η_{th}^0, and λ_∞^2 are given.

Then, in Eq. 5-3, the coefficient ζ of partition between propeller and jet is the only variable, and the maximum of the over-all efficiency η_0 occurs for $\zeta = \zeta_m$, so that

$$\zeta_m = \frac{\eta_p^2(\eta_{th}^0 + \delta\lambda_\infty^2) - (1 + \nu + \delta)\lambda_\infty^2}{\eta_p^2[\eta_{th}^0 - (1 + \nu)\lambda_\infty^2]} \tag{7-1}$$

Consider a turboprop burning kerosene [$\Phi_0 = (6.6 \text{ km/sec})^2$] and using no coolant ($\nu = 0$), with

$$\eta_{th}^0 = 0.4, \quad \eta_p = 0.85, \quad \delta = 60, \quad V_\infty = 0.2 \text{ km/sec (or } \lambda_\infty^2 = 0.00092)$$

Eq. 7-1 then gives $\zeta_m = 0.945$, that is, the optimum combination corresponds to the utilization by the propeller (with an efficiency of 0.85) of 94.5 per cent of the maximum mechanical power which could be developed by the engine considered. The corresponding maximum overall efficiency, from Eq. 5-3, is equal to 0.351, whereas for the pure jet ($\zeta = 0$) it would be only 0.209.

It is important to note that, if we vary ζ between its limits $\zeta = 0$ and $\zeta = 1$, the assumption of keeping η_{th}^0 constant does not correspond

[5] Editors note: Whenever the speed of an aircraft is so high that the energy of the fuel is not sufficient to accelerate the mass of fuel to the flight speed, drag results for turbojets as well as for turboprops as in the example given here.

exactly to the actual case, because η_{th}^0 has the tendency to diminish when we increase, to the profit of the propeller, the mechanical power developed by the expansion turbine. Eq. 5-1, 5-2, 5-3, and 5-4 permit, moreover, the evaluation of η_0 for the given characteristics in each particular case.

B,8. The Pure Turboprop. This is the case $\zeta = 1$, for which λ_∞ is limited (Art. 6) to λ^* defined by Eq. 6-1. In the domain $0 < \lambda_\infty <$

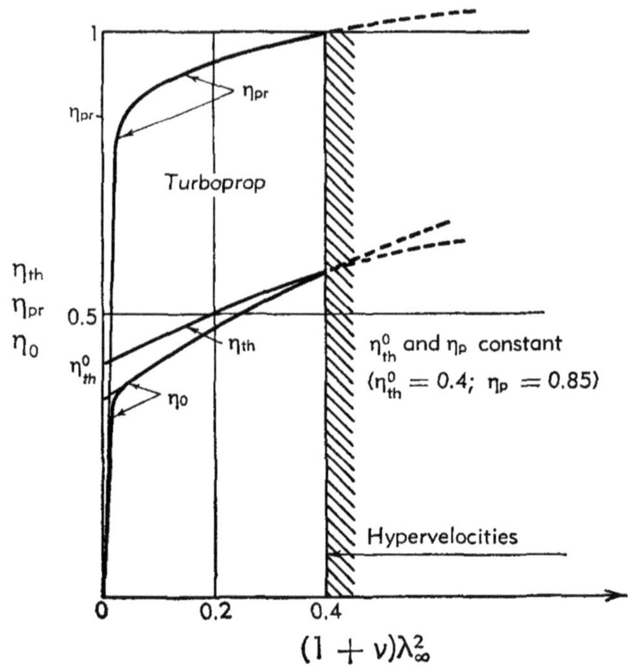

Fig. B,8.

$\sqrt{\eta_{th}^0/(1 + \nu)}$ we then have, from Eq. 5-1, 5-2, 5-3, and 5-4:

$$\eta_{th} = \frac{\eta_{th}^0 + (1 + \nu)\lambda_\infty^2}{1 + (1 + \nu)\lambda_\infty^2}, \quad \eta_\infty = \eta_{th}\eta_{pr}$$

When η_{th} and η_p remain constant, these efficiencies depend only on the parameter $(1 + \nu)\lambda_\infty^2$; their variation is represented as a function of this parameter in Fig. B,8, where we really are admitting that η_p drops very rapidly in the neighborhood of $\lambda_\infty = 0$, since the efficiency of a propeller can only be zero at the stationary condition.

In fact, for an actual power plant, η_p depends on λ_∞^2; nevertheless the propulsive efficiency η_{pr} always becomes unity for $(1 + \nu)\lambda_\infty^2 = \eta_{th}^0$, because the contribution from the propeller vanishes when the threshold of the hypervelocities is reached.

B,9. The Pure Air Flow Jet Engine and Pure Rocket. This is the case of $\zeta = 0$, with $\delta \neq 0$ for the air flow jet engine and $\delta = 0$ for the rocket. Eq. 5-1, 5-2, 5-3, and 5-4 give here

$$\eta_{th} = \frac{\eta_{th}^0 + (1+\nu)\lambda_\infty^2}{1 + (1+\nu)\lambda_\infty^2}, \quad \eta_{pr} = \frac{2\lambda_\infty^2}{\eta_{th}^0 + (1+\nu)\lambda_\infty^2}$$

$$\left[(1+\nu+\delta)\sqrt{\frac{\eta_{th}^0 + \delta\lambda_\infty^2}{(1+\nu+\delta)\lambda_\infty^2}} - \delta\right], \quad \eta_0 = \eta_{th}\eta_{pr} \tag{9-1}$$

When η_{th}^0, ν, and δ remain constant, these efficiencies depend only on λ_∞^2; η_{th} increases from η_{th}^0 to unity as a hyperbolic function of λ_∞^2 when the flight speed V_∞ increases from 0 to ∞. The propulsive efficiency η_{pr} always passes through a maximum, equal to unity,[6] when $\lambda_\infty = \lambda^*$, the threshold of the hypervelocities.

When λ tends toward infinity, η_{pr} tends toward the limit

$$(\eta_{pr})_{\lambda_\infty \to \infty} = \frac{2\delta}{1+\nu}\left[\sqrt{1 + \frac{1}{\delta}} - 1\right] \tag{9-2}$$

which is zero for the rocket ($\delta = 0$). For the air flow jet engine ($\delta > 0$) this limit remains finite, but it increases with δ and decreases with ν, as shown in Fig. B,9a.

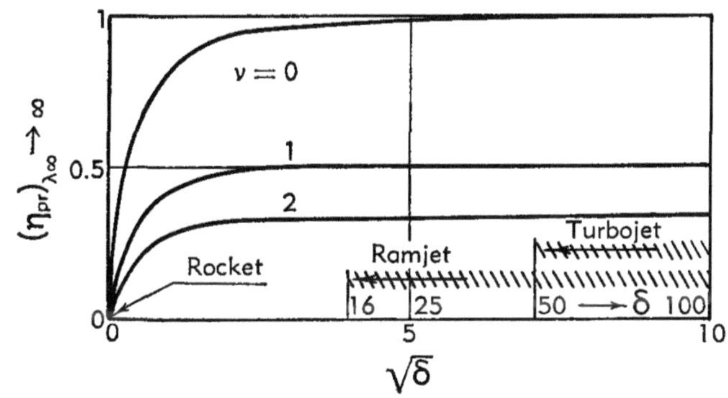

Fig. B,9a.

In practice and because of the relatively high values of δ which are used, the limit of η_{pr} for the ramjet as for the turbojet is very close to $1/(1+\nu)$, and therefore approximately equal to unity when the ramjet or turbojet does not consume any coolant ($\nu = 0$).

The over-all efficiency η_0 is zero for $\lambda_\infty = 0$, equal to $2\eta_{th}^0/(1 + \eta_{th}^0)$ for $\lambda_\infty = \lambda^*$, and tends toward the value given by Eq. 9-2 when λ_∞ increases indefinitely. For very high speeds, the over-all efficiency of the

[6] Between $\lambda_\infty = 0$ and $\lambda_\infty = \lambda^*$, η_{pr} would pass through a minimum in the case where we would simultaneously have $\delta < 0.25$ and $(1+\nu) > 38$. For a "jet" using a hydrocarbon of the kerosene type, this case is excluded because δ is at least of the order of 15 to 20 (for a ramjet) or 50 to 60 (for a turbojet).

air flow jet engine is therefore much greater than that of the rocket although ν is much lower.

As an example, Fig. B,9b shows the comparison of η_0 as a function of V_∞(km/sec) for a turbojet [$\Phi_0 = (6.6 \text{ km/sec})^2$, $\nu = 0$; $\eta_{th}^0 = 0.3$ or 0.6, and $\delta = 60$; $\eta_{th}^0 = 0.1$ or 0.3, and $\delta = 20$]; and for a rocket [$\Phi_0 = (2.2 \text{ km/sec})^2$, $\nu = 0$; $\eta_{th} = 0.3$ or 0.6; and $\delta = 0$]. This example merely shows the difference in the character of η_0 at low speeds and at very high speeds

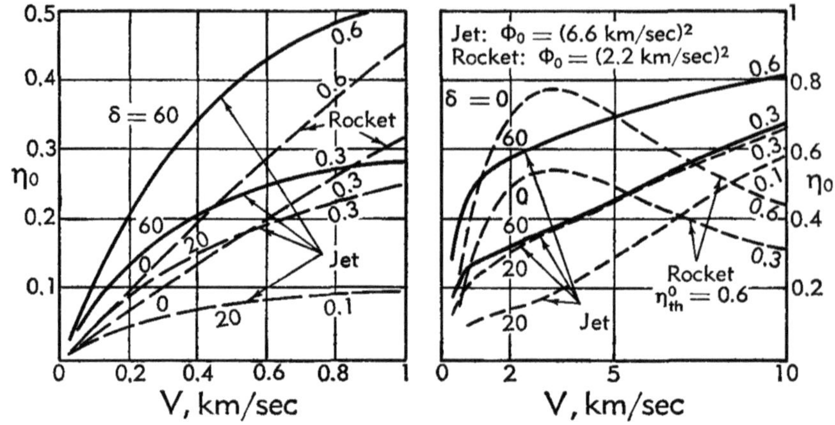

Fig. B,9b.

The air fuel ratio $\delta = 20$ corresponds pretty well to a ramjet, but the condition $\eta_{th}^0 = 0.3$ is far from realizable at low speeds for this type of jet.

B,10. Approximate Formulas for Low Speeds. The general formulas (Eq. 9-1) simplify considerably for low speeds, that is when we are able to consider

$$\frac{(1+\nu)\lambda_\infty^2}{\eta_{th}^0} = \frac{(1+\nu)V_\infty^2}{2\eta_{th}^0\Phi_0}$$

as practically negligible compared to unity, for example less than 0.05.

For a rocket, this condition corresponds to $V_\infty < 1$ km/sec if $\nu = 0$ and $\eta_{th}^0 \geqq 0.4$. For a jet with an air-fuel ratio, δ, more than or equal to 50 and for Φ_0 of the order of $(6.6 \text{ km/sec})^2$, this corresponds to a $V_\infty < 1.1$ km/sec if $\nu = 0$ and $\eta_{th}^0 \geqq 0.3$.

Under these conditions, Eq. 9-1 for the pure air flow jet engine become

$$\eta_{th} \cong \eta_{th}^0, \quad \eta_0 \cong 2\lambda_\infty[\sqrt{(1+\nu+\delta)(\eta_{th}^0 + \delta\lambda_\infty^2)} - \delta\tau_\infty], \quad \eta_{pr} \cong \frac{\eta_0}{\eta_{th}^0} \tag{10-1}$$

For the pure rocket ($\delta = 0$) they reduce to

$$\eta_{th} \cong \eta_{th}^0, \quad \eta_0 \cong 2\tau_\infty\sqrt{(1+\nu)\eta_{th}^0}, \quad \eta_{pr} \cong \frac{\eta_0}{\eta_{th}^0} \tag{10-2}$$

B,11. The Propulsive Jet Engine with Two Discharged Flows. For simplification, we shall not consider the turboprop hereafter, but consider only the pure direct reaction engine of the air flow jet engine type or of the rocket type.

First, let us remark that in the interior of this engine the flow of the active substance may divide itself into two or many partial flows. However, this circumstance does not alter the general formulas (Eq. 9-1), which only assume that the *discharge flow is unique* in the sense that the state f is practically uniform and identically defined in the various parts that can be contained in the exit section.

The other case is where, at the exit or downstream e or f, we can distinguish many distinctly different states, for the variance partial flows. This is said to be the case of an engine with *many ejected flows.*

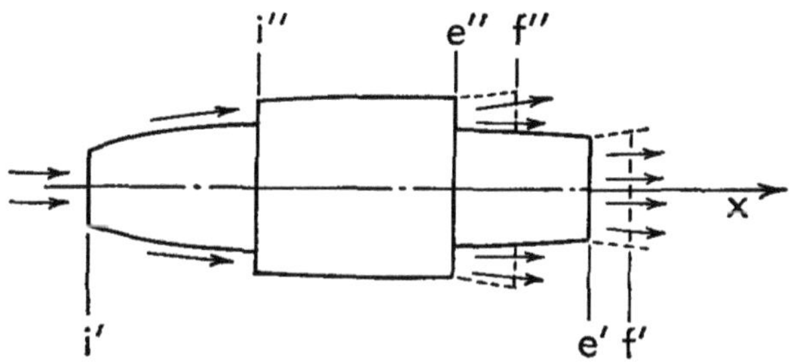

Fig. B,11.

The preceding definitions and general formulas apply to this case, provided we distinguish the average of E_{k_f} which enters the expression for the powers and the average of V_f which is to be considered in computing the thrusts. Therefore we no longer have the right to let $\overline{V}_f = \sqrt{2\overline{E}_{k_f}}$ as we have done above (Art. 5).

Let us carry out the necessary generalizations by restricting ourselves to the case of two *discharged flows,* which is generally sufficient, where we distinguish respectively, by *a single or double prime,* the quantities which refer to the primary or to the secondary flow.

The primary flow is chosen arbitrarily, provided it effectively consumes fuel ($m' \neq 0$), and we assume that a single type of fuel is being used, with a (specific) energetic value Φ_0 or a heating value $Q_0 = \Phi_0 - q_0$.

Note that all the air inlets operate at the same (specific) stagnation enthalpy $(h_0 + E_{k_\infty})$ of the upstream state ∞. Let us note also that each flow can have several air inlets and fuel or coolant injectors, which we distinguish by the numerical subscripts 1, 2, . . . , so that we have

$$\begin{cases} m'_{a,f,w} = (m'_{a,f,w})_1 + (m'_{a,f,w})_2 + \cdots \\ m''_{a,f,w} = (m''_{a,f,w})_1 + (m''_{a,f,w})_2 + \cdots \\ m_{a,f,w} = m'_{a,f,w} + m''_{a,f,w} \end{cases} \tag{11-1}$$

Finally, it is evident that the two flows can have many parts in common in the internal cycle, especially during the intake or compression of the air, and that these flows can exchange mechanical work or heat, for example, if the turbine run by the primary flow drives a compressor working on the secondary flow, or if there is present a heat exchanger transferring heat from one flow to the other.

Let us assume that there are distinct states f' and f'' for the two flows at a pressure p_∞ (and possibly different states e' and e''), which are sufficiently uniform so that we have separately

$$V_{f'} \cong \sqrt{2E_{\mathrm{k}f'}}, \quad V_{f''} \cong \sqrt{2E_{\mathrm{k}f''}} \tag{11-2}$$

Then, in order to simplify the formulas hereafter, by reference to the primary flow, in which m_f' is fundamentally positive, let us place

$$\delta' = \frac{m_\mathrm{a}'}{m_\mathrm{f}'}, \quad \nu' = \frac{m_\mathrm{w}'}{m_\mathrm{f}'}, \quad x = \frac{m_\mathrm{f}''}{m_\mathrm{f}'}, \quad \psi = \frac{m_\mathrm{w}''}{m_\mathrm{f}'}, \quad \mu = \frac{m_a'' + m_\mathrm{f}'' + m_\mathrm{w}''}{m_\mathrm{f}'(1 + \nu' + \delta')},$$

$$\alpha = \frac{E_{\mathrm{k}f''}}{E_{\mathrm{k}f'}} \tag{11-3}$$

Considering the operation under a fixed set of conditions and averaged over unit time, let us designate:

$m_\mathrm{f}' q_{i'f'}^{\prime e}$ = the heat transferred directly to the external atmosphere by the primary flow

$m_\mathrm{f}' q'$ = the heat transferred to the secondary flow by the primary flow (the reverse transfer is $-m_\mathrm{f}' q'$)

$m_\mathrm{f}' P'$ = the mechanical power furnished to the secondary flow by the primary flow (the reverse contribution is $-m_\mathrm{f}' P'$)

$m_\mathrm{f}' q_{f'}'$ = the sensible heat, of the primary mass flow in the state f'.

The previous unprimed notation is retained for the *over-all flow* (association of the primary flow and the secondary flow), consuming the quantity of fuel $m_\mathrm{f} = m_\mathrm{f}'(1 + x)$, and we assume that we know or can calculate from their appropriate definitions (Art. 4) the over-all quantities $\eta_{\mathrm{b}'} q_{if}^{e}$ from the definition (Eq. 4-14) of the global thermal efficiency at the stationary condition η_{th}^0. Then

$$\eta_{\mathrm{th}}^0 \Phi_0 = \eta_\mathrm{b} Q_0 - q_{if}^{e} - q_f \tag{11-4}$$

Considering the combustion in each flow (although combustion may not exist in the secondary flow) we can, from the definition (Eq. 4-4) applied separately to each flow, set up a corresponding combustion efficiency η_b' or η_b'', these two efficiencies being related moreover by

$$\eta_\mathrm{b}' + x\eta_\mathrm{b}'' = (1 + x)\eta_\mathrm{b}$$

If the secondary flow does not consume fuel, we evidently have $x = 0$, and η_b'' does not enter $\eta_\mathrm{b}' = \eta_\mathrm{b}$.

On the other hand, the proper thermodynamical evolution of the primary flow being assumed known from separate study, let us place

$$\eta'\Phi_0 = \eta'_b Q_0 - q'^{e}_{i'f'} - q'_{f'} - q' - P' \tag{11-5}$$

Then by means of a calculation that the reader can easily reestablish, we can express the ratio $\alpha = E_{k_{f''}}/E_{k_{f'}}$ by the formula

$$\alpha = \frac{(1 + x)\eta^0_{th} - \eta' + [\mu(1 + \nu' + \delta') - (x + \psi)]\lambda^2_\infty}{\mu(\eta' + \delta'\lambda^2_\infty)} \tag{11-6}$$

from which we can deduce the ratio of the average speeds $V_{f''}/V_{f'} = \sqrt{\alpha}$ which enters in the calculation of the kinetic thrust of each jet.

Finally, *the thermal efficiency in flight* η_{th} (combined efficiency) can be expressed by

$$\eta_{th} = \frac{(1 + x)\eta^0_{th} + (1 + \nu' + x + \psi)\lambda^2_\infty}{1 + x + (1 + \nu' + x + \psi)\lambda^2_\infty} \tag{11-7}$$

whereas the *over-all efficiency* η_0 (combined efficiency) can be expressed by:

$$\eta_0 = \frac{2\lambda^2_\infty}{1 + x + (1 + \nu' + x + \psi)\lambda^2_\infty} \left\{(1 + \mu\sqrt{\alpha})\sqrt{(1 + \nu' + \delta')\left(\frac{\eta'}{\lambda^2_\infty} + \delta'\right)} - [\delta' + \mu(1 + \nu' + \delta') - (x + \psi)]\right\} \tag{11-8}$$

while the propulsive efficiency η_{pr} (combined efficiency) remains defined, according to Eq. 4-9, by the ratio:

$$\eta_{pr} = \frac{\eta_0}{\eta_{th}} \tag{11-9}$$

Eq. 11-7, 11-8, and 11-9 furnish the generalizations we were looking for. When we let $x = \psi = \mu = 0$ and $\nu' = \nu$, $\delta' = \delta$, they reduce to Eq. 9-1, which formulas are convenient for the jet engine with a single discharged flow.

The above formulas are even applicable in the case of air flow jet engines or rockets containing two internal flows, but discharged into a mixed state by a single nozzle where we could possibly distinguish two states f' and f'' relative to each part of the mixture. These states are at the same pressure p_∞ and may or may not be at the same temperature and the same velocity.

Of course, the parameters η^0_{th} and η' expressed by Eq. 11-4 and 11-5 can only be calculated from a detailed thermodynamic study of the thermodynamical evolution of each of the two flows considered above. This point is taken up later (Art. 13).

B,12. The Ducted Fan. By means of a particular example, let us apply the preceding generalizations to a *ducted fan* (*propulsive compressor* or *turbo fan*) of the type composed of two homogeneous internal flows having their discharges in common.

In this engine (Fig. B,12a), the turbine T' of the primary flow is enlarged in order to drive the compressor C'' of the secondary flow. This compressor can have some stages in common with the compressor C' of the primary flow. Downstream of T' the two flows mix in order to be discharged by a single tailpipe N. Therefore we have two internal flows, with a portion of the evolution common to the inlet diffuser and the compressor, with a single ejected flow brought back to homogeneity by mixing before passing out of the end of the tailpipe.

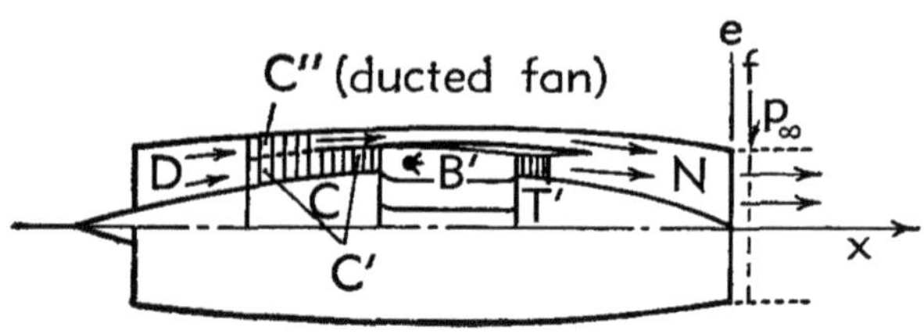

Fig. B,12a.

For simplification, we assume that:

1. The primary flow does not consume any coolant ($\nu = 0$) and the secondary flow consumes neither fuel ($x = 0$) nor coolant ($\psi = 0$).
2. The mixing of the two flows is isobaric and adiabatic, and a uniform mixture is obtained, that is $f' \equiv f'' \equiv f$.
3. The evolution of the secondary flow is completely reversible and adiabatic, that is isentropic, except in the mixing.
4. The evolution of the primary flow remains unaltered, whatever may be the relative magnitude μ of the secondary flow, until that stage of the turbine where the power necessary to compress the primary flow is reached. Downstream of this stage the primary flow is adiabatic and reversible until it mixes with the secondary flow.

As a result of these conditions, the additional stages of the turbine remove from the primary flow a certain amount of stagnation enthalpy,[7] this being the amount which the compressor transfers to the secondary flow and which the subsequent mixing restores to the final unique flow. Furthermore, and as it is easy to recognize, we have here

$$\eta'_b = \eta_b, \quad q'^e_{i'f'} \equiv q^e_{if}, \quad q'_{f'} = \frac{q_f}{1+\mu} \tag{12-1}$$

Finally, q_f is constant and equal to its value for the jet reduced to handling only the primary flow, and consequently deprived of a "ducted

[7] The (specific) stagnation enthalpy is the sum $h + E_k$.

fan." Then, taking into account Eq. 12-1 and $\alpha = 1$, Eq. 11-8 gives

$$\eta_0 = \frac{2\lambda_\infty^2}{1+\lambda_\infty^2}\left\{(1+\mu)\sqrt{(1+\delta')\left[\frac{\eta_{th}^0 + \mu\lambda_\infty^2}{(1+\mu)\lambda_\infty^2} + \delta'\right]} - [\delta' + \mu(1+\delta')]\right\} \quad (12\text{-}2)$$

For a turbojet using kerosene, with $\delta' = 60$ and $\Phi_0 = (6.6 \text{ km/sec})^2$, and for η_{th}^{0} constant and equal to 0.3, Fig. B,12b represents η_0 as a function of μ when V_∞ takes on the values 0.15, 0.30, and 0.50 km/sec. We see that the relative gain from the secondary dilution μ is greatly decreased as soon as the speed becomes of the order of 0.3 km/sec.

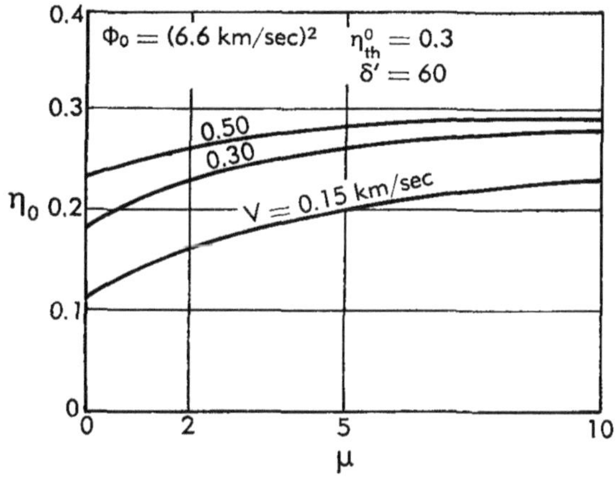

Fig. B,12b.

Actually the assumptions used in order to simplify the final expression (Eq. 12-2) for η_0 are too favorable to the "ducted fan," especially because we have assumed that the turbine and compressor introduced in this arrangement are perfect. The general formulas of Art. 11 permit, moreover, the exact evaluation of η_{th} and η_0 in each particular case by taking into account the individual characteristics of each.

B,13. The Turbojet with Bleed-off. Let us consider another example of the application of the general formulas, Eq. 11-4, 11-5, 11-6, 11-7, 11-8, and 11-9, where we proceed to a *detailed thermodynamic analysis* of the cycles of the two discharged flows, assuming each one homogeneous.

This example is the one of the *turbojet with bleed-off*, shown in Fig. B,13a, consisting of a turbojet from which a part of the air supply is drawn from the compressor in order to burn in the secondary burner B'' and to be discharged separately by the tailpipe N''. The compressor C'' of the secondary flow can consist of all or part of the stages of the primary flow compressor C'.

With reference to the preceding example we can also say that we are considering here a propulsive compressor C'', feeding a burner B'' whose exhaust is ejected separately without mixing with the primary flow. The particular advantage of this device is the possibility of heating the secondary air in the burner B'' to a very high temperature, as in a ramjet, because the secondary flow does not cross any stage of the turbine.

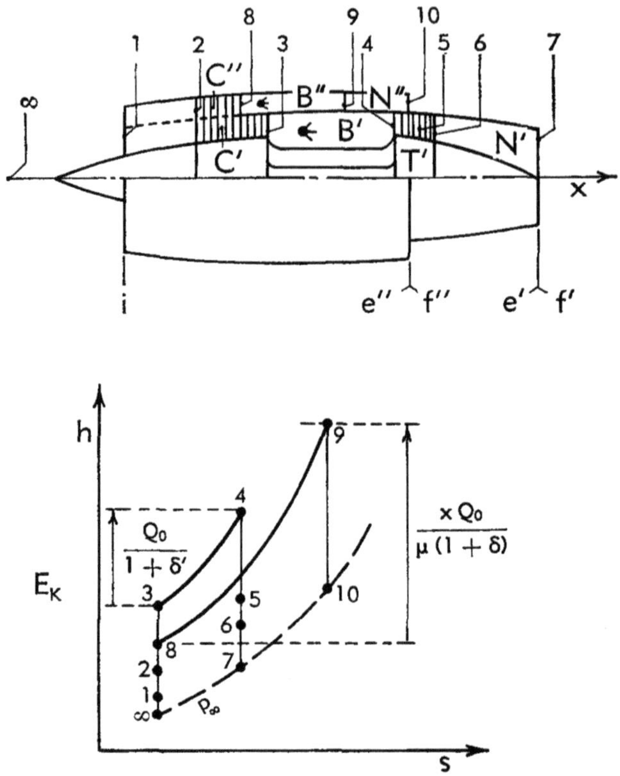

Fig. B,13a.

In order only to simplify the calculations in this schematic example, consider a relatively ideal engine by assuming hereafter that:

1. The engine does not consume any coolant ($\nu' = \psi = 0$) and the two combustions are complete ($\eta_b' = \eta_b'' = \eta_b = 1$).
2. All the transformations of the fluids are reversible and adiabatic, with the exception of the combustions which are not only complete and adiabatic, but also isobaric.
3. The (specific) kinetic energy E_k of the flow is negligible in all the states with numerical subscripts represented on the Mollier diagram, (s, h), of Fig. B,13a, with the exception of the states with subscripts ∞, 1, 7, and 10.

inducted air enters the inlet following various directions more or less oblique to the x axis. In such a case we would have to integrate the axial components of pressure and velocity V_i of the flow for each element of a better appropriate surface A_i, but for simplification we do not consider this particular case hereafter.

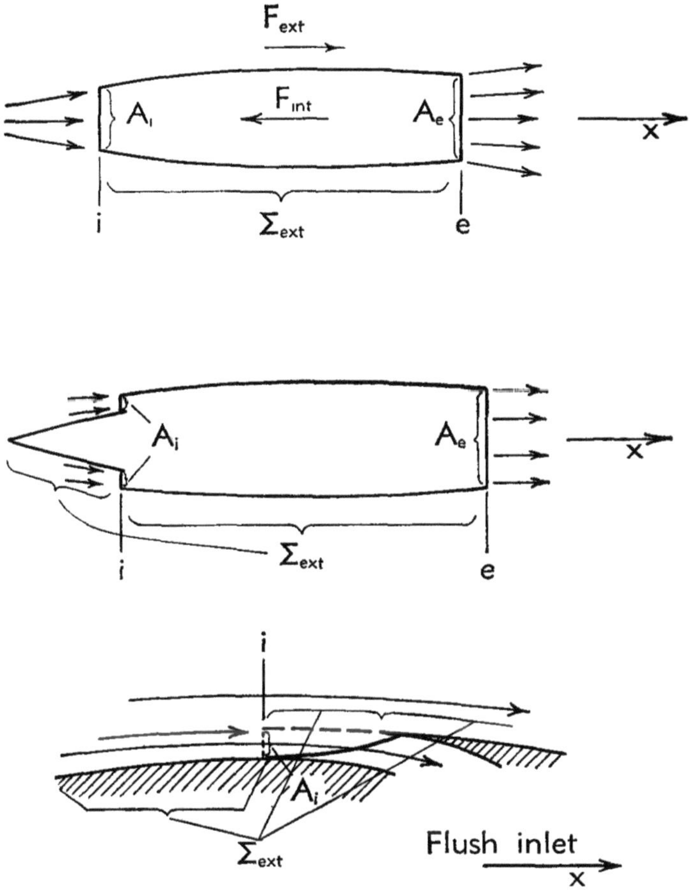

Fig. B,3a.

It is important to note, particularly as a result of the presence of shock waves in the neighborhood of the sections A_i and A_e, that the pressure of the internal flow in these sections is not always uniform or equal to the pressure p_∞ of the uniform atmosphere infinitely far upstream.

Therefore let us draw (Fig. B,3b), across the internal flow extended upstream of A_i and downstream of A_e, two sections A_∞ and A_f perpendicular to the x axis and where the pressure p_∞ exists, these sections being located as near as possible to A_i and A_e.

over-all efficiency

$$\eta_0 = \frac{2\lambda_\infty^2 a}{(1+\lambda_\infty^2)(c' + \xi\mu^* c'')} \left\{ \sqrt{y_\infty^2 - \xi(y_\infty^2 - 1)} - \left(1 - \frac{c'}{a}\right) + \xi\mu^* \left[\sqrt{\frac{\beta''}{b-1}\left(1 - \frac{1}{b\gamma''}\right)} - \left(1 - \frac{c''}{a}\right) \right] \right\} \tag{13-2}$$

Denoting by η_0^0 the over-all efficiency of the reference turbojet ($\mu = 0$) and the corresponding thrust by F^0, it is easy to calculate the ratios η_0/η_0^0 and F/F^0 as functions of $\xi(0 < \xi < 1)$ for the given values of V_∞ or λ_∞, a, γ', β', β'', and n''. This last factor represents the relative compression of the secondary flow with respect to that of the primary flow; the case $n'' = 1$ corresponds to a removal at the exit of the compressor C' of the reference turbojet.

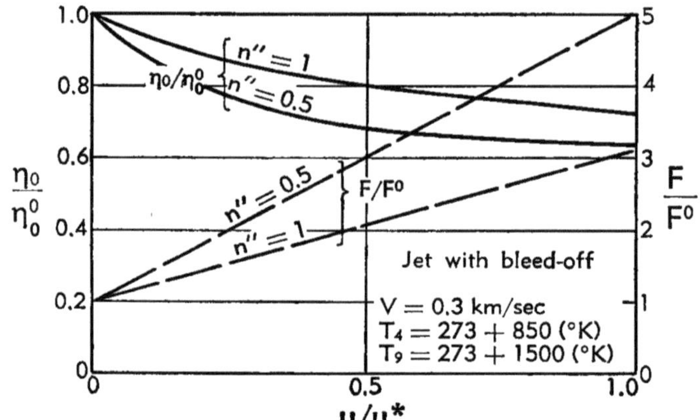

Fig. B,13b.

Consider the particular case $\Phi_0 = (6.6 \text{ km/sec})^2$, $T_\infty = 216.5$ (°K), $\gamma' = 1.87$ (the compressor C' having a pressure ratio equal to 6), $T_4 = 273 + 850$ (°K), $T_9 = 273 + 1500$ (°K) and $V_\infty = 0.35$ km/sec. The graph of Fig. B,13b shows, as a function of $\mu/\mu^* \equiv \xi$, the variation of η_0/η_0^0 and of F/F^0 depending on whether we choose $n'' = 0.5$ or $n'' = 1$.

We see that we are able to obtain a considerable increase in thrust, though at the cost of a certain increase in the maximum cross section due to the growth of the compressors and turbines, as well as to the addition of the burner B'', and also at the cost of a considerable drop in η_0.

Because the example treated is ideal, the thrust augmentation is overestimated and the drop in efficiency is underestimated in proportion to what both would be in an actual machine with all its imperfections.

B,14. Magnitude of the Thrust. The *magnitude* of the thrust can be defined in several ways by referring the thrust to a characteristic quantity of the engine, for example its weight or its maximum cross-sectional area.

The *gravimetric thrust* is the nondimensional ratio of the thrust of the engine to its weight. It is principally amenable to empirical formulation relative to this or that family of engines realized or studied. The inverse ratio is normally called "specific weight" of the jet engine.

On the other hand, the *frontal thrust* is the ratio of the total thrust to the maximum cross-sectional area, which is frequently capable of having a rather general expression, at least for jet engines.

Let us in fact assume that the cross-sectional area A_b of the primary burner or combustor (the cross section being occupied entirely or just in part by this combustor) is in a known ratio $\sigma_b = A_b/A$ to the maximum cross-sectional area A, and let us denote by the subscript b the average state of the flow entering the section A_b.

At that point, the portion of the fuel in the mass flow of the primary gaseous flow remains negligible, and this mass flow is $m'_f(\nu + \delta')$ where the part δ' of the coolant is generally taken into account because the coolant is generally injected upstream of the combustor.

We can consider the Mach number $M_b = V_b/a_b$ (where a is the speed of sound) to be characteristic of the average flow at the entrance of the burner, so that

$$m'_f(\nu' + \delta') = \rho_b A_b V_b = \rho_\infty a_\infty A M_b \sigma_b \frac{\rho_b a_b}{\rho_\infty a_\infty}$$

The thrust, conventionally represented by the kinetic thrust F_k, is for the case of two separately discharged flows (Art. 11)

$$F_k = m'_f(1 + \nu + \delta')(V_{f'} + \mu V_{f''}) - [\delta' + \mu(1 + \nu' + \delta') - (x + \psi)]V_\infty$$

The frontal magnitude of thrust $\Pi = F_k/A$ is expressed therefore as:

$$\Pi = \rho_b a_b M_b \sigma_b \left\{\frac{(1 + \nu' + \delta')(V_{f'} + \mu V_{f''}) - [\delta' + \mu(1 + \nu' + \delta') - (x + \psi)V_\infty]}{\nu' + \delta'}\right\} \tag{14-1}$$

If we take account of the expression (Eq. 11-8) for η_0, we can also express Π by

$$\Pi = \rho_b a_b M_b \sigma_b \frac{\eta_0[1 + x + (1 + \nu' + x + \psi)\lambda_\infty^2]}{\nu' + \delta'} \sqrt{\frac{\Phi_0}{2\lambda_\infty^2}}$$

For example, let us compare two air flow jets using kerosene [$\Phi_0 = (6.6 \text{ km/sec})^2$] with a simple flow ($x = \psi = \mu = 0$) without consuming coolant ($\nu' = 0$), and both flying at $V_\infty = 0.3$ km/sec. (This is $\lambda_\infty^2 = 0.00103$.) The first jet has a frontal magnitude of thrust Π_{RJ} and is a "ramjet" for which a good inlet diffuser furnished a certain ratio $\rho_b a_b/\rho_\infty a_\infty$ for a moderate value of the Mach number M_b. Accepting a fairly high combustion temperature, this ramjet is considered to operate with $\delta = 30$ and $\eta_0 = 0.1$.

The second jet has a frontal magnitude of thrust Π_{TJ} and is a "turbojet" using the same diffuser as the preceding ramjet; but superimposed on that initial compression is a mechanical compression of high efficiency which multiplies by 5.5 the preceding value of the ratio $\rho_b a_b/\rho_\infty a_\infty$ for the same fixed value of M_b. Furthermore, whereas the combustor occupies all the maximum cross section of the ramjet, it occupies only 60 per cent ($\sigma_b = 0.6$) in the case of the turbojet. Finally, the latter engine operates with $\delta = 55$ and $\eta_0 = 0.25$.

Eq. 14-2 then gives, for the ratio of the frontal magnitude of thrust of the turbojet to the ramjet,

$$\frac{\Pi_{TJ}}{\Pi_{RJ}} = 5.5 \times 0.6 \times \frac{0.25}{0.10} \times \frac{30}{55} = 4.5$$

Therefore, in the example considered where the maximum cross-sectional areas are equal, the turbojet furnishes a thrust equal to 4.5 times that of the ramjet.

B,15. Cited References and Bibliography.

Cited References

1. Lorin, R. Étude sur la propulsion des avions à grande vitesse. *Aerophile, Paris,* 82 (1908).
2. Lorin, R. Une expérience simple relative au propulseur à réaction directe. *Aerophile, Paris,* 1913.
3. Zwicky, F. Morphology of aerial propulsion. *Helv. Phys. Acta. 21,* 299 (1948).
4. Roy, M. *Thermodynamique des systèmes propulsifs à réaction,* Vol. 1. Dunod, Paris, 1947.

Bibliography

Buckingham, E. Jet propulsion for airplanes. *NACA Rept. 159,* 1923.

Cleveland Laboratory Staff. Performance and range of application of various types of aircraft propulsion systems. *NACA Tech. Note 1349,* 1947.

Crocco, G. A. Corpe aerodinamici a resistenza negativa. *Rend. accad. nazl. Lincei,* Rome, June 1931.

Jaumotte, A. Note sur les lois de la propulsion par réaction. *Rev. Universelle mines,* Brussels, 1950.

Katz, I. *Principles of Aircraft Propulsion Machinery,* Vol. 1. Pitman, New York, 1949.

Keenan, J. H., and Kaye, J. Survey of the calculated efficiencies of jet power plants. *J. Aeronaut. Sci. 14,* 437–450 (1947).

Reissner, H. Systematic analysis of thermal turbojet propulsion. *J. Aeronaut. Sci. 14,* 197–209 (1947).

Rinin, N. A. Propulsion a reazione senze utilizzazione delle 'aria esterna. *Volta High Speed Conference,* Rome, 1935.

Roy, M. La propulsion à jet avec utilisation de l'air extérieur. *Volta High Speed Conference,* Rome, 1935.

Roy, M. Recherches théoriques sur les systèmes motorpropulseurs à réaction. *Publs. sci. et tech. Ministère air France,* 1929.

Roy, M., Duban, P., and George, G. Turboréacteur à double flux. *Office natl. Etudes et Recherches aéronaut. 17,* 1948.

Roy, M., and George, G. Diagrammes universels pour l'étude du rendement global des réacteurs à simple flux. *Office natl. Etudes et Recherches aéronaut. 18,* 1948.
Silverstein, A. Research on aircraft propulsion systems. *J. Aeronaut. Sci. 16,* 197–222 (1949).
Steckin, B. S. Théorie du propulseur à réaction. *Tech. flotte aerienne russe,* Moscow, 1924.
Stemmer, J. *Le développement des fusées et des moteurs à réaction,* Vol. 3. Hoffmann, Zurich, 1944.
Zucrow, M. J. *Principles of Jet Propulsion and Gas Turbines.* Wiley, 1948.

SECTION C

THE TURBOJET ENGINE

C. A. MEYER

> "*We have long sought, and are seeking today, to ascertain whether there are in existence agents preferable to the vapor of water for developing the motive power of heat; whether atmospheric air, for example, would not present in this respect great advantage.*" —SADI CARNOT, 1824

In this section the methods used in the design and estimation of the performance of the turbojet engine are discussed in detail. Familiarity with the material in Vol. X and XI is assumed since a knowledge of this material is essential in order to obtain an understanding of the design and operation of the turbojet engine. This is so because it is convenient to consider the turbojet engine to be a gas turbine, the useful output of which is in the form of a high velocity gas jet rather than in the form of turbine shaft power. In this way we can apply the knowledge obtained in the study of the gas turbine to the study of the turbojet. The turbojet can also be considered to be a propelling device since it increases the velocity of the air it takes in so as to generate a reactive propelling thrust at its attachment to the aircraft. In this respect the turbojet is completely different from the stationary or land-based gas turbine which develops shaft power.

Although the gas turbine and the turbojet are similar in many ways, they also have many differences. For instance, the turbojet combustor must operate at the low densities encountered at high altitudes and, further, the turbojet must operate at the high blade loadings and stresses due to the high flight speeds. Since the turbojet engine must also fly, it is obvious that both its weight and size (frontal area) are more important than in the land-based gas turbine. Other requirements of the turbojet engine such as the short acceleration times required for carrier "wave off" operation, the electrical power requirements for operation of radar and wheels and flaps during landing, the compressor bleeding for cabin supercharging, and the large acceleration loads and inverted operation required of its bearings during flight maneuvers all contribute their share to the complication of the design and operation of the turbojet engine. It is also a fact that the selection of the design parameters for a turbojet engine is closely dependent on the airplane application. An interceptor airplane requires a powerful lightweight but somewhat less efficient engine, while a long range aircraft requires a more efficient and therefore possibly heavier engine.

Although the science of metallurgy and the methods of design and of fabrication of the turbojet engine are important in the determination of engine weight, no attempt is made to discuss these items since they are not considered to be within the scope of these volumes.

CHAPTER 1. BASIC TYPES AND SIMPLE CYCLE ANALYSIS

C,1. Description of Types. The turbojet engines shown in Fig. C,1 serve to illustrate the different types of components that have been used by the various manufacturers.

Engines 1, 3, 5, 6, 7, 8, 9, and 10 in Fig. C,1 have axial flow compressors C of various numbers of stages while engines 2 and 4 have double-flow and single-flow centrifugal compressors respectively.

The first British jet flight was powered with the W-1X engine, shown in Plate C,1a, which is of engine type 4 shown in Fig. C,1. Other centrifugal compressor engines are shown in Plate C,1b, and C,1c. Several pictures of axial flow compressor engines are shown in Plate C,1d, C,1e, C,1f, C,1g, C,1h, C,1i, C,1j, C,1k, C,1l, and C,1m.

Axial compressors allow higher efficiencies and lower frontal area at some sacrifice to weight, length, and operating stability [*1*].

Double-flow centrifugal compressors allow double the air flow and thus double the engine thrust for the same compressor diameter while introducing problems in supplying air to the inside runner.

Engines 1, 7, 8, and 10 in Fig. C,1 have a double annular combustion chamber (or burner) B, while engines 2, 3, 4, 5, and 6 each have several "can-type" combustion chambers. The annular combustion chamber is compact yet more difficult to develop than the multiple-can type where individual cans may be perfected separately.

Engines 1, 3, 7, 8, and 10 have controllable area exhaust nozzles while engines 2, 4, 5, 6, and 9 have fixed area exhaust nozzles N. The areas of the nozzles shown on engines 1, 8, and 10 are varied by positioning the center cone while the nozzles of engines 3 and 7 are composed of two "clam shells" or "eyelids" which are moved to vary the nozzle area. The use of a variable nozzle allows better part-load efficiency together with a shorter time to accelerate from idling to full thrust.

Engine 1 has a two-stage turbine while engines 2, 3, and 4 have single-stage turbines. The two-stage turbine allows a higher efficiency for the same or smaller diameter with some loss in allowable top temperature.

Engine 5 is called a ducted fan thrust augmenter [*2,3,4*]. In this engine

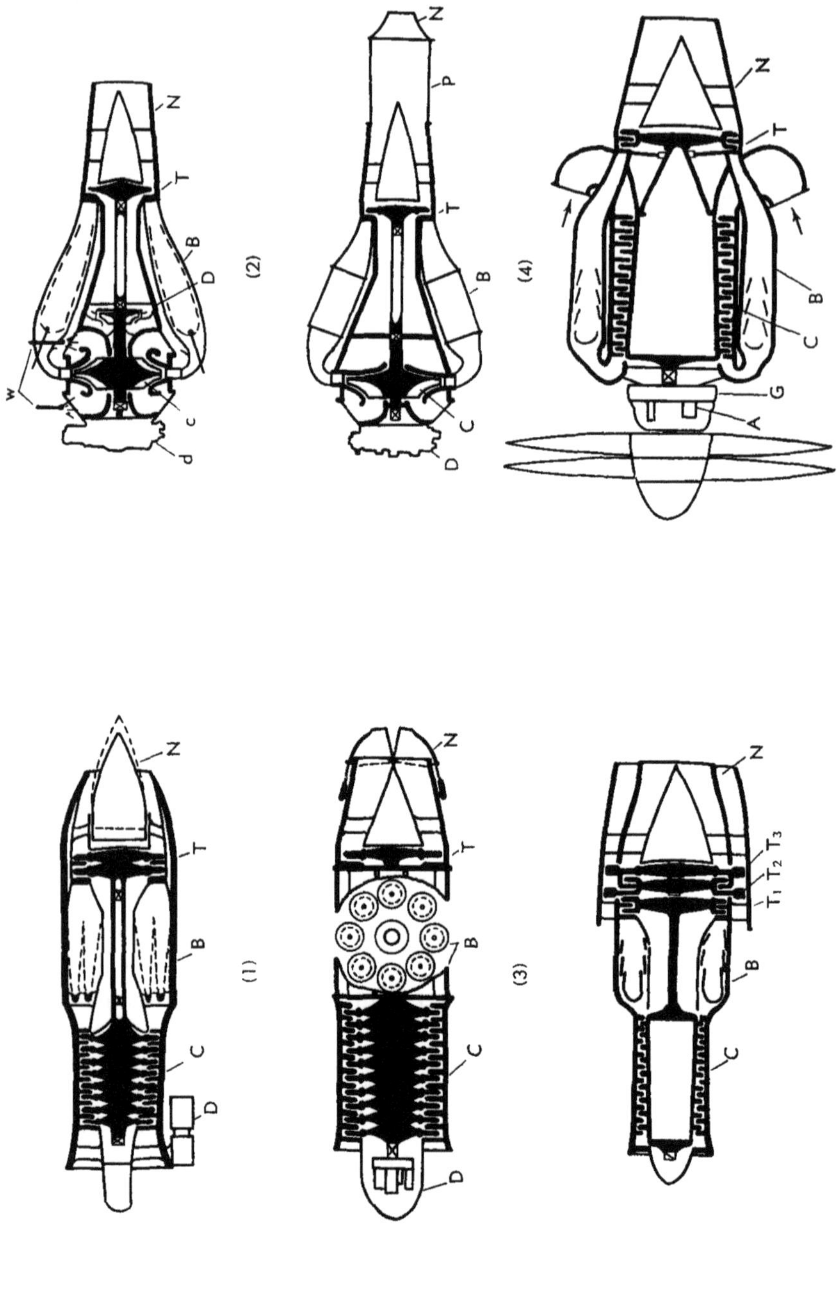
(1)
N
T
B
C
D
(2)
N
T
B
D
w
c
d
(3)
N
T
B
C
D
(4)
N
P
T
B
C
D
(5)
N
T_1 T_2 T_3
B
C
(6)
N
T
B
C
G
A

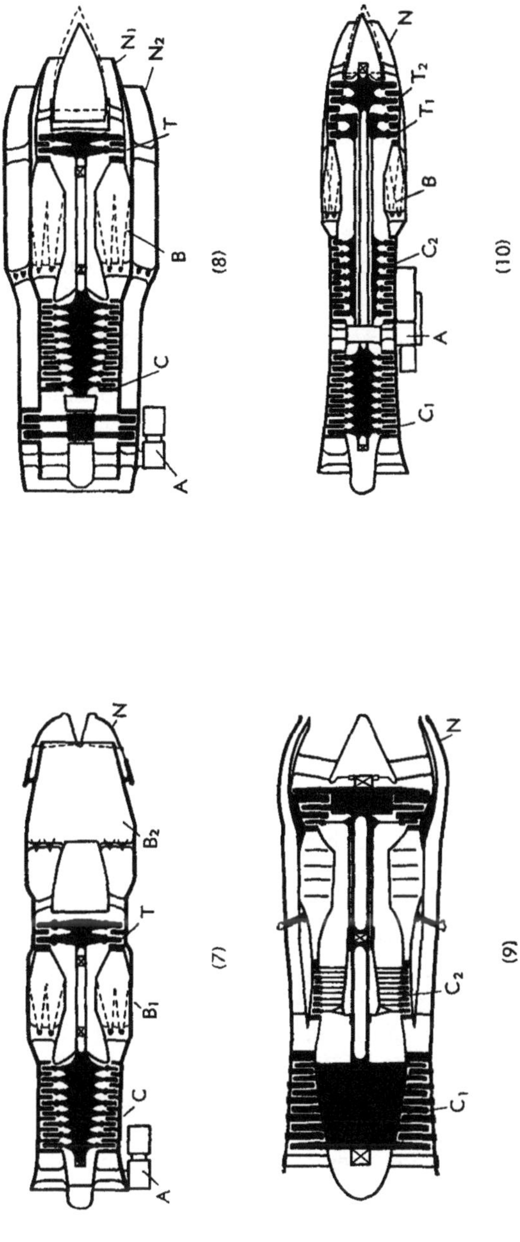

Fig. C,1. Engine types. (1) Turbojet, with axial compressor, two stage turbine annular combustion chamber, and center cone variable area nozzle. (2) Turbojet, with double entry centrifugal compressor, single stage turbine, can type combustion chamber, and a fixed area nozzle. (3) Turbojet, with axial compressor, single stage turbine can type combustion chamber, and clam shell variable area nozzle. (4) Turbojet, with single entry centrifugal compressor, single stage turbine, can combustion chamber, and a fixed nozzle area. (5) Ducted fan; axial compressor, three stage turbine. (6) Turboprop, with reverse flow axial compressor, can type combustion chamber and single turbine. (7) Turbojet with afterburner, axial flow compressor, two stage turbine annular combustion chamber and clam type variable area nozzle. (8) Bypass engine axial flow compressor, two stage turbine can type combustion chamber and central cone variable area nozzle. (9) Bypass engine axial flow compressor, two stage turbine annular combustion chamber. (10) Two spool turbojet axial flow compressor two stage turbines, annular combustion chamber central cone variable area nozzle.

two shrouded double-rotation propellers are driven directly by the turbine blade tips (avoiding a gear) in order to distribute the gas turbine output to more air, thus improving the wake or propulsive efficiency, and also to augment the take-off thrust (Sec. D).

Engine 6 shown in Fig. C,1 is a turboprop engine (Sec. D) which accomplishes the same or better improvement in wake efficiency and thrust as the thrust augmenter (engine 5) by using a gear and variable pitch double-rotation propeller. This engine has fewer mechanical problems than the thrust augmenter and has the additional flexibility of the variable pitch double-rotation propeller.

Engine 7 shown in Fig. C,1 is a turbojet engine with afterburning. The afterburner accomplishes somewhat the same results as a reheater in the gas turbine, i.e. it increases the output at the expense of considerable increase in fuel flow particularly when a regenerator is not used.

Engines 8 and 9 shown in Fig. C,1 are called ducted fan or bypass engines [*2,4*] and are quite similar to the turboprop except that a ducted fan is employed instead of a conventional propeller. The ducted fan has better high speed characteristics than a propeller, due to the fact that the air can be slowed down in the duct before the fan and the Mach number relative to the fan blades can be reduced, which is not possible with the conventional propeller. It is also possible to burn in the duct after the fan and thus obtain considerable augmentation in thrust at the expense of an increase in fuel flow. The Rolls-Royce Conway (Plate C,1j), which is of type 9 shown in Fig. C,1, avoids the gear through the use of a concentric shaft drive turbine [*5*].

Engine 10 of Fig. C,1 is a "double-spool" turbojet engine. It has a compressor which is divided in two, each portion of which is driven through concentric shafting by its own turbine. This arrangement is very helpful because it allows better matching of the front and back portions of the compressor since their respective rotational speeds can be varied independently. Engines designed for pressure ratios greater than 8 or 9 encounter compressor stalling difficulties in the front stages during starting as a result of the choking of the small flow areas at the rear of the compressor. The use of double-spool rotors alleviates this difficulty by allowing the air flow capacity of the rear half of the compressor to be augmented by increasing its speed relative to the front half of the compressor. The use of a double-spool compressor avoids the compromising of the design point performance in order to permit engine starting without stalling. The British Olympus Engine (Plate C,1l) is of this type.

Other devices have been used for correction of this starting difficulty, such as bleeding air from the middle stages of the compressor or varying the angle-setting of the compressor inlet guide vanes alone or in combination with angle changes of one or more of the front rows of stator vanes. The varying of the turbine nozzle vane settings has also proved useful in avoiding compressor blade stalling during the "off design" starting condition. Since the use of poorly designed variable angle vanes has been found to introduce large leakage losses, great care should be taken in designing them.

An extensive list of current turbojet engines together with their specifications is given in [6].

C,2. Thrust. In many applications of the turbojet engine it is important to accurately define thrust[1] so as to differentiate thrust forces

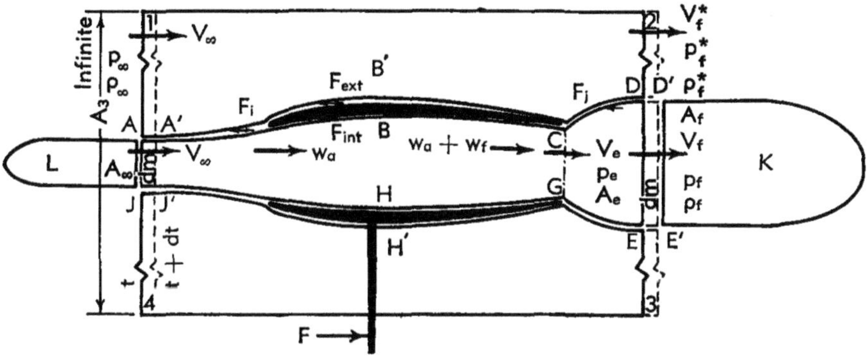

Fig. C,2a. Turbojet thrust and drag.

from nacelle drag forces. Also, a precise definition of thrust is helpful in cases where the static pressure is different after the engine than before the engine.

The thrust developed by a turbojet engine is most conveniently calculated by the application of Newton's laws. The following statement has been derived by applying Newton's laws to a group of bodies or particles [11].

"The resultant of forces (F) acting on the boundaries of any region is equal in magnitude and direction to the summation of the time rate of change of momentum of the particles or bodies within the region."

If one applies this law to the air passing through and around the turbojet engine, the net forces on the surfaces of the engine, and thus the thrust, can be calculated.

The above law is applied to a turbojet engine (see Fig. C,2a) where the region is assumed to surround the air which passes through the engine. The solid lines represent the region at time t while the dotted lines represent the region a short time later at time $t + dt$.

The momentum within the region is increased during the time dt by an amount $(dm/dt)V_f$ and is decreased by an amount $(dm/dt)V_\infty$, where dm is the mass flow during the time dt and V_f and V_∞ are velocities

[1] See Sec. B and [7,8,9,10] for further discussions on thrust definition.

as shown in Fig. C,2a. Notice that the momentum within the region $A'BCDEGHJ'$ remains the same during the time dt.

We then write the momentum equation (including the change in momentum of the fuel):

$$p_\infty A_\infty + F_i + F_{int} + F_j - p_f A_f = \frac{(w_a + w_f)}{g} V_f - \frac{w_a}{g} V_\infty \quad (2\text{-}1)$$

where F_i, F_{int}, and F_j represent the sum of the components of the forces in the flight direction on the indicated surfaces.

The drag B of a solid body shaped like $AB'DEH'J'$ and protected by cylindrical half bodies L and K which have zero drag [*11*, p. 118] can be obtained either as the sum of the forces on its surfaces[2] or by writing the momentum equation for the region 1 to 4 and 2 to 3 of Fig. C,2a containing the solid body.

$$p_\infty A_\infty + F_i + F_{ext} + F_j - p_f A_f = \int^{A_2} (\rho_\infty V_\infty^2 + p_\infty) dA - \int^{A_3} (\rho_f V_f^2 + p_f) dA \equiv B \quad (2\text{-}2)$$

For isentropic (adiabatic frictionless) flow of the air outside the engine, the integrals in Eq. 2-2 are zero since p_f is assumed equal to p_∞ and therefore $V_f = V_\infty$ and $\rho_f = \rho_\infty$.

Subtracting Eq. 2-2 from 2-1 one obtains the thrust force F.

$$F = F_{int} - F_{ext} = \frac{(w_a + w_f)}{g} V_f - \frac{w_a}{g} V_\infty - B \quad (2\text{-}3)$$

Since the dividing line between the engine thrust and airplane drag is a matter of definition, one could define B as nacelle (or airplane) drag and then the engine thrust would be the sum of the first two terms on the right-hand side of Eq. 2-3. The airplane drag would then include (Eq. 2-2) the forces F_i and F_j on the stream tubes at entrance and exit from the engine together with the forces on the end areas $p_\infty A_\infty$ and $p_f A_f$.

Where the stream tubes are in contact with the airplane the forces on the common surface can be neglected, since the forces are equal and opposite on the stream tube and on the airplane (Fig. C,2b).

There is a disadvantage in the above arbitrary division between airplane drag and engine thrust due to the practical difficulty which the airplane designer encounters in determining $F_j - p_f A_f$ which depends on the characteristics of the expanding (or contracting) jet. One can avoid this difficulty by first writing the momentum equation for the jet:

$$p_e A_e + F_j - p_f A_f = \left(\frac{w_a + w_f}{g}\right)(V_f - V_e) \quad (2\text{-}4)$$

Combine this equation with Eq. 2-3, eliminating the disturbing terms

[2] Notice that the forces on the end areas A_∞ and A_f are included.

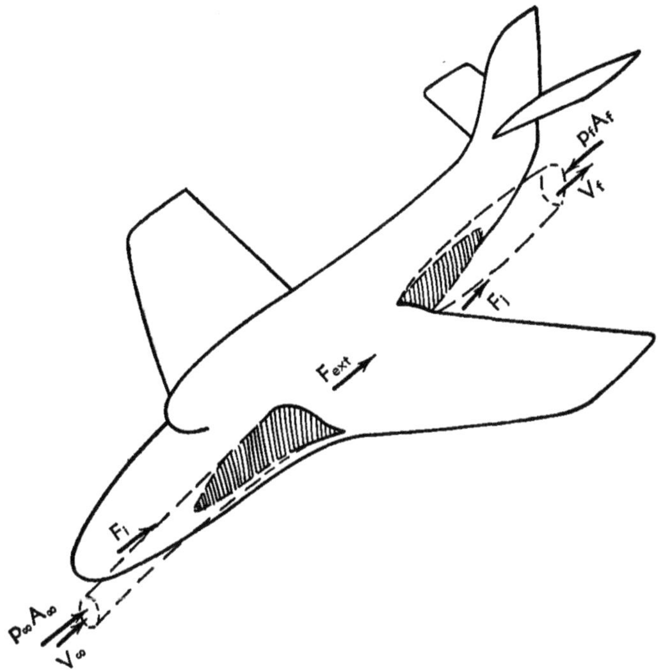

Fig. C,2b. Stream tube in contact with airplane.

$F_i - p_f A_f$; by adding and subtracting $p_f A_e$ we obtain

$$F = F_{int} - F_{ext} = \frac{(w_a + w_f)}{g} V_e - \frac{w_a V_\infty}{g} + (p_e - p_f)A_e - [p_\infty A_\infty + F_i + F_{ext} - p_f A_e] \qquad (2\text{-}5)$$

$$= (F_{net} - D)$$

If the drag is defined as the bracketed term in Eq. 2-5

$$D = [p_\infty A_\infty + F_i + F_{ext} - p_f A_e] \qquad (2\text{-}6)$$

The engine net thrust F_{net} is thereby defined as the sum of the first three terms on the right-hand side of Eq. 2-5.

$$F_{net} = \frac{(w_a + w_f)}{g} V_e - \frac{w_a}{g} V_\infty + (p_e - p_f)A_e \qquad (2\text{-}7)$$

Some interesting observations can be made regarding these arbitrary definitions of the engine net thrust equation (Eq. 2-7) and the airplane (or nacelle) drag equation (Eq. 2-6).

1. The drag as defined in Eq. 2-6 can be calculated independently of the jet expansion or contraction by the airplane designer.
2. The drag as defined is not zero for isentropic flow over the nacelle (see (4) below). However, the correct airplane force $F = (F_{int} - F_{ext})$

is obtained using Eq. 2-5 since no assumptions were made in deriving this equation.

3. The definition of net thrust (Eq. 2-7) assumes, according to Eq. 2-4, that the jet is either cylindrical ($F_j = 0,\ A_e = A_f$) or that atmospheric pressure surrounds the jet $F_j = p_f(A_f - A_e)$. Of course neither assumption is generally true.
4. The engine net thrust as defined by Eq. 2-6 is less than $[(w_a + w_f)/g]V_f - w_aV_\infty/g$, by an amount equal to $(F_j - p_fA_f + p_fA_e)$, while the drag as defined by Eq. 2-6 is less than B (Eq. 2-2) by the same amount.
5. Where the *exit* stream tube is in contact with the airplane, the forces on the airplane and the tube no longer cancel but must be included in the airplane drag. The forces should be considered as a part of F_{ext} (Eq. 2-6).

Summarizing, we have:

The engine gross thrust is given by

$$F_g = \frac{(w_a + w_f)}{g} V_e + (p_e - p_f)A_e \qquad (2\text{-}8)$$

The engine ram drag is given by

$$F_r = \frac{w_aV_\infty}{g} \qquad (2\text{-}9)$$

The engine net thrust is given by

$$F_{\text{net}} = F_g - F_r = \frac{(w_a + w_f)}{g} V_e + \frac{w_aV_\infty}{g} + (p_e - p_f)A_e \qquad (2\text{-}10)$$

The airplane (or nacelle) drag is given by

$$D = (p_\infty A_\infty + F_i + F_{\text{ext}} - p_fA_e) \qquad (2\text{-}11)$$

The resulting thrust on the airplane is given by

$$F = F_{\text{net}} - D \qquad (2\text{-}5)$$

In what follows we shall assume that the net thrust is given by Eq. 2-12, that there is no nacelle drag, and that the static pressure at the exit from the jet nozzle is the same as before the engine.

It has frequently happened that the measured thrust obtained during tests of turbojet engines in test chambers has failed to check those run in open test cells. The reason for the failure to check is often due to the fact that a small difference in pressure exists across the engine. This pressure difference should be accounted for as shown in the above equations.

It has often been suggested that the use of a large well-rounded, flow-measuring nozzle at the inlet to an engine results in too high a thrust measurement because of the favorable difference in pressure across the nozzle, but a careful application of the momentum law around the inlet duct shows this to be untrue. The flow nozzle actually causes a small loss in thrust due to its rather small fluid dynamic losses.

C,3. Definition of Terms.

Thermal efficiency. The thermal efficiency of a heat engine is generally defined as the ratio of the work output to the heat supplied [*12*].

$$\eta_{\mathrm{th}} = \frac{(V_j^2 - V_\infty^2)\dfrac{w_a}{2g}}{Q_l w_f} \equiv \frac{P_{\mathrm{th}}}{Q_l w_f} \equiv \frac{P_{\mathrm{th}}}{q_{\mathrm{add}}} \tag{3-1}$$

where Q_l is the lower heating value of the fuel.[3]

Wake or propulsive efficiency. The wake or propulsive efficiency is defined as the ratio of thrust power to net jet power [*12,13*]

$$\eta_{\mathrm{pr}} = \frac{F_{\mathrm{net}} V_\infty}{\dfrac{w_a}{2g}(V_j^2 - V_\infty^2)} = \frac{2}{1 + \dfrac{V_j}{V_\infty}} = \frac{1}{1 + \left(\dfrac{F_{\mathrm{net}}}{w_a}\right)\dfrac{g}{2V_\infty}} \tag{3-2}$$

Over-all efficiency. The over-all efficiency is defined as the ratio of thrust power to the fuel power [*2,14*]:

$$\eta_0 = \eta_{\mathrm{th}}\eta_{\mathrm{pr}} = \frac{F_{\mathrm{net}} V_\infty}{Q_l w_f} \tag{3-3}$$

Thrust specific fuel consumption. The thrust specific fuel consumption is the ratio of the fuel flow to the net thrust.

$$\frac{w_f}{F_{\mathrm{net}}} = \mathrm{tsfc} \tag{3-4}$$

This definition differs from the definition for shaft specific fuel consumption ssfc used for gas turbine or turboprop engines which is defined as the ratio of the fuel consumption to the shaft power. Notice Eq. 3-3 which shows that, for a given flight speed V_∞, the over-all efficiency is inversely proportional to the thrust specific fuel consumption. The thrust specific fuel consumption is particularly useful as a measure of perfection at standstill where the over-all efficiency is zero by definition.

Miles per pound. The miles per pound is defined as the ratio of the flight speed to the fuel flow

$$\frac{V_\infty}{w_f} = \mathrm{mpp} \tag{3-5}$$

The over-all efficiency is for a given net thrust F_{net} proportionate to the miles per pound.

[3] In the interest of simplicity, kinetic energy of the fuel has been omitted.

$$\eta_{th} = \eta_b \left\{ \frac{AE - B\left[1 + \left(1 + \frac{\Delta T_{ic}}{T_L}\right)(N-1)\right]}{AE + \delta\left[\frac{T_H}{T_L} - A - \left(1 + \frac{\Delta T_{ic}}{T_L}\right)(1+B)\right]} \right\} = \text{thermal efficiency}$$

$$A = \eta_t \frac{T_H}{T_L}\left[1 - \left\{\left(\frac{p_L}{p_H}\right)_c \left(\frac{p_H}{p_L}\right)_b \left(\frac{p_1 p_3}{p_2 p_4}\right)_{ex} \left(\frac{p_H}{p_L}\right)_r^{E-1}\right\}^{\frac{\gamma-1}{E\gamma}}\right]$$

$$B = \frac{1}{\eta_c}\left[\left\{\left(\frac{p_H}{p_L}\right)_c \left(\frac{p_H}{p_L}\right)_{ic}^{N-1}\right\}^{\frac{\gamma-1}{N\gamma}} - 1\right]$$

$$\delta = \frac{\text{temp. diff. in heat exchanger}}{\text{temp. range in heat exchanger}} = \frac{T_1 - T_4}{T_1 - T_3} = \frac{T_2 - T_3}{T_1 - T_3}$$

T_H = high temperature; T_L = inlet air low temperature

η_t = turbine efficiency; η_c = compressor efficiency

$\left(\frac{p_1 p_3}{p_2 p_4}\right)_{ex}$ = product of pressure ratios in heat exchanger

c_p = specific heat at constant pressure

$\gamma = c_p/c_v$ = ratio of specific heats; η_b = burner efficiency

E = no. of expansions; N = no. of compressions

$\left(\frac{p_H}{p_L}\right)_c$ = over-all pressure ratio across compressor(s)

$\left(\frac{p_H}{p_L}\right)_b$ = pressure ratio across burner

$\left(\frac{p_H}{p_L}\right)_{ic}$ = pressure ratio across intercooler(s)

$\left(\frac{p_H}{p_L}\right)_r$ = pressure ratio across reheater(s)

ΔT_{ic} = terminal temperature difference in intercooler(s)

Other Useful Data

$$q_{add} = \text{heat added} = c_p T_L \left\{AE + \delta\left[\frac{T_H}{T_L} - A - \left(1 + \frac{\Delta T_{ic}}{T_L}\right)(1+B)\right]\right\}$$

$$q_w = \text{heat to work} = c_p T_L \left\{AE - B\left[1 + \left(1 + \frac{\Delta T_{ic}}{T_L}\right)(N-1)\right]\right\}$$

Turbine temperature drop $T_H - T_L = T_L A$

Temperature after all but first stage of compression $T_3 = T_L(1+B)\left(1 + \frac{\Delta T_{ic}}{T_L}\right)$

Temperature after first stage of compression $T_{II} = T_L(1+B)$

Pressure ratio of each compression $= \left[\left(\frac{p_H}{p_L}\right)_c \left(\frac{p_H}{p_L}\right)_{ic}^{N-1}\right]^{1/N}$

Pressure ratio of each expansion $\left[\left(\frac{p_L}{p_H}\right)_c \left(\frac{p_H}{p_L}\right)_b \left(\frac{p_1 p_3}{p_2 p_4}\right)_{ex} \left(\frac{p_H}{p_L}\right)_r^{E-1}\right]^{1/E}$

Examples

Regeneration

(No reheat or intercool)

$N = 1$; $E = 1$; $\eta_t = 0.85$; $\eta_c = 0.80$; $\eta_b = 0.95$; $\delta = (T_1 - T_4)/(T_1 - T_3) = 0.10$

$\left(\frac{p_H}{p_L}\right)_b = 1.02$

$\left(\frac{p_1 p_3}{p_2 p_4}\right)_{ex} = 1.07$

$\left(\frac{p_H}{p_L}\right)_{ic} = 1.00$ (no intercooler)

$\left(\frac{p_H}{p_L}\right)_r = 1.00$ (no reheat)

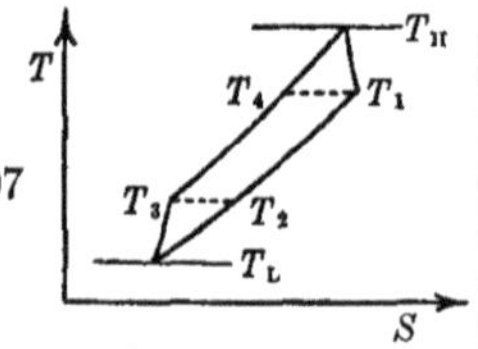

No regeneration

(1 reheat, 2 intercools)

$N = 3$; $E = 3$; $\eta_t = 0.85$ $\eta_c = 0.80$; $\eta_b = 0.95$; $\delta = 1.00$ (no regeneration)

$\left(\frac{p_H}{p_L}\right)_r = 1.03$

$\left(\frac{p_H}{p_L}\right)_{ic} = 1.02$

$\frac{\Delta T_{ic}}{T_L} = \frac{15°}{460 + 70°} = 0.0283$

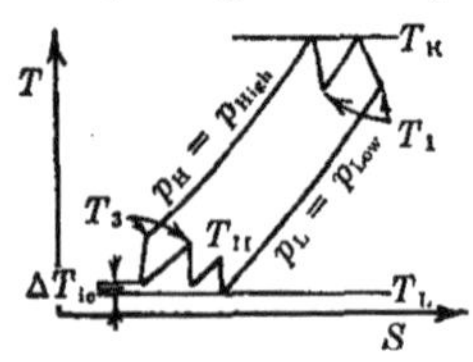

Fig. C,4a. Thermal efficiency of open gas turbine cycles.

C,4. Simple Cycle Calculation. The simplest kind of cycle calculation can be made by assuming constant air and gas thermodynamic properties. Thus, if we assume constant specific heats and constant gas constant, we can arrive at a simple expression for the efficiency and output (per pound of air) of the gas turbine cycle. In fact the author had occasion several years ago to develop such an expression for the thermal efficiency of a simple gas turbine cycle. Later, expressions giving

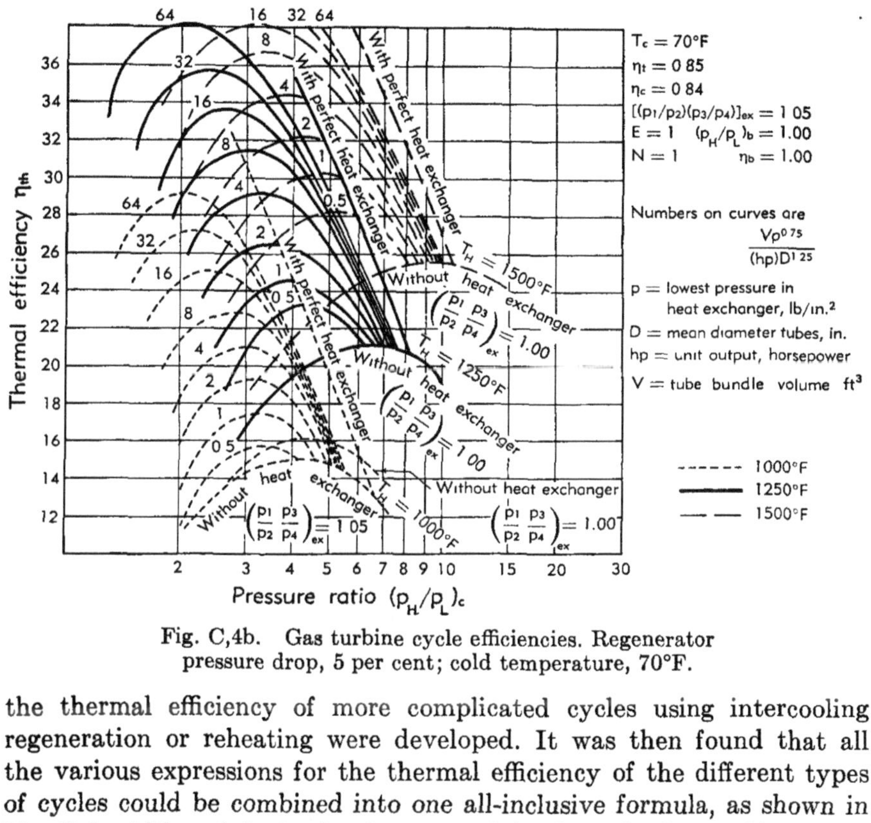

Fig. C,4b. Gas turbine cycle efficiencies. Regenerator pressure drop, 5 per cent; cold temperature, 70°F.

the thermal efficiency of more complicated cycles using intercooling regeneration or reheating were developed. It was then found that all the various expressions for the thermal efficiency of the different types of cycles could be combined into one all-inclusive formula, as shown in Fig. C,4a. Although in the development of this formula the specific heat of the combustion products was assumed equal to the specific heat of air and the additional mass of the fuel flow through the turbine was also neglected, the formula has been found to give sufficiently accurate results for most purposes. It is most accurate for indicating the effect, for instance, of changes of compressor or turbine efficiency or burner pressure drop on thermal efficiency.

Cycle calculations numbering into the hundreds have been carried out, covering a range of the variables contained in the formula of Fig. C,4a. For our present purposes the results shown in Fig. C,4b, C,4c, and C,4d are of interest. One can immediately see the effect of pressure

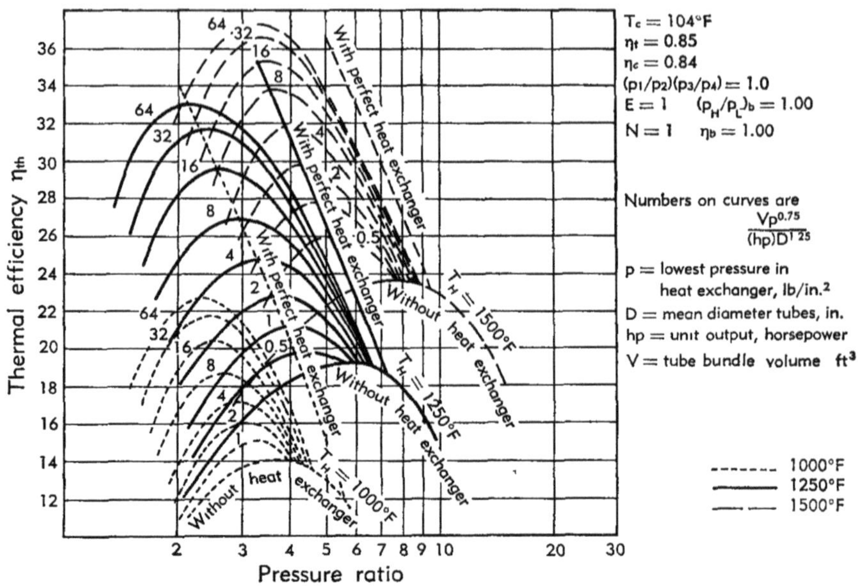

Fig. C,4c. Gas turbine cycle efficiencies. Regenerator pressure drop, 7 per cent; cold temperature, 104°F.

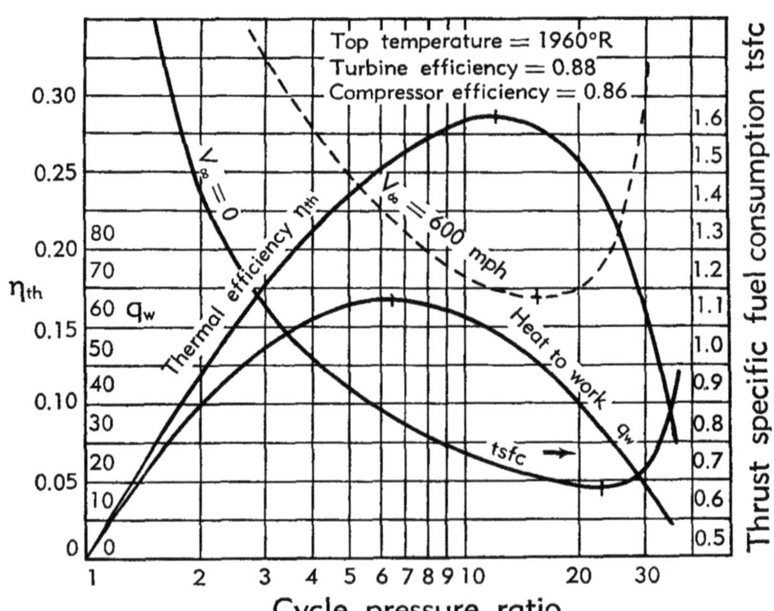

Fig. C,4d. Optimum cycle pressure ratios for thermal efficiency work output and specific fuel consumption.

ratio, top temperature, regenerator size, and ambient temperature on the cycle thermal efficiency. Curves of a similar type are also shown in [*12*, p. 439; *13*; *14*]. Although not of particular importance in our present discussion it is interesting to note that the parameters shown in Fig. C,4b and C,4c, giving regenerator volume per horsepower, have been calculated from the δ values by assuming a counterflow regenerator and optomizing the velocities on each side of the regenerator according to Keller [*14*]. Next, by means of Reynolds analogy [*15*, p. 158] between heat transfer and pressure loss, the volume of the regenerator tube bundle per unit output was minimized to give the values on the curves of Fig. C,4b and C,4c.

Of considerable interest is Fig. C,4d which shows for the simple cycle the values of work output in BTU per lb of air and also shows the values of thermal efficiency and thrust specific fuel consumption for standstill. Also shown is the thrust specific fuel consumption for a flight speed of 600 miles per hour.

A careful examination of Fig. C,4b, C,4c, and C,4d leads to the following conclusions:

1. There exists an optimum cycle pressure ratio which depends among other things on the type of cycle (regenerating or simple cycle), the component efficiencies, the top temperature, and what we desired to be optimum, i.e. the thermal efficiency η_{th}, the work output q_w, or the thrust specific fuel consumption. (See Fig. C,4d.)

   Since the cycle pressure ratio is equal to the product of the ram pressure ratio and the compressor pressure ratio, we should for the best desired result change the compressor ratio with flight conditions. Obviously this can only be approached in practice.
2. At sufficiently high cycle pressure ratios the thermal efficiency and cycle output fall to zero. This pressure ratio is higher for higher component efficiencies and for higher top temperatures. The turbojet engine at very high flight speeds can therefore experience excessively high cycle pressure ratios where the cycle output becomes small or zero. Further, at certain high speed flight conditions the compressor outlet temperature can equal the maximum allowable turbine inlet temperature, in which case it is impossible to burn any fuel without exceeding the allowable top temperature. The ramjet engine is more desirable under these flight conditions, since its compression is zero and the hot gases do not have to pass through a turbine and thus can be of higher temperatures.

   The comparative position of the turbojet engine with other propulsion means is shown in Fig. C,4e, where the role played by the cycle fuel efficiency and the propulsive efficiency in determining the over-all efficiency is clearly shown for the various propulsion means.

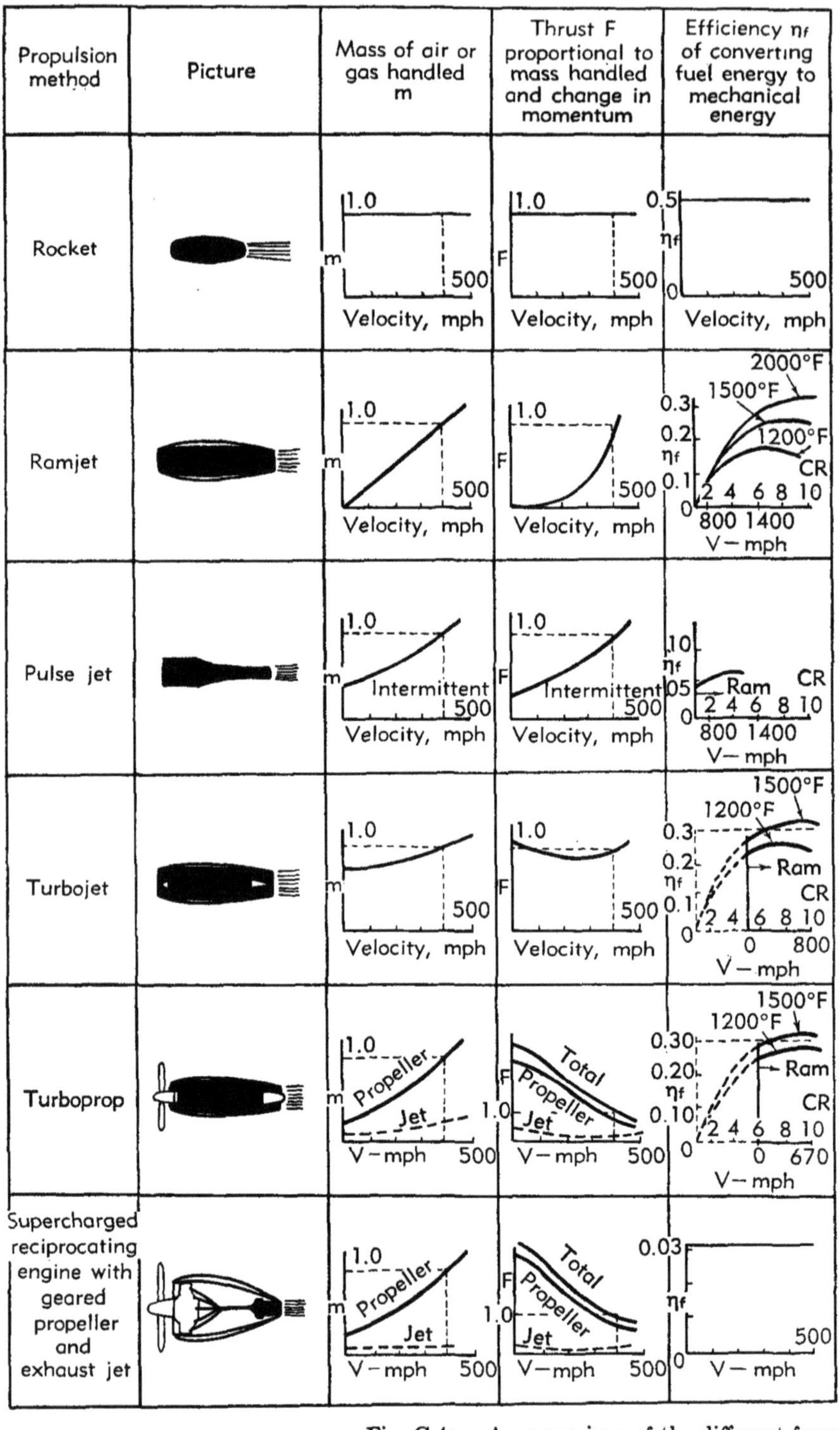

Fig. C,4e. A comparison of the different forms

Propulsive efficiency η_{pr}	Over-all efficiency	Relative frontal area (drag)	Relative weight of fuel for a given duration	Probable range of maximum flight speeds
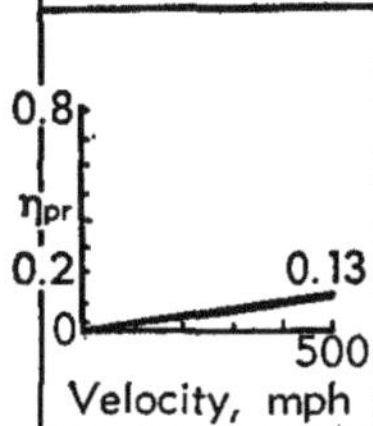	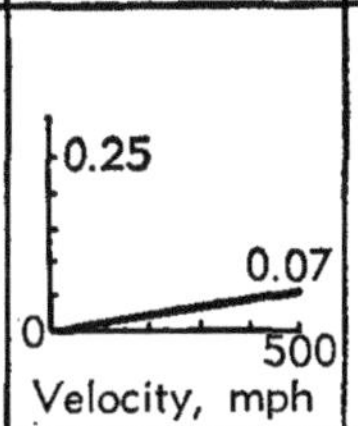			1. Above 600 mph 2. For aid in take-off 3. For flying bombs
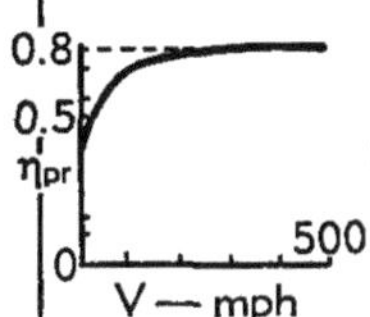	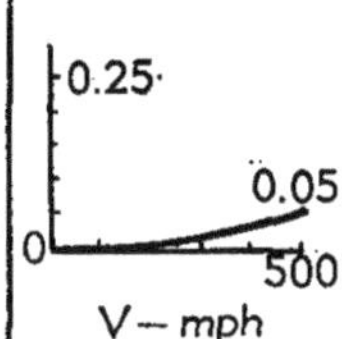			1. Above 500 mph 2. For flying bombs
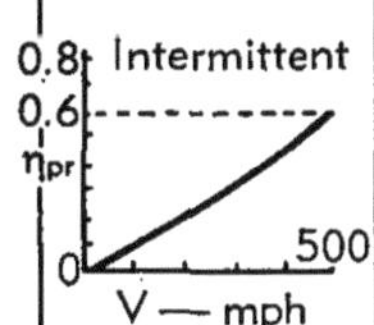	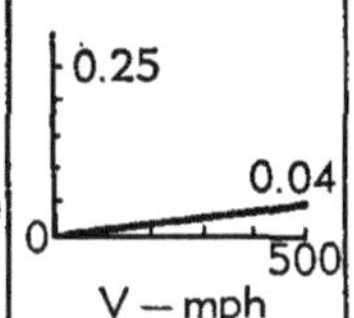			1. 300 to 600 mph 2. For flying bombs
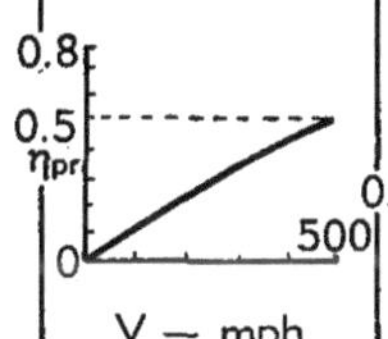	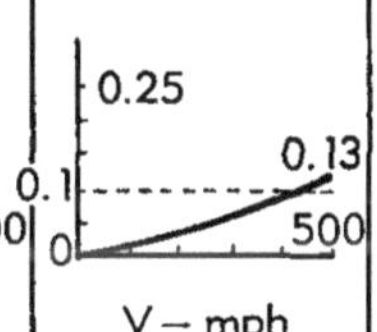			400 to 1500 mph
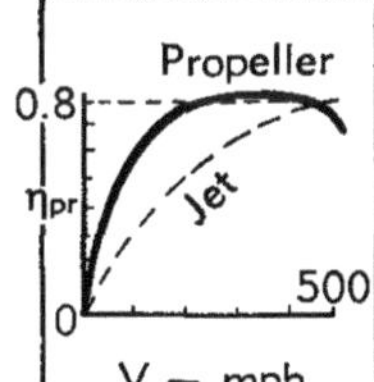	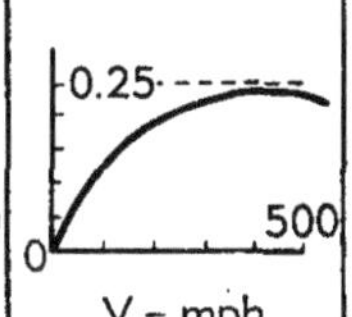			300 to 600 mph
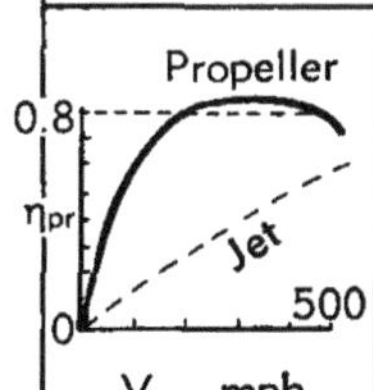	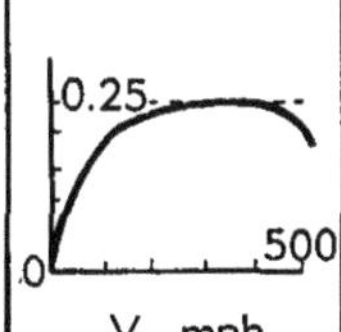			150 to 450 mph

of jet propulsion and reciprocating engines.

CHAPTER 2. ANALYSIS AND MATCHING OF COMPONENTS

C,5. Matching. The simple cycle calculation presented above has other shortcomings beyond the assumption of constant specific heats [*16,17*]. Some of these shortcomings are the following: (1) Single compression and expansion efficiencies are used whereas in fact the compression should be divided into two portions, the duct and the compressor efficiencies, which may not be equal. Likewise the expansion efficiency should be divided into the turbine and the exhaust nozzle efficiencies which also may not be equal. (2) The formula gives the efficiency in terms of the cycle pressure ratio, although in an actual engine the pressure ratio must be determined since it depends on the flight conditions, the flow characteristics of the compressor, and the fluid dynamic flow resistance of the turbine. (3) Nothing is mentioned regarding rotational speed or the amount of air taken in by the engine under varying flight conditions. (4) The component efficiencies must be assumed in using the above formula whereas the efficiencies of the components are actually dependent on their operating conditions, such as rotational speed, altitude flight speed, etc.

To properly account for the above items[4] we must estimate the performance of the individual components such as the inlet ducts, the compressor, the burner, the turbine, and the exhaust nozzle in terms of the variables which determine their performance. It is also necessary to combine the performance of the components for a selected altitude and flight speed in such a way as to satisfy the following requirements:

1. The rotational speeds of the turbine and compressor must be the same.
2. The turbine power must match the power required to drive the compressor, except for the small bearing loss.
3. The air flow passing through the compressor added to the small fuel flow must equal the gas flow through the turbine. Thus the turbine inlet pressure must be sufficient to overcome the fluid dynamic resistance to the passage of gas flow through the turbine.
4. The sum of the pressure rises due to the ram (flight speed) and to the compressor must equal the sum of the pressure drops due to the burner, turbine, and exhaust nozzle.

Once the above requirements have been satisfied, the jet velocity, thrust, and fuel flow are easily calculated.

C,6. Dimensional Analysis. If one is to carry out many cycle calculations satisfying the requirements given in Art. 5 for several altitudes and

[4] Refer to Fig. C,7w.

flight speeds, it is necessary to use dimensional analysis to shorten the work. By this means the number of variables can be reduced to a smaller number of dimensionless variables [*18*], thus considerably reducing the number of calculations.

The mathematical basis for dimensional analysis has been set down by Buckingham [*19,20,21,22,23*] in his "Pi" theorem, and the "principle of similitude" has been presented by Afanassjewa [*24*]. The underlying principle of the dimensional analysis set down by Buckingham and Afanassjewa is that equations which present the relationship between the variables connected with a physical phenomena must (if they are to be general) be independent of any arbitrary choice of the size of the standard units used to measure the variables. Thus the arbitrary selection of the foot or meter as a unit of length or the second as the unit of time should in no way influence an equation intended to generally describe the relationship between the variables connected with a given physical phenomena.

It is the author's view that we can considerably simplify the methods of dimensional analysis by taking care to avoid establishing any arbitrary standard units, which is easily accomplished by using the measurements of certain of the quantities connected with the given phenomena as standard units to measure the values of the remaining quantities. Thus in place of a standard unit of length, such as the foot or the meter, we may use any convenient length (e.g. the wing chord or a compressor diameter) associated with the phenomena we are studying.

We may, if we desire, use a density, a volume, an acoustic velocity, or the period of a vibration as a standard unit to measure the magnitude of other variables connected with the given phenomena.

With this simple, although not the less rigorous, view we can proceed to analyze the performance of the components of the turbojet engine and to combine these component performances to give the over-all performance of the engine in terms of dimensionless variables.

Referred quantities. In 1942 Bollay suggested simplifying the dimensionless quantities in order to make their use more convenient. If it is agreed that we are usually interested in one engine model at a time, we can eliminate the size factor from the dimensionless quantities. Further, we can refer all pressures and temperatures (or acoustic velocities) to their standard sea level static values. For example,

$$\text{the dimensionless flow} = \frac{w_a a_2^0}{g p_2^0 A_d} \tag{6-1}$$

becomes

$$\text{the referred flow} = \frac{w_a \sqrt{\theta_2^0}}{\delta_2^0} \tag{6-2}$$

The referred quantity[5]

1. has been arrived at by assuming that the acoustic velocity varies as the square root of the temperature. This is not too serious since we generally neglect the effect of the variation of γ and $\mathcal{R}$. This could be partially corrected by redefining θ as the ratio of acoustic velocities.
2. has dimension, i.e. the referred flow can be measured in, for instance, pounds per second, whereas the value of the dimensionless flow does not give one an idea of the engine size.
3. does not involve the question of which dimension was used as the characteristic size in the dimensionless quantity, which is the case, for instance, when one uses the Reynolds number.
4. is somewhat less general than the dimensionless number since the size factor has been eliminated.
5. represents the value of the particular variable while under standard static sea level pressure and temperature conditions.

In the following analysis we will use dimensionless quantities in order to be more general, although in practice the referred quantities are more desirable because of their simplicity.

C,7. Component and Engine Analysis Using Dimensionless Variables. During the following analysis[6] one should understand that losses of engine performance associated with, for instance, inlet ducting, exhaust ducting, compressor air bleeding, and power extraction from the auxiliary drive gear box are dependent on the airplane application of the engine. Thus the magnitude of the duct losses or the amount of air to be bled from the compressor or the amount of power to be extracted from the gear box for a given flight condition are not usually known by the engine designer. Also, the above requirements are generally different for the various airplane manufacturers. Therefore the turbojet engine manufacturer usually supplies to the airplane builder the "no-loss" performance of his engine together with means (to be described in detail later) for correcting the engine performance for the added losses introduced due to the airplane installation.

Other losses such as those due to nacelle cooling or due to the use of cooling air ejectors surrounding the jet nozzle are less susceptible of analysis and therefore require close cooperation between the engine and airplane designer. This is particularly true where a special jet nozzle (for instance one of a certain expansion ratio) is required by the particular airplane manufacturer.

[5] Previously called "corrected quantities."

[6] It has been suggested to use the expression "referred quantities" in place of the previously used "corrected quantities" in order to avoid the commonly used expression *corrected corrected quantities* when speaking, for instance, of quantities corrected for duct loss or compressor air bleeding.

The following discussion is given from the turbojet engine manufacturer's viewpoint, i.e. provision is made for introducing corrections for duct losses, the losses themselves not being discussed in too great detail.

The inlet duct. We are interested in obtaining for a given inlet duct the total pressure loss as a function of the flight velocity and the duct outlet velocity (the compressor inlet velocity).

Measuring the duct outlet total pressure in terms of the inlet total pressure and measuring the inlet and outlet velocities in terms of their respective acoustic velocities we have

$$\frac{p_2^0}{p_\infty^0} = \text{a function of } (M_\infty, M_2) \tag{7-1}$$

Similarly measuring the mass flow in terms of the duct outlet area, the acoustic velocity, and the total pressure we have

$$\frac{w_a a_2^0}{g p_2^0 A_d} = \text{a function of } (M_\infty, M_2) \tag{7-2}$$

It has been found to be more convenient when matching inlet duct performance with the engine performance to combine Eq. 7-1 and 7-2 by eliminating M_2 and thereby obtain for the duct performance

$$\frac{p_2^0}{p_\infty^0} = \text{a function of } \left(M_\infty, \frac{w_a a_2^0}{g p_2^0 A_d}\right) \tag{7-3}$$

In this way the flow through the duct and the flow into the engine can be matched, since the dimensionless flow rather than the Mach number M_2 is usually given in the compressor or engine performance.

In writing the relation (Eq. 7-3) we should bear in mind that we are considering a given duct which is characterized by a number of dimensionless geometric quantities (l_1/D, l_2/D, etc.), obtained by measuring all the linear dimensions l_1, l_2, etc. in terms of one dimension D taken as a standard unit. We are neglecting viscosity effects unless we introduce into the relation (Eq. 7-3) a dimensionless viscosity which could be a Reynolds number Re or could be the viscosity measured in terms of other convenient quantities, as in the discussion on the compressor.

Similarly, to be more general, we should include in Eq. 7-3 the ratio of specific heats γ. The omission of γ (or $\mathcal{R}$) in Eq. 7-3 signifies that its influence on the phenomena is negligible or is so assumed. In practice the influence of γ or Re usually cannot be calculated, and since there is a lack of test data on their effects we are forced by circumstances to neglect these items.

The relation (Eq. 7-3) can either be found experimentally [*25,26*] or estimated by calculation. A perfect duct, of course, will have

$$\frac{p_2^0}{p_\infty^0} = 1.00 \tag{7-4}$$

while any duct without heat addition or subtraction (which is usually the case) will have, independent of friction losses,

$$\frac{T_2^0}{T_\infty^0} = 1.00 \tag{7-5}$$

Turbojet engines are usually evaluated in test cells using large bell-mouthed inlet, air flow-measuring nozzles having very small losses (0.1 per cent total pressure loss). Obviously such a nozzle would not be efficient at high flight speeds where a rather sharp-lipped inlet is indicated in order to avoid loss of contact or compression shocks. This is particularly true where the contemplated flight speed is greater than the speed of sound [*27*]. The airplane designer must therefore compromise his inlet duct design to meet the high and low speed requirements. Very often the inlet duct is designed for high speed flight conditions and the ram doors are opened during take-off in order to provide additional inlet area to the engine, thus reducing the duct losses at low speed. Various means have been suggested to improve the inlet duct performance at supersonic speed. Oswatitsch [*28*] used one or more oblique shocks in place of a single plane shock to reduce inlet losses and other devices continue to be investigated in the laboratories [*29*].

Although a convergent-divergent diffuser is indicated for use at supersonic flight speed, test and theory [*30*] show that one has trouble in starting such diffusers due to the fact that a plane shock forms in front of the diffuser. This shock can be dislodged by opening up the diffuser throat, in which case its shock moves downstream into the diffuser past the throat where it is less intense and therefore introduces smaller losses. Theoretically the shock can be eliminated by again reducing the throat area to its original design value.

A very important consideration in ducting design is the maintaining of uniform velocity and pressure conditions over the duct cross section at the inlet to the compressor. This is true particularly of the axial flow compressor where the long inlet blades can be forced to vibrate and to fail in fatigue because of poor inlet flow conditions.

The reader is referred to the bibliography at the end of this section for a more detailed treatment of inlet ducting. Fig. C,7a shows some typical performance curves for inlet ducting.

The compressor. We are particularly interested in knowing the required driving power P_c of the compressor, the air flow w_a for a given rational speed n, and inlet conditions T_2^0 and p_2^0 for a given outlet pressure p_3^0.

For convenience we can replace the inlet temperature T_2^0 in our list of variables by the inlet acoustic velocity a_2^0 since the acoustic velocity is proportionate to the square root of the temperature. In so doing we

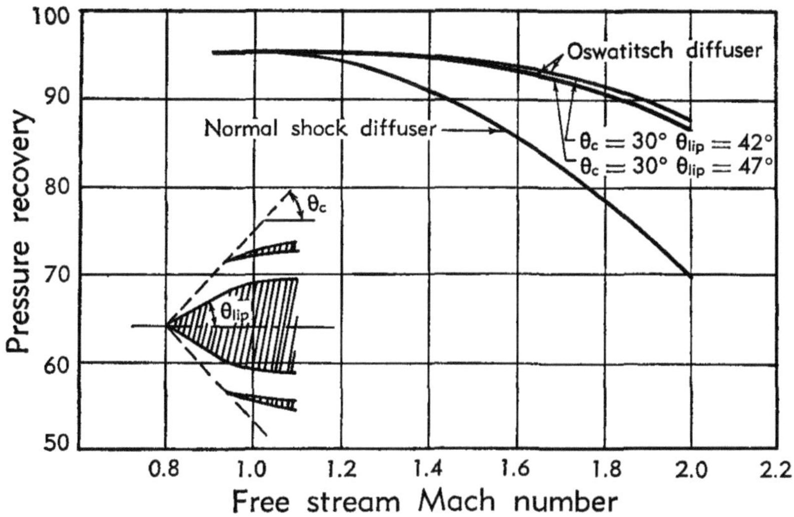

Fig. C,7a. Variation of pressure recovery with Mach number for supersonic inlets.

neglect the effects of changes in the ratio of specific heats γ and the gas constant $\mathcal{R}$, since

$$a_2^0 = \sqrt{\gamma g \mathcal{R} T_2^0} \tag{7-6}$$

The speed is measured in terms of the inlet velocity of sound and the diameter taken as standard units, thus obtaining the dimensionless speed:

$$\text{dimensionless speed} = \frac{nD_c}{a_2^0} \tag{7-7}$$

Similarly the outlet pressure is measured in terms of the inlet pressure taken as a standard unit:

$$\text{dimensionless outlet pressure} = \frac{p_3^0}{p_2^0} \tag{7-8}$$

The air mass flow and required driving power are most conveniently measured in terms of the inlet conditions and the size.

$$\text{dimensionless flow} = \frac{w_a a_2^0}{g p_2^0 D_c^2} \tag{7-9}$$

$$\text{dimensionless power} = \frac{P_c}{p_2^0 a_2^0 D_c^2} \tag{7-10}$$

Neglecting viscosity effects and variations of γ and c_p, we have for the compressor,

$$\text{dimensionless air flow } \frac{w_a a_2^0}{g p_2^0 D_c^2} = \text{a function of } \left(\frac{nD_c}{a_2^0}\right), \left(\frac{p_3^0}{p_2^0}\right) \tag{7-11}$$

$$\text{dimensionless power } \frac{P_c}{p_2^0 a_2^0 D_c^2} = \text{a function of } \left(\frac{nD_c}{a_2^0}\right), \left(\frac{p_3^0}{p_2^0}\right) \tag{7-12}$$

It is generally more convenient to use the compressor efficiency in place of the driving power since the efficiency has numerical values between zero and unity.

$$\eta_c = \frac{(h_3^0)_{is} - h_2^0}{h_3^0 - h_2^0} \tag{7-13}$$

The efficiency η_c is dimensionless and is also a function of the dimensionless speed and pressure ratio as is the dimensionless power. Thus,

$$\eta_c = \frac{[(h_3^0)_{is} - h_2^0] \dfrac{1}{(a_2^0)^2} \left(\dfrac{w_a a_2^0}{g p_2^0 D_c^2}\right)}{\left(\dfrac{P_c}{p_2^0 a_2^0 D_c^2}\right)} \tag{7-14}$$

Since

$$[(h_3^0)_{is} - h_2^0] \frac{1}{(a_2^0)^2} = \text{a function of } \left(\frac{p_3^0}{p_2^0}\right), \gamma, \left(\frac{\mathcal{R}}{c_p}\right) \tag{7-15}$$

Then

$$\eta_c = \text{a function of } \left(\frac{nD_c}{a_2^0}\right), \left(\frac{p_3^0}{p_2^0}\right), \gamma, \left(\frac{\mathcal{R}}{c_p}\right) \tag{7-16}$$

We must again remember that the above development is true for one compressor having a fixed geometry, i.e. we have for simplicity omitted (or assumed constant) the geometrical variables which are ratios of the various linear dimensions to one dimension taken as a reference. Also, the ratio of specific heats γ and the specific heat c_p will vary through the compressor as the temperature rises due to compression. We should therefore include in the above relations several dimensionless variables such as

$$\left(\frac{\gamma_{2.1}}{\gamma_2^0}\right), \left(\frac{\gamma_{2.2}}{\gamma_2^0}\right), \ldots, \left(\frac{c_{p2.1}}{c_{p_2}^0}\right), \left(\frac{c_{p2.2}}{c_{p_2}^0}\right), \text{ etc.}$$

which are used to express the change in γ and c_p through the compressor. Actually in practice we assume that these dimensionless ratios are not varied, i.e. the ratio of the value of γ or c_p at any stage of the compressor to their respective values at the inlet to the compressor is constant. Further we assume that the values of γ and c_p at the inlet to the compressor are constant. The uninitiated very often make the mistake of including the tested effects due to varying γ and c_p under the heading of "Reynolds number effects." We can easily imagine that angles of incidence to the blading through the compressor and thus the performance would be changed due to changes of γ and c_p.

If we desire to include the effect of viscosity we may introduce the well-known Reynolds number

$$Re = \frac{nD_c^2}{\nu} \tag{7-17}$$

which is the reciprocal of the kinematic viscosity measured in terms of the rotational speed and diameter. Or we may measure the absolute viscosity in terms of the total pressure p_2^0, size D_c, and total acoustic velocity a_2^0.

$$N_Q = \frac{p_2^0 D_c}{\mu a_2^0} = \text{condition number} = \frac{Re}{\gamma M} \tag{7-18}$$

This number which is Reynolds number divided by Mach number times γ has been used by the author for several years in accounting for viscosity effects in steam turbines. For want of a better name it has been called the "Q" number or "condition" number since it involves only the fluid conditions of pressure, viscosity, and acoustic velocity (together with the size D_c).

For a perfect gas the Q number is related to the ratio of molecular mean free path to the characteristic dimension. Since for a perfect gas the product μa_2^0 is proportional to the absolute temperature, the Q number is proportional to the pressure divided by the absolute temperature. For aircraft engines the Q number is therefore a function of the standard altitude conditions.

Similar statements apply to the variation of velocity throughout the compressor as given in the foregoing discussion on the variation of γ and c_p.

Considering the fact that the viscosity, the specific heat, and the ratio of specific heats all vary with temperature, and that their variations through the compressor depend on the dimensionless speed and pressure ratio, it appears that the effect of all these quantities could be accounted for most simply by introducing the compressor inlet temperature and pressure as variables in place of, for instance, Reynolds number and γ. This is particularly true since we are very seldom interested in working fluids other than air. Of course, we could use the referred pressure δ_2^0 and temperature θ_2^0 as the variables in place of p_2^0 and T_2^0.

It has generally been found that the results of altitude tests of complete engines when plotted in terms of referred (or dimensionless) quantities plot as one curve, independent of altitude. However, at extreme altitude this is not expected to remain true, particularly where combustion

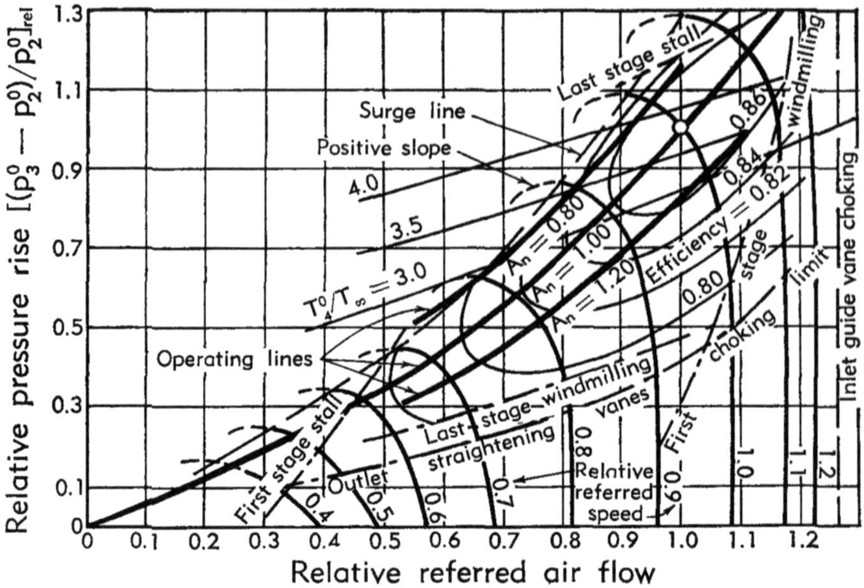

Fig. C,7b. Typical compressor performance map.

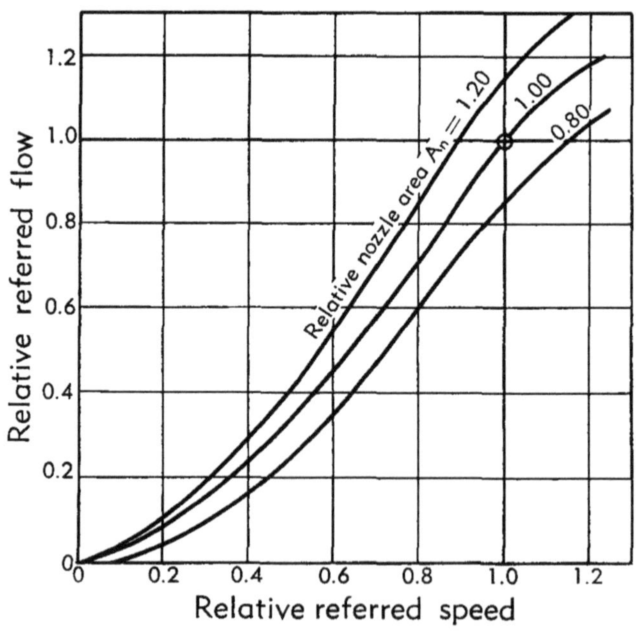

Fig. C,7c. Turbojet air flow for different engine rotational speed and exhaust nozzle areas.

"blow-out" occurs and the flame leaves the burner, passes through the turbine, and may remain in the exhaust duct or nozzle.

Plate C,7 shows the Reynolds number effect on the efficiency of a two-stage compressor, based on unpublished data obtained by W. R. New and H. B. Saldin.

A typical set of compressor performance curves is shown in Fig. C,7b and C,7c. Fig. C,7b is a combined plot of the relations (Eq. 7-11 and 7-16), while Fig. C,7c is a cross plot of Fig. C,7b.

Of interest in Fig. C,7b are the lines showing the relative aerodynamic loading [*31,32*] of the different stages in the compressor. Other effects

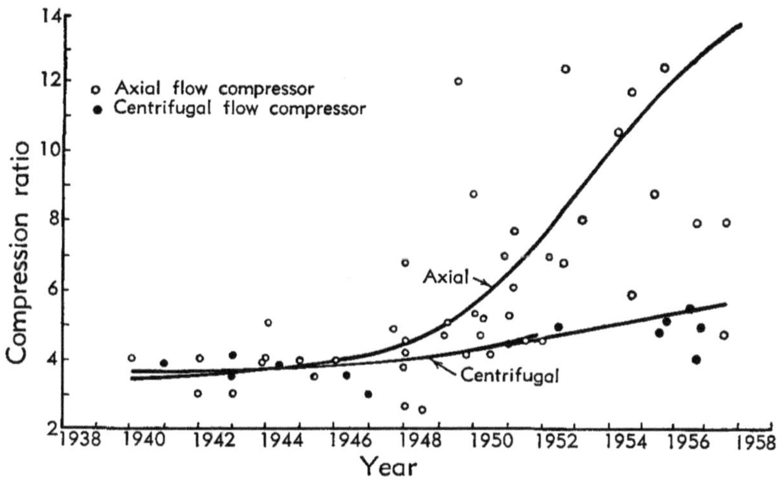

Fig. C,7d. Compression ratio of turbojets.

are indicated such as the effect of increasing the turbine temperature ratio T_4^0/T_∞, or of reducing the exhaust nozzle area, while the compressor is part of a complete turbojet engine.

The above relations hold for a single-spool compressor and the maximum pressure ratio for which a simple turbine engine can be built is somewhat limited[7] due to starting difficulties. When two or more spools are employed, there will be a set of curves for each unit. In this arrangement the inner and outer spindles run at different equilibrium rotational speeds during starting so that extreme (positive and negative) flow incidence angles on the front and back compressor blades are avoided.

The trends in compressor design have been shown by Driggs and Lancaster [*1*] who have assembled data for the various turbojet engine compressors. Typical compression ratios are shown in Fig. C,7d where single-spool engines approach 8 to 1.

The air flow per unit frontal area is shown in Fig. C,7e. Here the maximum value for the axial flow compressor is limited by critical flow

[7] Estimated by various authorities to be between 7 and 10 to 1.

in the inlet annulus (about 50 lb/sec/ft²) and by the ratio of annulus to frontal area. Axial flow compressors with low hub tip ratios at the inlet make the most use of the frontal area. Hub tip ratios of 0.5 or lower have been used in Fig. C,7f. The difficulty in successfully designing for low

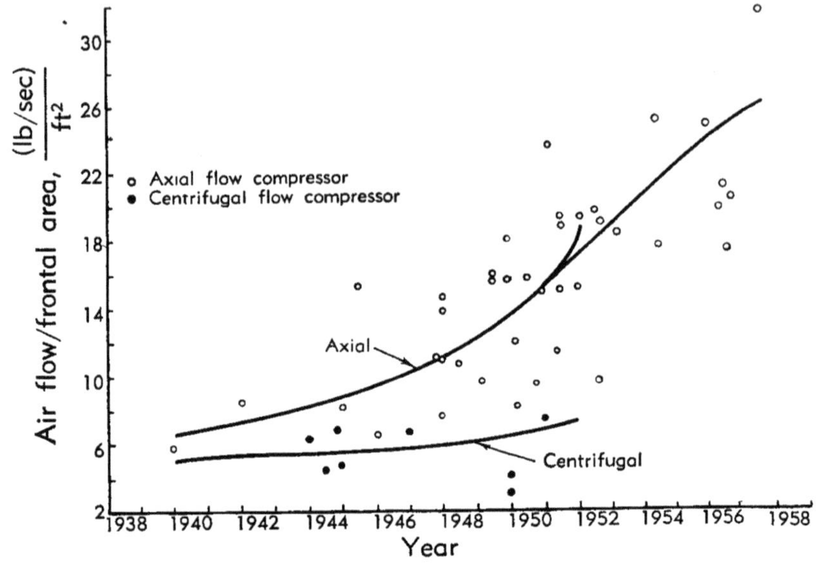

Fig. C,7e. Ratio of air flow to frontal area for turbojets.

Fig. C,7f. Inlet hub tip ratio of axial compressors.

hub tip ratios is attributed to the mechanical and fluid dynamic design of the first rotor blade. Centrifugal stresses for a given tip speed require, for the lower hub tip ratios, a greater ratio of blade hub-to-tip, cross-sectional area. As the blade hub cross-sectional area is increased with reducing hub tip ratio, it becomes increasingly difficult to maintain sufficient flow passages between the blades without increasing excessively the axial length of the blade. The lower hub diameter (lower wheel speeds) requires higher blade loadings if the same work is to be supplied to the air at the hub as at the tip. These more difficult stress and aerodynamic requirements generally result in a less efficient compressor. Greater flow per unit frontal area can be attained also by increasing the axial inflow

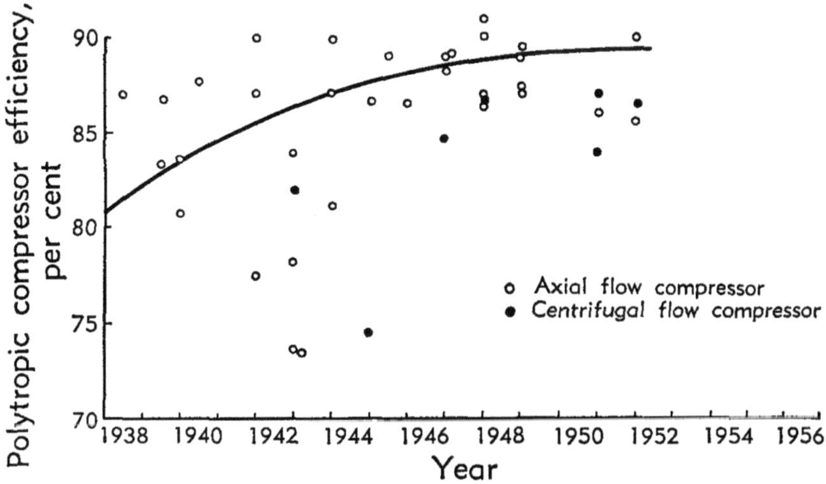

Fig. C,7g. Polytropic compressor efficiency for turbojets.

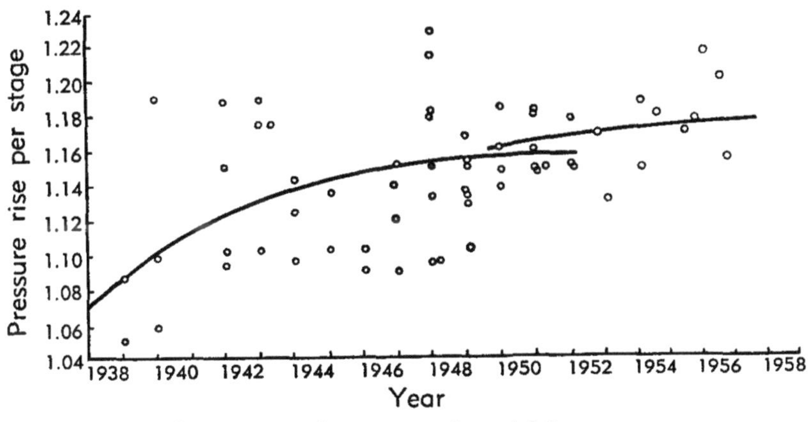

Fig. C,7h. Pressure rise per stage for axial flow compressors

velocity. However, this generally results in high Mach numbers relative to the rotating blade, which in turn results in losses in efficiency. This is particularly so at high altitudes where the ambient temperature is low, the acoustic velocity is low, and for a given rotational speed the Mach numbers are high.

The trend in flow per unit area of compressors must also be compared with the resulting change in efficiency (Fig. C,7g). Efficiencies approaching 90 per cent are being attained.

The efficiency of a compressor should also be compared to the work done per stage. Often compressors with greater pressure rise per stage save engine length at the expense of efficiency. The trend in stage pressure ratios is presented in Fig. C,7h where values between 8 and 23 per cent pressure rise per stage are shown. More correctly, the pressure rise per stage is dependent also on the wheel speed, which in turn is dependent on blade strength considerations. For this reason the enthalpy rise per stage divided by the energy corresponding to the mean (or tip) speed is a more accurate indication of stage pressure rise (performance).

The burner. The burner[8] is one of the simplest components of the turbojet engine when viewed from the manufacturing standpoint. However, the design of a burner is very complicated when examined from the theoretical standpoint for the following reasons:

1. The number of quantities which affect the operation of a burner are large (as shown below), so that even though we employ dimensional analysis to reduce the number of variables we still are left with an excessive number of dimensionless variables.
2. Very small changes in the geometry of a burner will make large differences in its performance. Thus the geometrical variables are large in number and also very important.
3. Burner performance is greatly influenced by the air inlet pressure and temperature, i.e., influenced by the Reynolds number or Q number, if you wish.
4. The burner must contain combustion at temperatures up to about 4000°R and yet must be mechanically durable when constructed of metals which begin to soften at half this temperature.

Some of the more obvious dimensionless variables which are connected with the performance of the burner and which have been devised by measuring some of the quantities in terms of others are:

$$\begin{aligned} &\text{The dimensionless air flow } \frac{w_a a_3^0}{g p_3^0 A_b} \\ &\text{The dimensionless pressure loss } \frac{p_\infty^0}{p_3^0} \text{ or } \frac{\Delta p_b}{p_3^0} \\ &\text{The dimensionless fuel flow } \frac{w_f}{w_a} \text{ or } \frac{w_f Q_p}{p a A_b} \\ &\text{The dimensionless temperature rise } \frac{T_4^0}{T_3^0} \end{aligned} \tag{7-19}$$

Other dimensionless variables such as those related to the properties of the fuel and those associated with the fuel injection system are more difficult to write down. For instance, the fuels generally used in turbojet

[8] Called "burner" instead of "combustor" to avoid subscript c conflict with compressor as in η_c, η_b.

engines are composed of various fractions which evaporate over a range of temperatures as shown in Fig. C,7i. Therefore it is difficult to set down one dimensionless number to express or measure this range of boiling points. Likewise if the latent heat of evaporization plays an important

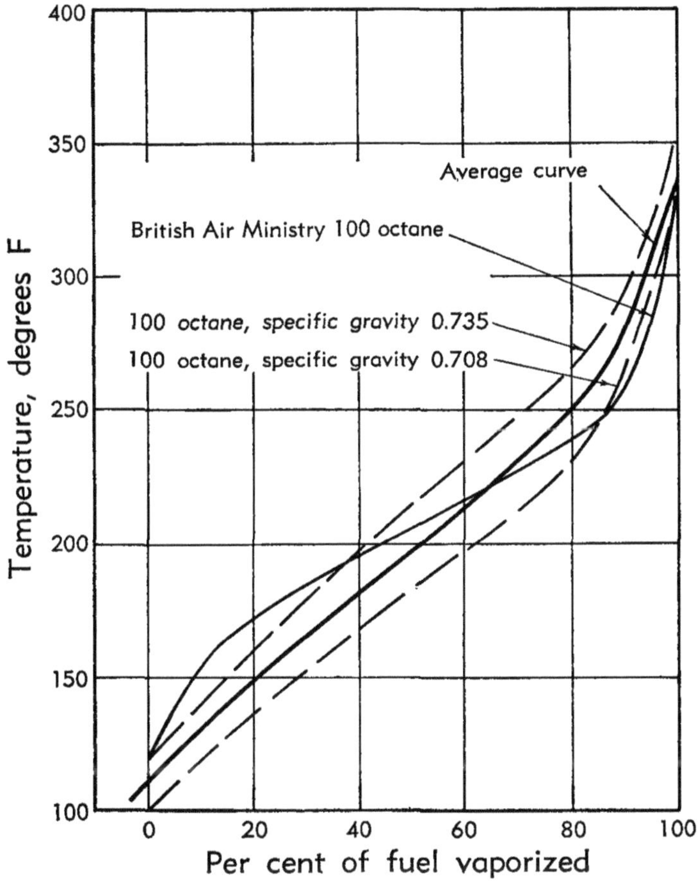

Fig. C,7i. Distillation curve of 100 octane gasoline.

role in the atomization and evaporization of the fuel in the burner it also has a range of values for the various fractions at different fuel temperatures. Therefore, the latent heat is difficult to express in terms of a single dimensionless number.

Some other quantities which have not been included in Eq. 7-19 and which could have an important bearing on the injection and vaporizing of the fuel are:

Δp = the fuel nozzle pressure drop
ρ_f = the fuel density

T_f, p_{v_f} = the fuel supply temperature and vapor pressure
c_{L_f}, c_{p_f} = the liquid and vapor specific heat of the fuel respectively
L_f = the average latent heat of the fuel
Q_l = the fuel lower heating value
μ_a = the air viscosity
S = the flame velocity
σ = the fuel surface tension
T_{ign} = ignition temperature
k_f, k_a = thermal conductivity of fuel and air

One can readily understand that although all of the quantities listed above may not be independent or important, those that are could lead

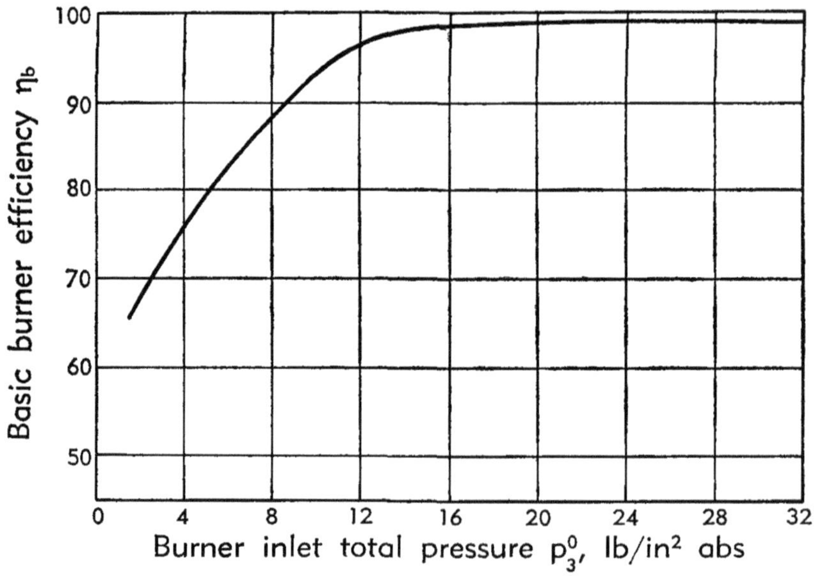

Fig. C,7j. Basic burner efficiency as affected by inlet pressure.

to possibly four or five additional dimensionless variables beyond those given in Eq. 7-19. This large number of dimensionless variables along with the large number of geometrical variables has made it impossible to completely test any given burner. In practice, combustor designers have generally been content to use the variables similar to those given in Eq. 7-19 together with the total pressure and temperature of the supply air. The use of Reynolds number (or Q number) in place of the pressure seems hardly warranted if one is forced by sheer number of variables to omit the other quantities listed above.

One approximate form of presenting the burner performance is shown in Fig. C,7j, C,7k, C,7l, and C,7m where correction factors are employed to correct the burner efficiency under one set of operating conditions

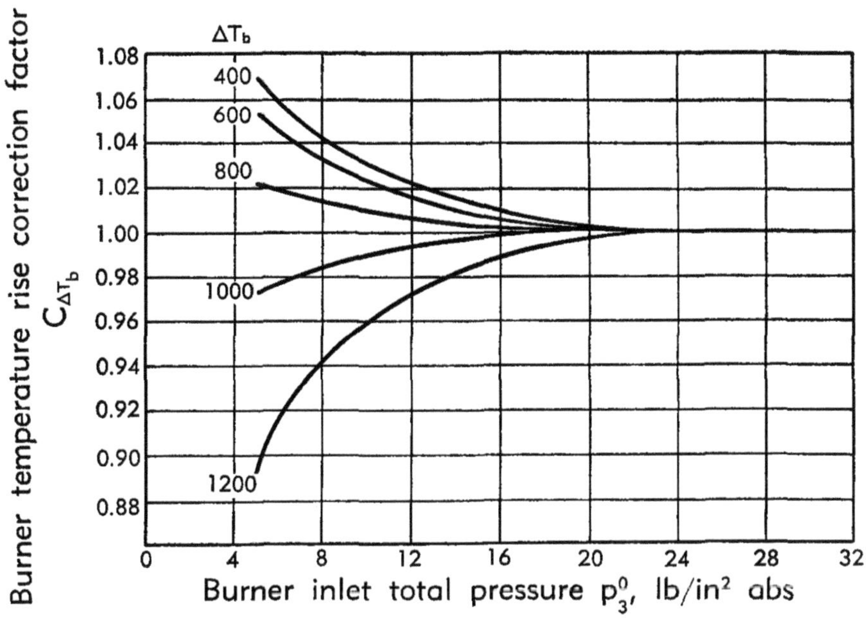

Fig. C,7k. Temperature rise correction factor for burner efficiency.

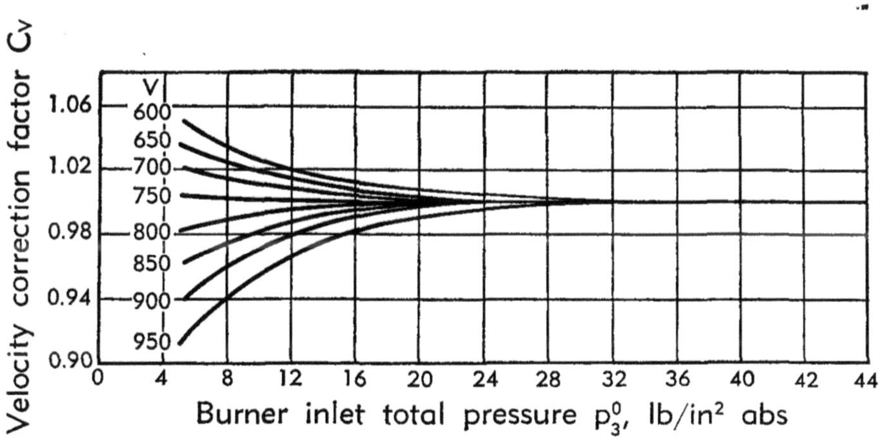

Fig. C,7l. Flow velocity correction factor for burner efficiency.

for changes in burner pressure level, burner temperature rise ΔT_b, air velocity, and air supply temperature. Admittedly such a presentation cannot be very general; however, in order to determine such a set of performance curves by test is quite a task. This task must be repeated if one should alter the geometry of the burner in order to improve its efficiency or pressure drop, which frequently happens. The burner is altered, improving its durability and thus invalidating all previous test results which are intended to furnish information over a wide range of dimensionless variables.

The burner efficiency η_b is conveniently related to the burner temperature rise ΔT_b, fuel air ratio w_f/w_a, fuel hydrogen-carbon ratio H/C, and gaseous specific heats by means of the following development.[9]

The energy equation for a perfect burner which uses the ideal fuel w_{f_i} is written

$$w_a(h_3^0 - h_{537}^0) + w_{f_i}[(Q_l)_f + (h_{f_{518.4}} - h_{f_{537}})] = (w_a + w_{f_i})(h_{4p}^0 - h_{537}^0) \quad (7\text{-}20)$$

where the lower fluid heating value is referred to 537°R (77°F) and where it is assumed that the fuel is supplied at the standard NACA sea level

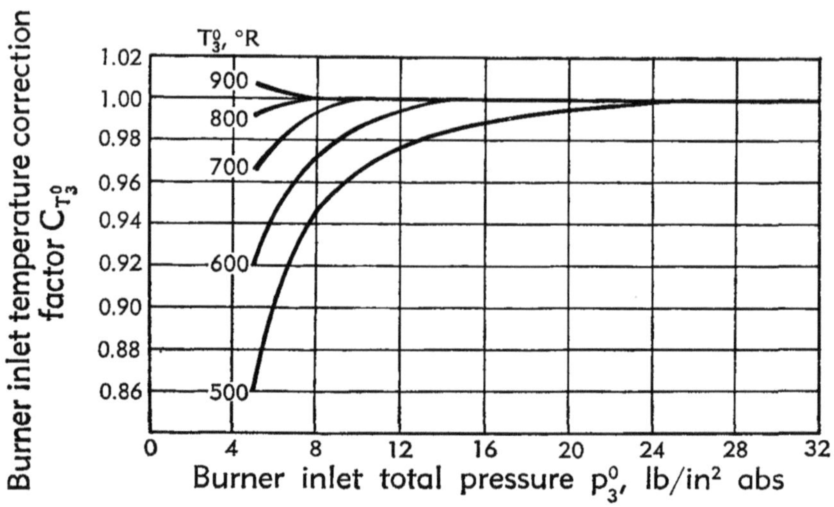

Fig. C,7m. Inlet temperature correction factor for burner efficiency.

temperature of 518.4°R. The enthalpy terms are given for convenience as differences between their 537°R values so that enthalpies referred to any base may be used. Eq. 7-20, after the addition of $+q_r$ to the right-hand side, is the defining equation for the heating value, the heating value being defined as the quantity of heat q_r removed in bringing the temperature of the products to the original temperature of the mixture (537°R). In this case the enthalpy difference terms in Eq. 7-20 are all zero.

The enthalpy of the products of combustions h_{4p}^0 at the exit of the burner is related [*41*] to the enthalpy h_4^0 of air at the same temperature by

$$h_{4p}^0 = \frac{1}{1 + \eta_b \dfrac{w_f}{w_a}} \left(h_4^0 + \eta_b \frac{w_f}{w_a} \lambda_4^0 \right) \quad (7\text{-}21)$$

[9] An alternate development is given by Withers in [*33*].

where

$$\lambda_4^0 = \frac{A_4^0 \frac{\mathrm{H}}{\mathrm{C}} + B_4^0}{\frac{\mathrm{H}}{\mathrm{C}} + 1}$$

$$A_4^0 = \frac{H_{\mathrm{H_2O}} - \frac{1}{2}H_{\mathrm{O_2}}}{2.016}; \quad B_4^0 = \frac{H_{\mathrm{CO_2}} - H_{\mathrm{O_2}}}{12}$$

$H_{\mathrm{H_2O}}, H_{\mathrm{O_2}}, H_{\mathrm{CO_2}}$ = molal enthalpies of water vapor, oxygen, and carbon dioxide respectively

The burner efficiency is conveniently defined in a way similar to the work of London [10] as the ratio of the ideal to the actual fuel consumption

$$\eta_b = \frac{w_{f_i}}{w_f} \tag{7-22}$$

Combining Eq. 7-20, 7-21, and 7-22 one obtains

$$\Delta T_b = c_b \eta_b \left(\frac{w_f}{w_a}\right) = T_4^0 - T_3^0 \tag{7-23}$$

where

$$c_b = \frac{Q_{\mathrm{eff}} - (\lambda_4^0 - \lambda_{537}^0)}{(h_{4a}^0 - h_{3a}^0)} \Delta T_b$$

$$= \text{a function of } T_3^0 \text{ and } \eta_b \frac{w_f}{w_a}$$

The effective heating value of the liquid fuel in terms of the liquid enthalpy and heating value at 537°R is:

$$Q_{\mathrm{eff}} = [(Q_l)_{f_{537}} + (h_f - h_{f_{537}})]$$

By means of Eq. 7-23 the burner temperature rise can now be easily calculated, once a curve of c_b is drawn as a function of burner inlet temperature T_3^0 and ideal fuel-air ratio $\eta_b(w_f/w_a)$.

A similar although slightly more complicated development for the afterburner leads to the following expression for afterburner temperature rise:

$$\Delta T_{ab} = c_{ab}\left(\eta_{ab} \frac{w_{f_{ab}}}{w_a}\right) - c_{1ab}\left(\eta_b \frac{w_f}{w_a}\right) \tag{7-24}$$

where

$$c_{ab} = \frac{Q_{\mathrm{eff}} - (\lambda_6^0 - \lambda_{537}^0)}{(h_{6a}^0 - h_{5a}^0)} \Delta T_{ab}$$

$$c_{1ab} = \frac{(\lambda_6^0 - \lambda_5^0)}{(h_{6a}^0 - h_{5a}^0)} \Delta T_{ab}$$

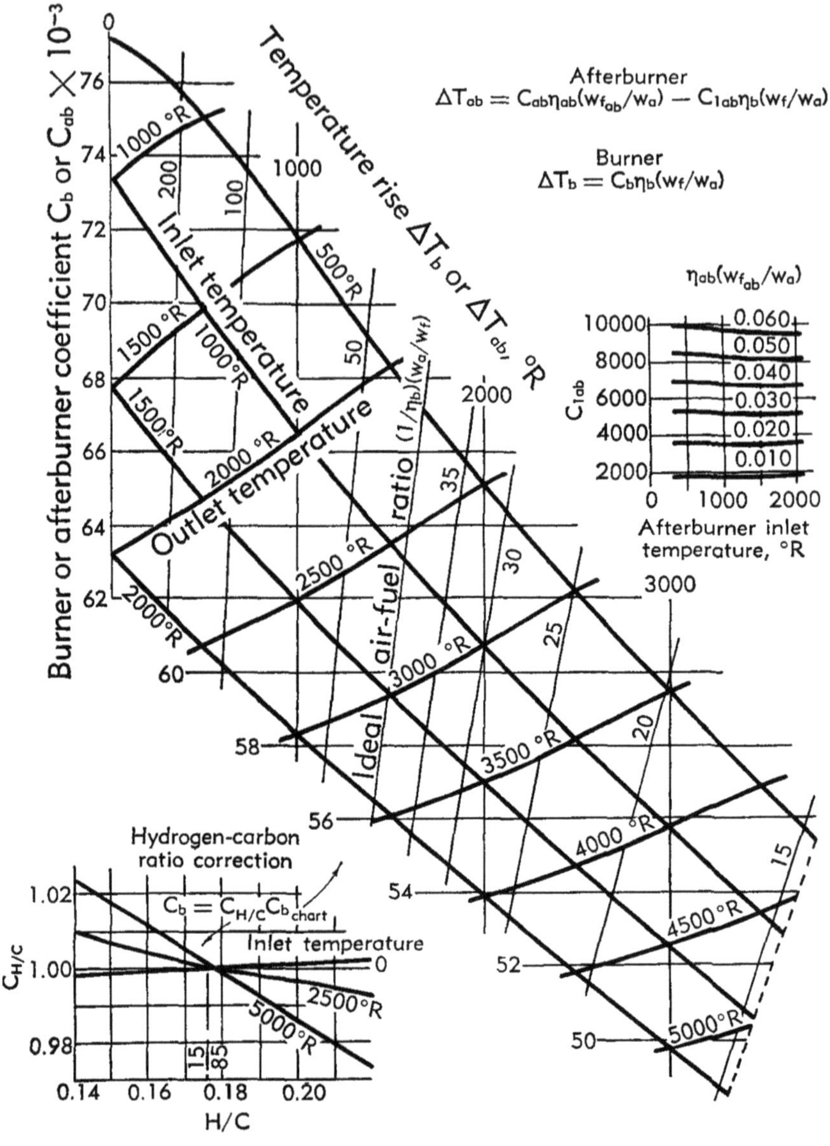

Fig. C,7n. Burner-afterburner chart.

Notice that c_{ab} is a similar expression to that for c_b in Eq. 7-23. Burner and afterburner curves are shown in Fig. C,7n from which the values of c_b, c_{ab}, and c_{1ab} are easily selected. Corrections for fuel hydrogen-carbon ratio and effective heating value (or fuel inlet temperature) are also given on this curve.

The above formulas can be altered to include dissociation by changing the expressions for the coefficients. So far it has not been found practicable to include the effect of dissociation in the above relation. This is due to the degree of uncertainty in the combustion efficiency itself, which depends on the accuracy of the measurement of the combustion temperature and due also to the uncertainty as to whether thermodynamic equilibrium is attained in the burner, after the turbine, or in the engine at all. The total pressure loss in a burner is most accurately found by test. A typical plot of the burner pressure loss is shown in Fig. C,7o.

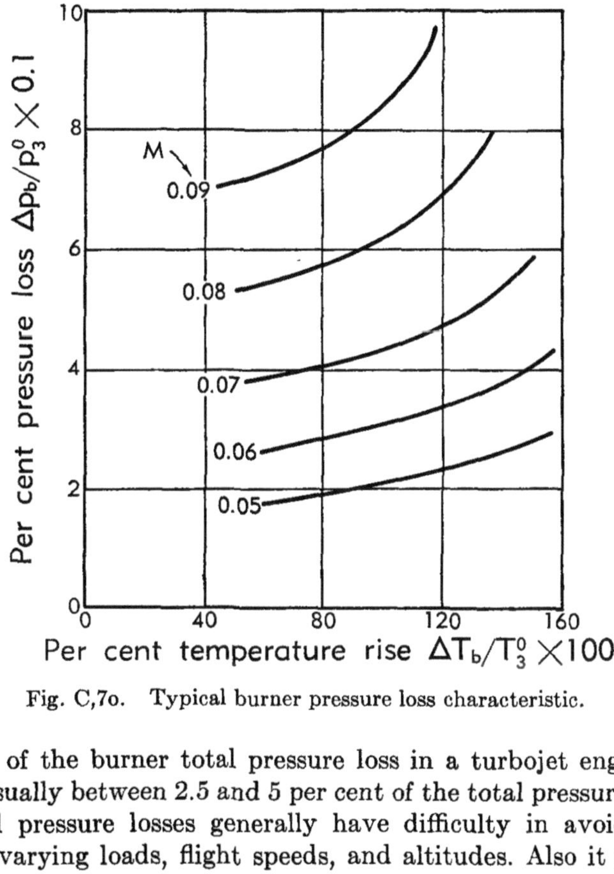

Fig. C,7o. Typical burner pressure loss characteristic.

The value of the burner total pressure loss in a turbojet engine at full thrust is usually between 2.5 and 5 per cent of the total pressure. Burners with small pressure losses generally have difficulty in avoiding "hot spots" at varying loads, flight speeds, and altitudes. Also it is difficult to obtain a durable burner when the pressure loss for air flow distribution is small.

A useful approximation to the burner pressure loss can be found by adding the burner pressure loss without burning (i.e. cold) to the so-called "momentum total pressure loss" for a constant area channel shown in Fig. C,7p. This curve has been calculated by Hawthorne and Cohen [*35*] by assuming heating without friction in a constant area channel and by writing the applicable continuity, energy, and momentum equations.

The operational altitude of turbojet engines as limited by combustor "blowout" is shown later in Fig. C,10b. As altitude is increased the density and therefore the efficiency of combustion is decreased until "blowout" occurs, i.e. the flame is blown out through the turbine and exhaust nozzle. The engine thrust and speed decrease.

Blowout can be overcome by higher flight speed which increases the combustion chamber pressure level. However, the aircraft may not be capable of this speed (see Fig. C,11b). The speed of the aircraft is thus limited to a narrow band between the maximum speed and the (lower) blowout speed.

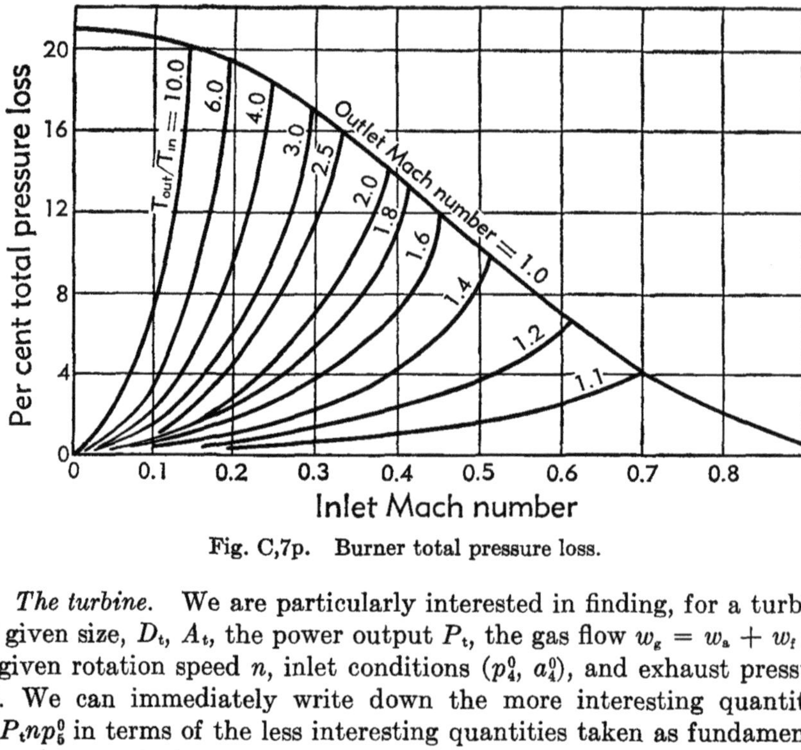

Fig. C,7p. Burner total pressure loss.

The turbine. We are particularly interested in finding, for a turbine of given size, D_t, A_t, the power output P_t, the gas flow $w_g = w_a + w_f$ for a given rotation speed n, inlet conditions (p_4^0, a_4^0), and exhaust pressure p_5^0. We can immediately write down the more interesting quantities $w_a P_t n p_5^0$ in terms of the less interesting quantities taken as fundamental units by a method similar to that used for the compressor. We thus obtain, neglecting viscosity, γ, and $\mathcal{R}/c_p$ effects, and assuming a constant set of geometrical ratios

$$\text{a function of} \quad \left(\frac{P_t}{p_4^0 a_4^0 A_t}\right), \left(\frac{nD_t}{a_4^0}\right), \left(\frac{p_5^0}{p_4^0}\right) = 0 \tag{7-25}$$

and

$$\text{a function of} \quad \left(\frac{w_g a_4^0}{g p_4^0 A_t}\right), \left(\frac{nD_t}{a_4^0}\right), \left(\frac{p_5^0}{p_4^0}\right) = 0 \tag{7-26}$$

It has been customary to use the turbine efficiency in place of dimensionless power[10] and also to use the velocity ratio in place of dimensionless speed in presenting turbine performance.

The efficiency is defined as

$$\eta_t = \left[\frac{h_4^0 - h_5^0}{h_4^0 - (h_5^0)_{is}}\right] \tag{7-27}$$

while the isentropic velocity ratio is

$$\nu_{is} = \sqrt{\frac{\sum U_m^2}{2gJ[h_4^0 - (h_5^0)_{is}]}} \tag{7-28}$$

where

$$U_m = \text{the mean wheel speed}$$

A typical set of efficiency curves is shown in Fig. C,7q for reaction, Rateau, and Curtis stages. An envelope curve is shown which indicates

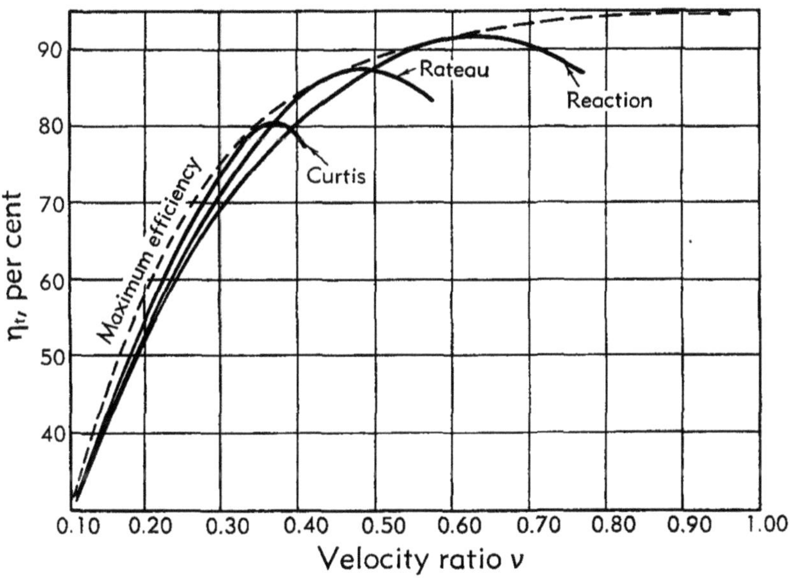

Fig. C,7q. Efficiency of different turbine types as a function of isentropic velocity ratio.

from past experience that, for low velocity ratios (large stage loadings), the best experimentally obtained efficiencies are lower than those for the higher velocity ratios. In general the turbine efficiencies are also considerably affected by the pressure ratio p_4^0/p_5^0 as shown in Fig. C,7r for a two-stage turbine. The variation of efficiency with pressure ratio and velocity ratio can be accurately obtained only by test. The variation of efficiency

[10] In using either power or efficiency we very often unknowingly discard some of the test information. For instance, starting torque cannot be obtained from either dimensionless power or efficiency (since these are zero at zero speed), but it can be found from a dimensionless torque.

is dependent among other things on the nozzle-blading area ratio, the leakage, the blade and nozzle shapes, the hub or tip taper, and the blade height variation through the turbine. A more detailed discussion of these variations is given in X,H.

A typical curve showing the variation of dimensionless gas flow represented by Eq. 7-26 for a two-stage turbine is shown in Fig. C,7s. Notice

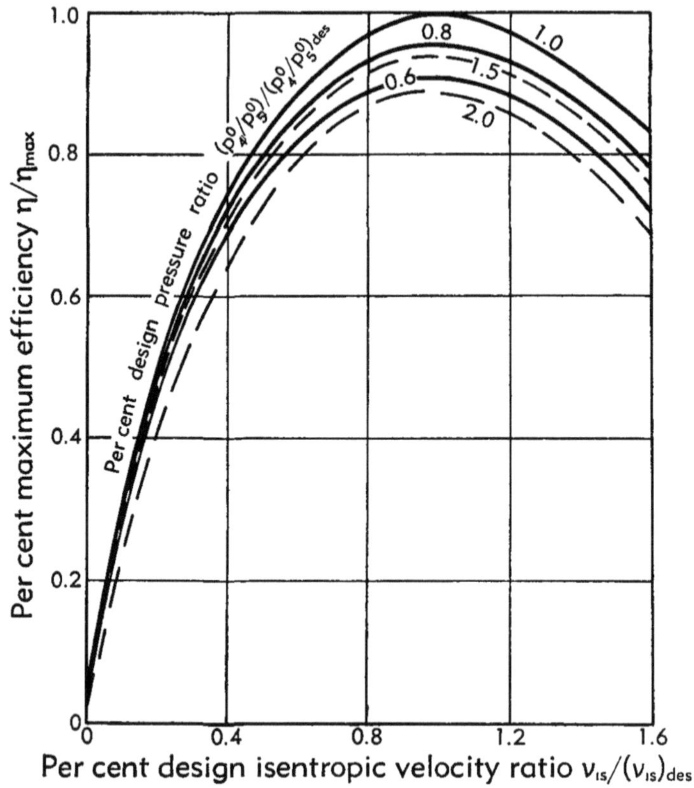

Fig. C,7r. Effect of pressure ratio on turbine efficiency.

the variation in dimensionless flow with speed for subcritical pressure ratios. These variations have been found to be due to the variations in the pressure drop across the rotor with different rotational speeds, which in turn cause variations in the nozzle back pressure and thus in the gas flow. Notice also that the maximum turbine flow may be somewhat less than the nozzle critical flow due to the successive restriction of the following stages. In general, good use is made of the fact that a large part of the operation of turbojet engines is in the region where the dimensionless flow is constant, i.e. on the nearly flat portion of the flow curve. In practice one frequently uses a tested turbine "effective area" $w_g a_4^0/g p_4^0$ in place of the dimensionless flow.

The exhaust (or propelling) nozzle. We are interested in the flow through and the gross thrust generated by a given nozzle in terms of its area, the pressure and temperature before the nozzle, and the pressure

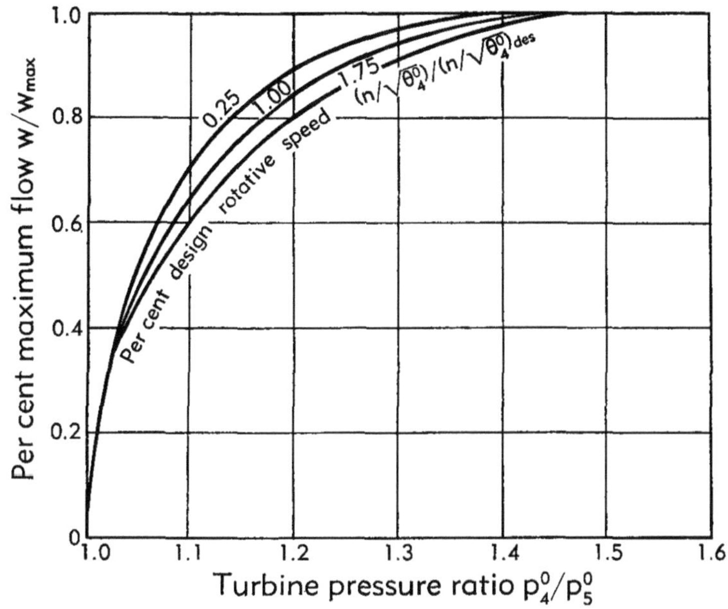

Fig. C,7s. Turbine flow variation with pressure ratio and rotational speed.

after the nozzle. Thus in dimensionless variables[11] we can write (neglecting Reynolds number and γ effects)

$$\text{a function of } \left(\frac{w_g a_6^0}{g p_6^0}\right), \left(\frac{p_6^0}{p_8}\right), A_n = 0 \tag{7-29}$$

$$\text{a function of } \left(\frac{F_g}{p_6^0 D_n}\right), \left(\frac{p_6^0}{p_8}\right), A_n = 0 \tag{7-30}$$

The above functions would be expected to be the same for nozzles of a given geometry except for Reynolds number effects. Of course the nozzle area A_n can be omitted in the above relations when we are dealing with fixed area nozzles.

A general chart[12] for use in the solution of flow problems is shown in Fig. C,7t *[36]*. The dimensionless velocity (Mach number), area, momentum, temperature, and force are given for a nozzle as a function of

[11] There is no advantage in nondimensionalizing A_n since we are usually interested in one engine size at a time. A_n/D_n^2 could be used. However, this would introduce the possibility of errors since we must consistently use the same D_n.

[12] Since this chart has general usefulness its symbols differ from those used in the remainder of this section.

pressure ratio. One can, for instance, enter the curve at a given flow per unit area, at a given Mach number, or at a given ratio of total to static pressure ($\eta = 1.00$) and solve for any of these quantities at a second station, given the area ratio, the pressure ratio across the section of channel, or the Mach number at the second station. The ratio of the dimensionless areas is of course the area ratio while the ratio of the pressure ratios along the bottom of the chart is the pressure ratio across the

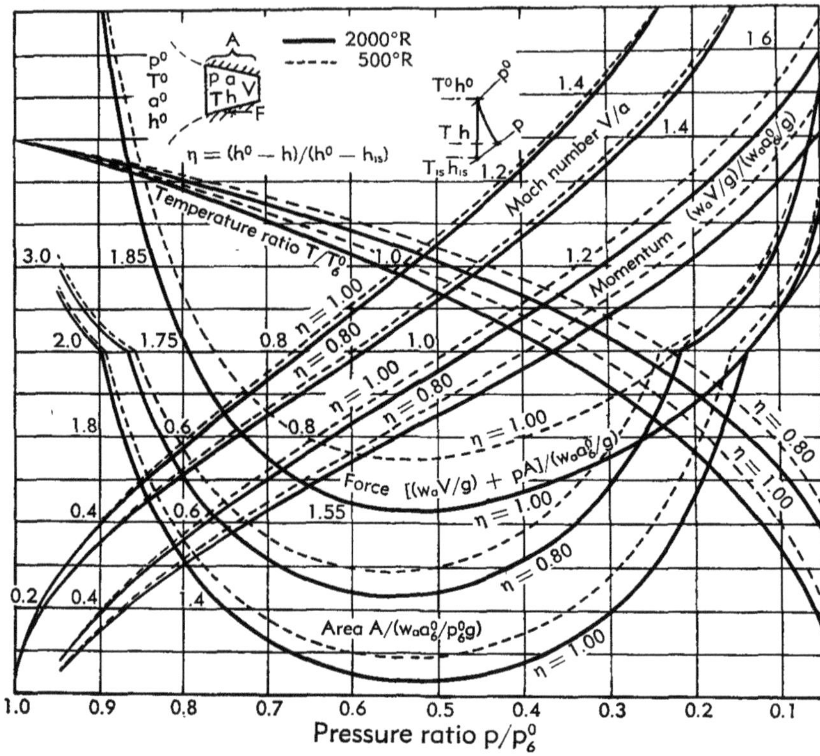

Fig. C,7t. Flow of air in nozzles.

channel section. Allowance for differences in specific heats and γ and expansion efficiency can be made by interpolating between the two temperature lines ($T_6^0 = 500°$; 2000°R) and the several efficiency lines. Subtraction of the values of the dimensionless force at two sections of a channel gives the dimensionless wall force between the two sections which may be required for stress calculations. Plane shocks can be estimated by noticing that the force $(w_a V/g) + pA$ and the flow per unit area are constant across the shock (the ratio of the dimensionless area is the ratio of the total pressure before and after the shock).

It is often the practice to use nozzle flow coefficients and thrust coefficients rather than the dimensionless flow or thrust, as stated earlier in this article. Also, a tested nozzle "effective area" $w_g a_6^0 / g p_6^0$ is frequently used.

It is interesting to review the definitions of the nozzle coefficients and to examine their interrelations. The nozzle thrust coefficient is defined as the ratio of the actual to the ideal thrust.[13] It is the quantity by which the ideal thrust is multiplied to obtain the actual thrust.

$$C_{F_n} = \frac{F_g}{(F_g)_{is}} = \frac{\int^{A_e} \frac{V_e dw_g}{g} + (p_e - p_f)dA}{(w_g V_f/g)_{is}} = \frac{\int^{A_e} [\rho V_e^2 + (p_e - p_f)]dA}{(w_g V_f/g)_{is}} \tag{7-31}$$

where subscripts refer to stations as shown in Fig. C,2a.

The nozzle velocity coefficient is defined as the ratio of the flow weighed average velocity to the ideal velocity.

$$C_{V_n} = \frac{\frac{1}{w_g}\int^{A_e} V dw_g}{(V_e)_{is}} = \frac{\frac{g}{w_g}\int^{A_e} \rho V^2 dA}{(V_e)_{is}} \tag{7-32}$$

Notice that this velocity coefficient is obtained only by means of a velocity and density survey over the nozzle area.

Another definition of the velocity coefficient C_{VF}, which is more useful than the above definition C_V, is based on obtaining a tested average velocity from the measured thrust and actual (not ideal) air flow rather than by surveying the test nozzle.

$$C_{VF_n} = \frac{\frac{F_g}{w_g/g}}{(V_e)_{is}} = \frac{\int^{A_e} [\rho V^2 + (p_e - p_f)]dA}{(V_e)_{is}(w_g/g)} \tag{7-33}$$

The thrust is easily calculated using this velocity coefficient and the actual air flow. Notice that when the nozzle flow is not expanding out the exit, $p_e = p_f$, $V_e = V_f$, and $C_V = C_{VF}$.

The nozzle flow coefficient is defined as the ratio of actual flow to ideal flow.

$$C_{w_n} = \frac{w_g}{(w_g)_{is}} = \frac{\int^{A_e} g\rho V dA}{(w_g)_{is}} = \frac{C_F}{C_{VF}} \tag{7-34}$$

The thrust coefficient is the product of the flow coefficient and the velocity coefficient C_{VF} obtained from the measured thrust. The energy

[13] One may question the arbitrarily chosen definition for ideal gross thrust $(F_g)_{is} = (w_g V_f/g)_{is}$ used in Eq. 7-31 because of neglect of the expansion losses. However, it is most convenient to define the ideal gross thrust in this way and to include the expansion losses in the thrust coefficient of the given nozzle.

coefficient (more often called the efficiency) is the ratio of the flow weighed average kinetic energy to the ideal kinetic energy.

$$C_{E_n} = \eta_n = \frac{E}{E_{is}} = \frac{\frac{1}{w_g}\int^{A_e} V^2 dw_g}{(V_e)^2_{is}} = \frac{\frac{g}{w_g}\int^{A_e} \rho V^3 dA}{(V_e)^2_{is}} \tag{7-35}$$

Frequently the efficiency $\eta_n = C_{E_n}$ is approximated as the square of the velocity coefficient

$$C_V^2 \cong C_E \tag{7-36}$$

One can readily understand that this involves the assumption that

$$\left(\int \rho V^2 dA\right)^2 \cong \frac{w_g}{g}\int \rho V^3 dA \cong \int \rho V dA \int \rho V^3 dA \tag{7-37}$$

which in turn assumes that the velocity and density are constant over the nozzle exit cross-sectional area.

The velocity coefficient is very often approximated as being equal to the flow coefficient. This involves the assumption that

$$\left(\int \rho V dA\right)^2 \cong \rho A \int^A \rho V^2 dA \tag{7-38}$$

which evidently assumes that the velocity and density are constant over the nozzle cross section.

Usually one employs the thrust coefficient C_F and the velocity coefficient C_V, and less frequently, the flow coefficient or energy coefficient, for the following reasons:

1. The thrust coefficient is easily obtained by test measurements of the nozzle area, the inlet and exit pressures, and the thrust. The thrust coefficient is also very useful for estimating the gross thrust during flight from the nozzle area and the inlet pressure and exit pressure.

$$C_{F_n} = \text{a function of } A_n,\ p_6^0,\ p_e,\ F_g,\ \gamma \tag{7-39}$$

2. The velocity coefficient C_{VF} also is easily obtained by test if the mass flow, thrust, nozzle inlet temperature, and nozzle pressure ratio are known.

$$C_{VF_n} = \text{a function of } F_g,\ w_g,\ \left(\frac{p_e}{p_6^0}\right),\ T_6^0 \tag{7-40}$$

The velocity coefficient thus requires the measurement of a mass flow and the nozzle inlet temperature rather than the measurement of the nozzle area as is required in order to obtain the thrust coefficient. Changes in nozzle dimensions (area) during operation thus will influence the thrust calculated from the thrust coefficient.

Mass flow and nozzle inlet temperature (together with nozzle pressure ratio) must be measured in flight if one is to calculate the gross thrust using the velocity coefficient. The mass flows can generally be obtained in flight from the known compressor characteristics and the referred speed.

3. The nozzle flow coefficient, also easily obtained by test, is principally of use in fixing nozzle area. However, the nozzle area is easily determined experimentally by cut and try methods to match the particular engine characteristic without recourse to the nozzle flow coefficients. The nozzle inlet temperature can be estimated if the nozzle flow coefficient, nozzle area, inlet and exit pressure, and the mass flow are known with sufficient accuracy.
4. The nozzle energy coefficient is generally of not too great use since the integral given in Eq. 7-35 must be obtained by surveying the nozzle exit. More frequently the energy coefficient, when needed, is approximately obtained as the square of the velocity coefficient (Eq. 7-36).

The theoretical thrust coefficient for frictionless nozzles of various expansion (area) ratios is shown in Fig. C,7u as a function of nozzle pressure ratio. As the pressure ratio increases across a frictionless expanding nozzle, the thrust coefficient at first remains constant (1.00) until sonic velocity is obtained in the throat. It then decreases due to the presence of an ever-enlarging plane shock in the expanding portion of the nozzle. As the pressure ratio is further increased, the thrust coefficient

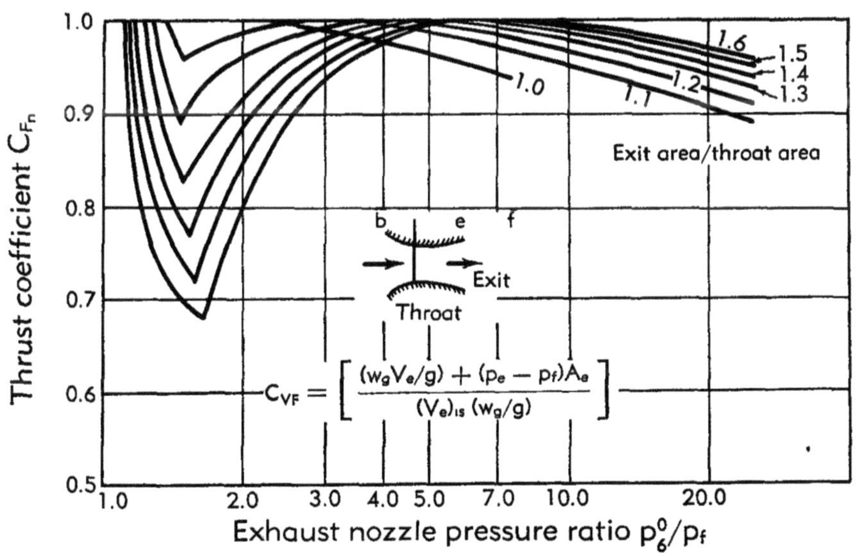

Fig. C,7u. Theoretical thrust coefficient of convergent-divergent jet nozzles.

next increases to unity due to the fact that the plane shock moves out of the nozzle, becomes more and more oblique, and less intense, and thus the shock losses steadily decrease to zero at the nozzle design pressure ratio. As the nozzle pressure ratio is still further increased, the flow expands out the nozzle exit and, although the flow is theoretically isentropic, the thrust coefficient steadily decreases due to the sidewise expansion of the jet.

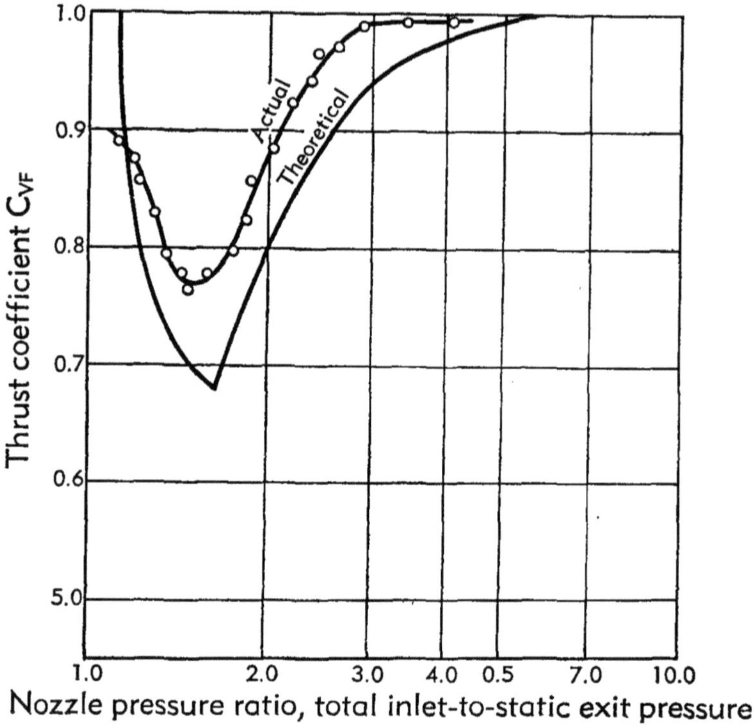

Fig. C,7v. Tested thrust coefficient of a convergent-divergent nozzle having a 1.60 area ratio.

The general characteristics shown in Fig. C,7u for frictionless nozzles are of course altered in actual nozzles where (1) friction is present, (2) the nozzle exit is conical rather than cylindrical as assumed in the calculation of Fig. C,7u, and (3) the shocks within the nozzle are often combinations of plane and oblique shocks rather than simply plane shocks as assumed in the calculation of Fig. C,7u.

The tested thrust coefficients of some typical nozzles are shown in Fig. C,7v. It is seen that the general shape of these curves agrees quite well with the theoretical curves of Fig. C,7u.

The gas generator. Internal performance. One can combine the compressor, burner, and turbine characteristics as estimated in the preceding sections to obtain the so-called "gas generator performance." In this view one imagines the combination of compressor, burner, and turbine to be a device which continuously compresses and heats the air passing through it. Accordingly, the temperature and pressure rise "generated" by the generator together with the rate of gas flow can be given dimensionlessly in terms of the dimensionless generator speed and top temperature taken as independent variables.

$$\frac{T_5^0}{T_2^0} = \text{a function of} \left(\frac{nD_c}{a_2^0}\right), \left(\frac{T_4^0}{T_2^0}\right) \tag{7-41}$$

$$\frac{p_5^0}{p_2^0} = \text{a function of} \left(\frac{nD_c}{a_2^0}\right), \left(\frac{T_4^0}{T_2^0}\right) \tag{7-42}$$

$$\frac{w_g a_5^0}{g p_5^0 A_5} = \text{a function of} \left(\frac{nD_c}{a_2^0}\right), \left(\frac{T_4^0}{T_2^0}\right) \tag{7-43}$$

Other dimensionless quantities such as the compressor ratio p_3^0/p_2^0, or temperature ratio T_3^0/T_2^0, or the dimensionless fuel flow $w_f Q_p / p_2^0 a_2^0 A_b$ are similarly dependent on the dimensionless speed and top temperature.[14]

The calculation of the gas generator performance can be considered as a part of the calculation of the performance of the complete turbojet engine, since by matching the performance of the inlet duct and the exhaust nozzle to the gas generator performance one obtains the performance of the complete engine. During the matching of the gas generator performance with the performance of the inlet duct and exhaust nozzle, one can assume either a fixed or variable exhaust nozzle having any type of engine control which is desired.

The complete engine. The matching of the performance of the engine components to obtain either the gas generator performance or the over-all engine performance is essentially a cut and try process. The method of combining the component performance to obtain the over-all engine performance, while identically satisfying the requirements already set down in Art. 5, can be followed more easily by referring to Fig. C,7w which outlines the procedure (see Art. 5).

In actual practice, the following facts, obtained as a result of experience, enable considerable reduction in the cut and try calculations.

1. The referred (or dimensionless) flow through the compressor is dependent only on the referred (or dimensionless speed) (Fig. C,7c). (The referred flow is really also dependent on the compressor pressure ratio.)
2. The turbine referred flow $w_g \sqrt{\theta_4^0/\delta_4^0}$, or effective area $w_g a_4^0 / g p_4^0$, is constant over a considerable part of the engine operating range (Fig. C,7s).

Assuming a referred (or dimensionless) compressor speed $n/\sqrt{\theta_2^0}$ one can immediately find the compressor referred flow from the compressor

[14] These relations are called "generator" or "internal" performance.

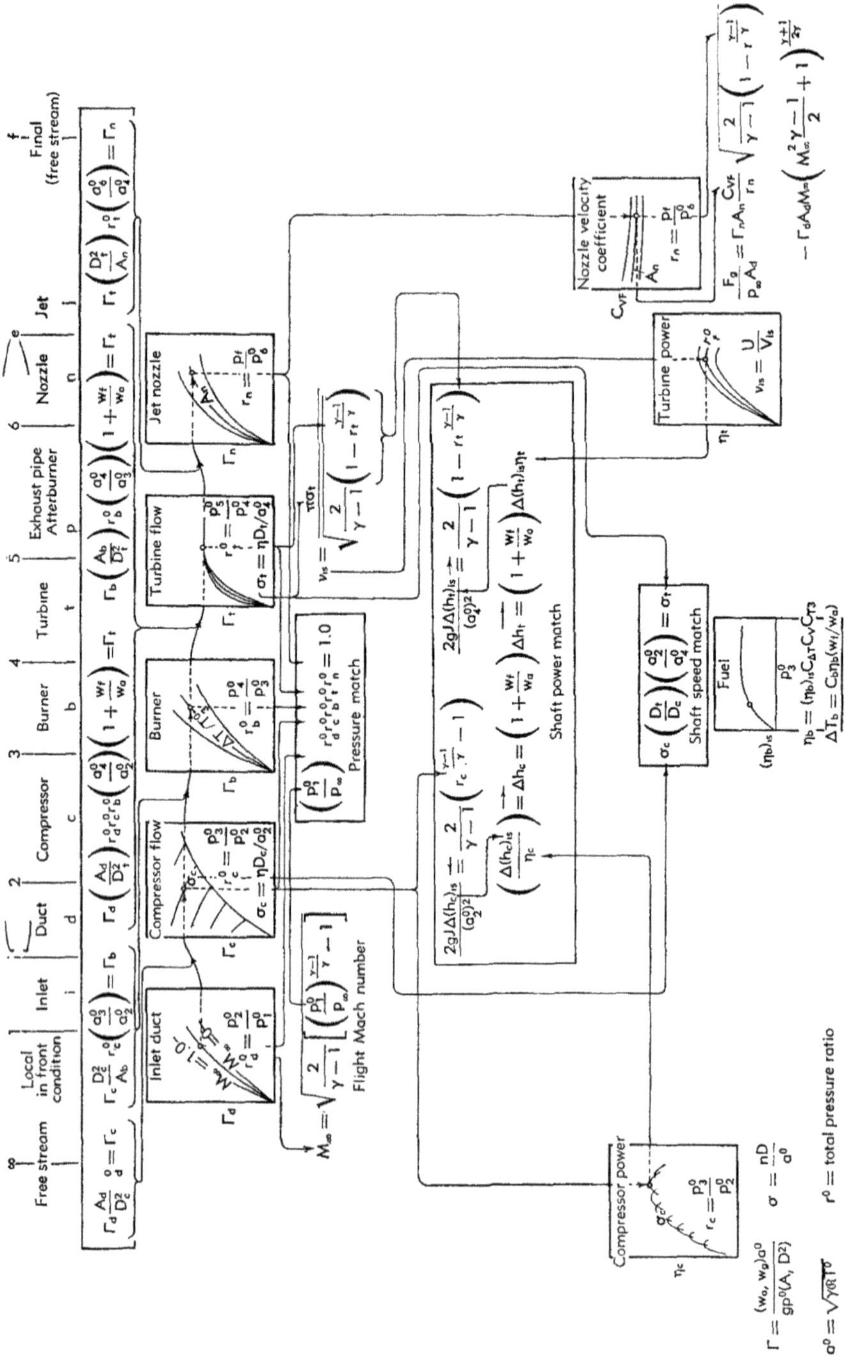

Fig. C,7w. Turbojet component matching diagram.

characteristic (Fig. C,7b). Knowing the constant value of the turbine referred flow $w_g \sqrt{\theta_4^0}/\delta_4^0$ (or effective area), together with the required top temperature T_4^0 (or θ_4^0), one can calculate δ_4^0/δ_2^0 or p_4^0/p_2^0. This is a first approximation to the compressor pressure ratio. Assuming a burner pressure loss (or pressure ratio) (p_3^0/p_4^0) one can obtain the compressor pressure ratio $p_3^0/p_2^0 = (p_4^0/p_2^0)(p_3^0/p_4^0)$.

The compressor efficiency can next be gotten from Fig. C,7b and thereby the compressor outlet temperature can be calculated. The burner inlet referred flow $w_a \sqrt{\theta_3^0}/\delta_3^0$ and the burner temperature rise ΔT_b can next be used together with Fig. C,7n to check the above assumed burner pressure ratio. Once this check is obtained the required turbine power (or enthalpy drop) can be gotten since it must be equal to the compressor power, except for the usually small bearing and auxiliary power losses. The turbine isentropic enthalpy drop $h_4^0 - (h_5^0)_{is} = (\Delta h^0)_{is} = \Delta h^0/\eta_t$ can next be estimated if we assume the turbine efficiency. The turbine-isentropic velocity ratio can be calculated using the known turbine (and compressor) speed and the turbine isentropic enthalpy drop. The turbine pressure ratio and isentropic velocity ratio together with Fig. C,7r yield the turbine efficiency which must check the value assumed above.

Fortunately the turbine velocity ratio does not vary greatly over the engine operating range so that it is easy to cut and try to obtain the turbine efficiency. The gas generator performance curves corresponding to Eq. 7-41 and 7-42 are easily calculated, once the turbine outlet pressure and temperature are known in terms of the compressor inlet pressure and temperature for the various assumed values of the compressor dimensionless speed.

If one now wishes to match a given fixed area jet nozzle to the generator performance in order to estimate the engine thrust, he must match the turbine outlet dimensionless flow to the jet nozzle inlet dimensionless flow. The nozzle pressure ratio should also equal p_5^0/p_2^0, if it is assumed that there are no inlet duct losses. This matching is accomplished most easily by overlaying the jet nozzle characteristics on the generator performance curve as shown in Fig. C,7x. Here the gas generator performance is plotted as a combination of Eq. 7-42 and 7-43 showing the dimensionless gas flow plotted against the log of the engine pressure ratio p_5^0/p_2^0 with parameter dimensionless speed. The log of the engine pressure ratio is used since one can easily obtain a match for any flight condition by sliding the nozzle characteristics to the right by an amount equal to the ram pressure ratio p_2^0/p_∞. The nozzle pressure ratio is equal to the generator pressure ratio multiplied by the ram pressure ratio.

$$\frac{p_5^0}{p_\infty} = \left(\frac{p_5^0}{p_2^0}\right)\left(\frac{p_2^0}{p_\infty}\right) \tag{7-44}$$

The intersection of the nozzle characteristic with the appropriate referred speed line sets the generator or engine total pressure ratio p_5^0/p_2^0 and therefore the temperature ratio, fuel flow, compression ratio, etc. of the engine. If one employs a variable area nozzle the solution is only slightly more complicated. For instance, if the control is such as to change the nozzle area with engine speed, one draws additional nozzle curves on Fig. C,7x and picks the appropriate speed and nozzle area as a match point. Other controls involving compressor outlet pressure, for instance,

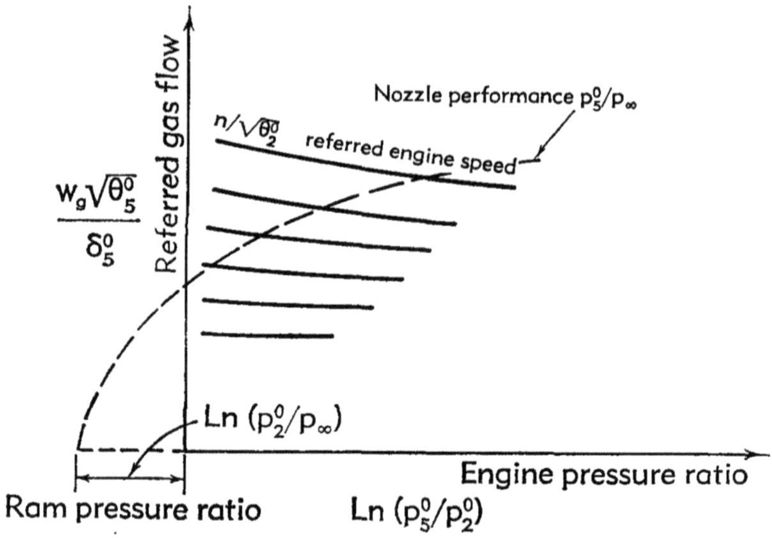

Fig. C,7x. Matching the jet nozzle characteristic with the engine gas generator performance.

can be solved by adding parameters of compressor pressure ratio to this figure.

Matching the two-spool engine. The matching of the components of a two-spool engine can be carried out in a similar fashion to that employed for the single-spool engine. Obviously the iterative process is more complicated due to the requirement of matching the speed, horsepower, compressor and turbine flows, and pressure ratios for the second shaft.

Here again it is helpful to remember certain points of experience which aid considerably in matching the components of the two-spool engine.

1. The referred flow of both compressors is dependent on their referred speed and only slightly influenced by their pressure ratios.
2. The referred flow of both turbines is essentially constant and slightly affected by their pressure ratios or speeds.

An assumed referred speed of the first compressor leads to an estimate of the referred flow. For an assumed top temperature the pressure before the first turbine can be estimated from its referred flow. Also the pressure before the second turbine can be estimated from its referred flow, once an assumption of the temperature before it is made. Thus the pressure ratio across the first turbine is approximately constant. The power output of the first turbine can now be estimated if its speed is assumed. Since the second compressor power must match the first turbine power one can iterate to find the center spool speed and the pressure before the second compressor. In order to do this the back pressure of the second compressor can be assumed to be the turbine inlet pressure plus an estimated burner pressure drop.

Once the pressure after the first compressor is found, the first compressor power can be found and matched with the second turbine power to find the pressure and temperature after the second turbine. This pressure and temperature, together with the gas flow, fixes the exhaust nozzle area.

If a match for a given nozzle area is required one must start over at the beginning with a different assumed speed and iterate until the required nozzle area is matched.

Fortunately the use of high speed electronic computers has practically eliminated the large amount of work required to carry out the iterative calculations outlined above.

CHAPTER 3. ENGINE PERFORMANCE, CONTROL, AND INSTALLATION

C,8. Performance Curves. A typical set of turbojet engine performance curves is sketched in Fig. C,8a. Several curves similar to Fig. C,8a are usually drawn for different standard altitudes and represent the ideal engine performance without duct losses, without air bleeding, and without power extraction from the gear box.

Loss correction factors. The engine performance for the case with inlet or exhaust duct losses, air bleeding, power extraction or nonstandard atmospheric temperature can be obtained by means of the correction equations (Table C,8a) using the correction factors shown in Fig. C,8b(1) to (10). The correction factors have been derived [*37*] by differentiating and combining the equations used in the cycle calculations. For instance, the effect of a one-per cent loss in engine inlet total pressure is traced through the engine to obtain successively its effect on air flow, compressor outlet pressure and temperature, fuel flow, top temperature, nozzle inlet pressure and temperature, jet velocity, gross thrust, and ram drag, and eventually the effect on net thrust is found. The correction factors themselves furnish a quick estimate of the loss of thrust and increased fuel consumption due to duct losses, air bleeding, and power extraction. The formulas for the correction factors are given in Table C,8b.

It should be noted that only the regions of interest on the performance curves (Fig. C,8a) need be corrected for duct losses, air bleeding, or power extraction. For instance only a short section of the military engine thrust curve near its intersection with the airplane drag polar needs to be corrected in order to obtain the airplane maximum flight speed. Similarly only the section of the curves in the region where the airplane drag polar is tangent to the "miles per pound" curves need be corrected if one is interested in obtaining the maximum airplane range.

Engine characteristics. Driggs and Lancaster [*1*] have assembled data on many of the most important characteristics of turbojet engines for the period 1937 to 1952. (The curves were extended by Lancaster in 1957.)

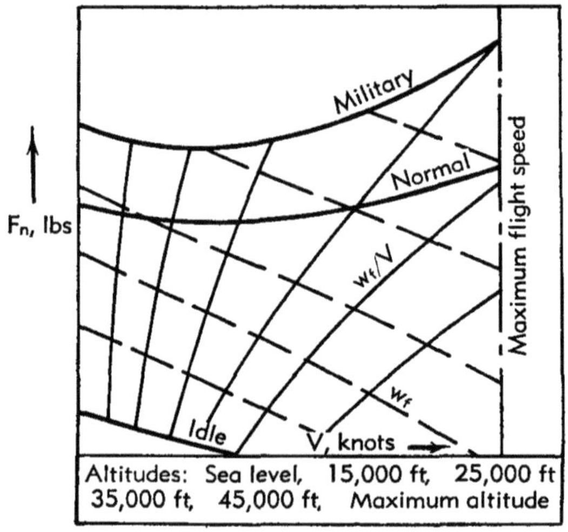

Fig. C,8a. A typical set of turbojet performance curves.

Their curves, reproduced here as Fig. C,8c to C,8h, show typical values of the important quantities such as thrust, specific fuel consumption, thrust per unit area, etc. The actual trend in improvement of these quantities is probably greater than shown by the curves due to the omission of data for the latest military engines.

The static military thrust of engines has been increasing at a rapid rate (Fig. C,8c). However, there appears to be a leveling off in the size of the centrifugal compressor type. This is possibly due to the difficulty in supplying larger quantities of air to the inner runner of the centrifugal compressor and also due to the larger drag of the centrifugal type, particularly when applied to high speed aircraft.

Axial flow engines of 18,000-lb thrust [*38*] are being developed.

The thrust of a turbojet engine varies as the square of its linear dimension, i.e. doubling the diameter increases the thrust by a factor of four. The weight of an engine, all other things being equal, is proportional

Table C,8a. Correction equations.

The per cent reduction in net thrust ($\Delta F_n/F_n$) or fuel flow ($\Delta w_f/w_f$) due to inlet duct losses, exhaust duct losses, bleeding air from compressor outlet, extracting power from the gearbox, and ambient temperature variation taken individually or in combination can be obtained for any altitude and flight speed by means of the following equations:

$$\text{(i)}\quad \frac{\Delta F_n}{F_n} = \underbrace{(1 + C_d)\left(\frac{\Delta p_2^0}{p_2^0}\right)_d}_{\text{inlet duct loss}} + \underbrace{C_d\left(\frac{\Delta p_5^0}{p_5^0}\right)_d}_{\text{exhaust duct loss}} + \underbrace{C_{bl}\left(\frac{w_{bl}}{w_a}\right)}_{\text{air bleed}} + \underbrace{C_{px}\left(\frac{(\text{hp})_{ext}}{(\text{hp})_{ref}}\right)}_{\text{power extraction}}$$
$$+ \underbrace{C_{T1}\left(\frac{\Delta T_\infty}{T_\infty}\right) + C_{T2}\left(\frac{\Delta w_a}{w_a}\right)_T}_{\text{ambient temperature variation}}$$

$$\text{(ii)}\quad \frac{\Delta w_f}{w_f} = \underbrace{\left(\frac{\Delta p_2^0}{p_2^0}\right)_d}_{\text{inlet}} + \underbrace{C_d'\left(\frac{\Delta p_5^0}{p_5^0}\right)_d}_{\text{exhaust}} + \underbrace{\left[1 + C_{bl}'\left(1 + \frac{w_{bl}}{w_a}\right)\right]\left(\frac{w_{bl}}{w_a}\right)}_{\text{air bleed}} + \underbrace{C_{px}'\left(\frac{\text{hp}_{ext}}{\text{hp}_{ref}}\right)}_{\text{power extraction}}$$

duct loss

$$+ \underbrace{C_{T1}'\left(1 + \frac{\Delta w_a}{w_a}\right)\left(\frac{\Delta T_\infty}{T_\infty}\right) + \left[-1 + C_{T2}'\left(1 + \frac{\Delta w_a}{w_a}\right)\right]\left(\frac{\Delta w_a}{w_a}\right)_T}_{\text{ambient temperature variation}}$$

$$\text{(iii)}\quad \underbrace{\left(\frac{\Delta w_a}{w_a}\right) = \frac{\Delta\Gamma}{\Gamma} - \bar{M}\left(\frac{\Delta T_\infty}{T_\infty}\right)}_{\text{ambient temperature variation}} \qquad \text{where} \quad \bar{M} = \frac{M_\infty^2 \dfrac{\gamma}{2} + \dfrac{1}{2}}{M_\infty^2 \dfrac{\gamma - 1}{2} + 1}$$

where

$\dfrac{\Delta p_2^0}{p_2^0}$	per cent loss in compressor inlet total pressure due to inlet duct loss
$\dfrac{\Delta p_5^0}{p_5^0}$	per cent loss in total pressure between turbine outlet and exhaust nozzle inlet
$\dfrac{w_{bl}}{w_a}$	per cent of total engine air flow bled from compressor outlet
$\dfrac{(\text{hp})_{ext}}{(\text{hp})_{ref}}$	per cent of reference horsepower extracted from the gearbox for aircraft accessories.
$(\text{hp})_{ref}$	reference horsepower arbitrarily taken as ten per cent of the sea level static military jet horsepower (given on performance curves)
$\dfrac{\Delta T_\infty}{T_\infty}$	per cent increase in ambient temperature over the standard NACA ambient temperature for the altitude concerned
$\left(\dfrac{\Delta w_a}{w_a}\right)_T$	per cent increase in actual engine air flow due to ambient temperature variation
$\dfrac{\Delta\Gamma}{\Gamma}$	per cent increase in referred engine air flow due to ambient temperature variation
C_d, C_{bl}, C_{px}, C_{T1}, C_{T2}	thrust correction factors for duct loss, air bleed, power extraction, and ambient temperature variation
C_d', C_{bl}', C_{px}', C_{T1}', C_{T2}'	fuel flow correction factors for duct loss, air bleed, power extraction, and ambient temperature variation

Table C,8b. Correction factors

	Inlet duct loss $\left(\frac{\Delta p_2^0}{p_2^0}\right)^*_d$	Exhaust duct loss $\left(\frac{\Delta p_5^0}{p_5^0}\right)^*_d$	Air bleed $\left(\frac{w_{bl}}{w_a}\right)$	Power extraction $\left[\frac{(hp)_{ext}}{(hp)_{ref}}\right]$
$\frac{\Delta F^*_{net}}{F_{net}}$	$1 + C_d = 1 + \frac{F_g\mathcal{R}(T_e)_{is}}{F_{net}2J(h_n)_{is}}$	$C_d = \frac{F_g\mathcal{R}(T_e)_{is}}{F_{net}2J(h_n)_{is}}$	$C_{bl} = \frac{F_g}{F_{net}}\left\{1 + \frac{\mathcal{R}(T_e)_{is}}{2J(h_n)_{is}} + \frac{1}{2(h_n)_{is}}\left[h_t - \frac{\mathcal{R}T_3^0}{J\eta_c}\right]\left[1 + \frac{(T_e)_{is}}{T_5^0}\left(\frac{1-\eta_t}{\eta_t}\right)\right]\right\}$	$C_{p_x} = \frac{F_g2545(hp)_{ref}}{F_{net}3600w_a2(h_n)_{is}}\left[1 + \frac{(T_e)_{is}}{T_5^0}\left(\frac{1-\eta_t}{\eta_t}\right)\right]$
$\frac{\Delta w_f^*}{w_f}$	$1 = 1$	$C'_d = 0$	$1 + C'_{bl}[1 + (w_{bl}/w_a)] = 1 + \frac{\mathcal{R}T_3^0}{J\eta_c h_b}\left[1 + \frac{w_{bl}}{w_a}\right]$	$C'_{p_x} = 0$
$\frac{\Delta F^*_{net}}{F_{net}}$	$1 + C_d = 1 + \frac{F_g\mathcal{R}(T_e)_{is}}{F_{net}2J(h_n)_{is}}$	$C_d = \frac{F_g\mathcal{R}(T_e)_{is}}{F_{net}2J(h_n)_{is}}$	$C_{bl} = \frac{F_g}{F_{net}}\left\{1 + \frac{\mathcal{R}(T_e)_{is}}{2J(h_n)_{is}} + \frac{B}{2(h_n)_{is}}\left[h_t - \frac{\mathcal{R}T_3^0}{J\eta_c}\right]\right\}$	$C_{p_x} = \frac{F_g2545(hp)_{ref}B}{F_{net}3600w_a2(h_n)_{is}}$
$\frac{\Delta w_f^*}{w_f}$	$1 = 1$	$C_d = 0$	$1 + C'_{bl}[1 + (w_{bl}/w_a)] = 1 - \frac{1}{h_b}\left[h_t - \frac{2\mathcal{R}T_3^0}{J\eta_c h_b}\right]\left[1 + \frac{w_{bl}}{w_a}\right]$	$C'_{p_x} = -\frac{2545(hp)_{ref}}{3600w_a h_b}$
$\frac{\Delta F^*_{net}}{F_{net}}$	$1 + C_d = 1 + \frac{F_g\mathcal{R}(T_e)_{is}}{F_{net}2J(h_n)_{is}}$	$C_d = \frac{F_g}{F_{net}2(h_n)_{is}}\left(\frac{\mathcal{R}(T_e)_{is}}{J} - C\right)$	$C_{bl} = \frac{F_g}{F_{net}}\left\{1 + \frac{\mathcal{R}(T_e)_{is}}{2J(h_n)_{is}} + \frac{1}{2(h_n)_{is}}\left[h_t - \frac{\mathcal{R}T_3^0}{J\eta_c}\right]\left[1 + \frac{(T_e)_{is}}{T_5^0}\left(\frac{1-\eta}{\eta}\right) + \frac{C}{c_pT_5^0}\left(\frac{1}{2} - \frac{\gamma}{\gamma - 1\eta_t}\right)\right]\right.$	$C_{p_x} = \frac{F_g2545(hp)_{ref}}{F_{net}3600w_a2(h_n)_{is}}\left[1 + \frac{(T_e)_{is}}{T_5^0}\left(\frac{1-\eta_t}{\eta_t}\right) + \frac{C}{c_pT_5^0}\left(\frac{1}{2} - \frac{\gamma}{\gamma - 1\eta_t}\right)\right]$
$\frac{\Delta w_f^*}{w_f}$	$1 = 1$	$C'_d = -\frac{D}{h_b}$	$1 + C'_{bl}[1 + (w_{bl}/w_a)] = 1 + \frac{1}{h_b}\left[\frac{\mathcal{R}T_3^0}{J\eta_c} + \frac{D}{c_pT_5^0}\left(h_t - \frac{\mathcal{R}T_3^0}{J\eta_c}\right)\left(\frac{1}{2} - \frac{\gamma}{\gamma - 1\eta_t}\right)\left(1 + \frac{w_{bl}}{w_a}\right)\right]$	$C'_{p_x} = \frac{2545(hp)_{ref}}{3600w_ac_pT_5^0h_b}\left(\frac{1}{2} - \frac{\gamma}{\gamma - 1\eta_t}\right)$

$$B = 1 + \frac{(T_e)_{is}}{T_5^0}\left(\frac{1-\eta_t}{\eta_t}\right) - \frac{1 - \frac{(T_e)_{is}}{2T_4^0}\left(\frac{\gamma+1}{\gamma}\right) - \frac{(\gamma-1)T_3^0}{2\gamma\eta_cT_4^0}\left[1 + \frac{(T_e)_{is}}{T_5^0}\left(\frac{1-\eta_t}{\eta_t}\right)\right]}{1 - \frac{(\gamma-1)T_3^0}{2\gamma\eta_cT_4^0}}$$

* The signs of the coefficients have been changed so that a decrease of these quantities is positive.

for three typical engine controls.

Ambient temperature variation $\left(\frac{\Delta T_\infty}{T_\infty}\right)$	$\left(\frac{\Delta w_a}{w_a}\right)_T$	Control
$C_{T1} = \frac{F_g c_p T_\infty}{F_{net} 2\eta_c (h_n)_{is}} \left[\frac{T_3^0}{T_\infty} - 1\right] \left[1 + \frac{(T_e)_{is}}{T_5^0}\left(\frac{1-\eta_t}{\eta_t}\right)\right]$	$C_{T2} = -1 - \frac{F_g}{F_{net} 2 (h_n)_{is}} \left\{\frac{\mathcal{R}(T_e)_{is}}{J} - \frac{\mathcal{R}T_3^0}{J\eta_c}\left[1 + \frac{(T_e)_{is}}{T_5^0}\left(\frac{1-\eta_t}{\eta_t}\right)\right]\right\}$	Constant turbine inlet temperature $\frac{\Delta T_4^0}{T_4^0} = 0$
$C'_{T1}[1 + (\Delta w_a/w_a)] = \frac{c_p T_\infty}{\eta_c h_b}\left[\frac{T_3^0}{T_\infty} - 1 + \eta_c\right]\left[1 + \frac{\Delta w_a}{w_a}\right]$	$-1 + C'_{T2}[1 + (\Delta w_a/w_a)] = -1 + \frac{\mathcal{R}T_3^0}{J\eta_c h_b}\left[1 + \frac{\Delta w_a}{w_a}\right]$	
$C_{T1} = \frac{F_g c_p T_\infty}{F_{net} 2\eta_c (h_n)_{is}}\left(\frac{T_3^0}{T_\infty} - 1\right) B$	$C_{T2} = -1 - \frac{F_g}{F_{net} 2 (h_n)_{is}}\left[\frac{\mathcal{R}(T_e)_{is}}{J} - \frac{\mathcal{R}T_3^0 B}{J\eta_c}\right]$	Constant turbine outlet temperature $\frac{\Delta T_5^0}{T_5^0} = 0$
$C'_{T1}[1 + (\Delta w_a/w_a)] = \frac{c_p T_\infty}{h_b}\left[1 + \frac{\Delta w_a}{w_a}\right]$	$-1 + C'_{T2}[1 + (\Delta w_a/w_a)] = -1$	
$C_{T1} = \frac{F_g c_p T_\infty}{F_{net} 2\eta_c (h_n)_{is}}\left[\frac{T_3^0}{T_\infty} - 1\right]\left[1 + \frac{(T_e)_{is}}{T_5^0}\left(\frac{1-\eta_t}{\eta_t}\right) + \frac{C}{c_p T_5^0}\left(\frac{1}{2} - \frac{\gamma}{\gamma - 1\eta_t}\right)\right]$	$C_{T2} = -1 - \frac{F_g}{F_{net} 2 (h_n)_{is}}\left\{\frac{\mathcal{R}(T_e)_{is}}{J} - \frac{\mathcal{R}T_3^0}{J\eta_c}\left[1 + \frac{(T_e)_{is}}{T_5^0}\left(\frac{1-\eta_t}{\eta_t}\right) + \frac{C}{c_p T_5^0}\left(\frac{1}{2} - \frac{\gamma}{\gamma - 1\eta_t}\right)\right]\right\}$	Constant exhaust nozzle area $\frac{\Delta A}{A} = 0$
$C'_{T1}[1 + (\Delta w_a/w_a)] = \frac{c_p T_\infty}{\eta_c h_b}\left\{\eta_c + \left[\frac{T_3^0}{T_\infty} - 1\right]\left[1 + \frac{D}{c_p T_5^0}\left(\frac{1}{2} - \frac{\gamma}{\gamma - 1\eta_t}\right)\right]\right\} \times \left[1 + \frac{\Delta w_a}{w_a}\right]$	$-1 + C_{T2}[1 + (\Delta w_a/w_a)] = -1 + \frac{\mathcal{R}T_3^0}{J\eta_c h_b}\left[1 + \frac{D}{c_p T_5^0} \times \left(\frac{1}{2} - \frac{\gamma}{\gamma - 1\eta_t}\right)\right]\left[1 + \frac{\Delta w_a}{w_a}\right]$	

$$C = \frac{c_p T_4^0 - \frac{\mathcal{R}(T_e)_{is}}{2J}\left(\frac{\gamma + 1}{\gamma - 1}\right) - \frac{\mathcal{R}T_3^0}{2J\eta_c}\left[1 + \frac{(T_e)_{is}}{T_5^0}\left(\frac{1-\eta_t}{\eta_t}\right)\right]}{\left[\frac{T_4^0}{T_5^0} - 1\right]\frac{\gamma + 1}{2(\gamma - 1)} + \frac{(\gamma - 1)T_3^0}{2\gamma\eta_c T_5^0}\left[\frac{1}{2} - \frac{\gamma}{\gamma - 1\eta_t}\right]}$$

$$D = \frac{c_p T_5^0 - \frac{\mathcal{R}T_3^0}{2J\eta_c}}{\left[\frac{T_4^0}{T_5^0} - 1\right]\frac{\gamma + 1}{2(\gamma - 1)} + \frac{(\gamma - 1)T_3^0}{2\gamma\eta_c T_5^0}\left[\frac{1}{2} - \frac{\gamma}{\gamma - 1\eta_t}\right]}$$

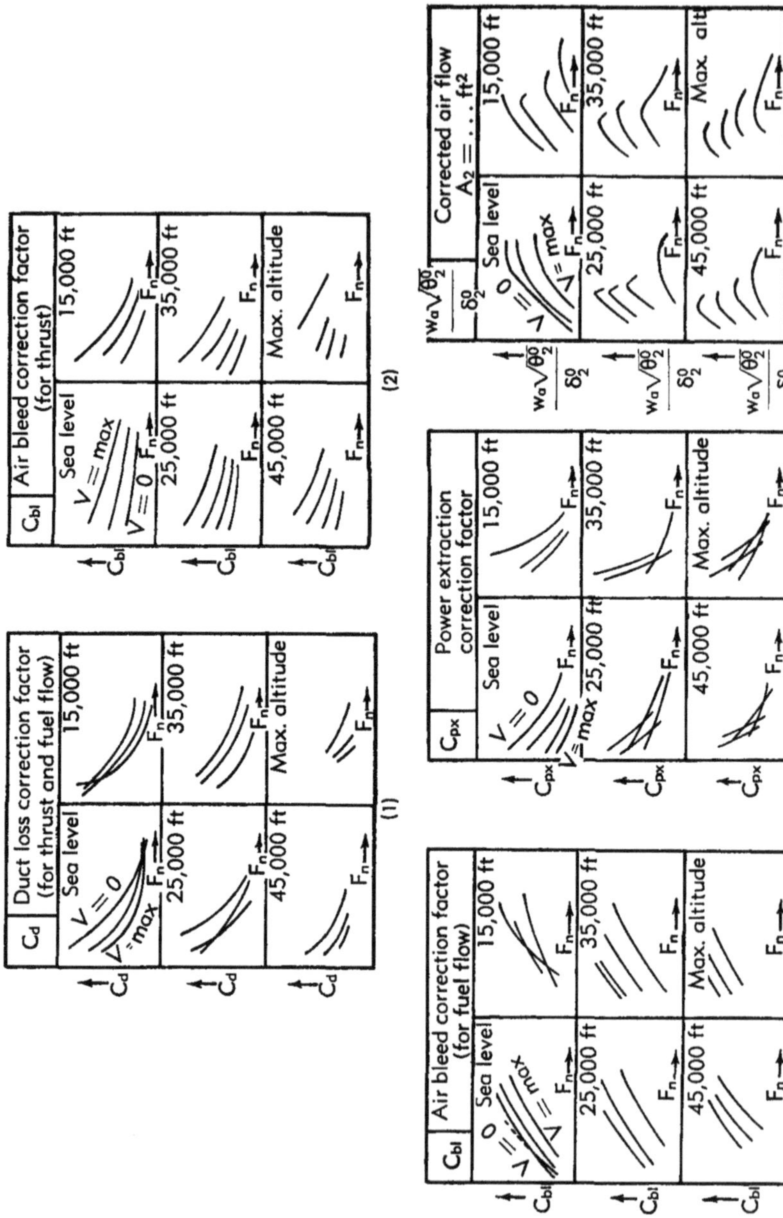
Duct loss correction factor (for thrust and fuel flow)
Air bleed correction factor (for thrust)
Air bleed correction factor (for fuel flow)
Power extraction correction factor
Corrected air flow
Sea level
15,000 ft
25,000 ft
35,000 ft
45,000 ft
Max. altitude
(1)
(2)
(3)
(4)
(5)

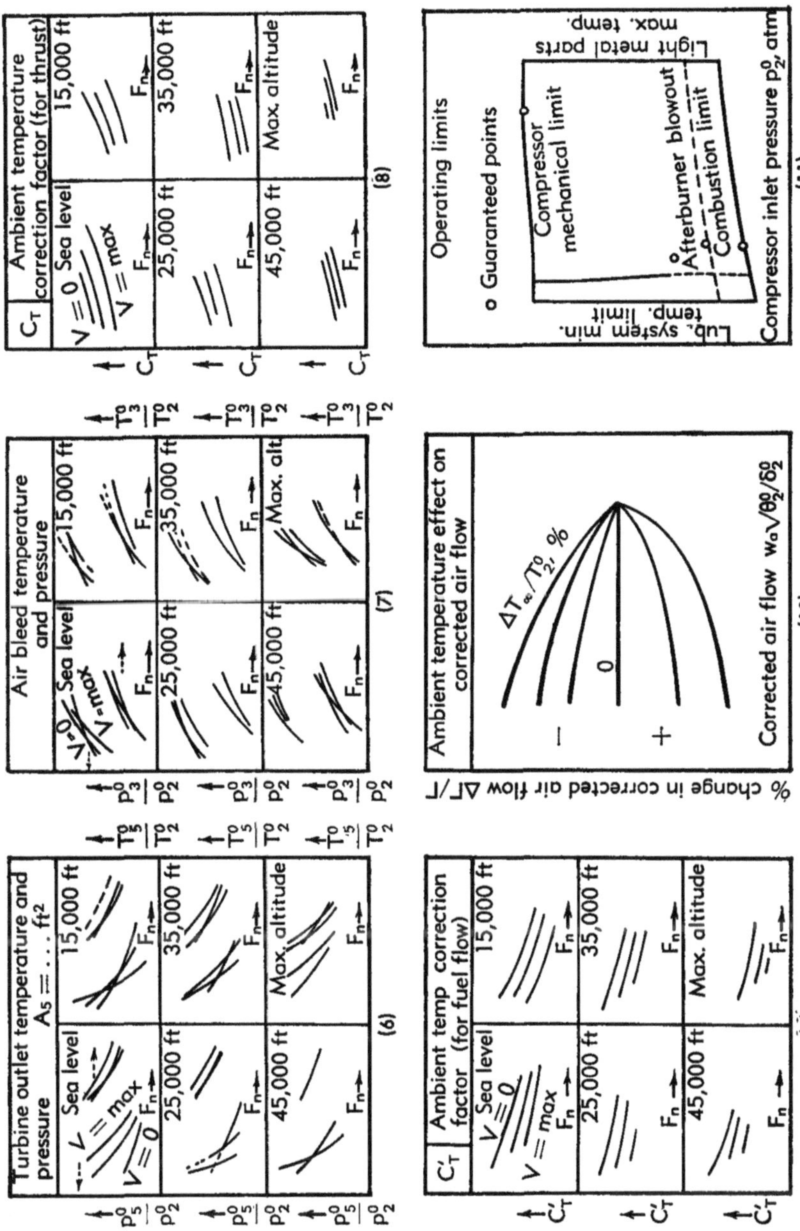

Fig. C,8b. Thrust correction curves.

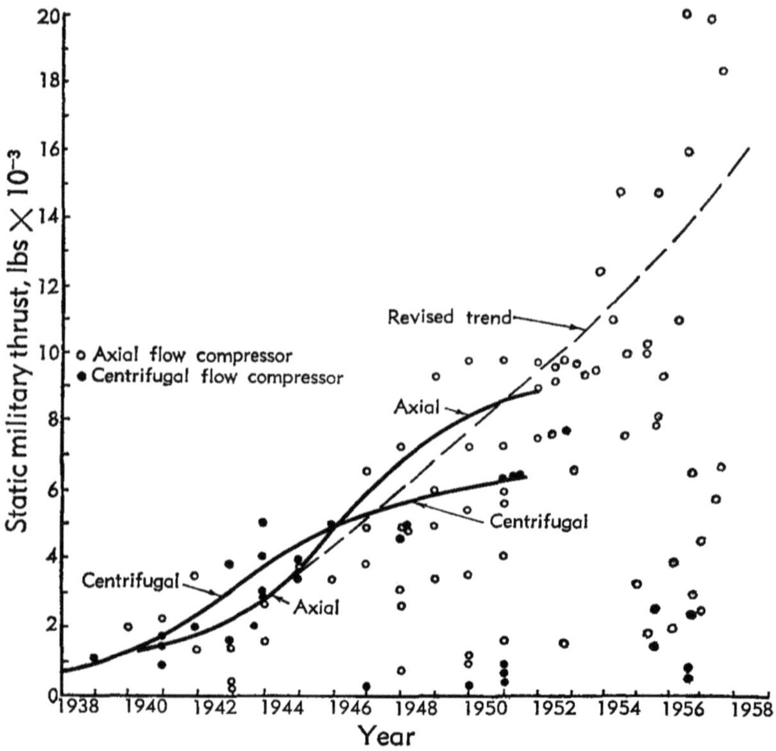

Fig. C,8c. Thrust of turbojets [1].

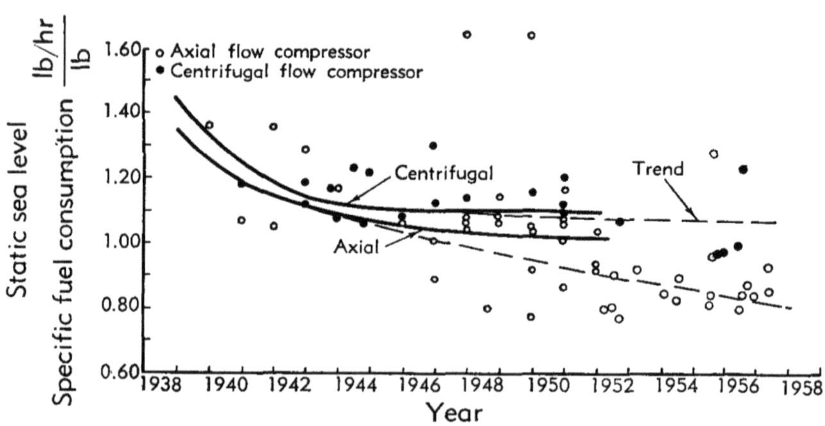

Fig. C,8d. Specific fuel consumption of turbojets [1].

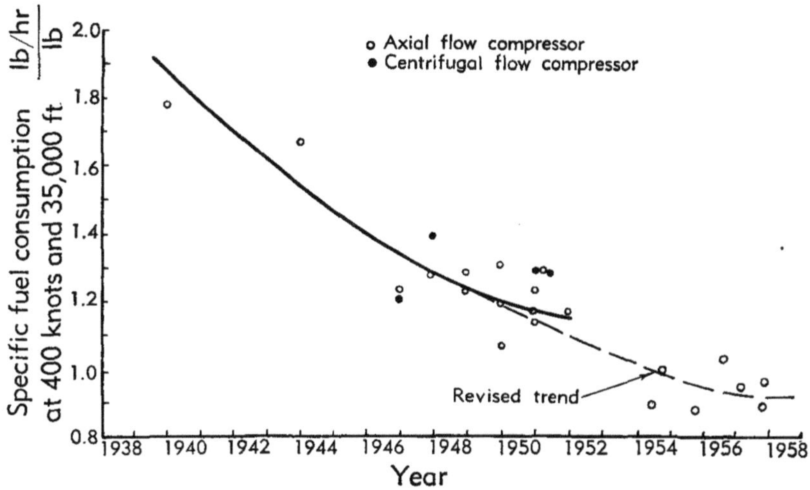

FIG. C,8e. Specific fuel consumption of turbojets at 400 knots and 35,000 ft. [1].

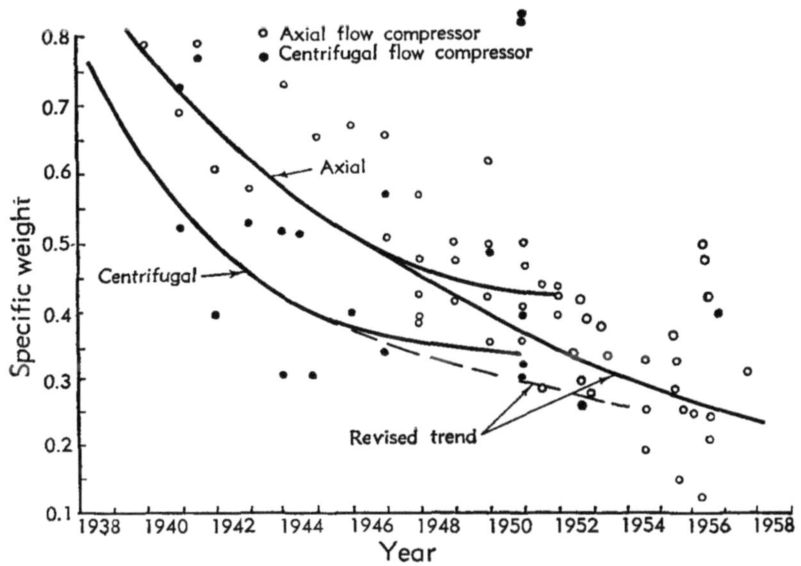

Fig. C,8f. Specific weight of turbojets [1].

to the cube of its linear dimension, that is, doubling its size causes the weight to increase eight times.

Thus, since the thrust of an engine varies as the square of its linear dimension and its weight varies as the cube of its linear dimension, one would expect to reach an optimum size giving minimum weight per pound of thrust. Technological advances in aerodynamics, strength, and weight control have in the past hidden the "square-cube law" so that larger engines are being designed with smaller specific weights (Fig. C,8f). Evidently there are other effects. The square-cube law assumes "all other things being equal" which is not exactly true. There are improvements in Reynolds number, and therefore efficiency, of the larger sizes, together

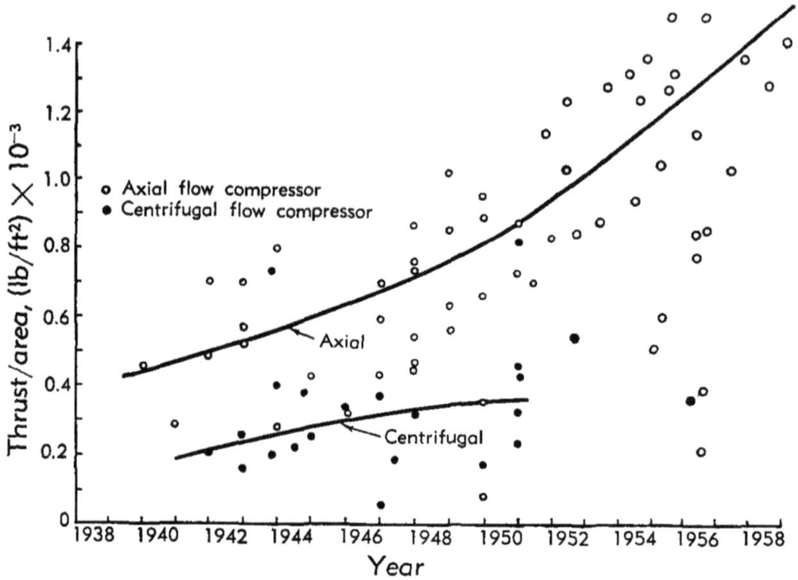

Fig. C,8g. Ratio of thrust to frontal area for turbojets [1].

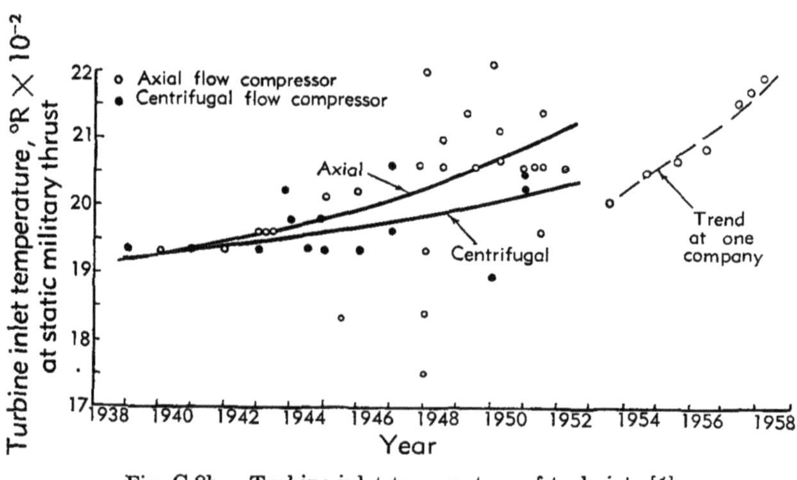

Fig. C,8h. Turbine inlet temperature of turbojets [1].

with reductives in relative dimensions such as blade tip clearances or sheet metal thickness, etc. which are established for purely manufacturing reasons.

The turbojet engine has not as yet run into a basic limit in size as did the piston engine with its cooling problems. The principal limit on turbojet engine size appears to be manufacturing limitations such as the size of satisfactory spindle forgings. Engine size may also be limited by the airplane size which has been the case for carrier aircraft.

The centrifugal compressor-type engine shows a definite advantage in specific weight over the axial flow type due mostly to savings in compressor weight. The axial flow type has advantages in both frontal area (Fig. C,8g) and specific fuel consumption (Fig. C,8d and C,8e). The latter advantage is due to the better efficiency (Fig. C,7g) of the axial flow compressor.

The ratio of thrust to frontal area (Fig. C,8g) is generally intended to show the relative drag of different engines. This is a hold-over from piston engine practice where the engine actually caused air drag. The air flow through the jet engine causes a net thrust on the engine mounts. Therefore the only drag associated with the engine is the nacelle drag (Eq. 2-2) which is ideally zero but which is probably dependent, not on the total engine frontal area, but on the annular frontal area outside the inlet stream tube and inside the maximum engine envelope. An attempt to indicate this is shown on the relative frontal areas (Fig. C,4e).

An important characteristic of the turbojet engine is the turbine inlet temperature which influences both the thermal efficiency and the output per pound of air swallowed by the engine. These in turn influence the specific thrust and specific fuel consumption. The trend in turbine inlet temperature is shown in Fig. C,8h where temperatures of 1700°F (2160°R) are shown. Turbine temperature and engine life are of course closely related. Some expendable engines for missiles are operated at very high temperatures due to their very short life.

C,9. Effect of Varying Humidity on Engine Performance. Tests of engines have shown that the effect of humidity on engine performance is generally small [*39*]. Some simplification is therefore justified in obtaining humidity correction factors. If one assumes that the change in humidity does not alter the engine component efficiencies and, further, that the Mach numbers and temperature ratios throughout the engine are held constant, one has:

For speed

$$\frac{n_m}{n} = \frac{a_m}{a_a} = \sqrt{\frac{\gamma_m \mathcal{R}_m}{\gamma_a \mathcal{R}_a}} \qquad \text{(assumed)} \tag{9-1}$$

For air flow

$$\frac{w_{a_m}}{w_a} = \sqrt{\frac{\gamma_m \mathcal{R}_a}{\gamma_a \mathcal{R}_m}} \tag{9-2}$$

For thrust

$$\frac{F_{g_m}}{F_g} = \frac{\gamma_m}{\gamma_a} \tag{9-3}$$

For fuel flow

$$\frac{w_{f_m}}{w_f} = \sqrt{\frac{\gamma_m \mathcal{R}_a}{\gamma_a \mathcal{R}_m} \frac{c_{p_m}}{c_{p_a}}} \tag{9-4}$$

where the subscripts $_m$ and $_a$ refer to moist and dry air respectively.

The above correction factors, devised by Samuels and Gale [*39*], have been found to agree quite well with test results. If one introduces the variation of the thermal properties with the change in specific humidity (Δm grains/lb of dry air), the above equations can be brought into more useful forms

$$\frac{\Delta n}{n} = \frac{\Delta a}{a} = +3.3 \times 10^{-5} \Delta m \tag{9-5}$$

$$\frac{\Delta w_a}{w_a} = -4.7 \times 10^{-5} \Delta m \tag{9-6}$$

$$\frac{\Delta F_{net}}{F_{net}} = \frac{\Delta F_g}{F_g} = -1.5 \times 10^{-5} \Delta m \tag{9-7}$$

$$\frac{\Delta w_f}{w_f} = +7.0 \times 10^{-5} \Delta m \tag{9-8}$$

The calculated variation of the maximum (saturation) specific humidity with altitude is shown in Fig. C,9 for standard NACA atmos-

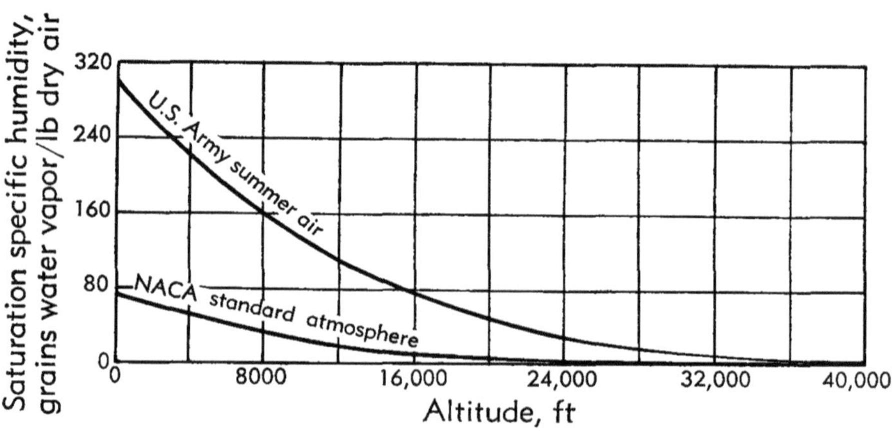

Fig. C,9. Saturation specific humidity for various altitudes.

phere and also for a U.S. Army hot day. One can see that for the worst case (Army hot-day sea level) we have (1) an increase of acoustic velocity and speed by 1 per cent, (2) a reduction of air flow by 1.4 per cent, (3) a reduction of thrust by 0.45 per cent, (4) an increase of fuel flow by 2.1 per cent, and (5) an increase of specific fuel by 2.65 per cent. One notices that the humidity corrections become small above an altitude of 15,000 feet.

Plate C,1a. W-1X engine. Sir Frank Whittle and associates with an early gas turbine in a disused foundary at Lutterworth. From film "Wonder Jet." (Courtesy of British Information Services.)

Plate C,1b. J33 engine. (Courtesy of General Electric Company.)

Plate C,1c. J48 engine, 7250-lb thrust. (Courtesy of Pratt and Whitney Aircraft.)

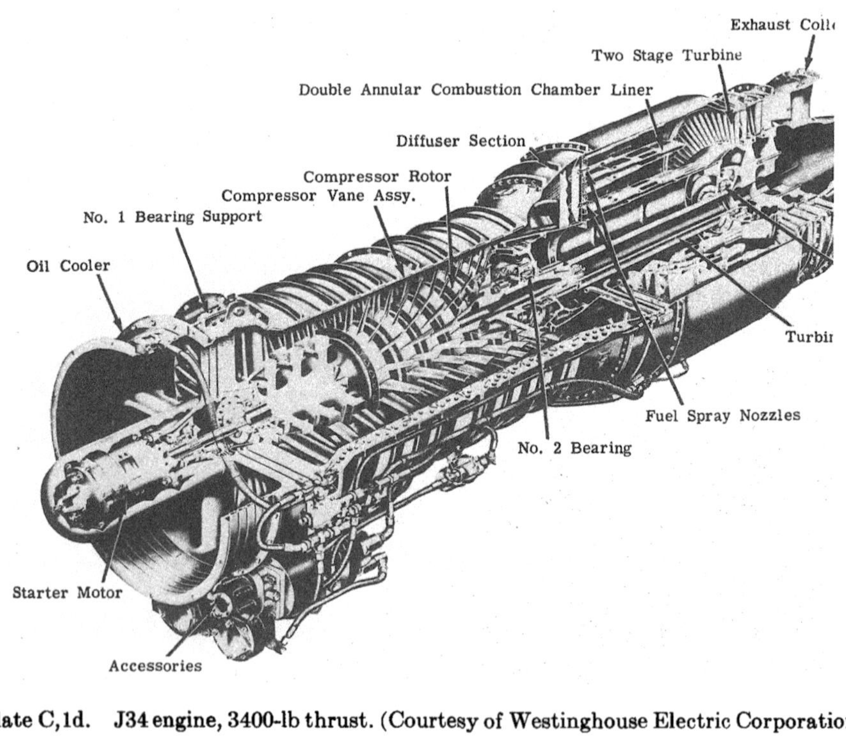

Plate C,1d. J34 engine, 3400-lb thrust. (Courtesy of Westinghouse Electric Corporation.)

Plate C,1e. ORENDA engine, 7000⁺-lb thrust. (Courtesy of A. V. Roe, Canada Limited.)

Plate C,1f. J71 engine, 10,000⁺-lb thrust. (Courtesy Allison Division, General Motors Corporation.)

Plate C,1g. J65 engine, 7200-lb thrust, British Armstrong Siddeley Sapphire. (Courtesy Curtiss-Wright Corporation.)

Plate C,1h. J73 engine, 9000-lb thrust. (Courtesy of General Electric Company.)

Plate C,1i. J75 engine, 10,000-lb thrust. (Courtesy of Pratt and Whitney Aircraft.)

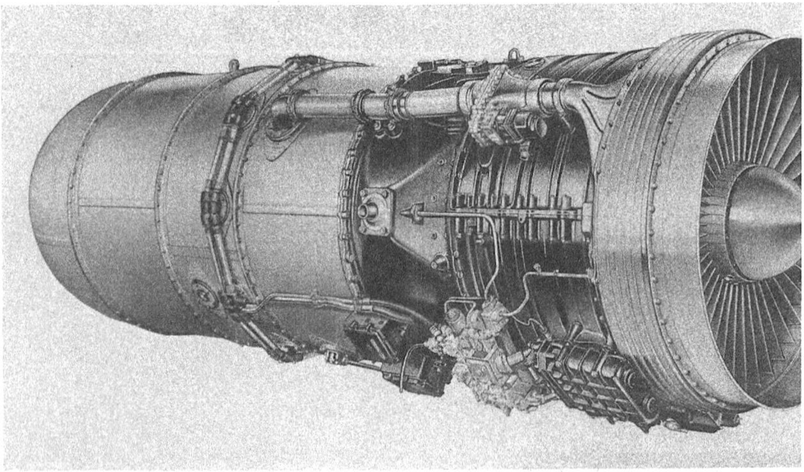

Plate C,1j. The Rolls-Royce "Conway" bypass turbojet, 13,000-lb thrust. (Courtesy of British Information Services.)

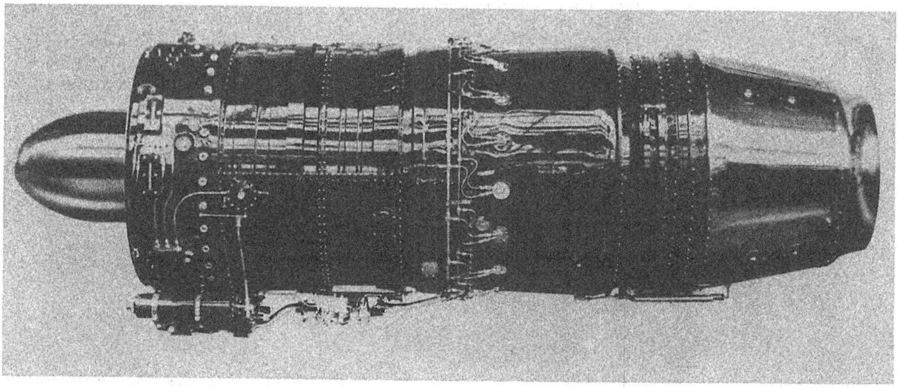

Plate C,1k. The DeHavilland "Gyron," 15,000-lb thrust. (Courtesy of British Information Services.)

Plate C,1l. The Bristol "Olympus" Turbojet, 9750-lb thrust. (Courtesy of British Information Services.)

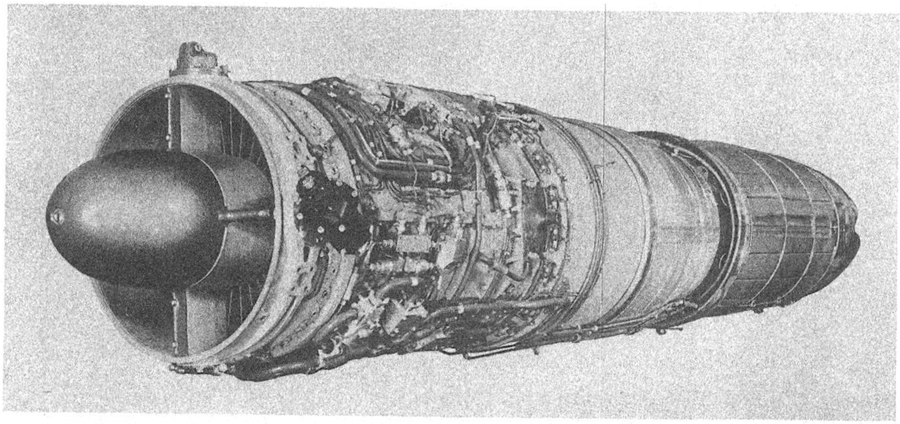

Plate C,1m. The French ATAR—"G" torbojet engine, 9700-lb thrust. (Courtesy of SNECMA.)

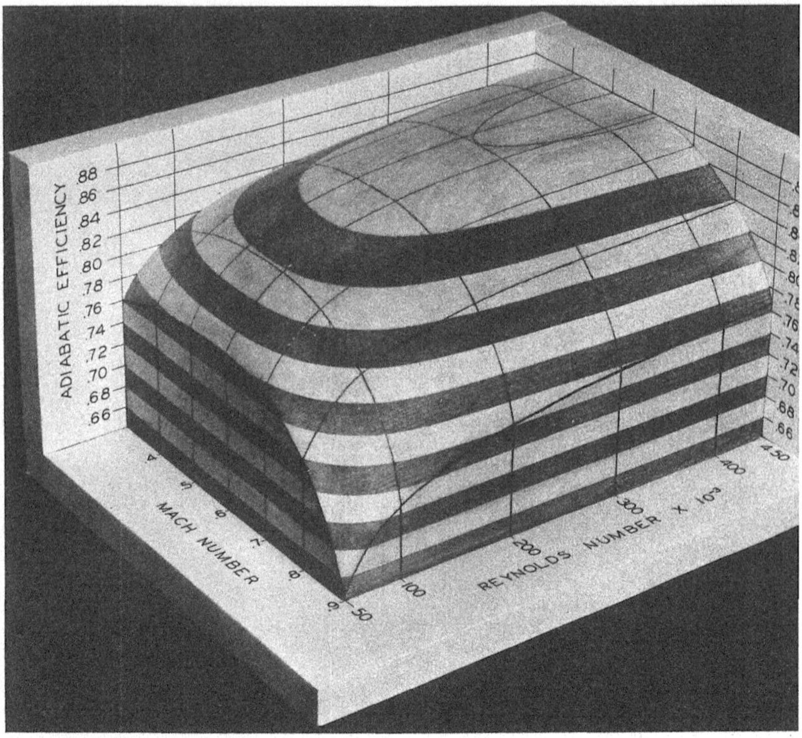

Plate C,7. Mach number and Reynolds number effects on compressor performance.

Plate C,24a. A J47 jet engine being readied for test in a sea level test cell. (Courtesy of General Electric Company.)

Plate C,24b. Sea level static test stand, National Adivsory Committee for Aeronautics.

Plate C,25b. Altitude test tank, National Advisory Committee for Aeronautics.

Plate C,25a. A J40 engine being readied for test in the altitude wind tunnel, National Advisory Committee for Aeronautics, Lewis Propulsion Laboratory, Cleveland, Ohio.

Plate C,25c. SW-24 Altitude test tank, National Advisory Committee for Aeronautics, Cleveland, Ohio.

Plate C,27a. Static test cell showing inlet silencer. (Courtesy of General Electric Company.)

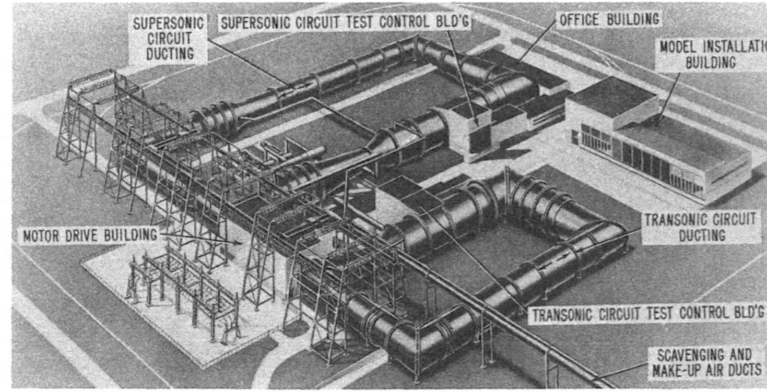

Plate C,27c. Arnold Engineering and Development Center. (Courtesy of U.S. Air Force.)

Plate C,27b. Facilities at Villaroche, France, showing many test cells.

Plate C,27d. Altitude tanks, AEDC. (Courtesy of U.S. Air Force.)

C,10. Effect of Variations of c_p, γ, and μ, Altitude Effects. Generally the effects of changes of c_p and γ on engine performance can be approximately allowed for by calculating the engine performance, using data similar to that given by Keenan and Kaye [*40*] for air. One should remember, however, that the appropriate values of c_p and γ for combustion products should be used in the burner, turbine, and exhaust nozzle calculations. Actually experience has shown that there is very

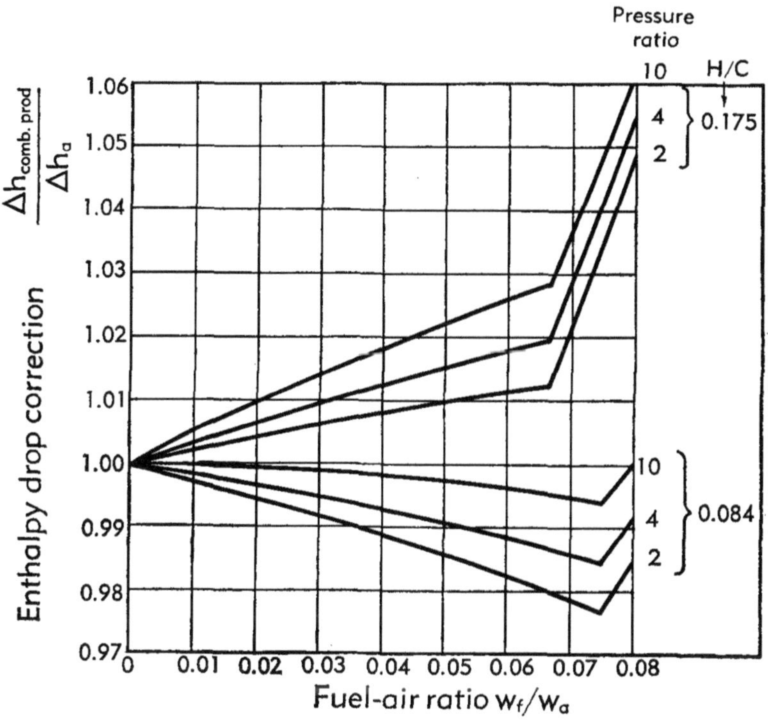

Fig. C,10a. Enthalpy drop correction factor for various air-fuel, hydrogen-carbon ratios and pressure ratios.

little error introduced in the calculated enthalpy drop across the turbine or exhaust nozzle if one uses the properties of air instead of the properties of combustion products. Fig. C,10a shows the ratio of enthalpy drop for combustion products compared to that for air [*41*] to be very small for the usual fuel (hydrogen-carbon) ratios (15/85). This is evidently due to compensating effects of changes in c_p and γ. Although for a given pressure ratio the difference in the value of γ between air and combustion products *leads to large differences in turbine temperature ratio, the difference* in the specific heats c_p compensates and we obtain essentially the same enthalpy drop using air or combustion products. However, this is definitely not the case for the burner calculations where large errors in burner temperature rise result if the appropriate specific heat for the products of combustion is not used. A convenient method of allowing for this difference in specific heats is given in Art. 7 under *The burner*.

One should also bear in mind the fact that changes in c_p and γ cause small changes in densities and velocities throughout the compressor, combustor, turbine, and jet nozzle. Therefore one would expect variations in flow incidence angle to the compressor and turbine blades which in turn would affect the compressor and turbine efficiencies. It is impossible at present to estimate these changes in efficiencies because of changes in c_p and γ. Performance changes due to changes in air viscosity are generally of such a complicated nature that they must be found empirically.

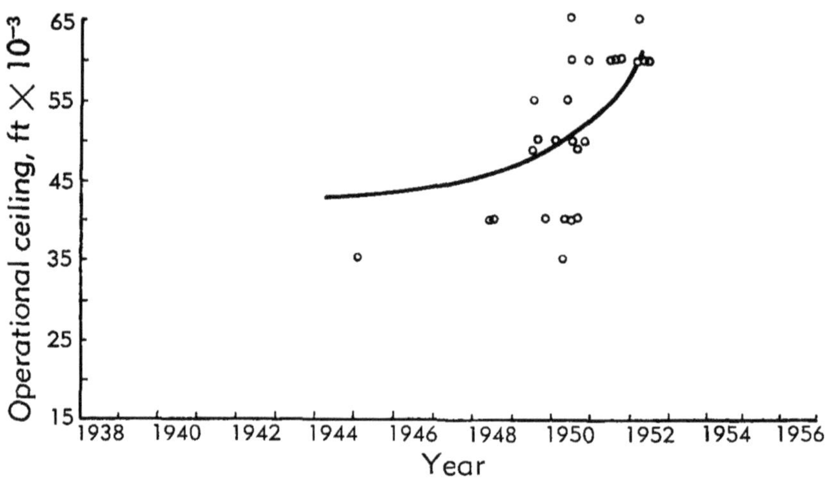

Fig. C,10b. Operational ceiling of turbojets [1].

Other effects are present, such as the differential expansion of the casing, due to high or low ambient temperatures which cause changes in compressor and turbine tip clearances.

Usually it is the practice to lump the effects on performance of c_p, γ, μ, and the differential expansion, etc. under one heading called "altitude effects." This is convenient since the performance curves (Fig. C,8a) are drawn separately for each of several altitudes. The altitude effects are found when the tested performance at several altitudes fails to plot on a single curve when using referred or dimensionless variables. Altitude effects have been found to vary between engines but rarely are present below 25,000 ft altitude; other engines have not shown altitude effects up to 50,000 ft altitude. The first altitude effect to show itself is usually an excessive fuel consumption due to falling burner efficiencies at the high altitude (at low pressures or low Reynolds numbers). At the higher altitudes effects due to Reynolds number on the compressor and turbines begin to appear.

The trend in operational ceiling of turbojets [1] is shown in Fig. C,10b. There is a continuing research effort to improve combustors to permit operation up to 70,000 ft altitude.

The operation of afterburners at altitude is even more difficult than the engine combustor due to the lower level of density as a result of the lower pressure and higher temperature at the afterburner inlet. The altitude limit of the afterburner is generally from ten to twenty thousand feet lower than the basic engine.

C,11. Starting, Windmilling, Ignition, and Acceleration. Due to the poor efficiencies of the engine components and the pressure ratio at low rotational speeds, the turbojet engine is not self-sustaining at speeds below about twenty per cent of full speed. In starting, turbojet engines must therefore be cranked up by the starter to at least twenty per cent of full speed. The required starting horsepower varies for the different engines depending on the size, design pressure ratio, required starting time, etc. Generally the electrical starting power is about four-to-five horsepower for each 1000 pounds of engine rated thrust.

The starting torque is composed of (1) air windage, (2) rotor inertia, and (3) mechanical friction.

Dimensional analysis shows that the air windage torque T_w is related to the speed n, size D, acoustic velocity a, and air density ρ_a as follows

$$\frac{T_w}{\rho_a D^5 n^2} = \text{a function of } \left(\frac{nD}{a}\right) \cong \text{const} \tag{11-1}$$

Generally, since the starting speeds are low and the effect of Mach number is small, the function on the right side of Eq. 11-1 is a constant, and therefore the windage torque is proportional to the density, the square of the speed, and the fifth power of the size D.

If one assumes that a constant acceleration is provided by the starting device, the inertia torque T_i is then related to starting time t, the average rotor material density ρ_m, the ratio φ of starting speed to full speed, the size D, and the full speed n as follows:

$$\frac{T_i t}{\rho_m D^5 n \varphi} = \text{const} \tag{11-2}$$

The mechanical bearing loss and the auxiliary driving gear losses generally vary as the square of the speed, and are greatly influenced by oil temperature and therefore by ambient temperature. The mechanical and gear losses can increase as much as a hundred per cent in going to the extremely low (−67°F) temperatures.

Summarizing we then have:

1. The windage horsepower (proportional to $T_w \times n$) varies directly as the air density ρ_a, the fifth power of the size D, and the cube of the speed n.
2. The inertia horsepower varies inversely as the required starting time t, directly as the fifth power of the size D, and as the square of the speed n.
3. The friction horsepower generally varies as the square of the speed n and in a complicated manner with ambient or oil temperature.

Once the spindle of the turbojet engine is rotated by the starter to a sufficient speed, fuel and ignition must be supplied so that the

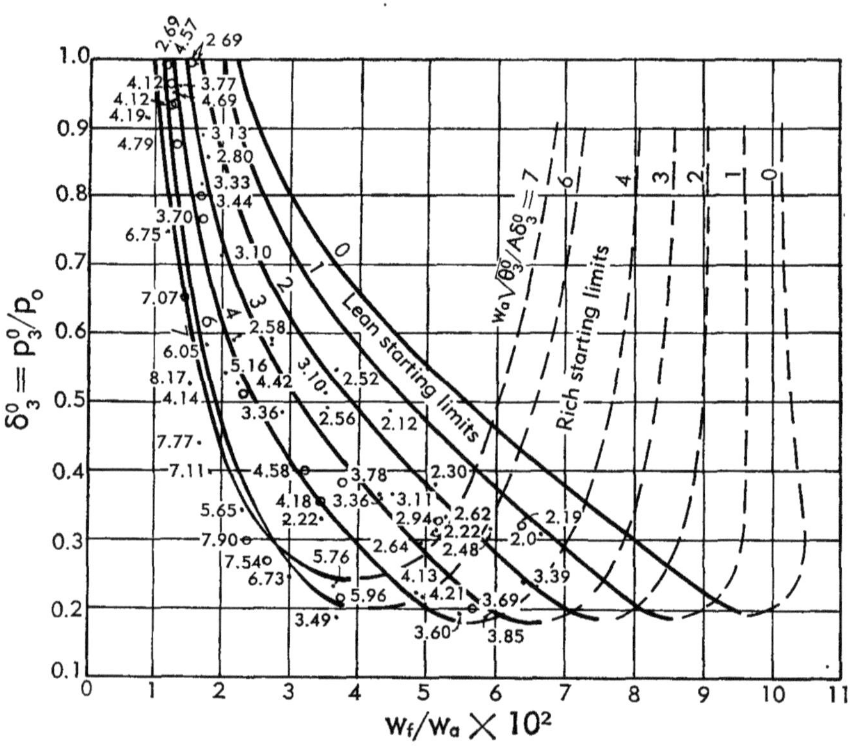

Fig. C,11a. Turbojet engine ram starting limits for various combustion inlet pressures, fuel-air ratios, and burner referred air flow.

engine itself will quickly assist the starter, even though the speeds and top temperatures may be too low for the engine to be self-sustaining. Although ignition is generally easy to initiate at sea level, great difficulty is experienced at the higher altitudes due to the low air density and temperature. A typical set of burner ignition limiting curves is shown in

Fig. C,11a.[15] Notice that there is an optimum fuel-air ratio required for each Mach number.

If one contemplates starting the engine by windmilling under altitude flight conditions, the curve (Fig. C,11a) must be matched with the pressure level and the flow Mach number existing in the combustion chamber

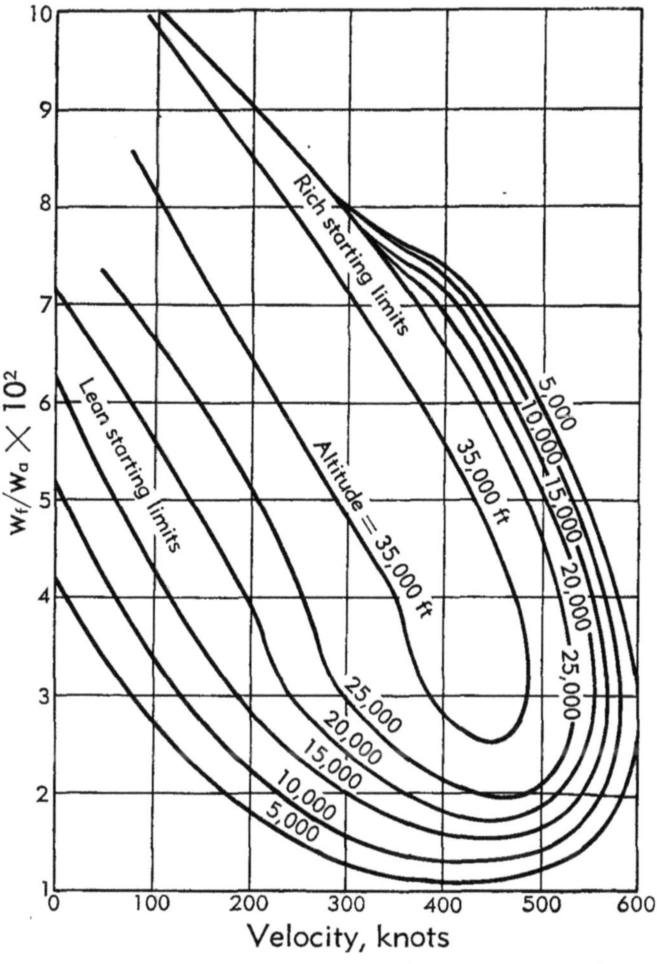

Fig. C,11b. Turbojet ram starting limits as a function of fuel-air ratio, altitude, and flight speed.

under the windmilling condition. The results of one such matching study are sketched in Fig. C,11b. This curve is interesting because it shows that for best effect the fuel rate for air starting should be varied with flight speed.

Windmilling of a turbojet engine during flight is useful as a starting

[15] Compiled by D. T. Browning.

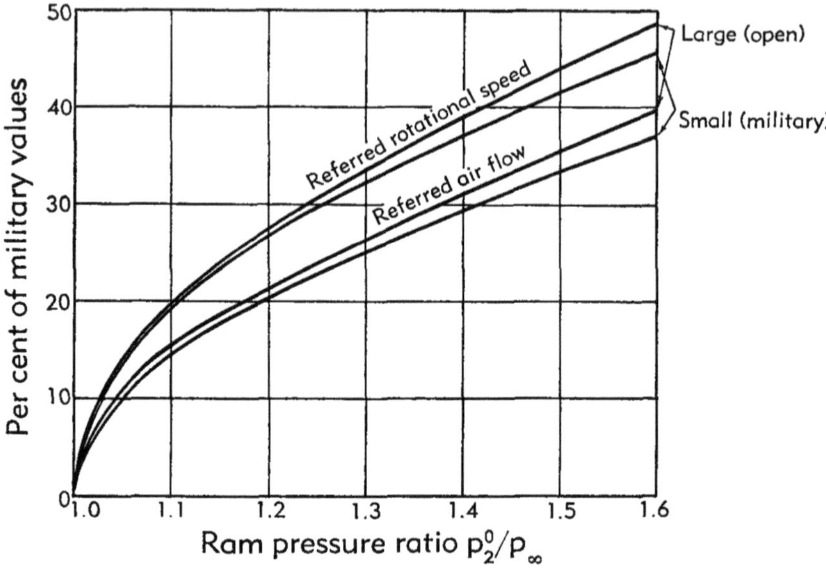

Fig. C,11c. Windmilling speed and referred air flow of a typical turbojet engine. For two exhaust nozzle sizes.

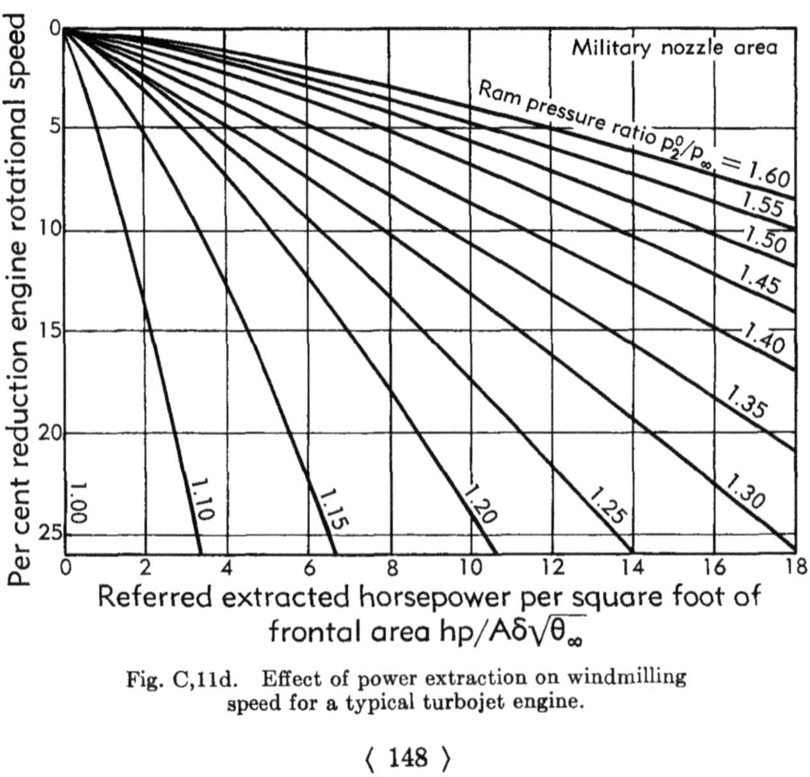

Fig. C,11d. Effect of power extraction on windmilling speed for a typical turbojet engine.

means as indicated above and is also useful during "let down" prior to landing where at minimum thrust one is required to extract electrical power from the engine generator for operation of wheels, flaps, radar, etc. Also it is often required during windmilling to bleed air from the compressor for anti-icing, instruments, auxiliary power, etc.

The windmilling performance of a typical turbojet engine is shown in terms of referred quantities in Fig. C,11c, C,11d, C,11e, and C,11f.

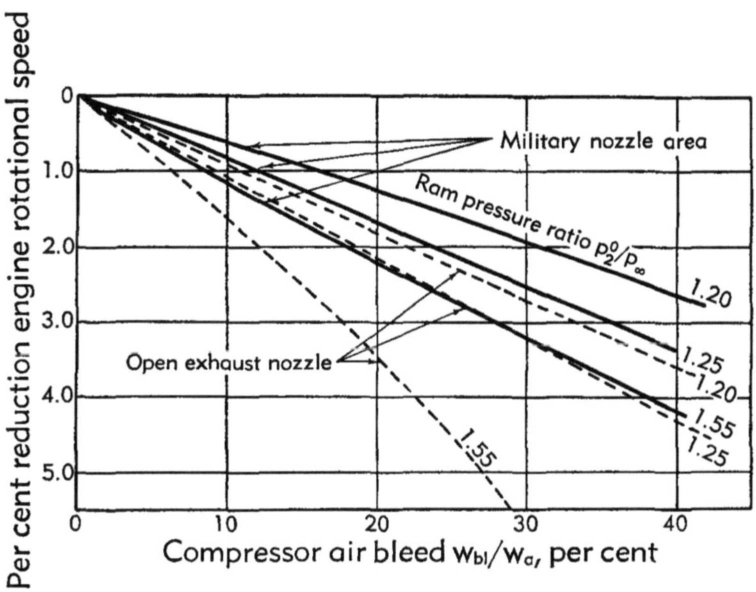

Fig. C,11e. Effect of air bleeding on windmilling speed of a typical turbojet engine.

Fig. C,11c shows the variation of the referred speed and referred air flow as fractions of the full load values for different ram pressure ratios and exhaust nozzle areas. Curves of Fig. C,11d and C,11e are typical curves showing the reduction in engine rotational speed (required for electrical frequency considerations) as a function of the extracted power and the amount of air bled. A typical curve showing the drag of a turbojet engine during windmilling is shown in Fig. C,11f.

The time required for a turbojet engine to accelerate from idling to full thrust is very important, particularly during emergencies when landing, such as occur during a carrier "wave off." The time required to accelerate, as may be suspected, depends considerably on the altitude since the accelerating torques depend (like the engine thrust) on the air pressure, while the rotor inertia is the same at sea level or high altitude. This is shown to be true by means of a dimensional analysis. We assume during the analysis that the temperature during acceleration is

[16] Compiled by H. S. Jordan.

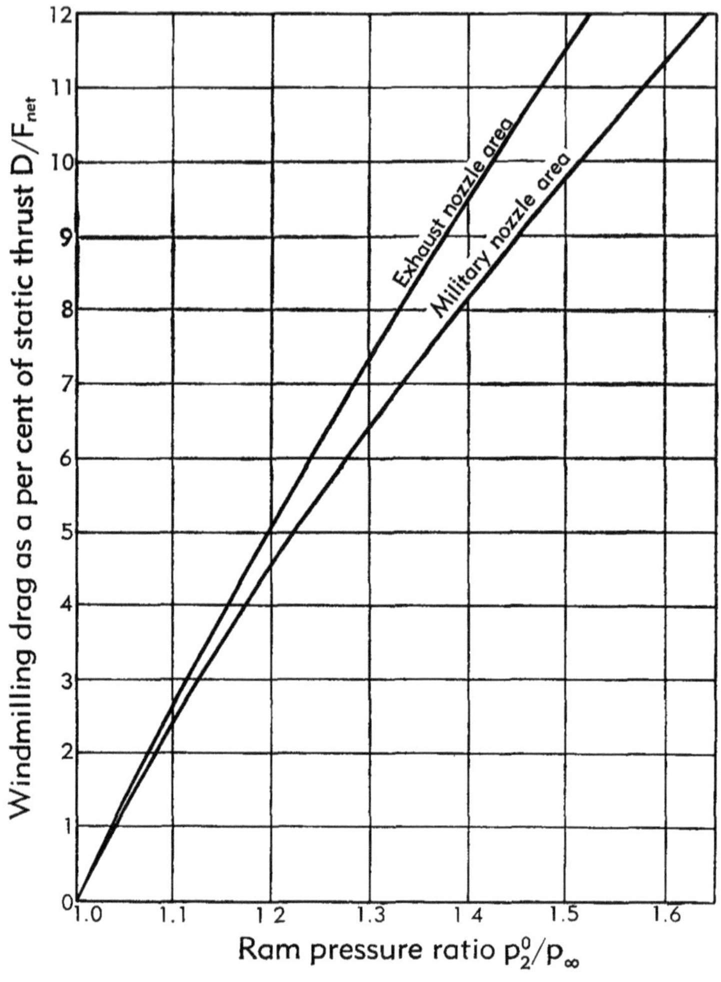

Fig. C,11f. Windmilling drag of a turbojet engine.

dimensionless, i.e.

$$\frac{T_4^0}{\theta_2^0} = \text{a function of } \left(\frac{n}{\sqrt{\theta_2^0}}\right) \tag{11-3}$$

By making this assumption[17] we avoid introducing additional variables connected with the engine control and thereby simplify the analysis. The result of such an analysis gives a function of

$$\left(\frac{tp_\infty D^4}{Ia}, \frac{n_1 D}{a_2^0}, \frac{n_2 D}{a_2^0}, \frac{p_2^0}{p_\infty}, Re, \frac{ta}{D}\right) = 0 \tag{11-4}$$

[17] Another way of stating the assumption is that the control during acceleration is such as to regulate the fuel supply so that we move along a single line (for instance close to the stall line) on the dimensionless compressor characteristic (Fig. C,7b).

where t is the acceleration time, I is the rotor moment of inertia, and n_1, n_2 are the speeds at the beginning and end of the acceleration. Evidently the last dimensionless quantity in Eq. 11-4 is only important if the acceleration time t is so short compared to D/a that the pressures throughout the engine during acceleration are different from their steady state values.

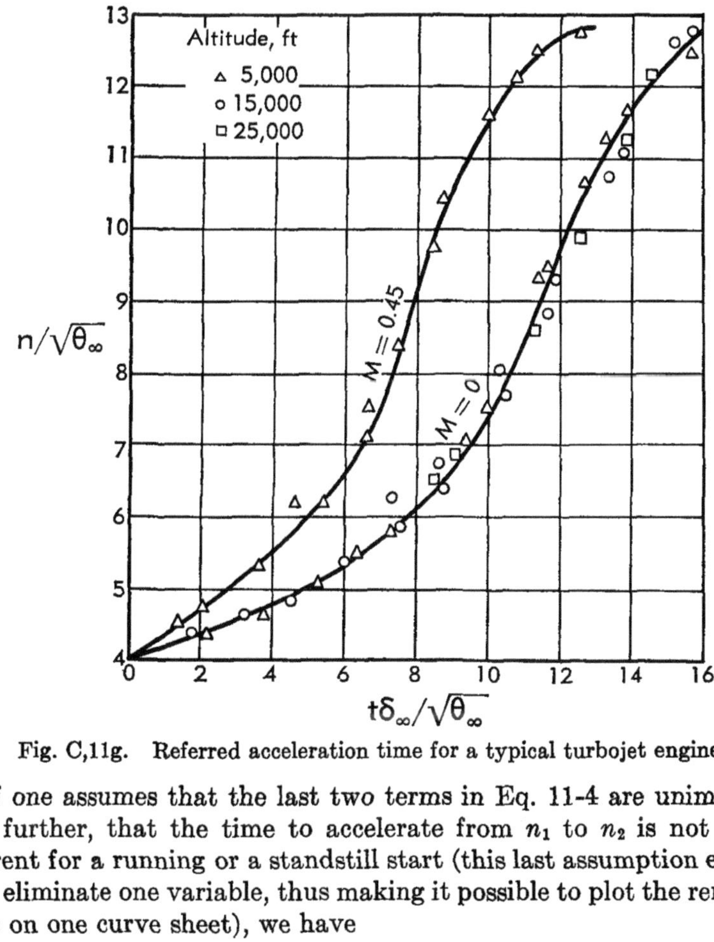

Fig. C,11g. Referred acceleration time for a typical turbojet engine.

If one assumes that the last two terms in Eq. 11-4 are unimportant and, further, that the time to accelerate from n_1 to n_2 is not greatly different for a running or a standstill start (this last assumption enabling us to eliminate one variable, thus making it possible to plot the remaining three on one curve sheet), we have

$$\text{a function of } \left(\frac{tp_\infty D^4}{Ia}\right), \left(\frac{nD}{a}\right), \left(\frac{p_2^0}{p_\infty}\right) = 0 \tag{11-5}$$

We can conveniently convert these quantities to referred quantities and obtain

$$\text{a function of } \left(\frac{t\delta_\infty}{\sqrt{\theta_\infty}}\right), \left(\frac{n}{\sqrt{\theta_\infty}}\right), \left(\frac{p_2^0}{p_\infty} \text{ or } M\right) = 0 \tag{11-6}$$

A typical plot of such a function is shown in Fig. C,11g for two flight Mach numbers. This curve has been drawn by sliding to the right or left the time scale for several different test curves until the points coincided. Some of the scatter of the points is due to the assumption regarding standstill or running starts. This curve shows that the acceleration time for a given flight Mach number between any two *referred* speeds varies inversely as the pressure level and directly as the square root of the absolute temperature. Since $t\delta_\infty/\sqrt{\theta_\infty}$ has a given value between any two referred speeds, the time will be directly proportional to $\sqrt{\theta_\infty}/\delta_\infty$.

C,12. The Variable Area Jet Nozzle. The variable area jet nozzle first came into being as an aid in starting and acceleration to avoid compressor stall. Later other advantages became evident. Some of the more important advantages and disadvantages for using a variable area jet nozzle are as follows:

Advantages.

1. Avoids compressor stall during starting and acceleration.
2. Reduces acceleration time.
3. Improves low thrust specific fuel consumption.
4. Reduces loss in take-off thrust on unusually warm days.
5. Avoids excessive or too low maximum temperatures under varying flight conditions.
6. Requires no nozzle size adjustments due to variations between engines or to airplane duct losses.

Disadvantages.

1. Mechanical complexity.
2. Requires "no leakage" seal for high efficiency.
3. At low thrust (cruise) the air flow is high and thus the duct losses are generally large.
4. It is difficult to make a practical variable area converging-diverging nozzle.

An explanation of some of the reasons for the above advantages and disadvantages would be helpful.

One can reduce the thrust of a turbojet engine from its full load value by steadily opening the jet nozzle and reducing the top temperature T_4^0 (reducing the fuel flow), while maintaining the rotative speed constant. This process can be continued until the jet or exhaust nozzle is open to its limiting amount. In this way we maintain a high air flow while reducing greatly the top temperature, fuel flow, and compressor pressure ratio. The compressor operating point B (Fig. C,12a) corresponding to this condition is on the maximum speed line at a low pressure ratio, far removed from the stall line (see Fig. C,7c). Since the nozzle is opened a maximum amount at this point B, the thrust can only be reduced further by simultaneously lowering the top temperature and rotational speed which carries us along the operating line *BC* (Fig. C,12a). This operation during reduction of thrust (or its reverse acceleration) avoids the compressor stall line. Therefore, for two reasons this method of varying thrust is helpful:

1. It permits large amounts of fuel to be added during acceleration along the line *CB* (Fig. C,12a).
2. The thrust increase along *BA* (Fig. C,12a) can be accomplished in a very short time due to the constant speed "no inertia" process where only the top temperature and nozzle area need be changed rapidly.

The variable area exhaust nozzle reduces the specific fuel consumption at part thrust by enabling an optimum matching of the efficiencies of the

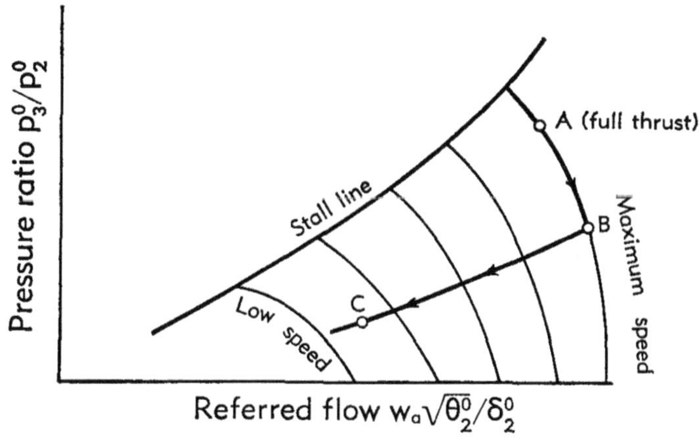

Fig. C,12a. Operating line on the compressor map of a variable area jet nozzle turbojet engine.

compressor and turbine. This is more clearly indicated by Fig. C,12b, which shows typical performance curves for a turbojet engine using various fixed area exhaust nozzles. One can see that for optimum performance the nozzle area (and rotational speed) must be varied with engine thrust. One would expect the variation of optimum speed, exhaust nozzle area, and temperature with thrust to be involved due to the fact that the cycle pressure ratio, the air flow, the component efficiencies, and the burner pressure losses all vary in a complicated manner with these variables. One generally finds that different optimum variations between rotational speed and nozzle area are indicated at altitude and at sea level. Thus the control schedule between speed temperature and exhaust nozzle area must be compromised to approach optimum conditions at the required altitude and at sea level.

The advantage in improved efficiency at part thrust due to using a variable area nozzle has been found in some instances to reduce with increasing altitude. In fact if the rotational speed is maintained high

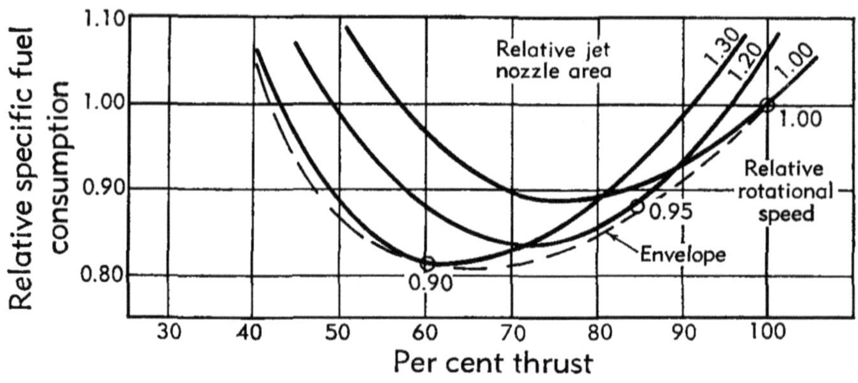

Fig. C,12b. The effect of using a variable jet nozzle on minimum specific fuel consumption.

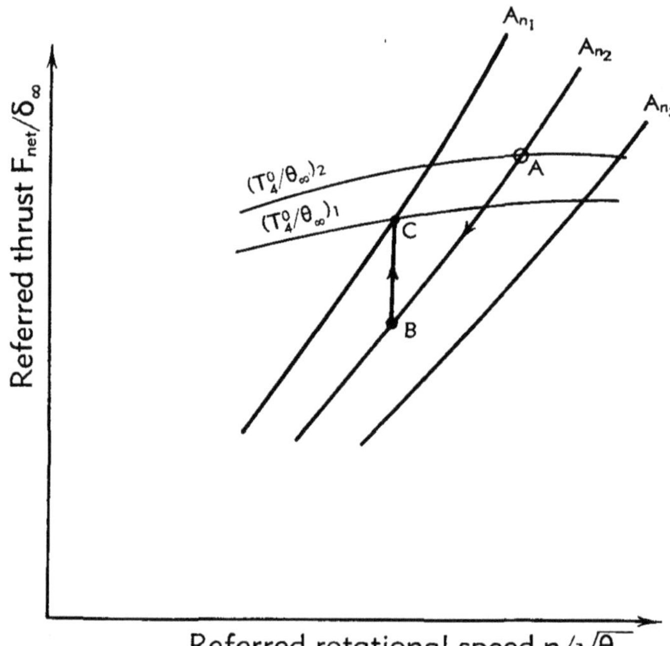

Fig. C,12c. The effect of using a variable jet nozzle in reducing hot day thrust loss.

at part thrust (for efficiency and acceleration reasons) the airflow is also high. This means that any duct losses are magnified due to the fact that these losses vary approximately as the square of the air flow. This duct-loss effect can sometimes overcome the advantage of better component matching and higher cycle pressure ratio.

On excessively warm days the referred rotational speed $n/\sqrt{\theta_\infty}$, air flow and pressure ratio, and therefore the engine thrust, are considerably reduced for a turbojet engine with a fixed area nozzle A to B (Fig. C,12c). By means of a variable area exhaust nozzle the nozzle area can be reduced so that the maximum temperature and rotational speed coincide (point C) and the loss in thrust can be partially restored. Fig. C,12d is an interesting plot obtained by dividing the referred top temperature T_4^0/θ_2^0 by the square of the referred speed $n/\sqrt{\theta_2^0}$ and plotting this ratio against the referred speed. The ordinate T_4^0/n^2 is then independent of temperature θ_2^0, and for limiting temperature T_4^0 and speed n it has a constant value. The vertical line represents the rated speed under standard conditions. Notice that as θ_2^0 is decreased (cold day), i.e. as $n/\sqrt{\theta_2^0}$

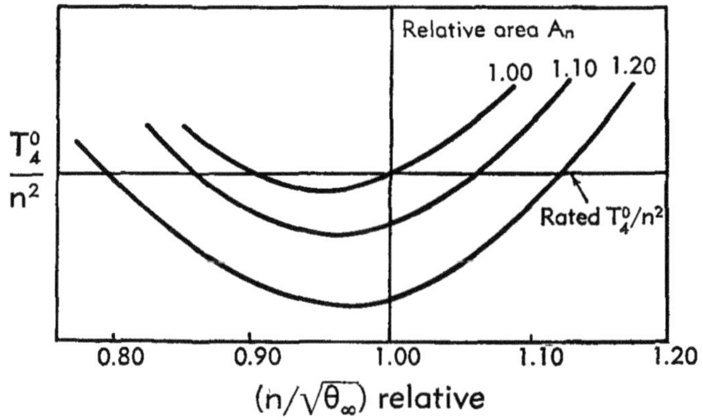

Fig. C,12d. The effect of variable jet nozzle area in preventing overtemperaturing on hot or cold days.

is increased, we approach the limiting value of T_4^0/n^2. Likewise if θ_2^0 is increased (hot day or high flight speeds), $n/\sqrt{\theta_2^0}$ is decreased, and again we approach the limiting value of T_4^0/n^2. This typical curve shows why we must operate a fixed nozzle engine below the limiting temperature or speed on a standard day (i.e. we must derate the engine) in order to insure not exceeding the allowed top temperature on cold or hot days or at high flight speeds.

In summary one could say that even discounting the gain in performance due to the use of a variable nozzle, the advantages of ease in starting (particularly for high compression ratio machines), together with the maintaining of thrust on hot days, are sufficient reasons for the use of variable exhaust nozzles.

C,13. Controls. As an introduction to the problems associated with the control of turbojet engines, it is interesting to examine carefully the following typical set of design requirements [*42*].

1. The control should require the operation by the pilot of only a single control lever.

2. The control should be sensitive at all altitude and flight speeds to permit no difficulty in, for instance, formation flying.
3. The control should be stable and accurate under all ambient temperatures and pressures under all altitudes and flight speeds on hot and cold days.
4. The control should enable an acceleration in a short time without overtemperaturing the engine, stalling the compressor, or causing the combustor to "blow out."
5. The control should enable the engine to idle stably at all altitudes and flight speeds.
6. The control should "fail safe," i.e. if the control fails it should not jeopardize the safety of the aircraft.
7. The control should be lightweight.
8. The control should function when using fuel with adverse properties such as dirty fuel, or fuel of high or low viscosity on cold or hot days, or fuel containing water which might freeze.
9. The controls must function under specified angular and linear accelerations and must operate while the engine is inverted in order not to restrict the flight maneuverability of the airplane.
10. The thrust should vary approximately in a linear manner with the throttle position.
11. For solid injection burners the fuel nozzle pressure drop must be sufficient at high altitude to give good fuel atomization.

Controls for turbojet engines can generally be divided into the following three kinds: (1) for engines with fixed area exhaust nozzles, (2) for engines with variable area exhaust nozzles, and (3) afterburner controls.

Fixed exhaust nozzle controls. Since for an engine with a fixed area exhaust nozzle the engine speed and the thrust are dependent on the fuel flow rate, one need only control the fuel flow to control the engine. However, due to the great range of fuel flows required (35 to 1 between high speed, sea level, cold day and high altitude, hot day) one might possibly use either of the following two systems:

1. A control which regulates the fuel flow by means of the throttle position with a compensation by an aneroid barometer (a barostat) for changes in ambient pressure in order to utilize the full throttle travel at altitude.
2. A control which is isocronous and which regulates the fuel flow by setting the per cent of full speed by means of the throttle position.

The latter type of control is to be preferred over the former, due to the disadvantage that scheduling requires considerable knowledge of the altitude effects on the engine fuel flow rate.

In Fig. C,13a, which shows the relationship between engine speed and throttle position used in the early days, one notices the sensitivity to thrust at high speed. Fig. C,13b shows a more adequate arrangement; however, here the control is sensitive to engine speed at low thrust. Fig. C,13c shows the introduction of a flat spot for idling in which the throttle is permitted to move several degrees without any response in rpm or thrust, thus eliminating the speed sensitivity at idling.

There are several arrangements of controls which are automatic and which permit engine acceleration without compressor stall, overtemperaturing, and without causing burner blowout at altitude. One such system

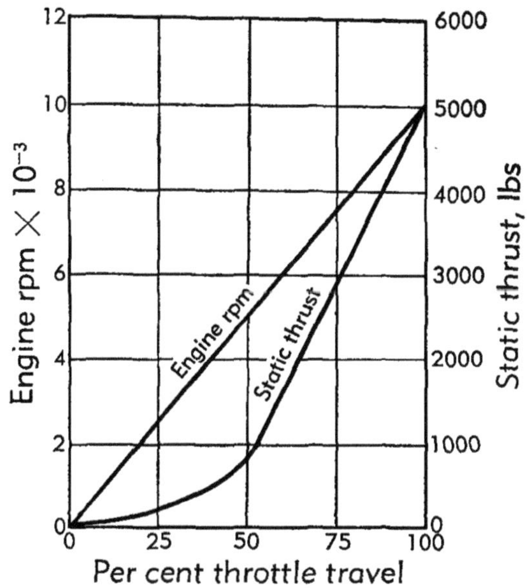

Fig. C,13a. Early engine control characteristic.

controls the rate of fuel addition by means of the rate of fuel pressure rise in the fuel control. This system is compensated for altitude. Another system employs a dashpot connection between the throttle and the fuel controller so that, regardless of the rate at which the throttle is moved, the fuel will be fed to the engine at a scheduled rate.

Generally, the above types of controls have the disadvantage that they do not enable the attainment of the maximum acceleration rate. A control which regulates fuel by means of a measured temperature is to be preferred in this connection. However, the response rate of the thermocouple must be faster than the turbine blades if the control is to be a success.

Controls for variable area exhaust nozzles. The advantages of using a variable area exhaust nozzle have already been enumerated in Art. 13. The control of the exhaust nozzle area is preferably such as to maintain the best fuel economy, as shown for instance in Fig. C,13b for all altitudes

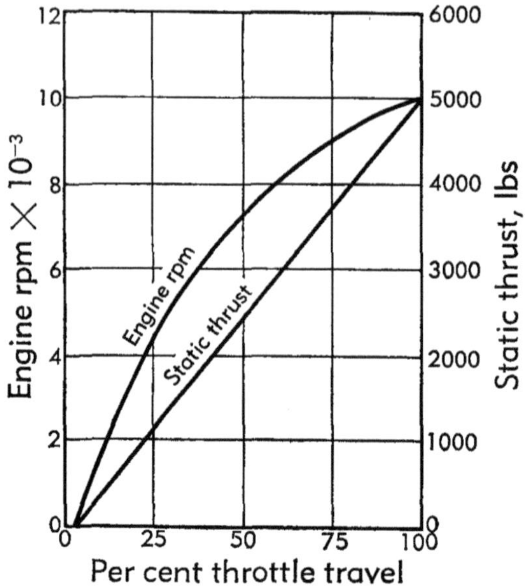

Fig. C,13b. Later control giving a linear variation of thrust with throttle travel.

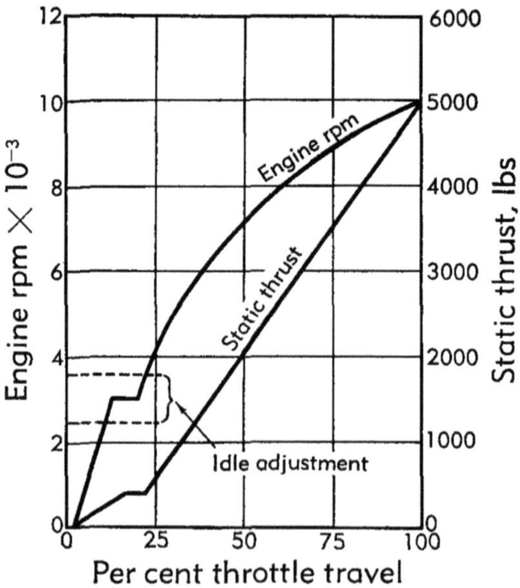

Fig. C,13c. Control with idling flat spot to reduce speed sensitivity.

and flight speeds. For a typical engine one generally has to compromise the control to favor some altitude and flight speed since it is generally difficult to maintain optimum conditions for all altitudes and speeds, due to Reynolds number effects, etc.

Usually one will find that for best performance the thrust is reduced from its maximum value by maintaining (or slightly reducing) the rotational speed while the nozzle area is increased and the fuel flow and top temperature are reduced. Thrust is reduced in this manner until the maximum area of the exhaust nozzle is reached, after which the thrust

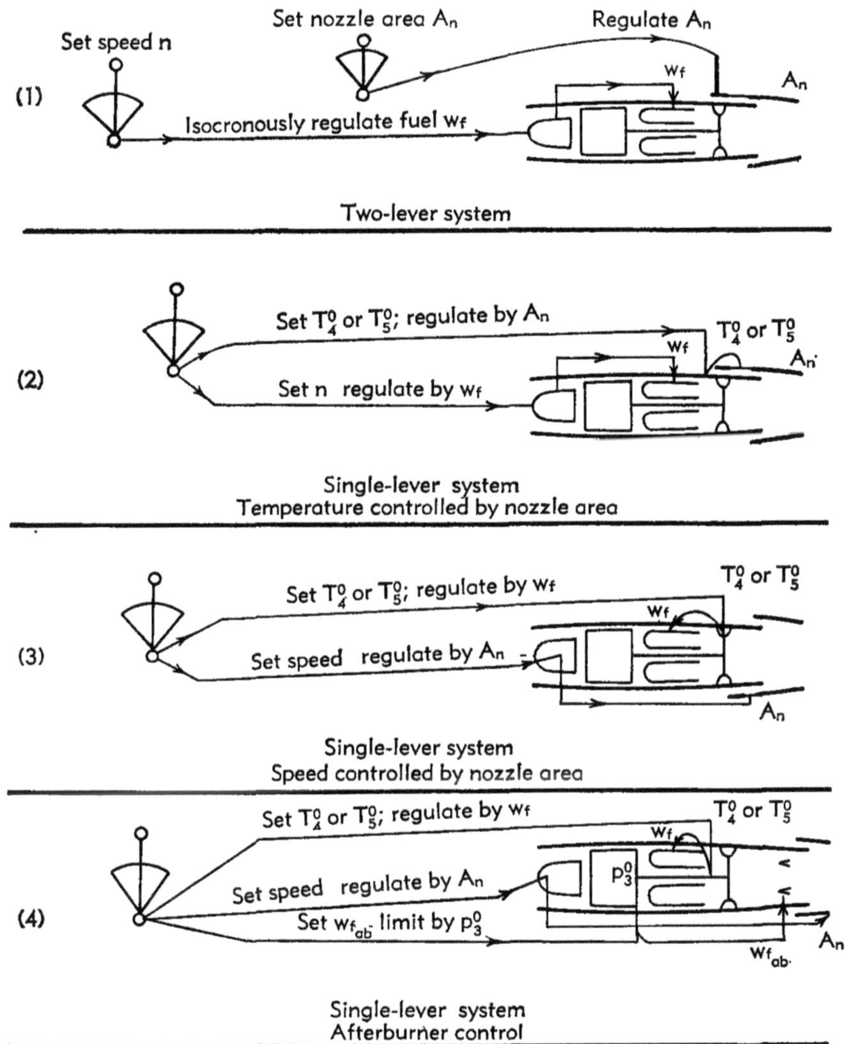

Fig. C,13d. Various types of turbojet engine controls.

is further reduced as in a fixed area engine, by further reduction of the fuel flow which causes a reduction in rotational speed.

The simplest method (Fig. C,13d(1)) of accomplishing the above control would be by means of two separate levers, one for setting speed which could regulate the engine fuel flow, the other for controlling exhaust nozzle area. A further simplification, of course, would be to use a two-position exhaust nozzle in place of a modulated one.

The disadvantages of the two-lever system are: (1) the operational complexity of having two levers, and (2) the difficulty of the pilot in optimizing the lever settings for various flight and atmospheric conditions.

A method of correcting these disadvantages is sketched in Fig. C,13d(2) in which a single lever is used to set both speed n and, for instance, turbine outlet temperature T_4^0. The speed is controlled by regulation of the fuel flow while the temperature T_4^0 is maintained by regulating the exhaust nozzle area. This system, although simple from the pilot's standpoint, has been found to control poorly during rapid changes in thrust, principally due to the high response rate required of the temperature measuring device.

A more improved type of control is shown in Fig. C,13d(3) which is a reversed version of the system shown in Fig. C,13d(2). Here the exhaust nozzle is positioned to regulate the speed while the fuel flow is regulated to maintain the temperature T_5^0.

All of the above types of controls function similarly to those for a fixed exhaust nozzle engine, once the exhaust nozzle reaches its maximum opened position.

The variable area controls are generally more difficult to develop from the standpoint of stability and safety. Slow response to sudden load changes can cause destructive overspeeds or overtemperaturing.

Controls for afterburners. Although there are many possible types of controls for afterburners, only one type is described here because of space limitations. This typical control is based on the following assumptions regarding conditions of operation: (1) While the afterburner is in use the engine will be operating at full military conditions of speed and temperature. (2) The effective turbine nozzle area (or referred flow $w_g \sqrt{\theta_4^0}/\delta_4^0$) is constant. This assumes that the turbine is choked. (3) The percentage pressure loss across the burner is small.

The above assumptions which are closely true in practice enable one to devise an afterburner control (Fig. C,13d(4)) which is simply an addition to the engine control (Fig. C,13d(3)). The maximum fuel flow to the afterburner is limited proportionally to the measured compressor outlet pressure P_3^0.

This is possible since the turbine inlet temperature and θ_4^0 is constant (assumption 1); therefore, since the referred flow is constant (assumption 2), the gas flow w_g is proportionate to the turbine inlet pressure P_4^0 or δ_4^0. With slight error as a result of assumption 3, the gas flow is assumed proportionate to the compressor outlet pressure. Therefore the fuel gas ratio to the afterburner can be limited by restricting the fuel flow proportionately to the gas flow, which in turn is approximately proportionate to the compressor outlet pressure. Thus, with this type of control, when the throttle is advanced past the full military condition for the engine, fuel is fed to the afterburner and ignited. The maximum amount of fuel that can be supplied to the afterburner is limited by the compressor outlet pressure. The engine conditions of speed and temperature are maintained by the usual engine controls and are supposedly unaffected by the afterburner operation.

C,14. High Speed Flight. In estimating the performance of turbojet engines at high flight speeds and various altitudes, many problems arise which require solution, a few of which follow:

1. The effect of high flight speed on the performance of engine components.
2. The effect of high flight Mach number on the inlet diffuser and exhaust nozzle.
3. Thermal cycle effects causing changes in thrust and fuel consumption.
4. The variation of top temperature with flight conditions introducing a control problem.

Of course, the answers to most of these problems as they affect over-all performance of a given turbojet engine can be found by wind tunnel tests duplicating high altitude and flight speed conditions. However, such over-all tests of a given engine usually do not lead to better designs unless the effects of flight speed and altitude on the component parts are also determined.

Mach number effects on engine components. The turbojet engine compressor is the component of the engine most affected by high internal Mach numbers. It is well known that airfoils, when operated at flight speeds approaching the velocity of sound, suffer losses due to compression shocks induced when the local velocities near the foil reach or exceed the local velocity of sound. Similarly, a compressor blade grid suffers losses when the air velocities relative to the blade approach the velocity of sound. All airfoils and compressor blade grids can be characterized by a critical Mach number. The critical Mach number is the Mach number relative to the blade which will cause sonic velocities near the blade surface. Critical Mach numbers of blade grids are usually lower than those of single foils due to the induced velocities caused by neighboring blades of the grid.

High Mach numbers relative to the compressor blades thus induce losses which result in losses in compressor efficiency and cause a loss in over-all engine performance.

Let us now inquire into the variation of Mach number relative to the compressor blades under different engine-operating conditions.

The static performance of a given turbojet engine can be presented by

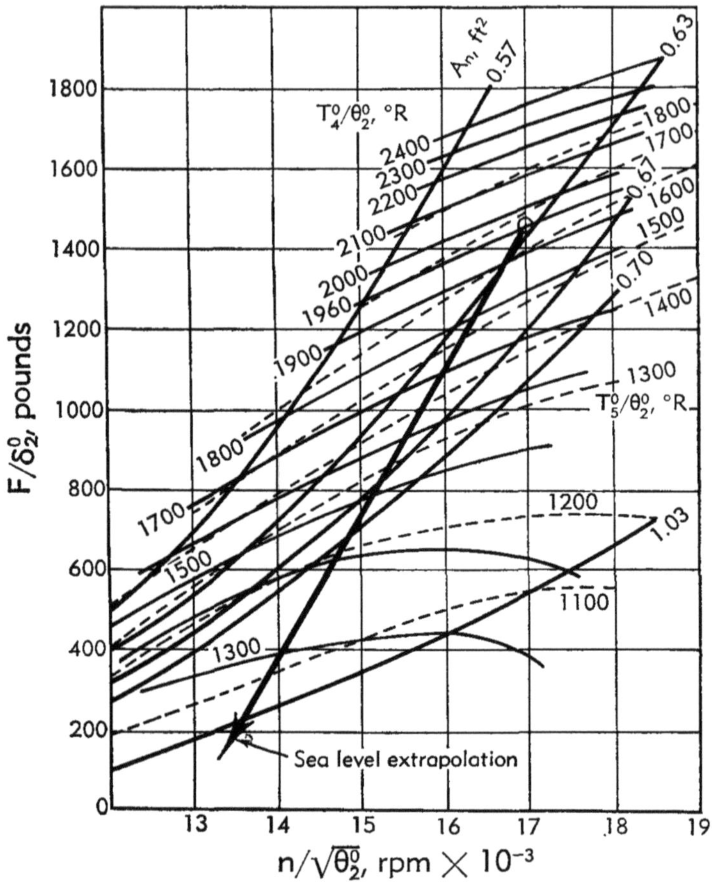

Fig. C,14a. Typical performance of a turbojet engine showing referred thrust, top temperature, and tailpipe temperature for various jet nozzle areas and referred rotational speeds.

the following relations, provided Reynolds number effects are neglected.

$$\text{a function of } \left(\frac{F_{\text{net}}}{\delta_2^0}\right), \left(\frac{n}{\sqrt{\theta_2^0}}\right) A_e = 0 \tag{14-1}$$

$$\text{a function of } \left(\frac{T_4^0}{\theta_2^0}\right), \left(\frac{n}{\sqrt{\theta_2^0}}\right), A_e = 0 \tag{14-2}$$

$$\text{a function of } \left(\frac{w_a \sqrt{\theta_2^0}}{\delta_2^0}\right), \left(\frac{n}{\sqrt{\theta_2^0}}\right), A_e = 0 \tag{14-3}$$

These relations are given for a typical turbojet engine in Fig. C,14a and C,14b. The foregoing relations imply that for a given turbojet engine, if the exhaust nozzle area A_e and the referred rotational speed $n/\sqrt{\theta_2^0}$

are fixed, the referred thrust F_{net}/δ_2^0, the referred top temperature T_4^0/θ_2^0, and the referred air flow $w_a \sqrt{\theta_2^0}/\delta_2^0$ are also fixed. Other relations similar to the foregoing could also be written showing that, once the referred speed and exhaust nozzle area are fixed, all the Mach numbers, pressure ratios, and temperature ratios throughout the engine are also fixed. This implies that all the velocity triangles throughout the engine are also fixed.

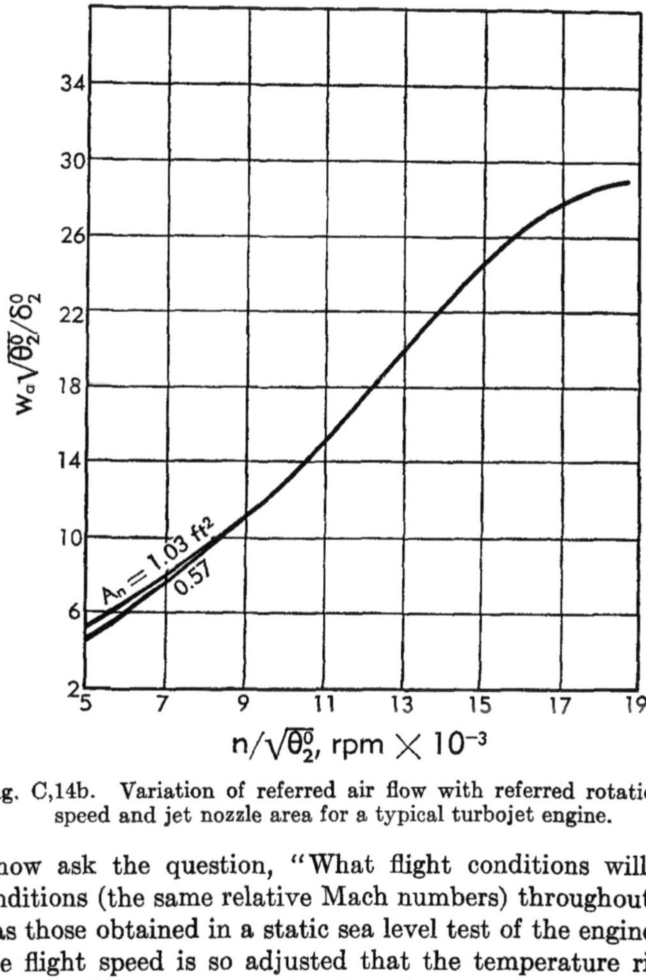

Fig. C,14b. Variation of referred air flow with referred rotational speed and jet nozzle area for a typical turbojet engine.

We now ask the question, "What flight conditions will give the same conditions (the same relative Mach numbers) throughout the compressor as those obtained in a static sea level test of the engine?"

If the flight speed is so adjusted that the temperature rise due to ram (flight speed) compensates for the reduction in temperature due to increased altitude, we can thereby maintain the compressor inlet total temperature T_2^0 at its static sea level value. Thus for a given actual rotational speed n, the referred rotational speed $n/\sqrt{\theta_2^0}$ for the adjusted flight condition would be identical to the sea level static value. Further, for a given limiting total top temperature T_4^0, the referred top temperature T_4^0/θ_2^0 would be identical to its sea level value.

The pressure ratios, temperature ratios, and Mach numbers throughout the unit could only be maintained at their sea level values, provided that the pressure ratio p_5^0/p_2^0 were also adjusted to its sea level value. The adjustment of this pressure ratio can be accomplished by adjusting suitably the exhaust nozzle area. Thus by adjusting the compressor inlet temperature to its static sea level value by increasing the flight speed with altitude and by adjusting suitably the exhaust nozzle, all temperature ratios, pressure ratios, velocity triangles, and Mach numbers

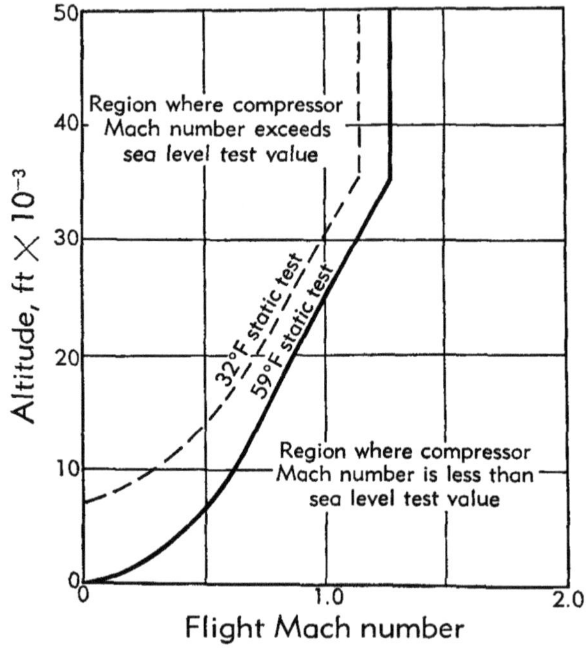

Fig. C,14c. Curves showing flight conditions at which the compressor Mach number is identical to the value obtained in a static sea level test.

throughout the engine are maintained identical to their sea level static values.

The solid curve in Fig. C,14c was drawn using the foregoing reasoning to show the flight conditions at rated top temperature and rotational speed, which give conditions (Mach numbers) within the engine similar to those of a static sea level test at 59°F ambient temperature. A static test at 32°F ambient temperature gives conditions (Mach numbers) within the engine the same as those corresponding to flight conditions shown by the dotted curve in this figure.

Thus we see that, insofar as Mach number troubles within the engine are concerned, they occur at high altitudes and low flight speeds. Thus Mach numbers within the engine are less than those occurring in a sea level static test for flight conditions below the curve of Fig. C,14c. This is true independently of the ram recovery efficiency, because the ram temperature recovery (excluding heat losses) is always 100 per cent.

Compressors for jet engines which are to operate efficiently at high altitudes and low flight speeds must be designed with Mach numbers relative to the blades which are not excessive at 12 per cent overspeed on the ground. The effect of this increase in referred speed differs with various compressor types and designs. The lower the designed relative Mach number, compared to the critical Mach number of the compressor

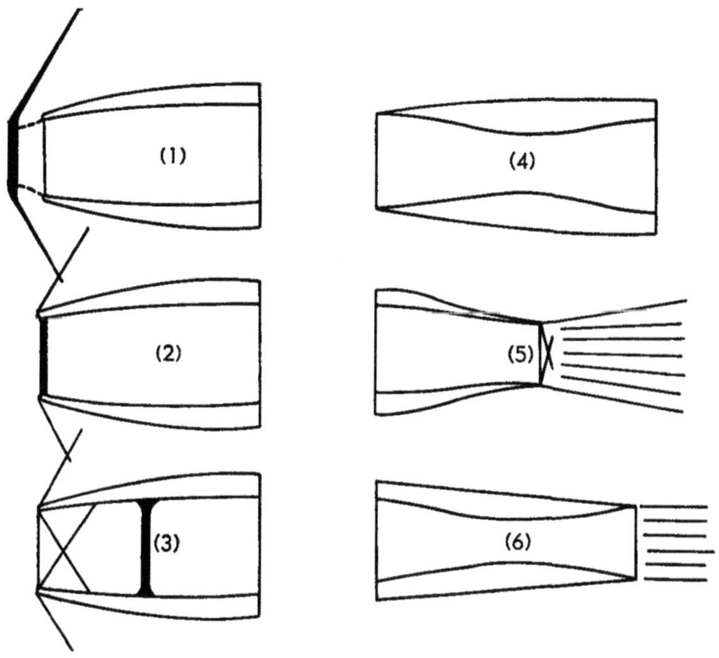

Fig. C,14d. Operation of inlet diffuser and exhaust nozzles at high flight Mach number. (1), (2), (3): Successive positions of a plane shock in a divergent inlet diffuser. (4): Theoretical supersonic inlet diffuser. (5): Expansion from a convergent jet nozzle. (6): Convergent-divergent jet nozzle for use with superacoustic jets.

blade grid, the less will be the loss due to overspeed operation of the compressor. Some factors affecting the critical Mach number of the blades are blade loading, camber, thickness, pitch, and stagger angle.

The compressor can be designed for low relative Mach numbers. However, this leads to larger engine sizes for a given output (smaller flow per unit area). Another possible solution is to reduce actual engine rotational speed at the high altitudes. This leads to reduced output at altitude, and thus also to larger sizes per unit output. We should therefore always be careful to compare engine output at altitude with engine size.

The basic problem, so far as axial compressors are concerned, is to develop compressor blade grids which will operate efficiently at high relative Mach numbers and which will give acceptable outputs. Compressor blade grid tests at low inlet temperature, or possibly tests with one of the freon gases (which have low acoustic velocities), are indicated research methods for solving this high altitude flow, flight speed problem.

Inlet diffuser and exhaust nozzle. The following discussion is intended to present possible ways of allowing for variations in the performance of the inlet diffuser and exhaust nozzle at high flight speeds.

The operation of the normal inlet diffuser at a flight Mach number above unity would be approximately as shown in Fig. C,14d. Fig. C,14d(1)

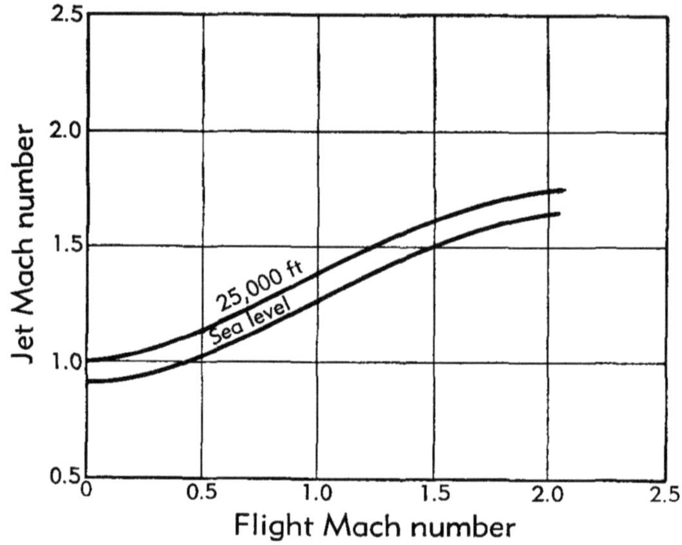

Fig. C,14e. Curve showing variation of jet Mach number with flight Mach number for typical turbojet engine.

shows the conditions without flow into the jet unit. A plane shock evidently occurs out in front of the diffuser, the shock becoming more and more near the edge of the diffuser and tapering off in intensity until it becomes a Mach line far from the diffuser. The flow after passing through the shock passes around the inlet to the diffuser. The admission of a small amount of flow into the diffuser (by reducing the diffuser outlet pressure) causes the shock to move toward the diffuser inlet, and the air after passing through the shock diffuses into the inlet and spreads out further in the diffuser.

Fig. C,14d(2) shows the conditions if the diffuser outlet pressure were further reduced and adjusted to cause the shock to occur in the diffuser inlet.

Fig. C,14d(3) shows the conditions if the diffuser outlet pressure is still further reduced. The shock is evidently drawn into the diffuser and increases in intensity. Shocks and Mach lines are also caused by the leading edge of the diffuser. The more or less plane shock inside the diffuser causes separation and poor diffusion after the shock. After the shock enters the diffuser inlet, the air flow evidently remains constant due to the established conditions at the inlet.

It has been suggested that the theoretical superacoustic diffuser be used as shown in Fig. C,14d(4), thus entirely avoiding compression shocks. However, studies of this type of diffuser show that it is very critical in operation and also that shocks can be minimized only for particular operating conditions.

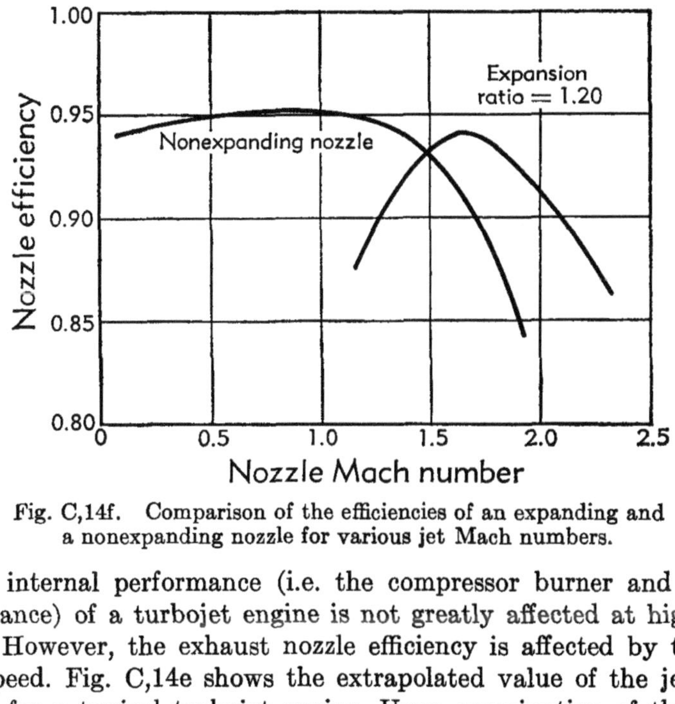

Fig. C,14f. Comparison of the efficiencies of an expanding and a nonexpanding nozzle for various jet Mach numbers.

The internal performance (i.e. the compressor burner and turbine performance) of a turbojet engine is not greatly affected at high flight speeds. However, the exhaust nozzle efficiency is affected by the high flight speed. Fig. C,14e shows the extrapolated value of the jet Mach number for a typical turbojet engine. Upon examination of this figure we find that since the jet Mach number is usually greater than unity, the flow explodes out of the normal exhaust nozzle, as shown in Fig. C,14d(5). We might expect this explosion to cause losses in nozzle efficiency at the higher flight Mach numbers. A suggested correction for this difficulty is to use an expanding exhaust nozzle, as shown in Fig. C,14d(6). However, past experience has given nozzle efficiencies somewhat as shown in Fig. C,14f. It is noticed that a nonexpanding nozzle gives good performance up to jet Mach numbers of about 1.50, which is due to the relatively small amount of explosion required at this Mach number. The improvement in using an expanding nozzle of area ratio 1.20 is substantial only for jet Mach numbers of the order of 1.80.

Returning to Fig. C,14e, we see that the use of an expanding exhaust nozzle will probably not be required until flight Mach numbers of about 1.50 are contemplated. The losses in the expanding nozzle at low flight speed due to compression shocks are also a determining factor in the use of this type of exhaust nozzle.

Cycle effects. Fig. C,14g shows the results of extrapolating the static performance of a typical turbojet engine to high flight Mach numbers.

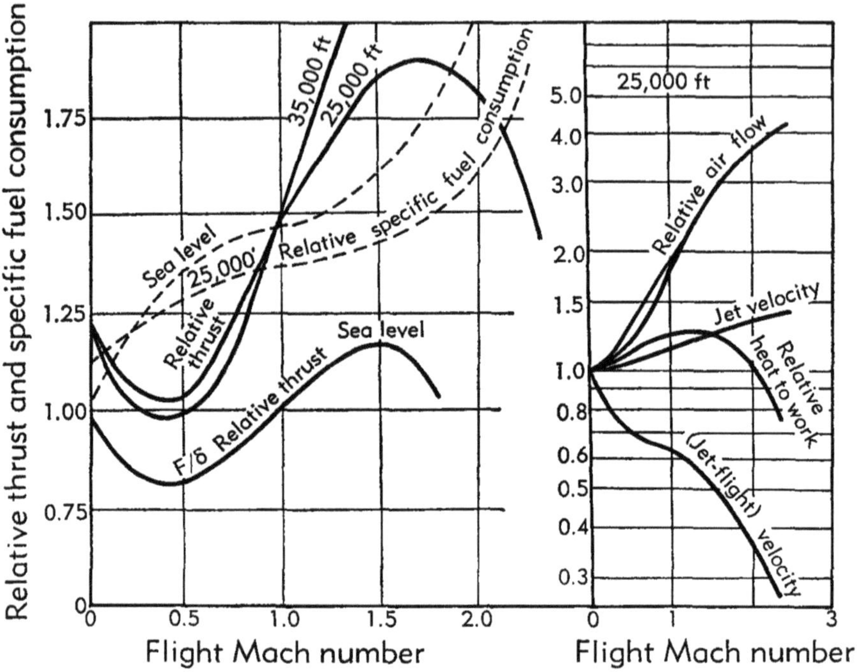

Fig. C,14g. Curve showing relative variations of net thrust, specific fuel consumption, air flow, jet velocity, and heat to work with flight Mach number.

Losses associated with a plane shock at the inlet to the diffuser were assumed in making the extrapolation. The heavy line in Fig. C,14a gives the locus of operating points for the sea level extrapolation shown in Fig. C,14g. Note the reduction in referred rotational speed $n/\sqrt{\theta_2^0}$ and referred top temperature T_4^0/θ_2^0 in going to high flight speeds at sea level. This is due to the increase in compressor inlet total temperature T_2^0, resulting from ram. It is interesting to note the odd shape of the thrust curve shown in Fig. C,14g, and to inquire into the reasons for this peculiar shape. The following factors combine to determine the shape of the thrust curve:

1. The actual air flow for a given referred rpm increases with the compressor inlet total pressure p_2^0, and inversely as the square root of the total temperature T_2^0. Also, for a given actual rpm, the referred rpm

is reduced at increasing flight speed, causing a reduction in referred air flow according to the engine characteristic curve (Fig. C,14b). Fig. C,14g shows the relative increase in air flow due to the combined factors cited.

2. The work output per pound of air q_w first rises and then falls with increasing flight speed (Fig. C,14g). This is explained most easily by referring to Fig. C,4d which shows the results of a cycle calculation. It is noticed that, for an engine designed for a compression ratio of 4, increasing the flight speed increases the cycle pressure ratio and thus causes first an increase in output per pound and, at still higher flight speeds (higher cycle pressure ratios), a reduction in output.
3. The jet velocity continues to increase with increasing flight speed, even though the cycle output per pound of air begins to decrease at high flight speeds.
4. The difference between jet and flight velocity decreases with flight speeds, somewhat offsetting the large increase in air flow. The net thrust then varies proportionally to the product of the air flow and the difference between jet and flight speed, thus accounting for the odd shape of the thrust curve.
5. At the higher flight Mach numbers the shock losses become considerable.

A study of Fig. C,4d shows some of the known facts regarding turbojet engines. For example, the selection of the over-all cycle pressure ratio depends on the purpose to be served by the engine. If we are interested in maximum output for a given size of engine, a cycle pressure ratio of 6 to 7 would appear best. If over-all efficiency is important, the cycle pressure ratio would be selected to give a minimum specific fuel consumption (pressure ratio 16 to 23 depending upon the flight speed).

Top temperature control. Another important problem encountered at high flight speed (and also at high altitude) is the control of the top temperature of the turbojet engine. As already shown in Art. 13, a fixed exhaust area engine can exceed the allowable top temperature at high flight speeds due to reduction of the referred speed $n/\sqrt{\theta_2^0}$ (see Fig. C,13d).

Eq. 14-2 indicates the relation between the referred top temperature, the referred rpm, and the exhaust nozzle area for the static case. If we consider operation at different flight speeds, the referred flight speed $V_\infty/\sqrt{\theta_2^0}$ would have to be included as a variable in Eq. 14-2. Now, if we introduce a new variable by dividing the referred top temperature by the square of the referred rotational speed to have

$$\left(\frac{T_4^0}{\theta_2^0}\right)\Big/\left(\frac{n}{\sqrt{\theta_2^0}}\right)^2 = \frac{T_4^0}{n^2} = \text{a function of } \left(\frac{n}{\sqrt{\theta_2^0}}\right), \left(\frac{V}{\sqrt{\theta_2^0}}\right), A_n \quad (14\text{-}4)$$

In order to operate the turbojet engine at rated top temperature T_4^0 and rated speed n, the quantity T_4^0/n^2 has a fixed value independent of flight speed or altitude. Since the referred engine speed varies with flight speed and altitude, Eq. 14-4 shows us that the exhaust nozzle area A_n will generally have to vary in order to maintain T_4^0 constant. It is true that for some particular engines the required exhaust nozzle area variation is negligible and a fixed exhaust nozzle can be used. Unfortunately, this is generally not the case.

C,15. Second Law Analysis. Of particular interest is the distribution of the losses throughout the various elements of a turbojet engine. An analysis based on the second law of thermodynamics [*43*] is ideal for clearly pointing out the losses in their true perspective. The changes in enthalpy h or available energy $A = h - T_0 s$ throughout a fictitious engine are shown in Table C,15. After a careful examination of this table one notices:

1. The loss in available energy in the combustor is 25.36 per cent. This loss is due to the low temperature at which the combustor air is supplied to the burner and the top temperature limit (assumed) of 1500°F. This loss could at best be reduced to 20 per cent since the Carnot efficiency for the assumed limit of 1500°F is 0.80.
2. The sum total of the losses due to the compressor, incomplete combustion, the burner pressure loss, the turbine, and the jet nozzle is equal to only 8.43 per cent.
3. The available energy loss in the exhaust is 20.76 per cent which together with the burner loss of 25.36 per cent make up major losses of available energy.

C,16. Installation Problems. The optimum solution of the many problems and requirements that arise when a turbojet engine is installed in an airplane is a prerequisite for a successful aircraft [*44*]. Some of the more important requirements are discussed from both the airplane designer's and the engine designer's points of view.

A jet engine installation should have a minimum of inlet and exhaust ducting losses. The inlet duct should have low total pressure loss over the complete speed range. According to Karanik [*45*] a 10 per cent loss in duct efficiency causes a 7 per cent loss in sea level rate of climb. Also, the time needed to climb to 40,000 ft is increased 15 per cent.

An ejector surrounding the jet nozzle should suck sufficient cooling air around the engine and through the nacelle so as not to overheat the air frame. This ejector should operate at all flight speeds (including standstill) and for all exhaust nozzle positions, causing a minimum loss of thrust or increase in drag.

Table C,15. Second law analysis of a turbojet engine.

Engine characteristics
Compression ratio 16:1; burner pressure ratio $p_4^0/p_3^0 = 0.98$
Efficiencies: $\eta_c = 0.86$; $\eta_b = 0.98$; $\eta_t = 0.90$; $\eta_n = 0.95$
Temperatures: $T_4^0 = 1500°F$; ambient $= -67°F = 393°R$
Altitude = 35,000 ft; flight Mach number = 0.85

Item		BTU/lb of air	Available energy loss	Per cent
Fuel energy to engine		252.66		100.00
Loss due to incomplete combustion		5.06		2.00
Net energy to air	247.60	247.60		
Resulting increase in available energy	183.57			
Difference	64.03		64.03	25.36*
Loss due to burner Δp_b			0.53	0.21
Reduction in available energy in turbine	156.56			
Turbine output	151.90			
Loss of available energy	4.66		4.66	1.84
Reduction in available energy through nozzle	131.67			
Kinetic energy of jet	128.70			
Loss of available energy in nozzle	2.97		2.97	1.18
Compressor input	151.90			
Increase available energy in compressor	143.80			
Loss of available energy in compressor	8.10		8.10	3.20
Available energy in exhaust			52.44	20.76
Enthalpy in exhaust		132.73		
Kinetic energy jet	128.70			
Kinetic energy ram	13.83			
Engine int output	114.87	114.87	114.87	45.45
Wake kinetic energy loss		58.00	58.00	23.10
Thrust output		56.87	56.87	22.35

* A Carnot engine would lose only 20 per cent.

Provisions should be made to supply auxiliary power to the airplane. According to McCarthy [*45*] an engine having 8000-lb thrust should supply 100 horsepower or more in the forms of pneumatic (high and low pressure), electrical, and hydraulic power. A variety of take-off pads to accommodate mechanical extension shafts should be provided.

Provisions should be made for bleeding up to 10 per cent of the engine air flow through any one of several bleed bosses on the compressor housing.

All engine components such as fuel pumps, oil pumps, filters, tachometers, and generators should be mounted as compactly as possible on the engine while still maintaining minimum frontal area.

The fuel system requires great care in designing. For instance, at the present high rates of climb the fuel boils, and losses due to evaporation become appreciable. Tank pressurization requires heavy tanks to stand 6 lb/in.^{2} These tanks cannot be self-sealing since this is limited to only 2 lb/in.^{2} Combinations of refrigeration and pressurization may be the answer.

The starting power requirement is another problem. A 5000-lb thrust engine requires from 20 to 30 hp while a 10,000-lb thrust engine requires from 25 to 50 horsepower. Electrical starting of this capacity is too large and heavy. Pneumatic or explosive means are usually lighter.

The jet engine should be capable of standing the high positive and negative g loads and angular velocities encountered in flight without the deflections and gyroscopic effects causing blade rubs or other failures.

The lubrication system should be capable of operation while the engine is inverted. One method is to arrange to change over the system by automatic valves when the engine is inverted. Only for a short period during change-over is the lubrication cut off.

Flow meters and thrust meters should be installed to aid the pilot and a second emergency fuel system should be supplied.

The engine should be mounted in the airplane so that no airframe stresses are created by the engine and vice versa.

The engine should be easily removed from the aircraft with a minimum of disconnections.

The engine should be capable of operating under icing conditions, i.e. heating or other means should be applied to the inlet duct to prevent clogging with ice. Generally an engine will consume ice in small pieces without damage. However, if large pieces of ice are allowed to accumulate and break off, serious damage results. A more detailed treatment of the icing problem is given in [*46*].

A removable inlet screen should cover the inlet duct to prevent damage by large stones picked up from the ground.

Military aircraft engines should be capable of being pierced by bullets through the less vital sheet metal parts of the engine, such as the combustor afterburner or exhaust duct, without causing damage to the aircraft due to the leaking of hot gases.

Many of the installation requirements mentioned above require close coordination between the aircraft and engine builder. The use of jet pods attached through a strut to the aircraft wing requires the engine designer to bury the engine auxiliaries in a smooth low drag nacelle or use auxiliary power plants located in the fuselage of the aircraft.

CHAPTER 4. THRUST AUGMENTATION

C,17. Introduction. Augmentation of the turbojet engine has taken many forms, some of which are rather easily developed into practical means while others will become useful devices only after many years of development effort. Fortunately it is always quite easy to carry out thermodynamic calculations of the various methods of augmentation and to roughly estimate the relative advantage to be gained by using the ideas which have been proposed. Before examining more closely the basic principals behind the different methods of thrust augmentation, it is interesting to compare the relative thrusts and specific fuel consumptions of several augmentation methods in order to indicate the thermodynamic advantage of each method. Table C,17 lists the relative thrusts and specific fuel consumptions of different augmentation means, based on a consistent set of assumed component efficiencies. Although the table lists the relative thrusts and liquid consumption for static sea level conditions, analysis indicates that considerably larger augmentation ratios are found at the higher flight speeds while smaller values are found at increasing altitudes. One can understand from Table C,17 that large augmentations are theoretically possible at the expense of engine complexity and increases in engine weight.

Table C,17. Comparison of estimates of performance of various turbojet engines. Sea level static conditions. Based on thermodynamic calculations.

	Relative thrust	*Relative specific liquid consumption*
Turbojet engine	1.00	1.00
Burner water injection	1.25	6.00
Compressor water injection	1.30	3.00
Afterburning	1.45	2.30
Afterburning plus compressor water injection	1.60	5.00
Compressor bleed and burn plus water injection in burner	2.00	13.6
Ducted fan (or bypass engine)	2.10	0.65
Ducted fan with burning in outer duct	2.80	2.35
Ducted fan with burning in outer duct and with afterburning	3.30	2.90

C,18. Liquid Injection.

Compressor liquid injection. The injection of liquid into the compressor inlet is intended to cool the mixture by evaporation of the liquid (water, alcohol, and ammonia have been used) [*47,48*]. The resulting decrease in temperature increases the density of the mixture, and therefore reduces the required compression work and increases the mass flow. The work required per unit mass to adiabatically compress and deliver a gas is given by the enthalpy difference

$$\Delta h = \int dh = \int \left(\frac{\gamma}{\gamma - 1}\right)\left(\frac{n - 1}{n}\right) v dp = \int \frac{1}{\eta_c} v dp \qquad (18\text{-}1)$$

where n is the polytropic exponent and η_c is the polytropic compression efficiency.

Notice that any reduction in specific volume v proportionally reduces the compression work for a given pressure difference. The reduced compressor work and increased mass flow both contribute to an increase in engine thrust at the expense of greater liquid consumption. For best results the liquid should be evaporated prior to its entrance into the compressor so that the air is saturated at the start of compression. As compression proceeds and the mixture temperature is increased, the relative humidity decreases, enabling more liquid to be evaporated. For this reason one injects more liquid than required to saturate the air at the entrance to the compressor.

The process within a compressor is very complex, the unevaporated fluid is separated centrifugally, the mixture is cooled, and, in the case of water injection, the average molecular weight of the mixture is reduced, water vapor having a molecular weight of 18 compared to 29.8 for air. The difference in the thermodynamic properties between air and the mixture causes changes in air flow and axial velocities within the compressor, which in turn change the flow incidence angles relative to the compressor blades, thus introducing changes in compressor efficiency and pressure ratio. As a result of the complex nature of the fluid injection process the only reliable means of determining the extent of augmentation is by testing. Actually one finds that considerably more liquid is required than is theoretically indicated and also somewhat less augmentation is obtained.

Burner liquid injection. The advantage of introducing liquid into the combustion chamber is that the work required to compress and inject one pound of a dense liquid is much less than that required to compress one pound of the less dense (bulkier) air. Eq. 18-1 is useful in demonstrating this point. The specific volume of water is only about one fifth of one per cent of the average volume of the air being compressed.

Thus by evaporating a liquid in the combustion chamber we obtain a large volume of vapor which requires a negligibly small amount of compression work at the expense of additional fuel needed to evaporate the liquid. In the cases where the liquid injected is combustible (for instance, ammonia) a partial saving is made in the additional fuel required to evaporate the liquid. During water injection the control problem can be greatly simplified if a mixture of alcohol and water is injected so that the combustion of the alcohol will supply just sufficient heat to evaporate the water. Where the liquid injected is burned, possibly poisonous products of combustion might result, as in the case when ammonia is injected.

When a large amount of liquid is injected into the combustion chamber and evaporated, the pressure before the turbine must rise since the flow resistance of the turbine is substantially fixed. This increase in pressure at the compressor outlet can induce the compressor to surge or stall.

The limit of augmentation by this means is somewhat dependent on the difference between the compressor, operating-pressure ratio, and the maximum (stall) pressure ratio. A compressor having little margin in pressure ratio (too few stages) does not allow much liquid injection in the combustion chamber. The liquid consumption and augmentation obtained is greatly dependent on the surge margin designed into the compressor and on the stability of the combustor. Accurate values can only be obtained by testing. One finds that actual test values of the augmentation ratio and the ratio of specific liquid consumption (1.20 and 7.00 respectively) fall short of the analytical values listed in Table C,17.

C,19. Afterburning. The basic principle[18] in the use of afterburning to augment the thrust of a turbojet engine is also demonstrated by again writing Eq. 18-1 for the case with nozzle expansion

$$\Delta h = \int dh = \int \eta_n v dp \tag{19-1}$$

where η_n is the nozzle efficiency. The gases leave the turbine in a turbojet engine at several atmospheric pressures. By increasing the bulk (the specific volume v) by heating these gases one can proportionally increase the energy obtainable during their expansion back to atmospheric pressure. In fact the operation of the turbojet engine itself is easily explained on this basis. After compression of the air, heating enables one to increase the bulk of the gases available for expansion back to atmospheric pressure. In this way we are able to extract in the turbine sufficient energy from the gases to pay for the work of the compressor and to have sufficient energy left to develop the high speed jet.

Since the specific volume of the gases at a given pressure is proportional to the absolute temperature, the expansion energy available in the nozzle is approximately proportional to the absolute temperature

[18] A detailed discussion of afterburners is given in [*49*].

of the gases at the afterburner outlet. Therefore, doubling the afterburning absolute temperature approximately doubles the energy available in the jet. Since, however, the jet velocity varies as the square root of the enthalpy drop across the nozzle, the jet velocity is increased only by a factor of $\sqrt{2}$ for the case where the afterburner doubles the turbine outlet temperature. All of the above reasoning assumes no pressure loss in the afterburner due to flame holder diffusers, etc. One can estimate the effect of change of temperature rise and pressure loss in an afterburner on the jet velocity by differentiating the energy equation (assuming the gases have constant thermal properties):

$$\frac{dV_j}{V_j} = \frac{1}{2}\frac{dT_6^0}{T_6^0} + C\frac{dp_6^0}{p_6^0} \tag{19-2}$$

where

$$C = \frac{\frac{\gamma - 1}{2\gamma}}{\left(\frac{p_6^0}{p_j}\right)^{\frac{\gamma-1}{\gamma}} - 1}$$

= the percentage loss in jet velocity due to a per cent loss in p_6^0.

The value of C is shown in Fig. C,19 as a function of the nozzle pressure ratio.

An examination of Fig. C,19 and Eq. 19-2 shows that a per cent loss in p_6^0 at a nozzle pressure ratio of 2.4 ($C = 0.5$) is about equivalent to a per cent loss in T_6^0. Assuming a value of $T_6^0 = 4000°R$, each one per cent loss in p_6^0 (say 0.35 lb/in.^{2} in 2.4 atmospheres) nullifies 40°F of the afterburner temperature rise as far as thrust augmentation is concerned.

The problems associated with afterburning are basically similar to those encountered in the engine combustor (Art. 7). Since the pressure level in the afterburner is lower than in the engine burner a more difficult combustion problem is encountered in the afterburner at altitude. The limiting altitude at which an afterburner will light off and burn is, as might be suspected, dependent on the gas velocity through the afterburner. Larger (usually heavier) burners allow higher altitude operation. Actual values of augmentation compare favorably with the analytical values listed in Table C,17 [*42,50*].

The testing of afterburners has brought out many practical and analytical difficulties. It has been found impossible to divorce the afterburner performance from the jet nozzle performance. This difficulty arises because burning generally takes place in the jet nozzle so that the nozzle cannot be correctly assumed to be an adiabatic device. Therefore the definition of afterburner nozzle efficiency has little significance if based on an ideal frictionless adiabatic process. Further, an exceptional afterburner combustion efficiency (if it could be accurately evaluated) is of little value if it is obtained at the expense of a large pressure drop. The difficulty of measuring the top temperature accurately is next to impossible, so that an efficiency based on a top temperature measurement should be avoided. Accordingly it has been found convenient by the

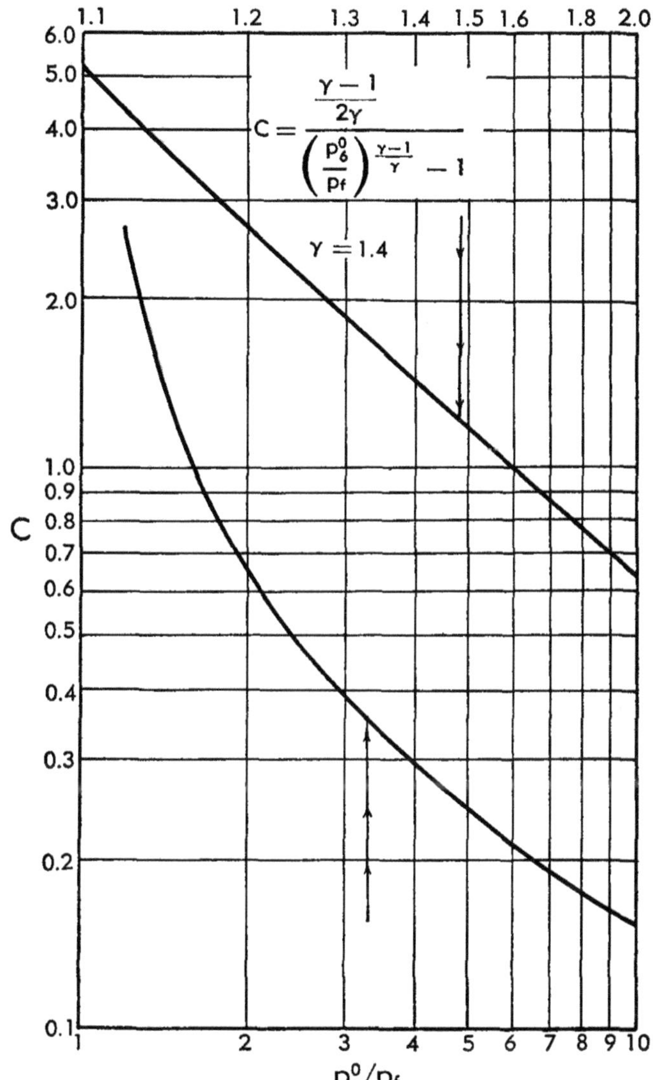

Fig. C,19.

author to define an afterburning efficiency covering the afterburner and nozzle combined which can readily and accurately be evaluated by experiment.

One notices that, for an engine developing little or no pressure ratio

p_5^0/p_∞, even large amounts of afterburning do not augment the thrust. Accordingly a useful definition of afterburner efficiency is the ratio of the actual increase in jet energy to the ideal increase in jet energy, based on the measured afterburner fuel flow and the measured gross thrust.

$$\eta_{ab} = \frac{(1 + w_{f_{ab}}/w_g)V_{j_{ab}}^2 - V_j'^2}{(1 + w_{f_{ab}}/w_g)V_{j_{ab}}'^2 - V_j'^2} \tag{19-3}$$

where $w_{f_{ab}}/w_g$ = afterburner fuel-gas ratio

$V_{j_{ab}} = F_{g_{ab}} \times g/w_{g_t}$ = calculated jet velocity while afterburning based on measured thrust and total gas flow w_{g_t}.

$V_{j_{ab}}'$ = ideal calculated afterburner jet velocity based on ideal top temperature corresponding to perfect burning and zero afterburner pressure drop.

V_j' = the ideal calculated jet velocity of the engine alone based on the engine temperature T_5^0 and the engine (nozzle) pressure ratio p_5^0/p_∞, assuming isentropic nozzle expansion.

The above definition of afterburner efficiency enables a rather accurate and consistent test evaluation of a given afterburner and nozzle combination based on the measured engine pressure ratio p_5^0/p_∞, the turbine outlet temperature T_5^0, the afterburning gross thrust $F_{g_{ab}}$, the total gas flow w_g, and the afterburner fuel flow $w_{f_{ab}}$.

The afterburner efficiency defined by Eq. 19-3 would be expected to depend, among other things, on the pressure ratio available to the afterburner, the Mach number at the inlet to the afterburner, and the pressure level in the afterburner. One would expect a large carefully designed afterburner to have an efficiency as defined by Eq. 19-3 of only about 80 per cent, since this efficiency includes the nozzle friction as well as the combustion and flame holder friction losses.

Needless to say, the pressure loss while not afterburning is an important characteristic of any afterburner. There is generally a diffuser following the turbine to reduce the gas velocities to enable burning on the flame holders, which are of many and varied types depending on the experience of the designer. Sometimes a double outer wall is used to permit cooling of the outermost wall to prevent heating of the airplane structure. These devices together with the fuel nozzle system all contribute to the nonburning pressure losses. Depending on the relative diameter (which fixes the through-flow velocity), the length (which influences the completeness of combustion), and the general excellence of design, the nonburning pressure losses might vary from 10 to 15 per cent to as little as 2 per cent. It should be noted, however, that an afterburner having a small nonburning pressure loss might thereby be limited to a small amount of augmentation due to the small size of its flame holders. It has been suggested [7] that flame holders can be eliminated by injecting fuel before the turbine nozzles and allowing the mixture to burn after the turbine.

The selection of the size, weight, and augmentation of the afterburner depends to a large extent on the airplane application. The take-off thrust augmentation, the required duration of augmentation during combat, the added weight of the afterburner, and the added fuel load, tankage, etc. must enter into the selection of the afterburner. A small lightweight afterburner having small nonburning losses and a small maximum augmentation may in some applications prove superior to a large heavy afterburner with large augmentation and large nonburning pressure losses.

C,20. Afterburning with Compressor Water Injection. The combination of afterburning with compressor water injection enables (1) an increased mass flow due to cooling the air prior to its entrance into the compressor, (2) a reduction in compression work due to the cooling by the injected water, (3) a larger engine pressure ratio p_5^0/p_∞ due to the smaller required compression work, and (4) a larger afterburning augmentation due to the larger mass flow and the larger engine pressure ratio p_5^0/p_∞.

The afterburning is limited, however, due to the influence on the combustion process of the water injection in the compressor. The limit of afterburning with water injection is largely dependent on the conservatism used in the design of the afterburner. The values given in Table C,17 are therefore rough estimates based on an assumed limit of afterburning.

C,21. Compressor Air Bleed and Burn with Water Injection in Burner. This idea is based on the fact that any air bled from the engine, after the compressor, can be heated by combustion to a very high (stoichiometric) temperature and passed through a separate jet nozzle. The top temperature of this air is not limited since, unlike the engine air, it does not have to pass through a turbine. The amount of air which is bled in this manner is, of course, not available to the turbine for developing power to drive the compressor. If, however, water is injected into the combustion chamber and evaporated, it could replace the bled air, insofar as the turbine gas flow requirements are concerned. Ideally all the excess air is normally used to cool the combustion products in the engine burner before the turbine can be bled, but actually an excess of air must be supplied to the burner in order to insure a high combustion efficiency. Approximately one third of the compressed air can be bled and heated to a very high temperature and passed through the auxiliary nozzle. The increased thrust comes from the auxiliary jet while the engine thrust remains about the same as for the case without bleeding. Water injection can be employed in the auxiliary jet; however, calculations show that there is little advantage since the same auxiliary jet thrust is obtained when the air is heated to a high temperature or when water is injected, the top temperature reduced, and the mass flow increased.

The bleed and burn operation as outlined above is capable ideally of increasing thrust about 100 per cent (Table C,17). This can only be accomplished practically with a control system capable of (1) opening the auxiliary jet nozzle, (2) igniting the auxiliary burner, (3) water injecting in the main engine burner, (4) injecting excess fuel in the engine burner to heat the injected water, and (5) controlling the jet nozzle and/or fuel flow to maintain engine speed. One must understand that such a control system is not simple.

C,22. Ducted Fan and Bypass Engines. The wake or propulsion efficiency (Art. 3) of a turbojet engine is generally low, due to the high velocity jet leaving a large amount of kinetic energy in the backwash behind the engine. The use of a geared propeller to distribute the output energy of the engine to a large mass of air and thus to improve the wake efficiency has a disadvantage due to the large losses associated with the high Mach number relative to the tips of the propeller, particularly at high flight speeds. The ducted fan (Fig. C,1(5)) is arranged so that in place of a conventional propeller we reduce the air velocity in a diffusing duct prior to entering the fan and thus gain the advantage of a better propulsion efficiency without the consequent propeller Mach number losses.

Typical axial flow jet engines already operate at the maximum allowable tip speeds and swallow air flows which are close to choking (the maximum obtainable). Evidently the only way in which we can arrange to have more air swallowed by an engine in the interest of improving propulsive efficiency is to reduce the rotational speed and increase the tip diameter. This is because of the limit on tip speed due to stress and also to fluid dynamic compressibility effects at the blade tips. The rotational speed must be lowered with each increase in compressor inlet tip diameter in the same manner as one does in building larger engines.

It may be recalled from dimensional reasoning that, other things being equal (such as relative blade Mach numbers, etc.), the air flow and therefore the thrust is proportional to the square of the linear dimensions of a jet engine. A bypass engine Fig. C,1(8) might be imagined as a small turbojet engine which drives the front end of the compressor of a larger engine through a reduction gear. Since a large portion of the output of the engine goes to the fan rather than directly to the jet it should be understood that the turbine will have to be enlarged, possibly by adding another stage. The gear may also be avoided by having a separate turbine drive the first compressor through a concentric shaft.

This type of engine combination thus handles a large amount of air, only a portion of which goes through the engine. Once the basic concept of the bypass engine is understood, other questions soon present themselves. What is the optimum portion of the air handled by the compressor which should go through the engine? What is the best compression ratio of the secondary air stream compared to the engine compressor ratio?

As might be expected, the answers to these questions depend on (1) the attainable component efficiencies of the compressor, burner, turbine, etc., (2) the engine weight, and (3) the application, i.e. the required range and performance of the aircraft.

The estimates listed in Table C,17 should be considered as rough estimates since they are based on thermodynamic cycle calculations and do not consider the practical difficulties, for instance, burning in the outer duct at the same time that the basic engine is afterburning. The design problem associated with the wall between the engine afterburner and outer burning duct would be tremendous. However, the gains theoretically possible with the bypass engine appear very attractive [*51*] particularly since one has in the same engine the choice of either low specific fuel consumption or large augmentations at the expense of much poorer specific fuel consumptions.

The ducted fan without afterburning or duct burning would be expected to perform in a way somewhat similar to the ducted fan thrust augmenter (Fig. C,1(5)) which has its fan directly connected to the tip of the turbine blades, thus avoiding the gear with its additional weight and losses. The thrust augmenter has not met with great acceptance since it must be mounted far back on the airplane wing to allow clearance for the ducted fan. This position is unfavorable when viewed from the wing flutter standpoint because the center of mass is too far to the rear.

CHAPTER 5. COORDINATION OF DESIGN

C,23. Design Problems. The methods used to design the compressor, burner, and turbine have been treated in detail in Vol. XI. However, there are many other problems [*52*] associated with the integration and manufacturing of these components into a complete and satisfactory turbojet engine. A discussion of a few of the problems together with some of the means employed for their solution may be of interest.

Optimization. The design of any turbo-engine can be carried out to varying degrees of optimization. For example, we may optimize the cycle pressure ratio to obtain either the least specific fuel consumption or the greatest specific thrust (Fig. C,4d). Again, we may be interested in engine weight, in which case estimates of engine component weights for various air flows, efficiencies, and pressure ratios must be made. Optimizing specific fuel consumption or engine weight are usually only steps toward obtaining the greatest flight range or maximum speed. A better optimization can be obtained if the complete problem including the airplane is studied.

For instance, given a certain tactical problem for a military aircraft, i.e. range, load, etc., it may be required to optimize the engine flow, pressure ratio, top temperature, etc. to obtain the highest flight speed or the greatest rate of climb at a specified altitude. Today it is possible through the use of electronic computers to carry out thousands upon thousands of optimization calculations to find the best design parameters for a new engine in combination with an airplane for a specific task. Obviously the proposed application of the engine changes considerably the value of the design parameter. An engine to drive a missile may be completely different from one to be used in a long range aircraft. For example, the ratio of engine-to-fuel weight is much smaller for the long range airplane. If the engine weight is only ten per cent of the fuel weight, a one per cent reduction in fuel consumption justifies a ten per cent increase in engine weight. Any increase in weight less than ten per cent results in a net improvement. Actually the saving in airplane weight due to tankage, wings, landing gear, etc. justifies a further increase in engine weight for the same fuel weight saving. Therefore if a reliable optimization of engine design parameters is to be obtained one should consider all the effects, including the effect of the engine size and the air flow on nacelle drag.

Axial compressor design. The continued improvement in compressor air flow per unit inlet area, the increase in pressure rise per stage, the improvement in efficiency, and the reduction of weight all contribute to better designs independently of the application. Higher compressor flows are obtained by higher axial velocities or by lower hub tip ratios. Higher velocities mean high relative Mach numbers at the blade tips which generally leads to lower efficiencies due to shock losses. Lower hub diameters mean wider or thicker blades at the root and heavier disks to support the longer heavier blades. The lower wheel speed and thick blades at the hub make it very difficult to put into the air the same amount of work as at the blade tip, without a loss in efficiency. The selection of the type of compressor to be used for a given design is thus complicated since it depends on how far we have progressed in maintaining high blade efficiency and strength while increasing the Mach number and the blade thickness, and reducing the hub tip ratio.

In the interest of ease of manufacturing it has often been the practice in the past to use the same blade (cast or forged) cut off at different lengths for the different rotating rows [*53*]. Other designs use two or more different blade forgings. Generally this ease of manufacturing is purchased at the expense of a greater engine length and weight since the blades and disks in the latter stages are wider and heavier than they need to be. Due to the generally large production of engines and blades, and since for production reasons several sets of dies must be made for the same blade, it is now convenient to have each blade row a separate forging or casting with each row reduced to its minimum width and weight.

Either constant hub or constant tip diameters have also been used for ease of manufacturing. Here difficulties have occurred in the design of the diffuser transition from compressor outlet to the combustor, particularly when the compressor outlet annulus is at a different diameter than that required for the combustor. Small hub diameters at the compressor outlet increase blade heights and therefore reduce leakage losses. In some designs the exit tip diameters have been kept as small as possible to give the engine a "wasp-waisted" appearance and thus to provide space around the engine for the auxiliary drive which is placed near the compressor outlet. However, the lower outlet compressor diameter means more stages due to the reduced wheel speeds.

Savings in engine weight can be made by using only two bearings instead of three for a single spindle engine. In this case, bearings are placed one before the compressor and one after the turbine. With this arrangement one finds it difficult, because of thermal deflections and shaft bending, to maintain small tip clearances at the compressor outlet where the blade heights are smallest. This then requires a compromise between engine weight and performance.

As the referred speed of an engine is increased due to flight at altitude there is a tendency for the air flow distribution at the compressor outlet to become skewed, i.e. the flow per unit area is greater at the tip than at the hub. This causes maldistribution of air to the burner and thus results in hot spots on the turbine nozzles. The prediction of the compressor outlet flow distribution at "off-design" conditions is next to impossible and therefore flow mixers or other means at the compressor outlet have been employed in some engines to solve these problems. The use of greater pressure drops for flow distribution in the burner is one way of correcting for poor compressor outlet flow distribution. However, this is not attractive because of the poorer over-all engine performance.

One of the toughest problems [*54*] in the development of the turbojet engine is to so design the compressor that it will start easily and quickly without surging and will accelerate quickly to military speed. The compressor should also not have a high speed stall at altitude due to the higher pressure level and lower velocity at compressor outlet. Some compressors have required motoring the engine to 60 per cent of full speed before the front stages unstall. A good compressor should unstall the front stages one at a time rather than have a number of stages unstall at once. Many times the stalling of a compressor is cured by changing the inlet guide vane-setting. However this often reduces air flow and therefore reduces engine thrust. The problem is to obtain maximum air flow and efficiency with minimum acceleration time without stalling at all altitudes.

The combustor. In order to obtain the best over-all engine performance while keeping in mind the stress and creep of the turbine blades, it has been found useful to optimize the radial temperature variation at the inlet to the turbine. The gas temperature is designed to be higher at the blade tips where the stress is lower and to be lower at the hub where stresses are the greatest. The maintaining of this optimum temperature distribution is difficult, particularly at extreme altitude near blowout where the combustion efficiency is deteriorating and the flow distribution from the compressor may be poor due to the "off-design" compressor conditions and to the lower Reynolds number.

The combustor cross-sectional area may be thought of as being made up of the primary zone where velocities are kept low and areas large, and where close to stoichiometric combustion takes place. Lower velocities, i.e. larger flow areas, contribute to stable combustion. The remaining cooling air passes around the primary zone and requires a larger or smaller cross-sectional area, dependent on its velocity which in turn sets a part of the pressure losses. The remaining pressure losses take place in the primary and mixing zones. One can see that any combustor design is a compromise between combustor altitude limit, combustor pressure loss, and therefore engine performance and airplane or nacelle drag and engine weight. Hence the best design of the combustor is closely dependent on the engine application.

The turbine. The allowable operating temperature of a turbine is often specified to have a radial variation with a higher temperature at the tip than at the hub diameter. In order to avoid blade failures one must check the radial and circumferential variation of temperature which is a difficult thing to do since it requires traversing the turbine inlet. The turbine temperature limit is generally set by specifying the total amount of creep (say 1 per cent creep in 1000 hours). Creep beyond this amount causes the blade tips to rub under design acceleration loading. The blade tips may be ground off at overhaul to restore the clearance.

A tabulation of the stress rupture values for various gas turbine alloys has been compiled by the U.S. Bureau of Ships [*55*] and is shown in Table C,23.

Any maldistribution of temperature at the turbine inlet is possibly caused by poor fuel or air distribution in the combustor which may be traced to poor outlet flow from the compressor and sometimes to poor inlet duct conditions.

Another important item in the turbine design is the outlet area. In many of the first turbine designs the turbine exhaust areas were not as large as they might be. The thought then was that the gases would later be accelerated in the nozzle so that the turbine exit axial velocities could be high, leading to short blades. However, this is not the case. During starting and acceleration it is an advantage to arrange to let the turbine develop as much power as it can. This can only be done if the exhaust duct and nozzle act as a diffuser, thereby reducing the pressure at the turbine outlet, so that the developed energy can be used to drive the turbine rather than be expended in a high velocity jet. Considerable savings in starter power and weight are a result. A turbine with low outlet area will choke sooner under starting conditions and thus limits the power available to drive the compressor. Other advantages of having a large turbine exit area are: (1) The velocities through the turbine under design conditions are lower, leading to lower blading losses. (2) Any exhaust ducting losses are reduced due to lower velocities. (3) The lower velocities at the turbine exit are more adaptable to afterburners.

Larger turbine exit areas mean longer blades and less favorable flow conditions at the blade roots. The larger required blade cross-sectional areas at the hub occur where the Mach number is highest and the blade pressure drop (reaction) is low. Also, since the tip diameter is generally held constant to avoid changes in clearance due to axial differential thermal expansions, the increase in blade height toward the exhaust is obtained by a reduction of the hub diameter. This often causes excessive divergence in the blade path which has been known to cause serious losses in efficiency.

The amount of swirl leaving the turbine should be minimized since it has been known to cause losses in the exhaust nozzle or in the diffuser ahead of the afterburner. Straightening vanes after the turbine have been used to insure zero swirl conditions entering the jet nozzle or afterburner. The use of straightening vanes after the turbine can be imagined as a half a turbine stage. When a single-stage turbine is overloaded, exit swirl usually results. Similarly if a two-stage turbine is overloaded a set of straightening canes may be helpful in improving turbine and nozzle or afterburner efficiency. In any case straightening vanes insure zero swirl conditions for the nozzle or afterburner over a wide range of loads.

Turbine cooling. The cooling of turbine blades or nozzle vanes may be employed either to conserve strategic materials or to improve engine performance while using the same amount of critical material. The effects of turbine cooling may be divided into: (1) changes in turbine losses or efficiency, (2) changes in work output of the turbine, and (3) changes in size or weight of the turbine.

Studies of the effects of blade cooling have been made by Hawthorne [*56*] which indicate that the reduction in efficiency is small compared to the loss of work output. Tests of a small water-cooled turbine by Alpert and Grey [*57*] confirm these findings.

A simplified understanding of these effects may be obtained by notic-

Table C,23. Gas turbine alloy stress rupture values.

Alloy	Condition	Nominal composition, per cent									
		C	Cr	Ni	Co	Mo	W	Cb	Ti	Fe	Other
AISI 302	F*	0.14	18.	9.						Bal	
AISI 310	F	0.25	25.	21.						Bal	
AISI 316	F	0.10	17.	12.		2.5				Bal	
AISI 321	F	0.08	18.	9.5					$5 \times C$	Bal	
AISI 347	F	0.08	18.	10.5				$10 \times C$		Bal	
Inconel	F	0.05	14.	78.						7.	
Inconel X	F	0.05	15.	73.				0.70	2.5	7.	Al0.9
Inconel W	F	0.05	15.	73.					2.5	7.	Al0.9
Nimonic 80A	F	0.04	20.	75.					2.4	0.5	Al1.2
Nimonic 90	F	0.04	20.	55.	20.				2.4	0.5	Al1.2
Waspaloy	F	0.07	20.	Bal	14.	3.			2.5	2.0	Al1.2
Hastelloy B	E	0.05		65.		29.				5.0	
Hastelloy C	E	0.1	14.5	57.		17.	4.5			5.5	
Refractalloy 26	F	0.03	18.	37.	20.	3.			3.	Bal	Al0.3
Refractalloy 70	C	0.1	20.	20.	30.	8.	4.			Bal	
Vitallium (H.S. 21)	C	0.2	28.	2.	62.	6.				2.	
422-19 (H.S. 30)	C	0.4	25.	16.	52.	6.				2.	
M.I.T. "J" Alloy	C	0.7	23.	6.	60.	6.				2.3	Ta 2.
M.I.T. 100 NT-2	C	1.	20.	30.	20.	3.	2.2			Bal	Ta 2.
X-63 (G.E.)	C	0.4	25.	10.	55.	6.				3.6	
X-40 (H.S. 31)	C	0.5	25.	10.	55.		7.			2.	
61 (H.S. 23)	C	0.4	24.		68.		6.			2.	
6059 (H.S. 27)	C	0.5	26.	33.	34.	6.				2.	
19-9 DL	F	0.3	19.	9.		1.3	1.2	0.4	0.2	Bal	
19-9 W-Mo	F	0.12	19.	8.5		0.35	1.25	0.5	0.35	Bal	
S-590	F	0.5	20.	20.	20.	4.	4.	4.		Bal	
S-816	F	0.4	20.	20.	44.	3.5	4.	4.		4.	
L-251	C	0.4	18.5	10.	53.		14.5			2.	B 0.06
L-605	F	0.15	20.	10.	Bal		15.			2.	
HE 2048	C	0.4	26.	30.	15.	4.	2.			Bal	B 0.15
N-155 (Med. C)	E	0.35	20.	20.	20.	3.	2.5	1.		Bal	N0.1
I-1360 (G.E.)	C	0.1	10.	Bal		5.		2.		6.4	Al6. B0.5
GT-45	F	0.08	17.	14.		3.		0.5	0.3	Bal	Cu3.
H.S. 88	F	0.15	12.	15.		2.	0.5		0.6	Bal	B0.1
Timken (16-25-6)	F	0.1	16.	25.		6.				Bal	N0.15
Discaloy 24	F	0.03	13.	26.		3.			2.3	Bal	Al0.6
Tinidur (A-286)	F	0.08	15.	26.		1.3			1.9	Bal	V 0.3 Al0.35
K-42-B	F	0.06	18.	42.	22.				2.6	Bal	Al0.6
G-18-B	F	0.4	13.	13.	10.	2.	2.5	3.		Bal	

* C, cast; F, forged; E, either cast or forged.

Stress rupture, psi × 10⁻³											
1200°F		1350°F		1500°F		1600°F		1700°F		1800°F	
100 hr	1000 hr	100 hr	1000 hr	100 hr	1000 hr	100 hr	1000 hr	100 hr	1000 hr	100 hr	1000 hr
23.	16.5		7.5		3.5						
	17.		8.7		4.5		3.		2.4		1.95
	25.		14.0		7.		4.		2.3		1.3
29.5	17.5		7.6		3.7		2.75				
	17.0		9.0		4.4						
22.	14.5	10.5	6.8	5.7	3.7	4.2	2.7			2.5	1.55
80.	68.	50.	40.	28.	18.	17.					
63.	49.	45.	32.	22.5	11.						
63.	49.	45.	32.	22.5	11.						
76.	64.	50.	38.	28.	18.	17.					
		52.5	40.	28.	20.	16.					
58.	33.2	34.	25.	17.	11.3	11.7	8.3				
				14.1	10.7						
74.		48.		29.	18.						
				19.	15.	12.5	9.6				
54.	43.	36.5	28.0	22.	16.	18.4	14.5	13.1	10.5	8.8	6.4
		47.	36.3	30.4	23.5	19.0	14.7	14.7	11.5	10.1	7.1
		52.5	40.5	34.8	28.2	26.4	21.	17.2	12.3	12.8	9.3
				30.5	22.8	26.1	19.3	18.2	13.1	13.6	9.9
				26.8	20.2						
55.	46.	44.8	34.	28.	23.2	21.3	18.5	17.3	14.4	11.3	9.8
58.	47.	35.2	27.	26.7	20.5	17.1	11.8	13.5	9.4	8.7	5.6
55.	46.	36.4	30.	24.	18.8	16.	12.	12.	8.6	9.2	7.
60.	50.	35.	20.5	13.3	10.						
54.5	42.	28.5		13.5							
52.	40.	31.5	23.5	20.	15.5	14.0	10.5	9.2	6.5	5.5	3.4
64.	49.2	40.	29.5	27.	21.1	14.8	10.	9.8	6.4	5.3	3.1
72.	59.5			28.1	22.8						
45.				24.	19.					7.2	4.5
		36.	28.	22.	16.	17.5	13.			9.5	
60.	45.5	36.	26.	24.5	18.6	15.5	11.4	7.6	4.7	4.9	2.75
				45.	32.5						
46.1	39.	27.	22.4	15.	12.						
43.	36.	25.	19.3	12.1	9.2						
57.	42.	28.	19.	13.3	9.						
56.	46.	30.5	17.								
65.5	45.	34.5	21.3	13.	8.						
62.	39.5	36.5	26.5	22.5	15.						
52.	37.	28.5	20.5	16.	11.						

ing that the ideal turbine work is equal to the integral of the isentropic enthalpy drop.

$$h_4^0 - (h_5^0)_{is} = \int (dh)_{is} = \int v dp^0 \tag{23-1}$$

where v is the specific volume, h^0 the total enthalpy, and p^0 the total pressure.

The actual turbine work is found by multiplying the isentropic enthalpy drop by the expansion efficiency η.

$$W = \eta[h_4^0 - (h_5^0)_{is}] = \int \eta v dp^0 \tag{23-2}$$

An examination of this equation shows:

1. The differential work is proportional to the product of the efficiency η, the specific volume v, and the pressure drop.
2. The efficiency may be affected by cooling due to changes of the density or viscosity in the cooled boundary layer or by flow angle changes due to density changes which affect the axial velocity. These effects are usually small [*51*,*58*].
3. The change in average specific volume due to cooling during the expansion is the principal effect. If the turbine nozzle is uncooled and the complete pressure drop is taken across the nozzle there is no loss in the integral equation (Eq. 23-2).

We can imagine that, due to blade cooling, the gases are cooled by the blade which results in a smaller increase in specific volume for a given increase in top temperature than for the uncooled blade. The lost work due to blade cooling would be

$$\Delta W = \int \eta \Delta v dp^0 \tag{23-3}$$

where Δv is the reduction in average specific volume due to blade cooling.

The jet nozzle and ejector. The problem of designing an ejector to surround the jet nozzle for the purpose of nacelle cooling is complicated by the fact that any change in the nozzle position influences the ejector performance [*59*]. The problem would be much simpler if the nozzle area were varied by means of a movable center cone; thus the outside of the nozzle and therefore the ejector configuration would remain fixed. However, this is not a very good solution since it is difficult to use an adjustable center cone nozzle with an afterburner, due to the high temperature in the center of the nozzle. Further, it is difficult to obtain sufficient area variation for afterburning with a variable center cone.

The problem of designing a movable, yet leakproof, nozzle is a difficult one. In the case of the eyelid-type nozzle, sealing must be maintained around the complete circumference, and a blanket of cool air must be provided to avoid thermal distortions and leakage which may cause flowback and heating of the nacelle.

Thrust reversal. Turbojet engines have been at a disadvantage since their inception because of their inability to reverse thrust. Although many different types of thrust reversers have been tested successfully, not many have reached the production stage. The development of the turbojet engine should not be considered complete until a practical thrust reverser is found.

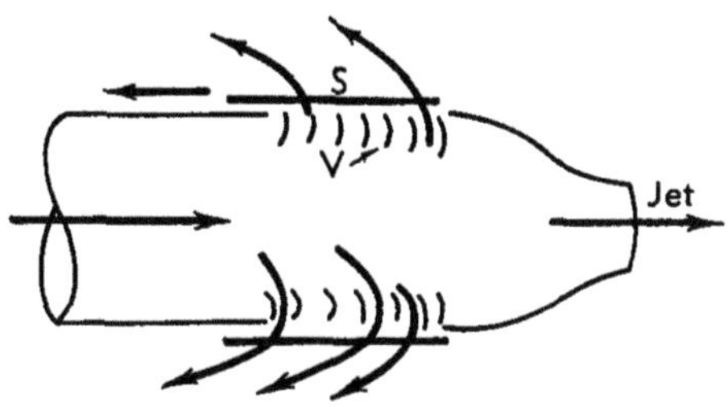

Fig. C,23a. Thrust reverser, Westinghouse.

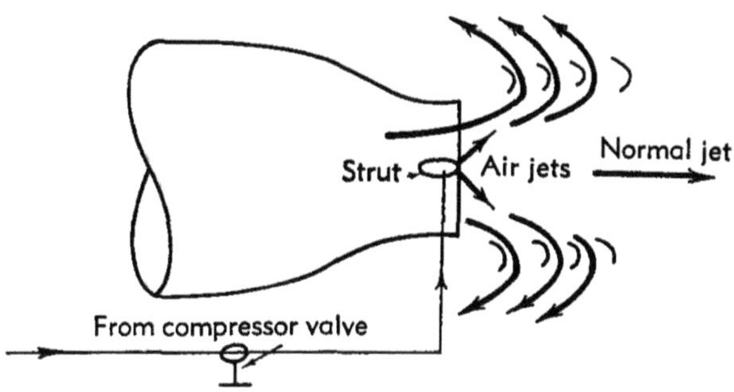

Fig. C,23b. Thrust reverser, SNECMA.

One type of reverser (Fig. C,23a), successfully tested several years ago at Westinghouse, had a series of circular turning vanes *V* covered by a sliding shield *S*. When the shield was moved to uncover the vanes the high pressure gases escaped and were directed forward, thus reversing the thrust.

Another type of reverser successfully demonstrated by SNECMA in France (Fig. C,23b) uses a jet of compressor air to deflect the main jet through reversing vanes outside the normal jet stream. Complete thrust reversal is obtained in a matter of seconds by this means.

CHAPTER 6. TESTING

C,24. Types of Tests. The size and cost of test facilities required to adequately develop a turbojet engine has grown tremendously during the past decade. The first turbojet engines were tested in simple cells with measurements confined to the principal variables, such as thrust, fuel consumption, rotational speed and air flow (Plate C,1a). The continuing increase in engine size and the more difficult performance and operational specification which must be met has been responsible for the construction of large and expensive testing facilities all over the world. Originally the only means of determining engine performance under altitude conditions was by flight test where instrumentation and test personnel were limited and testing was hazardous. Some of the first tests were run by mounting the engine in the bomb bay of a bomber or under the fuselage of a propeller-driven fighter plane. Today there are several elaborate test facilities located throughout the world which enable the most powerful engines to be tested under extreme altitude conditions. A large part of the cost of these facilities is due to the refrigeration and evacuating equipment required to condition the great quantities of air handled by the engine. Also, considerable investment in test instrumentation is required to solve the many problems arising in the development of an engine. Problems of starting, ignition, acceleration, combustion hot spots, compressor surging, blade vibration, combustion blowout, nacelle cooling, anti-icing, to name a few, all require facilities and much instrumentation for their solutions.

Engine testing can be divided into four principal categories:

1. *Development testing* which is carried out during the development of the engine. The engine components having been previously tested in component test rigs, the complete engine is tested for over-all performance giving thrust air flow and specific fuel consumption, etc., together with operational characteristics such as compressor stall, acceleration time, control stability, overheating, combustor hot spots, etc. These tests may be run under static (no ram) sea level conditions in a test cell as shown in Plates C,24a or C,24b. Anti-icing methods have been tested in refrigerated test tanks. More natural icing conditions have been tested in cells on top of Mt. Washington.

2. *Type test or endurance tests.* Before a new engine type is accepted for military or civilian use it is generally required to pass an intensive test to demonstrate its ability to stand up for long periods of time under conditions simulating those found in actual operation. The engine is usually subjected to long periods of full thrust followed by periods of intermittent full and idling thrust in order to show any defects due to rapid heating and cooling. After these tests the engine is usually disassembled and the parts inspected for any signs of excessive wear or failure. These tests are usually run in static sea level test cells similar to Plates C,24a or C,24b.

3. *Flight tests.* These tests demonstrate the ability of the engine to operate satisfactorily in the airplane. Many difficulties may be uncovered

during these tests. For instance, poor flow distribution in the inlet duct may be encountered during some flight altitudes such as during landing or take-off. The air supply for nacelle cooling may not be adequate at standstill requiring ejector modification.

4. *Acceptance tests* are tests run on production engines prior to shipment. These tests are generally of short duration since they are run on well-proven engines. Usually the engine must demonstrate the contract guaranteed thrust, specific fuel, and oil consumption together with certain operational requirements such as the acceleration time. During these tests the engine is usually calibrated, i.e. its performance is logged and final adjustments are made. In addition to complete engine tests there is also production testing of components and accessories. According to Zaehringer [*58*] one manufacturer utilizes 75 production cells, costing five million dollars, for jet-accessory production testing. These cells permit testing from −80°F to 1000°F and simulated altitudes to 75,000 ft.

C,25. Test Methods. Test methods have generally been developed and improved along with the engines. As an example, one of the first jet engine altitude test facilities available in America is at NACA's Clevcland laboratory (see Plate C,25a). Before the introduction of the jet engine, complete piston engine nacelles were tested here by mounting them on a wing panel which was placed across the tunnel and on which the thrust could be measured by means of weighing devices situated beneath the tunnel and connected to the ends of the wing panel. The huge closed circuit tunnel could be evacuated and refrigerated to simulate altitudes up to 55,000 ft while large fans simulated air speeds up to a Mach number of unity. Adapted to jet engine testing, the first improvement was to collect the hot exhaust gases in a pipe downstream of the engine jet nozzle so that they could be sprayed with water and removed from the tunnel, thus reducing the refrigeration load. Next, a direct connected pipe to the engine inlet insured cooler air reaching the engine inlet, thus simulating higher altitudes quickly without waiting to cool the whole tunnel. Since the air flowing around the engine contributed little to the test except by nacelle cooling, the type of tank test facility shown in Plates C,25b and C,25c was used when more facilities were added.

The use of a tank-type facility introduces some difficulties. A particularly troublesome item is the measurement or simulation of the proper conditions surrounding the jet nozzle. Unless the tank diameter is large compared to the jet diameter, considerable variation in static pressure is found in the region of the jet. In one case (at the Philadelphia Naval Base) the proper ram conditions could only be adjusted by adjusting the internal engine pressure ratio and ignoring the various static pressures along the tanks near the jet nozzle. This difficulty can be avoided if the ratio of tank diameter to jet diameter is equal to 10 or larger. Another difficulty is the measurement of thrust which is influenced by the pressure variation on the surfaces within the tank. In order to simulate ram conditions a partition is usually arranged to divide the tank in two. Labyrinth-type seals between the partition and the engine prevent excessive leakage past the partition. Ram conditions are simulated by arranging a pressure difference across the partition. Any pressure differences on the tank end walls must be accounted for in calculating the engine thrust.

Thrust measurements. Thrust measurements have been made using many different methods, among which are bell crank and weigh scales, hydraulic or pneumatic pistons, straingauges or electric load cells. In flight, gross thrust has been indicated on a meter by means of a combination of pressure transducers and electric circuits that electrically calculate the thrust from the jet nozzle pressure ratio, which was measured by means of differential transformer straingauges. In flight, straingauges or load cells on the engine mounts have been used. Engine thrusts have been checked in the airplane while on the ground by measuring the force on a perforated cylinder which is arranged to catch the jet blast and turn it at right angles to the engine axis. In test cells one must be careful that fuel and control lines do not influence the thrust measurement. Even if flexible hose is used for the fuel supply line it is well to have the thrust stand move little or not at all to eliminate errors due to outside connections. Flexible hose connections usually stiffen when fluid pressure is applied.

The static thrust of the installed engine has also been measured by means of a thrust-measuring device which holds the airplane wheels. Some sources of error in thrust measurement that should be avoided are:

1. In a test cell or nacelle, a large amount of air flowing around the engine can cause a drag on the engine and accessories, particularly if the cell is small.
2. Large amounts of cooling air flowing around the engine in a test cell have momentum changes which influence the measured thrust.
3. If the engine and cell-cooling air do not enter the test cell at right angles to the engine axis, an error in measured thrust occurs due to the momentum of the entering air along the engine axis.
4. If a small test cell is used it is possible for the cell-cooling air to cause a pressure difference before and after the engine which must be combined with the measured thrust.

Fuel flow. Fuel flow has also been measured in several ways, among which are tank and weigh scale, volumetric tank, rotometer, propeller, and piston disk or bellows displacement meters. Ultrasonic meters based on sound transmission through the moving fluid and magnetic meters [*60*], based on the voltage generated in the moving fluid, are also used. The first three methods have the disadvantage of not being applicable to flight tests. Rotating impellers on which the driving torque is measured have also been employed to measure the mass flow rate directly.

Fuel flow totalizers are also an essential flight test instrument, particularly when testing variable flight conditions such as during climb tests.

Air flow. Air flow is generally measured in test cells by means of a bellmouth standard flow nozzle mounted directly on the engine as shown in Plates C,24a and C,24b. In this way only the air flow which enters the engine is measured. In a setup such as shown in Plate C,25c, the cell-cooling air or leakage air past the tank partition seals is not measured.

In the test arrangement shown in Plate C,25a, the flow is measured by means of a measuring nozzle in the air supply pipe. Air leakage past a slip joint must be allowed for. In flight the air flow can be found if the compressor characteristic is known from previous calibrations. Measurement of compressor speed, inlet total temperature and pressure, and pressure ratio are easily made in flight, enabling an estimate of air flow.

Temperature measurements. Temperature measurements throughout a turbojet engine are satisfactorily made using thermocouples. Difficulty is generally encountered in measuring the high temperatures before or after the turbine. Temperature measurements can be of two kinds: those which are required for thermodynamic calculating, such as in calculating the compressor or turbine efficiency or those which are used to control the engine to prevent overheating. The temperature survey probes shown in Plate C,25a are for the purpose of determining the temperature distribution (hot spots) at the outlet of the burner under high altitude conditions. Obviously this equipment is impractical in the airplane. Usually several thermocouples are grouped into a harness in parallel to closely read an average turbine outlet temperature for control purposes. Actually this device may not give a very accurate temperature reading; however, it is rugged and consistent enough for control purposes.

During developmental tests the author used a calibrated jet nozzle to determine a thermodynamic turbine outlet temperature. The air flow was measured by the flow nozzle at engine inlet. Since the flow through the jet nozzle is proportional to the total pressure and the area, and inversely proportional to the square root of the jet nozzle inlet temperature, the total temperature in the exhaust can be found. This method gives an effective jet nozzle temperature even though the temperature is not constant over the flow cross-sectional area entering the jet nozzle. The top temperature can be estimated by adding the compressor temperature rise to the turbine outlet temperature since the turbine and compressor work can be assumed equal. Suitable allowance can be easily made for the larger flow in the turbine due to the fuel added and also for the difference in specific heat of the combustion products and air. Accurate temperature measurement in afterburners is very difficult to obtain, due to the fact that they are the highest in the engine and are generally and intentionally not uniform in the interest of keeping the walls and the variable jet nozzle cool. The temperature measurement can be avoided by defining a combined afterburner jet nozzle efficiency as explained in the section on afterburners.

Rotational speed. There is generally no difficulty in measuring speed using a tachometer generator driven by the auxiliary gear box. The generator signal is also easily adaptable for use in electronic engine control systems.

Tip clearance. An accurate knowledge of turbine and compressor tip clearance is essential to the development of turbojet engines. Quick shutdowns at low ambient temperatures may cause the turbine to seize, due to the sudden chilling of the turbine housing. Large tip clearances could of course be used with a resulting loss in efficiency. A better solution of the problem is to keep the housing hot by ventilating with hot combustion gases.

Tip clearance measurements in the compressor are useful in insuring that the clearances while running are small and uniform (i.e. the rotor is not eccentric). A few different types of tip clearance instruments have been devised. One simple and inexpensive type [*31*] uses a pneumatic piston to force a piece of $\frac{1}{8}''$ copper tubing into the machine. The tubing is ground off by the blade tips and is then withdrawn by relieving the air pressure. Removing the device from the engine and again energizing the pneumatic piston, the amount of the tube protruding leads to a measure of the tip clearance. More elaborate devices have been used which continuously and remotely measure tip clearance by automatically moving a probe into the engine until an electric contact is made with the blade, causing the probe to be withdrawn. This process is continuously repeated and a continuous reading is obtained.

Blade vibration. Blade vibration has been measured by straingauges baked on the rotor and stator blades of the compressor. Stalling-induced vibrations and their frequencies are easily indicated by this means. Electrical contact to the gauges on the rotor is accomplished through slip rings on the shaft. Very often blade stall vibrations have been indicated by this means when no other evidence of stall is present, such as pressure fluctuation or noise.

Traversing. In order to find detailed information about the flow in the compressor it is helpful to traverse between the rotating and stator rows to determine flow angle, velocity (total and static pressure and temperature). This can be done with manually or remotely operated traversers [*31*] which accurately position a probe to any desired depth or angular position. This information is very helpful in indicating which stage is stalling during quick acceleration in the static sea level cell or in the altitude chamber.

Position indicators. Differential transformers are generally used to indicate position, such as exhaust nozzle and adjustable compressor vane positions.

C,26. Data Reduction. The data read during early engine flight tests were limited at best to the amount the pilot could jot down on a knee pad. Later photopanels were installed to photograph an instrument panel situated, for instance, in the nose of the plane. This method, although far superior to the pilot's reading the instruments, had the disadvantage of the film development and reading which consumed considerable time and effort. Next the data was transmitted directly to the ground by telemetering. Telemetering has the disadvantage of short range. Much more data is taken today [*61*] using magnetic tape recorders which use segmented commutators that switch to several tracks on a magnetic tape recorder. In this manner it is possible to read over 400 instruments. The magnetic tape readings can be converted, in a digitizer, to digital signals which are placed on a second magnetic tape which can be fed to an electronic computer. The computer then calculates the results and types them or plots them on automatic X-Y recorders. This type of data reduction is also used in static test cells or altitude tanks. A large volume of data is thus taken and calculated automatically and plotted without human hands.

Testing of engine controls and accessories has been carried out using an electronic engine simulator or analogue which is arranged to feed the engine characteristics to the actual controls, thus checking the control or accessory without an engine. The complete airplane has also been simulated in this manner [*62*].

C,27. Test Facilities. One of the most difficult problems associated with the testing of turbojet engines has been the noise problem. All test cells must be equipped with noise suppressors at the air inlet and at the jet exhaust; otherwise complaints will be received from distances up to a mile. The noise level can generally be reduced to an acceptable amount by means of sound-absorbing material and wavy inlet ducts as shown in the airplane test facility in Fig. C,27. The exhaust gases are usually sprayed with water to protect the soundproofing material. A typical air inlet silencing system is shown in Plate C,27a.

As a matter of experience it is generally found that the high-pitched sounds are rather easily removed by soundproofing materials. The low frequency rumble appears to be the most difficult to suppress. Huge and expensive tuned resonators arranged to form a wave trap are sometimes required to successfully reduce these low frequency, high energy sounds.

Wind tunnel or altitude tank test facilities obviously do not have a noise problem.

There are many fine engine test facilities located in many countries throughout the world. For example, a Swedish laboratory is described in [*63*]. The extensive test facilities at Villaroche, France, are shown in Plate C,27b. One of the largest engine test facilities is at the Arnold Engineering Development Center at Tullahoma, Tennessee (Plate C,27c). This facility allows the testing of complete engine nacelles. There are two wind tunnels, a transonic ($M = 1.6$) and a supersonic ($M = 1.5$ to 4.9) tunnel. The test sections ($16 \times 16 \times 40$) for each tunnel are removable so that complete engine setups can be changed quickly. A

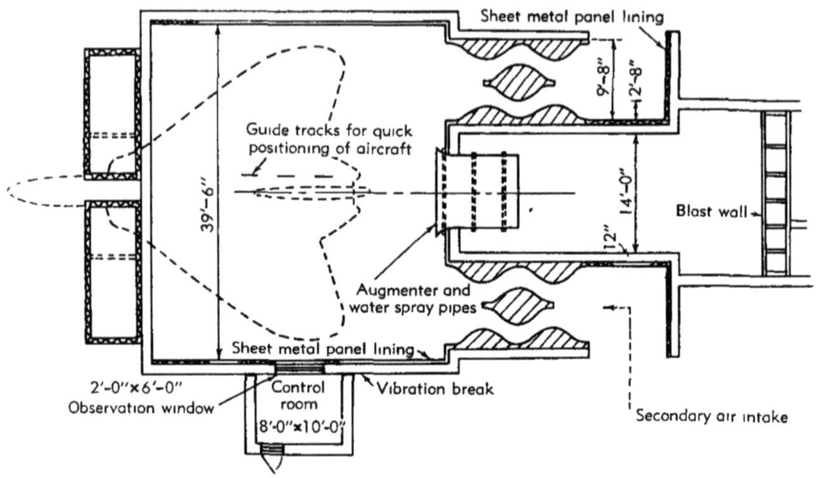

Fig. C,27. Run-up muffler. (Courtesy Koppers Company, Inc.)

three-stage compressor supplies air for the transonic tunnel while an eighteen-stage compressor supplies air for the supersonic tunnel. The compressors are 575 feet long and supply 13 million cubic feet of air per minute. A total of 216,000 hp in four motors drive the compressors. Altitudes up to 100,000 feet can be simulated.

An altitude tank facility (Plate C,27d) containing three tanks provides test sections 12.25 feet in diameter with adjustable lengths of from 19 to 78 feet. An air flow of 225 pounds per sec is supplied by four 5000 horsepower centrifugal compressors at temperatures from -120 to $+650°$F. The air is dried to 1.1 grains per pound. Exhausters of 40,000 hp enable the evacuation of the 3000°F exhaust gases after they are sprayed with water.

C,28. Cited References.

1. Driggs, I. H., and Lancaster, O. E. *Trans. Am. Soc. Mech. Engrs. 75*, 217, 234 (1953).
2. Godsey, F. W., and Young, L. A. *Gas Turbines for Aircraft.* McGraw-Hill, 1949.
3. Finch, V. C. *Jet Propulsion.* National Press, Millbrae, Calif., 1948.

4. Scheuter, K. R. *Engr's. Digest 4*, 151–157 (1947).
5. Lovesey, A. C. *Aviation Age 26*, 28 (1956).
6. Table of engines. *Aviation Age 26*, B-14 (1956).
7. Durham, E. P. *J. Aeronaut. Sci. 17*, 425–428 (1950).
8. Jakobbson, B. *J. Roy. Aeronaut. Soc. 55*, 226 (1951).
9. Reissner, H. *J. Aeronaut. Sci. 14*, 197 (1947).
10. London, A. L. *Trans. Am. Soc. Mech. Engrs. 1*, 234 (1948).
11. Prandtl, L., and Tietjens, O. G. *Applied Hydro- and Aeromechanics.* McGraw-Hill, 1934.
12. Keenan, J. H., and Kaye, J. *J. Aeronaut. Sci. 14*, 442 (1947).
13. Adams, H. T. *Elements of Internal Combustion Turbine Theory.* Cambridge Univ. Press, 1949.
14. Keller, C. *U.S. Patent 2,172,910*, Sept. 1939.
15. McAdams, W. H. *Heat Transmission.* McGraw-Hill, 1933.
16. Goldstein, A. W., and Alpert, S., et al. *NACA Rept. 928*, 1948.
17. Sanders, J. C., and Chapin, E. C. *NACA Rept. 987*, 1950.
18. Bridgeman, P. W. *Dimensional Analysis.* Yale Univ. Press, 1931.
19. Buckingham, E. *Phys. Rev. 4*, 346 (1914).
20. Buckingham, E. *Phil. Mag. 48*, 141 (1924).
21. Buckingham, E. *J. Wash. Acad. Sci. 4*, 347 (1914).
22. Buckingham, E. *Nature 96*, 208 (1915).
23. Buckingham, E. *Trans. Am. Soc. Mech. Engrs. 37*, 263–296 (1915).
24. Afanassjewa, T. E. *Phil. Mag. 1*, 257 (1926).
25. Roberts, H. E., and Langtry, B. D. *J. Aeronaut. Sci. 17*, 429–435, 456 (1950).
26. Brajnikoff, G. B. *NACA Tech. Note 2697*, 1952.
27. Meyer, C. A. *Trans. Am. Soc. Mech. Engrs. 69*, 237 (1947).
28. Oswatitsch, K. *J. Roy. Aeronaut. Soc. 50*, 921 (1946).
29. Thomas, A. N. *Aviation Age 26*, 24 (1956).
30. Kantrowitz, A., and Donaldson, C. duP. *NACA Research Mem. L5D20*, 1945.
31. Meyer, C. A., and Benedict, R. P. *Trans. Am. Soc. Mech. Engrs. 74*, 1327–1336 (1952).
32. Salzman, F. *Schweiz. Bauzit. 124*, 13 (1944).
33. Constant, H. *Gas Turbines and Their Problems.* Todd Reference Library, London, 1949.
34. Schmitt, H. E. *Aero. Eng. Rev. 9*, (*12*), 18–24 (1950).
35. Hawthorne, W. R., and Cohen, M. *Roy. Aircraft Establishment Rept. E3997*, 1945.
36. Meyer, C. A. *Trans. Am. Soc. Mech. Engrs. 72*, 465 (1950).
37. Meyer, C. A., and Faught, H. F. *Aero. Eng. Rev. 10*, (*1*), 33–41 (1951).
38. Hunter, F. S. *American Aviation 20*, 57 (1956).
39. Samuels, J. C., and Gale, B. M. *NACA Tech. Note 2119*, 1950.
40. Keenan, J. H., and Kaye, J. *Thermodynamic Properties of Air and Combustion Products.* Wiley, 1949.
41. Pinkel, B., and Turner, R. L. *NACA Wartime Rept. E23*, 1945.
42. Cody, C. S. *Am. Soc. Mech. Engrs. Paper 48-AV-3*, 1948.
43. Keenan, J. H. *Thermodynamics.* Wiley, 1945.
44. Luskin, H., and Klein, H. *Am. Soc. Mech. Engrs. Paper 50-A-102*, 1950.
45. Karanik, J., McCarthy, J. S., Benedict, M. C., and Gerdan, D. *Symposium, Aero. Eng. Rev. 9*, 28–47 (1950).
46. Fraser, D., Pettit, K. G., and Bowler, H. *Aeronaut. Eng. Rev. 11*, 44, 1952.
47. Wilcox, C. E., and Trout, M. *NACA Rept. 1008*, 1951.
48. Ellisman, C. *NACA Research Mem. E7G29*, 1947.
49. Cleveland, F. A. *J. Aeronaut. Sci. 15*, 305–311 (1948).
50. Wilkinson, P. H. *Aero Digest 60*, 17–18 (1950).
51. Pearson, H. *Aviation Age 26*, 32 (1956).
52. Hawthorne, W. R. *SAE Trans. 54*, 347–357 (1946).
53. Constant, H., Howell, A. R., and Reeman, J. Development of the British gas turbine jet unit. *Inst. Mech. Engrs., War Emergency Issue 12.* Republished *Am. Soc. Mech. Engrs.*, Jan. 1947.

54. Novak, R. A. *Aviation Age 26*, B-10 (1956).
55. Hafer, A. A. *Trans. Am. Soc. Mech. Engrs. 75*, 127 (1953).
56. Hawthorne, W. R. *Am. Soc. Mech. Engrs. Paper 55-A-186*, 1955; *55-A-191*, 1955.
57. Alpert, S., and Grey, R. E. *Am. Soc. Mech. Engrs. Paper 55-A-202*, 1955.
58. Zaehringer, A. J. *Jet Propulsion 26*, 506–507 (1956).
59. Cortright, E. M., Jr. *Aeronaut. Eng. Rev. 15*, 59 (1956).
60. Munch, R. H. *Ind. and Eng. Chem. 44*, 83A (1952).
61. Granholm, J. W. *Aviation Age 26*, G5 (1956).
62. Evans, W. L. *Aviation Age 26*, (*3*), 42 (1956).
63. Walter, L. *Industrial Laboratories 7* (*2*), 25 (1956).

SECTION D

THE TURBOPROP ENGINE

I. H. DRIGGS
O. E. LANCASTER

D,1. Introduction. The basic principles[1] involved in generating the potential power of a turbojet and a turboprop are identical. They each have a compressor, a burner, and a turbine which runs the compressor. This common part of each produces energy in the form of a hot gas at a high pressure. The only difference in the engines is the manner in which the available energy is converted to useful thrust. As previously explained, in a turbojet the energy is converted to thrust by expanding the gases through a nozzle to form a jet at high velocity. By directing this jet backward the reaction gives a forward component of thrust. In a turboprop, part of the potential energy is removed by additional turbine stages that drive a propeller which activates a large mass of air; the residual energy is then used to produce a jet stream similar to that of a turbojet, but because of the lower energy level the jet velocity is much lower.

For convenience of thought in comparing the two engines, the part that is common to the turbojet and the turboprop is usually referred to as the hot gas generator, or power generator, and the means for converting the energy to thrust power is called the propulsive system. The hot gas generator is analogous to the steam boiler on a steam engine, in that each produces available energy stored in a gaseous medium at a high temperature and pressure.

Fig. D,1a gives a schematic comparison of the turbojet and the turboprop. It shows the common part, the gas generator, and the two different propulsive systems. In actuality, the turbine which runs the propeller and the turbine which runs the compressor may or may not be combined mechanically to form one multistage unit. When the two are mechanically coupled, the engine is called a fixed turbine machine. In this case Fig. D,1a is slightly fictitious because all stages operate to drive the one shaft which propels both the compressor and turbine, and the division of power absorbed by each depends upon the operating condition.

[1] The opinions or assertions contained herein are the private ones of the authors and are not to be construed as official or reflecting the views of the Navy Department or the Naval Establishment at large.

When there are actually two turbines, as depicted by the figure, the second one is attached to the propeller by a concentric shaft passing through the shaft of the gas generator. In this case the two turbines operate independently; the first always supplies the power to the compressor, and the second supplies the power to the propeller. A unit which drives only the propeller is called a free turbine. There is still another possibility; the first turbine may drive only part of the compressor, the high compression stages, and the second drives the propeller

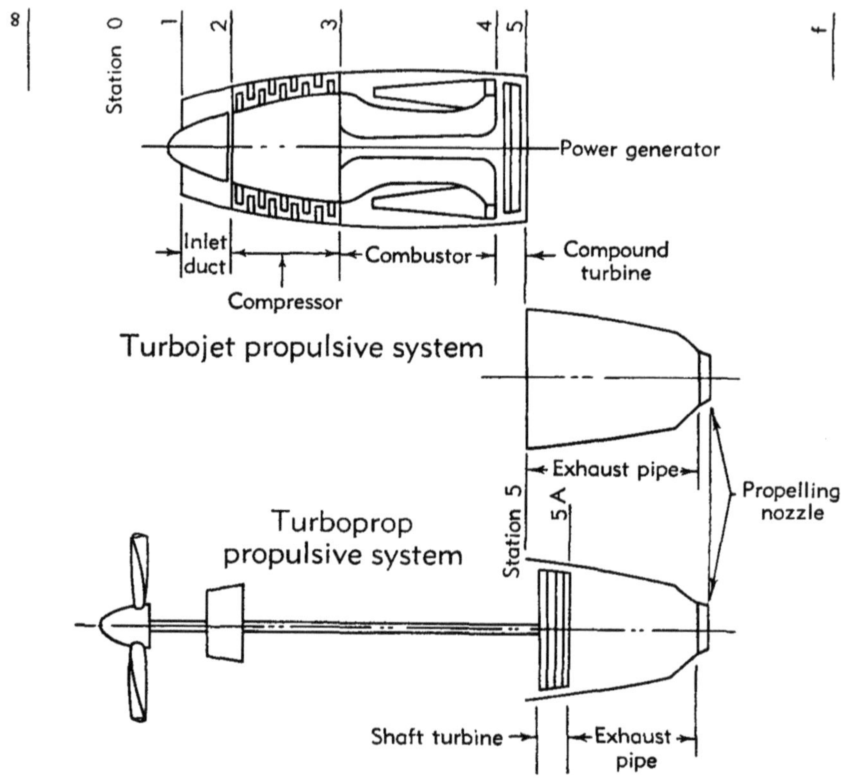

Fig. D,1a.

and the low compression stages of the compressor. In this case the design is called a coupled turbine as well as a two-spool engine. In this sense it is similar to a two-spool turbojet and theoretically they both could be multispool units.

Why a turboprop? It is natural to ask, "Why the additional complication?" Why replace a simple light exhaust pipe (an annular piece of metal), the propulsive system of the turbojet, by a turbine, a shaft, a gear box, a heavy propeller, and an exhaust pipe (the propulsive system of the turboprop)? Not only is the system more complicated, it is much heavier, so why do it? The answer is straightforward. The latter system can convert more of the energy of the gas generator to useful work or thrust; that is, for a moving aircraft more of the available gas energy is given to the aircraft and less is left in the directed stream of gas, and, for stationary aircraft, where no energy is given to the airplane, more thrust is developed. This results from the fact that the energy of a gas is proportional to the product of the mass of the gas and the square of the velocity, whereas the work or thrust delivered to the airplane is proportional to the product of the mass and the velocity change.

The useful energy produced by a gas generator can accelerate the mass of air m, which passes through the engine, to a velocity V_j relative to the airplane. This same energy could also be used to give a larger quantity of air M, of which m is only a part, a velocity V relative to the airplane. Thus, in the ideal case, the energy equivalence gives the equation[2]

$$\frac{m}{2}(V_j^2 - V_\infty^2) = \frac{M}{2}(V^2 - V_\infty^2) \tag{1-1}$$

where V_∞ is the flight velocity. The net thrust energy imparted to the airplane is the difference between this fixed energy and the amount of energy left in the slipstream, that is

$$FV_\infty = \frac{M}{2}(V^2 - V_\infty^2) - \frac{M}{2}(V - V_\infty)^2 \tag{1-2}$$

or

$$F = M(V - V_\infty) \tag{1-3}$$

where F is pounds of thrust. It follows from the differentiation of Eq. 1-1 and 1-3 that

$$\frac{dF}{dM} = \frac{(V - V_\infty)^2}{2V} \tag{1-4}$$

This derivative is positive so long as there is a rearward velocity of the slipstream relative to the surrounding air. Therefore, for a given amount of energy, the greater the mass of air activated and the smaller its change in velocity, the greater the thrust. The turboprop by means of its propeller achieves this very thing. It imparts a small velocity to a large mass of air whereas a jet activates only the small mass of air which passes through the engine. This is illustrated in Fig. D,1b. The ideal upper limit of the ratio of the thrust of a turboprop to that of a turbojet

[2] Here the mass of the fuel is neglected. In the more general case treated by Roy in Sec. B, this mass is included and then when the flight velocity is high enough so that the energy obtained from burning fuel is insufficient to give the mass of the fuel a velocity equal to the flight velocity, a turboprop or turbojet would produce drag. The flight velocities here are sufficiently low so that this is not a consideration. Other factors, such as efficiency, are more important than the omission of fuel flow.

is

$$\frac{1}{2}\left(\frac{V_j}{V_\infty} + 1\right)$$

which for positive thrust is always greater than one, and the ratio may be quite large depending upon the magnitude of the given energy per pound of gas in the gas generator and the flight speed. Although these ideal conditions cannot be reached, they establish a goal. Eq. 1-4 shows that the increase of thrust with mass activated is less pronounced the nearer the

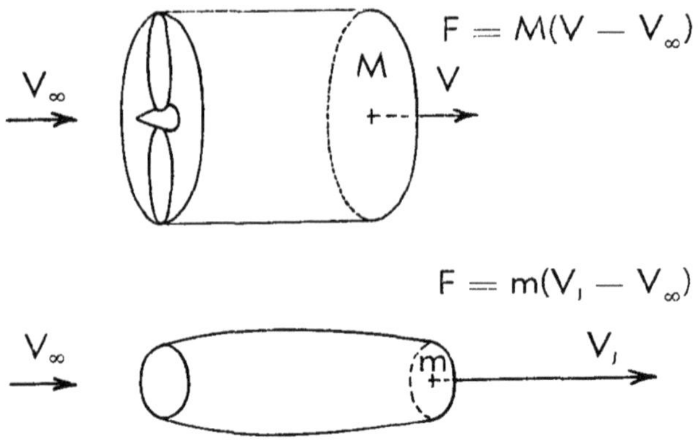

Fig. D,1b.

limit. Thus, from a practical point of view, it is possible to limit the propeller size without seriously reducing the thrust.

D,2. Minimum Propeller Efficiency Required. Although air propellers have been designed which are highly efficient, certain losses are necessary. In order to compare more exactly the propulsive systems of the two engines, it is necessary to take into consideration these losses as well as others due to mechanical friction. The thrust horsepower of a turboprop is given by the following relation:

$$\text{thp} = \frac{1}{550g}\{[w_a + w_f][\eta_p\eta_g\eta_t gJ(1-\alpha)h_g + V_\infty\sqrt{2gJ\eta_j\alpha h_g}] - V_\infty^2 w_a\} \quad (2\text{-}1)$$

where V_∞ is the flight speed in feet per second, w_a is the pounds of engine air per second, w_f is the pounds of fuel per second, h_g is the ideal useful generator enthalpy in BTU per pound of gas, α is the per cent of this energy which is used to form the jet velocity, and the remainder $1 - \alpha$ is used to turn the turbine and the propeller. This latter quantity is reduced by multiplying it by the efficiency of the turbine η_t, the efficiency of the gears in the gear box η_g, and the efficiency of the propeller η_p, to give useful energy; J, g, and 550 are the conversion factors which convert the thermal units to horsepower. The jet energy is reduced by the jet efficiency η_j.

It can readily be shown by differentiation that the thrust horsepower has its maximum value, provided the efficiencies of the components are assumed to be independent of the division of power, when

$$\alpha = \frac{\eta_j V_\infty^2}{2gJh_g(\eta_p \eta_g \eta_t)^2} \tag{2-2}$$

Thus, the most economical division of power between the shaft and the jet depends upon the flight velocity, the amount of energy per pound of air,

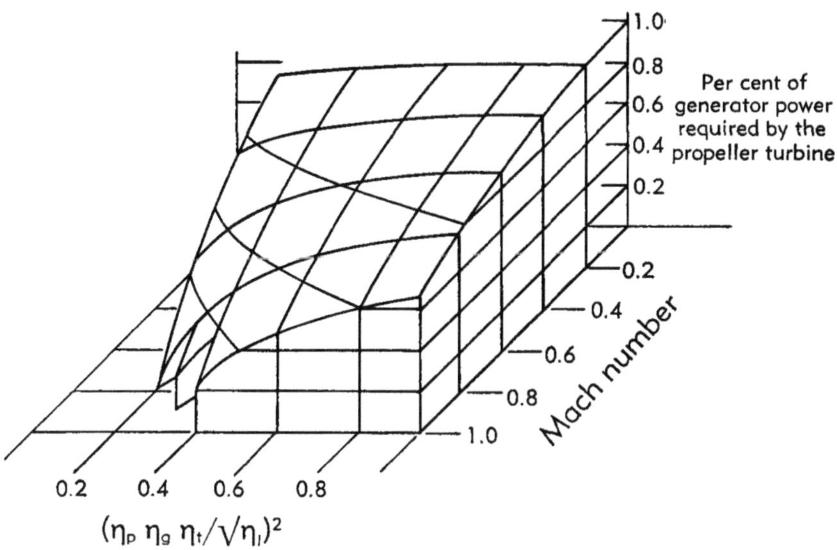

Fig. D,2. Per cent of generator power required by the propeller turbine for maximum thrust horsepower. $h_g = 80$.

and the efficiency of the turbine, the propeller, the gears, and the jet. Since $2gJh_g$ is the square of the maximum jet velocity which can be obtained from the given energy and which is always much greater than the square of the flight velocity, at full power, then unless the square of the product of the three efficiencies is very small, the amount of energy removed in the jet nozzle should likewise be small. This relationship is illustrated in Fig. D,2. In the case of a free turbine this optimum division of power can be nearly realized. The power absorbed by the propeller can be adjusted by varying its pitch; however, the efficiencies of the turbine and propeller are both functions of the power delivered. When the engine has a fixed turbine, the optimum division of power may be set for one operating point, and, with a fixed area nozzle, the division of power is then determined everywhere.

The substitution of the value of α from Eq. 2-2 in Eq. 2-1 gives the value of the maximum possible thrust power for a turboprop. When the small fuel term is ignored, the result is

$$(\text{thp})_{\text{max}} = \frac{w_{\text{a}}}{550g}\left[\eta_{\text{p}}\eta_{\text{g}}\eta_{\text{t}}gJh_{\text{g}} + V_{\infty}^2\left(\frac{\eta_{\text{j}}}{2\eta_{\text{p}}\eta_{\text{g}}\eta_{\text{t}}} - 1\right)\right] \tag{2-3}$$

This value may be compared with the maximum power that can be delivered by a jet without a propeller. The thrust horsepower of a turbojet is given by the relation

$$(\text{thp})_{\text{j}} = \frac{(w_{\text{a}} + w_{\text{f}})\sqrt{2gJh_{\text{g}}\eta_{\text{j}}V_{\infty}^2} - w_{\text{a}}V_{\infty}^2}{550g} \tag{2-4}$$

The differentiation of this equation with respect to V_{∞}, when the small fuel term is again neglected, shows that the maximum value of the thrust power occurs when

$$V_{\infty} = \tfrac{1}{2}\sqrt{2gJh_{\text{g}}\eta_{\text{j}}}$$

or since

$$V_{\text{j}} = \sqrt{2gJh_{\text{g}}\eta_{\text{j}}}$$

$$V_{\infty} = \tfrac{1}{2}V_{\text{j}} \tag{2-5}$$

and the maximum power of the jet engine is

$$(\text{thp})_{\text{j}_{\text{max}}} = \frac{w_{\text{a}}}{550g}\left(\tfrac{1}{2}gJh_{\text{g}}\eta_{\text{j}}\right) \tag{2-6}$$

A relation between the efficiencies of the propeller, the turbine, the gears, and the nozzle which will make the maximum powers of the turboprop and the turbojet engines equal may be determined by equating the right-hand members of Eq. 2-3 and 2-6. Thus, the maximum possible powers will be equal when

$$\eta_{\text{p}}\eta_{\text{g}}\eta_{\text{t}} = \tfrac{1}{2}\eta_{\text{j}}$$

For representative values for η_{g}, η_{t}, and η_{j} of 0.98, 0.87, and 0.95 respectively, the propeller efficiency is 55.5 per cent. Thus, for propeller efficiencies of 56 per cent or above, the thrust power of the turboprop is greater than the thrust power of the jet. These statements hold for the optimum condition for both the propeller and the jet; that is, when the energy distribution is given by Eq. 2-2 and the velocity of the hot gases out of the propelling nozzle of the turbojet is twice the flight speed as defined by Eq. 2-5. At other than these discrete points for each engine type this solution is no longer valid. It is evident that as the flight speed increases it may become increasingly difficult to maintain the relationship expressed by Eq. 2-2 and 2-5; either the propeller efficiency may become lower than the limit set by the former, or it may become impossible to meet the condition of jet velocity imposed by the latter equation. However, it is obviously easier to satisfy the condition (Eq. 2-2) than Eq. 2-5, for in the former there is one more degree of freedom, one which is independent of flight speed, regardless of the gas power. Fig. D,2 illustrates that for one practical energy level a propeller is advantageous even at efficiencies much lower than 56 per cent.

The analysis above shows that it should be possible (provided η_p is equal to or greater than 56 per cent) to obtain more thrust power from a given hot gas generator if the propulsive unit is a propeller rather than a jet. This does not answer the whole problem, however. The weight of the additional turbine, shaft, gear box, and propeller greatly increases the power plant weight, and unless the saving in fuel during cruise or loiter is sufficient to overcome the initial handicap, a larger and heavier aircraft will result. This larger plane will have higher drag and thus may not be as fast as an aircraft designed around a turbojet. Another fact usually not considered when comparing propulsive systems is the great destabilizing effect of a tractor propeller. This reduction of stability, then, will require larger tail surfaces in order to provide proper stability. It is therefore evident that a simple comparison between the turboprop and turbojet is not a sufficient basis for a definite decision as to the ultimate merits of one or the other. A series of aircraft must be designed around each engine type and the results compared before firm decisions can be reached. The discussion above has shown that such a thorough study may demonstrate superiority for the propeller in comparison with the jet and at least should serve to encourage further research and study rather than a cavalier dismissal of the propeller, even for very high speed aircraft.

It follows from Eq. 2-1 and 2-4 that as the value of the total enthalpy h_g is increased the turboprop engine, operating at any chosen flight speed, experiences more rapid increases in thrust than a turbojet. The thrust power of a turboprop increases linearly with the total enthalpy, whereas the thrust power of a turbojet increases linearly with the square root of the total enthalpy. Since the useful enthalpy is increased by greater cycle temperatures and higher compression ratios, either together or singly, it is evident that the importance of developments to improve either of these items will be greater for the propellered unit than for the jet.

D,3. Propeller Performance. There are two basic types of propellers: the usual design currently in use on piston-engined aircraft and a unit enclosed in a shroud or duct. Either, of course, may have any number of blades, any diameter, or any number of blade rows. As long as the propelling device acts upon a greater mass of air than that passing through the turbine, it is a propeller. It may run at the same rotating speed as the turbine or may be driven through a gear train so as to operate at a lower speed. Both of these basic types may be designed so that the blade angles of the propeller remain fixed or change, either automatically or by some controlled or predetermined pitch schedule. The ducted fan and the by-pass engine mentioned by Meyer (Sec. C) come under the general classification of a turboprop.

A simplified method for analyzing propellers of the normal type which has seen considerable use both in this country and abroad is presented in [*1*] and the references contained therein. In essence the propeller is considered as an airfoil operating in a spiral path, and then the lift and drag forces are resolved into directions parallel and normal to the axis of rotation. By an appropriate integration factor the thrust and power coefficients are calculated from the airfoil characteristics at 0.7 of the propeller radius. This method is advantageous in that the airfoil characteristics can be found from wind tunnel or other tests upon rather simple designs and then extended by the theory to cover changes in blade width and number as well as to tandem counter-rotating banks. By means of these methods the performance of constant blade angle, constant rpm, or controlled angle propellers can be estimated with a sufficient degree of accuracy. In the original discussion of the method [*1*] the following equations and their derivations are given:

$$C_F = \frac{1.5\sigma C_L}{\cos\phi}\left(1 - \frac{C_D}{C_L}\tan\phi\right) \tag{3-1}$$

$$C_P = \frac{3.3\sigma C_L}{\cos\phi}\left(\frac{C_D}{C_L} + \tan\phi\right) \tag{3-2}$$

$$\eta_P = \frac{1 - \dfrac{C_D}{C_L}\tan\phi}{1 + \dfrac{C_D}{C_L}\cot\phi} \tag{3-3}$$

$$\frac{C_{D_i}}{C_L} = \sin\alpha_i \tag{3-4}$$

$$\frac{C_{D_i}}{C_L} = \frac{\sigma C_L \cos\alpha_i}{4K\sin(\phi + \alpha_i)} \tag{3-5}$$

$$\frac{C_D}{C_{L_i}} = \frac{C_{D_p}}{C_L} + \frac{C_{D_i}}{C_L} \tag{3-6}$$

A combination of Eq. 3-1 and 3-2 yields

$$\frac{C_D}{C_L} = \frac{0.455C_P - C_F\tan\phi}{C_F + 0.455C_P\tan\phi}$$

where the values are evaluated at the 0.7 propeller radius.

K Goldstein correction factor for a finite number of blades
ϕ advance angle, $\tan\phi = V_\infty/0.7D\pi n$

n rotational speed of the propeller, revolutions per sec
D diameter of the propeller, ft
V_∞ flight velocity, ft/sec
C_{D_i} drag coefficient due to lift
C_{D_p} profile drag coefficient. This term includes all the drag except that attributed by theory to induced drag.
C_D total drag coefficient
C_F thrust coefficient $= F/\rho n^2 D^4$
C_P power coefficient $= P/\rho n^3 D^5$
F thrust, pounds
P shaft power, ft lb/sec
ρ density of the ambient air, (lb $\times$ sec^2)/ft^4
α_i downwash angle
η_p propeller efficiency
σ solidity of the propeller

In the previous article an approximate value for η_p was estimated at which the thrust power from a given hot gas generator would be

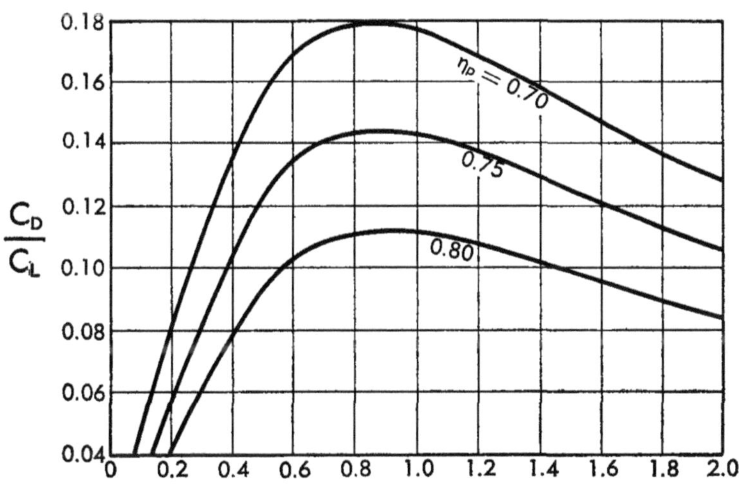

Fig. D,3a. Drag-lift ratios for various propeller efficiencies.

identical whether jet or propeller propulsion was employed. In case η_p should be smaller than this value, at the proper flight speed a turbojet would be superior. Therefore a higher value would be necessary before a propeller would be considered. It is assumed that η_p must equal 0.70 to be of interest. From Eq. 3-3 the value of C_D/C_L required to give any prescribed efficiency can be obtained as a function of the advance ratio. Fig. D,3a shows these values required for propeller efficiencies of 0.7, 0.75, and 0.8. It is evident from these curves that tan ϕ should be around

0.8 to 0.9 in order to provide the maximum margin for the ratio C_D/C_L. However, to more fully appreciate whether the required values are possible, it is desirable to consider an example. Wind tunnel test data supply values for C_P and C_F from which C_D/C_L can be calculated for the

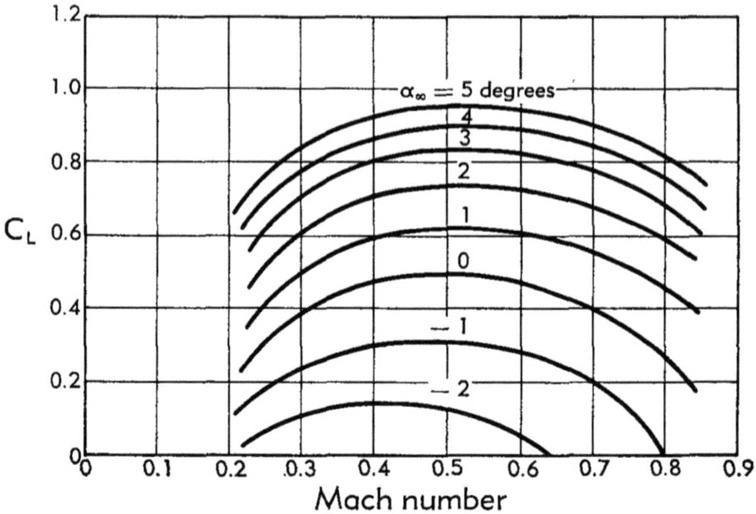

Fig. D,3b. Lift coefficient from tests of a propeller.

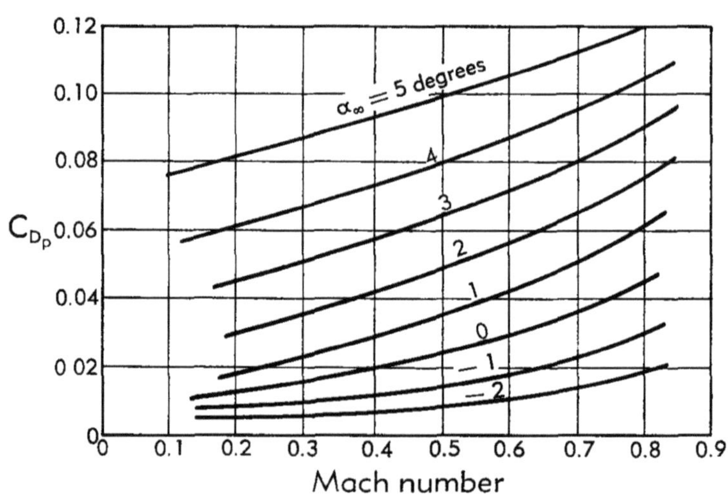

Fig. D,3c. Profile drag coefficient from tests of a propeller.

propeller tested. Then σC_L may be obtained from either Eq. 3-1 or 3-2 and thence α_i and C_{D_i}/C_L from Eq. 3-4 and 3-5, and finally C_{D_p} from Eq. 3-6. Since the blade-setting angle is always given in wind tunnel tests, the true angle, at infinite aspect ratio, between the airfoil at the 0.7 radius and the relative air stream can be determined from the relation

$$\alpha_\infty = \theta - \phi - \alpha_i$$

where θ is the blade-setting angle and α_∞ is the angle equivalent to infinite aspect ratio. The calculated lift coefficient and the profile drag

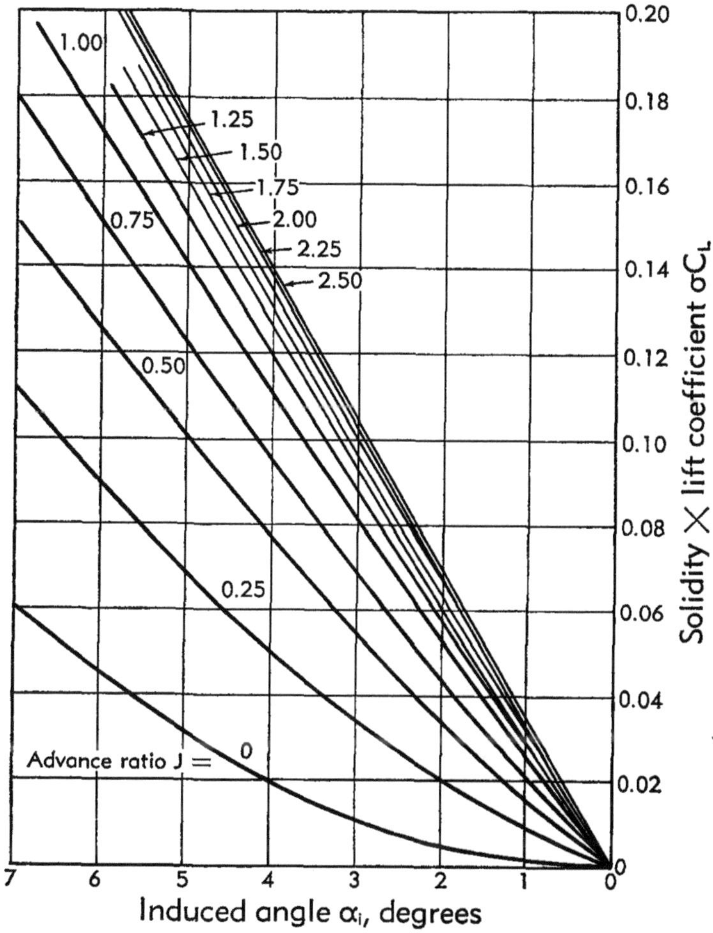

Fig. D,3d. Lift coefficient vs. induced angle of attack for a four-bladed single-rotating propeller.

coefficients can be plotted versus α_∞ for additional use. As explained in [*1*], the values may be employed to predict the performance of other propellers with different diameters, solidities, cambers, and number of blades. Fig. D,3b and D,3c illustrate data upon a propeller blade which might be considered suitable for modern high speed aircraft. These curves were determined by the methods outlined above and described more fully in [*1*].

Now let us see if it is possible to obtain a propeller efficiency of at least seven-tenths by a propeller with characteristics given in Fig. D,3b and D,3c. Assume $V_\infty/nD = 2$, $\sigma = 0.2$, $C_L = 0.6$, $M_\infty = 0.7$, and a four-bladed propeller. Now from Fig. D,3d which gives σC_L versus α_i for a four-bladed propeller, $\alpha_i = 3\frac{1}{2}°$. Hence $C_{D_i}/C_L = \sin 3\frac{1}{2}° = 0.061$. From Fig. D,3b and D,3c, $C_{D_p} = 0.0555$. Thus $C_{D_p}/C_L = 0.0925$, whence $C_D/C_L = 0.0925 + 0.061 = 0.1535$. For $V_\infty/nD = 2$, $\tan \phi = 0.91$. Therefore from Fig. D,3a, the efficiency would be about 74 per cent.

Other propellers with different planforms have been tested which yield much higher efficiencies than the one illustrated here. Thus, not only in theory but also in actual tests, it has been demonstrated that the turboprop is thermodynamically a more efficient propulsive power system.

D,4. Performance Analysis.

Generator performance. In order to get a background for the analysis of the various types of turboprop engines it is desirable to consider briefly

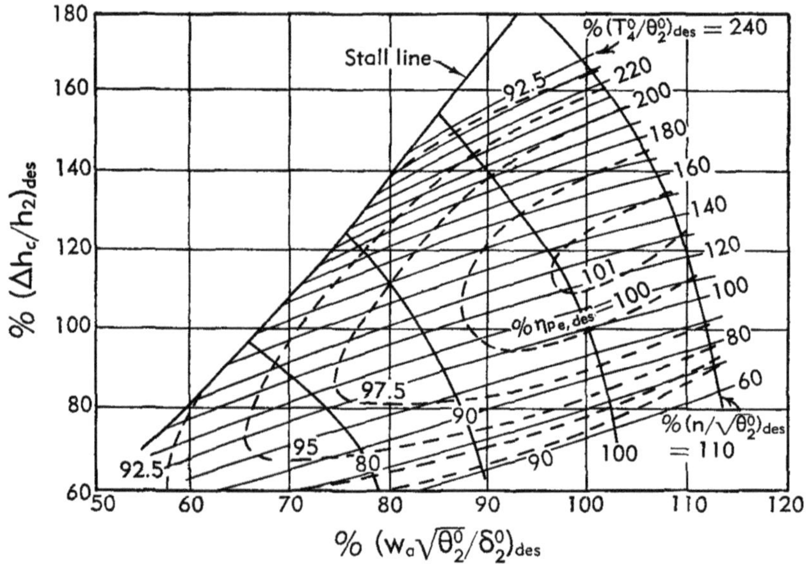

Fig. D,4a. Generator internal solution. $(\Delta h_c/h_2)_{des} = 0.50$.

the performance of the hot gas generator. The performance of any generator is defined by the internal solution which balances the rotational speed, and the air mass flow of the compressor and turbine, the turbine and compressor energy output and input respectively, and the compressor outlet pressure with the turbine inlet pressure for various cycle temperatures. The solution may be obtained by the analytical method described in XI,L or by the graphical method described in [2]. As explained in

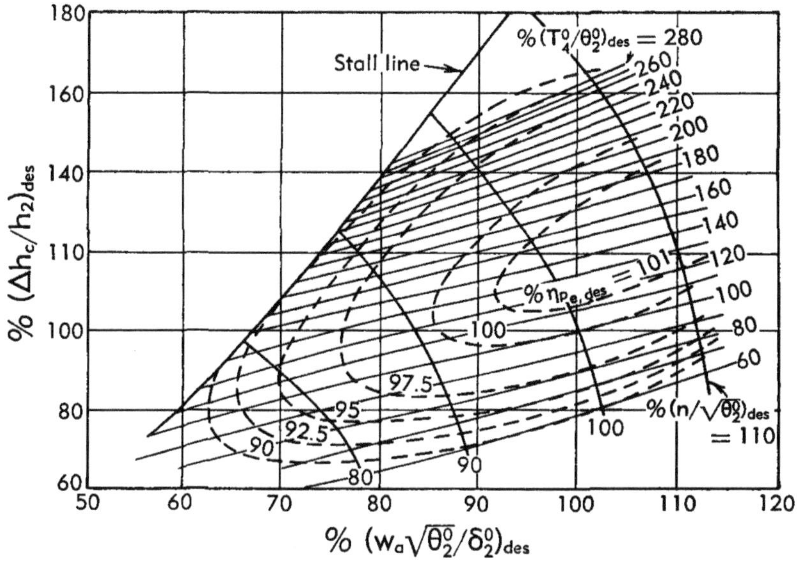

Fig. D,4b. Generator internal solution. $(\Delta h_c/h_2)_{des} = 0.75$.

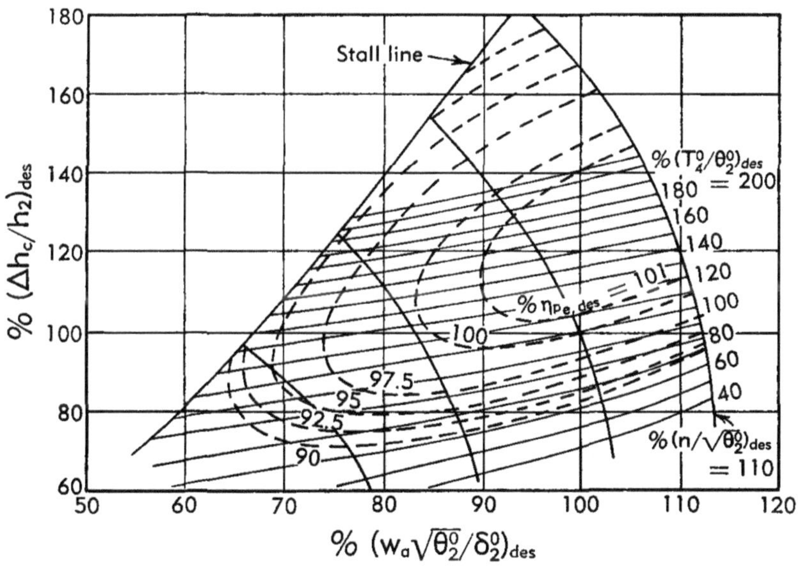

Fig. D,4c. Generator internal solution. $(\Delta h_c/h_2)_{des} = 1.00$.

both of these references the results can be expressed in a simplified form by the use of dimensionless parameters. The conditions required to maintain a steady state are determined, once two of the generalized variables are prescribed, for example the ratio of the top operating temperature to the compressor inlet temperature and the rotational

Mach number of the compressor. Fig. D,4a, D,4b, D,4c and D,4d give the internal solution for four different gas generators whose design values of $\Delta h_c/h_2$ are 0.5, 0.75, 1.00, and 1.25 respectively, where Δh_c is the enthalpy rise per pound of air through the compressor and h_2 is the enthalpy of the air entering the compressor. These graphs give the compressor enthalpy, the mass flow, the temperature ratio, the rotational speed, and the engine efficiency at all possible operating points. The quantities are plotted in terms of their design values, that is, $\% X_{des}$ means the value of X divided by the value of X at the design point.

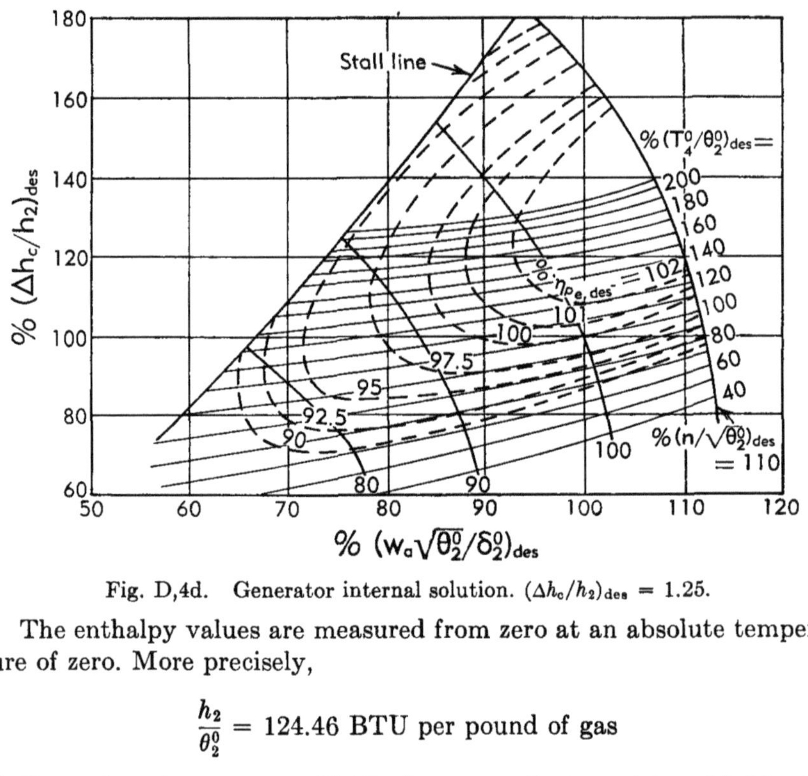

Fig. D,4d. Generator internal solution. $(\Delta h_c/h_2)_{des} = 1.25$.

The enthalpy values are measured from zero at an absolute temperature of zero. More precisely,

$$\frac{h_2}{\theta_2^0} = 124.46 \text{ BTU per pound of gas}$$

For a given temperature T_4^0/θ_2^0, the energy that must be added by the fuel is obtained from

$$\frac{\Delta h_b}{\theta_2^0} = \frac{h_4}{\theta_2^0} - \frac{\Delta h_c}{\theta_2^0} - \frac{h_2}{\theta_2^0} \tag{4-1}$$

where the value h_4/θ_2^0 corresponds to the value of T_4^0/θ_2^0 and the specific heat of the combustion products.

The value of the ideal useful generator energy Δh_{gc} at each of the points in the figures can be readily obtained, when the value at the design point is known, from either the table of gas characteristics or the solution graph of Fig. D,4e. The additional ideal generator energy result-

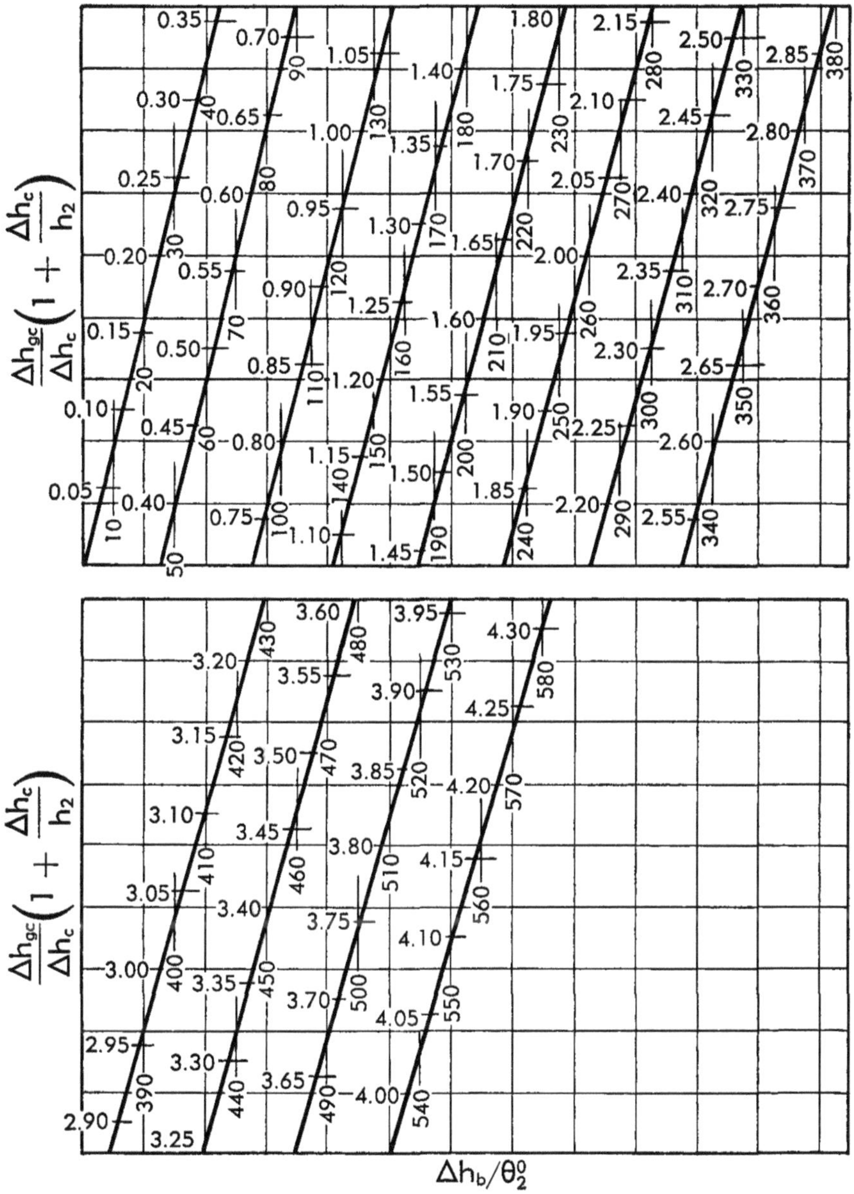

Fig. D,4e. Generator enthalpy of ideal turbocompressor cycle.

ing from the effects of flight velocity can be determined from the gas tables or read directly from Fig. D,4f [3].

In order to obtain the actual useful energy of the generator it is necessary to have a knowledge of the pressure losses (or efficiency) through the compressor, the combustor, and the turbine. The quantity

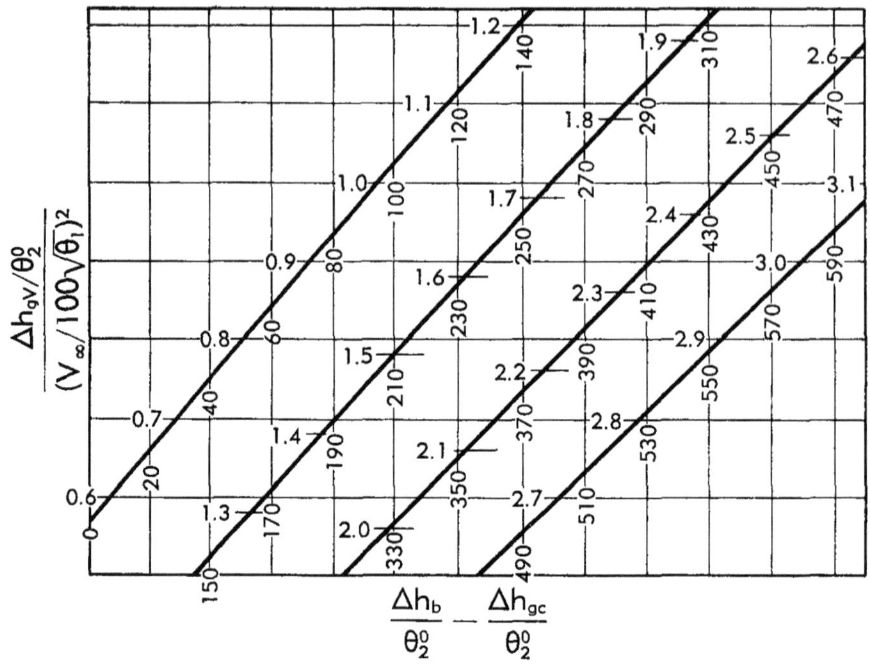

Fig. D,4f. Expansion enthalpy of ideal velocity cycle.

% $\eta_{p_{e,des}}$ in the figures is a measure of the efficiency of the engine in terms of these pressure losses.

$$\eta_{p_e} = 1 - \frac{\Delta s_{tot}}{c_{p_a} \ln\left(1 + \frac{\Delta h_c}{h_2}\right)} \tag{4-2}$$

where Δs_{tot} is the total increase in entropy, namely

$$\Delta s_{tot} = \Delta s_c + \Delta s_b + \Delta s_t \tag{4-3}$$

and Δs_c, Δs_b, and Δs_t are the increase in entropy due to pressure losses in the compressor, the burner, and the turbine, respectively.

$$\Delta s_c = c_{p_a}(1 - \eta_{p_c}) \ln\left(1 + \frac{\Delta h_c}{h_2}\right) \tag{4-4}$$

$$\Delta s_b = -0.06854 \ln\left(1 - \frac{\Delta p_b}{p_2}\right) \tag{4-5}$$

$$\Delta s_t = c_{p_g}\left(1 - \frac{1}{\eta_{p_t}}\right) \ln\left(1 - \frac{\Delta h_t}{h_4}\right) \tag{4-6}$$

These increases in entropy and the actual available ideal useful generator enthalpy are illustrated in Fig. D,4g.

The actual available gas enthalpy Δh_{gc_a} is obtained by subtracting the unavailable energy from the ideal values. The results corresponding to Fig. D,4a to D,4d are shown in Fig. D,4h, D,4i, D,4j, and D,4k, respectively.

Now the energy per pound of air that is available for driving a propeller is the difference between the total available Δh_{gc_a} and the amount

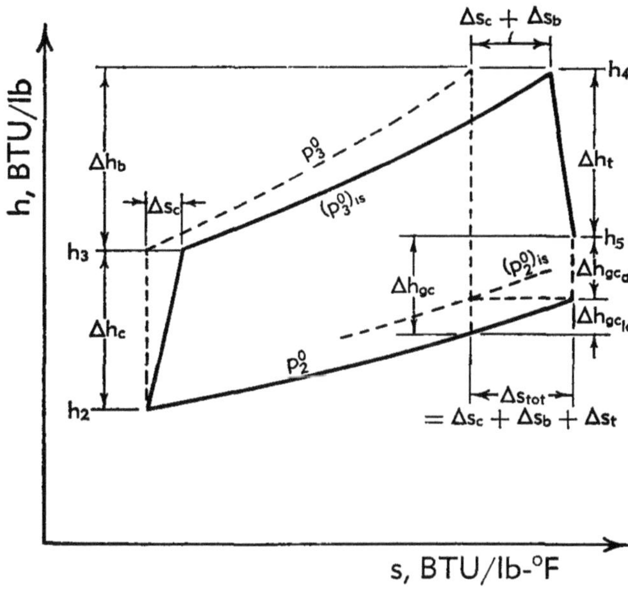

Fig. D,4g. Gas turbine cycle with losses.

of energy required to produce the flow out of the exhaust nozzle, station *e*. From the energy equation

$$V_j = \sqrt{2gJ\eta_j\Delta h_j} \tag{4-7}$$

and from the continuity equation

$$\frac{w_a}{g} = \rho_e A_e V_j \tag{4-8}$$

where ρ_e is the gas density at the exit and A_e is the exit area. Thus

$$\frac{w_a\sqrt{\theta_e}}{\delta_e} = K\sqrt{2gJ\eta_j\frac{\Delta h_j}{\theta_e}} \tag{4-9}$$

and

$$K = \frac{gA_e p_0}{RT_0}$$

where the subscript $_0$ refers to standard sea level conditions. Or, in terms

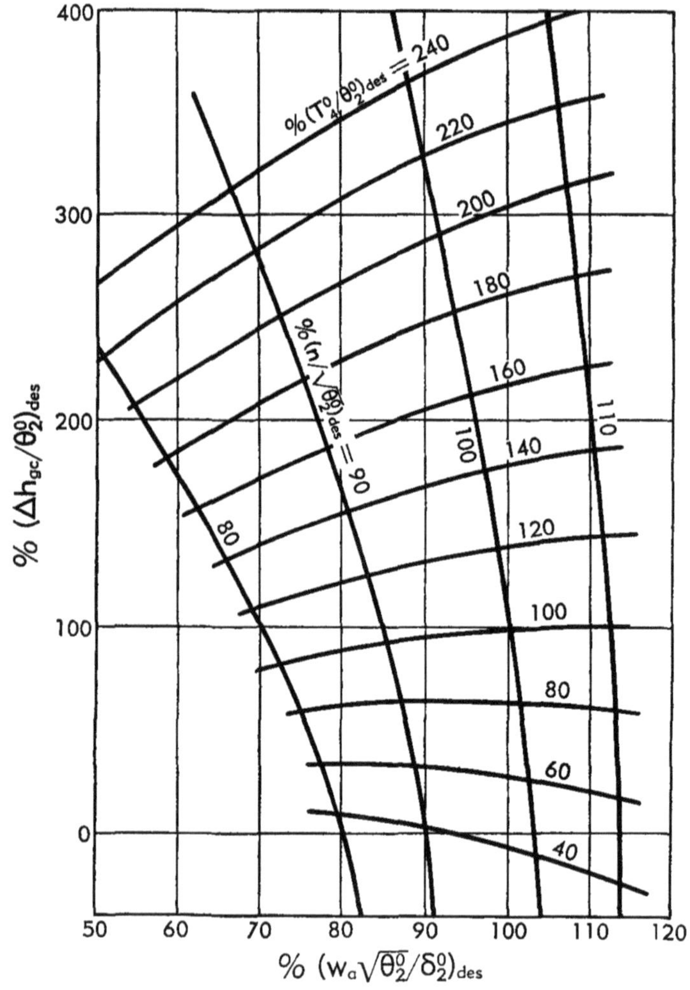

Fig. D,4h. Useful enthalpy. $(\Delta h_c/h_2)_{des} = 0.50$.

of the compressor inlet conditions,

$$\frac{w_a\sqrt{\theta_2^0}}{\delta_2^0} = K\sqrt{2gJ\eta_j\frac{\Delta h_j}{\theta_e}}\left(\frac{\delta_e}{\delta_2^0}\sqrt{\frac{\theta_2^0}{\theta_e}}\right) \tag{4-10}$$

Now it follows from Eq. 2-2 that, for the maximum thrust power,

$$V_j = \frac{n_j V_\infty}{\eta_p\eta_g\eta_t} \tag{4-11}$$

Thus

$$\sqrt{2gJ\eta_j\frac{\Delta h_j}{\theta_2^0}} = \frac{\eta_j\dfrac{V_\infty}{\sqrt{\theta_\infty}}}{\sqrt{\theta_2^0/\theta_\infty}\;\eta_p\eta_g\eta_t} \tag{4-12}$$

that is, the amount of energy required for the jet is determined when the four efficiencies are known. The area of the nozzle can then be determined from Eq. 4-10. Since the expansion is to ambient pressure, $\delta_e/\delta_\infty = 1$. The value of the outlet temperature can be found precisely

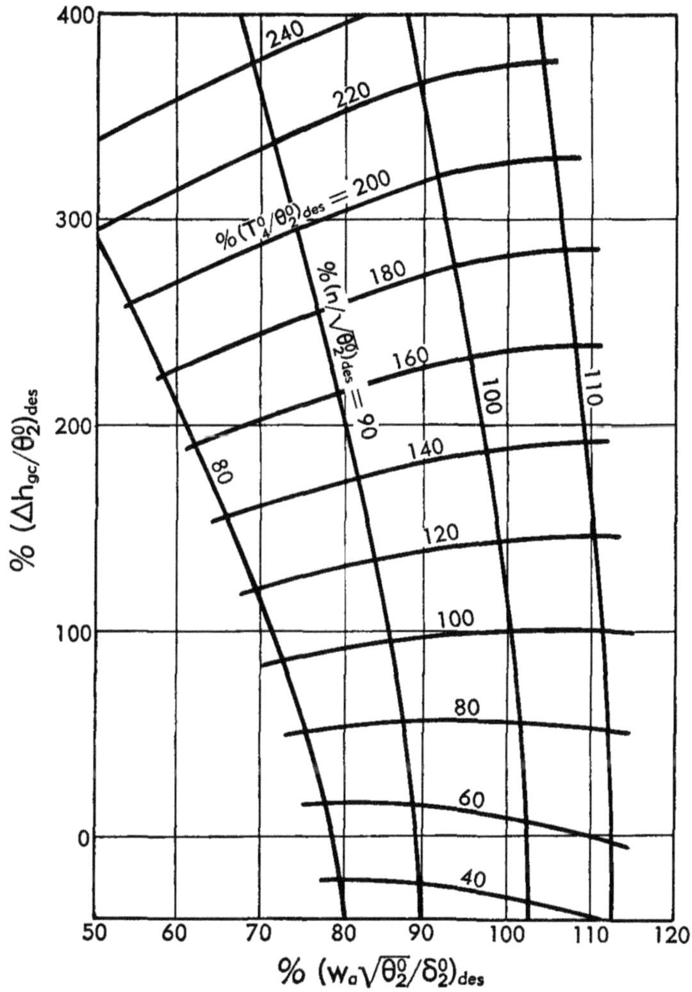

Fig. D,4i. Useful enthalpy. $(\Delta h_c/h_2)_{des} = 0.75$.

only by iteration or by a graphical means employing the *h-s* tables or charts, since the increase in entropy through the additional turbine tends to increase this temperature. Also this expansion takes place through two elements—the turbine and propelling nozzle, each with different efficiencies. However, little error is introduced if the total expansion is assumed to take place with an efficiency equal to that of the turbine.

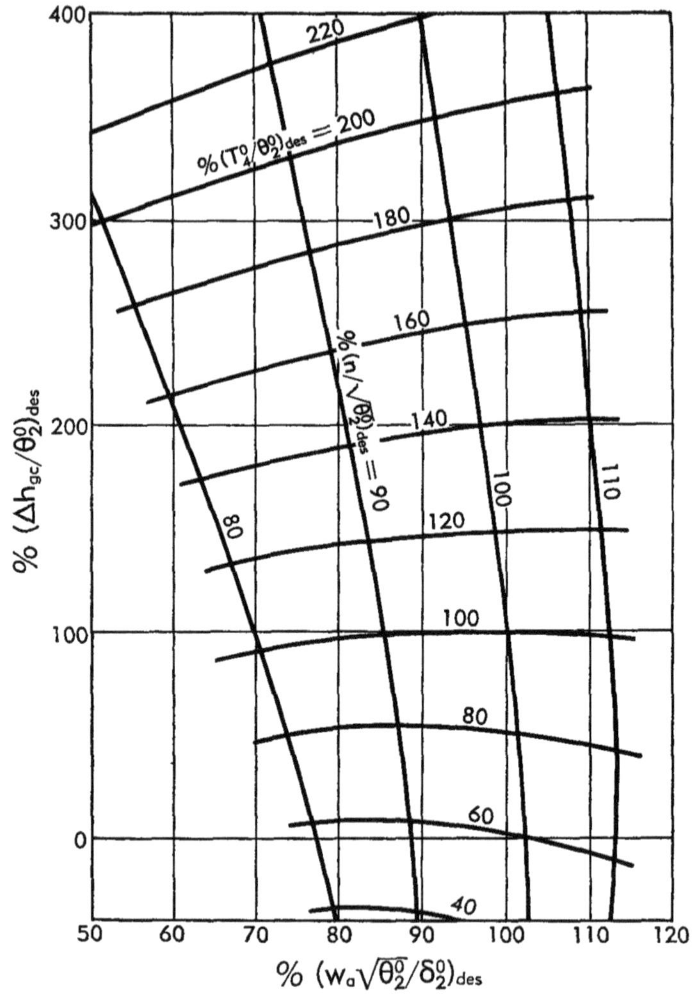

Fig. D,4j. Useful enthalpy. $(\Delta h_o/h_2)_{des} = 1.00$.

Under this assumption

$$\frac{T_e}{T_2} = \frac{h_4/\theta_2^0}{518.4c_{p_g}(h_4/h_e)^{\eta_{p_t}}} \tag{4-13}$$

where h_e is the ideal exit enthalpy, viz.

$$\frac{h_e}{\theta_2^0} = \frac{h_4}{\theta_2^0} - \frac{\Delta h_o}{\theta_2^0} - \frac{\Delta h_{gc}}{\theta_2^0} \tag{4-14}$$

It would be possible to compute the area required for each operating point. Thus, by means of a variable area nozzle, it would always be possible to obtain the split in energy which produces the maximum thrust horsepower at each engine operating point. Usually the engines are designed for one fixed area. This area may be arbitrarily selected or chosen so as to give the maximum thrust horsepower at a particular point.

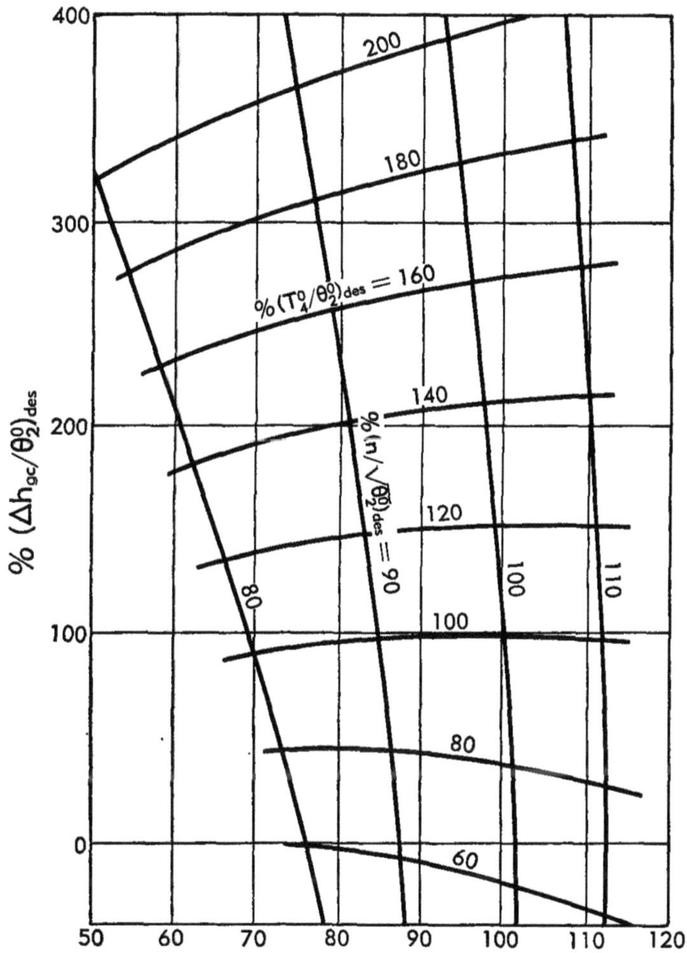

Fig. D,4k. Useful enthalpy. $(\Delta h_0/h_2)_{des} = 1.25$.

Once the area is fixed the energy per pound of gas required to exhaust the total amount of gas is computed from Eq. 4-10. The data from which the preceding graphs were computed indicate that the turbine efficiency is nearly constant in the usual operating range. This fact allows the use of a constant value of η_{p_t} in Eq. 4-14.

The enthalpy that can be employed to drive the propeller and its turbine is obtained by subtracting the Δh_j from the total,

$$\frac{\Delta h_{\text{shaft}}}{\theta_2^0} = \frac{\Delta h_{\text{gc}_a}}{\theta_2^0} - \frac{\Delta h_j}{\theta_2^0} \tag{4-15}$$

Fig. D,4l illustrates a typical solution for available shaft enthalpy for an engine with fixed area nozzle. The curves at the left give the energy for one air speed. The curves at the right give a means for obtaining corrected values for other speeds. Fig. D,4m gives the enthalpy which is used to produce the jet thrust.

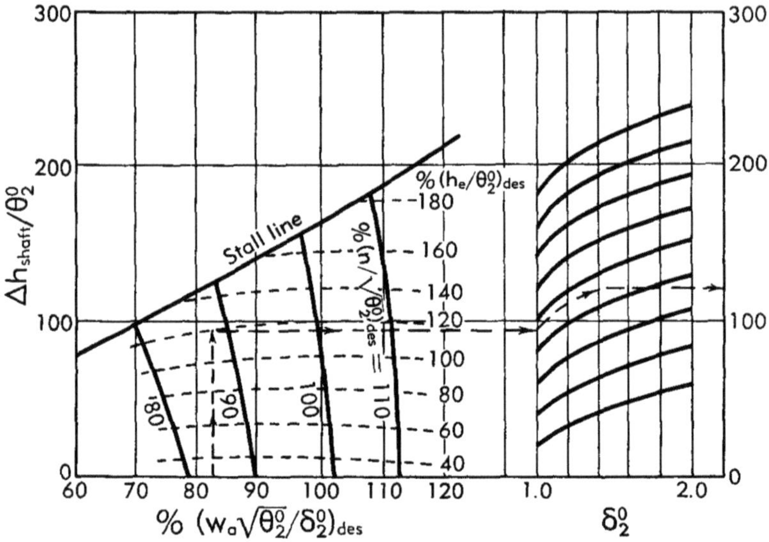

Fig. D,4l. Enthalpy for shaft power. $(\Delta h_c/h_2)_{des} = 0.75$.

It will be noted that the enthalpy available at any speed is a function of two quantities. Any two of the four quantities can be chosen at random. For example, a rather wide latitude exists in the choice of rpm and fuel flow for any power output. This freedom of operation is of great importance in the design of the control system.

Fixed and free turbines. So far the discussion holds whether the turbine is fixed or free. The difference comes in the design of the turbine. In obtaining the internal solution it was assumed that the turbine was designed to match the compressor. In case the engine is to have a free turbine, this would be the proper design for the turbine. The second turbine would then be designed to swallow the air flow of the generator and at the same time remove the enthalpy given in Fig. D,4l. The contours on the figures are for the rotational speed of the compressor. The rotational speed of the turbine need not be connected in any way to these values.

If the engine is to be of a fixed turbine type, then the turbine must be designed to deliver the power to both the compressor and the turbine. This is done by adding stages or by changing the blade camber, thus increasing the lift coefficients and turning angles. The rotational speed, as well as the mass flow, must now be the same as for the generator turbine. Only if the energy required to run the compressor is proportional to the shaft energy will the turbine efficiency employed in getting the generator performance be correct. As explained in XI,L this is true only for a fixed pitch propeller. In other words, this would be true only along

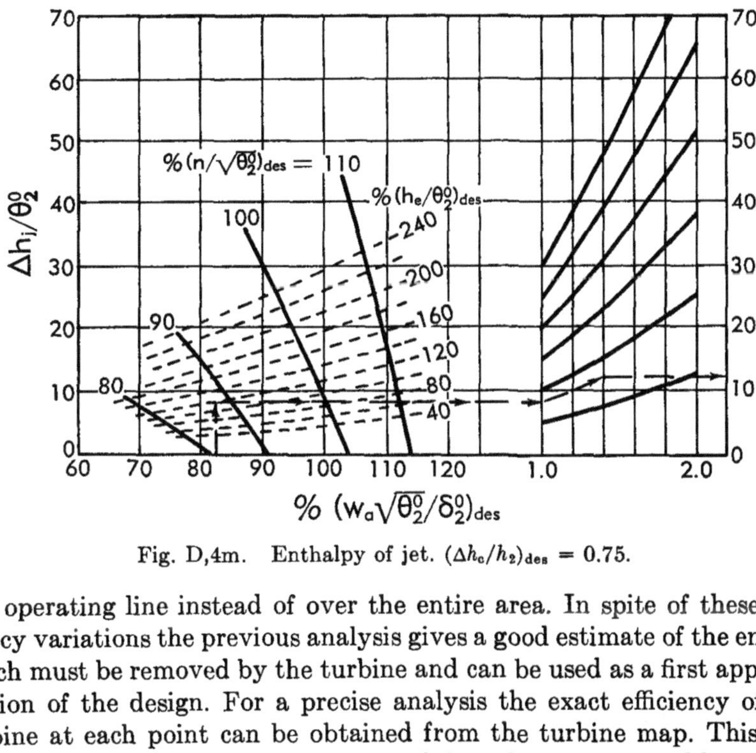

Fig. D,4m. Enthalpy of jet. $(\Delta h_c/h_2)_{des} = 0.75$.

one operating line instead of over the entire area. In spite of these efficiency variations the previous analysis gives a good estimate of the energy which must be removed by the turbine and can be used as a first approximation of the design. For a precise analysis the exact efficiency of the turbine at each point can be obtained from the turbine map. This efficiency in turn affects the temperature of the exhaust gases and hence the energy employed by the jet. The effect on the jet energy is small. However, a reduction in the turbine efficiency over that assumed in the previous analysis reduces the shaft energy considerably since the total power of the turbine is so large—about three times the shaft power.

The free turbine has an extra degree of freedom, that is, the rotational speed may be different from that of the engine. For a given level of energy and mass flow from the generator, a variation in rotational speed of an unchoked turbine would necessarily produce a change in the shaft power delivered. Nevertheless, this might be highly desirable because the propeller efficiency is a function of the rotational speed. In particular, in helicopter applications, it is necessary to change the rotational speed for different maneuvers. Hence the free turbine is desirable.

Coupled turbine. In a coupled turbine engine the inner high compressor and turbine behave like a hot gas generator and the performance

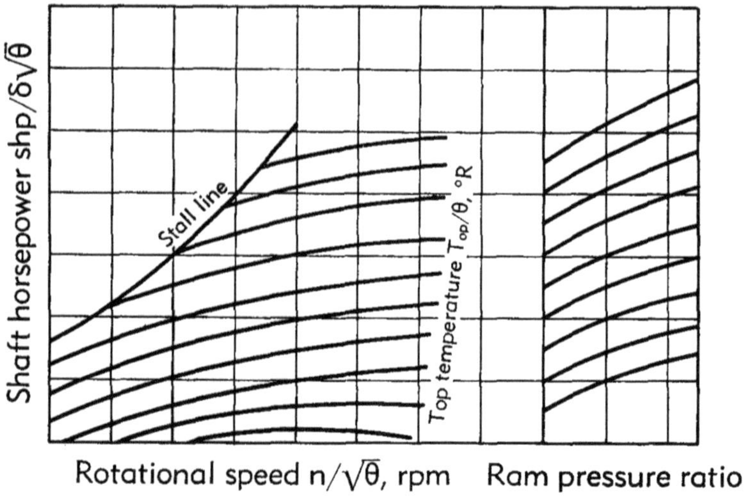

Fig. D,4n. Shaft horsepower of a turboprop.

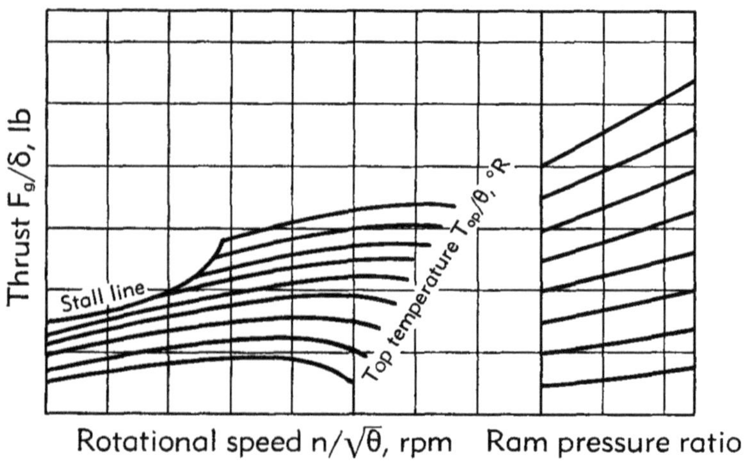

Fig. D,4o. Jet thrust of a turboprop.

of the inner unit can be represented precisely like that of any other gas generator. The entering air flow is prescribed by the low compression compressor; this and the fuel flow (top operating temperature) determine the operation of the unit. The performance of the entire unit is obtained from computations for each air flow and rotational speed of the low compressor for various operating temperatures of the high compression spool. This can be done by assuming values from the low compression map and then proceeding step by step through the engine to the exhaust.

Turboprop performance. Once an engine has been designed and the performance completely calculated, then it is usually plotted in terms of shaft horsepower and jet thrust versus the engine rotational speed for various operating temperatures, where corrections for changes in velocity are made approximately as they were in Fig. D,41 and D,4m. Examples of such plots are given in Fig. D,4n and D,4o. Once a control

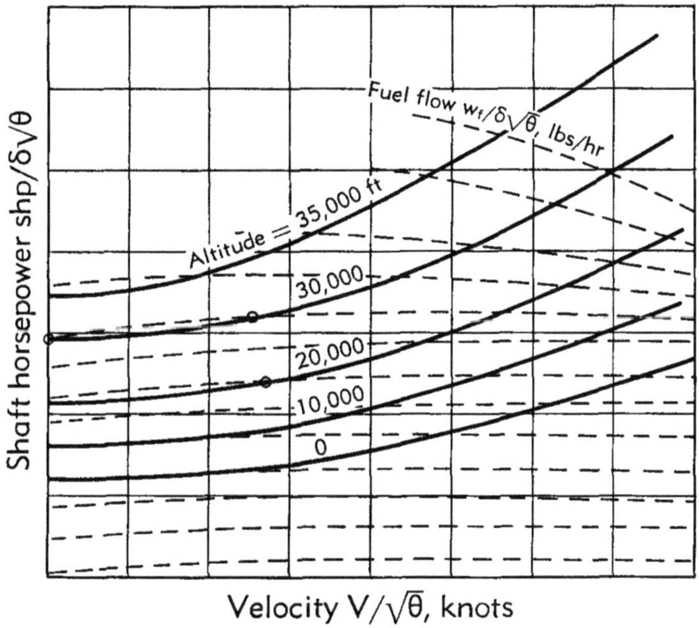

Fig. D,4p. Shaft horsepower of a turboprop for a single-lever control.

schedule is determined which leaves only one degree of freedom, the shaft power and jet thrust are then plotted versus flight velocity with contours of rotational speed and fuel flow. See Fig. D,4p and D,4q. In these figures the upper bounds to the thrust and power are given for several altitudes as determined by ambient temperatures according to the standard atmosphere. Although the performance does not quite generalize, it usually can be given to a high degree of accuracy upon one chart, provided that a correction for fuel flow with altitude is given.

For preliminary airplane performance it is convenient to combine the shaft power and the jet power into one single term called the equivalent shaft horsepower. This is a power which when multiplied by the propeller efficiency gives a close approximation to the thrust horsepower. Statically,

$$\text{eshp} = \text{shp} + \frac{F}{2.5} \tag{4-16}$$

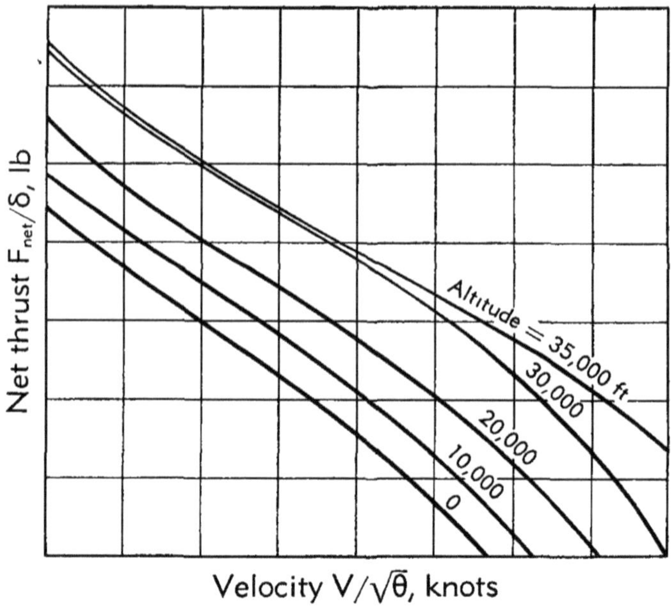

Fig. D,4q. Jet thrust of a turboprop for a single-lever control.

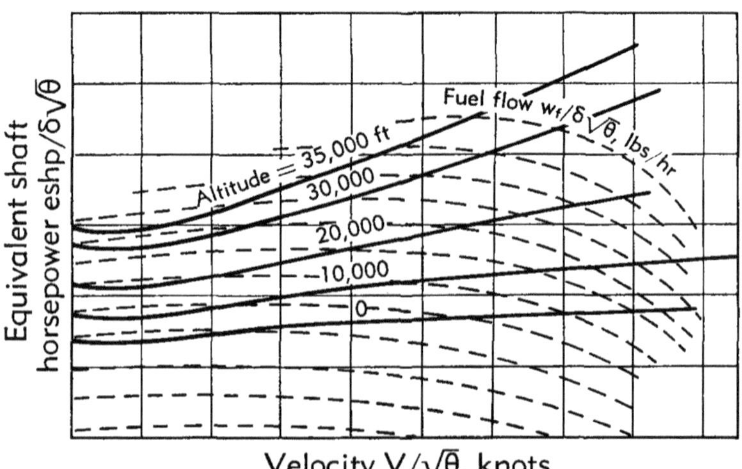

Fig. D,4r. Equivalent shaft horsepower of a turboprop for a single-lever control.

At flight velocity

$$\text{eshp} = \text{shp} + \frac{FV_\infty}{\eta_p 550} \tag{4-17}$$

where η_p is a propeller efficiency usually taken as 0.85 and V_∞ is the flight velocity in feet per second. See Fig. D,4r.

Power augmentation. Even though a turboprop is the most powerful of engines, there are times when an augmentation of thrust would be desirable. Because of the low pressure ratio across the jet the potentialities of afterburning are very small, in fact so small that it would probably be cheaper weightwise to build a slightly larger engine.

Water and alcohol injection at the compressor[3] inlet is a second possible means of augmentation. Actually this can be used with fixed turbine turboprops to augment the thrust by about the same percentage as that for the turbojets (10 to 20 per cent). This may be enough to supply the extra needed for a tropical temperature take-off. However, this scheme is again not very fruitful for a free turbine. This results from the fact that the main increase in power comes from the saving in compressing the air at a lower temperature, that is, less power is required to run the compressor turbine, and this saving is not fully transferred to the shaft turbine.

D,5. Propeller Turbine Design. When one is confronted with the problem of designing a turboprop to meet certain specifications, there are certain logical steps to be taken:

1. Select a compression ratio.
2. Choose the gas turbine temperature for cruise.
3. State the equivalent horsepower per unit engine.
4. Assume a split of about 85 per cent and 15 per cent of the energy for shaft power and jet power respectively.
5. Assume a polytropic efficiency of the compressor and turbine, say 0.9 for each.
6. From these assumptions, get a first approximation of the mass flow.
7. Obtain a compressor map, i.e. the performance of the compressor throughout the operating range.
8. Design a turbine which fits the compressor for size which can absorb the power at the compressor rpm. If a suitable turbine cannot be made, then it is necessary to modify the compressor.
9. Choose a propeller and reduction gear ratio.
10. After determining the propeller efficiency at cruise, go back and determine the final split in energy in accordance with Eq. 2-2.
11. Determine the exhaust nozzle size which will give this area.
12. Select the control system.

Many studies have been made which show the effect of compression ratio and temperature upon the cycle efficiency. In general the higher the compression ratio, with the associated higher temperature, the more efficient the cycle. Since in the case of a turboprop the propulsive effi-

[3] Injection at the burner inlet also produces an augmentation. This augmentation must result from the increase in mass flow resulting from the reduced back pressure produced by the cooling effect of the evaporation.

ciency is independent of the generator power per pound of gas, it follows that the higher the compression ratio and temperature the better the specific fuel consumption. Ideally, infinite compression ratio is best. However, with realizable compressor and turbine efficiencies there is a finite limit, and this limit is lower the poorer the component efficiencies. The nearer this limit is approached the smaller the rate of improvement. Moreover, the higher the compression ratio the heavier the engine. Therefore some practical value of the compression ratio must be chosen which balances the benefits of a low specific fuel consumption and a low specific weight as determined from the proposed use of the engine. Current engines range in compression ratio from 4 to 12 and values as high as 20 may be desirable.

This choice of compression ratio determines the number of spools in the engine. For compression ratios less than some value between 7 or 8, a single-spool engine would be the lightest, and above this value a two-spool engine would perhaps be lighter. Engines with more than two spools appear to be too complicated to build because of the bearing lubrication problems. Naturally the exact dividing line between the single- and multiple-spool units is not clear cut. Opinion seems to vary on this from place to place. Some believe a good single-spool compressor can be built up to a compression ratio of at least 14. In order to accomplish this, perhaps it would be necessary to use both interstage bleed and variable inlet guide vanes for starting purposes, whereas for compression ratios of less than 10, interstage bleed appears to be sufficient. There is little doubt about the ability to produce a single-spool compressor with a high compression ratio and a good efficiency at the design point, but the question is really how well will it perform at points off the design point which result from variations of altitude and ram. Two spools should flatten out the compressor peak. If the impracticality of three shafts, as mentioned before, is correct, then the compression ratio for a free turbine engine is necessarily limited to that which can be produced by a single-spool compressor. By the same reasoning the compression ratio for a fixed turbine is also limited. Hence when one selects the compression ratio, he must at the same time decide whether the turbine will be coupled or of the fixed or free turbine type. Although current engines have low compression ratios, future designs will probably have compression ratios of 12 or above and consequently will be of the coupled turbine type. These two-spool engines have most of the desirable characteristics of the free turbine and yet their disadvantages when compared with the fixed turbine single-spool engines are not so pronounced. The starting power, the windmilling drag, and the accelerations are all intermediate to the other two. The first decision required in the case of a two-spool engine is the division of compression between the low compression and the high compression compressors. These decisions are based upon fabrication and other practical considerations other than the computation of performance to obtain the optimum value. There appears to be agreement on the division by one British and one American manufacturer. However, another American manufacturer feels that a higher compression ratio for the inner free unit is desirable. It should be pointed out that, with the exception of one of these two-spool studies, the gas generators were basically designed for turbojets and then converted to turboprops. A design for a turboprop generator from the outset might yield different results.

The magnitude of the compression ratio will be influenced by the decision concerning the use of a heat exchanger. A low compression ratio with a heat exchanger may be more efficient than a high compression ratio without one. The higher the compression ratio the smaller the effect of a heat exchanger.

The operating gas temperature should be selected for the cruise condition. It should be taken as high as possible for continuous operation. With a turbojet engine excessive temperatures are not desirable because, the higher the generator energy per pound of air, the higher the jet velocity and the lower the propulsive efficiency. However, with a turboprop the propulsive efficiency is essentially constant, and when the gas temperatures are high enough to give a slight reduction in cycle efficiency, the lower weight of the engine will offset this loss for most practical operation ranges. This is surely true in military aircraft where weight is a primary variable. The ability to obtain a high temperature is so important for a turboprop that high temperature metals and methods of cooling are well worth the investigation to determine their effectiveness.

The shaft horsepower, or better the equivalent shaft horsepower, of the unit will depend upon the application. The engines, however, must remain somewhere within the practical limits. For an axial flow compressor the lower limit appears to be around 1000 eshp static sea level rating. Even units of this size suffer adverse Reynolds number effects. For cruise purposes it is desirable to operate the engines at normal power and consequently much thought has been given to the idea of developing nests of multiples of a small engine. In operation, units could be cut off when not needed and by such a scheme a low specific fuel consumption could be obtained. A second consideration is the weight of the units per shaft horsepower. Very small or extremely large engines have a higher specific weight than intermediate size engines. (See Fig. D,5.) So, from the consideration of weights alone, there is a minimum size which is profitable to build the engines even if they can be nested to form multiple units. In the last analysis there are many practical considerations which may determine the size. Such things as manufacturing limits, gas-bending loads, vibrations, and Reynolds number may be the determining factor. The question of Reynolds number is so vital in the design of turboprops that it deserves further elaboration. As a result of the large amount of power which can be developed per pound of air, the last stage compressor blades of a small or medium-powered engine are so short and the associated chord lengths so low that the blades operate at medium Reynolds numbers even at sea level. At high altitudes there is a severe loss in efficiency. On an early design this condition was so bad that an engine of 1200 eshp would not operate at 35,000 feet. Although better designs can be made now, this may still limit the lower size. A compressor blade should have a chord of at least 0.6 inch. For these small blades very low manufacturing tolerances must be maintained. At such small radii, performance differences can be observed for tip

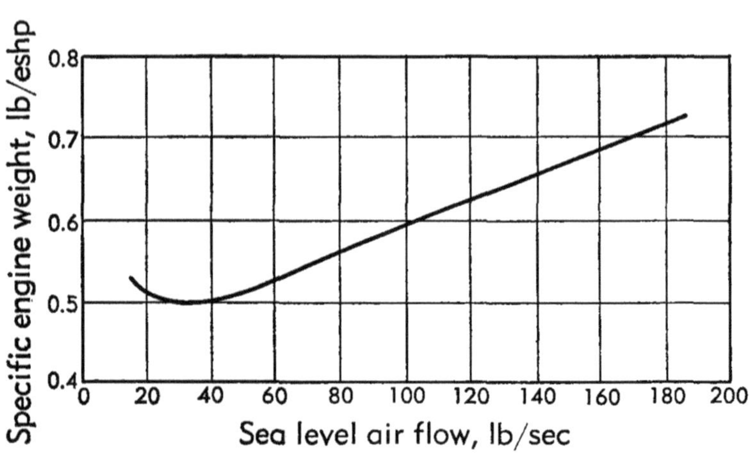

Fig. D,5. Specific weight of turboprop engines.

radius changes of 0.001 inch. Moreover, the small aspect ratio of the blades affect the compression efficiency adversely.

The upper size limit appears to be determined by the facilities for forging large castings. Possibly 250 pounds of static sea level air per second is the largest size to consider. On the other hand, the limitations of the amount of power that can be absorbed by a practical propeller may possibly limit the size to 180 pounds of air per second or less.

After the power size has been determined, the next step is to calculate the equivalent shaft horsepower per pound of air for the assumed compression ratio, operating temperature, efficiencies, and energy split between the propeller and jet. This can be done by the methods outlined in XI,L or by means of the Mollier charts where the entropy increases are determined by the assumed efficiencies. The quotient of the desired power and the power per pound of air gives a first approximation of the air flow.

Since the design of the compressor is the same as for a turbojet, it is not considered here. Once it has been designed, the compressor map must be obtained either from direct test or from the generalization of other tests.

The next step, the design of the turbine, is perhaps the most important for the turboprop. Because of the tremendous amount of power delivered by the turbine, it is of primary importance to have a turbine with high efficiency. In the design of a turbojet the efficiency of the turbine is about equally as important as the efficiency of the compressor, whereas in the design of a turboprop the efficiency of the turbine is about twice as important as the efficiency of the compressor [*4*]. As a rule of thumb, a one per cent increase in turbine efficiency produces a three per cent increase in power and a three per cent decrease in specific fuel consumption. There has been a tendency to sacrifice efficiency for a reduced number of turbine stages to reduce weight, and yet the weight of additional turbine stages is small compared to that of the gear box and propeller and the possible savings in fuel if the turbine efficiency can be raised.

One of the first criteria for the design of a turbine is to insure that it will give satisfactory operation over the entire range of conditions. The pressure drop across the turbine varies widely. The lowest pressure ratio across the turbine occurs when the airplane is loitering at low altitude, while the maximum values occur at maximum speed and maximum power flights at altitudes above 35,000 feet. The latter values may be twice as much as the former. If the exhaust nozzle area is to be fixed, the turbine must be designed to absorb the power at the highest compression ratio without exceeding the limiting load for any of the turbine stages. The turbine then will operate for the other extreme condition, even though there is a reduced efficiency. It appears that a small pressure drop per stage gives best performance. However, practical consideration forces the use of a small number of stages. Satisfactory turbine stages have been designed for a temperature drop of from 150 to 250°F. These values are a function of the entering temperature. The first stage can yield a higher temperature drop than the successive stages. Moreover, it is the last stages that are more adversely affected by the changes in pressure ratio. The details of the turbine design are given in X,H. However, it must be said here that more should be known about turbine design. Why lose 3 per cent in efficiency? Some turbines are operating at a peak efficiency of 81 per cent, whereas it appears possible to design a turbine with an efficiency of 91 per cent.

The choice of the propeller depends upon the use of the engine. For high static thrust the propeller should have camber. At high speed it should not have camber. At the static condition it should have a high rotational speed, while the higher the flight speed the lower the rotational speed, etc. The camber, thickness, diameter, pitch, planform, and solidity are determined by the aircraft mission. It is desirable also to choose the gear box best suited for the mission. Unfortunately the cost of development of gear boxes has been so great that usually only one reduction ratio has been built and it has been hoped that the propeller characteristics could be chosen so as to give a satisfactory propulsive system. This difficulty may have been caused by lack of experience in designing gears to absorb large powers at high revolutions, but as experience is gained these difficulties may be eliminated and additional gears to provide a series of reduction ratios may become available.

D,6. Controls. One of the major problems which is unique to turboprops is that of the controls. The operation of the engine has one more degree of freedom than the turbojet. Even with fixed jet nozzle area, as previously explained, there are two independent variables, that is, two quantities must be specified before the engine has a unique operating point. This in itself should not make the controls too difficult, for the two variables could be and usually are mechanically coupled so that there is only one independent control lever as in the turbojet. The difficulty comes in the attempt to capitalize upon these two variables to obtain performance which could not be achieved without the two degrees of freedom. As a result, the sensitivity of the controls must be more delicate than those employed on other types of engines. The fundamental quantity to be controlled is the thrust to the aircraft in accordance with its needs. Moreover, it is desirable to achieve this thrust with the least fuel consumption and from the pilot's point of view it is almost a necessity that this be accomplished by means of one control lever. Yet, all operation points must be within the physical limitations of the engine. Thus the sensitivity of the controls arises from two kinds of factors: those resulting from specifications which attempt to get the most output from the engine and those inherent in the turboprop power plant which must be met in order to insure safe flight. Here are some of the goals to attain:

1. Negative horsepower in landing approaches so as to reduce the landing speed.
2. Reverse pitch of the propeller during ground roll so as to shorten landing distance.
3. Avoid excessive asymmetrical thrust in multiple engine aircraft.
4. Eliminate windmilling drag of the engine and propeller when there is an engine failure.
5. Rapid engine acceleration.
6. Obtain maximum power without exceeding physical limitations of engine.
7. Maintain a given power level without oscillations.
8. Protect the engine against inadvertent types of failure in propeller pitch.

9. Protect gear box against excessive torque changes.
10. Burn dirty variable density fuel.
11. Use the least amount of this fuel for a given thrust.

All of these goals can be met, and all of them are desirable. As indicated at the beginning of this section, some of the requirements are essential while others are imposed by decisions and are not absolutely necessary. The question arises as to what is the cost in complexity and weight to obtain them. Is it really worthwhile? This can only be answered after a complete analysis of the system. The British have flown turboprops with essentially the same type of propeller controls which were developed for reciprocating engines. Even without 1 and 2 above, the landing conditions would be better than for turbojet aircraft. If 7 were relaxed, there would be a loss of power at the extreme physical limits which probably could be sacrificed at least during early operations, and it is believed that condition 10 is too severe for this stage of development. Requirement 3 results from the high powers delivered by turboprops. At high speed, a small change in pitch angle of a propeller produces a large variation in thrust and would impair handling aircraft characteristics. (This may require the pitch stops to be controlled to within plus or minus one-fourth degree.) The rapid engine acceleration 5 is required for the wave-off condition in carrier operation. To meet certain of these specifications, because of the high inertia of turboprops, it has been necessary to develop a pitch rate of change of the propeller of as high as 84° per second. Yet, on the other hand, a change of 3° at 500 knots gives a torque change which may be excessive for the gear box. These last two requirements are somewhat incompatible and hence give an added complication to the controls. Moreover, the high rate of pitch change leads to instability. The asymmetrical force on a multiple engine airplane is extremely large when a turboprop fails. This results from two sources, first, the power loss from the engine not operating (which is *large* per nacelle) and second, the *large* amount of drag required to drive the propeller and the turbine when the propeller is windmilling. It is felt that this excessive drag produced by windmilling after engine failure has been the cause of several accidents. The negative drag from a turboprop can be as much as 20,000 pounds, much greater than can be obtained from a reciprocating system. It may be so great that it cannot be counterbalanced by the aircraft controls. This condition has led to the development of devices which increase the propeller pitch toward feathering or decouple the engine when a reverse torque is indicated.

In order to more fully appreciate these problems, a few of them are discussed here in more detail. First, there seems to be a logical choice for the fundamental control variables. A satisfactory method of metering the fuel to jet engines as a function of air flow has been developed already.

A method for regulating the engine speed by governing the propeller pitch is well known from the systems used on piston engines. So, it appears natural to assume initially that the two independent variables should be taken as fuel flow and engine speed. Now, from all possible operating conditions determined from the internal solution, curves of specific fuel consumption can be plotted for constant fuel flow for various engine speeds and an envelope curve can be obtained which will give the points for least fuel consumption. The points on this curve are the desired operating values. On Fig. D,6a these points are the maximum values of the power for a fixed fuel flow.

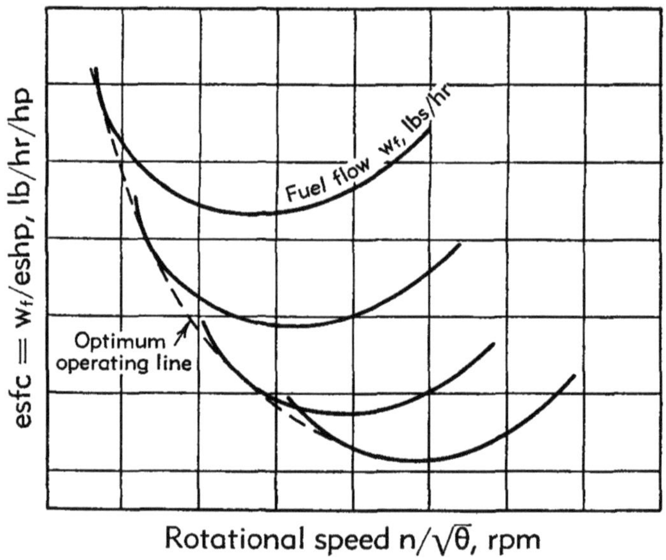

Fig. D,6a. Equivalent specific fuel consumption of a turboprop.

Physical limits. Before developing the controls, it is necessary to be sure the physical limits of the machine are not exceeded; namely, maximum temperature, maximum rotational speed, compressor surge, burner blowout, excessive flame lengths (at low densities), and intake pressure (at high Mach number). Because of the creep rate and the yield strength of the turbine blades, the centrifugal stresses must not exceed a given quantity which is a function of the air density. Thus there is a maximum flight velocity at any given altitude. As already mentioned, the compressor produces an unstable flow in certain areas. Not only does the compressor function improperly, but once the oscillations are encountered, the variations in air flow may be so great that there will be excessive heating of the turbine. The danger of encountering surge appears during accelerations, and if this occurs the engine fails to accelerate. Both the high and low fuel blowout limits may be experienced. The occurrence of the latter is likely during decelerations and the former at high altitude operation. Even before there is a blowout limit in this low density operation area, the flames may continue through the turbine to such an extent that the gases are hotter in the rotor blades than in the combustion chamber and hence cannot be tolerated. Before one can design a control system, he must consider these limitations and be sure that they are not exceeded. It is not difficult to schedule the engine so that the operating line for stable operation will lie within the six limiting conditions. But during the transition from one operating point to another, there is great danger of exceeding these limits.

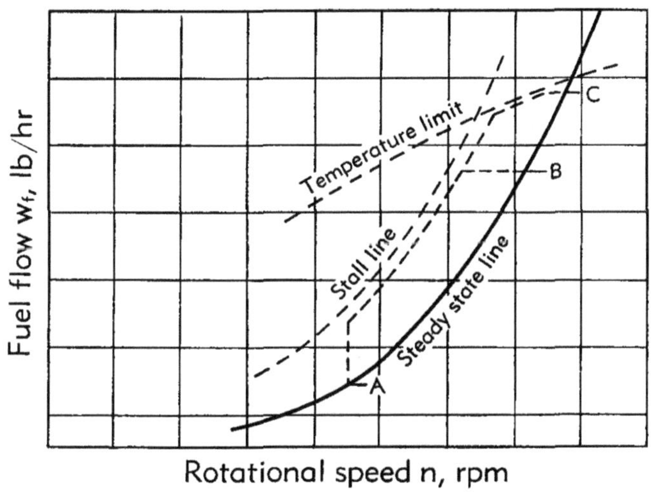

Fig. D,6b. Admissible increment of fuel for acceleration.

If an engine is capable of rapid changes in power, the equilibrium running points cannot ever be too close to the limit lines. There must be a sufficient range of admissible values in order to produce the changes. This is illustrated in Fig. D,6b. If an engine is operating at point A and it is desired to accelerate to point B, the total increment in fuel cannot be added without a change in engine speed, or otherwise two-engine limits the surge line and the maximum temperature would be exceeded. This problem is not as critical for turboprops as for turbojets because there is a way of increasing the engine speed without increasing the fuel flow. One would say the solution is to increase rpm first. This is not a satisfactory solution because of the large powers required to accelerate the compressor and turbine. So, even here the limitations are real. Fig. D,6c gives a plot of useful power versus rpm for contours of equal fuel flow. The stall line and the temperature limit line are again clearly marked. Also, a possible operating line passing through the maximum operating temperature and rpm is indicated. It can be seen, because of the flatness of the fuel flow curves, that the operating line could be varied within wide limits without seriously affecting the economy. Moreover, it can be seen that the useful operating range of the engine is limited to the upper half of the speed range. This results from the fact that below this point the compression ratio is insufficient to permit the development of a useful amount of power. It is also obvious that if the engine operation is to remain within the safe zone near maximum power a pilot cannot be given independent controls of the two variables. A relationship between the fuel flow and the rpm must be fixed.

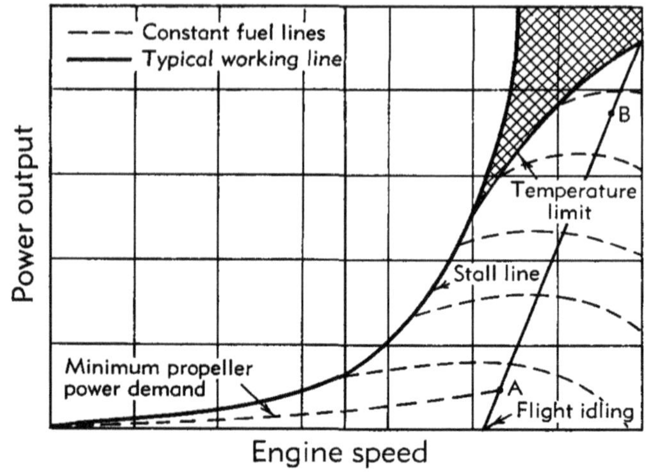

Fig. D,6c. Power output with contours of constant fuel flow.

Rotational Speed. Before considering a control system which will provide this relationship, it is necessary to appreciate the functioning of the speed control or constant speed unit and the way in which engine speed is controlled by the propeller pitch angle. The greater the pitch of a propeller the more power required to rotate it, and vice versa. For a given power and aircraft speed the pitch angle of a constant speed propeller determines the rotational speed. If either the power or the forward speed are increased at a fixed rotational speed, a greater propeller pitch is required, and vice versa. If the power and forward speed are held constant, an increase in rotational speed is obtained by shifting to a finer pitch and a reduced speed by coarsening the propeller blade pitch. A constant speed unit is a governor, controlling propeller pitch in such a way as to maintain the rotational speed at a preselected value. Thus, if the rotational speed is above the selected value, the governor unit causes the propeller pitch to increase until the speed is reduced to the selected value. Similarly if the speed is below the selected value, the unit causes a reduction in the propeller pitch until the rpm is increased to the selected value. Fig. D,6d and D,6e present the operations pic-

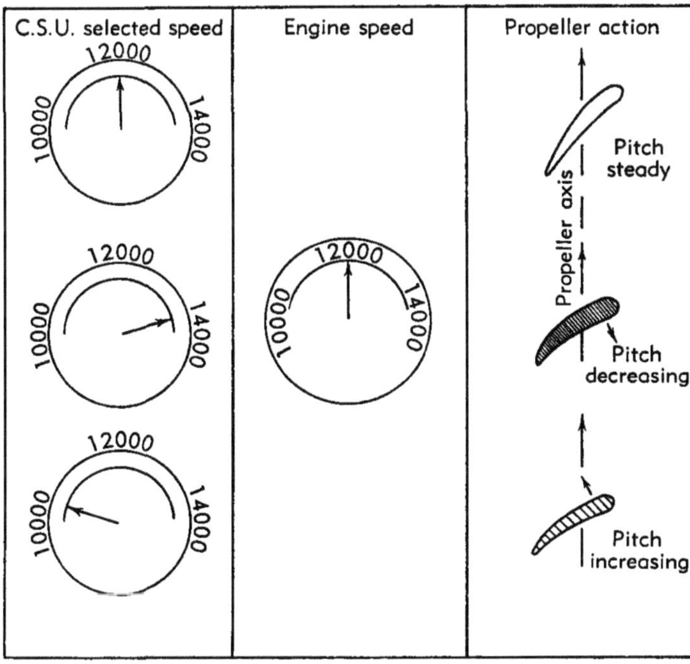

Fig. D,6d. Pitch angle changes with selected rpm.

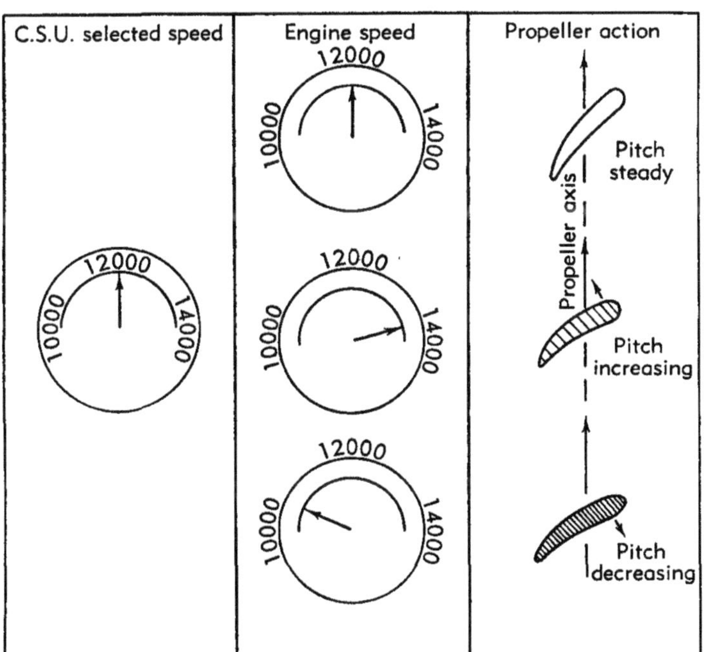

Fig. D,6e. Pitch angle changes for constant rpm.

torially. Fig. D,6d indicates the changes in pitch angle as a result of altering the selected rpm from the steady state conditions; while D,6e shows the changes in pitch in the event that the engine rpm varies from the selected speed.

It should be noted that the governor unit only changes the propeller pitch when there is a difference between the actual rotational speed and the selected speed. Finally, it is necessary to understand that the rate at which the propeller will change its pitch depends on the actual

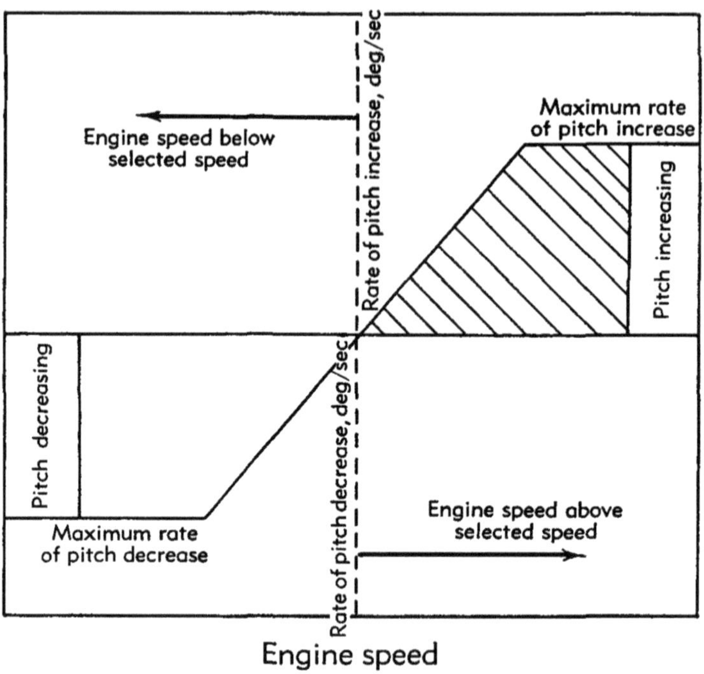

Fig. D,6f. Rate of pitch angle change.

difference in value between the engine speed and the desired speed. The greater the speed difference the higher the rate of pitch change until a maximum rate is achieved (Fig. D,6f). From these figures, it should be clear how the engine or propeller speed governor works.

Single-lever control system. Now a simple fuel-rpm relationship can be established by coupling a turbojet fuel system, which meters fuel in accordance with the air flow weight, and a constant speed governor to a single lever in the cockpit. (See Fig. D,6g.) By adjusting the relative lengths of the levers fitted to the fuel control and the speed control units, any operating line can be obtained. Such a system operates satisfactorily provided that the pilot's control lever is moved very slowly from one position to another. However, if the pilot's control lever is moved rapidly to a higher power, the fuel supply to the engine is increased and at the same time the selected speed is increased. Due to the magnitude of inertia of the engine and propeller the engine speed does not follow the selected speed instantaneously; instead, it gives a result more like Fig. D,6h or D,6i. At first the engine speed is below the selected speed and the propeller moves toward fine pitch. This causes the engine speed to increase

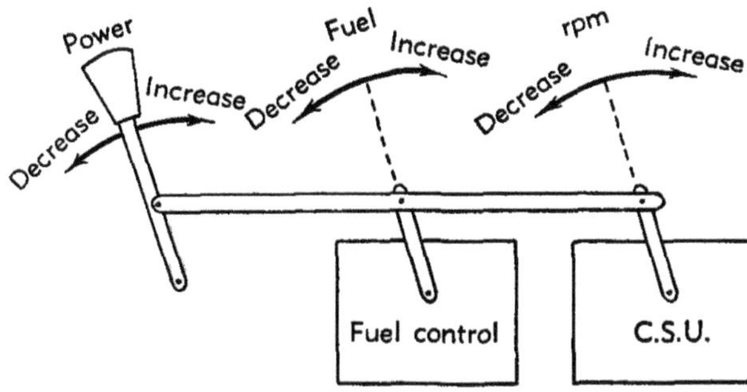

Fig. D,6g. Single-lever control.

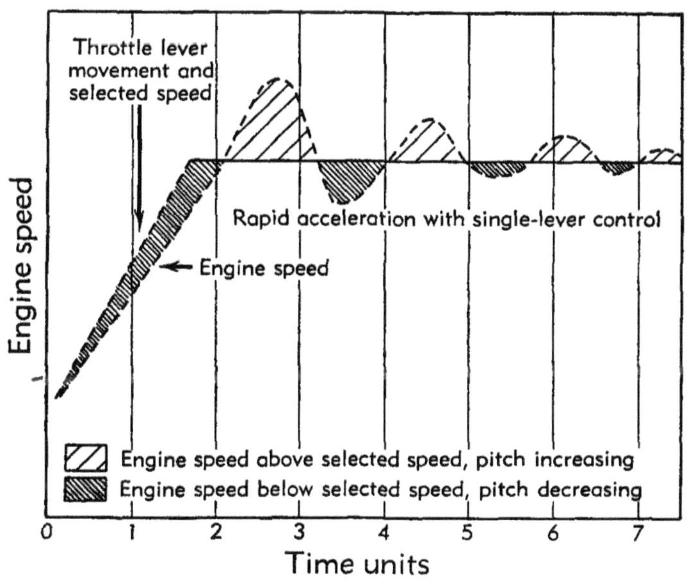

Fig. D,6h. Response to rapid acceleration.

more rapidly until it is above the selected speed when the propeller pitch starts to coarsen for the first time. This system suffers from three major disadvantages: First, the initial response to the opening of the throttle is in the wrong sense because the propeller pitch and thrust momentarily decrease instead of increasing; second, there is an appreciable lag before the thrust does increase; and third, some time elapses before the engine stabilizes and runs steadily at the new selected conditions. On the rapid reduction of power there are similar disadvantages. These defects were avoided by Armstrong Siddeley Motors, Limited, by incorporating a hydraulic lever servo between the fuel flow control and the constant speed unit [*5*]. This hydraulic servo (see Fig. D,6j) provides a differential rate of

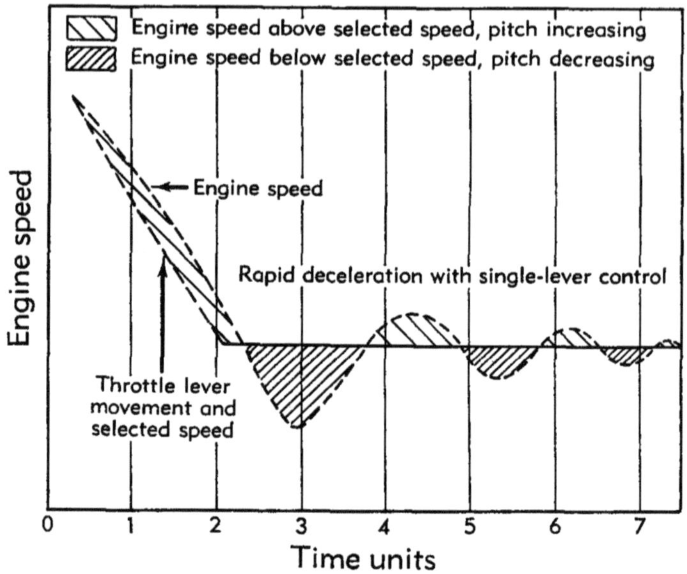

Fig. D,6i. Response to rapid deceleration.

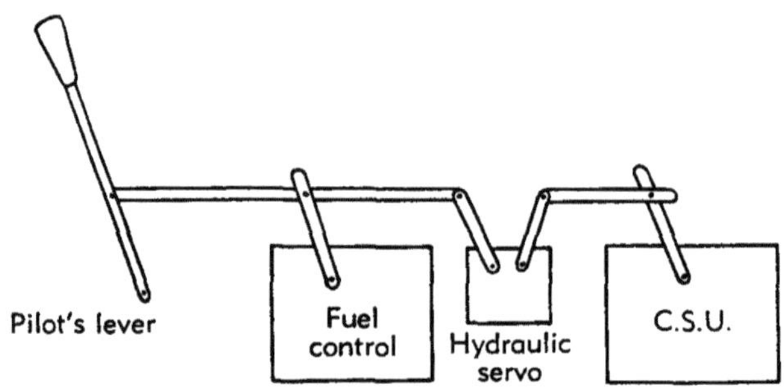

Fig. D,6j. Single-lever control with hydraulic servo.

movement between the fuel control and the speed selector, thereby causing the selected rpm to lag behind the movement of the pilot's lever. The response is illustrated in Fig. D,6k and D,6l. For a normal acceleration of about five seconds the immediate effect is to increase the propeller pitch and thrust. This results from the fact that selected rpm is lower than the actual speed, causing the pitch to coarsen progressively, thus increasing the thrust progressively and rapidly. Similar improvements are obtained on deceleration.

Plate D,7a. Turboprop reduction gear. (Courtesy of Pratt and Whitney Aircraft.)

Plate D,7b. Turboprop reduction gear. (Courtesy of Pratt and Whitney Aircraft.)

Plate D,8a. Propeller designed for 5000 to 7500 shaft horsepower. Diameter 18 ft; thickness ratio 0.055 at $0.75R$; blade width 18 in.; blade activity factor 146; total weight 1051 lb. (Courtesy of Curtiss-Wright Aircraft Corporation.)

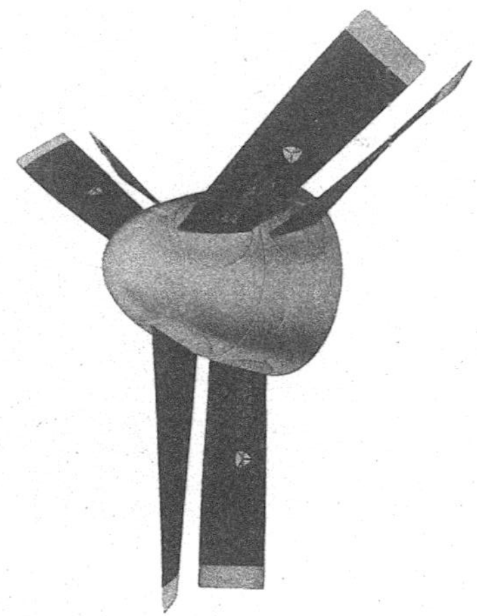

Plate D,8b. A dual propeller designed for 5000 to 7500 shaft horsepower. Diameter 14 ft; thickness ratio 0.050 at $75R$; blade width 18.4 in.; activity factor 168; total weight 1575 lb. (Courtesy of Aeroproducts Operation, Allison Division, General Motors Corporation.)

Plate D,8c. A propeller designed for 8000 to 12,000 shaft horsepower. Diameter 15 ft; thickness ratio 0.0449 at $0.75R$; blade width 23.5 in.; activity factor 200; total weight 1927 lb. (Courtesy of Curtiss-Wright Corporation, Propeller Division.)

Plate D,9a. Dynamometer used for shaft horsepower absorption. (Courtesy of Pratt and Whitney Aircraft.)

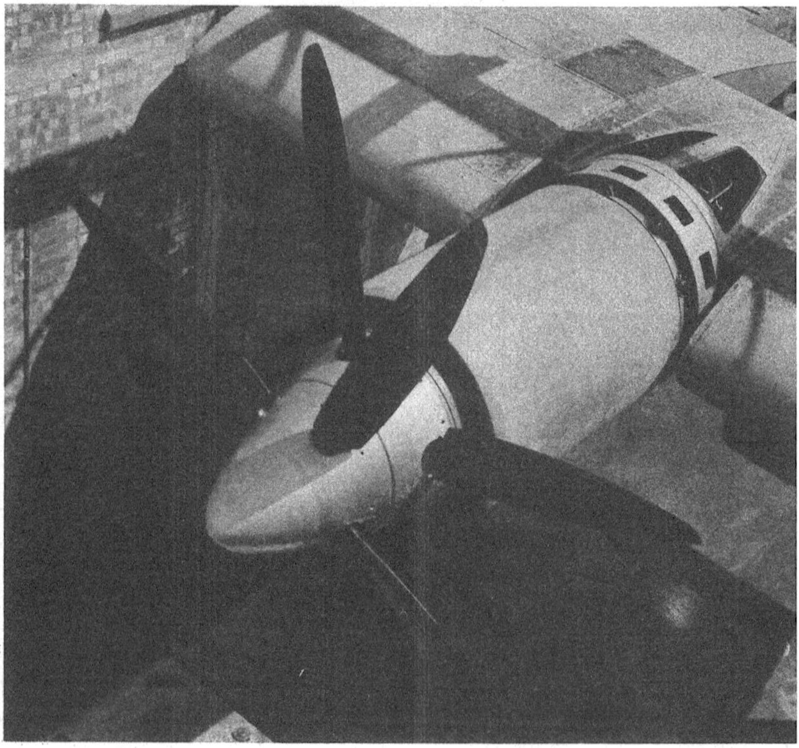

Plate D,9b. Python turboprop in a test bed. (Courtesy of Armstrong Siddeley Limited.)

Plate D,9c. The T-34 engine equipped with a Hamilton standard propeller ready for operation in a test cell. (Courtesy of Pratt and Whitney Aircraft.)

Plate D,9d. Installation of a Mamba in the nose of a Lancaster for flight test purposes. (Courtesy of Armstrong Siddeley Limited.)

Plate D,10a. An E-type ducted spinner. (Courtesy of Curtiss-Wright Corporation, Propeller Division.)

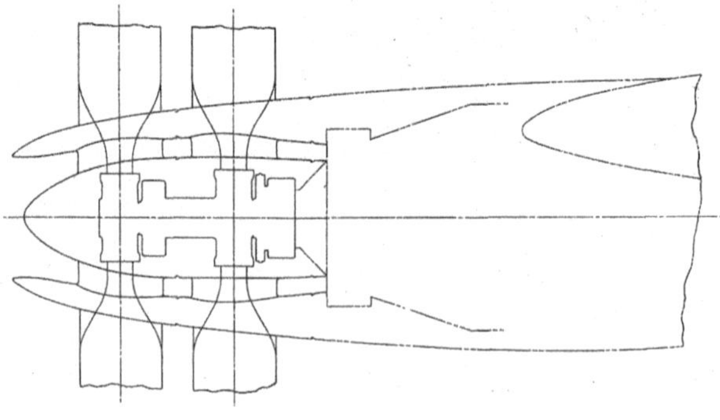

Plate D,10b. A typical type-E cowl for a dual rotation propeller on a turboprop engine installation. (Courtesy of Curtiss-Wright Corporation, Propeller Division.)

Plate D,10c. Installation of the Mamba in the Apollo Aircraft. (Courtesy of Armstrong Siddeley Limited.)

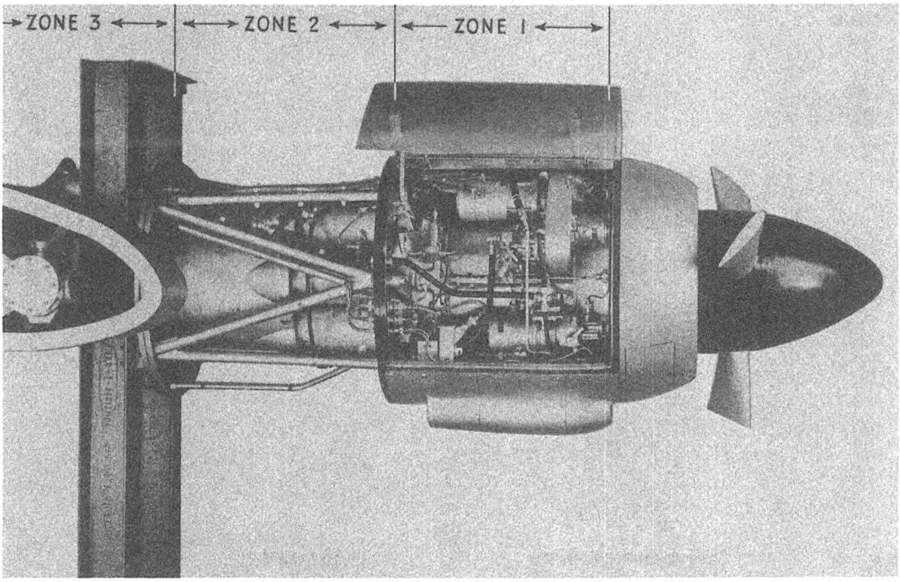

Plate D,10d. Installation of the Mamba in the Apollo Aircraft. (Courtesy of Armstrong Siddeley Limited.)

Plate D,11a. The Proteus.

Plate D,11b. The Dart.

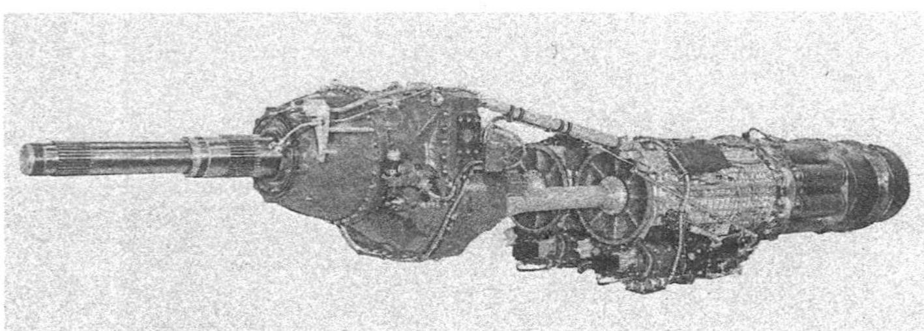

Plate D,11c. The T-40. (Courtesy of the Allison Division, General Motors Corporation.)

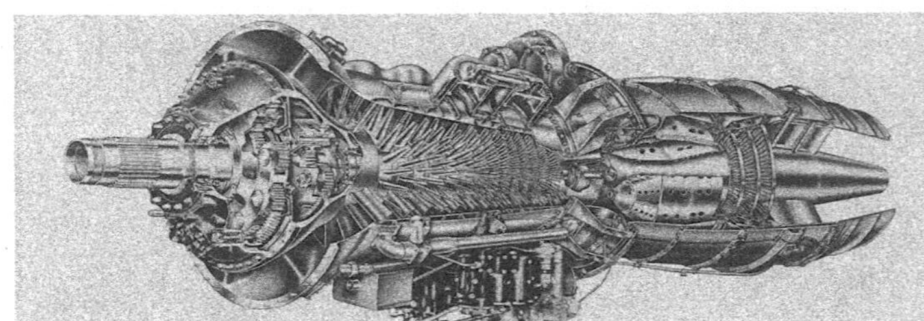

Plate D,11d. The T-34. (Courtesy of Pratt and Whitney Aircraft.)

Plate D,11e. The British Princess Flying Boat. Long range flying boat for military or civilian use. Powered by 4 coupled Proteus and 2 single Proteus engines with total 37,800 eshp; span of 219 feet; wing area 5000 square feet; gross weight of 315,000 lb; maximum pay load of 40,000 pounds; cruise speed about 380 miles per hour; maximum range of 5500 miles. It can transport 200 military troops for a shorter range. (Courtesy of the British Information Services.)

Plate D,11f. U. S. Navy Flying Boat XP5Y-1. Powered by 4 T-40 engines which deliver 22,000 eshp; span 146 feet; gross weight of 60 tons; maximum speed of more than 350 miles per hour; normal range of about 3000 miles. It is the first of a series of new design ultra-fast seaplanes being developed by the Navy. (Courtesy of Consolidated Vultee Aircraft Corporation.)

Plate D,11g. The Vickers Viscount 701. New medium-sized British turboprop airliner; carries 40 passengers; powered by 4 Darts with a total of 6000 eshp; span 94 ft; wing area 963 square feet; gross weight 56,000 lb; cruise speed of about 300 miles per hour and normal range of about 1000 miles. (Courtesy of the British Information Services.)

Plate D,11h. British Britannia. Long range commercial turboprop aircraft for 50 to 100 passengers; propelled by 4 Proteus engines with shaft horsepower of about 16,000 eshp; span 140 feet; wing area 2055 square feet; gross weight 140,000 lb; cruising speed about 360 miles per hour; maximum range 5600 miles. (Courtesy of the British Information Services.)

Plate D,11i. U. S. Navy Skyshark A2D-1. Single-seat, carrier-based attack aircraft, propelled by one Allison T-40. (Courtesy of the United States Navy.)

Plate D,11j. U. S. Navy Savage XA2J-1. Carrier-based, three-seat combat aircraft, propelled by 2 Allison T-40's. (Courtesy of North American Aviation, Incorporated.)

Plate D,11k. The British Gannet. Three-seat, carrier-based, anti-submarine aircraft, powered by an Armstrong Siddeley Double Mamba. (Courtesy of the British Information Services.)

Plate D,11l. The British Wyvern. Single-seat, carrier-based naval fighter, powered by the Armstrong Siddeley Python. (Courtesy of the British Information Services.)

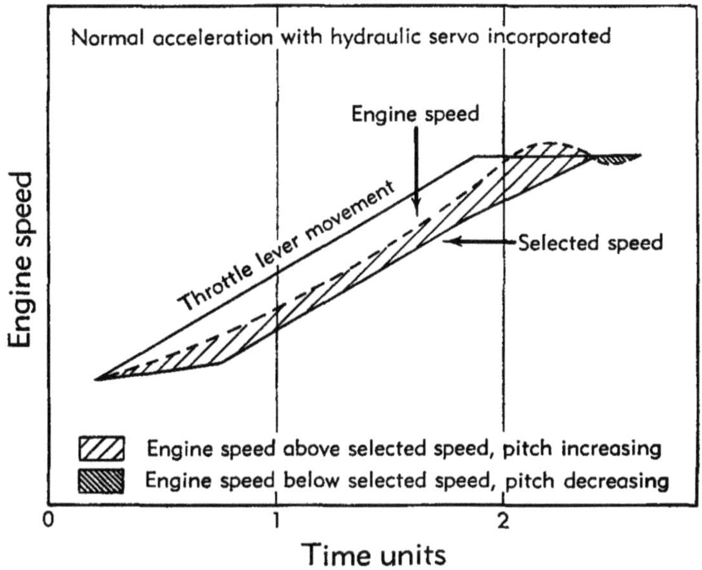

Fig. D,6k. Accelerations with single-lever hydraulic servo control.

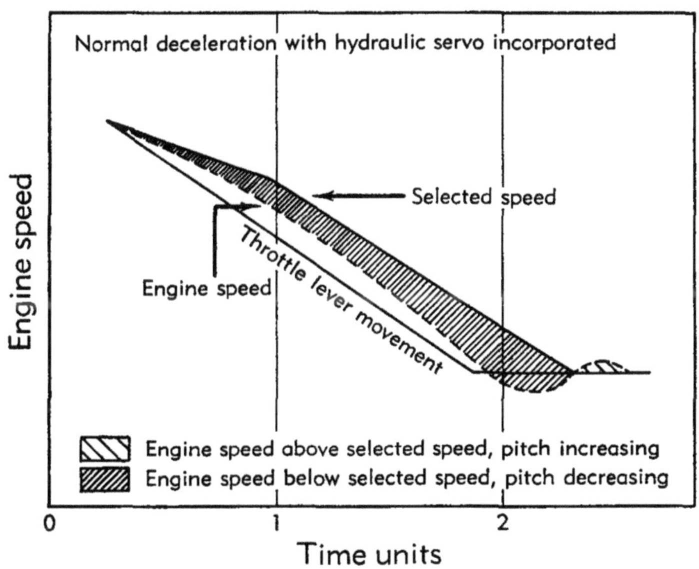

Fig. D,6l. Deceleration with single-lever hydraulic servo control.

When the turboprop is used in carrier aircraft, immediate response is necessary. The thrust should drop immediately to zero when the pilot receives the cut signal, and at the same time the engine must be capable of achieving full thrust very rapidly in the event of a wave-off. Thus, time for acceleration must be less than five seconds, possibly as low as

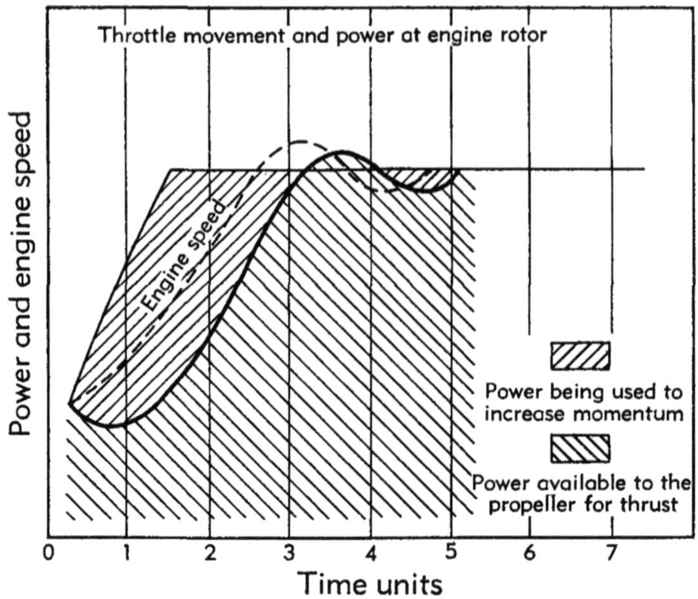

Fig. D,6m. Power division during accelerations.

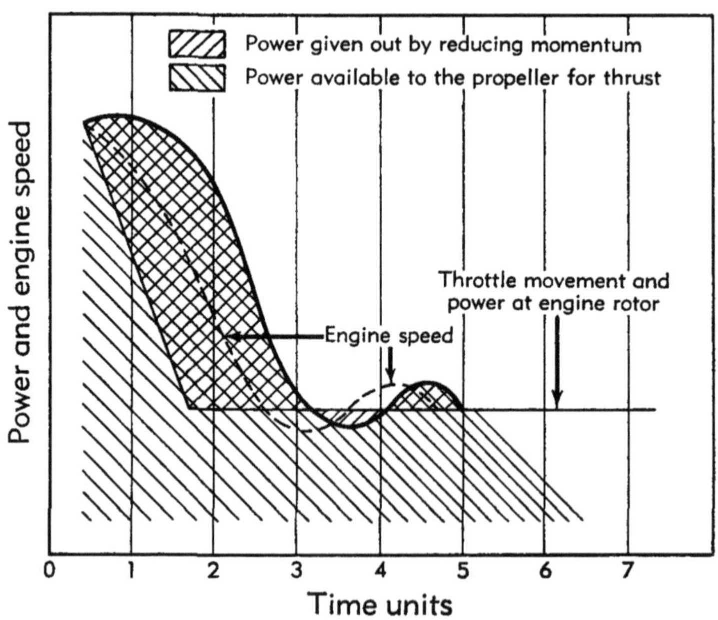

Fig. D,6n. Power division during deceleration.

1.5 seconds. Whether a proposed system can meet these requirements can be ascertained from a plot of the power delivered by the turbine and the power required to increase the inertia of the system as a function of time. The difference between the two quantities is the power available for producing thrust. Fig. D,6m and D,6n give plots for the above proposed system for acceleration and deceleration. Such plots clearly show that the change in inertia required to change the speed of the engine produces a considerable time delay which could be avoided if the engine were scheduled for constant speed. Analyses such as these

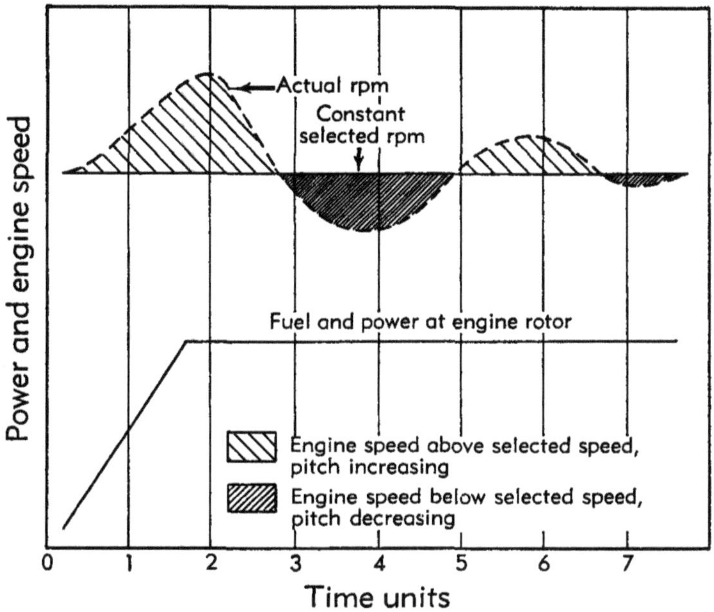

Fig. D,6o. Increasing power with constant selected rpm.

have led to the adoption of constant speed controls for at least two of the fixed turbine turboprops. Others have adopted nearly constant speed operating lines.

Even with a constant speed operating line it is not possible to leave the selector set at one speed. For when fuel is first added, the engine rpm will increase before the pitch will change since an increment in speed is necessary to activate the system. Thus the inertia delays are not removed (see Fig. D,6o). To minimize the change in rpm, it is necessary to change the pitch of the propeller at the same time the fuel flow is changed. That is, on acceleration it is necessary to temporarily select a lower rpm in order to provide the incremental difference between the actual and the selected rpm so as to prevent an actual change in rpm (see Fig. D,6p). When such a system is perfected, the increased power of the turbine is immediately available for thrust. Such a system permits the change from flight-idle to full power in about two seconds. Hence, it appears to be best for carrier-based operations to have an operating line of constant speed from flight-idle up to maximum power. For land-based operations this acceleration requirement need not be so stringent.

Below flight-idle the operating line is determined by other considerations. There is usually a prescribed blade angle setting associated with power lever position. As the lever is shifted downward from flight-idle to ground-idle the pitch is changed from some positive angle to a small

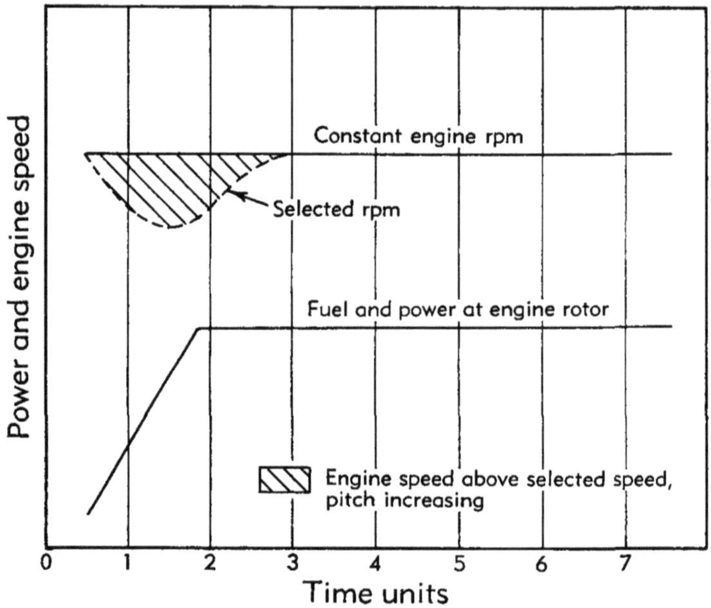

Fig. D,6p. Selected rpm for constant rpm during acceleration.

negative angle. A further change of the lever may give full reverse pitch. From flight-idle to ground-idle the fuel and engine speed are first reduced and may be reduced throughout, depending upon the negative pitch of the propeller, but from ground-idle to full reverse the fuel and rpm are increased. Throughout this range the fuel flow is determined by the propeller requirements (see Fig. D,6q).

As a result of the decision for constant speed or near constant speed of a turboprop, in order to obtain this rapid change in power, the pitch change mechanism must operate at a much faster rate than is necessary for a reciprocating engine. The resulting difficulties associated with this high pitch change have been previously mentioned.

Innumerable devices could be included to insure protection of the engine and pilot in case of failures. As previously mentioned, the windmilling drag of a turboprop is about twice as great as that of a reciprocating engine. This results from the fact that the low pitch propeller stop for turboprops is from 8° to 12°, whereas the fine pitch stop for conventional piston engines is about 25° or more. The low value for the fine pitch stop results from the high rotational speeds even at partial power. This windmilling drag may be as much as 60 per cent of the maximum power of the unit. Thus, upon engine failure the thrust reduction per nacelle may be 1.6 times the thrust power of the engine. This produces an asymmetrical force which may be too large to be counterbalanced by

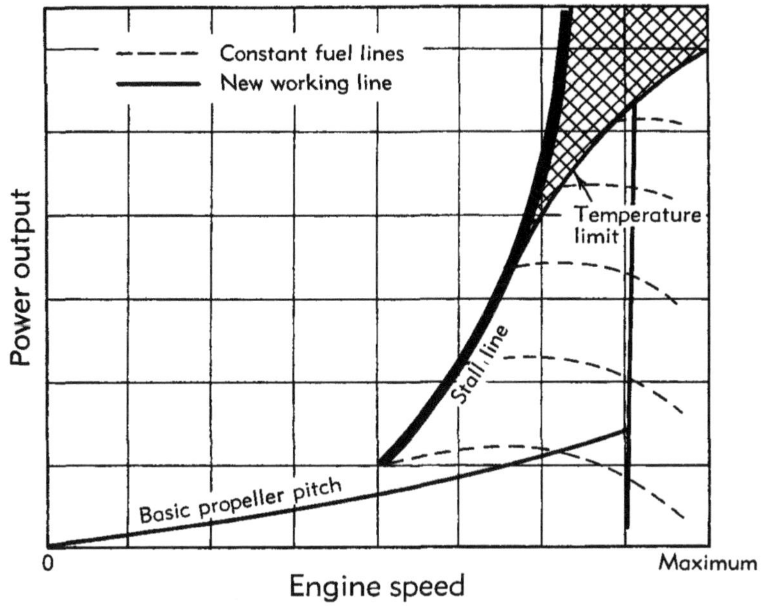

Fig. 6q. Operating line for constant speed turboprop.

the airplane control surfaces. Hence, a device which immediately eliminates this drag on engine failure is considered important. The present devices are based upon a method of increasing the propeller pitch when reverse torque is encountered. One difficulty with such systems is the avoidance of propeller feathering as a result of power variations or turbulence. In the case of mechanically coupled engines it is necessary to decouple an engine which fails. Should the propeller pitch mechanism fail there are two dangers: (1) if the propeller goes into flat pitch, the engine may over-speed and throw the turbine or compressor blades and (2) if it goes into coarse pitch, the turbine blades may be over-heated.

All of the above discussions have centered around a fixed turbine. Many of the problems are the same for free and coupled turbines. In the case of free turbines the compressor speed is controlled by the position of the fuel throttle. The barometric pressure control unit permits a substantially constant engine speed at one throttle setting regardless of the altitude and flight velocity. The propeller and power turbine speed are determined by the pitch of the propeller, which may be fixed, have a finite number of stops, or be controlled by a constant speed governor. For take-off it is desirable to employ a high propeller rotational speed and at cruise speed a medium speed depending upon the tip Mach number limitation. Since the speed of the compressor turbine, and hence the generator power of the engine, are determined by the fuel flow, in a free turbine the operating line cannot be a constant speed curve. The speed of the engine must reduce with power. It is true that the airplane thrust can be changed by varying the pitch of the propeller but this variation is too small to permit constant speed over the entire range. Moreover, if it could be done, it would not be economical. Because of the possibility of leaving the propellers at fine pitch while accelerating the compressor, it is possible to accelerate the free turbine engines fairly rapidly anyway. On the other hand, the free turbine enhances the runaway problem should the propeller go into fine pitch. Yet, should an engine fail, there is less windmilling drag since the propeller would turn only the power turbine. For a given propeller and reduction gear box, the relation between the propeller speed and the compressor turbine speed can be determined in order to obtain the best specific fuel consumption. In the case of the Bristol Proteus, the most economical points of operation are where the two turbines operate at about the same rpm.

Comparison of fixed and free turbine engines. At this point, it is convenient to summarize the characteristics of the fixed and free turbine by giving the advantages of each.

Advantages of a free turbine over a fixed turbine:

1. Safer in case of power failure, since there is less windmilling drag.
2. Simpler controls.
3. More degrees of freedom on propeller speed. There are many aircraft missions in which it would be desirable to have a two or more speed gear box to achieve the best propeller performance. The free turbine gives the equivalent of a gear box with a continuum of gear ratios.
4. Because of (3) it is the "only" satisfactory gas turbine for a helicopter.
5. Low starting powers since the propeller need not be accelerated while starting the gas generator.
6. Faster rate of stopping of propeller after engine is shut off.

Advantages of a fixed turbine over a free turbine:

1. Mechanically less complicated, since there is only one shaft.
2. Better turbine performance. There is no penalty in performance because of improper distribution of load between the two turbines. In the free turbine engines the compressor turbine must run the compressor regardless of the power, and the propeller turbine must run

the propeller. At many operating points one turbine may be overloaded while the other is underloaded, and as a result, the efficiency of each is less than that which could be obtained with a single turbine.

3. Simpler to design. Developmentally the design of a free turbine must be more accurate, since it is more difficult to change it.
4. More aircraft braking effect on landing.
5. More rapid acceleration.
6. Possible to augment thrust by water and alcohol injection for tropical take-offs.
7. Compression ratios may be higher. (This is somewhat controversial.)

The coupled turbine is an intermediate between the fixed and the free turbine, and hence has some advantages of both.

D,7. Gear Box and Accessories. The propeller, turbine, and the compressor are aerodynamically similar machines. The diameter of a propeller must be greater than the diameter of the engine. Thus the most efficient rotational speed for the engine is not the best rotational speed for the propeller. An even more important consideration is the structural aspect. If the engine blades are operating at a high rotational speed, and the propeller is operating at the same speed, the centrifugal loads on the blades will be so great that it would not be feasible to construct them. Therefore, if a propeller is to have a significantly larger diameter than the engine, then it is necessary to have a gear box which reduces the engine speed to a desirable propeller speed. (The ducted fan, because of the small diameter, has been able to operate without a reduction gear.) The speed ratio of the reduction gears will depend upon the diameter of the propeller and the amount of power to be absorbed. When turboprops were initially considered, it was assumed that the propeller blades would have to rotate slowly enough to keep the blade tip Mach number of the order of magnitude of 1, that is, about the same order of magnitude as that of the turbine blades. Moreover, in order to absorb the power, it was assumed that the propeller blade diameter would have to be from 7 to 12 times as great as that of the gas generator unit. Hence it would be necessary to develop a reduction gear box with a speed ratio of from 7 to 12. More recently it has been found that it is possible to design propellers which operate at supersonic speeds and still maintain a fairly good efficiency. With these propellers, the reduction gear ratio may be reduced.

The development of reduction gears was not a new problem in principle. However, the magnitude of the powers to be absorbed was beyond that of previous designs. Destructive vibrations caused by tooth clearances and deflections due to high loads caused great difficulty: scuffing of the teeth, fatigue failure of small parts and the cases themselves, and breakdown of the oil film. It became necessary to use gear designs that run with small clearance; to design forms that can withstand high tooth pressures; and to lubricate by means of an oil fog instead of the usual oil baths. It is probable that helical gears with stub teeth will provide solutions to these problems.

The current reduction gears are usually either spur or compounded planetary gear types or a combination of both. The best of these reduction gears run about 0.1 pound per shaft horsepower for a reduction ratio of about 10 to 1. (Plates D,7a and D,7b illustrate one such set of gears.) Gear trains have been produced for both single and counter-rotating propellers. In the latter case, in addition to reducing the speed of rotation, it is necessary to provide equal and opposite rotation to concentric shafts. Besides the main reduction gear, there are gears for power take-off pads for driving all the auxiliary equipment of both the airplane and engine. The fuel pump, lubricating oil pressure and scavenger pump, oil metering pump, centrifugal breather, and generator are usually driven by the compressor shaft, whereas the propeller control unit, the propeller tachometer and synchronizing altimeter, and the torquemeter pump are driven from the low speed side of the reduction gear. The current practice is to have a take-off pad to run the aircraft accessories which is separate and distinct from the pads to run the engine.

Starting is one of the basic problems of turboprops. There is a minimum rotational speed before an engine can run under its own power. Hence it is necessary to supply some external power to accelerate the engine rotor and the propeller to this speed. This power is used to accelerate the propeller and turbine and to pump air through the engine. The energy required to accelerate the rotor and propeller is independent of the time, while the power to pump the air is a function of the amount of air as well as the pressure head, and hence is a function of time. The burners may be lit almost as soon as the engine starts to rotate, and this hot air supplies almost enough turbine power to pump the air. Hence the external energy for starting the turboprop is almost independent of time. Therefore the size of the starter motor may vary over a fairly wide range. The upper size of the starter is determined by the torque carrying capacity of the starter drive, and the lower size is limited by the time that the turbine blades may be subjected to excessive high temperatures created by burning during acceleration. Perhaps the time for acceleration should not exceed 20 seconds. The motor may be of any type, but the electric motor is the most desirable. The starter is connected to the engine through a gear train. In the case of mechanically coupled dual units there is usually a declutching system which provides for starting either engine without rotating the propeller.

All engines are equipped with a brake which prevents engine and propeller rotation of an inoperative engine whether parked or in flight.

This brake is also employed to decelerate and stop the propeller. In flight the brakes do not give sufficient torque to stop the propeller unless it is first feathered. In the newer coupled turbine designs, it is anticipated that two brakes will be employed, one on each shaft.

D,8. Turboprop Propellers. Before the advent of turboprops, propellers of 20 feet in diameter had been built and operated, but no propeller had been designed to absorb as much as 6000 shp. Now there is a demand for a propeller for over 15,000 shp. The power absorption ability of a propeller depends upon the diameter, the rotational speed, and the solidity or blade area. It rises with the fifth power of the first, the third power of the second, and the first power of the third. A propeller with a diameter much greater than 20 feet would not be feasible for aircraft use. Moreover, increasing the diameter and/or rotational speed increases the Mach number of the blade sections until they operate in a compressible flow region. Increasing the blade area at least within limits does not give any adverse effects. Thus one is led directly to either wide blades or a large number of blades, the latter of which are prohibited by the space around the shank unless the blades are placed on tandem hubs either rotating in the same or in counter directions. In view of these limitations, it is logical to investigate carefully the losses encountered at the higher rotational speed. When all sections of the blades are operating at supersonic speed, the performance of the propeller may be quite accurately calculated by strip theory methods. Results of such calculations indicated that the propeller efficiency of very thin blades can remain as high as 80 per cent. The efficiency of a propeller operating in the intermediate range with some subsonic and some supersonic sections is not as easily predicted but by inferences it appears that the efficiency of a properly designed propeller may be almost as high even in this region. The simple analysis given previously (Art. 3) indicates that for constant C_D/C_L values, tan ϕ should be about 1 for the most efficient operation. This would mean that the rotational speed of the mean chord would be equal to the flight speed. Hence, for supersonic flight, the rotational speed should be supersonic at the mean chord, which would give a relative speed of about 1.4 times as much. Of course, the value of C_D/C_L is not independent of speed in the compressible flow range because of the drag rise. The actual speed of rotation will be a function of this drag rise which tends to reduce the value of ϕ for the highest efficiency and hence the actual rotational speed. In order to hold the drag to a minimum the blade sections should be as thin as possible. Thickness ratios of 2 per cent at the 0.9 radius are contemplated for supersonic propellers. Aerodynamically, the problem is clear. Structurally it is difficult. In summary, the initial considerations of magnitude of power and efficiency at the design point indicate that the propeller blades should be wide and very thin and the rotational speed should be high. Plates D,8a, D,8b, and D,8c are pictures of propellers designed for medium, medium large, and large power turboprops, respectively.

At high speeds, the central portion of the propeller near the hub gives a negative thrust coefficient. Hence it is necessary to prevent flow through this area, which is done by surrounding the hub by a spinner. Such a spinner must be faired into the nacelle or fuselage to give correct aerodynamical flow. An illustration of such a spinner is given in Plate D,9b. The spinner should cover from 0.2 to 0.33 of the radius depending upon the design and operational speed of the propeller. The ducted inlet mentioned later under installation is a satisfactory form for the spinner.

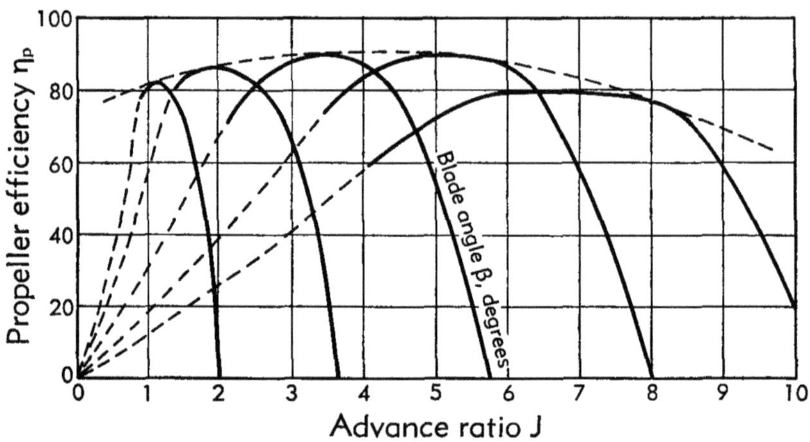

Fig. D,8a. Propeller efficiency, as function of blade angle and advance ratio.

The constant speed control system which appeared to be the logical answer for the fixed turbine from the standpoint of engine acceleration requires that the propellers operate efficiently over the widest possible range of speeds and powers. A conventional plot of propeller efficiency versus J is given in Fig. D,8a. For a constant rotational speed and constant flight velocity, J is constant; thus variations in power require changes in β. A glance at the figure shows that large changes in power at constant flight velocity produce large variations in propeller efficiency. On the other hand, if the rotational speed were reduced with power, the efficiency of the propeller might be maintained near the envelope. Further observation of the figure shows that, unless the shaft power of a constant speed engine changes rapidly enough with flight velocity to change the pitch angle significantly, the propeller efficiency may be extremely low in some regions. Fig. D,8b shows a plot of the actual efficiency obtained as a result of a variation of power with velocity in accordance with the power requirements for level flight. The loss in efficiency resulting from the constant speed operation is less than might have been expected for this particular case.

With a free turbine the rpm can be varied with speed, in which case the propeller efficiency could be made to remain on the envelope of values. However, this is usually not accomplished. Ordinarily a few pitch settings give a satisfactory performance. In throttling the engine, the speed of both the propeller and the engine should be reduced for best efficiency.

The retention of the blades in the hub is not a new problem except for the tremendous forces which are associated with a high rotational

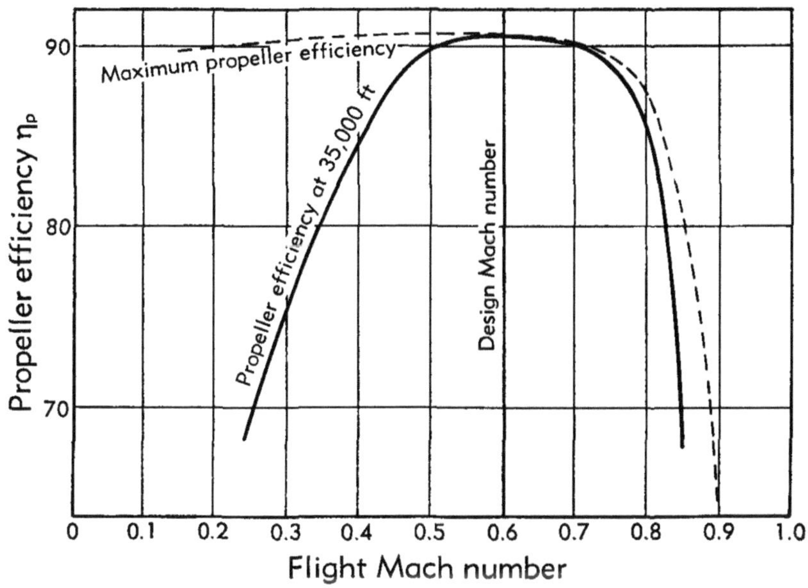

Fig. D,8b. Propeller efficiency in flight at 35,000 ft.

speed. For a supersonic propeller the blade loads on the hub may be as much as 900,000 lb. When, as usually assumed, a pitch change is required, the blade must be free to turn in its retainer about the centerline. This requires a cylindrical shank and the propeller should be mounted in a frictionless bearing. Even then the centrifugal twisting moment for a transonic propeller is of the order of 125,000 in.-lb per blade. Moreover, the pitch-changing mechanism must be an integral part of the hub. There are three basic types of these mechanisms, classified by the motivating power: mechanical, electrical, and hydraulic.

The structural problems for turboprop propellers are accentuated by the requirement for thinner blades. In addition to the stresses resulting from the rotational speed, engine torque, and thrust, it is necessary to make allowance for certain dynamic stresses, notably the $I \times P$

excitations and flutter. A high torsional rigidity and modulus is required. Research is still in progress to find strong metals with these characteristics which also have satisfactory fabrication qualities. Any method may be employed to obtain the proper shape; the blades may be extruded, welded, or brazed just so long as uniformity of structure can be obtained. The uniformity is necessary to produce a light blade. At the present time titanium appears to be the metal with the greatest promise. It has all of the characteristics to make it attractive for blades, except that it has a relatively low modulus, which for the torsional rigidity makes the blades thicker than steel. This increase in thickness decreases the efficiency.

The propeller designers feel that it is possible to build a propeller for 15,000 shp. This, however, is not an easy task, and it may not be possible to do it within a practical weight range. While on the subject of weights it is appropriate to note that the weight of the propeller and gear box on the current designs almost equals the weight of the gas turbine generators. In fact, these weights are in a ratio of about 5 to 7 and in some cases they are in a ratio of 7 to 5. Hence it can be seen that if the weight of a turboprop is the fundamental variable which determines their usefulness, or if the weight is what is keeping them out of operation, then the two most fruitful places to look for weight saving are in the propeller and the gear box. Moreover, there is about twice as much chance for success when working with the propeller, because it is about twice as heavy as the gear box. At present, the weights are about 0.2 and 0.1, respectively, per shaft horsepower. With most optimistic assumptions for hollow blades of new high strength metal operating at supersonic speeds it may be possible to make the propeller as light as 0.12 pounds per shaft horsepower. At the same time, because of the possibility of lower speed ratios, there may be a saving in the reduction gear weight. Should the sum of the two be reduced to about 0.2 pounds per shaft horsepower, then the total specific weights of the propulsive system would be reduced by about 8 or 9 per cent. Such a reduction would make the turboprop a competitor of the turbojet in many more military operations.

D,9. Engine Testing. The testing of turboprops is in many ways like the testing of turbojets and, in other ways, like the testing of reciprocating aircraft engines. The measurements required for determining the operation of the compressor, the burner, and the turbine are exactly the same as those for turbojets. The measurement of the shaft power is different from that for testing piston engines only in the magnitude of the power to be measured.

The testing program is divided into two phases, static tests on a test bench and flight tests. The bench tests may be divided into five types:

experimental, developmental, qualification, modification, and acceptance testing.

It is difficult to draw the line between experimental and developmental testing. Together they include all tests on the components and the engine which are necessary to understand and produce an operating engine. Qualification tests are those which demonstrate that the engine developed will deliver a certain prescribed performance over a period of time without any apparent change in the engine. Modification tests refer to tests to determine the effects of changes of certain components which are deemed necessary after the engines have passed their qualification tests. The time of operation for these tests may be less than that for the qualification, depending upon the nature of the changes, and it is not necessary to duplicate tests of components which were not modified. The acceptance tests are those run on production engines to insure a standard performance.

Experimental and developmental tests. Because of the advantages of high compression ratios and high temperature, and the tremendous amount of power removed by the turbine in turboprop engines, it is essential in the development of these engines that a thorough knowledge of the fundamentals be obtained. Unless the efficiency of the compressor can be maintained at a reasonably high level, the gain in cycle efficiency of high compression ratio will be lost. Moreover, for starting and off-design operation, a complete knowledge of the flow becomes more important the higher the compression ratio. So here it is necessary to determine the complete performance of a single-stage compressor over the entire flow range, i.e. it is necessary to study the flow in the unstable operation regions (Fig. D,9a). Some of the stages may be operating in this unstable region A, some in the stable region B, while others are operating in the other extreme region C of decreasing pressure. This condition frequently occurs when starting. By complete understanding of a single stage over the entire area, it is possible to determine the amount and location of bleed air which will permit the best starts, and the information will also be valuable in helping to match the stages to give better off-design operation. The methods for obtaining the data are discussed in X,D. For optimum development it would be desirable to obtain a complete map of the turbine as well as that of the compressor. However, the amount of power absorption is so great (about three times the shaft power of the engine) that at present no facility is equipped to obtain the complete data on turbines for the large engines.

In bench testing a turboprop the usual measurements are the compressor inlet and outlet total pressure and temperature; the turbine outlet temperature and pressures; the rotational speed; the flow, pressures, and temperatures of the oil; the fuel flow; and the shaft horsepower. In addition, many test rigs measure the turbine inlet temperature.

D · *THE TURBOPROP ENGINE*

The testing of the turboprop differs from that of the turbojet in many ways. First is the requirement for measuring the shaft horsepower. This is done by a dynamometer, a torquemeter built in the engine reduction gear, or by an external torquemeter. An indication of the magnitude of the power to be determined can be obtained from Plate D,9a which shows three large eddy current dynamometers coupled in series with one electric dynamometer which can absorb 800 horsepower or can be used for driving up to 650 horsepower. A second problem which is unique to turboprops is the method of measuring temperature and stress on the rotating parts. Since the leads from the straingauges and thermocouples cannot be taken out through the front, it is necessary to bring them out

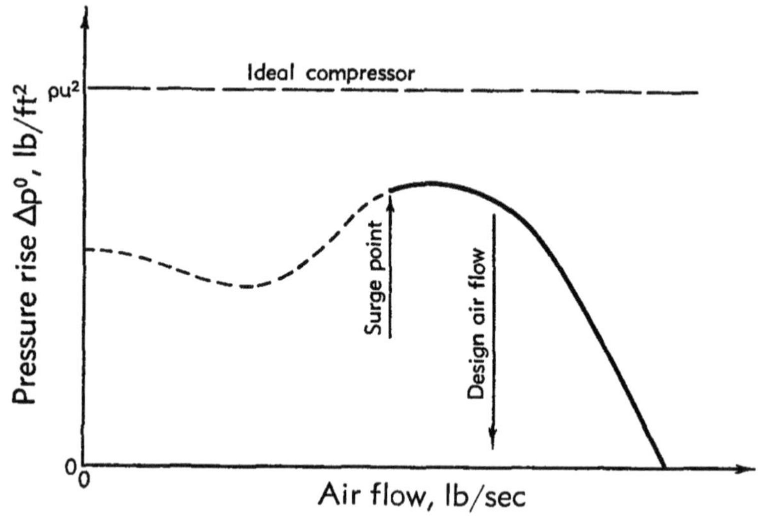

Fig. D,9a. Tested pressure rise and air flow of a single-stage compressor for 1 rpm.

the turbine end by means of slip rings inside the tail cone. The high temperatures involved make the measurements more complicated. Third, the thrust of the jet stream is not measured directly and must be computed. This can be done from pressure and temperature measurements by a rake at the exhaust nozzle and a measurement of the air flow. Since this jet thrust is such a small part of the total thrust, the accuracy of the measurements may be relaxed somewhat. In fact, in some test cells the jet thrust is estimated without any rake measurements in the exhaust jet. Fourth, when it is desired to test the engine under ram or altitude conditions with the propeller on the shaft, normal facilities will not yield a supply of air sufficient to permit the propeller to operate in the simulated environment. Hence the complete performance of the engine cannot be obtained without flight tests. Ram testing at high powers is necessary to test the gear box, yet testing these high powers without the propeller air at the same conditions may produce disastrous vibrations. Fifth, it is more desirable to measure the turbine inlet temperature since a completely satisfactory estimate cannot be obtained from measurements of turbine outlet temperature. Lastly, a large test cell is necessary when the propeller is attached and care must be taken to insure that the propeller does not develop adverse air flow patterns.

After the bench testing the complete engine is transferred to a hanger test bed. Plates D,9b and D,9c show two installations in hanger test beds.

Qualification tests. The first step in a qualification test on an engine is to calibrate the engine, that is, prior to connecting the propeller, the power delivered for various settings of engine speed and gas temperature is measured in a brake test bed. The readings are made after the engine has been allowed to operate at these conditions for a few minutes, possibly five. Next, the controls are set as nearly as possible to give the two conditions at the various power settings, ground-idle, flight-idle, maximum continuous power (normal power), maximum intermediate power (military power), and maximum take-off power. Then the propeller is connected to the shaft and the complete propeller engine is calibrated and the controls adjusted if necessary. Then the engine is ready to begin its 150-hour endurance qualification test. This total time is broken down into a number of short periods, during each of which various operating conditions are tested. During the 150-hour period, there will be a specified number of hot and cold starts and some false starts. A cold start is a start after 2 or more hours of natural cooling of a stationary engine. A hot start is a start within 15 minutes after shutting down. A false start is simulated by elimination of the ignition. The engine is accelerated from ground-idle to take-off power and after a short operating time decelerates back to ground-idle. The times for the moving of the throttle are limited to, say, one second, and the times for the accelerations and decelerations are measured. There will be a prescribed number of hours at a series of percentages of continuous rated power, at maximum intermediate power, at maximum take-off power, at flight-idle, at ground-idle, and a certain number of changes to reverse the pitch of the propeller. All stops are made by means of the propeller brake. After completion of the 150-hour tests the engine should be completely disassembled and all parts examined for wear and distortion. If there are no excessive conditions, the engine has passed its qualification test.

Acceptance tests. After an engine has passed its qualification test and is put into production, each engine is subjected to an acceptance test: This is a test of about 2-hours duration with a certain prescribed time at each of the operating conditions mentioned in the qualification tests. After this time, the engine is stripped for inspection and, on completion of the rebuilding a final run of the same nature is made. In case any engine fails to meet the specification, defective parts may be replaced and additional tests may be run. However, if an engine fails to meet the specification after a prescribed number of hours, it is rejected.

Short method for correcting test results to standard conditions. When one is running tests on the engine, it is highly desirable to know whether

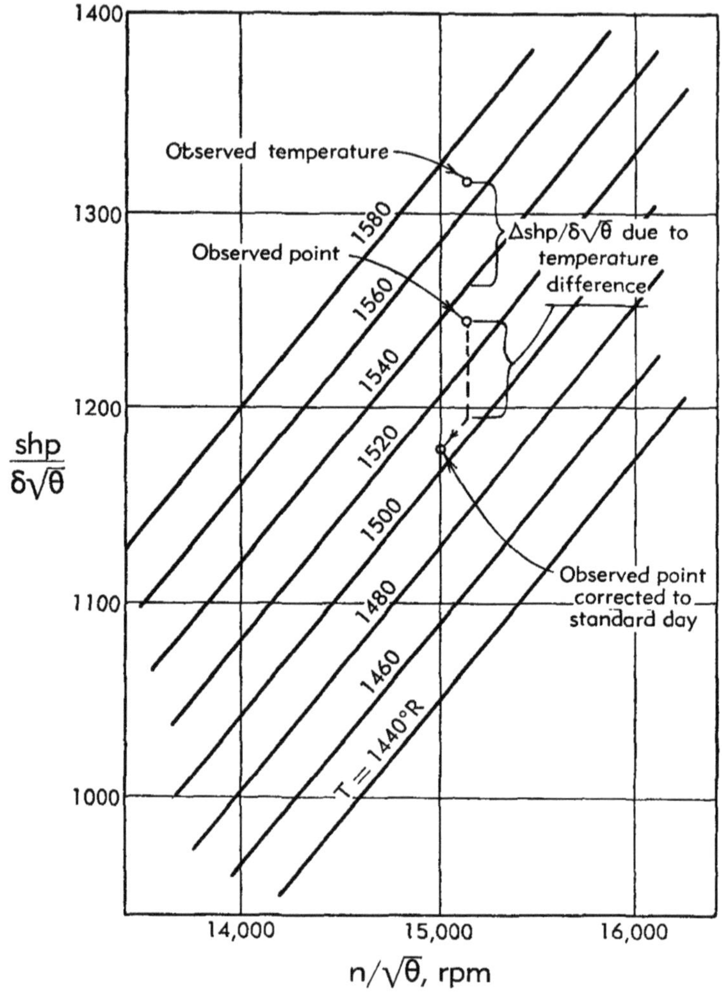

Fig. D,9b. Chart for short method of correcting tests to standard conditions.

the current run is consistent with other tests and predictions. This should be known before the test is completed and the engine removed from the test bed. The atmospheric conditions are usually not standard. Hence it is desirable to have a quick method of converting the test data to standard atmospheric conditions. Fig. D,9b illustrates a method for doing this. On this grid, contours giving the variation of generalized shaft horsepower versus rpm for constant exhaust jet pipe temperature are plotted from estimates of the performance. Now when a test point is obtained at a given tailpipe temperature and rpm, the corrected value at these conditions on a standard day can be obtained by computing the corrected shaft power and temperature for the tested conditions. Then the results may be converted to the standard conditions by linear interpolation or extrapolation along the constant jet pipe temperature contours and the constant rpm lines. For example, assume that the observed shp is 1200 at an observed rpm of 15,000 and an observed jet pipe temperature of 1540°R, intake temperature 510°R, and barometer 14.3 lb/in.[2] The corrected shaft horsepower is

$$1200\left(\frac{14.7}{14.3}\right)\sqrt{\frac{518.4}{510}} = 1245$$

The corrected jet pipe temperature is

$$1540\left(\frac{518.4}{510}\right) = 1565°\text{R}$$

Corrected rpm

$$15{,}000\sqrt{\frac{518.4}{510}} = 15{,}130$$

Now from the chart the change in power per degree temperature is 2 and the change in power per rpm is 0.125. Thus the power for a standard day with a rotation speed of 15,000 and a tailpipe temperature of 1540 would be approximately

$$\begin{aligned} &= 1245 - 2(1565 - 1540) - 0.125(15{,}130 - 15{,}000) \\ &= 1179 \end{aligned}$$

The results can also be obtained geometrically as illustrated in Fig. D,9b.

Flight test. All of the leading aircraft companies have found flight testing an essential part of the development of engines. In this country the engines have been installed in the nose of a B-17, and in England, the Lancaster has been employed (see Plate D,9d). An engine may be qualified for flight test purposes by 50 hours of bench testing. As indicated earlier, flight testing is necessary since the high speed flight conditions cannot always be simulated. It is particularly essential in evaluating the basic control system. It is also important in checking the lubrication system, the emergency controls, starts in flight, and the altitude performance.

D,10. Installation. One of the main arguments in favor of the gas turbine engine has been the fact that it admits a cleaner airplane with a small frontal area. Although the turboprop is not as satisfactory from this point of view as a turbojet, it is much better than a reciprocating engine. Nevertheless there are several problems which are characteristic of turboprop installations. First among these is the design of the inlet ducts. A feature of gas turbine engines is the large quantity of air that they inhale in flight, and in order to obtain a satisfactory thermal efficiency, this air must enter at high pressure. With a turbojet it is possible to design an inlet duct with a low pressure loss. However, with a turboprop the problem is much more difficult. Since the air intakes are behind the propeller, it is usually necessary for the inlet air to pass through the propeller near the hub. With a conventional spinner, the losses may be as great as 12 to 15 per cent of the total free stream pressure. This situation can be greatly improved by spinner platforms (see Plate D,9b). Blade juncture platforms which cover the blade hub juncture can be inserted about the propeller shank and will improve the situation.

Perhaps the best solution to the problem is the design of a ducted spinner. In this type of duct the outer engine cowls are brought forward over and in front of the propeller blades, forming an annulus through which the inlet air flows (see Plates D,10a and D,10b). The part surrounding the propeller blades, called a spinner, is allowed to rotate with the propeller, and inside it the round portions of the shanks of the propeller blades are faired in with streamlined rotating cuffs set at the correct incidence angle. Aerodynamically, it appears possible to design such a spinner which would actually yield an air pressure to the compressor inlet greater than the free stream pressure. However, this would give the correct flow at only one point and any departure from this point would tend to produce less favorable conditions. Actual tests indicate a value of about 95 per cent of the total pressure can be obtained. Sometimes an under cowl scoop-type duct is employed.

Although the ducted spinner appears to be the best from the pressure recovery standpoint, it has some disadvantages. It is much heavier than other designs and is more difficult to construct and maintain in service. It is desirable to provide space for a spinner afterbody, which will permit variations of spinner diameters (required by airplane designs) and still have a smooth flow of the air from the propeller spinner to the air inlet. In order to insure that there are no discontinuities in the inlet passage, the engine and the cowl must be rigidly connected.

On the reliability side, it is necessary to make the inlet ducts and all early parts of a rugged construction in order to prevent parts from breaking off or shaking loose which would permit particles to pass through the engine, notching or destroying the compressor or turbine blades. The small particles that produce notches may produce failure at a later date and hence may be almost as hazardous as a large particle which produces immediate failure. (Although it is not in an exactly coherent spot in this discussion, it might be well to say that operations which permit the propeller of an airplane to pick up particles that may enter a following plane should be avoided.)

Another problem associated with the air inlets is that of preventing ice formation. Ice may form and choke off the engine, or large particles of ice may break off and go through the engine causing damage to the compressor blades. Attention should be directed toward antiicing rather than deicing. This can be done by circulating compressor discharge air through the inlet guide vanes or by injecting hot gases into the air intake, whenever icing conditions are likely to occur.

Not only is the induction of the air a design problem, but the disposal of the hot exhaust gases is also troublesome for the airplane design engineer. In this respect, however, the turboprop is better than the turbojet since there is less gas, it is cooler, and the ducts need not be straight because of the low velocity of the turboprop. The ducts emerge from the fuselage or nacelle at an angle.

The exit requirements for the gases coupled with the long engine lengths require that in wing-mounting the engines be underslung or overhung over the main spars of the wing unless a pierced or banjo spar is used. When they are mounted in the fuselage, the cockpit must lie above the exhaust ducts, which give the plane a "broken back" appearance. In this respect the turboprops are worse than turbojets because the inlets ducts for the latter may be aft of the canopy.

It was first thought that the installation of the turboprops would be simpler than that for reciprocating engines. It seemed possible, because of the smooth running engine, to eliminate the flexible mounts. Also, the need for external cooling air is eliminated. Now there appears to be some question about elimination of the flexible mounts since many of the airplane designs have vibration troubles. With the large propeller, which is required to absorb the power, and the light engine it is quite likely that the vibrations produced will not be damped. In fact, it is quite possible for the engine to be mounted so as to obtain resonance between the propeller vibrations and the natural vibration of the wings. Conditions near such a resonance point are almost sure to produce vibration fatigue of some part. An installation which damps out these vibrations is therefore essential.

As a result of the tendency to make the smaller diameter nacelles which are permitted with the turboprops, the problem of spacing the accessories in the nacelles becomes acute. Of course, the aircraft designer could return to the large nacelles, but then he would lose one of the advantages of this type of engine. The power pads, the control system, the oil, and the oil pumps all give protrusions which destroy the symmetry and from the point of view of the aircraft are a nuisance.

Fire walls must be provided to separate the hot portions of the engine from the airplane structures and care must be taken to insure that the fuel and oil pipes are routed so that, in case of a failure, the fluids will not come in contact with the hot areas. Moreover, the fumes should be prevented from getting into a location where they can be ignited by a leak in the hot gas chambers of the engine. In nacelle installation, there is the additional problem of insuring that the propeller be located sufficiently far in front of the wings for uniform flow, and far enough away from the fuselage to avoid excessive pressure from propeller tips.

Typical power plant installation. A compact installation is illustrated by Lindsey [*6*] in his description of the Mamba in the Apollo Aircraft. This installation was worked out to occupy the minimum space and diameter. The power plant is a completely self-contained unit and fits into a cowling with a maximum diameter of 31 inches. The power plant is divided into three zones by stainless steel fume and fireproof bulkheads. The compressor zone (zone 1) contains all engine accessories, including fuel pumps, oil pumps, constant speed unit, feathering pump, oil tank, and cooler. It is ventilated and the air temperature inside this zone will not exceed 50°C. The combustion chamber zone (zone 2) includes the six combustion chambers, the turbine feed manifold, and the turbine itself. The chambers are covered with a light stainless steel cylindrical shield which matches up at the turbine end with the outer cowling, and the annular space between this shield and the outer cowling is ventilated to keep the temperature down to 100°C. All pipes and electrical services to zone 1 (see Plates D,10c and D,10d) pass through this annular space and are adequately protected. The electrical leads to the feathering pump are fed through a separately vented duct. The jet pipe compartment (zone 3) aft of the turbine houses the exhaust cone and tailpipe and, in addition, the ignition booster coils and the aircraft accessory boxes. This compartment merges into the wing structure, and the temperatures in the compartment are expected to be low.

Weight and space were saved by making the nose cowl immediately aft of the propeller blades an oil tank, in which welded construction was used to eliminate the possibility of compressor damage caused by broken rivets passing through the compressor. The entire oil system is in zone 1. Oil temperatures are automatically controlled by an electrically actuated flap governed by a thermostat, and transmitters are fitted to give visual flap indication to the pilot. The engine mounting is of welded tube construction for lightness, and, although of rigid design, some work is now being done on rubber mountings. Only two controls are provided, one for throttle control and the other for operation of the isolator value which is interconnected mechanically with the override lever for mechanical feathering. The controls are housed in light gauge steel tubes through zone 2. The drive from the engine accessories casing is taken from the engine through zones 1 and 2 to a bevel box above the exhaust cone jet pipe joint. The drive is then bifurcated and taken downward and outward at 45° to subsidiary boxes on which are fitted cabin blowers, generators, and hydraulic pumps.

A three-blade 10-foot diameter de Havilland hydromatic propeller with dual blades is fitted onto the shaft. Two thirds of the blade length is fitted with electric deicing shoes which are energized by means of a hub generator fitted in the rear of the propeller. The hubs of the propeller are submerged into a streamline spinner. The air passes over this spinner into the cowl inlet duct.

The propeller is arranged with the usual constant speed control and can be fully feathered. It is also suitable for use with power braking. The complete power plant weighs 1400 pounds of which 400 pounds is attributed to the propeller and its auxiliary equipment.

This installation of the Mamba in the Apollo is one of the better installations and it clearly illustrates some of the problems and a solution. Other installations have the ducted spinner which from an aerodynamic point of view is an improvement over this one.

D,11. Past, Present, and Future Roles of Turboprops in Aircraft Propulsion. Prior to or possibly concurrently with the first turbojets, the possible advantage of the turboprop was recognized. In spite of the additional complications and weights resulting from the extra turbine stages, the gear box and the propeller, the high possible propulsive efficiencies made the idea so attractive that projects were soon initiated for the construction of turboprops. The Armstrong Siddeley Motors, Limited, in England, used the ASX axial flow turbojet as the basis of the Python turboprop. This engine first ran in March, 1945. The Rolls-Royce, Limited, of England, made an adaptation of their Derwent turbojet to form an experimental turboprop, called the Trent. It became the first turboprop to propel an aircraft, in September, 1945. The Bristol Aeroplane Company, Limited, England, was the first to concentrate upon the development of a turboprop, per se, their first unit being the Theseus I. It had a combined compressor consisting of eight axial stages followed by one centrifugal stage. The compressor was driven by a two-stage turbine and the propeller was driven by a single-stage free turbine by means of a shaft which extended forward through the hollow compressor rotor shaft. This unit was unique in that it had a heat exchanger; that is, the hot exhaust gases and the air leaving the compressor were passed through adjacent tubes in a matrix so that the hot exhaust gases transferred some heat to the compressed air, thereby reducing the amount of fuel required in the burner and thus reducing the specific fuel consumption. This unit and the T-36, built by the Chrysler Corporation, are the only units which have been built with a heat exchanger, and neither is now in production.

At about the same time as the British developments, the General

Electric Company in America designed and produced the TG-100. This engine made its maiden flight on December 21, 1945, when it propelled the XP-81. Actually the first turboprop designed in the United States was the turbodyne of the Northrop Aircraft Company. The commitment for the production of this engine began in 1941, and the engine was wrecked in the test bed some four years later. Both of these engines were of the fixed turbine type.

Subsequent to these early developments in England, the Bristol Company produced a larger unit, the Proteus (Plate D,11a), and later developed a supercharged turboprop, the Orion. Rolls-Royce produced the Clyde, the Dart (Plate D,11b) and the Tyne; Armstrong Siddeley produced the Python II, the Mamba I, and the Mamba II; and the Napier Company produced the NAIAD and the Eland. In addition, many of these single units have been mechanically coupled to form engines with twice the power; namely, the Mamba, the Proteus, and the NAIAD. In America, the Allison Division of General Motors produced the T-38 and a mechanically coupled pair of these units called the T-40 (Plate D,11c), and then developed higher power versions of both, called the T-56 and T-54 respectively. The Pratt and Whitney Aircraft Division of United Aircraft Corporation has produced the T-34 (Plate D,11d) and later worked on two other units, the T-52 and the T-57. The Wright Aeronautical Company produced and tested the T-35, which no longer exists, and did development work on the T-47 and the T-49.

Russia has developed the largest turboprop of them all. The French have produced a medium-size axial flow turboprop and a series of small centrifugal flow turboprop units which developed from 225 to 350 shaft horsepower. These latter units utilize only 4.4 to 6.8 pounds of engine air per second. Also in the small turboprop group is an American engine produced by the Boeing Aircraft Company.

A fairly complete summary of the characteristics of present turboprops is given in [7]; however, no specific consumptions are given. Table D,11a gives a briefer summary of the engine facts, but it does include some specific fuel consumptions. Also, this table indicates the first date at which the performance was obtained. It includes two ducted fans, since in the interpretation of this section they are really turboprops with a fixed pitch shrouded propeller. The performance values in the table are for static sea level conditions, in terms of the static equivalent shaft horsepower eshp, which has been previously defined. The eshp is used merely for the convenience of combining the two quantities into a single one for comparison purposes. Since most of the power is in the shaft, there is less error in converting jet thrust to equivalent shaft power than there would be to convert the shaft power to thrust. In the table w_f denotes pounds of fuel per hour, W the weight of the engine without propeller but including gear box, A the maximum frontal area, and L is the length of the unit.

Table D,11a. Turboprops.

Name	Year	eshp	$\frac{w_t}{\text{eshp}}$	$\frac{W}{\text{eshp}}$	$\frac{\text{eshp}}{A}$	$\frac{\text{eshp}}{L}$	Cruise sfc
British							
Theseus 1	1945	2075	0.940	0.980	159.0	235.0	0.81
Theseus 21	1947	2142		0.880	163.0	242.0	0.80
Theseus 11	1948	2425		0.760	185.0	352.0	0.80
Theseus 502	1950	2540		0.870	160.0	368.0	0.77
Proteus I	1947	3885					
Proteus II	1949	3510		0.830	433.0	422.0	0.51
Coupled Proteus II	1949	7040		1.020	347.0	505.0	0.51
Proteus 705	1952	3780		0.70	467.0		0.62
Proteus 755	1953	4150		0.68	488.0		0.48
Orion I	1955	5150	0.64	0.60	541.0	551.0	0.395
Cotswold	1946	1425		0.53	279.0	196.0	0.58
Trent I	1945	1062		1.175			
Clyde	1946	3460	0.618	0.810			
Dart I	1948	1125	0.830	0.890	139.0	143.0	
Dart 504	1951	1515	0.83	0.610	187.0	139.0	
Dart 505	1951	1540	0.75	0.67	195.0	189.0	
Dart 7	1954	2105	0.67	0.68	270.0	258.0	0.58
Dart 8	1955	2500	0.64				
Tyne 1	1955	4695		0.43	512.0	563.0	0.41
Mamba I	1944	1136	0.765	0.66	284.0	170.0	0.72
Mamba II	1949	1435		0.61	312.0	224.0	0.72
Double Mamba II	1949	2855	0.78	0.70	266.0	338.0	0.72
Python I	1945	4130	0.765	0.72	328.0	364.0	0.58
Python II	1949	4560		0.65			
NAIAD I	1948	1590		0.69	372.0	187.0	
Coupled NAIAD	1950	3150		0.71	315.0	367.0	
Eland 1-3	1953	3000	0.64	0.78	422.0	300.0	
Eland 4	1955	4000	0.56	0.45	579.0	400.0	0.46
Eland 5	1955	4200	0.56	0.43	593.0	417.0	
Gazelle I	1955	1260		0.56	225.0	216.0	0.70
French							
TB-1000	1950	1520		0.70	370.0	170.0	0.68
TB-1000A	1952	2000	0.81	0.50	488.0		
Artouste II	1952	396	1.07	0.73	137.0	71.0	
Artouste III		594		0.54	247.0	123.0	0.65
Bastan		741		0.53	309.0	146.0	0.67
Marcadau	1952	400	1.07	0.87	146.0	81.0	
Turmo II	1953	396	1.07	0.81	107.0	97.0	
Turmo III	1954	786		0.67	150.0	143.0	0.72
American							
T31 TG-100	1945	2440		0.820	325.0	259.0	
T38-A	1948	2720	0.580	0.455	682.0	244.0	0.56
T40-A	1948	5580	0.580	0.465	631.0	386.0	0.56
T34-P	1950	5206	0.610	0.456	635.0	454.0	0.62
T34-P-2	1951	5700	0.615	0.495	576.0	392.0	0.55

Table D,11a (continued)

Name	Year	eshp	$\frac{w_t}{\text{eshp}}$	$\frac{W}{\text{eshp}}$	$\frac{\text{eshp}}{A}$	$\frac{\text{eshp}}{L}$	Cruise sfc
T37-NA-4 Turbodyne	1947	10,000	0.615	0.700	1000.0	782.0	
T35-W-1	1947	5500	0.591	0.625	501.0	681.0	
T47-W	1951	8500					
T49-W	1952	12,000		0.526			
T50-BO	1948	220		0.640	100.0	630.0	
T51-T-1	1956	280		0.91	200.0	91.0	0.99
T51-T-3	1956	425		0.55	192.0	142.0	0.97
T53-L-1	1953	825		0.56	282.0	206.0	0.71
T54-A	1953	7500					
T55-L	1955	1800		0.417			
T56-A	1953	3750	0.54	0.47	625.0	310.0	0.46
T57-P	1955	15,000					
T58-GE	1955	1024	0.68	0.32	732.0	208.0	0.64
T59-W							
T-60-BO	1956	400					
T-61-A	1956						
T-62-S	1956						
Russian							
M 022	1950	5000		0.84	538.0	272.0	0.66
M 028	1952	8190		0.807	650.0	350.0	0.70
Type K	1953	11,800		0.43	1060.0	600.0	0.57
British Ducted Fans							
F-3		1720		1.32*	117.0	147.0	
F-5		1938		1.18*	112.0	159.0	
Bypass engines							
Conway	1951	5200	1.75	0.63*	542.0	482.0	
Soulor	1957	282					

* Includes propeller weight.

Many of the engines were experimental, others failed to find a purchaser and hence ceased to exist. At the present time, in the United States, five turboprops have remained in production to some extent and at least three others may be produced. In England, there are still four, not counting the coupled units, in production and at least one other is contemplated.

There have been twenty aircraft fabricated which were designed around turboprop engines. (This does not include flying test beds which are modifications of existing aircraft.) Of these twenty, six went into production. These planes include large sea planes, large commercial land planes, and Naval fighters and attack aircraft. They are illustrated in Plates D,11e, D,11f, D,11g, D,11h, D,11i, D,11j, D,11k, and D,11l. Other turboprop transport aircraft are being built.

On July 29th, 1950, the British European Airways inaugurated the world's first scheduled commercial air service of a turboprop aeroplane between London and Paris. This was with a single aircraft, the original British Viscount prototype powered by four Darts. In 26 days, the plane flew 137 hours and 57 minutes and carried 1815 passengers. This exploratory operation provided evidence of the potential attractiveness of the turboprop aircraft both from the point of view of the passengers and crew. At the same time, it helped to clarify many of the operational problems.

On April 19, 1953, the British European Airways established the first regular commercial airline with a turboprop. The Viscounts are now flown on regular service routes from London to Cyprus, Istanbul, Geneva, Zurich, and Stockholm. According to Masfield, in the *British European Airways* Magazine, "During the first 1000 hours of scheduled commercial operation, the BEA Viscounts have operated for an average sector distance of 625 statute miles, achieved a block speed (doors shut to doors open) of 221 miles per hour, including all aerodrome taxiing and delays. In the first 1000 hours of operation, Viscounts have flown 221,000 revenue miles, carried 7700 passengers and 77 tons of mail and freight and have earned 240,000 pounds (sterling) of revenue. They have achieved a regularity of operation of 99.3 per cent.

"On the cost side, the break-even load factor on the 625 mile average stage has worked out at 62 per cent on total costs, including interest on capital and the full weight of airline overheads. The achieved load factor has been 76 per cent.

"Thus, during their first 1000 hours of operation, BEA's turboprop Viscounts have made an over-all profit for the Corporation of 45,000 pounds (sterling) in carrying an average load on each flight of 32 passengers and 800 pounds of mail and freight. Not bad as a start.

"In this 1000 hours of flying the Rolls-Royce Dart propeller-turbine engines have made some 3200 million revolutions (3.2 billion in American parlance) and have been remarkably free of troubles."

On June 26, 1955, the Capital Airlines established the first regular commercial turboprop air line in the United States. They employed Viscount planes equipped with improved Dart engines. Within a year they were operating twenty of these planes on eight different routes. Another version of the same plane was also put into commercial use in Canada.

The American Airlines have ordered 35 Lockheed Electra turboprop airplanes powered by Allison T-56 engines.

Previous to the experience of the Darts, employed in the Viscounts described above, all aircraft engines employed in civilian airlines had been developed to a reliable state by military operation. There are many people in the United States who say that the commercial air business cannot afford to pay for the operational time required to perfect a new engine. This may be true in general but the above experience gives some contradictory evidence. It is true, however, that this engine was designed and tested with a view to military application, and the series of proofing flights designed to test the aircraft and its auxiliary equipment in simulated scheduled operations were carried out under an agreement with the Ministry of Supply. Nevertheless, several of the British engine designers are very optimistic about the place of turboprops in the commercial cargo field and are designing engines solely for that purpose. In addition to the routes by the British European Airlines, the British Overseas Airways Corporation plans to establish a scheduled line to South America and Australia. They have contracted for the Bristol Britannia airplanes which are powered by the Proteus.

In order to better understand the optimistic feeling it is well to look at the airplane requirements. Of the many characteristics of engines, the most important ones from the standpoint of aviation are the thrust or power of the unit, the specific fuel consumption, the specific weight, and the thrust per unit frontal area and per unit size. A large thrust is important for take-off and high velocity. A low specific fuel consumption is required for long range endurance. A light engine permits smaller airplanes, or greater range, or more payload, or a combination of all three factors. A small frontal area admits the possibility of a cleaner airplane with less drag, thus requiring less thrust for a given performance. In some designs, space considerations are important, hence the smaller the better.

The turboprop produces more power per unit nacelle frontal area and volume than any other engine (i.e. at low speed), and the units may be built in any size above a certain minimum. On this point, turbine engines have an advantage over a reciprocating engine.

Table D,11b. The subscript $_c$ refers to the values for cruise conditions. W_p denotes the weight of the propeller, W denotes the weight of the engine plus gear box, w_f denotes pounds of fuel per hour, and eshp denotes equivalent shaft horsepower, as previously defined.

	Static sea level		400 knots at 35,000 ft	
	$\frac{w_f}{\text{eshp}}$	$\frac{W + W_p}{\text{eshp}}$	$\frac{w_{fc}}{(\text{eshp})_c}$	$\frac{W + W_p}{\text{eshp}}$
Reciprocating engine	0.73	1.360	0.42	3.30
Turboprop (current)	0.56	0.600	0.44	1.63
Turbojet, axial (current)	2.55	1.050	0.83	0.80
Turbojet, centrifugal (current)	2.75	0.825	0.90	0.65

Table D,11b throws some light upon two other fundamental properties, the specific fuel consumption and the specific weight. At take-off power (static sea level) the turboprop has a definite advantage over both the turbojet and the reciprocating engine, and at a high cruise speed of 400 knots at 35,000 feet it is almost as good as a reciprocating engine on specific fuel consumption and is far superior on specific weight. On the other hand, it is superior to the turbojet in fuel consumption, but is much heavier. These comparisons make the future of turboprops look bright. Moreover, it is always expected that there will be improvements

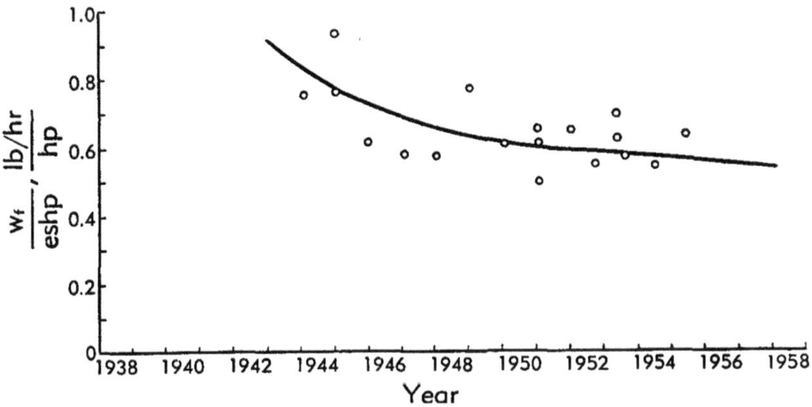

Fig. D,11a. Specific fuel consumption of turboprops. Static sea level military power.

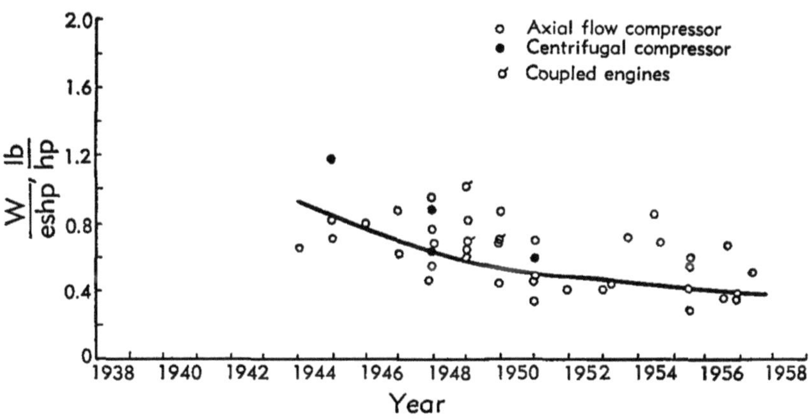

Fig. D,11b. Specific weight of turboprops.

and refinements over a period of years which will make them even more promising. Fig. D,11a to D,11d show the points corresponding to the time the various engines passed their experimental runs. These graphs indicate what improvements have been made and also indicate what improvements might be expected. The curves are all for static sea level conditions. Corresponding curves for flight speed indicate that in the very near future it should be possible to obtain specific fuel consumptions as low as 0.36. The points and curves in Fig. D,11b indicate that a specific weight of less than 0.4 has been obtained. This figure, which is much less than the number 0.6 given in Table D,11b above, does not include the propeller weight. This emphasizes the tremendous weight of the propeller. Actually, in recent design, the propeller and gear box account for about $\frac{5}{12}$ or more of the entire weight of the propulsive system, so that it may be emphasized again that, in looking forward to improvements,

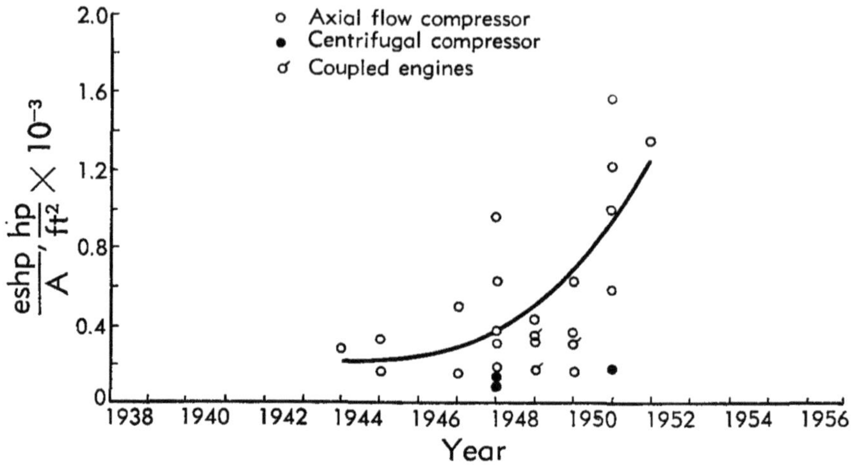

Fig. D,11c. Ratio of equivalent horsepower to frontal area for turboprops.

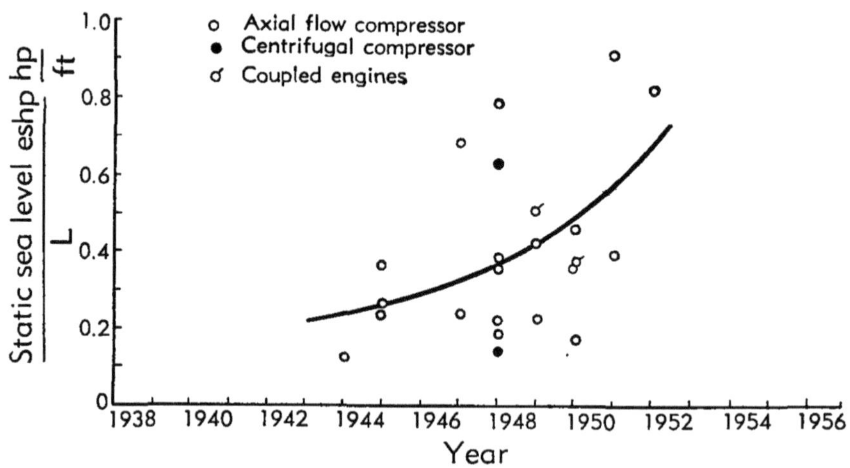

Fig. D,11d. Ratio of equivalent shaft horsepower to length for turboprops.

the most fertile area will be in methods of reducing the weights of the propeller and gear box.

With only the performance obtained in current developments, the turboprop appears attractive for large cargo and transport planes required to fly long distances at moderate speed. The greatest question in the minds of the operators is whether the engines are reliable. Are the advantages pronounced enough to warrant the risk? The greatest current need for turboprop development is more flight time in service aircraft. Only under these conditions can the development difficulties be isolated and remedied. There definitely is a place for the turboprop airplane of the military cargo and transport type. Also, there may be a place for the best of these engines in combat planes. But before these engines can seek their place, they will have to have enough service operation to demonstrate their reliability. Because of the high speeds required and relative short time at cruise, the turboprop will probably not be able to compete with the turbojet engine for combat aircraft. This is true unless the propeller and gear box weights are greatly reduced, say to about two thirds of their present weight, or some different division of power between the propeller and compressor is employed. Only two divisions of power have been employed in design studies of military aircraft, viz. the ducted fan and the turboprop with about 90 per cent of the useful energy going into the propeller. Perhaps for high speed applications an intermediate value, say 50 per cent division of energy, would be more desirable than either.

In conclusion, it may be said that the turboprop shows great promise for commercial and military transport operations, and appears quite promising for helicopter application. However, the development will likely be slow and over a long period of time.

D,12. Cited References and Bibliography.

Cited References

1. Driggs, I. H. *J. Aeronaut. Sci. 5*, 337 (1938).
2. Driggs, I. H. *SAE Quart. Trans. 2*, 596 (1948).
3. Driggs, I. H., and Lancaster, O. E. *Gas Turbines for Aircraft.* Ronald Press, 1955.
4. Lancaster, O. E. Combustion gas turbines for airplanes. *NavAer ADR Rept. R-39*, Oct. 1944.
5. The problem of control of propeller turbine engines. Issued by *Flight Section Armstrong Siddeley Motors, Limited*, June 1950.
6. Lindsey, W. H. *J. Roy. Aeronaut. Soc. 53*, 137 (1949).
7. . *Interavia 8*, 72 (1953).

Bibliography

Ainley, D. G. *Proc. I of Mech. Eng. 159*, 230 (1948).
Brady, G. W. Propellers for high powers and transonic speeds. *Anglo-American Aeronautical Conference*, Sept. 4–7, 1951.
Driggs, I. H., and Lancaster, O. E. *Trans. Am. Soc. Mech. Engrs. 75*, 217 (1953).
Gorton, R. E., and Miller, B. E. *SAE J. 60*, 60 (1952).
Killip, E. L. *Interavia 8*, 81 (1953).
Treseder, R. C., Brooks, M., and Kessler, J. R. *Aero. Eng. Rev. 11*, 1 (1952).

SECTION E

THE RAMJET ENGINE

CHAPTER 1. INTRODUCTION

DEMARQUIS DALE WYATT

E,1. Description of Engine. The ramjet engine, in common with all jet engines, produces propulsive power by increasing the momentum of the working fluid through some form of heat release so that the momentum of the exiting jet exceeds the momentum of the entering air stream. In contrast to other air-breathing jet engines the working cycle is accomplished without mechanical compression of the working fluid and without intermittent semiclosed combustion cycles; consequently the ramjet is mechanically the least complicated air-breathing engine yet devised.

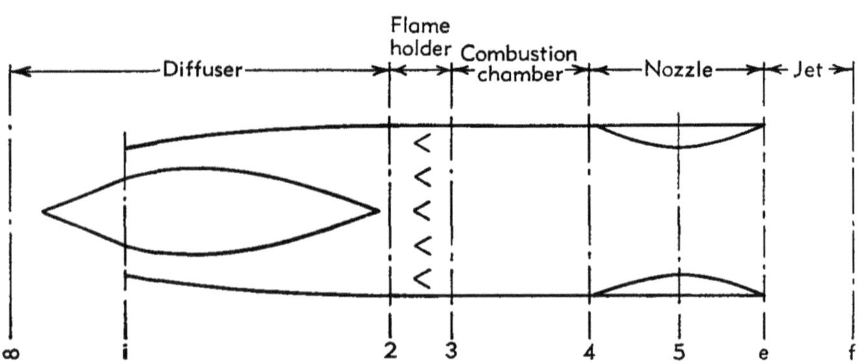

Fig. E,1. Schematic of ramjet engine.

The basic elements of the engine can be represented schematically as in Fig. E,1. The charge air is inducted to the combustion or heat release chamber through a diffuser, so-called because the entering air stream is generally decelerated from the forward speed of the engine to a lower combustion entrance velocity. In engines employing liquid or gaseous fuels for the energy addition to the charge air, the fuel is injected through orifices or nozzles at some point in the diffuser process. The location of the point of fuel injection depends upon the desired degree of atomization and vaporization of the fuel and the desired mixture pattern of the fuel and charge air at the entrance to the combustion chamber.

At the end of the diffuser passage is located a flame holder, a turbulence-generating device offering shelter to the subsequent combustion flame. The fuel-air mixture entering the combustion chamber is ignited in the sheltered regions of the flame holder by some independent ignition source. In most engines this ignitor is only required to function at the onset of combustion, the combustion process being self-propagating for continuous operation.

The combustion of the fuel-air mixture is ideally completed in the combustion chamber, normally a constant area passage. (In ramjets utilizing an external heat source such as a nuclear reactor, the fuel-injection, flame-holding combustion process is replaced by a heat exchanger from which the uncontaminated charge air is discharged at an elevated temperature.)

The products of combustion leaving the combustion chamber are, in general, at a higher pressure than the free stream air surrounding the engine. The exhaust nozzle is used to expand the flow and thereby to convert the pressure into velocity. The flow velocities after combustion are subsonic; hence an initial contraction of the flow area is necessary. This contraction must be sufficient to either reduce the internal pressure to the ambient value or to accelerate the internal flow to sonic velocity. At low design flight speeds, ambient pressure is achieved with unchoked flow and the nozzle terminates with the convergence. At high flight speeds the internal flow chokes at pressures above ambient and for maximum thrust realization a diverging passage is added downstream of the throat in which the flow continues to expand to supersonic velocity until ambient pressure is attained.

E,2. Engine Cycle. The ideal ramjet engine operates on a Brayton cycle which is characterized by constant pressure combustion and complete expansion of the products of combustion to the initial or ambient pressure. An enthalpy-entropy diagram illustrating the engine cycle is shown in Fig. E,2.

In the ideal engine, the air at ambient pressure enters the engine at station i (Fig. E,1) and is isentropically compressed through an enthalpy rise which is proportional to the square of the flight velocity. Fuel is added after diffusion and the heat of combustion increases the enthalpy at constant pressure to the end of the combustion chamber, station 4. The products of combustion are isentropically expanded to ambient pressure resulting in a discharge velocity which is proportional to the square root of the enthalpy change in the nozzle. The cycle is closed through an external heat rejection at constant pressure.

In the actual engine cycle as contrasted with the ideal cycle, pressure losses are encountered at practically every step in the flow processes. As shown in Fig. E,2, the diffusion process from station ∞ to 2 is considered to be adiabatic and hence has the same enthalpy rise as the ideal engine; however, inevitable diffusion pressure losses result in a lower pressure after diffusion than with the ideal engine.

The flame holder may be considered to operate in a region of constant enthalpy where the pressure in the cycle is further reduced as a result of flame holder drag (station 2 to 3). The heat addition in the combustion chamber is also accompanied by an inevitable pressure drag in the actual engine since the combustion process is not carried out at zero velocity.

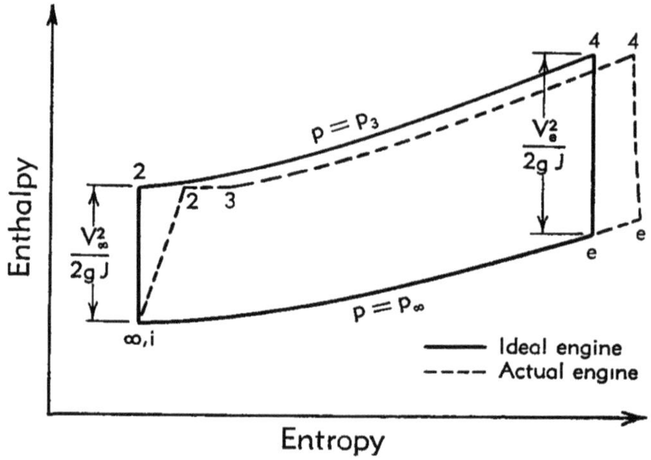

Fig. E,2. Ramjet cycle diagram.

Thus, through a multiplication of pressure losses the final temperature (enthalpy) after combustion (station 4) is reached with a lower pressure in the actual engine than with the ideal engine. As a result the enthalpy change available to create discharge velocity as a result of flow expansion to ambient pressure is reduced, as compared to the ideal engine.

In addition to the pressure losses incurred ahead of the nozzle, additional losses generally arise in the nozzle of the actual engine which further reduce the potential discharge velocity by increasing the entropy and hence increasing the enthalpy at ambient pressure.

E,3. Probable Applications of Ramjet. The fundamental limitation of the ramjet engine as a propulsive device is obvious from a study of Fig. E,2. The engine will not yield an internal thrust force until the flight speed is high enough to produce a diffuser pressure rise that is sufficiently great to exceed the pressure losses in the flame holder, combustion chamber, and exhaust nozzle. An effective propulsive thrust force will not be obtained until the flight speed is enough greater than this minimum value for the internal thrust to compensate the drag directly chargeable to the engine installation. A yet higher speed must be obtained before the propulsive thrust is great enough to overcome the drag of the vehicle in which the engine is installed and thus to qualify the engine as a useful propulsive device.

The minimum flight speed at which the ramjet can qualify as a propulsion system for aircraft depends upon the degree to which the internal pressure losses and the installation drag are minimized and upon the drag of the propelled airframe. In general it has been found that the minimum speed application of the ramjet is not less than about 400 mph. (This speed is the relative speed of the engine and not necessarily of the airframe. Thus, ramjet engines, tip-mounted on the rotors of helicopters, have proved to be adequate propulsive devices at zero forward airplane speed.) The efficiency of the engine is very low at these minimum speeds, however, and applications of interest at subsonic speeds are confined to installations in which the attractions of light weight, mechanical simplicity, and low cost outweigh considerations of fuel consumption.

As the engine flight speed is increased the thrust increases as a function of the square of the flight speed (see Art. 4) in the same order as the airframe drag. Thus a ramjet engine, properly designed, is capable of producing the very high thrusts necessary for supersonic propulsion. The engine efficiency, and hence the specific fuel consumption, ideally improve as flight speed increases so that the principal range of application of the ramjet as a competitive power plant appears to lie in the propulsion of supersonic vehicles.

The inability of the engine to produce a useful propulsive force at zero or low flight speeds necessitates the use of an auxiliary power plant to initially accelerate the vehicle to the required take-over speed of the ramjet. Such auxiliary power is also required for controlled landing of the vehicle. As a consequence the ramjet engine is not well-suited for conventional aircraft applications (with the exception of the helicopter) and the principal application of the engine appears to be further restricted to missiles or other similar vehicles of a one-flight expendable nature.

E,4. Important Performance Parameters.

Thrust. Ramjet engine thrust is defined as the net change in total momentum as the net working-fluid passes through the engine. In accordance with conventional aerodynamic definitions of drag, the pressure terms in the total momentum expressions are taken relative to the ambient pressure. Referring to Fig. E,1,

$$F = m_e V_e + A_e(p_e - p_\infty) - m_\infty V_\infty \tag{4-1}$$

For the special case where the nozzle discharge pressure is ambient, the thrust equation reduces to

$$F = m_e V_e - m_\infty V_\infty \tag{4-2}$$

It should be noted that this conventional thrust definition does not yield a thrust force that is identical to the axial component of the pressure integral over the internal engine surfaces, except when the total momentum at the diffuser intake (station i) is the same as the total momentum in the free stream (station ∞).

The general thrust equation (Eq. 4-1) can be simplified by regrouping and factoring terms, and by employing the mass flow equation $m = \rho VA$ to give

$$F = \rho_\infty V_\infty^2 A_\infty \left(\mu \frac{V_e}{V_\infty} - 1\right) + A_e p_\infty \left(\frac{p_e}{p_\infty} - 1\right) \tag{4-3}$$

where $$\mu = \frac{m_e}{m_\infty} = 1 + f$$

The specific size of the engine, as implied by A_∞, and the ambient pressure p_∞ can be generalized by converting the thrust to a thrust coefficient. In accordance with the conventional definition of drag coefficient, the thrust coefficient is defined as

$$C_F = \frac{F}{\frac{1}{2}\rho_\infty V_\infty^2 A} \tag{4-4}$$

where the choice of reference area A is arbitrary. It is frequently useful to base the engine thrust coefficient on the free stream area of the air flow tube entering the engine A_∞, so that

$$C_{F_\infty} = 2\left(\mu \frac{V_e}{V_\infty} - 1\right) + 2\frac{A_e}{A_\infty}\frac{p_\infty}{\rho_\infty V_\infty^2}\left(\frac{p_e}{p_\infty} - 1\right) \tag{4-5}$$

From the perfect gas relation $p/\rho = \mathcal{R}T$ and the defining relation for Mach number, $M^2 = V^2/\gamma\mathcal{R}T$, Eq. 4-5 can be simplified to

$$C_{F_\infty} = 2\left(\mu \frac{V_e}{V_\infty} - 1\right) + \frac{2}{\gamma_\infty M_\infty^2}\frac{A_e}{A_\infty}\left(\frac{p_e}{p_\infty} - 1\right) \tag{4-6}$$

The thrust coefficient from Eq. 4-6 can be readily converted into a coefficient based on any chosen area A_x by the simple relation

$$C_{F_x} = C_{F_\infty}\frac{A_\infty}{A_x} \tag{4-7}$$

Specific fuel consumption. The specific fuel consumption sfc is the fuel rate in pounds per hour per pound of engine thrust, or

$$\text{sfc} = \frac{w_f}{F} \tag{4-8}$$

where w_f is the fuel rate in pounds per hour. From the relation

$$w_f = 3600fw_a = 3600fg\rho_\infty V_\infty A_\infty \tag{4-9}$$

where f is the fuel-air ratio, and Eq. 4-4, the specific fuel consumption can be related to the thrust coefficient C_{F_∞} as

$$\text{sfc} = \frac{7200gf}{C_{F_\infty} V_\infty} \tag{4-10}$$

Specific impulse. The specific impulse is the thrust per unit weight of fuel consumed per second. From Eq. 4-8 to 4-10, this becomes

$$I_{sp} = \frac{3600F}{w_f} = \frac{C_{F_\infty} V_\infty}{2gf} \tag{4-11}$$

Engine efficiency. The engine efficiency η_0 represents the fraction of the heat released in the combustion process that is converted into work, and is a useful parameter for comparing the range possibilities of various engine designs and operating conditions.

The work output of the engine is FV, ft-lb/sec. The heat input into the combustor per second is $w_f H_f/3600$ where H_f is the lower heating value of the fuel in BTU/lb. The efficiency is, therefore,

$$\eta_0 = \frac{3600FV_\infty}{w_f H_f J} \tag{4-12}$$

From the previously developed relations the engine efficiency in terms of thrust coefficient becomes

$$\eta_0 = \frac{C_{F_\infty} V_\infty^2}{2gJfH_f} \tag{4-13}$$

CHAPTER 2. CALCULATION OF STATE CONDITIONS

DEMARQUIS DALE WYATT

E,5. Introduction. In the design of a ramjet engine and for some phases of performance calculations it is necessary to know the pressures, temperatures, velocities, and flow areas at each point in the engine. Various analytical methods of varying degrees of accuracy are in common usage for these calculations. The more rigorous methods are generally the most complicated in application and since in many instances a high degree of computational accuracy is unwarranted because of the low order of precise knowledge of the details of some of the flow processes, the less rigorous but more easily applied analytical methods have a definite utility.

It is almost universal practice to make the basic assumption that the flow at any given station in the engine is one-dimensional, or invariant across the passage. Anticipated or experimentally measured deviations of the real flow from this one-dimensional model are generally accounted for by the introduction of empirical correction factors. This approach to the problem is warranted considering the present level of understanding of the complex flow patterns and interactions in the actual flow processes.

E,6. Bases for Different Analytical Methods. The analytical methods to be discussed are primarily differentiated by the exactness with which they evaluate the total energy content of the fluid. In all the methods the perfect gas law, $p = \rho \mathcal{R} T$, is assumed to be the equation of state for both the charge air and the products of combustion.

The energy equation may be expressed as

$$h_a + \frac{V_a^2}{2gJ} = h_b + \frac{V_b^2}{2gJ} - Q \tag{6-1}$$

which for adiabatic processes ($Q = 0$) reduces to

$$h + \frac{V^2}{2gJ} = \text{const} \tag{6-2}$$

For a perfect gas, Eq. 6-2 may be written as

$$c_p T + \frac{V^2}{2gJ} = c_p^0 T^0 \tag{6-3}$$

or

$$\frac{V^2}{2gJ} = \int_T^{T^0} c_p dT \tag{6-4}$$

Enthalpy method. In the most exact method of calculating states in the ramjet cycle, the energy equation (Eq. 6-3) is solved using values of specific heat of the form

$$c_p = K_0 + K_1 T + K_2 T^2 + K_3 T^3$$

and the enthalpy $h = \int c_p dT$ is evaluated above some arbitrary base temperature. In application, this method requires the use of tables or charts such as those of [*1,2,3,4,5*] in which enthalpy is presented as a function of temperature.

Average specific heat method. As a first approximation, the specific heat in the right-hand member of Eq. 6-4 can be considered to have a constant average value $\bar{c}_p$, such that

$$\frac{V^2}{2gJ} = \bar{c}_p \int_T^{T^0} dT = \bar{c}_p (T^0 - T) \tag{6-5}$$

A detailed computational technique based upon this method is presented in [6].

Arbitrary specific heat method. Analysis and computations are considerably reduced if no effort is made to determine an average specific heat based on the actual temperatures of the cycle, but rather to select arbitrary specific heats which approximately fall in the proper range. In the application of this widely used method it is customary to assume a value of c_p such that γ, $\{c_p/[c_p - (\mathfrak{R})/gJ)]\}$ equals 1.4 for all processes prior to combustion and has some lesser constant value (generally 1.3) after combustion.

Constant specific heat method. This is the easiest, though least accurate, of the analytical methods. In the method it is assumed that the gas has specific heats throughout the engine such that $\gamma = 1.400$ regardless of the temperature or constituent properties of the fluid.

E,7. Diffuser Calculations.

Enthalpy method. The total energy in the free stream, per unit mass, is

$$h^0_\infty = h_\infty + \frac{V^2_\infty}{2gJ} \tag{7-1}$$

where

$$V_\infty = M_\infty a_\infty = M_\infty \sqrt{\gamma_\infty \mathfrak{R}_\infty T_\infty} \tag{7-2}$$

The variation of γ with temperature for air is shown in Fig. E,7a.

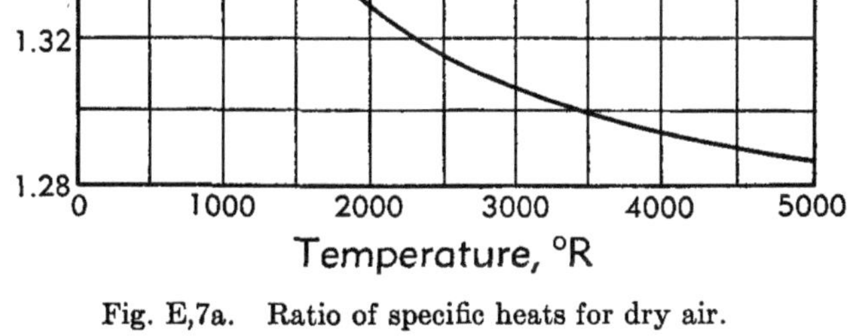

Fig. E,7a. Ratio of specific heats for dry air.

The diffusion process is adiabatic (VII,E) so that $h^0_2 = h^0_\infty$. In air tables, such as those of [1], enthalpy is presented as a function of temperature so that $T^0_\infty = T^0_2$ can be determined; these tables also record values of relative pressure p_r as a function of temperature, and show that along an isentropic the ratio of the relative pressures corresponds to the ratio of the absolute pressures. Therefore

$$p_\infty^0 = p_\infty \frac{p_{r_\infty}^0}{p_{r_\infty}} \tag{7-3}$$

Knowing the diffuser pressure recovery p_2^0/p_∞^0, the total pressure at the end of diffusion is

$$p_2^0 = p_\infty^0 \frac{p_2^0}{p_\infty^0} \tag{7-4}$$

When the diffuser discharge velocity is specified, the static enthalpy after diffusion is

$$h_2 = h_2^0 - \frac{V_2^2}{2gJ} \tag{7-5}$$

from which the static temperature T_2 can be determined.

When only the diffuser discharge Mach number M_2 is specified or known, neither h_2 nor V_2 can be directly calculated since both are a function of the unknown temperature T_2. In this event an iteration process is required to solve the simultaneous relations

$$V_2 = \sqrt{2gJ(h_2^0 - h_2)} \quad \text{and} \quad V_2 = M_2 \sqrt{\gamma_2 \mathcal{R}_2 T_2} \tag{7-6}$$

Once T_2 is determined, the static pressure after diffusion is given by the relation

$$p_2 = p_2^0 \frac{p_{r_2}}{p_{r_2}^0} \tag{7-7}$$

The flow area at the diffuser discharge is determined from the conservation relation

$$A_2 = A_\infty \frac{p_\infty V_\infty T_2}{p_2 V_2 T_\infty} \tag{7-8}$$

Average specific heat method. The approximate form of the energy equation (Eq. 6-5) can be written as

$$T_\infty^0 - T_\infty = \frac{V_\infty^2}{2gJ\bar{c}_{p_\infty}} \tag{7-9}$$

Since $$\bar{c}_{p_\infty} = \frac{\bar{\gamma}_\infty}{\bar{\gamma}_\infty - 1} \mathcal{R}_\infty$$

and $$V_\infty^2 = M_\infty^2 \gamma_\infty \mathcal{R}_\infty T_\infty$$

$$\frac{T_\infty^0}{T_\infty} = 1 + \frac{\bar{\gamma}_\infty - 1}{2} M_\infty^2 \frac{\gamma_\infty}{\bar{\gamma}_\infty} \tag{7-10}$$

In order to put Eq. 7-10 into the form of the conventional isentropic equations based on Mach number, an effective specific heat ratio $\gamma_{\text{eff}_\infty}$

can be defined as

$$\gamma_{\text{eff}_\infty} = 1 + \frac{\gamma_\infty}{\bar{\gamma}_\infty}(\bar{\gamma}_\infty - 1) \tag{7-11}$$

so that

$$\frac{T_\infty^0}{T_\infty} = 1 + \frac{\bar{\gamma}_{\text{eff}_\infty} - 1}{2} M_\infty^2 \tag{7-12}$$

In addition to the known static temperature T_∞, from which γ_∞ is determined (Fig. E,7a), a knowledge of the stagnation temperature T_∞^0 is required to establish $\gamma_{\text{eff}_\infty}$ since $\bar{\gamma}_\infty = (\gamma_\infty + \gamma_\infty^0)/2$. An iteration process is therefore necessary to establish compatible values of γ_∞^0 and T_∞^0. The total temperature after diffusion will, of course, be equal to the free stream total temperature from the assumption of adiabaticity across the diffuser.

The free stream total pressure can be calculated from the ambient static pressure, and the free stream static and total temperature by the isentropic one-dimensional flow equation

$$p_\infty^0 = p_\infty \left(\frac{T_\infty^0}{T_\infty}\right)^{\frac{\gamma_{\text{eff}_\infty}}{\gamma_{\text{eff}_\infty} - 1}} \tag{7-13}$$

so that, knowing the diffuser pressure recovery p_2^0/p_∞^0, the total pressure after diffusion becomes

$$p_2^0 = p_\infty^0 \frac{p_2^0}{p_\infty^0} \tag{7-14}$$

For all practical diffuser discharge Mach numbers for ramjet engine operation, the static temperature after diffusion T_2 will be sufficiently close to the total temperature that $\gamma_2 \cong \gamma_2^0$. Hence the remaining flow properties at diffuser discharge can be calculated as

$$T_2 = T_2^0 \left(1 + \frac{\gamma_2^0 - 1}{2} M_2^2\right)^{-1} \tag{7-15}$$

$$p_2 = p_2^0 \left(1 + \frac{\gamma_2^0 - 1}{2} M_2^2\right)^{-\frac{\gamma_2^0}{\gamma_2^0 - 1}} \tag{7-16}$$

$$V_2 = M_2 \sqrt{\gamma_2^0 \mathcal{R}_2 T_2} \tag{7-17}$$

If a diffuser discharge velocity is specified instead of a Mach number, the discharge Mach number M_2 can be determined by combining Eq. 7-15 and 7-17 to give

$$M_2^2 = \frac{\dfrac{V_2^2}{\gamma_2^0 \mathcal{R}_2 T_2^0}}{1 + \dfrac{\gamma_2^0 - 1}{2}\left(\dfrac{V_2^2}{\gamma_2^0 \mathcal{R}_2 T_2^0}\right)} \tag{7-18}$$

As in the enthalpy method, the diffuser discharge area is given by the equation

$$A_2 = A_\infty \frac{p_\infty V_\infty T_2}{p_2 V_2 T_\infty} \tag{7-19}$$

Arbitrary specific heat method. In applying this method to the calculation of diffuser discharge conditions it is assumed that the specific heat ratio γ remains constant at 1.4 throughout the diffusion process. The applicable equations thus have the same form as those in the average specific heat method but are simplified.

$$T_\infty^0 = T_2^0 = T_\infty(1 + 0.2M_\infty^2) \tag{7-20}$$

$$p_\infty^0 = p_\infty(1 + 0.2M_\infty^2)^{3.5} \tag{7-21}$$

$$p_2^0 = p_\infty^0 \frac{p_2^0}{p_\infty^0} \tag{7-22}$$

$$T_2 = T_2^0(1 + 0.2M_2^2)^{-1} \tag{7-23}$$

$$p_2 = p_2^0(1 + 0.2M_2^2)^{-3.5} \tag{7-24}$$

$$V_2 = 49.02M_2\sqrt{T_2} \tag{7-25}$$

The discharge area can be calculated from Eq. 7-19. It is frequently convenient, however, to utilize another method where tables of functions of M are available. The isentropic area ratio A^*/A denotes the flow area contraction necessary to bring the flow at Mach number M to sonic velocity along an isentropic path. The isentropic area ratio is related to Mach number by

$$\frac{A^*}{A} = M\left(\frac{\frac{\gamma+1}{2}}{1 + \frac{\gamma-1}{2}M^2}\right)^{\frac{\gamma+1}{2(\gamma-1)}} = \frac{1.728M}{(1 + 0.2M^2)^3} \tag{7-26}$$

From the mass conservation equation,

$$A_2 = A_\infty \frac{p_\infty^0}{p_2^0} \frac{\frac{A^*}{A_\infty}}{\frac{A^*}{A_2}} \tag{7-27}$$

Constant specific heat method. Since the ratio of the specific heats is always 1.4 in this method of calculation, Eq. 7-20 through 7-27 apply for the diffusion process.

Comparison of various methods of calculating diffuser discharge flow conditions. The relative accuracies of the several methods of computing diffuser discharge flow conditions are compared in Fig. E,7b and E,7c

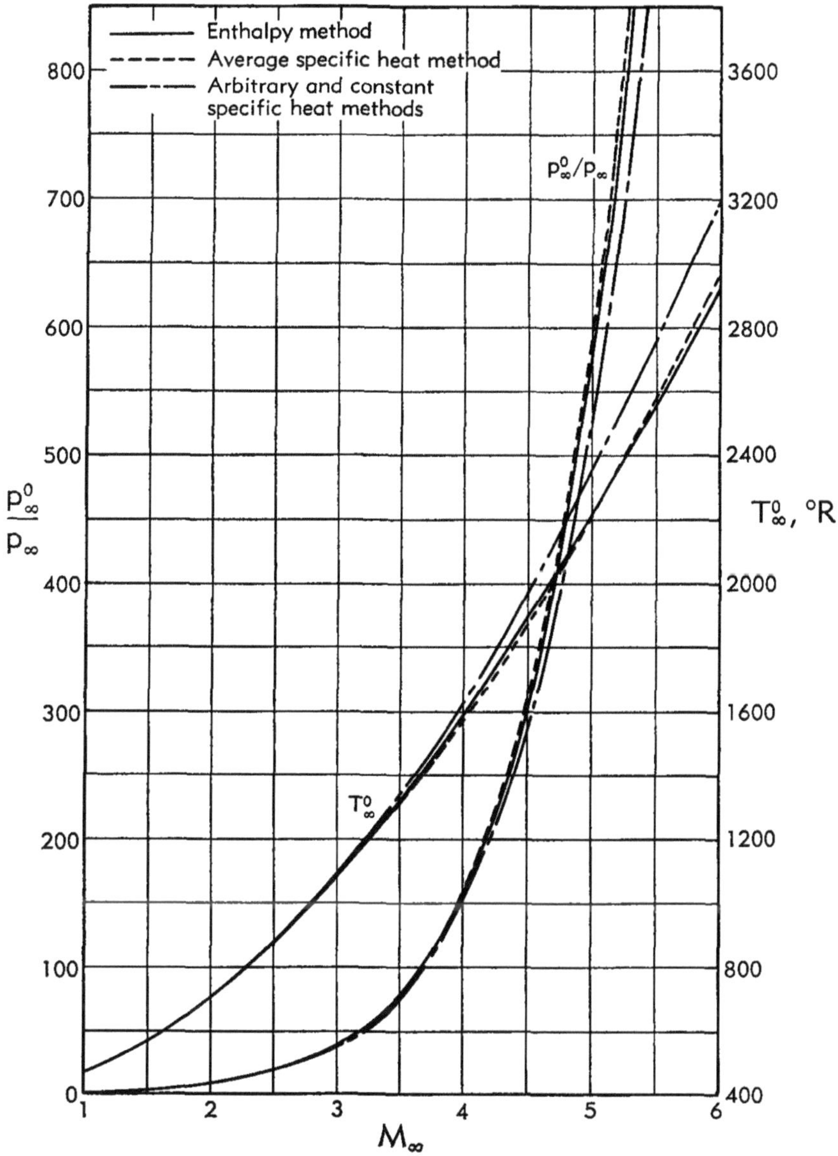

Fig. E,7b. Free stream stagnation conditions calculated by different methods. $T_\infty = 392°R$.

for an arbitrarily selected set of flow conditions. In Fig. E,7b are the free stream stagnation temperatures and the ratio of stagnation to ambient static pressure for flight over a range of Mach numbers in the NACA standard tropopause ($T_\infty = 392°R$).

The calculations by the enthalpy method may be taken as accurate.

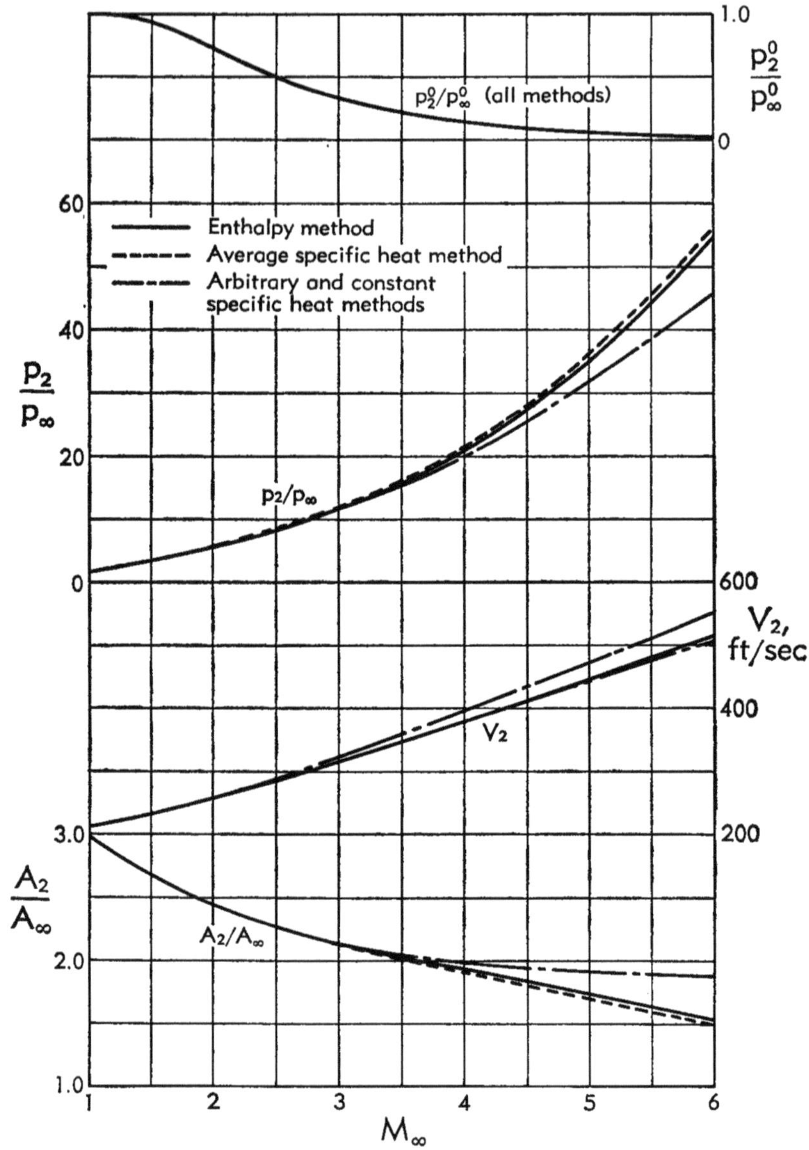

Fig. E,7c. Flow conditions at diffuser discharge by different methods of calculation. $M_2 = 0.20$, $T_\infty = 392°R$.

In general the average specific heat method yields very acceptable results. The stagnation temperature is accurate within approximately one per cent and the pressure ratio within 2.5 per cent. The arbitrary specific heat and constant specific heat methods (which are identical for the diffuser calculations) closely predict the accurate values for Mach numbers up to about 3.0. At higher Mach numbers, however, the stagnation temperature is progressively overestimated, with a maximum error of 280°R at Mach number 6. The stagnation to static pressure ratio is consistently underestimated for Mach numbers above 3.0 with maximum errors reaching as high as 18 per cent.

A constant diffuser discharge Mach number of 0.20 was arbitrarily assumed for calculations of the ratio of discharge static pressure to ambient static pressure, the discharge velocity and the ratio of discharge flow area to free stream flow area presented in Fig. E,7c. In order to give the results of a measure of quantitative significance, a diffuser pressure recovery schedule corresponding to normal shock diffusion (see VII,E) was assumed as indicated at the top of the figure.

Again referring to the enthalpy method of calculation as the standard of accuracy, the average specific heat method is apparently adequate. The arbitrary specific heat and constant specific heat methods are subject to the inherent errors arising from the calculation of free stream stagnation conditions and hence underestimate the discharge static pressures at the higher Mach numbers and overestimate the discharge velocities and flow area ratios.

E,8. Flame Holder Calculations. Any attempt to obtain extreme accuracy in the calculation of flame holder flow effects is somewhat fruitless inasmuch as the flame holder functions successfully by virtue of the local distortions it introduces into the flow, hence basically negating the principle of one-dimensional flow. Detailed considerations of the state of the fuel, which is generally injected at or upstream of the flame holder and which may have an indeterminant degree of evaporation across the flame holder, are consequently not warranted and for simplicity it is assumed for most analyses that the fuel is effectively injected downstream of the flame holder. Under this assumption the mass flow and total energy existing at the diffuser discharge must be conserved across the flame holder.

Momentum is not conserved across the flame holder and this momentum change, or flame holder drag, must be experimentally evaluated. The drag is generally defined by a pressure loss coefficient, $(p_2^0 - p_3^0)/q_2 = K$, which is largely independent of the diffuser discharge Mach number in the practical range of interest. In [7] it has been shown that it is convenient and sufficiently accurate to include within the flame holder pressure loss coefficient the friction losses occurring further downstream in the combustion chamber.

Enthalpy Method. The flame holder pressure loss coefficient can be converted to a flame holder pressure recovery

$$\frac{p_3^0}{p_2^0} = 1 - K\frac{q_2}{p_2^0} \tag{8-1}$$

Since mass flow is assumed to be conserved across the flame holder, it follows that for a constant area duct

$$\frac{p_3}{p_2} = \frac{V_2 T_3}{V_3 T_2} \tag{8-2}$$

The static pressure ratio on the left side of Eq. 8-2 can be expressed as

$$\frac{p_3}{p_2} = \frac{p_3}{p_3^0}\frac{p_2^0}{p_2}\frac{p_3^0}{p_2^0} \tag{8-3}$$

so that Eq. 8-2 can be written as

$$\frac{p_3^0}{p_3}\frac{T_3}{V_3} = \frac{p_3^0}{p_2^0}\frac{p_2^0}{p_2}\frac{T_2}{V_2} \tag{8-4}$$

The left-hand member of Eq. 8-4 involving downstream conditions can thus be evaluated as a whole from known conditions ahead of the flame holder and the flame holder pressure recovery given by Eq. 8-1. The pressure terms in the left-hand member can be eliminated as variables through the relation

$$\frac{p_3^0}{p_3} = \frac{p_{r_3}^0}{p_{r_3}} \tag{8-5}$$

since the reference pressures are functions of the total and static temperatures after the flame holder.

The remaining unknowns in the left-hand member of Eq. 8-4, T_3 and V_3, are also related by the energy equation

$$h_3^0 = h_2^0 = h_3 + \frac{V_3^2}{2gJ} \tag{8-6}$$

Thus, through the use of air tables such as those in [1] and either a graphical solution or an iteration process, T_3 and V_3 can be evaluated.

Average specific heat method. From the compressible flow relations, Eq. 8-1, defining the flame holder pressure recovery, becomes

$$\frac{p_3^0}{p_2^0} = 1 - \frac{K}{2}\frac{\gamma_2^0 M_2^2}{\left(1 + \frac{\gamma_2^0 - 1}{2}M_2^2\right)^{\frac{\gamma_2^0}{\gamma_2^0 - 1}}} \tag{8-7}$$

Mass conservation in a constant area duct gives

$$\gamma_2^0 p_2 \frac{M_2^2}{V_2} = \gamma_3^0 p_3 \frac{M_3^2}{V_3} \tag{8-8}$$

or, assuming specific heat changes across the flame holder to be negligible,

$$\frac{p_3}{p_2} = \frac{M_2}{M_3}\sqrt{\frac{T_3}{T_2}} \tag{8-9}$$

From the assumption of conserved total energy across the flame holder, Eq. 8-9 may be written

$$\frac{p_3}{p_2} = \frac{M_2}{M_3}\left(\frac{1 + \frac{\gamma_2^0 - 1}{2} M_2^2}{1 + \frac{\gamma_2^0 - 1}{2} M_3^2}\right)^{\frac{1}{2}} \tag{8-10}$$

The pressure ratio on the left-hand side of Eq. 8-10 can be expressed as

$$\frac{p_3}{p_2} = \frac{p_3}{p_3^0}\frac{p_2^0}{p_2}\frac{p_3^0}{p_2^0} = \frac{p_3^0}{p_2^0}\left(\frac{1 + \frac{\gamma_2^0 - 1}{2} M_2^2}{1 + \frac{\gamma_2^0 - 1}{2} M_3^2}\right)^{\frac{\gamma_2^0}{\gamma_2^0 - 1}} \tag{8-11}$$

so that

$$\frac{p_3^0}{p_2^0} = \frac{M_2}{M_3}\left(\frac{1 + \frac{\gamma_2^0 - 1}{2} M_3^2}{1 + \frac{\gamma_2^0 - 1}{2} M_2^2}\right)^{\frac{\gamma_2^0 + 1}{2(\gamma_2^0 - 1)}} \tag{8-12}$$

M_3 can therefore be determined from a simultaneous solution of Eq. 8-7 and 8-12. The velocity V_3, downstream of the flame holder, is

$$V_3 = M_3\sqrt{\gamma_3^0 \mathcal{R}_3 T_3} \tag{8-13}$$

Arbitrary specific heat method. Inasmuch as the flame holder is analytically assumed to be ahead of the combustion zone, a specific heat ratio of 1.4 is assumed for all flame holder calculations. The pertinent equations above are thereby simplified to

$$\frac{p_3^0}{p_2^0} = 1 - 0.7K\frac{M_2^2}{(1 + 0.2M_2^2)^{3.5}} \tag{8-14}$$

and

$$\frac{p_3^0}{p_2^0} = \frac{M_2}{M_3}\left(\frac{1 + 0.2M_3^2}{1 + 0.2M_2^2}\right)^3 \tag{8-15}$$

Constant specific heat method. Eq. 8-14 and 8-15 apply for this method, since $\gamma = 1.40$.

Comparison of methods. The flame holder total pressure ratio and the Mach number after the flame holder are indicated on Fig. E,8 as a function of the diffuser discharge Mach number for a flame holder having a pressure loss coefficient of 2. In order to apply the enthalpy method it was assumed that the total temperature was 3000°R and for the average specific heat method a ratio of specific heats of 1.3 was selected.

In the range of diffuser discharge Mach numbers of practical interest ($0.10 < M_2 < 0.25$) all of the methods of calculation give essentially identical results for the downstream Mach number. In the same range

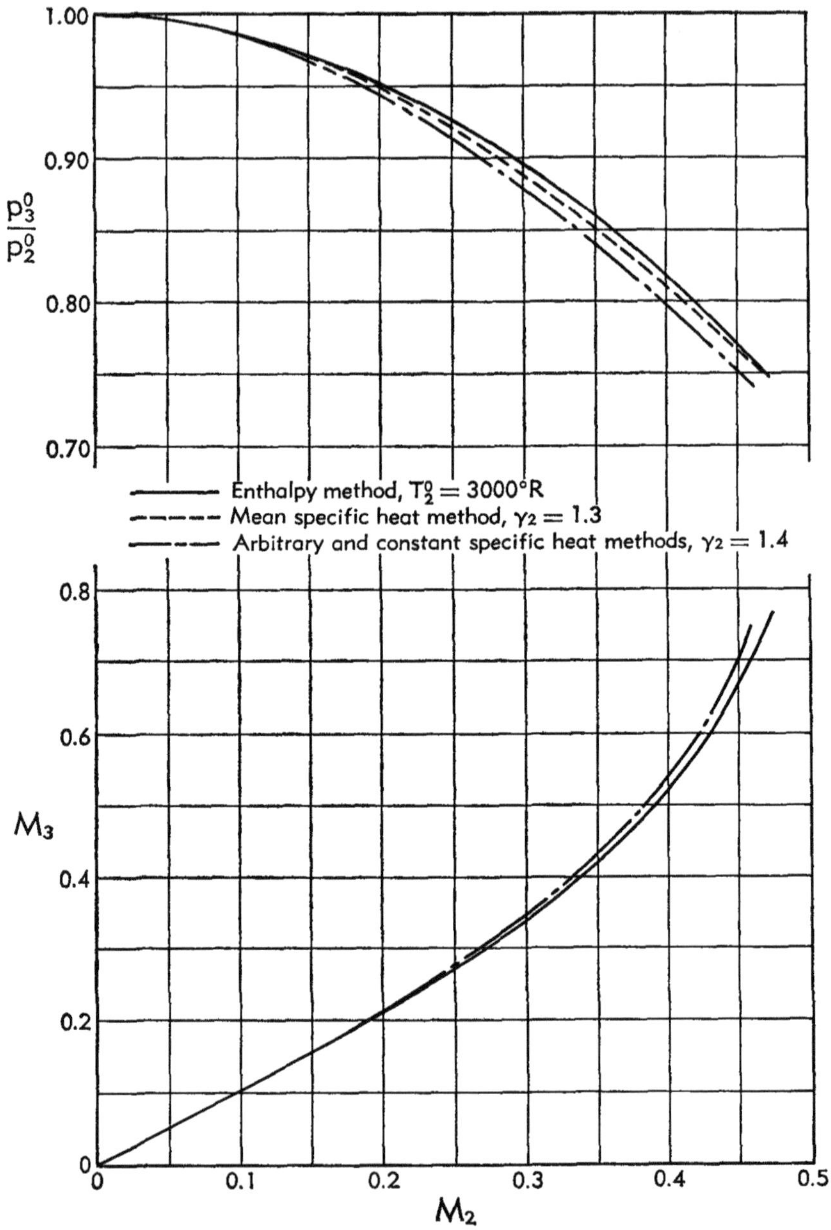

Fig. E,8. Flame holder flow effects calculated by different methods.

the constant specific heat method yields flame holder pressure losses as much as one per cent greater than those calculated by the enthalpy method, while the average specific heat method halves the error.

The total temperature assumed in calculating Fig. E,8 is extreme. Lower temperatures, more representative of the bulk of ramjet calculations, would reduce the spread in the calculations.

E,9. Combustor Calculations. The ramjet combustion chamber is generally a constant area passage and will be so considered herein. The combustion process will be assumed to include the injection of fuel (generally as a liquid) and ideal complete combustion-forming end products of carbon dioxide and water. Incomplete combustion will be accounted for by assuming an effective reduction in the heat of combustion of the fuel.

The general energy equation across the combustion chamber is

$$h_4^0 = h_3^0 + fh_f + Q_c \tag{9-1}$$

where f is the fuel-air ratio, h_f is the enthalpy of the fuel in its liquid state, and Q_c represents the heat addition to the working fluid from the combustion process.

The heat addition Q_c may be calculated as

$$Q_c = fH_f\eta_b \tag{9-2}$$

where H_f is the lower heating value of the fuel and η_b is the combustion efficiency. For liquid hydrocarbon fuels the lower heating values in BTU/lb of fuel is adequately represented by [*3*],

$$H_f = 15935 + 15800\,\frac{\mathrm{H}}{\mathrm{C}} \tag{9-3}$$

The hydrogen-carbon ratio H/C for a fuel represented by C_nH_m is

$$\frac{\mathrm{H}}{\mathrm{C}} = \frac{1.008m}{12.01n} \tag{9-4}$$

The enthalpy of the liquid fuel h_f may frequently be neglected without serious error. When retained in Eq. 9-1 this value is generally determined by assuming a constant specific heat of 0.5, so that

$$h_f = 0.5(T - T_{\mathrm{ref}}) \tag{9-5}$$

where T is the temperature of the fuel at injection and T_{ref} is the base temperature used for enthalpy determination.

The enthalpies before and after combustion must be based on a common unit, generally a pound of mixture. When referred to a mixture base, Eq. 9-1 becomes

$$h_4^0 = \frac{h_3^0 + fh_f + fH_f\eta_b}{\mu} \tag{9-6}$$

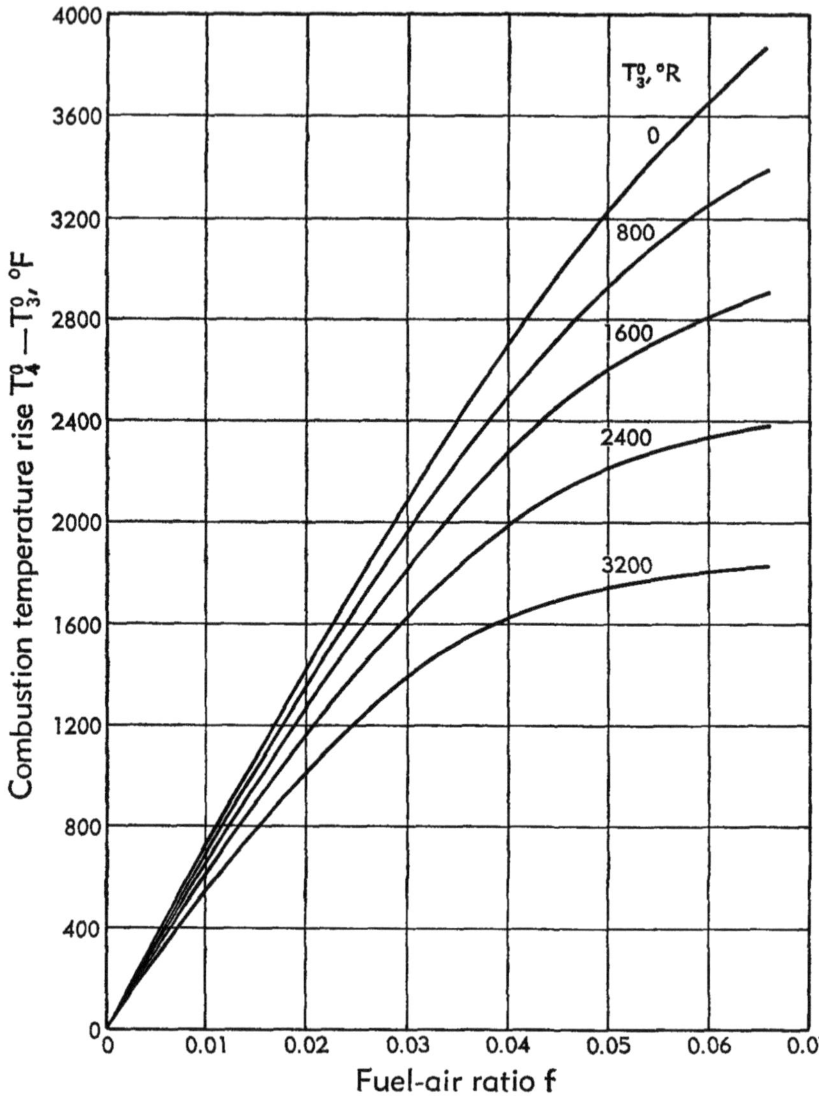

Fig. E,9a. Theoretical temperature rise from combustion of *n*-octane (C_8H_{18}) and air. (From [*6*].)

The elements of the foregoing simplified analysis have been applied to determine the temperature rise across a combustor for various hydrocarbon fuels by a number of investigators. Fig. E,9a, taken from [*6*], illustrates the working form of these results for a specific fuel.

In addition to the energy equation, the combustion process is defined by the mass flow equation

$$m_4 = \mu m_3 \tag{9-7}$$

and the momentum equation

$$m_4V_4 + A_3p_4 = m_3V_3 + A_3p_3 \tag{9-8}$$

Enthalpy method. Assuming that the products of combustion satisfy the perfect gas equation, it follows from Eq. 9-7 that

$$\frac{p_4}{p_3} = \mu \frac{V_3}{V_4}\frac{T_4}{T_3}\frac{\mathcal{R}_4}{\mathcal{R}_3} \tag{9-9}$$

The momentum balance represented by Eq. 9-8 can be rewritten as

$$\frac{p_4}{p_3} = \frac{\dfrac{V_3^2}{\mathcal{R}_3T_3} + 1}{\dfrac{V_4^2}{\mathcal{R}_4T_4} + 1} \tag{9-10}$$

Elimination of the pressure terms between Eq. 9-9 and 9-10 and separation of the states before and after combustion yields

$$V_4 + \frac{\mathcal{R}_4T_4}{V_4} = \frac{1}{\mu}\left(V_3 + \frac{\mathcal{R}_3T_3}{V_3}\right) \tag{9-11}$$

The right-hand member of Eq. 9-11 can be calculated from known conditions after the flame holder. The gas constant $\mathcal{R}_4$ of the combustion products can be determined by the relation [*8*]

$$\mathcal{R}_4 = \frac{1716.2 + \dfrac{12334f\left(\dfrac{\mathrm{H}}{\mathrm{C}}\right)}{1 + \left(\dfrac{\mathrm{H}}{\mathrm{C}}\right)}}{\mu} \tag{9-12}$$

so that only the temperature and velocity after combustion remain as unknowns in Eq. 9-11.

The total energy at the end of combustion can be expressed as

$$h_4^0 = h_4 + \frac{V_4^2}{2gJ} \tag{9-13}$$

so that a second equation containing V_4 and T_4 (indirectly) is available. Eq. 9-11 and 9-13 can be solved simultaneously either by iteration or by graphical solution to determine V_4 and T_4 and the static pressure after combustion, p_4, can then be evaluated from either Eq. 9-9 or 9-10.

The total pressure after combustion is determined from the relation

$$p_4^0 = p_4 \frac{p_{r_4}^0}{p_{r_4}} \tag{9-14}$$

where the reference pressures are taken from enthalpy tables or charts at the static and total temperatures after combustion.

The Mach number of the flow after combustion is

$$M_4 = \frac{V_4}{\sqrt{\gamma_4 \mathcal{R}_4 T_4}} \tag{9-15}$$

so that the ratio of specific heats of the products of combustion must be known. Fig. E,9b presents the calculated value of γ_4 from [6] for n-octane fuel combusted with air.

Average specific heat method. Using the Mach number relations, Eq. 9-9 derived from the mass flow relation across the combustor, can be

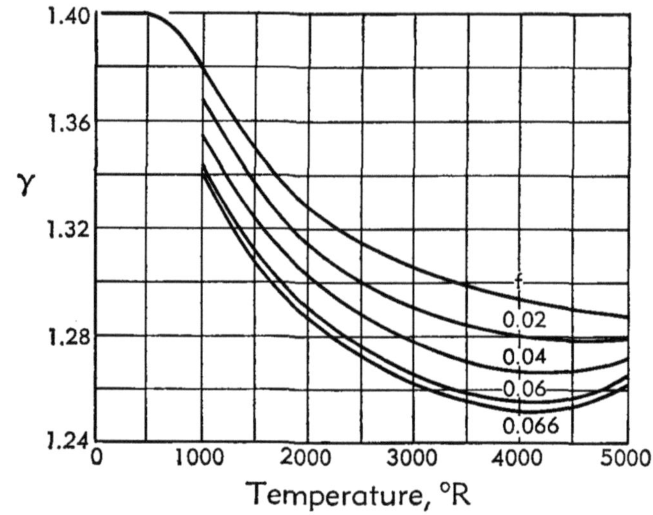

Fig. E,9b. Specific heat ratios for products of combustion of n-octane (C_8H_{18}) and air [6].

written as

$$\frac{p_4}{p_3} = \mu \frac{M_3}{M_4} \sqrt{\frac{\gamma_3 \mathcal{R}_4 T_4}{\gamma_4 \mathcal{R}_3 T_3}} \tag{9-16}$$

Utilizing the relations

$$\frac{T_4}{T_3} = \frac{\dfrac{T_3^0}{T_3}}{\dfrac{T_4^0}{T_4}} \frac{T_4^0}{T_3^0}$$

and

$$\frac{T^0}{T} = 1 + \frac{\gamma_{\text{eff}} - 1}{2} M^2$$

Eq. 9-16 becomes

$$\frac{p_4}{p_3} = \sqrt{\frac{\gamma_3}{\gamma_4}} \mu \sqrt{\frac{\mathcal{R}_4 T_4^0}{\mathcal{R}_3 T_3^0}} \frac{M_3}{M_4} \left(\frac{1 + \dfrac{\gamma_{\text{eff}_3} - 1}{2} M_3^2}{1 + \dfrac{\gamma_{\text{eff}_4} - 1}{2} M_4^2} \right)^{\frac{1}{2}} \tag{9-17}$$

The momentum balance across the combustion zone given by Eq. 9-10 can be written as

$$\frac{p_4}{p_3} = \frac{\gamma_3 M_3^2 + 1}{\gamma_4 M_4^2 + 1} \tag{9-18}$$

Elimination of the pressure terms between Eq. 9-17 and 9-18 and separation of the variables gives

$$\frac{\sqrt{\gamma_4}\, M_4 \left(1 + \frac{\gamma_{\text{eff}_4} - 1}{2} M_4^2\right)^{\frac{1}{2}}}{\gamma_4 M_4^2 + 1} = \mu \sqrt{\frac{\mathcal{R}_4 T_4^0}{\mathcal{R}_3 T_3^0}} \frac{\sqrt{\gamma_3}\, M_3 \left(1 + \frac{\gamma_{\text{eff}_3} - 1}{2} M_3^2\right)^{\frac{1}{2}}}{\gamma_3 M_3^2 + 1} \tag{9-19}$$

from which M_4 can be determined for known conditions downstream of the flame holder and the heat addition.

In a rigorous sense the exact value of M_4 depends on the unknown static temperature T_4, which must be used with the total temperature T_4 and specific heat sources such as Fig. E,9b to determine T_4 and γ_{eff_4}. From a practical viewpoint, however, the difference between γ_4 and γ_4^0 will be small, since M_4 has a limiting value of unity and the lowest value of T_4 will be on the order of $0.85T_4^0$. Hence sufficient accuracy is retained in the calculations if it is assumed that $\gamma_4 = \gamma_4^0 = \gamma_{\text{eff}_4}$. It has previously been shown that the concurrent assumption that $\gamma_3 = \gamma_3^0 = \gamma_{\text{eff}_3}$ is valid for conditions immediately prior to combustion. Eq. 9-19 can therefore be simplified to

$$\frac{\sqrt{\gamma_4^0}\, M_4 \left(1 + \frac{\gamma_4^0 - 1}{2} M_4^2\right)^{\frac{1}{2}}}{\gamma_4^0 M_4 + 1} = \mu \sqrt{\frac{\mathcal{R}_4 T_4^0}{\mathcal{R}_3 T_3^0}} \frac{\sqrt{\gamma_3^0}\, M_3 \left(1 + \frac{\gamma_3^0 - 1}{2} M_3^2\right)^{\frac{1}{2}}}{\gamma_3^0 M_3^2 + 1} \tag{9-20}$$

The static temperature after combustion is

$$T_4 = \frac{T_4^0}{1 + \frac{\gamma_4^0 - 1}{2} M_4^2} \tag{9-21}$$

and the static pressure p_4 can be evaluated from Eq. 9-18.

The total pressure ratio across the combustion zone is determined from combining Eq. 9-18 and the identity

$$\frac{p_4}{p_3} \equiv \frac{\frac{p_3^0}{p_3}\, \frac{p_4^0}{p_3^0}}{\frac{p_4^0}{p_4}} = \frac{p_4^0}{p_3^0} \frac{\left(1 + \frac{\gamma_3^0 - 1}{2} M_3^2\right)^{\frac{\gamma_3^0}{\gamma_3^0 - 1}}}{\left(1 + \frac{\gamma_4^0 - 1}{2} M_4^2\right)^{\frac{\gamma_4^0}{\gamma_4^0 - 1}}} \tag{9-22}$$

to give

$$\frac{p_4^0}{p_3^0} = \frac{\gamma_3^0 M_3^2 + 1}{\gamma_4^0 M_4^2 + 1} \frac{\left(1 + \frac{\gamma_4 - 1}{2} M_4^2\right)^{\frac{\gamma_4^0}{\gamma_4^0 - 1}}}{\left(1 + \frac{\gamma_3^0 - 1}{2} M_3^2\right)^{\frac{\gamma_3^0}{\gamma_3^0 - 1}}} \tag{9-23}$$

Arbitrary specific heat method. The analytical approach is identical to that of the average specific heat method; however, the ratio of specific heats after combustion is arbitrarily selected. For most calculations a value of $\gamma_4 = 1.3$ is used and the principal equations reduce to

$$\frac{\sqrt{1.3}\, M_4(1 + 0.15M_4^2)^{\frac{1}{2}}}{1.3M_4^2 + 1} = \mu \sqrt{\frac{\mathfrak{R}_4 T_4^0}{\mathfrak{R}_3 T_3^0}} \frac{\sqrt{1.4}\, M_3(1 + 0.2M_3^2)^{\frac{1}{2}}}{1.4M_3^2 + 1} \tag{9-24}$$

$$\frac{p_4}{p_3} = \frac{1.4M_3^2 + 1}{1.3M_4^2 + 1} \tag{9-25}$$

$$T_4 = \frac{T_4^0}{1 + 0.15M_4^2} \tag{9-26}$$

$$\frac{p_4^0}{p_3^0} = \frac{1.4M_3^2 + 1}{1.3M_4^2 + 1} \frac{(1 + 0.15M_4^2)^{4.33}}{(1 + 0.2M_3^2)^{3.5}} \tag{9-27}$$

Constant specific heat method. The change in specific heat ratio and gas constant accompanying the combustion of fuel and air at the elevated temperatures is ignored. Consequently the principal equations in the discussion of the average specific heat method are simplified to

$$\frac{M_4(1 + 0.2M_4^2)^{\frac{1}{2}}}{1.4M_4^2 + 1} = \mu \sqrt{\frac{T_4^0}{T_3^0}} \frac{M_3(1 + 0.2M_3^2)^{\frac{1}{2}}}{1.4M_3^2 + 1} \tag{9-28}$$

$$\frac{p_4}{p_3} = \frac{1.4M_3^2 + 1}{1.4M_4^2 + 1} \tag{9-29}$$

$$T_4 = \frac{T_4^0}{1 + 0.2M_4^2} \tag{9-30}$$

$$\frac{p_4^0}{p_3^0} = \frac{1.4M_3^2 + 1}{1.4M_4^2 + 1} \left(\frac{1 + 0.2M_4^2}{1 + 0.2M_3^2}\right)^{3.5} \tag{9-31}$$

Comparison of methods. Typical total pressure ratios across the combustion zone and the Mach numbers after combustion are presented on Fig. E,9c as a function of the Mach number at the start of combustion. The total temperature of the air at the start of the combustion process was arbitrarily assumed to be 1000°R for these calculations. Two fuel-air ratios are indicated. The final gas temperatures of 2111°R for $f = 0.0166$ (25 per cent of stoichiometric) and 3078°R for $f = 0.0332$ were used in all the methods of calculation.

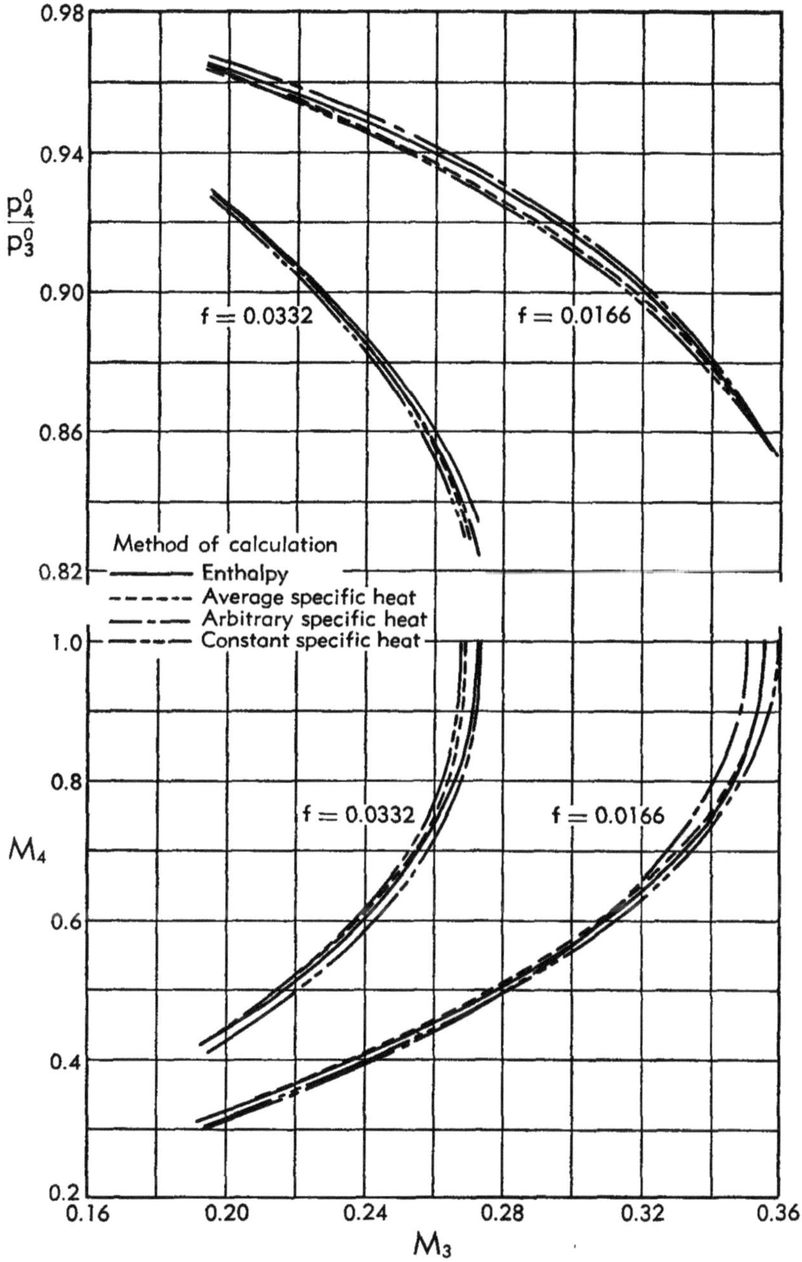

Fig. E,9c. Flow variations across combustor by different methods of calculation. $T_3^0 = 1000°R$.

Combustion total pressure ratios calculated by the various methods differed by less than one per cent throughout the range of variables, indicating the insensitivity of this parameter to the ratio of specific heats. The combustor discharge Mach number shows a greater variation. The average specific heat method shows excellent agreement with the enthalpy method, calculating over most of the range indicated, and, in the range of combustion inlet Mach numbers of practical interest, the arbitrary specific heat method is acceptably close.

E,10. Nozzle Calculations. The products of combustion approaching the exhaust nozzle will be at or below sonic velocity and, in general, will have a pressure greater than ambient. In order to maintain these flow conditions it is necessary that the nozzle either discharge at ambient pressure or discharge with sonic or supersonic velocity so that the internal flow is insensitive to the external pressure. Thus, as a general rule, the exhaust nozzle must initially converge. If the internal pressure is decreased to ambient pressure with subsonic velocity in the nozzle the nozzle process is complete. If sonic velocity is attained at pressures above ambient the nozzle may be reexpanded to further reduce the internal pressures to the ambient level prior to discharge. In the latter case the complete nozzle configuration is referred to as convergent-divergent (see VII,E).

Although maximum engine thrust is ideally realized with a nozzle that discharges at ambient pressure, the amount of area reexpansion is, in some cases, so large that the nozzle discharge area becomes the maximum frontal area of the engine. It is frequent practice in such cases to terminate the nozzle expansion at some previously established maximum area, such as the combustion chamber area. The nozzle flow will, therefore, be only partially expanded and will discharge at a supersonic velocity but at a pressure greater than ambient. Thus nozzles may be classified as sonic discharge (no expansion), partial expansion, or complete expansion nozzles.

Regardless of the amount of expansion, mass continuity is maintained so that $m_4 = m_5 = m_e$. If heat transfer through the nozzle is neglected the total energy is also conserved, or $h_4^0 = h_5^0 = h_e^0$. Entropy will not, in general, be conserved, however, because of friction losses in the nozzle. For design purposes the specification of friction losses as total pressure ratios leads to the greatest ease of calculating and this practice is utilized in the subsequent analyses of nozzle flow.

Enthalpy method. For most operating conditions of practical interest for the ramjet engine, the nozzle throat is choked. For this case

$$M_5^2 = \frac{V_5^2}{\gamma_5 \mathcal{R}_5 T_5} = 1 \tag{10-1}$$

In the energy equation, therefore,

$$h_5^0 = h_5 + \frac{\gamma_5 \mathcal{R}_5 T_5}{2gJ} = h_4^0 \tag{10-2}$$

Knowing the total enthalpy of the products of combustion, h_4, Eq. 10-2 can be solved for T_5 and Eq. 10-1 for V_5.

From the mass continuity equation,

$$\frac{A_5}{A_4} = \frac{1}{C_d} \frac{p_4}{p_5} \frac{\mathcal{R}_5 T_5}{V_5} \frac{V_4}{\mathcal{R}_4 T_4} \tag{10-3}$$

where C_d is the discharge coefficient for the nozzle throat. This coefficient is required to account for the vena-contracta effect of the real flow and the magnitude depends on the convergent contour of the nozzle.

The pressure terms in Eq. 10-3 may be written as

$$\frac{p_4}{p_5} = \frac{p_4}{p_4^0} \frac{p_5^0}{p_5} \frac{p_4^0}{p_5^0} = \frac{p_{r_4}}{p_{r_4}^0} \frac{p_{r_5}^0}{p_{r_5}} \frac{p_4^0}{p_5^0} \tag{10-4}$$

so that

$$\frac{A_5}{A_4} = \frac{1}{C_d} \frac{p_{r_4}}{p_{r_4}^0} \frac{V_4}{\mathcal{R}_4 T_4} \frac{p_{r_5}^0}{p_{r_5}} \frac{\mathcal{R}_5 T_5}{V_5} \frac{p_4^0}{p_5^0} \tag{10-5}$$

All the terms on the right-hand side of Eq. 10-5 are known from the previously calculated temperatures, since the reference pressures are a function of temperature

If the nozzle throat is not choked when ambient pressure is attained, $p_5 = p_\infty$ and

$$p_{r_5} = p_{r_5}^0 \frac{p_\infty}{p_4^0} \frac{p_4^0}{p_5^0} \tag{10-6}$$

From the static enthalpy at the nozzle discharge corresponding to p_{r_5}

$$V_5 = \sqrt{2gJ(h_4^0 - h_5)} \tag{10-7}$$

The nozzle discharge area is determined by Eq. 10-5.

When the internal flow pressures are sufficiently large that the nozzle chokes at a pressure level above ambient and a divergent nozzle section is desirable to realize additional thrust, the amount of divergence may be specified in several ways. If the nozzle expansion ratio is specified by previous geometric considerations (i.e. $A_e = A_3$, for example), the discharge conditions are obtained from the mass conservation equation

$$\frac{p_e V_e}{\mathcal{R}_e T_e} = \frac{p_5 V_5}{\mathcal{R}_5 T_5} \frac{A_5}{A_e} \tag{10-8}$$

which can be written as

$$\frac{p_{r_\infty}}{p^0_{r_e}} \frac{V_e}{\mathcal{R}_e T_e} = \frac{p_{r_5}}{p^0_{r_5}} \frac{V_5}{\mathcal{R}_5 T_5} \frac{p^0_5}{p^0_e} \frac{A_5}{A_e} \tag{10-9}$$

The energy equation for the expansion region is

$$h^0_e = h_e + \frac{V^2_e}{2gJ} = h^0_5 \tag{10-10}$$

The left-hand member of Eq. 10-9 can be evaluated as a whole from known conditions at the throat, the specified nozzle expansion ratio, and the total pressure ratio across the expansion section. By a simultaneous solution of Eq. 10-9 and 10-10 the discharge velocity and static temperature are determined. The discharge pressure is evaluated from the relation

$$p_e = p^0_e \frac{p_{r_e}}{p^0_{r_e}} \tag{10-11}$$

The amount of nozzle divergence may be indirectly specified by specifying the nozzle discharge pressure. For maximum thrust the discharge pressure is ambient so that $p_e = p_\infty$. The discharge temperature is obtained from the relative discharge pressure p_{r_e} where

$$p_{r_e} = p^0_{r_e} \frac{p_\infty}{p^0_5} \frac{p^0_5}{p^0_e} \tag{10-12}$$

and the discharge velocity from the relation

$$V_e = \sqrt{2gJ(h^0_e - h_e)} \tag{10-13}$$

The nozzle expansion ratio is determined from Eq. 10-8.

Average specific heat method. The mass continuity equation may be written as

$$\frac{A_5}{A_4} = \frac{1}{C_d} \frac{p_4}{p_5} \sqrt{\frac{\gamma_4 \mathcal{R}_5 T_5}{\gamma_5 \mathcal{R}_4 T_4}} \frac{M_4}{M_5} \tag{10-14}$$

The pressure terms in Eq. 10-14 may be written as

$$\frac{p_4}{p_5} = \frac{p_4}{p^0_4} \frac{p^0_5}{p_5} \frac{p^0_4}{p^0_5} = \frac{p^0_4}{p^0_5} \frac{\left(1 + \frac{\gamma_{\text{eff}_5} - 1}{2} M^2_5\right)^{\frac{\gamma_{\text{eff}_5}}{\gamma_{\text{eff}_5} - 1}}}{\left(1 + \frac{\gamma_{\text{eff}_4} - 1}{2} M^2_4\right)^{\frac{\gamma_{\text{eff}_4}}{\gamma_{\text{eff}_4} - 1}}} \tag{10-15}$$

The temperature term becomes

$$\frac{T_5}{T_4} = \frac{T_5}{T^0_5} \frac{T^0_4}{T_4} = \frac{1 + \frac{\gamma_{\text{eff}_4} - 1}{2} M^2_4}{1 + \frac{\gamma_{\text{eff}_5} - 1}{2} M^2_5} \tag{10-16}$$

so that

$$\frac{A_5}{A_4} = \frac{1}{C_d}\sqrt{\frac{\gamma_4 \mathcal{R}_5 p_4^0}{\gamma_5 \mathcal{R}_4 p_5^0}}\frac{M_4}{M_5}\frac{\left(1 + \frac{\gamma_{\text{eff}_5} - 1}{2} M_5^2\right)^{\frac{\gamma_{\text{eff}_5}+1}{2(\gamma_{\text{eff}_5}-1)}}}{\left(1 + \frac{\gamma_{\text{eff}_4} - 1}{2} M_4^2\right)^{\frac{\gamma_{\text{eff}_4}+1}{2(\gamma_{\text{eff}_4}-1)}}} \tag{10-17}$$

When the nozzle throat is choked, $M_5 = 1$ and Eq. 10-17 simplifies to

$$\frac{A_5}{A_4} = \frac{1}{C_d}\sqrt{\frac{\gamma_4 \mathcal{R}_5 p_4^0}{\gamma_5 \mathcal{R}_4 p_5^0}}\, M_4 \frac{\left(\frac{\gamma_{\text{eff}_5} + 1}{2}\right)^{\frac{\gamma_{\text{eff}_5}+1}{2(\gamma_{\text{eff}_5}-1)}}}{\left(1 + \frac{\gamma_{\text{eff}_4} - 1}{2} M_4^2\right)^{\frac{\gamma_{\text{eff}_4}+1}{2(\gamma_{\text{eff}_4}-1)}}} \tag{10-18}$$

As previously noted, sufficient accuracy is retained if it is assumed that $\gamma_5^0 = \gamma_5 = \gamma_{\text{eff}_5}^0$. Since $T_5^0 = T_4^0$, $\gamma_5^0 = \gamma_4^0$ and Eq. 10-18 may be further simplified to

$$\frac{A_5}{A_4} = \frac{1}{C_d}\frac{p_4^0}{p_5^0} M_4 \left(\frac{\frac{\gamma_4^0 + 1}{2}}{1 + \frac{\gamma_5^0 - 1}{2} M_4^2}\right)^{\frac{\gamma_4^0+1}{2(\gamma_4^0-1)}} \tag{10-19}$$

When the nozzle throat is not choked, the Mach number at which the throat pressure equals ambient pressure is

$$M_5 = \left\{\frac{2}{\gamma_4^0 - 1}\left[\left(\frac{p_5^0}{p_\infty}\right)^{\frac{\gamma_4^0-1}{\gamma_4^0}} - 1\right]\right\}^{\frac{1}{2}} \tag{10-20}$$

The nozzle contraction ratio is then calculated from Eq. 10-17, the static temperature at the throat is

$$T_5 = \frac{T_4^0}{1 + \frac{\gamma_4^0 - 1}{2} M_5^2} \tag{10-21}$$

and the throat velocity becomes

$$V_5 = M_5 \sqrt{\gamma_5 \mathcal{R}_5 T_5} \tag{10-22}$$

If the nozzle is reexpanded and the expansion ratio is specified, an equation similar to Eq. 10-18 results, which when transposed is

$$\frac{\left(1 + \frac{\gamma_{\text{eff}_e} - 1}{2} M_e^2\right)^{\frac{\gamma_{\text{eff}_e}+1}{2(\gamma_{\text{eff}_e}-1)}}}{M_e} = \sqrt{\frac{\gamma_e \mathcal{R}_5}{\gamma_5 \mathcal{R}_e}}\frac{A_e}{A_5}\frac{p_e^0}{p_5^0}\left(\frac{\gamma_5^0 + 1}{2}\right)^{\frac{\gamma_5^0+1}{2(\gamma_5^0-1)}} \tag{10-23}$$

In the expanded flow γ_e cannot be assumed equal to γ_e^0 if accuracy is desired. Hence it is necessary to solve Eq. 10-23 by iteration. Initially Eq. 10-23 is solved with $\gamma_e = \gamma_e^0 = \gamma_{\text{eff}_e}$ and a value of M_e is established.

With this M_e the static temperature T_e is determined from

$$T_e = \frac{T_e^0}{1 + \frac{\gamma_{\text{eff}_e} - 1}{2} M_e^2} \tag{10-24}$$

and a corresponding value of γ_e is used to determine γ_{eff_e} according to

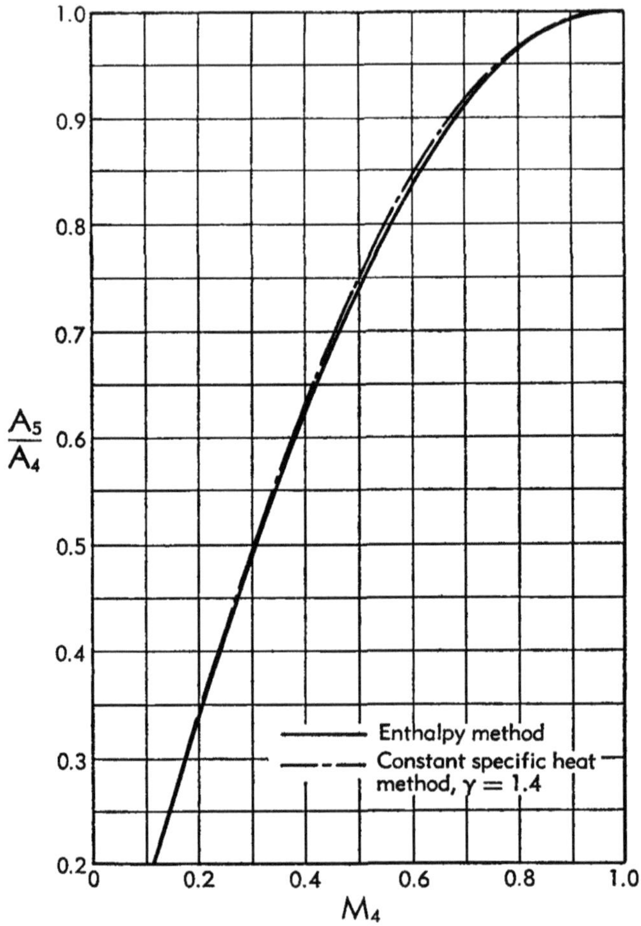

Fig. E,10a. Nozzle convergence area ratios. $T_4^0 = 3000°\text{R}, f = 0.0166, p_5^0 = p_4^0$.

Eq. 7-11. Eq. 10-23 is resolved using the new values of γ_e and γ_{eff_e} and the process is continued until T_e converges.

The static pressure at the discharge station is given by

$$p_e = p_5^0 \frac{p_e^0}{p_5^0} \left(1 + \frac{\gamma_{\text{eff}_e} - 1}{2} M_e^2\right)^{-\frac{\gamma_{\text{eff}_e}}{\gamma_{\text{eff}_e} - 1}} \tag{10-25}$$

The inverse problem of determining the required flow expansion to produce a selected discharge pressure p_e requires the solution of Eq. 10-25 for M_e, the solution of Eq. 10-24 for T_e, and an iteration until T_e converges. Eq. 10-23 can then be solved for A_5/A_e.

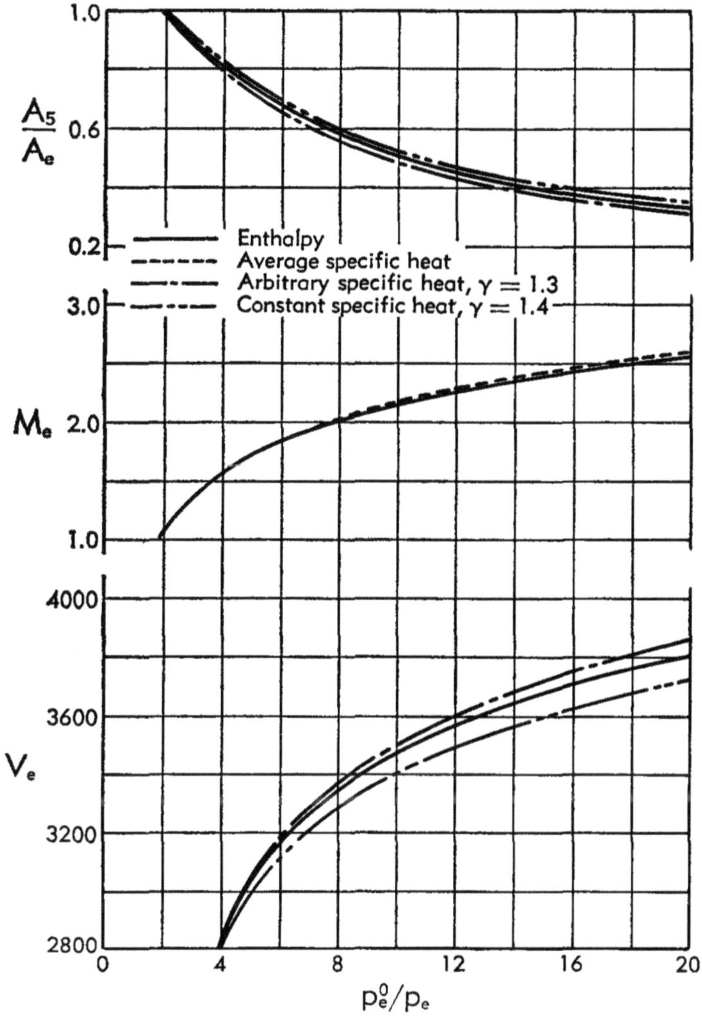

Fig. E,10b. Conditions at discharge of completely expanded nozzles. $T_6^0 = 2000°R$, $f = 0.0166$, $p_e^0 = p_5^0$.

The discharge velocity is given by

$$V_e = M_e \sqrt{\gamma_e \mathcal{R}_e T_e} \tag{10-26}$$

Arbitrary specific heat method. The nozzle flow process being downstream of the combustion zone, the specific heat ratio is taken as constant at $\gamma = 1.3$. The equations above apply, using $\gamma = 1.3$ throughout.

Constant specific heat method. The equations of the average specific heat method are used with a constant value of $\gamma = 1.4$ for all flow conditions.

Comparison of methods. Nozzle convergence ratios A_5/A_4 are indicated in Fig. E,10a as a function of combustor discharge Mach number M_4. Only the enthalpy method and constant specific heat method calculations are shown since the other two methods yield results that fall between the two. It is apparent that the calculation of A_5/A_4 is relatively insensitive to the assumption of specific heat ratio, the maximum error between the methods shown being less than one per cent.

When the nozzle is reexpanded and the discharge Mach numbers exceed unity, the calculations become more sensitive. Fig. E,10b indicates flow discharge conditions for complete nozzle expansion over a range of nozzle pressure ratios, p_e^0/p_e. The average specific heat method shows close agreement with the enthalpy method for all quantities of interest. The other two methods closely predict the discharge Mach number and are not presented in the figure because of the close coincidence. For the particular operating conditions, however, the arbitrary specific heat method ($\gamma = 1.3$) overestimates the discharge velocity and the required discharge flow area. The constant specific heat method underestimates the discharge velocity and flow area.

CHAPTER 3. THEORETICAL ENGINE PERFORMANCE

DEMARQUIS DALE WYATT

E,11. Introduction. The important performance parameters of the ramjet engine were presented in Art. 4. In the design of a particular engine it is the designer's goal to optimize one or more of these parameters while satisfying other considerations such as minimum weight, structural integrity, and minimum complexity. From a design viewpoint there may be wide possibilities in the selection of components to satisfy the latter considerations. Thus it is necessary to evaluate the effect of component changes on the engine performance in order to arrive at the optimum engine design.

Although the ramjet engine is relatively uncomplicated as compared to other air-breathing engines, there do exist enough variables within the free choice of the designer to preclude optimization of components as entities. Consequently component evaluations must be made in terms of over-all engine performances. A complete evaluation of component effects on engine performance should include an analysis of the engine for all anticipated operating conditions. Desirable component trends can be delineated, however, from an analysis of the engine at the principal operating, or design, point.

E,12. On Design Performance Parameters. The ultimate utility of a given engine design can only be evaluated by its demonstrated ability to accomplish the necessary flight purpose of the vehicle in which it is installed. For general analytical purposes, however, some more general figures of merit must be utilized. This figure of merit should embrace, insofar as possible, all of the primary and secondary considerations affecting the choice of engine design. A method of analysis which leads to quantities that are indicative of the comparative range capabilities of a vehicle of fixed gross weight with different engine designs appears to satisfy these requirements.

The Breguet flight path is frequently employed to evaluate the range possibilities of various engine-airframe combinations. This path assumes flight at a constant lift-drag ratio and velocity and hence requires that the evaluation be based on properties of the vehicle after flight boost is terminated and design speed is attained.

The Breguet range equation may be written as

$$\text{Range} = V_\infty I_{sp} \frac{L}{D} \ln \frac{W_0}{W_e} \tag{12-1}$$

where L/D is the lift-drag ratio of the vehicle and W_0 and W_e are the initial and final weights of the vehicle respectively. For comparisons of various flight speeds, V_∞ can be eliminated from Eq. 12-1 by substituting from Eq. 4-11 and 4-13 to give

$$\text{Range} = JH_f\eta_0 \frac{L}{D} \ln \frac{W_0}{W_e} \tag{12-2}$$

Eq. 12-2 may be interpreted as incorporating the heating efficiency of the fuel as measured by the heat content per unit weight H_f, the propulsive efficiency of the engine η_0, the aerodynamic efficiency of the vehicle as measured by the lift-drag ratio L/D, and the structural efficiency of the vehicle as measured by $\ln W_0/W_e$.

The engine efficiency η_0 can be directly evaluated for various ramjet designs. The other terms in the range equation cannot be directly evaluated from a consideration of the power plant alone, however, and thus the effects of alternate engine design possibilities on these terms can only be qualitatively estimated.

The lift-drag ratio of the vehicle is, of course, influenced by the drag of the engine. When the engine is submerged in some portion of the airframe, the drag solely arising from the engine is difficult to generalize, however, and there appears to be no wholly adequate representative engine parameter for this purpose. If the engine is comparatively isolated from the rest of the airframe, on the other hand, as in a nacelle-type installation, the engine drag can be calculated. In this case it is convenient to redefine the engine efficiency from Eq. 4-13 as

$$\eta_0' = \frac{(C_{F_\infty} - C_{D_\infty})V_\infty^2}{2gJfH_t} \tag{12-3}$$

so that the lift-drag ratio in Eq. 12-2 represents the drag of the vehicle exclusive of the engine installation. To the extent that aerodynamic interferences between the semi-isolated engine and the airframe can be neglected, the engine efficiency η_0' therefore represents comparative range capabilities of the vehicle, except for engine weight considerations.

The empty weight of the vehicle and hence the term ln (W_0/W_e) for a fixed gross weight, will depend on engine weight, which in turn will be proportional to engine size in some manner. It has been pointed out in Art. 4 that engine size can be generalized through the use of a thrust coefficient, so that for a specific application the size will be inversely proportional to the thrust coefficient.

If the thrust coefficient is based on the free stream tube area of the flow entering the engine, the engine size variation represented by changes in thrust coefficient will be the inlet area variation. The inlet area may not be indicative of the engine weight, however, so that variations in C_{F_∞} do not necessarily reflect as significant changes in ln (W_0/W_e). Since the major portion of the engine weight is likely to be determined by the combustion chamber and associated auxiliary equipment, thrust coefficients based on the combustion chamber area, C_{F_2}, are more likely to reflect engine weight differences, even though in many cases such coefficients do not correspond to thrusts per unit maximum frontal area of the engines.

E,13. Generalized Thrust Coefficient. The thrust coefficient based on the free stream tube area of the air flow passing through the engine is shown in Eq. 4-6 to be

$$C_{F_\infty} = 2\left(\mu \frac{V_e}{V_\infty} - 1\right) + \frac{2}{\gamma M_\infty^2}\frac{A_e}{A_\infty}\left(\frac{p_e}{p_\infty} - 1\right) \tag{13-1}$$

When the nozzle discharges at ambient pressure the last term in Eq. 13-1 vanishes. For this condition a regrouping of terms and an expression of velocity by the equivalent Mach number relation yields

$$\frac{C_{F_\infty} + 2}{\mu} = 2\frac{M_e a_e}{M_\infty a_\infty} \tag{13-2}$$

By using the identity,

$$\frac{a_e}{a_\infty} \equiv \frac{a_e}{a_e^0}\frac{a_\infty^0}{a_\infty}\frac{a_e^0}{a_\infty^0}$$

the compressible flow relations,

$$a = \sqrt{\gamma \mathcal{R} T}$$

$$\frac{T^0}{T} = 1 + \frac{\gamma - 1}{2} M^2$$

and the simplifying assumption that the gas properties and specific heat ratios are constant throughout the engine, there results

$$\frac{C_{F_\infty} + 2}{\mu \left(\frac{T_e^0}{T_\infty^0}\right)^{\frac{1}{2}}} = 2 \frac{\left(1 + \frac{\gamma - 1}{2} M_\infty^2\right)^{\frac{1}{2}}}{M_\infty} \frac{M_e}{\left(1 + \frac{\gamma - 1}{2} M_e^2\right)^{\frac{1}{2}}} \tag{13-3}$$

In deriving Eq. 13-2 it was assumed that the nozzle discharges at ambient static pressure ($p_e = p_\infty$). It therefore follows that

$$\frac{p_e^0}{\left(1 + \frac{\gamma - 1}{2} M_e^2\right)^{\frac{\gamma}{\gamma - 1}}} = \frac{p_\infty^0}{\left(1 + \frac{\gamma - 1}{2} M_\infty^2\right)^{\frac{\gamma}{\gamma - 1}}} \tag{13-4}$$

After substituting from Eq. 13-4 into 13-3 for the discharge Mach number M_e, and simplifying, there is obtained

$$\frac{C_{F_\infty} + 2}{\mu \left(\frac{T_e^0}{T_\infty^0}\right)^{\frac{1}{2}}} = \frac{2}{M_\infty} \left\{ M_\infty^2 + \frac{2}{\gamma - 1} \left[1 - \left(\frac{p_\infty^0}{p_e^0}\right)^{\frac{\gamma - 1}{\gamma}} \right] \right\}^{\frac{1}{2}} \tag{13-5}$$

Eq. 13-5 shows that the thrust characteristics of a ramjet engine depend on the over-all total pressure ratio, the total temperature ratio, and the flight Mach number. This dependency is graphically shown in Fig. E,13a for a value of $\gamma = 1.4$. It must be emphasized that only qualitative significance can be attached to thrust calculations from Eq. 13-5 and Fig. E,13a because of the simplifying assumptions regarding the constant state of the gas through the engine. The general trends obtained from the generalized thrust coefficient are valuable, however, in the evaluation of the causes of thrust changes calculated by more precise methods for systematic ranges of important variables.

For example, Fig. E,13a illustrates the relatively diminishing influence of moderate total pressure losses in the engine cycle as flight Mach number is increased. For a constant total temperature ratio across the engine, the thrust per pound of air flow handled is very sensitive to pressure losses at subsonic and low supersonic flight speeds. At higher Mach numbers, however, appreciable pressure losses must be encountered for substantial thrust reductions below the ideal value to occur.

This diminishing importance of moderate pressure losses as flight

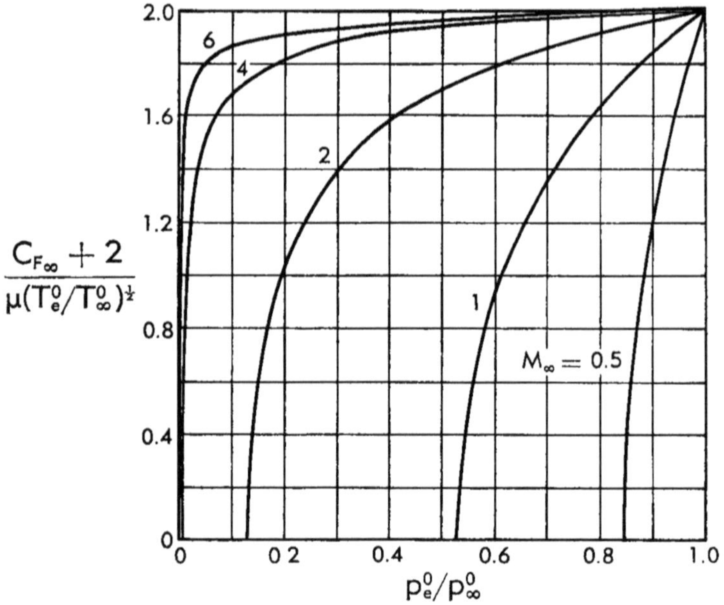

Fig. E,13a. Generalized thrust coefficient.

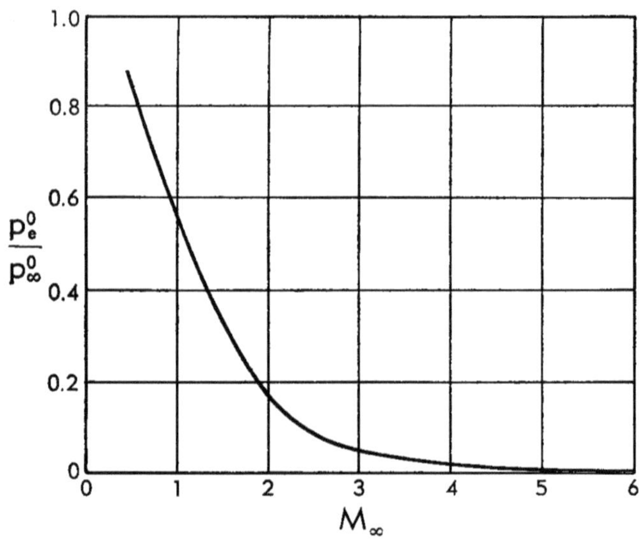

Fig. E,13b. Over-all pressure ratio yielding zero thrust with stoichiometric combustion.

speed is increased is further illustrated in Fig. E,13b, which shows the lowest over-all total pressure ratio for which the ramjet engine will produce a net internal thrust with stoichiometric combustion of a typical hydrocarbon fuel ($f = 0.0678$). At a flight Mach number of 0.5 the engine will suffer an internal drag if the over-all losses exceed approximately 15 per cent of the free stream total pressure. From Fig. E,9c it is apparent by extrapolation that these losses would be exceeded in the combustion process alone, thus illustrating the deficiencies of the ramjet at low speeds. At a flight Mach number of 6, on the other hand, the internal thrust force does not drop to zero until almost the entire free stream total pressure is dissipated through various losses within the engine cycle.

It has been shown in Art. 4 that the several parameters for evaluating ramjet engine efficiency (specific fuel consumption, specific impulse, and engine efficiency) are all directly or inversely proportional to the thrust coefficient defined by Eq. 13-5. Hence Fig. E,13a shows that the engine efficiency is markedly affected by pressure losses of any magnitude at subsonic and low supersonic speeds, but only by very large pressure losses in the higher speed range.

E,14. Effects of Fuel-Air Ratio and Flight Speed. A general performance map for on-design ramjet engine operation over a range of flight Mach numbers and fuel-air ratios is presented in Fig. E,14a. The fuel-air ratios for this and succeeding calculations have been divided by the stoichiometric value ($f = 0.0678$) to yield an equivalence ratio E_R, which indicates the fraction of stoichiometric fuel involved in the combustion process. The equivalence ratios shown correspond to the effective values of fuel-air ratio; that is, the theoretical values which would yield the assumed heat release. These effective fuel-air ratios are related to the actual fuel-air ratios by the relation

$$f_{\text{eff}} = \eta_b f \tag{14-1}$$

Thus the apparent engine efficiency on the ordinate of Fig. E,14a must be multiplied by the combustion efficiency to yield a true engine efficiency.

The calculations of this and succeeding performance curves have been performed using the enthalpy method. The air tables of [*1*] were used for all calculations, it being assumed for simplicity that the products of combustion retained the properties of air. Spot checks by more accurate techniques using the gas charts of [*5*] indicate that negligible error was introduced by this procedure.

The curves of Fig. E,14a correspond to a constant combustion chamber inlet Mach number, completely expanded exhaust nozzle without loss, and a constant diffuser efficiency, η_d, for all conditions. The diffuser efficiency is derived in [*9*] and VII,E.

As might be expected, Fig. E,14a shows that at any flight Mach number the thrust of the ramjet engine increases continuously as the equivalence ratio increases. The general level of engine efficiency increases as flight speed is increased for the assumed diffuser efficiency schedule. At the lower flight speeds, the maximum engine efficiency is achieved with the lowest values of equivalence ratio. At Mach numbers of 4 and above, the maximum efficiency condition is shifted in the direction of higher equivalence ratios.

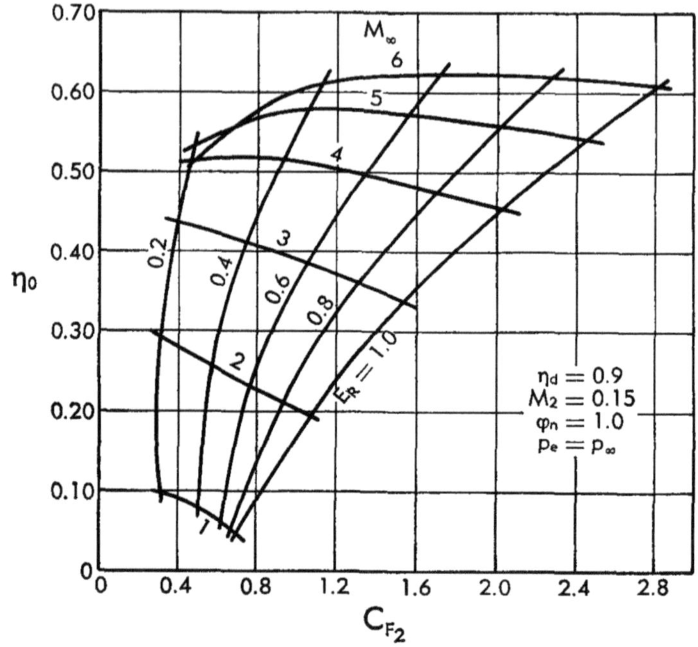

Fig. E,14a. Engine performance map.

The latter effect is also illustrated by Fig. E,14b which indicates the variation in specific fuel consumption with flight Mach number for a range of equivalence ratios. In addition to the shift in optimum equivalence ratio as flight speeds are increased, these calculations also show that minimum specific fuel consumptions are encountered in the speed range from 2.3 to 3.5, depending on equivalence ratio, in contrast to the generally constant increase in engine efficiency with increasing flight speed through Mach number 6. This incongruity is due to the small increase in specific fuel consumption at higher than optimum speeds as compared with the velocity change of the vehicle. As indicated in Art. 12, the engine efficiency offers a better indication of the range potentialities of the engine.

It is not possible from an examination of Fig. E,14a to categorically delineate conditions of optimum engine operation at any particular flight

Mach number. At a Mach number of 3, for example, the engine having an equivalence ratio of 0.2 is 30 per cent more efficient than an engine operating with an equivalence ratio of 1.0, but for a given installation would be almost four times as large as the high equivalence ratio engine. If engine weight is not critical for the installation, the greatest range for a given gross vehicle weight, or the least gross weight for a required

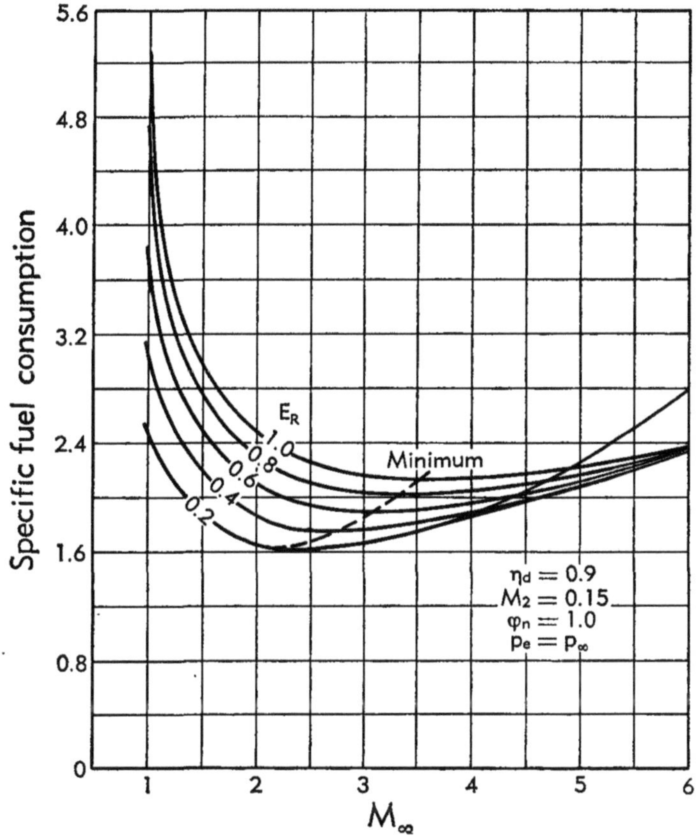

Fig. E,14b. Specific fuel consumption.

range, would be achieved with the larger, more efficient engine. To the extent that engine weight significantly affects the quantity ln (W_0/W_e), the optimum engine condition would shift to the higher equivalence ratios.

E,15. Influence of Diffuser Pressure Recovery. Except at subsonic and low supersonic flight speeds the major source of pressure losses in the ramjet engines is likely to be the diffuser. As shown in VII,E, the pressure recoveries for simple diffuser designs rapidly deteriorate as flight speed is increased and complex diffuser geometries must be employed to

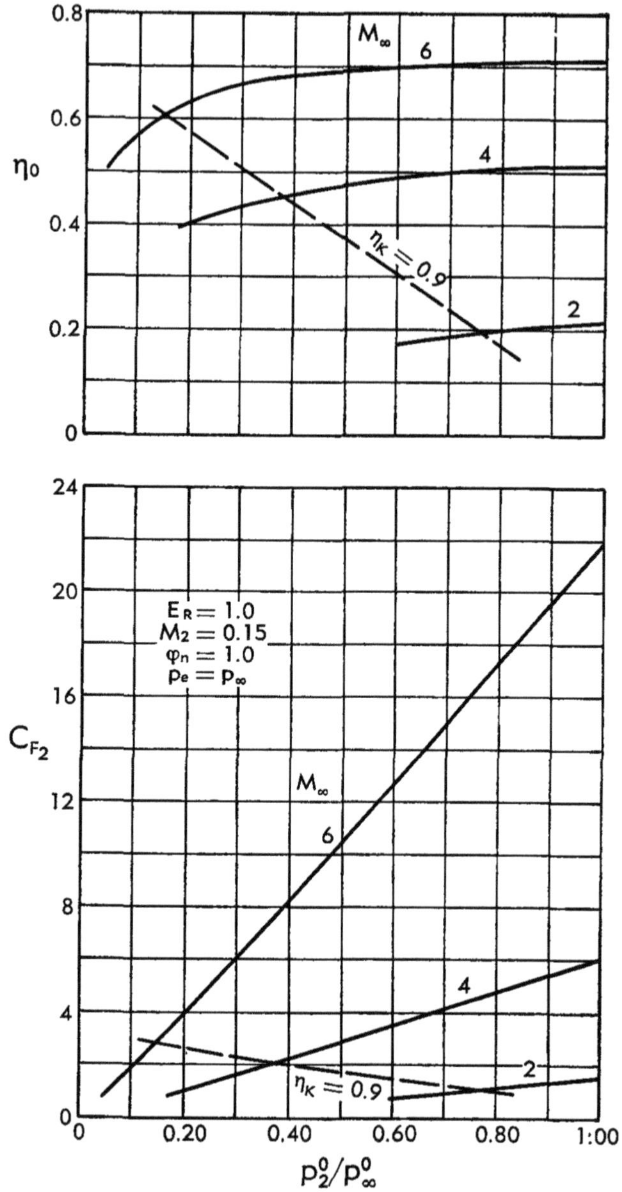

Fig. E,15. Effect of diffuser pressure recovery on ramjet performance.

attain higher levels of diffuser performance at Mach numbers above 2.5. Fig. E,15 indicates that very high pressure recoveries at the higher flight speeds are principally important only where engine size and weight are primary factors, however.

Consider the effects of a 40-per cent loss in free stream total pressure for the conditions indicated. At a flight Mach number of 2, the efficiency is 18 per cent lower than for isentropic diffusion. At Mach number 4 this loss in efficiency is reduced to 6 per cent and at Mach number 6 to only 2 per cent. This trend is qualitatively predictable from the generalized thrust coefficient of Fig. E,13a.

At each flight speed, the engine with 60-per cent diffuser pressure recovery will have about twice the frontal area of an engine with an isentropic diffuser, however. Consequently, design conditions may exist in which the attainment of high pressure recovery assumes greater importance. (The pressure recovery has another effect not considered at this point. The combustion pressure level is proportional to diffuser pressure recovery; consequently altitude limits are effectively increased for a given combustion efficiency as pressure recovery increases.)

The pressure recoveries corresponding to a diffuser kinetic energy efficiency of 0.9 are indicated in Fig. E,15. This efficiency schedule has been used in most of the computations in this article as being representative of diffuser pressure recoveries attainable without excessively elaborate diffusion geometry. The resultant efficiencies and thrust coefficients at the higher Mach number are low enough to suggest that improved pressure recoveries above this level would be warranted even without consideration of engine weight.

E,16. Influence of Diffuser Discharge Mach Numbers. The diffuser discharge Mach number influences the thrust per pound of air flow, and consequently the propulsive efficiency of the ramjet through its influence on the total pressure losses across the flame holder and the total pressure losses in the combustion chamber. In addition, the diffuser discharge Mach number affects the ratio of combustor area to free stream tube area for a fixed diffuser pressure recovery, and hence influences the thrust coefficient based on the combustor area. Typical representations of these effects are shown in Fig. E,16a for a range of flight Mach numbers.

Over the range of diffuser discharge Mach numbers from 0.05 to the value required to sustain choking in the combustion chamber without nozzle contraction, the changes in flame holder and combustion pressure losses are significant. At $M_2 = 0.05$, for example, the pressure losses are only from 0.5 to 1.5 per cent of the total pressure after diffusion, whereas with choking in the combustor the losses range from 17 to 23 per cent depending on flight Mach number and the consequent total temperature ratio obtainable from a constant equivalence ratio (see Fig. E,9a). As can be deduced from the generalized thrust coefficient (Art. 13) the effect on C_{F_∞} and hence η_0 is less pronounced. At a Mach number of 2, operation with straight pipe choking is 10 per cent less efficient than at $M_2 = 0.05$. At a flight Mach number of 6, the decrement in efficiency due to high diffuser discharge Mach numbers is only about

2 per cent. As previously discussed, the effects of diffuser discharge Mach number on engine size are much more pronounced, with the influence of M_2 increasing as flight speed is increased.

In addition to the effects on internal flow parameters, the diffuser discharge Mach number, through its effect on the required flow area

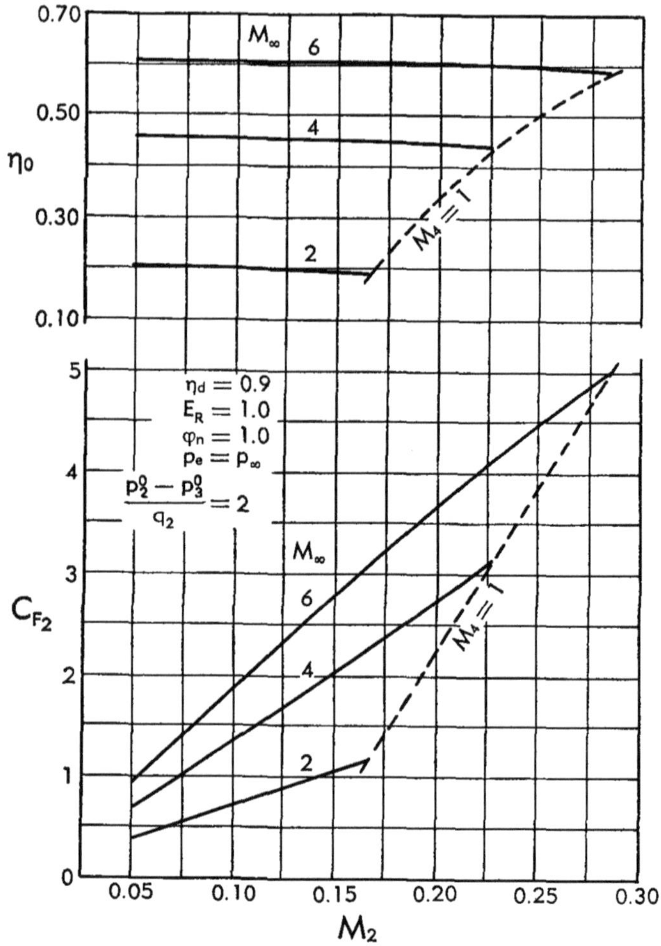

Fig. E,16a. Typical effects of diffuser discharge Mach number.

ratio between the inlet and the combustor, may change the engine drag. The effects of considering these drag changes for an isolated engine are illustrated in Fig. E,16b (these calculations are based on slightly different diffuser pressure recoveries and equivalence ratios than those of Fig. E,16a; hence no direct comparison of the two figures should be attempted). The drags in this example have been estimated by assuming the external geometry to consist of conical and cylindrical segments of reasonable length-diameter ratios and using pressure drag coefficients from [*10*] with friction drag coefficients corresponding to fully turbulent compressible flow.

In contrast to the performance analysis without drag consideration in which the maximum engine efficiency is found to occur at the lowest

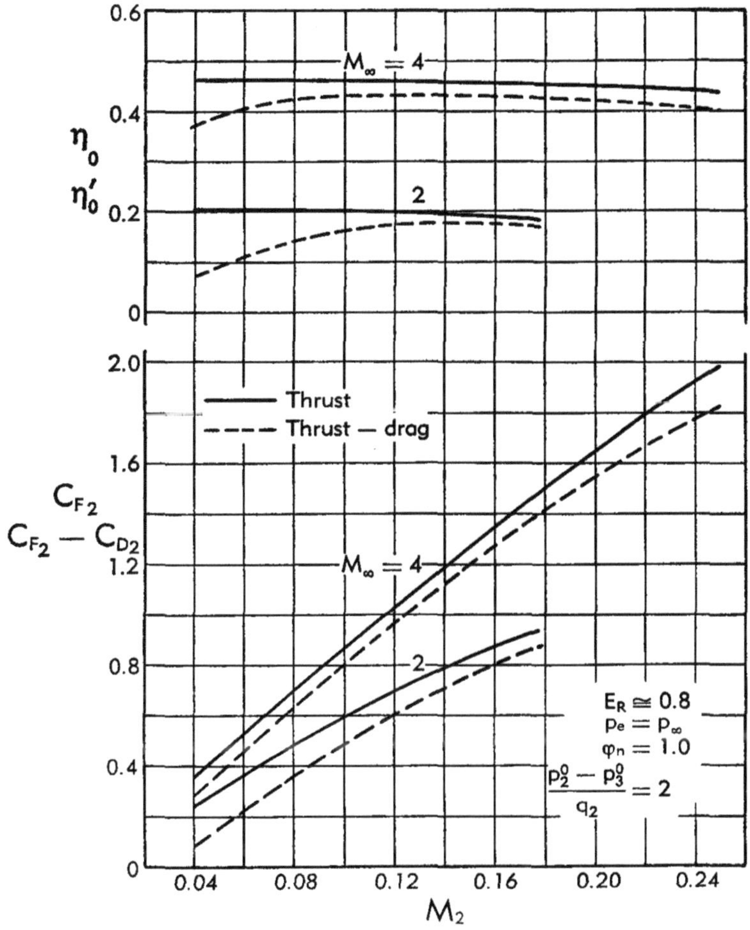

Fig. E,16b. Drag effect on diffuser discharge Mach number selection.

values of diffuser discharge Mach number, the inclusion of reasonable engine drag values causes the propulsive efficiency to optimize at diffuser discharge Mach numbers in the range from 0.12 to 0.16 at flight Mach numbers of both 2 and 4.

It is significant that the maximum engine efficiency at Mach number 4, with drag considered, occurs at a diffuser discharge Mach number that is low enough to cause the combustor area to be larger than the inlet area, so that diffuser pressure drag is not entirely eliminated. This result stems from the balance between thrust per pound of air flow and drag. The maximum efficiency occurs at the diffuser discharge Mach number at which the rate of change of thrust due to increasing internal pressure losses exceeds the rate of change of drag due to decreasing project frontal area.

E,17. Effects of Flame Holder Pressure Loss Coefficient. From the preceding discussion on the slight effects of diffuser discharge Mach number on engine efficiency it can be inferred that the effect of the flame holder pressure loss coefficient at a fixed diffuser discharge Mach number

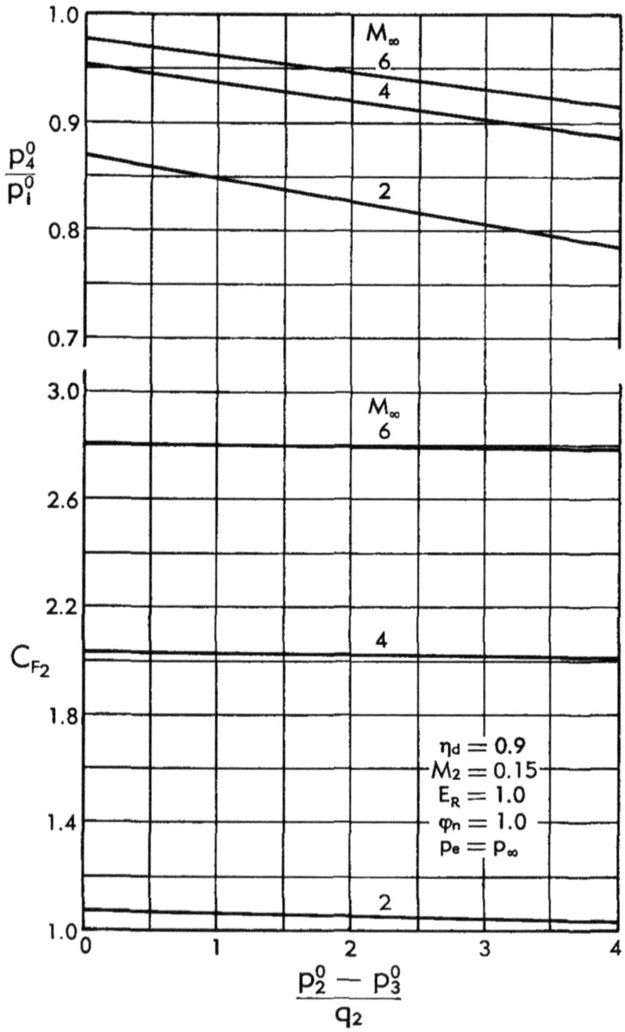

Fig. E,17. Effects of flame holder pressure drop.

is also small. This is verified by the calculations shown in Fig. E,17 for a range of flame holder pressure loss coefficients and flight speeds.

Although increases in the loss coefficient from 0 to 4 have an appreciable effect on the total pressure ratio across the flame holder and combustion zone (the combustion losses are influenced by the flame holder through the change in Mach number at the start of combustion), these losses do not result in significant thrust variations for otherwise constant ramjet conditions. At a flight Mach number of 2, the thrust change in going from a flame holder having no pressure loss to one that has a loss coefficient of 4 results in a thrust change of less than 5 per cent. Inasmuch as the diffuser geometry is unaltered by the changes in flame holder loss characteristics, the change in C_{F_∞} and η_0 is, percentagewise, identical to the change in C_{F_2}.

E,18. Nozzle Effects. The exhaust nozzle can exert a powerful influence on the performance of a ramjet engine if the nozzle does not adequately expand the exhaust gas or if it produces large pressure losses. The importance of adequate expansion can be inferred from the generalized thrust coefficient of Art. 13 through the concept of an effective total pressure loss due to underexpansion.

A nozzle terminating at station e (see Fig. E,1) with a discharge pressure greater than ambient pressure could hypothetically be extended so that through some form of unspecified pressure losses the discharge pressure could be reduced to the ambient value with no change of momentum between station e and the new discharge station f. The ramjet engine, extended to station f, would then be susceptible to the analysis of Art. 13.

The hypothetical nozzle extension will have to satisfy the mass, momentum, and energy conservation equations so that

$$m_f V_f = m_e V_e + A_e(p_e - p_\infty) \tag{18-1}$$

$$m_f = m_e = \rho_e V_e A_e = \frac{p_e V_e A_e}{\mathfrak{R} T_e} \tag{18-2}$$

and $T_f^0 = T_e^0$, for simplification. Combining Eq. 18-1 and 18-2 there results

$$\frac{V_e V_f}{\mathfrak{R} T_e} = \frac{V_e^2}{\mathfrak{R} T_e} + 1 - \frac{p_\infty}{p_e} \tag{18-3}$$

By converting the velocities to Mach numbers

$$\gamma M_e M_f \left(\frac{T_f}{T_e}\right)^{\frac{1}{2}} = \gamma M_e^2 + 1 - \frac{p_\infty}{p_e} \tag{18-4}$$

From conservation of energy,

$$\frac{T_f}{T_e} = \frac{T_f}{T_f^0}\frac{T_e^0}{T_e} = \frac{1 + \frac{\gamma - 1}{2} M_e^2}{1 + \frac{\gamma - 1}{2} M_f^2} \tag{18-5}$$

The last term in Eq. 18-4 can be written

$$\frac{p_\infty}{p_e} = \frac{p_\infty}{p_e^0}\left(1 + \frac{\gamma - 1}{2} M_e^2\right)^{\frac{\gamma}{\gamma - 1}} \tag{18-6}$$

After substituting Eq. 18-5 and 18-6 into Eq. 18-4 and simplifying, there results

$$\frac{M_f}{\left(1 + \frac{\gamma - 1}{2} M_f^2\right)^{\frac{1}{2}}} = \frac{1 + \gamma M_e^2 - \frac{p_\infty}{p_e^0}\left(1 + \frac{\gamma - 1}{2} M_e^2\right)^{\frac{\gamma}{\gamma - 1}}}{\gamma M_e\left(1 + \frac{\gamma - 1}{2} M_e^2\right)^{\frac{1}{2}}} \tag{18-7}$$

for which M_f can be determined as a function of the nozzle pressure ratio p_∞/p_e and the nozzle discharge Mach number M_e.

The apparent total pressure loss in the hypothetical nozzle extension is

$$\frac{p_f^0}{p_e^0} = \frac{p_f^0}{p_f}\frac{p_f}{p_e^0} = \frac{p_\infty}{p_e^0}\left(1 + \frac{\gamma - 1}{2} M_f^2\right)^{\frac{\gamma}{\gamma - 1}} \tag{18-8}$$

which can be solved from Eq. 18-7.

The apparent total pressure ratio calculated from Eq. 18-7 and 18-8 for $\gamma = 1.3$ is shown in Fig. E,18a as a function of the nozzle discharge Mach number for a range of nozzle pressure ratios. It is obvious that as the nozzle pressure ratio, p_∞/p_e^0, decreases (corresponding to increasing flight speed) the necessity for adequate nozzle reexpansion is increased. This is emphasized by Fig. E,18b which presents the effective nozzle total pressure ratio for a convergent ($M_e = 1$) exhaust nozzle.

If the flow process ahead of the exhaust nozzle could be effected without pressure losses, the nozzle pressure ratio p_∞/p_e^0 would be a minimum for each flight Mach number, and the effective nozzle losses would be a maximum. At a Mach number of 4, the use of a convergent instead of a completely expanded exhaust nozzle would cause an effective pressure loss of approximately 93.5 per cent insofar as the complete engine cycle corresponding to the analysis of Art. 13 is concerned. In other words, the thrust coefficient based on the free stream area of the air flow, and consequently the engine efficiency, for an engine operating without internal pressure loss but with a convergent exhaust nozzle, would be no greater than for another engine having a completely expanded exhaust nozzle but with cumulative internal pressure losses amounting to 93.5 per cent of the free stream total pressure.

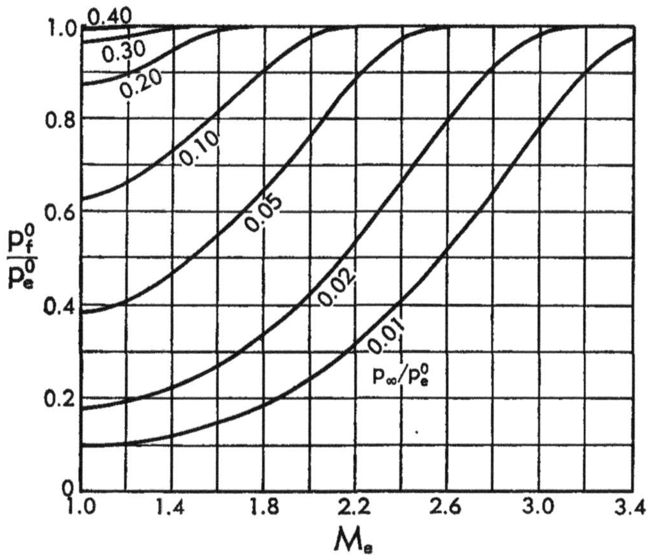

Fig. E,18a. Effective total pressure ratio from incomplete nozzle expansion. $\gamma = 1.3$.

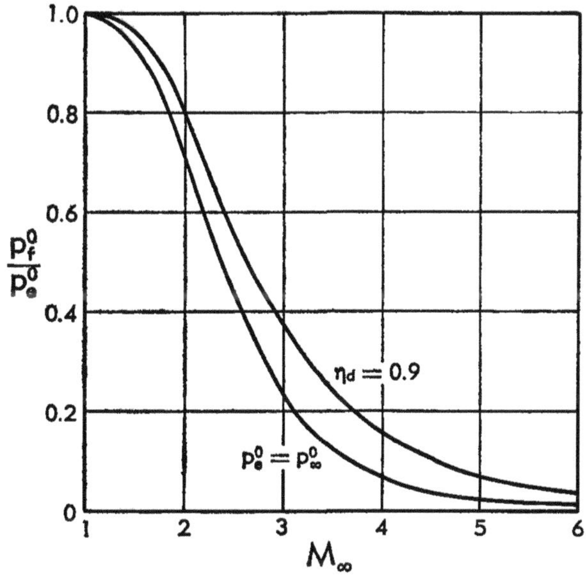

Fig. E,18b. Effective nozzle total pressure ratio with convergent nozzle.

The presence of pressure losses in the engine tends to reduce the nozzle pressure ratio p_∞/p_e^0 and hence to reduce the apparent pressure loss due to underexpansion. This is illustrated in Fig. E,18b by the higher effective nozzle total pressure ratio corresponding to a diffuser efficiency of 0.9. However, the change in effective nozzle losses due to internal pressure losses is not sufficient to markedly change the requirements for correct nozzle expansion.

The quantitative influence of nozzle expansion on ramjet performance is illustrated for a flight Mach number of 4 in Fig. E,18c. On the upper portion of the figure is shown the engine efficiency-thrust coefficient map

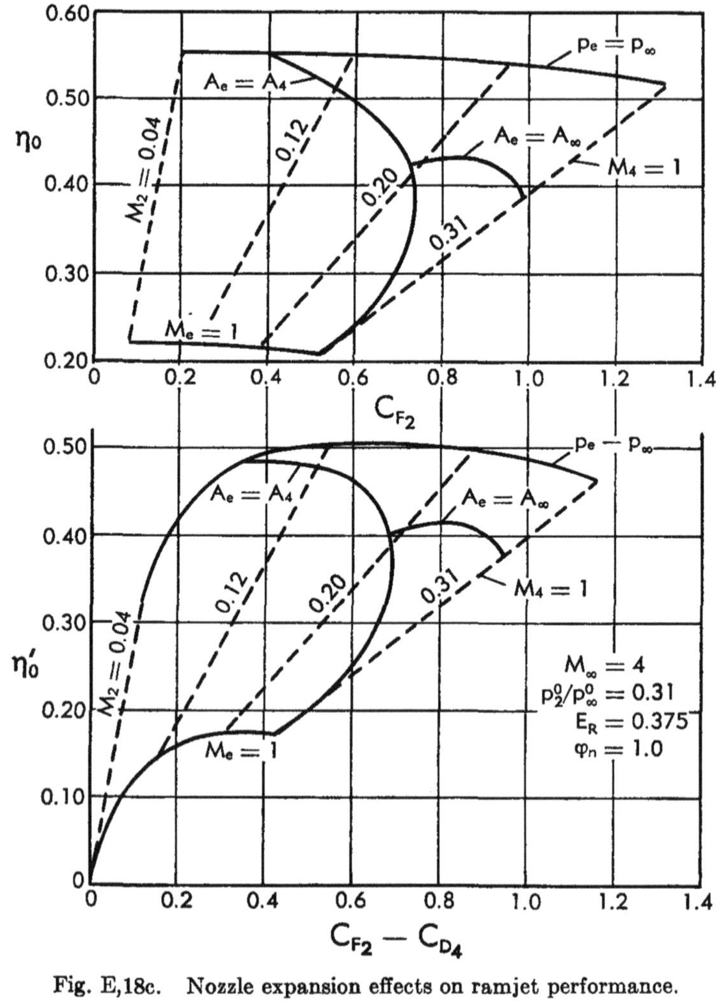

Fig. E,18c. Nozzle expansion effects on ramjet performance.

for a range of diffuser discharge Mach numbers and nozzle expansions. A similar map on the lower part of the figure takes the engine drag for an isolated engine into account.

The maps are bounded on the lower limit by a convergent nozzle characteristic, and on the upper limit by a completely expanded nozzle. Also shown are the performance lines for nozzle expansion to the combustion chamber area ($A_e = A_4$) and to the inlet area ($A_e = A_\infty$) for the range of diffuser discharge Mach numbers in which the inlet area exceeds the combustion chamber area.

Neglecting the effect of drag, the upper portion of Fig. E,18c shows that maximum efficiency is obtained with the lowest diffuser discharge Mach number and with the completely expanded exhaust nozzle, and that maximum thrust is obtained with choking in the combustion chamber ($M_4 = 1$) and a nozzle expanding to ambient pressure. The use of a convergent exhaust nozzle would reduce both the efficiency and the maximum thrust about 60 per cent.

Consideration of the influence of engine drag shows the same trends with diffuser discharge Mach number previously discussed in Art. 16, namely that the optimum Mach number with complete expansion is increased to about 0.16 from an efficiency viewpoint. For a convergent nozzle both the optimum efficiency and thrust are attained with the highest possible diffuser discharge Mach number. The lowest engine drag is obtained with $A_e = A_4$ up to diffuser discharge Mach numbers of about 0.19 and with $A_e = A_\infty$ for higher values of M_2. It is noteworthy that neither maximum thrust minus drag nor maximum efficiency is obtained with this minimum drag. Furthermore, as mentioned in Art. 16, the highest engine efficiency is obtained with a diffuser discharge Mach number of 0.16, which results in a combustion chamber larger than the inlet so that a finite inlet drag as well as a nozzle drag is present.

In addition to performance losses due to incomplete nozzle expansion, other losses may occur because of friction and shock losses within the nozzle. There are many ways of expressing these losses; one of the easiest to evaluate experimentally is a nozzle coefficient φ_n, defined as the ratio of the actual nozzle thrust to the ideal nozzle thrust for a completely expanded nozzle operating at the same nozzle pressure ratio. When the actual nozzle is completely expanded, φ_n becomes the ratio of the actual jet velocity to the ideal jet velocity.

From the development of the thrust coefficient equations in Art. 4 and 13 it can be shown that the thrust coefficient based on the free stream tube area of the flow entering the engine, C_{F_∞}, is the difference between the jet thrust coefficient based on the same area and a free stream momentum coefficient $mV_\infty/q_\infty A_\infty$ which has a constant value of 2, or

$$C_{F_\infty} = C_{J_\infty} - 2 \tag{18-9}$$

Consequently it can be shown that the thrust coefficient with nozzle losses is related to the coefficient without losses, $(C_{F_\infty})_i$, by the equation

$$C_{F_\infty} = \varphi_n[(C_{F_\infty})_i + 2] - 2 \tag{18-10}$$

and that the coefficient based on the combustor area is, similarly,

$$C_{F_2} = \varphi_n[(C_{F_\infty})_i + 2]\frac{A_\infty}{A_2} - 2\frac{A_\infty}{A_2} \tag{18-11}$$

Thus, for constant engine operating conditions, the thrust coefficient varies linearly with φ_n at a slope determined by $(C_{F_\infty})_i$ and A_∞/A_2. Typical variations are shown in Fig. E,18d for several flight Mach numbers and arbitrary engine operating conditions. In general C_{F_∞} will decrease as the flight Mach number is increased at a constant fuel-air ratio because of decreasing total temperature ratio across the engine.

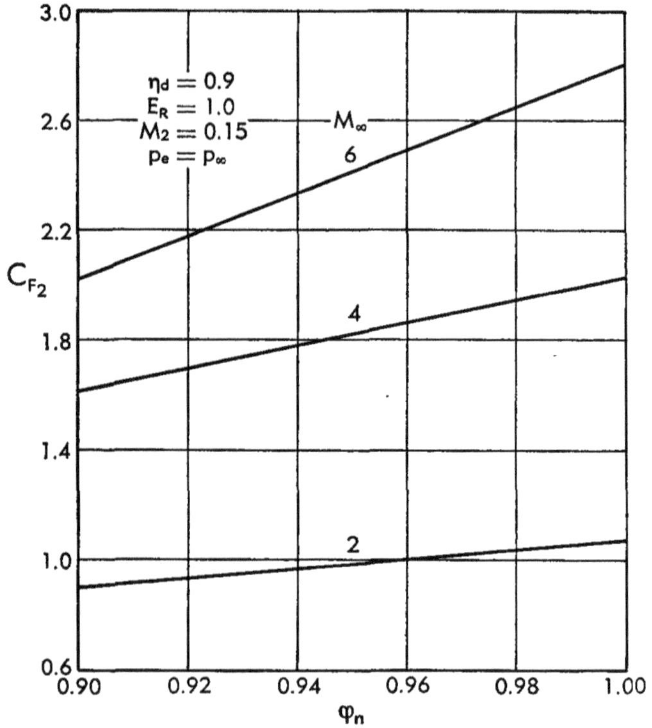

Fig. E,18d. Effect of nozzle losses.

Consequently the adverse effect of nozzle losses increases as flight speed is increased. Inasmuch as the changes in nozzle losses do not affect the diffuser conditions, the percentage changes in C_{F_2} in Fig. E,18d due to nozzle losses are numerically equal to engine efficiency changes.

E,19. Off-Design Operation. When the engine design is fixed to yield maximum engine performance in terms of either thrust or propulsive efficiency for one operating condition, the performance at other operating conditions will be less than optimum to a degree that depends on the amount of control variables incorporated in the engine. Normally, fuel-flow control can be provided for some desired schedule of operation (see Chap. 4), but the inclusion of variable inlet and exhaust nozzle geometries, though desirable for some purpose, results in a high degree of complexity, both in the mechanical design of the engines and in the control system.

A comparison of ramjet engine operation at off-design speeds with fixed and with variable exhaust nozzles is indicated in Fig. E,19a. The design point for each engine has been assumed to be at a flight Mach number of 2 with stoichiometric fuel-air ratio. It has been assumed that a

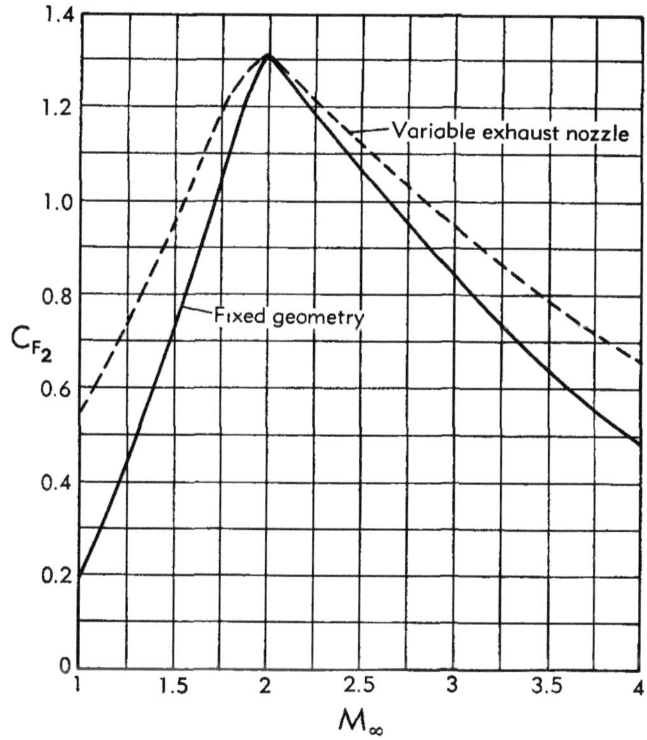

Fig. E,19a. Typical performance of ramjet engine at off-design speeds. $M_\infty = 2$.

fuel control system is incorporated which maintains this fuel-air ratio at all flight speeds.

With the fixed geometry engine the thrust coefficient decreases rapidly at flight speeds both above and below the design speed. When the variable exhaust nozzle is employed the change in thrust coefficient with speed changes is reduced. The reasons for these engine behaviors can be deduced from Fig. E,19b, which shows the variation of several significant internal flow parameters with flight speed.

With the fixed geometry engine the Mach number at the downstream end of the combustion chamber remains essentially constant (see Eq. 10-18). Since the total temperature ratio across the combustion chamber varies with flight Mach number for a constant fuel-air ratio because of the variation in combustor inlet stagnation temperature (Fig. E,9a), the diffuser discharge Mach number also varies with M_∞ (see Eq. 9-19, for example). In order to satisfy this nozzle-dictated diffuser discharge

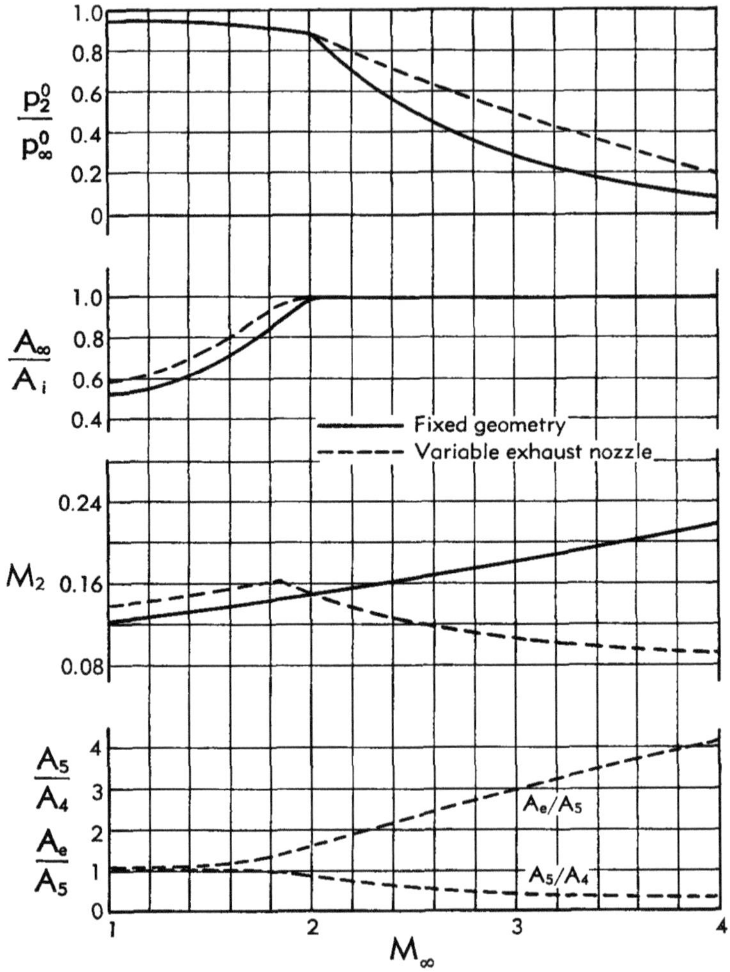

Fig. E,19b. Variation of flow parameters at off-design speed. Design $M_\infty = 2$.

Mach number, the diffuser flow must be supercritical; i.e. the terminal shock wave regresses into the diffuser passage and the diffuser inlet intercepts a full stream tube of air ($A_\infty = A_i$). Under this operating condition the resultant diffuser pressure recovery is given by Eq. 7-27 as a function of the flight and diffuser discharge Mach numbers and the diffuser area ratio A_2/A_i. The diffuser recovery decreases rapidly as the flight speed is increased and in conjunction with the decreasing total temperature ratio causes the decrease in thrust coefficient shown in Fig. E,19a.

At flight speeds less than $M_\infty = 2$ with the fixed geometry engine, the diffuser discharge Mach number continues to decrease as a result of the nozzle constriction and the diffuser flow becomes subcritical, i.e. the diffuser mass flow is less than the maximum possible. From the assumed subcritical pressure recovery schedule indicated in Fig. E,19b the area ratio A_∞/A_2 required to satisfy the diffuser discharge Mach number at a given flight speed, and consequently the mass flow ratio A_∞/A_i can be calculated. The rapid decrease in thrust of the fixed geometry engine shown in Fig. E,19a is primarily a result of the reduced engine air flow at below-design speeds. (The spilled air flow also causes an additive drag to be incurred (VII,E) which, for the purposes of Fig. E,19a has been subtracted from the calculated thrust to yield the net thrust shown. Though not an inappreciable term, the additive drag does not alter the basic trends of the thrust at below-design speeds and hence is not shown separately.)

If a variable exhaust nozzle is employed, the deleterious constricting effects of the nozzle can be largely eliminated. At above-design speeds, for example, the exhaust nozzle throat can be reduced to maintain critical diffuser flow, or maximum pressure recovery without flow spillage. The resultant pressure recovery for fixed inlet geometry is shown in Fig. E,19b. (It has been assumed in the calculations of the variable exhaust nozzle engine that the nozzle outlet area is simultaneously adjusted to give complete nozzle expansion. The necessary variations in throat area and outlet area are indicated in Fig. E,19b.) The increased performance of the variable nozzle engine at above-design speeds is due to the higher operating pressure recovery of the cycle.

At below-design flight speeds, the nozzle throat would continue to be increased as flight speed is decreased in order to maintain maximum air flow through the engine. With a design diffuser discharge Mach number of 0.15 at flight Mach number 2, however, the combustion chamber would choke at a flight Mach number of 1.8, so that further nozzle control could not be effected and the engine would operate as a fixed geometry engine at lower flight speeds, though at a higher mass flow ratio than the other engine. The increased performance of the variable geometry engine at Mach numbers less than 2 is largely the result of this increased air flow since the pressure recovery of the diffuser has been assumed independent of the subcritical operating condition.

At the present level of ramjet development, the potentially beneficial effects of variable engine geometry indicated in Fig. E,19a are generally sacrificed in favor of simplicity and resultant reliability. Consequently the ramjet engine is largely a constant flight speed engine.

CHAPTER 4. RAMJET CONTROL SYSTEMS

DEMARQUIS DALE WYATT

E,20. Introduction. The control system for a ramjet engine must perform two basic functions: (1) maintain desired engine performance throughout the flight of the vehicle, and (2) minimize departure from desired performance during transients resulting from external disturbances acting on the engine-airframe system. This second element, which exists for all engine-airframe systems, assumes greater importance for the ramjet engine because the engine performance is greatly affected by flight velocity.

It is difficult to discuss the ramjet engine control problem in general terms because different criteria for the engine and the control system design must be applied for each different application. There are, however, certain overlapping considerations for widely different applications. The present discussion will be confined to these common considerations.

E,21. Performance of Fixed Geometry Engines. In Art. 19 the off-design-speed performance of a ramjet engine over an extended range of flight speeds was briefly considered. The variation of several internal flow and performance quantities pertinent to the engine control problem will now be considered in greater detail for a smaller range of flight speeds corresponding to potential speed excursions during externally imposed transients. For this purpose, calculations have been made for a fixed geometry ramjet engine designed to cruise at a flight Mach number of 3.0 with a diffuser discharge Mach number of 0.199 and an engine total temperature ratio of 2.2. For simplicity in the drag calculations the exhaust nozzle has been assumed to be convergent-divergent with the expansion terminating at an area equal to the combustion chamber area, although as pointed out in Art. 18 the resultant performance is not the maximum possible with the engine.

The variation of the thrust-minus-drag coefficient for an isolated engine (corresponding to the net thrust available for propulsion of the vehicle insofar as the engine drag can be isolated from the airframe drag), is shown in Fig. E,21a as a function of the effective fuel flow rate to the engine for a range of flight Mach numbers near 3.0. (The effective fuel rate $\eta_b w_f$ in pounds per hour has been divided by $A_2 p_\infty$ to generalize the effects of engine size and altitude in the tropopause region of the atmosphere.) At any Mach number, when the fuel rate is reduced below the level required to maintain critical inlet flow, the diffuser goes into supercritical operation with a strong internal shock system and a consequent reduced pressure recovery. The combined effects of the lowered pressure recovery and the reduced engine temperature ratio cause a rapid reduction in thrust. On the other hand, when the fuel rate is increased above the value required for critical diffuser operation, the diffuser is forced into subcritical operation with bow shock spillage and the attendant high additive drag. As a consequence, although the engine operates with a high temperature ratio it does so with a reduced air flow and an increased drag, so that the net thrust increases only slightly for large increases in fuel flow.

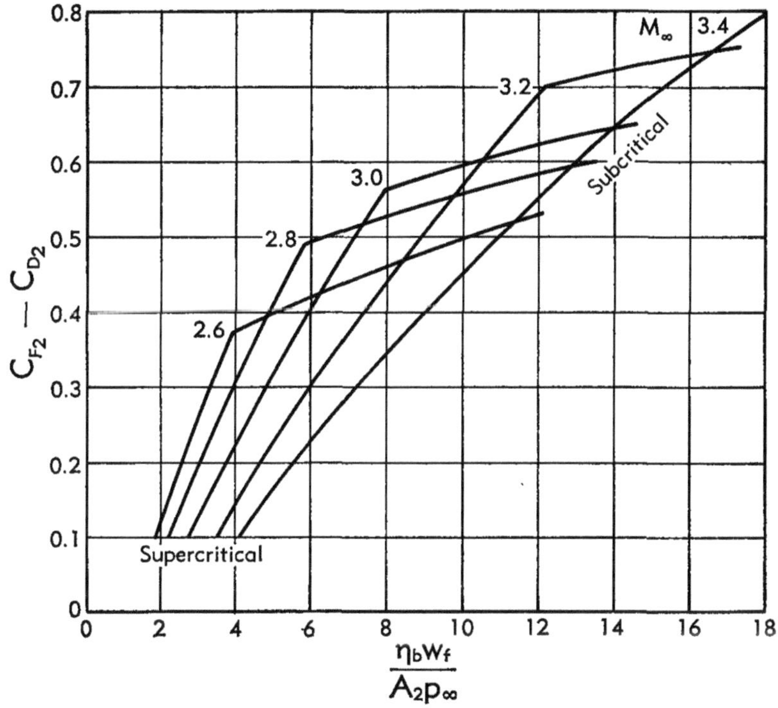

Fig. E,21a. Thrust variation with fuel flow.

Corresponding variations in the specific fuel consumption of the engine as a function of the effective fuel rate are shown in Fig. E,21b. Minimum specific fuel consumption, and hence maximum engine efficiency, occurs with critical inlet flow at each Mach number. The specific fuel consumption increases rapidly when the fuel flow forces the engine into subcritical operation. It also increases for supercritical diffuser operation, but at a lower initial rate so that for the first increments of supercritical operation there is only a slight reduction in engine efficiency.

Inasmuch as the attainment of a particular level of thrust output is the primary requisite for the engine, the specific fuel consumption has a dependent relation in the control problem. This dependency is illustrated by Fig. E,21c in which the specific fuel consumption is presented as a

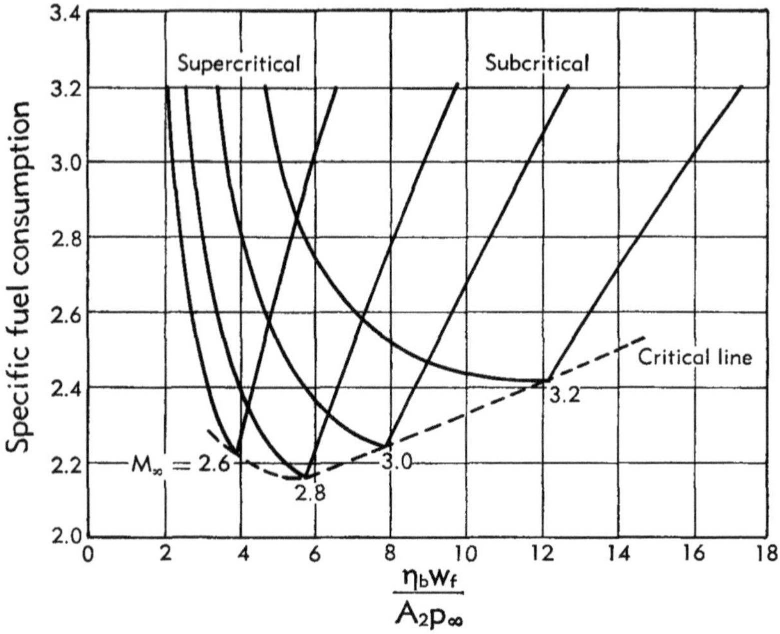

Fig. E,21b. Specific fuel consumption variation with fuel flow.

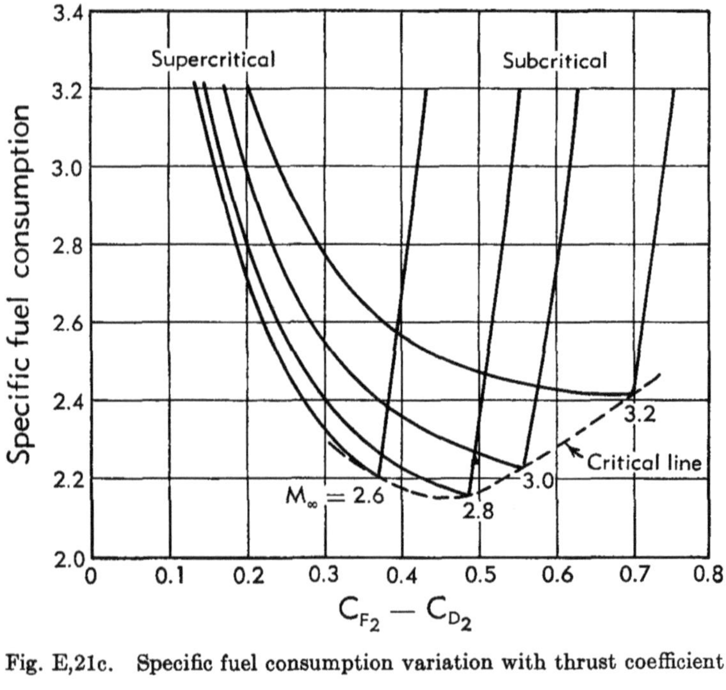

Fig. E,21c. Specific fuel consumption variation with thrust coefficient.

function of the net thrust. If, at any flight Mach number, the flight requirement demands greater thrust than can be developed with critical inlet flow, the specific fuel consumption must increase markedly. Thrust requirements that are only moderately lower than the critical diffuser flow values can be satisfied, however, with only small increases in specific fuel consumption.

E,22. Control Considerations for Fixed Geometry Engines. It is the function of the engine control to provide the proper engine output

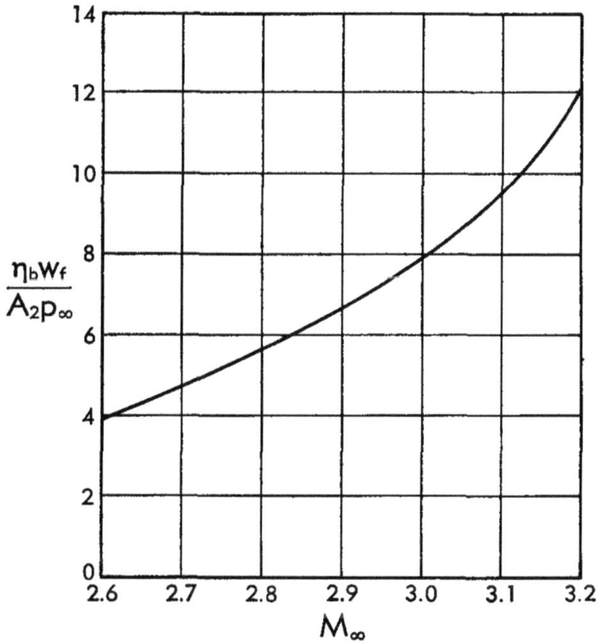

Fig. E,22. Fuel flow required for critical engine operation over a range of flight speeds.

for the particular requirements of the given flight plan. From a gross weight or range consideration it is normally desirable for the engine to provide the thrust at maximum efficiency (minimum specific fuel consumption) and for the ramjet engine this represents a most difficult control requirement. As indicated in Fig. E,21b the condition of maximum efficiency occurs at the critical diffuser condition. Thus a possible control requirement is to maintain the diffuser at its critical point. In general, one of the fundamental control considerations of the supersonic ramjet engine is the determination of the method of manipulation of the input variable to obtain this satisfactory diffuser and hence engine performance.

The fixed geometry ramjet has but one input variable which can be manipulated to obtain a given operating point; this variable is the fuel flow. The effective fuel flow required for critical diffuser operation is shown in Fig. E,22 as a function of flight Mach number. The fuel flow for this mode of engine operation (which results in minimum specific fuel consumption) increases exponentially with flight speed and varies threefold over the narrow range of flight speeds shown. One way to maintain this mode of engine operation would be to schedule the corrected fuel flow as a function of flight Mach number according to Fig. E,22. Since the ordinate of this figure contains the combustion efficiency, however, such a control scheme would only be as good as the prior knowledge of the combustion efficiency variations and the ability to meter the fuel accordingly. Therefore such a control can offer no positive assurance of proper engine operation.

It is advantageous to utilize some engine variable which is independent of the combustion efficiency, but which is directly indicative of the engine operating condition and the engine thrust output to control the fuel flow to the engine. Two engine parameters which satisfy this requirement are the ratio of diffuser discharge total pressure to ambient static pressure (diffuser pressure ratio) and the diffuser discharge Mach number. (For control purposes it is much easier to measure the diffuser pressure ratio p_2^0/p_∞ than the pressure recovery p_2^0/p_∞^0, hence the use of the former quantity for control purposes.)

E,23. Control of Diffuser Pressure Ratio for Fixed Geometry Engine. The variation of the net thrust with diffuser pressure ratio for the assumed engine is shown in Fig. E,23a for the range of flight speeds considered. The critical diffuser line indicated on the figure corresponds to the most efficient engine operating conditions and may be taken to represent the desired pressure ratio schedule for the engine to follow. A block diagram of a possible control system utilizing the control relation between diffuser pressure ratio and flight Mach number is shown in Fig. E,23b.

The system of Fig. E,23b is a typical closed loop control acting upon the requirements of a scheduled reference value of the controlled variable. Flight Mach number is sensed and this signal is fed into a scheduler, the output of which is the required value of the pressure ratio corresponding to the critical line in Fig. E,23a. The pressure ratio existing in the engine is measured and compared with the required value; and any error or difference between these values acts through a fuel control computer and serves to vary the fuel flow in such a manner as to bring the error to zero, thus attaining the desired operating point.

The system described is advantageous because it automatically corrects for the effect of pressure altitude on engine performance and is not influenced by variations in combustion efficiency. The system has a limiting feature, however, which may constitute a severe disadvantage. This limiting feature is a result of the diffuser characteristic illustrated in Fig. E,23c, where diffuser pressure ratio is plotted as a function of effective fuel flow at constant flight Mach numbers. In the supercritical

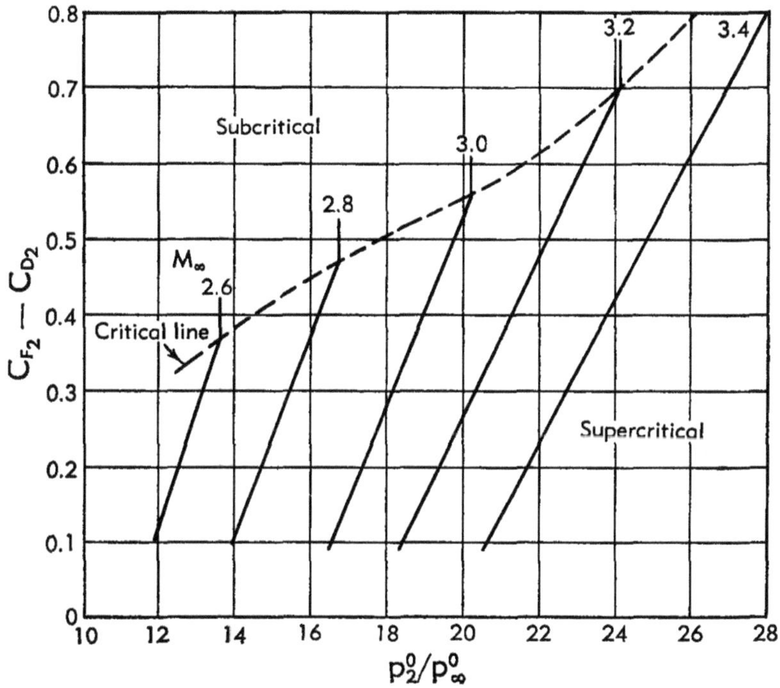

Fig. E,23a. Thrust variation with diffuser pressure ratio.

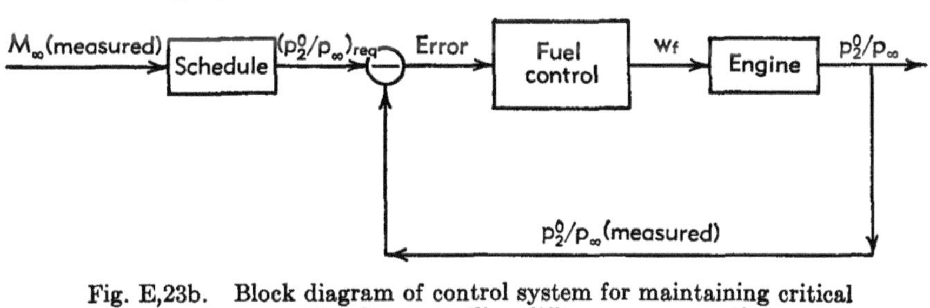

Fig. E,23b. Block diagram of control system for maintaining critical engine operation by controlling diffuser pressure ratio.

region of diffuser operation an increase in fuel flow results in an increase in pressure ratio; hence the engine gain is positive and the negative feedback in the control system provides the proper control action. At the critical diffuser point there is a discontinuity in the slope of the diffuser pressure ratio curve, the nature of the discontinuity being a function of the particular diffuser characteristics. In general there are two possible behaviors of pressure ratio in the subcritical region; the pressure ratio may have a constant value as assumed in these calculations or it may decrease with increasing fuel flow with possible attendant pulsing. In either event it is possible for the control to increase fuel flow continuously in an attempt to raise the pressure ratio. In the case of the decreasing pressure ratio characteristic, this results from the change in size of engine gain which results in positive feedback in the control. For the flat characteristic, a schedule calibration error calling for an unattainable

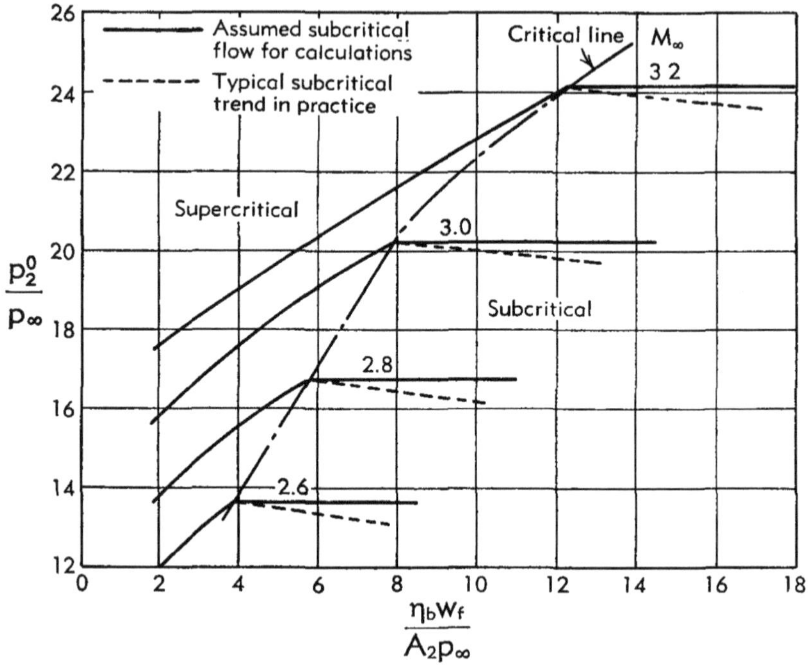

Fig. E,23c. Assumed variation of diffuser pressure ratio with effective fuel flow.

pressure ratio could initiate a continuously increasing fuel flow. Therefore, for satisfactory operation of this control scheme a maximum fuel flow limit must be incorporated in the system.

E,24. Control of Diffuser Discharge Mach Number for Fixed Geometry Engine. The use of diffuser discharge Mach numbers to control the engine fuel flow overcomes the principal disadvantage of the preceding system. The control system using this parameter is essentially the same as that shown in Fig. E,23b for the pressure ratio system, the difference being that diffuser discharge Mach number is measured and the output of the scheduler is the required value of this parameter, according to the critical line shown in Fig. E,24a. This system possesses all of the advantages of the system operating on diffuser pressure ratio in that it compensates for changes in pressure due to altitude and is unaffected by variations in combustion efficiency. In addition, as shown by

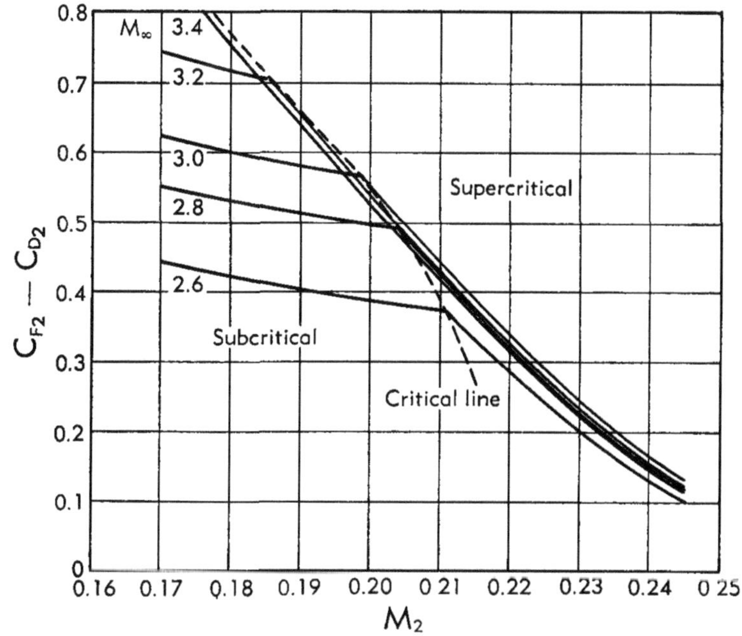

Fig. E,24a. Thrust variation with diffuser discharge velocity.

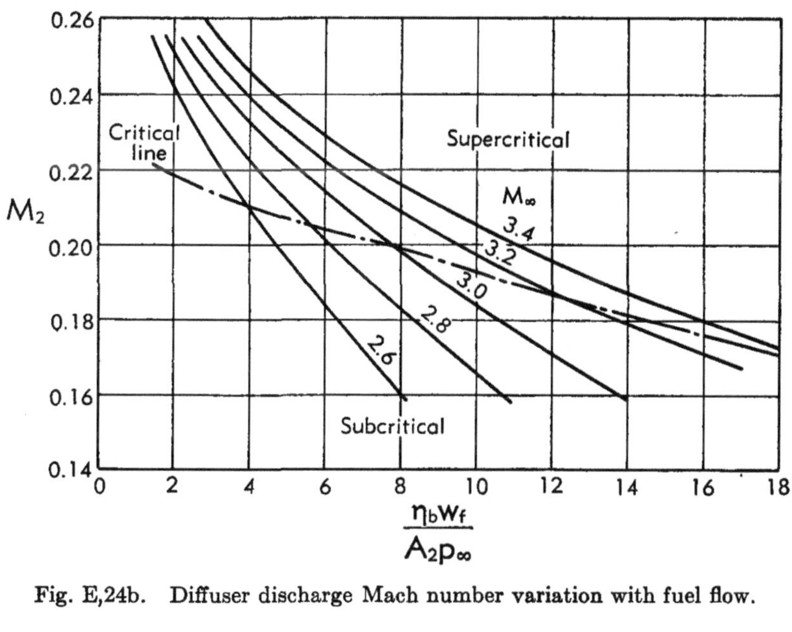

Fig. E,24b. Diffuser discharge Mach number variation with fuel flow.

Fig. E,24b the relation between diffuser discharge Mach number and effective fuel flow is continuous, so that a continuous increase of fuel flow cannot be caused by inadvertent subcritical operation. Of course, this system could result in subcritical engine operation in the event that the schedule was in error and called for a lower value of diffuser discharge Mach number than that accompanying critical flow. In this event the control would continue to maintain the subcritical condition provided that this was a stable engine operating point.

The chief disadvantage of the diffuser discharge Mach number control system would appear to be the small order of variation of the controlled parameter, the variation of M_2 from 0.16 to 0.25 being sufficient to cover almost the complete thrust range at a given flight Mach number (Fig. E,24a). This low range would require the development of highly accurate sensing devices to measure the diffuser discharge Mach number.

It should be noted that the control systems discussed above can be used to attain modes of engine operation other than critical-point operation, since at any flight Mach number there exist unique relations among diffuser pressure ratio, diffuser discharge Mach number, and engine thrust. The attainment of any desired mode of engine operation within the capabilities of the ramjet merely requires the use of a different control schedule relation in the systems presented. The present discussion has been limited to critical-point operation because, as previously stated, this represents not only the most efficient mode of engine operation but also presents the greatest problems in control operation.

E,25. Performance of Variable Geometry Engine. As pointed out in Art. 19 the ramjet engine is essentially a constant flight speed device that cannot operate too far from its design point without being overly large or inefficient. This handicap is somewhat lessened if the geometry of the engine can be varied. In particular, variation of the exhaust nozzle can provide a large range of thrusts at a given flight speed and hence can provide excess thrust for acceleration and maneuver.

The performance map of an engine incorporating a variable area exhaust nozzle is shown in Fig. E,25 for a flight Mach number of 2.4. (It should be noted that this engine is a different configuration than that previously considered.) Only the critical and supercritical operating regions are shown on this map.

The critical line forming the left-hand boundary of the engine map corresponds to the maximum diffuser pressure ratios considered attainable with this configuration. The upper boundary represents the maximum value of heat release attainable with the fuel considered; the lower boundary indicates the minimum fuel-air ratio at which combustion can be maintained. The right-hand boundary on the performance map corresponds to the maximum diffuser discharge Mach number for which combustion can be maintained. The values shown are all arbitrarily selected for illustrative purposes and for an actual installation depend on the characteristics of the particular engine design and the fuel.

It is evident from Fig. E,25 that variations in thrust can most efficiently be accomplished through manipulation of the exhaust nozzle area to maintain engine operation on the critical line. This is to be expected from previous discussions.

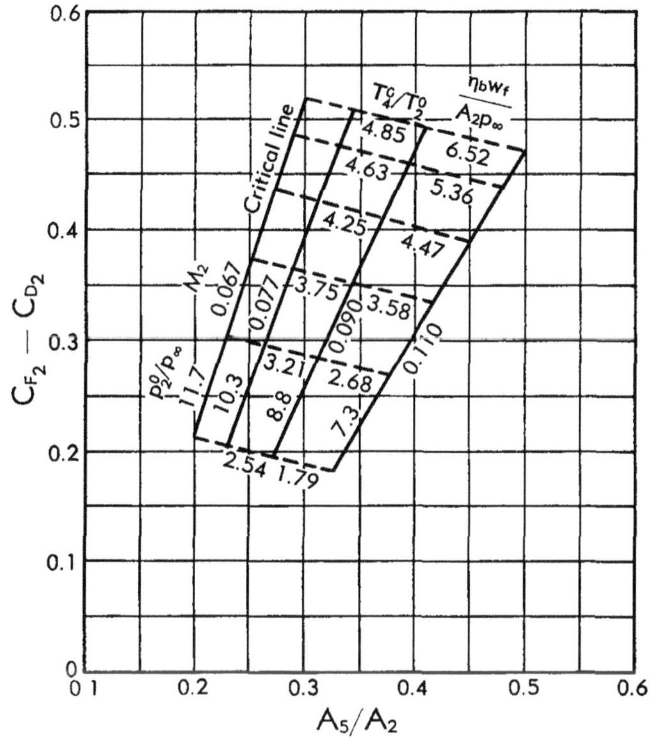

Fig. E,25. Ramjet performance with variable area exhaust nozzle. $M_\infty = 2.4$.

E,26. Control Considerations for Variable Geometry Engines.

The addition of the variable exhaust nozzle to the ramjet has in effect added another degree of freedom to the system; that is, the independent variables are now two in number, the fuel flow and the throat area. However, for operation at maximum efficiency a schedule of pressure ratio or diffuser discharge Mach number still provides this mode of operation as in the case of the fixed geometry engine. The variable area can be used to obtain the desired thrust level with minimum specific fuel consumption. The controls discussed in Art. 23 and 24 can be directly applied to the variable geometry engine to attain critical diffuser conditions with the nozzle area scheduled to give the required thrust level or to prevent diffuser pulsing by opening when pulsing is sensed.

For operation at other than the critical diffuser conditions it is possible to schedule a different value of pressure ratio or diffuser discharge Mach number. Such a procedure would be required if the thrust required for a given flight condition were below the value that could be obtained at maximum efficiency.

E,27. Effects of Flight Plan on Controls for Fixed Geometry Engine. The considerations discussed above are concerned only with

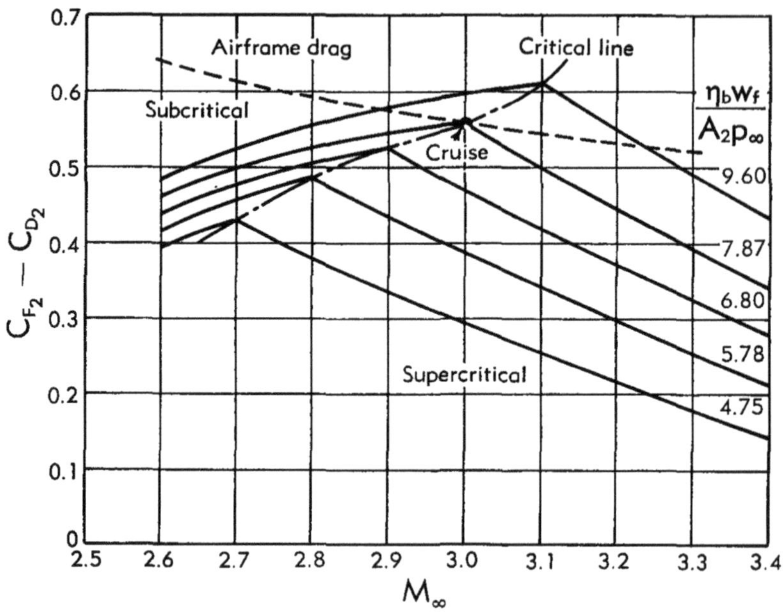

Fig. E,27a. Flight map for a fixed geometry engine.

the attainment of a given mode of engine operation. It was observed that the fixed and variable geometry engines could utilize similar control systems to obtain a desired engine operating point. When the engine is considered in relation to the airframe it is to propel, however, the control requirements and systems are no longer the same for the two engine types.

The operating map for a fixed geometry ramjet engine is shown in Fig. E,27a, where the net thrust is plotted as a function of flight Mach number for several values of corrected fuel flow. Superimposed on the figure is a curve representing the drag of a typical airframe at a constant angle of attack and essentially constant altitude. The drag curve has been chosen to give a design cruise point of the engine-airframe combination at a Mach number of 3.0 with critical diffuser flow.

If the engine controls are selected to cause the engine to operate at a constant corrected fuel flow, the configuration is partially unstable in flight velocity. Any external disturbance acting to decrease the flight speed would place the system in a region where the drag would exceed the thrust and continuous deceleration would result. On the other hand, a disturbance acting to increase flight velocity would also place the system in a region where the drag would exceed the thrust and the deceleration force would be a stabilizing effect tending to restore the system to cruise velocity. If the engine control were designed to maintain critical diffuser flow the engine thrust line would follow the critical line on Fig. E,27a and the system would be unstable for either form of velocity disturbance.

Two obvious methods for stabilizing flight velocity present themselves. The first is to utilize available aerodynamic and gravitational forces to restore the aircraft to the desired flight condition. If, for example, the aircraft is below cruising speed due to a transient disturbance the angle of attack may be decreased. This change in angle of attack would place the aircraft in a shallow dive, thus increasing the velocity. The second method of velocity control would be to vary engine thrust through fuel flow control in order to provide a restoring force to the system. For an engine designed to cruise at critical diffuser conditions this cannot be done safely, however, since an increase in fuel flow would force the diffuser into subcritical operation and potential inlet flow instability. (The increase in thrust from subcritical operation is small in any event and long response times would result.) It would therefore appear necessary to design the engine for supercritical operation at the cruise condition in order to obtain the necessary thrust margin for flight velocity stabilization. Such a design would, of course, result in a larger engine to propel a given airframe and cruise operation at a reduced efficiency as compared to the engine designed to cruise at critical diffuser conditions.

A block diagram of a possible control system for velocity stabilization by the first method is shown in Fig. E,27b. Flight Mach number is sensed and compared with the required Mach number. The error, or difference, between these values acts on an angle-of-attack control to vary the angle of attack in such a manner as to reduce the error to zero. The control would act to increase the angle of attack in an overspeed condition and to decrease it for an underspeed condition. The engine control plays a secondary role in this system. Essentially it should be designed to maintain the maximum safe value of thrust during underspeed conditions and to avoid blowout of the engine during overspeed excursions. As such the control can be a critical-point control of the type discussed previously with a fuel flow limiting device incorporated for overspeed conditions.

The response of an aircraft under the proposed control action would be a damped oscillation in both altitude and velocity with a frequency approximately that of the phugoid motion of the airframe. The initial excursion in altitude due to the velocity disturbance might be quite large, depending on the characteristics of the airframe and the control gain. The response times would be comparatively long since the available damping is rather small. Such a response is characteristic of aircraft velocity responses at high altitudes, because such systems are characterized by high inertia and operate in a rarefied atmosphere with attendant small restoring forces.

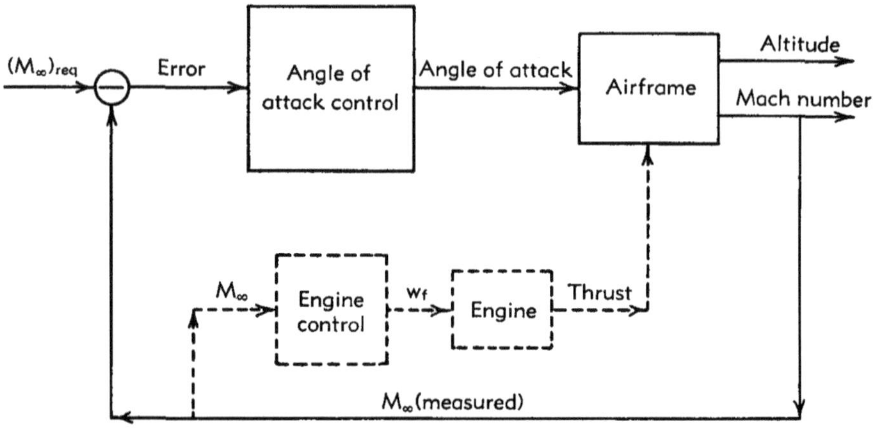

Fig. E,27b. Block diagram of control system for stabilizing flight velocity through angle of attack.

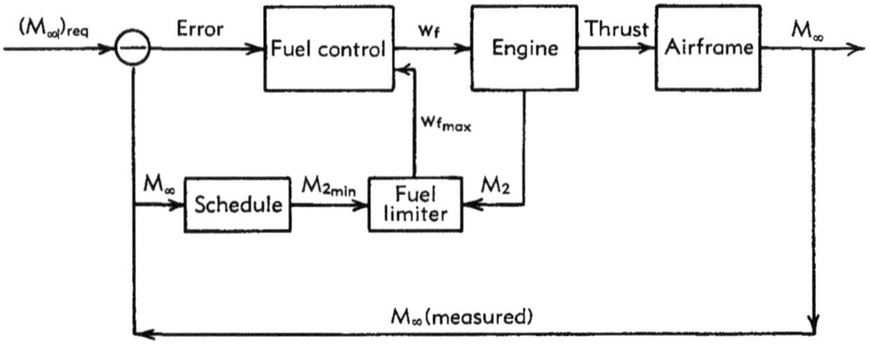

Fig. E,27c. Block diagram of control system for stabilizing flight velocity through engine thrust, supercritical engine operation at cruise point.

A block diagram of a control system using engine thrust variations to stabilize flight velocity is shown on Fig. E,27c. Flight Mach number is sensed and compared with the required value. The error signal acts through a fuel control to change the engine thrust to restore the error to zero. The use of high control gains for such a system, that is, large changes in fuel flow for a unit error in flight velocity, can result in undesirable subcritical engine operation during a transient. In order to prevent such a condition a limiting device may be incorporated in the system to prevent excessive fuel flow. Such a device would be the system considered in Art. 24 for attaining critical engine operation by scheduling diffuser discharge Mach number with flight velocity. This system can be designed to operate as a limiting device rather than as a primary fuel control and would then limit fuel flow to safe values during the underspeed conditions.

Under such control action (i.e., thrust control with limited action, which can only be applied for the case of supercritical cruise operation), the velocity response is almost linear with a slight superimposed oscillation. Altitude would vary slightly during the transient, but the magnitudes involved are of no great importance. The time of response is quite similar to that required to damp the oscillations in the angle-of-attack control system.

E,28. Effects of Flight Plan on Controls for a Variable Geometry Engine. The ramjet with a variable area exhaust nozzle has a great

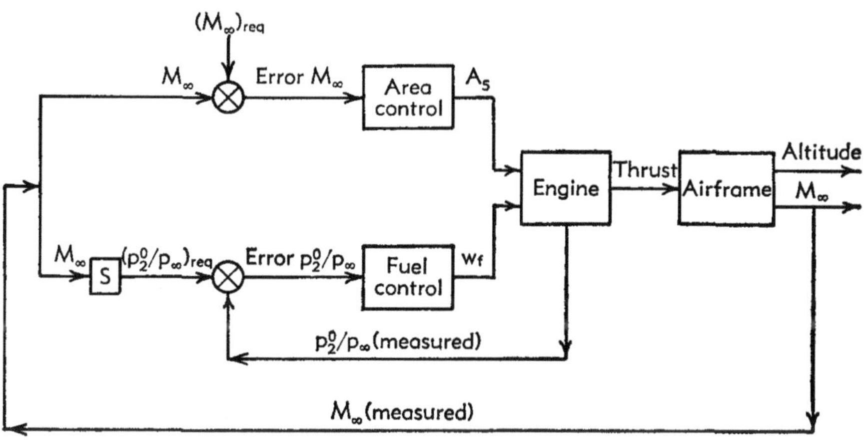

Fig. E,28. Block diagram of flight velocity control system for ramjet engine with variable area exhaust nozzle.

advantage over the fixed geometry engine with regard to the control of flight velocity. The engine may be operated at maximum efficiency during cruise and, through the agency of the variable area nozzle, may have a large available thrust margin to control flight velocity without operating in the subcritical region. Referring to Fig. E,25, the maximum combustion temperature forms the upper boundary to available thrust, and since this is not the most efficient cruise condition an aircraft would normally have an engine installation operating below this thrust level for cruise, leaving the remaining thrust for control purposes and maneuver.

A block diagram of a possible control system for controlling flight speed with a variable geometry engine is shown in Fig. E,28. Flight Mach number is sensed and the signal is sent to two points. The first point is a scheduler where the output sets the desired level of diffuser pressure ratio or discharge Mach number, as the case may be, which in turn acts through the fuel control to obtain this value. This part of the control thus acts to maintain critical engine operation. The second point to which the flight Mach number signal is sent is an error detector wherein it is compared with the required flight speed. The error signal in this circuit actuates the nozzle area servo which increases the exit area for underspeed conditions and decreases it for overspeed conditions. Increasing the area at constant fuel flow lowers the diffuser pressure ratio which causes an increase in fuel flow, thus increasing thrust. Thus, in effect, this system results in control action along a series of critical lines and represents a more efficient mode of engine operation and controls. Care must be exercised in the design of such a system to prevent the interaction of the two control loops from causing a control instability, a possible condition in multiloop engine control systems.

The response of a system acting under such control action will be quite similar to that of the thrust control scheme for the fixed geometry engine, with the difference that larger loop gains are possible with the present system and hence faster responses are to be expected.

E,29. General Remarks on Engine Control. The control schemes discussed herein are based on analytical considerations of the characteristics of ramjet engines. The actual design and construction of the necessary equipment and control components represents a task of great complexity. Systems such as those discussed require sensing devices of great precision and sensitivity, fuel systems with good response characteristics, stability, etc. In addition, the internal dynamics of ramjet engine operation must be well known in order to prevent instabilities from arising when engine and control systems are joined. Such things as the noise levels and the frequencies involved within the engine and the dynamics of diffuser pulsing must be evaluated prior to design of the control components, for these engine characteristics could very easily rule out some of the control schemes proposed. Such things as variations of engine performance with angle of attack must also be determined, for this may necessitate additional control elements if the engine is greatly affected by this variable. In essence, the control schemes presented are feasible, but the proof of their applicability depends upon the evaluation of the dynamics of performance of the ramjet engine which will in turn establish the requirements for the dynamic response characteristics of the control computer and servo elements.

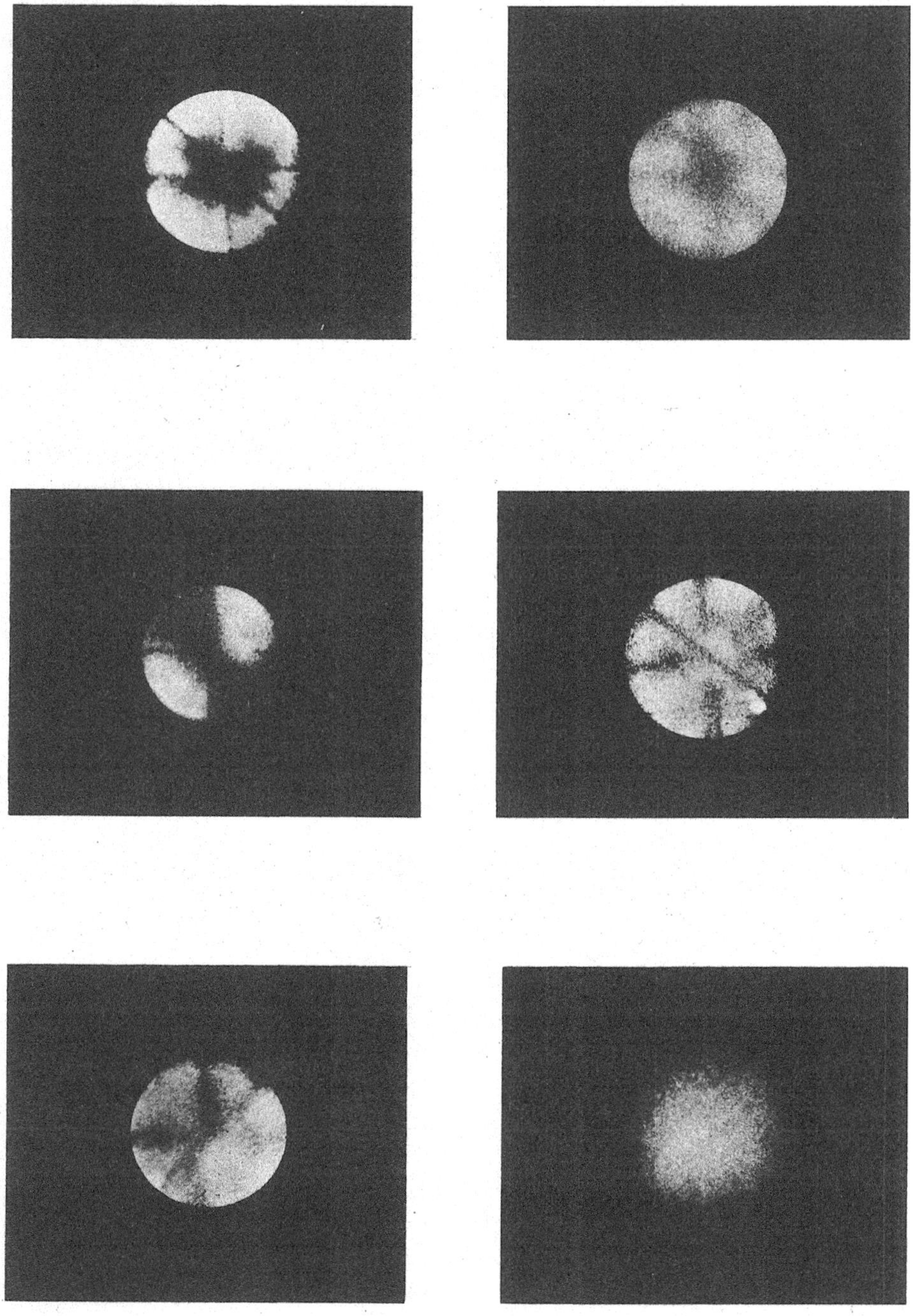

Plate E,32. Typical combustion photographs of ramjet engines taken through a periscope.

Plate E,33a. View of test section of free jet facility with jet diffuser around engine inlet cowl.

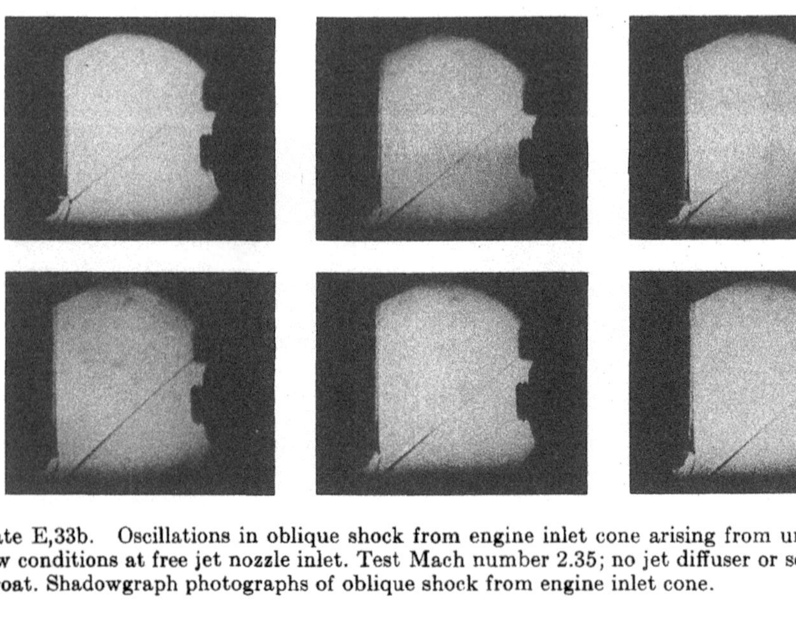

Plate E,33b. Oscillations in oblique shock from engine inlet cone arising from uneven flow conditions at free jet nozzle inlet. Test Mach number 2.35; no jet diffuser or second throat. Shadowgraph photographs of oblique shock from engine inlet cone.

Plate E,33c. Semicircular free jet nozzle installed in test chamber.

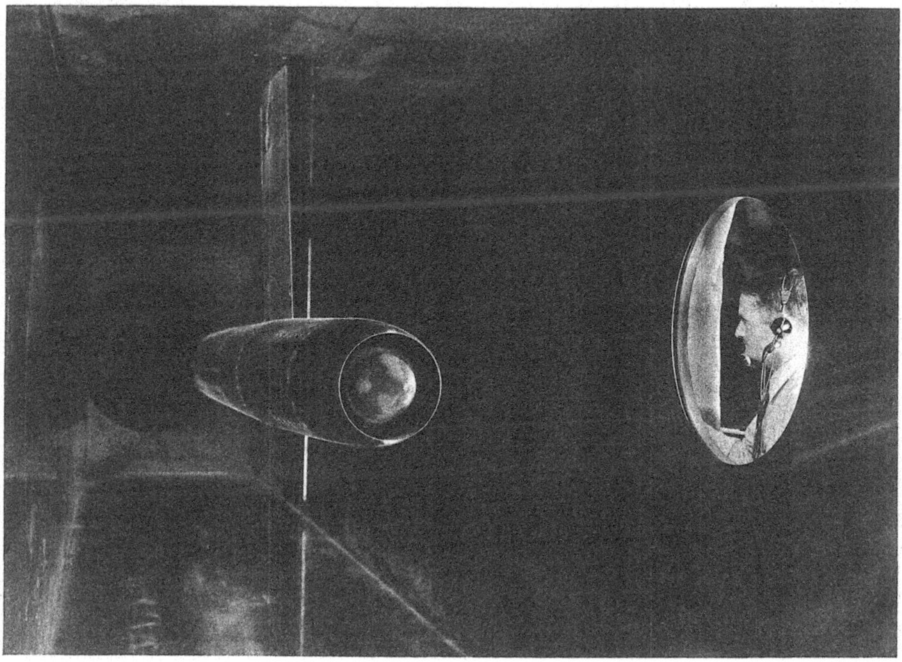

Plate E,34a. Ramjet engine mounted on a strut in test section of supersonic wind tunnel.

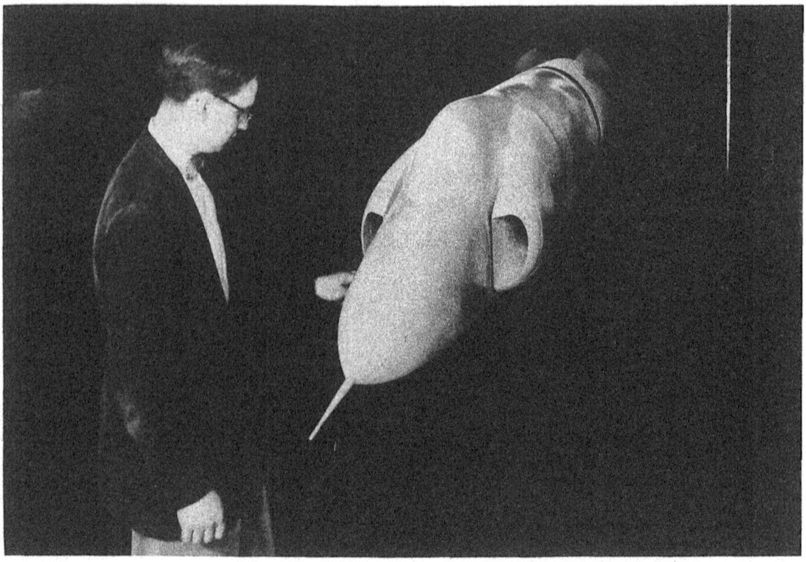

Plate E,34b. Model for inlet diffuser investigation mounted on sting in supersonic wind tunnel.

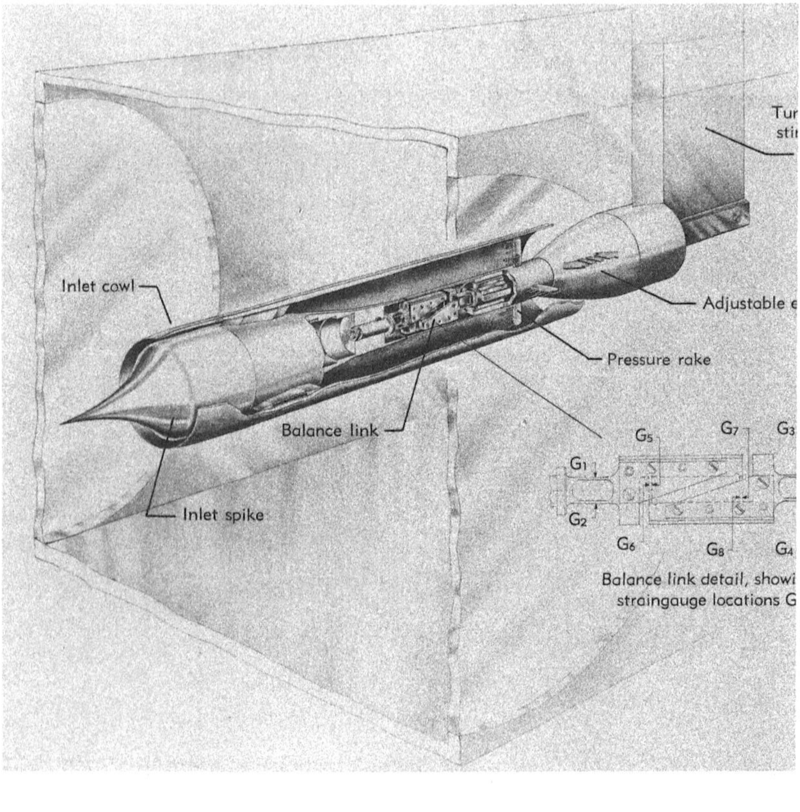

Plate E,34c. Schematic drawing of a sting-mounted model with an internal straingauge force balance.

Plate E,35a. Ground-launched ramjet test vehicle and booster on launcher.

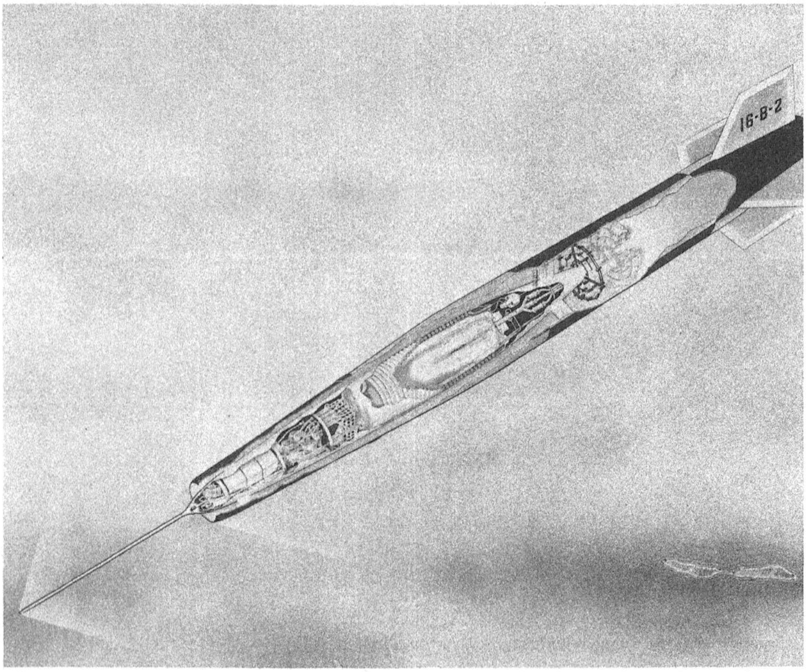

Plate E,35b. Air-launched ramjet unit. Top, ramjet unit mounted beneath center wing panel of carrier airplane; bottom, cutaway view of ramjet unit during free flight.

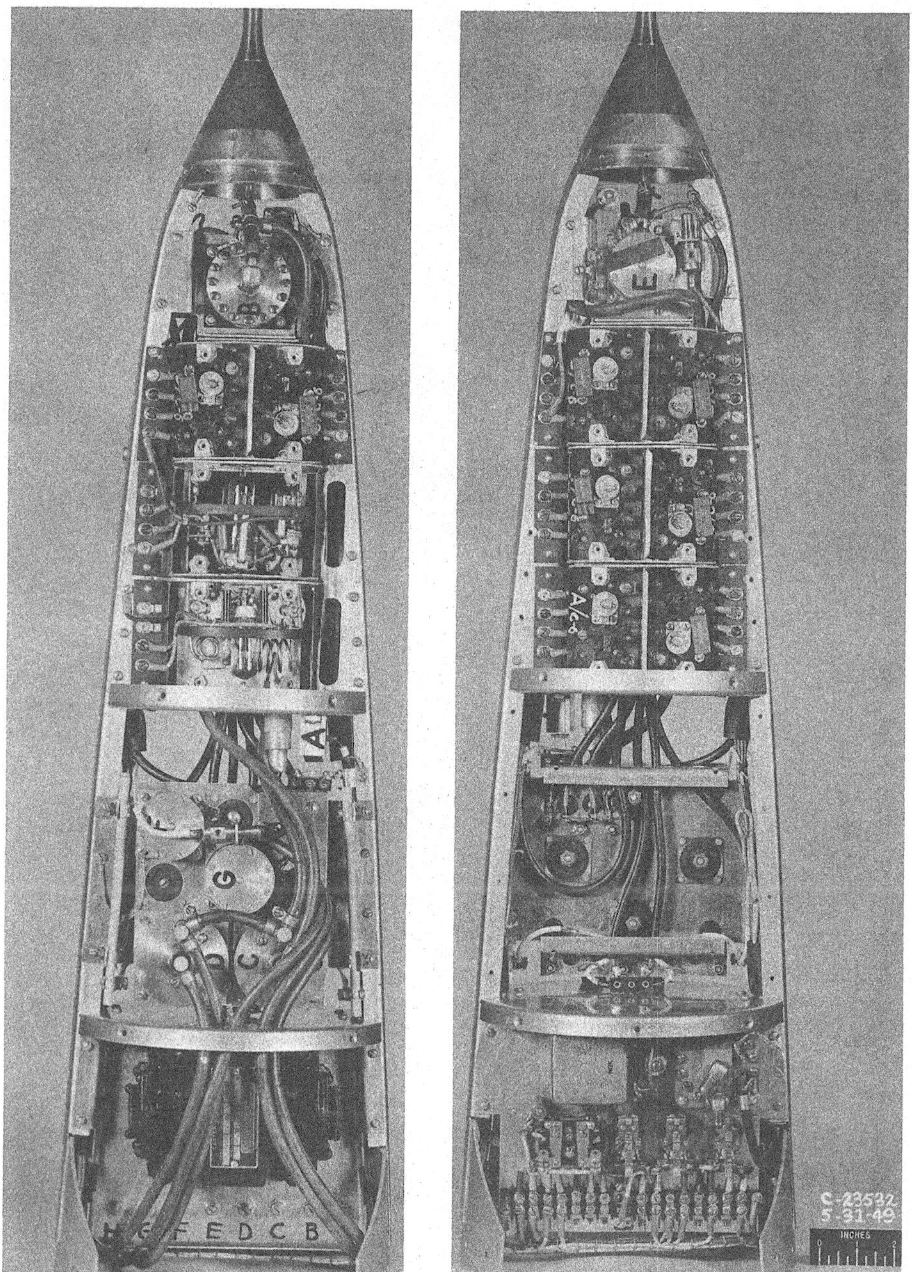

Plate E,35c. Typical eight-channel telemeter nose section.

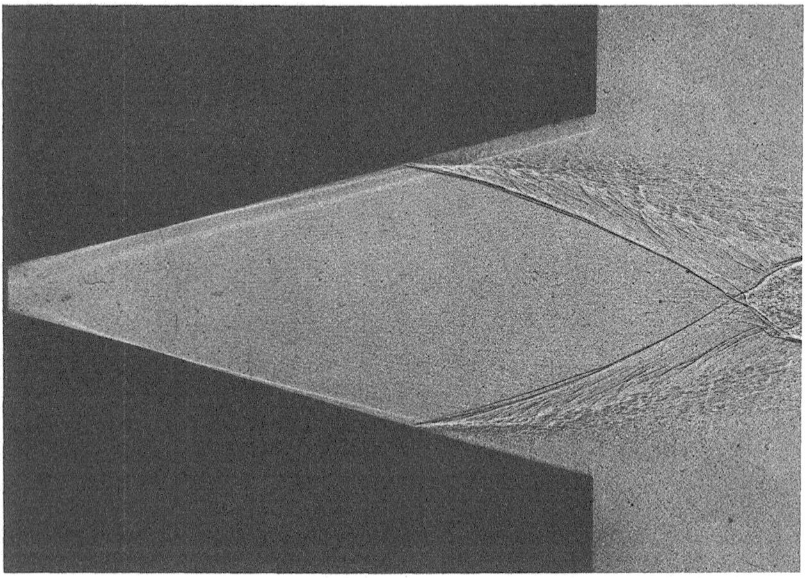

Plate G,3. Spark shadowgraph of flow in overexpanded nozzle.

CHAPTER 5. RAMJET TEST FACILITIES

BRUCE T. LUNDIN

E,30. Introduction. Although the idea of ramjet propulsion was patented by the French engineer René Lorin as early as 1913, the field remained nearly dormant for over 30 years. It was not until 1944, in fact, that serious work on the ramjet engine was started, principally at the Applied Physics Laboratory of The Johns Hopkins University and at the NACA. This relatively long period of time between the conception of the idea and the start of serious work was, of course, largely a matter of awaiting the practical visualization of the high flight speeds that are necessary for efficient engine operation.

The actual development of the engine into a useful power plant, however, also had to await the availability of large and specialized test facilities. The history of the development of the ramjet engine has been and, in fact, will continue to be determined to a very large degree by the quality and capacity of test facilities. This close dependence of progress in the field on test equipment thus warrants a brief discussion of the technical characteristics of these specialized facilities and some of the associated experimental procedures.

Many factors combine to make successful ramjet-engine development so dependent on test facilities. The principle of ram compression, upon which the engine operates, introduces a high degree of mechanical simplicity that belies the truly complex nature of the aerodynamic and combustion processes involved. Obtaining the efficient compression and expansion necessary for good engine performance requires proper design of diffusers and nozzles for three-dimensional supersonic, mixed, and subsonic flow processes. Combustion must also occur, both efficiently and with low pressure loss at high rates of heat release under difficult environmental conditions and frequently over wide ranges of operation. The closely coupled nature of the various components of the engine also severely limits the amount of useful work that can be done on isolated components. The combustor, in particular, is highly influenced by the nature of the flow leaving the diffuser; the diffuser, in turn, may be strongly influenced by the flow field about the body in which it is installed.

Added to these difficulties of component interaction and complexity is that of scale. The importance of Reynolds number effects on the aerodynamic processes involved and the present lack of knowledge for scaling combustors requires that most experimental investigations into ramjet performance be carried out in large scale models and, ultimately, with full scale hardware. Thus the very principle of operation of the ramjet engine that introduces a striking simplicity in both basic concept and mechanical design has, paradoxically, also introduced complex aerodynamic and combustion processes that require complete, large scale engines for successful research and development.

Dominant in the above considerations is the fact that the ramjet engine, operating as it does by ram compression, cannot pump its own supply of air. Unique among all present aircraft power plants in this regard, it therefore can be operated only in actual flight or in large air-blowing facilities. Being essentially a supersonic power plant, these facilities for ramjet research must have what may be considered a supersonic character. Large air flow capacity, high pressure ratio, and high air temperatures must be the basic characteristics of ramjet facilities.

The type of information required, and hence the nature of the test facility, may vary widely, depending on the objective of the investigation. For a performance evaluation of an engine, measurement of thrust output and fuel consumption, together with such operating limits as inlet stability and combustor blowout, are obvious requirements. Other factors, such as durability and starting characteristics, are also frequently necessary to a performance evaluation. Research investigations aimed at contributing to engine design or development usually involve much additional information on detailed flow or combustion processes. This information may include, for example, the growth and stability of boundary layers, internal air flow distribution, fuel-air ratio distribution within the operating combustor, detailed surveys of wall static pressures, and knowledge of shock movement within a diffuser during transient operation.

In view of the wide range of problems involved, many types of test facility have been developed for ramjet investigations. These include, briefly, facilities for isolated exhaust nozzle investigations, facilities for combustor investigations, free jet facilities for complete engine investigations, supersonic wind tunnels for simulation of external as well as internal flow conditions (for either inlet diffuser or complete engine investigations), and, finally, flight test facilities. Each of these facilities obviously possesses certain advantages as well as disadvantages, depending upon the requirements of the investigation. The purposes and limitations of these five types of facility, as well as their technical characteristics and associated experimental procedures, are described in the following articles of this chapter.

The free jet method of testing is treated in somewhat greater detail than the four other methods because it is relatively new, quite complex, and has not been widely discussed in the unclassified literature. The supersonic wind tunnel and flight-testing techniques, on the other hand, are only briefly reviewed because of their thorough treatment elsewhere. The discussion of instrumentation and data reduction methods that is included in the articles on the diffuser, the nozzle, and combustor testing are, of course, common to all methods of testing.

In addition to the equipment comprising the test engine or component installation that is discussed herein, a great deal of auxiliary plant equipment is also required for most of the methods of testing. This equipment includes the compressors, exhausters, heaters, dryers, exhaust gas coolers, and associated control and piping systems that provide the air and exhaust services at the test installation. This part of the central laboratory, however, is not discussed herein. Many different types and arrangements of this equipment may be used with equal effectiveness and the choice of design is usually dictated as much by the nature of existing equipment that may be available, by the contemplated method of operation, and by the environment of the site as by purely technical considerations.

E,31. Exhaust Nozzle Facilities. The principal items of information required from investigations of exhaust nozzles are the thrust characteristics and the air flow handling capacity. Static pressure distributions along the wall of the nozzle are also frequently measured to provide additional information on the internal flow processes.

The thrust characteristics of a nozzle are usually defined by a thrust coefficient C_{F_n}, which is the ratio of the actual jet thrust produced by the nozzle at a given pressure ratio to the thrust ideally available by isentropic expansion of the same air flow over the specified pressure ratio. Thus,

$$C_{F_n} = \frac{F_j}{F_i} \tag{31-1}$$

The air flow capacity of a nozzle is determined as the flow per unit of nozzle throat area w_a/A_5 corrected to some standard nozzle inlet conditions, as follows:

$$\text{Air flow parameter} = \frac{w_a\sqrt{\theta}}{A_5\delta} \tag{31-2}$$

For choked flow (nozzle pressure ratios greater than critical) this air flow parameter has a theoretical value of 49.5 lb/sec-ft^2. Comparison of the actual air flow parameter with this theoretical value thus provides a flow coefficient.

Test equipment. To determine the above characteristics, measurements are required of the actual thrust produced, the air flow through the nozzle, the nozzle-inlet total pressure and temperature, the ambient or exhaust static pressure, and the throat area. A typical nozzle test rig in which these various quantities may be measured is shown in Fig. E,31a. The test nozzle is installed on a mounting pipe located inside a test chamber. The test chamber is connected to central laboratory compressed air and exhaust equipment that provides the desired pressure ratio across the nozzle. In addition to flow capacity requirements, the inlet air must be dried or heated sufficiently to prevent condensation shocks from forming within the nozzle.

The jet thrust of the test nozzle is determined from a complete force balance on the nozzle and mounting pipe. Many different arrangements of or types of thrust stands may, of course, be used to provide this force balance. A very high degree of accuracy is, however, necessary, which requires careful attention to the action of various external pressure forces; special flexure members and labyrinth seals are therefore preferred to conventional bearings and slip joints. In the setup shown, the mounting pipe is freely suspended by four flexure rods connected to a bed plate. The pressure forces, both external and internal, are transmitted from the bed plate through a flexure-plate-supported bell crank and linkage to a balanced air-pressure diaphragm force-measuring cell. The pair of labyrinth seals maintain the pressure difference across the nozzle. These seals are installed in a reduced-area section to reduce the magnitude of the inlet pressure force and to increase the air velocity at the flow-measuring station. The vent between the two seals decreases the pressure difference across the second labyrinth and prevents a pressure gradient on the outside of the diffuser section because of an air blast from the seal.

Calculation methods. For the case of a plug nozzle illustrated in Fig. E,31a, a complete force balance on the system is

$$\frac{w_a}{g}V_1 + A_1p_1 - p_{bm}(A_1 - A_L) + p_\infty(A_4 - A_L) - F_d$$
$$- p_\infty(A_4 - A_e - A_{p_e}) - \frac{w_a}{g}V_e - \int_0^{A_{p_e}} p\,dA_p - A_ep_e = 0 \quad (31\text{-}3)$$

where

$$\frac{w_a}{g}V_1 + A_1p_1 = \text{total inlet momentum}$$

$$\frac{w_a}{g}V_e + A_ep_e = \text{total outlet momentum}$$

$$p_{bm}(A_1 - A_L) = \text{pressure force on the bellmouth inlet}$$

$$p_\infty(A_4 - A_L) = \text{pressure force on the diffuser}$$

$$p_\infty(A_4 - A_e - A_{p_e}) = \text{pressure force on the nozzle cowl}$$

$$\int_0^{A_{p_e}} p\,dA_p = \text{pressure force on the plug extension}$$

Eq. 31-3 states, essentially, that the difference between the inlet and exit momentum is equal to the pressure forces on the system. Because

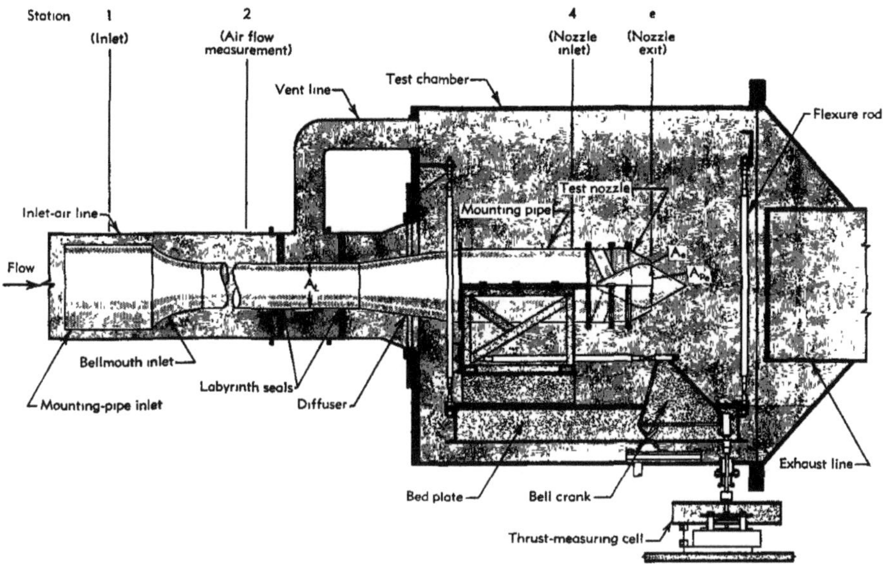

Fig. E,31a. Schematic drawing of thrust stand for exhaust nozzles.

the desired jet thrust F_j is defined as

$$F_j = \frac{w_a}{g} V_e + A_e(p_e - p_\infty) + \int_0^{A_{p_e}} p\,dA_p - A_{p_e}p_\infty \qquad (31\text{-}4)$$

it may be calculated from the following equation:

$$F_j = \frac{w_a}{g} V_1 + p_1A_1 - p_{bm}(A_1 - A_L) - A_Lp_\infty - F_d \qquad (31\text{-}5)$$

The isentropic, or ideally available, thrust required for the determination of the thrust coefficient is computed from

$$F_i = w_a\sqrt{\frac{2\gamma}{\gamma - 1}\frac{\mathfrak{R}T_4}{g}\left[1 - \left(\frac{p_\infty}{p_4^0}\right)^{\frac{\gamma-1}{\gamma}}\right]} \qquad (31\text{-}6)$$

The air flow w_a is determined by integrating the total and static pressures across the air flow-measuring station (station 2, Fig. E,31a) in accordance with the following equation:

$$w_a = p_2A_2\sqrt{\frac{2\gamma g}{(\gamma - 1)\mathfrak{R}T_2^0}\left(\frac{p_2^0}{p_2}\right)^{\frac{\gamma-1}{\gamma}}\left[\left(\frac{p_2^0}{p_2}\right)^{\frac{\gamma-1}{\gamma}} - 1\right]} \qquad (31\text{-}7)$$

The inlet velocity V_1 may be similarly obtained from pressure and temperature measurements at station 1.

Instrumentation. Numerous pressure and temperature measurements are required for the various terms in the preceding equations. These include total and static pressures at stations 1 and 2, total pressures and temperatures at the nozzle inlet (station 4), exhaust ambient pressure, and static pressure surveys on the outside wall of the bellmouth and on the surface of the plug of the nozzle. Static pressures are usually measured both on the wall and in the main stream. To provide accurate results, the flow at each station is extensively surveyed by several instrument rakes, with 30 total pressure tubes, 14 static pressure probes, and 12 total temperature thermocouples being typical. Skin or metal temperatures are also required to permit the various geometric flow areas to be corrected for thermal expansion.

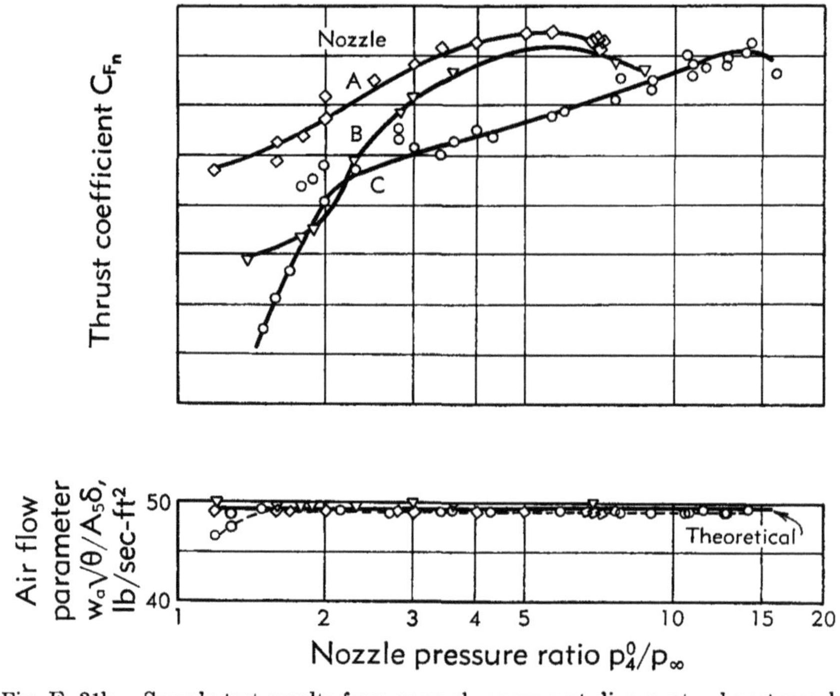

Fig E, 31b. Sample test results from several convergent-divergent exhaust nozzles.

Typical test results. The thrust coefficient and air flow parameter are usually plotted against the nozzle pressure ratio as shown in Fig. E,31b. Data are presented for three convergent-divergent nozzles having different expansion ratios and, hence, different design pressure ratios. The peak thrust coefficient occurs at approximately the design pressure ratio. The performance at lower pressure ratios (overexpanded nozzle) differs appreciably from what would be expected from one-dimensional flow theory because the higher back pressure propagates forward into the nozzle through the boundary layer. The air flow parameter of a nozzle is constant as long as the throat is choked. It is apparent in Fig. E,31b that the three different nozzles become unchoked at different pressure ratios.

A typical static pressure distribution along the wall of a convergent-divergent nozzle is shown in Fig. E,31c for several pressure ratios. Inspection of these data, together with associated theoretical distributions, reveals the regions of flow separation within the nozzle and provides

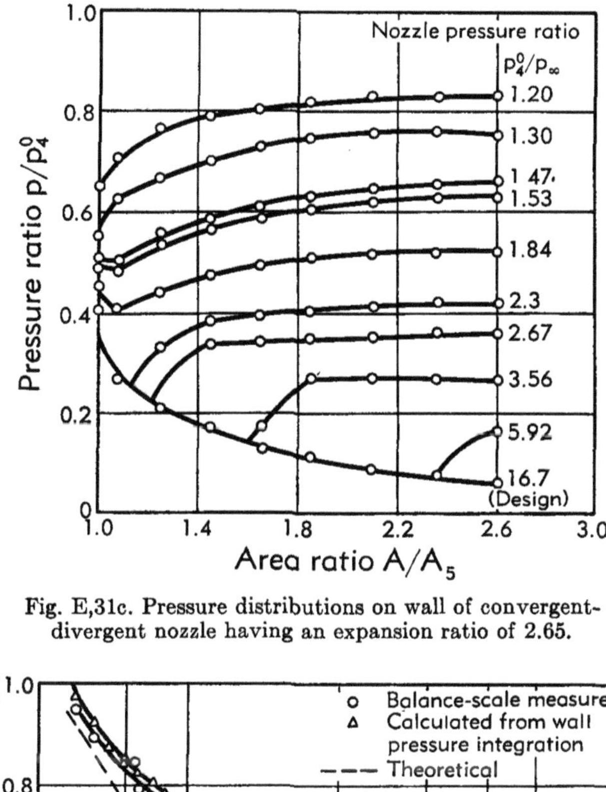

Fig. E,31c. Pressure distributions on wall of convergent-divergent nozzle having an expansion ratio of 2.65.

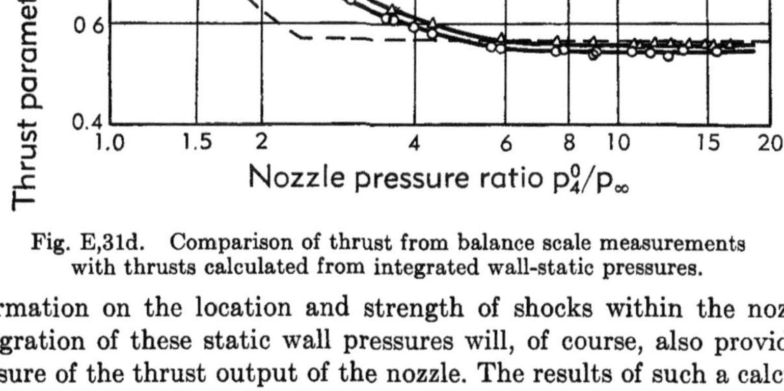

Fig. E,31d. Comparison of thrust from balance scale measurements with thrusts calculated from integrated wall-static pressures.

information on the location and strength of shocks within the nozzle. Integration of these static wall pressures will, of course, also provide a measure of the thrust output of the nozzle. The results of such a calculation for a range of nozzle pressure ratios is shown in Fig. E,31d. Also shown in this figure is the actual thrust as determined from force measurements and the thrust predicted from one-dimensional flow theory. The difference between the theoretical thrust and that computed from static wall pressures represents the losses due to departure from theoretical expansion. The effect of internal skin friction losses may be evaluated from a comparison of the actual thrust with the thrust computed from static wall pressures.

E,32. Combustor Test Facilities. Basic research into the fundamental processes of combustion is usually carried out in isolated test combustors that are connected directly to air-supply piping. Typical of such research are investigations into the mechanisms of turbulent flame spreading and of fuel droplet evaporation and diffusion. In contrast to such fundamental research, performance evaluations or development programs on combustors for ramjet engines are concerned primarily with over-all performance and operational characteristics. Among these characteristics are combustion efficiency, pressure drop, stability limits, lean and rich burning limits, starting characteristics, and durability. Transient performance characteristics of a complete combustor are also necessary to engine control studies. This type of an investigation, aimed at the development or performance evaluation of a ramjet combustor, requires a closer simulation of engine flow conditions than is possible with a completely isolated combustor. At least part of a representative inlet diffuser must be provided at the combustor inlet and some sort of an exit restriction or exhaust nozzle is required at the combustor outlet.

Test equipment. Requiring essentially a complete engine, tests of ramjet combustors are usually conducted in an altitude test chamber. The test chamber is, of course, connected to air service equipment that is capable of providing the desired pressure, temperature, and flow rate conditions at the engine inlet and outlet.

Two types of installation, differing in the degree to which the inlet diffuser of the engine is simulated or reproduced, have been commonly used. These two types of installation are schematically illustrated in Fig. E,32a. In both setups, the engine (or test article) is mounted on a thrust platform or bedplate inside the test chamber. A front bulkhead separates the inlet air from the exhaust section of the chamber and thus permits the desired pressure ratio to be maintained across the engine. A rear bulkhead is usually necessary to prevent recirculation of hot exhaust gases into the engine compartment.

For the subsonic inlet setup, shown in Fig. E,32a, top, only the subsonic part of the engine inlet diffuser is provided. Connected to this annular diffuser is a subsonic bellmouth inlet. The flow between the test chamber and the combustor is thus entirely subsonic. The exhaust nozzle is usually maintained choked for the dual purpose of isolating the combustor from any resonant coupling effects with the exhaust system and to permit certain simplifications in data reduction. The desired average values of temperature, pressure, and velocity at the combustor inlet may be maintained for various fuel-air ratios by control of the inlet air conditions and exhaust nozzle area. The flow profiles entering the combustor may, however, differ widely from actual flight conditions because the shock system that exists within the diffuser of an engine during supersonic flight is not reproduced. Adjustment of these flow profiles may,

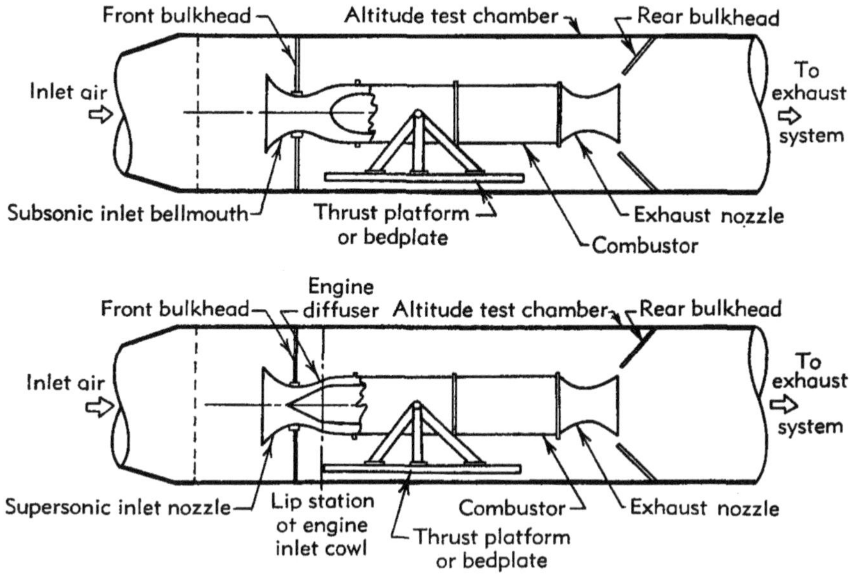

Fig. E,32a. Schematic diagram of direct-connect setup for tests of ramjet engine combustor. Top, subsonic inlet setup; bottom, ducted nozzle setup.

of course, be accomplished by screens or vanes in the inlet diffuser. Such flow control devices, however, would not only have to be tailored by cut-and-try methods to match previously established profiles, but also varied for each test point.

The ducted nozzle setup, shown in Fig. E,32a, bottom, provides a closer simulation of the inlet shock system at, of course, the expense of somewhat greater complexity of equipment and a higher facility pressure-ratio requirement. The internal geometry of the test engine is similar to that of the flight version except for the installation of a bellmouthed, convergent-divergent nozzle at the engine inlet. This nozzle, in conjunction with the cone of the engine diffuser, accelerates the air from stagnation conditions in the test chamber to a supersonic velocity. The nozzle is designed so the Mach number at the cowl-lip station is identical to what it is in flight under supercritical engine operating conditions (normal shock swallowed). The total pressure at the bellmouth inlet is set at the value that would occur in flight behind the oblique shock off the cone. This arrangement therefore permits simulation of the Mach number, pressure, and temperature conditions that occur downstream of the lip with the free stream inlet of a flight configuration and serves to position the normal shock in the subsonic diffuser at approximately the same location it would assume in flight. Exact simulation of this internal shock system is probably not obtained, however, because of the influence of the boundary layer on the inlet bellmouth and the non-representative boundary layer on the inlet cone of the test engine. A second limitation of this method of installation is that subcritical engine operation, in which the normal shock is expelled and air spills over the cowl, cannot be simulated.

Calculation methods. Many of the combustion characteristics of interest such as stability limits, lean or rich blowout limits, and pressure drop may be measured more or less directly without involved data reduction procedures. The air flow is determined by flow orifices in the supply piping or, more frequently, by pressure and temperature surveys at some station in the engine inlet (Eq. 31-7). Determination of the combustor outlet temperature and, hence, the combustion efficiency present greater difficulties because these temperatures are usually beyond the range of thermocouples. Many different methods of indirectly determining outlet temperature or enthalpy have been used. Among these are flow continuity at the exhaust nozzle throat (Eq. 31-6), exhaust calorimeters, gas sampling, and various optical methods. Accurate determination of temperature from thrust measurement is difficult because the required thrust is usually a small difference between two large numbers (the scale reading and the pressure force on the front bulkhead seal). Gas sampling and optical methods not only require rather elaborate equipment but, being essentially point readings, necessitate extensive surveys to yield average values. With the exhaust calorimeter, the engine exhaust gas is passed through a large, sealed chamber into which water is sprayed. This water reduces the temperature of the gases to values that may be accurately measured by thermocouples. By measuring the gas flow, the water flow, the inlet water temperature, the outlet temperature of the mixture, and any external heat losses, a heat balance provides the enthalpy of the engine exhaust gases. Although this method is straightforward and accurate, the rather large and cumbersome size of the equipment required limits its use primarily to small scale engines.

The most successful method for determining the exhaust temperature of large engines is by flow continuity through the nozzle throat. A convenient form of the continuity equation for this purpose is

$$T_5^0 = \left(\frac{p_5^0 A_5 C_d}{w_g}\right)^2 \frac{\gamma g}{\mathcal{R}} \left(\frac{2}{\gamma + 1}\right)^{\frac{\gamma+1}{\gamma-1}} \tag{32-1}$$

This equation assumes that the exhaust nozzle is choked. If it is not, an equation of the following form must be used:

$$T_5^0 = \left(\frac{p_5 A_5 C_d}{w_g}\right)^2 \frac{2g}{\mathcal{R}} \frac{\gamma}{\gamma - 1} \left(\frac{p_5^0}{p_5}\right)^{\frac{\gamma-1}{\gamma}} \left[\left(\frac{p_5^0}{p_5}\right)^{\frac{\gamma-1}{\gamma}} - 1\right] \tag{32-2}$$

The area blocked by the pressure rakes is accounted for in A_5 and the flow coefficient C_d is determined from cold flow calibrations. An accurate determination of C_d is necessary since an error of 1 per cent in C_d will result in approximately a 2-per cent error in T_5^0 and as much as a 4- or 5-per cent error in combustion efficiency. Because the value of γ is dependent upon the gas temperature, an iterative process, which converges rapidly, is used.

Many different definitions of combustion efficiency have been used, each with its own advantages and disadvantages. These definitions are (1) the ratio of actual enthalpy rise to heat input in the fuel, (2) the ratio of the ideal fuel flow for the actual temperature rise to the actual fuel flow, (3) the ratio of the actual temperature rise to the ideal temperature rise for the fuel flow, and (4) the ratio of the actual- to ideal-enthalpy rise based on the corresponding temperature rises. It will be recognized that methods (3) and (4) give values of efficiency that are appreciably greater than those computed by methods (1) and (2) at fuel-air ratios above stoichiometric, but the four methods do not differ greatly at lower fuel-air ratios.

One additional method, which is obtained directly from flow continuity, is particularly suited to ramjet engines and worthy of mention. In this method, the combustion efficiency η is defined as

$$\eta = \frac{f_{\text{ideal}}}{f_{\text{actual}}} \tag{32-3}$$

where f_{ideal} is the ideal fuel-air ratio that would have produced the same combustor exit total pressure p_5^0 as was measured for the burning conditions under consideration. Thus, the efficiency is related only to the combustor exit total pressure, obviating the measurement or determination of the gas temperature.

For an engine operating supercritically, the air flow at a given inlet pressure is the same for both burning and nonburning conditions which will be designated in what follows by the subscripts $_h$ and $_c$ respectively. The ideal fuel-air ratio may therefore be determined by this condition of equal air flows as follows:

$$w_a = (\rho_5 C_d A_5 V_5)_c = \frac{(\rho_5 C_d A_5 V_5)_h}{1 + f} \tag{32-4}$$

By use of the equation of state, converting static pressure and temperature to total values, and some manipulation, the following equation relating the total pressure at the exhaust nozzle under burning to non-burning conditions is obtained.

$$\frac{(p_5^0)_h}{(p_5^0)_o} = \sqrt{\frac{T_5^0}{T_\infty^0}}\,(1 + f)\,\sqrt{\frac{\left[\left(\frac{\gamma + 1}{2}\right)^{\frac{\gamma+1}{\gamma-1}}\left(\frac{\mathcal{R}}{\gamma}\right)\right]_h}{\left[\left(\frac{\gamma + 1}{2}\right)^{\frac{\gamma+1}{\gamma-1}}\left(\frac{\mathcal{R}}{\gamma}\right)\right]_o}} \tag{32-5}$$

The pressure ratio $(p_5^0)_h/(p_5^0)_o$ may now be evaluated for various ideal fuel-air ratios by using theoretical combustion charts to find $(T_5^0)_h$. These data may then be plotted as f_{ideal} against $(p_5^0)_h/(p_5^0)_o$. With this plot, the ideal fuel-air ratio f_{ideal} may be obtained for each measured value of $(p_5^0)_h/(p_5^0)_o$ and the combustion efficiency of Eq. 32-3 computed.

It should be noted that this combustion efficiency is not a chemical efficiency such as a heat-balance or enthalpy-rise method would indicate. Instead, this efficiency is more representative of over-all engine performance because it indicates how effectively the fuel is being used to provide thrust rather than how completely it is being burned.

Instrumentation. The instrumentation required may vary widely depending upon the nature of the investigation. Typical of special combustor instrumentation are fuel-air ratio sampling probes within the combustion zone, high frequency pressure pickups for combustion instability measurements, and burner skin thermocouples. If the air-flow profiles within the diffuser are distorted, extensive pressure surveys at various stations across the duct are necessary. These surveys may involve at least six radial rakes of eight total pressure probes, eight static pressure probes, and three thermocouples each. Important to all performance investigations is a water-cooled tail rake across the exhaust nozzle throat. This rake usually spans the diameter of the nozzle and contains 10 or 12 total pressure probes located on centers of equal areas. Rotameters, calibrated orifices, and rotating-vane flowmeters have all been successfully used to measure fuel flows.

An important instrument for all combustor testing is a periscope with its line of sight directly into the exhaust nozzle along the axis of the engine. This periscope is used both as an operating guide for engine operation and to determine the extent and nature of combustion under various conditions of operation. It is usually composed of conventional optical components arranged to throw an image on a ground-glass screen for visual observation and into a camera for permanent records. The periscope head is water-jacketed and a blast of air is maintained over the face of the lens both for cooling and to prevent the accumulation of dirt.

Some typical photographs taken through a periscope are presented in Plate E,32. The location and extent of combustion with the various burner configurations and arrangements of fuel injection used are clearly evident in the photographs. The dark radial lines are various instrument rakes.

Validity of direct-connect testing. Because the direct-connect method of testing does not completely reproduce the diffuser flow conditions that occur in flight, the validity of the results obtained has long been under question. As far as the performance of the combustor is concerned, which is the primary objective of this method of testing, the principal factor that would influence the validity of the results is the flow profile at the

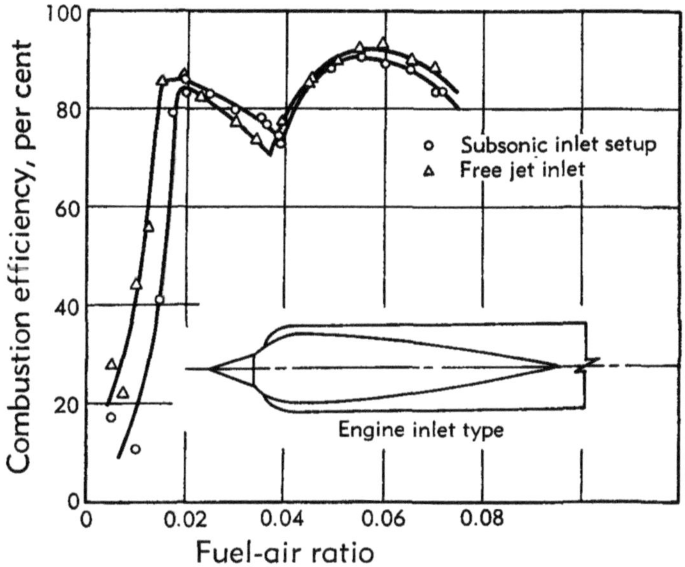

Fig. E,32b. Comparison of combustion efficiency of ramjet engine in subsonic inlet setup and free jet facility. Engine with symmetrical nose inlet at zero angle of attack. Free jet operation; supercritical inlet at Mach number 3.0.

combustor inlet. If fairly uniform flow profiles may be expected in flight, the results should be valid; however, if nonuniform profiles are to be expected, the combustion performance may differ from what would occur in flight.

Unfortunately, data are not available to compare the performance of a given combustor with both the subsonic inlet setup and the ducted nozzle setup. Some data are available, however, that compare the performance of a combustor with a subsonic inlet setup and with the correct inlet conditions from a free jet setup (Art. 33). This comparison is shown in Fig. E,32b for an engine with a symmetrical nose inlet at zero angle of attack. The diffuser of the free jet setup is operating supercritically. The remarkably close agreement of the combustion efficiencies, within the range of normal data scatter, testifies to the validity of a subsonic inlet for these particular conditions of operation.

This good agreement between the two methods of testing does not

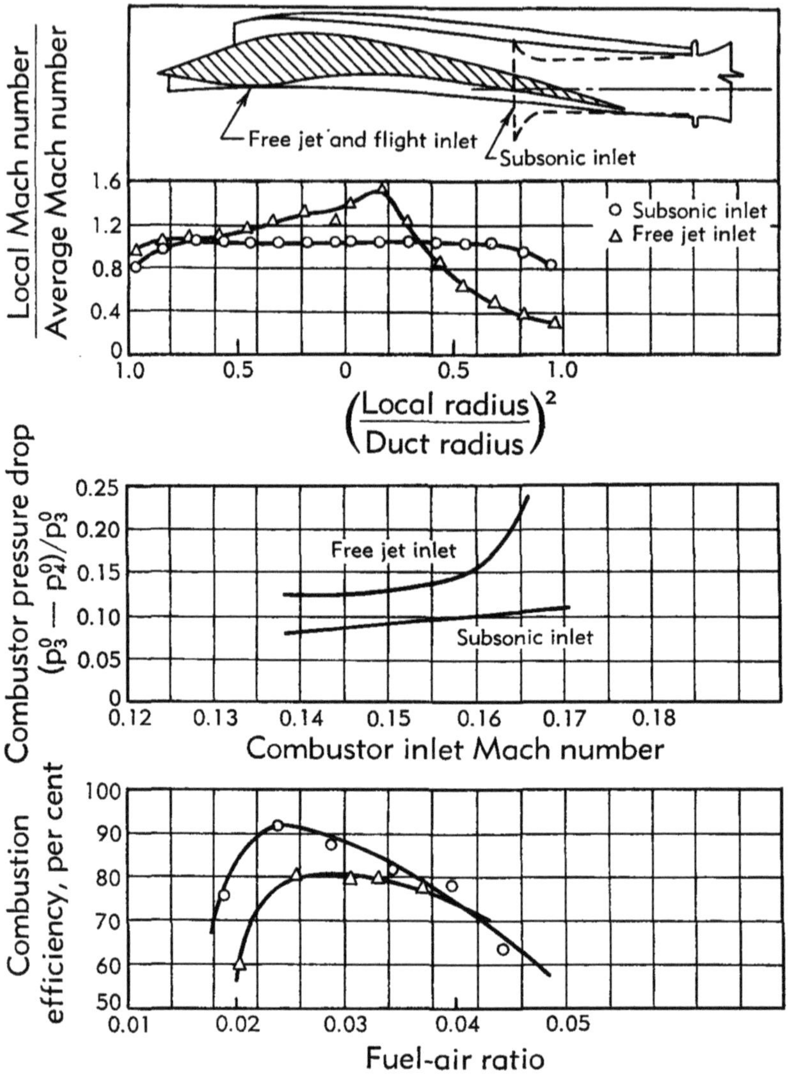

Fig. E,32c. Comparison of combustion characteristics of ramjet engine in subsonic inlet setup and free jet facility. Engine with asymmetric side inlet at 3° angle of attack. Free jet operation; supercritical inlet at Mach number 2.75. Top, combustor inlet flow profiles; middle, pressure drop; bottom, combustion efficiency.

exist, however, for more complex inlet arrangements at angle of attack. Data from a side inlet configuration are presented in Fig. E,32c, in which the combustor inlet flow profiles, the combustor pressure drop, and the combustion efficiency are compared. The flow profiles for the free jet conditions are considerably more distorted than for the subsonic inlet which, in turn, results in both a higher pressure drop and a lower combustion efficiency over most of the range of fuel-air ratios. Severe problems of upstream burning and burner-shell overheating also occurred in the free jet setup, but were absent with the subsonic inlet.

In spite of this loss of accurate reproduction of flow conditions with the subsonic inlet, much useful work of a preliminary nature may, of course, often be accomplished under these conditions. It is the final performance evaluation of an engine and detailed combustor development that loses its validity. It should also be noted that this comparison is for the simple subsonic inlet, in which no effort is made to reproduce the supersonic portions of the flight inlet. It is quite likely that the ducted nozzle type of inlet setup would provide much closer simulation of actual flight conditions.

E,33. Free Jet Facilities. With the free jet method of testing, the engine inlet is submerged in a supersonic jet of air in which the Mach

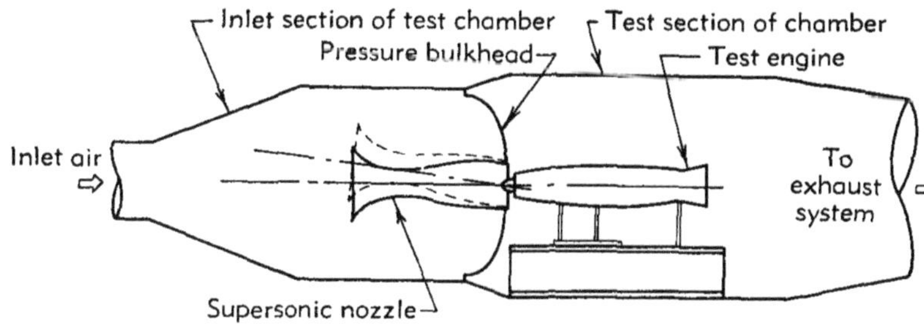

Fig. E,33a. Schematic diagram of free jet test facility for ramjet engine.

number, pressure, temperature, and angle-of-attack conditions of actual flight are reproduced. In this manner, a good simulation of the internal flow conditions in the inlet diffuser, and hence throughout the engine, may be obtained. No attempt is made to simulate external flow conditions around the outside of the engine.

General description. A schematic diagram of a simple form of a free jet facility is presented in Fig. E,33a. Both the supersonic jet nozzle and the engine being tested are shown installed inside a test chamber. Air at stagnation conditions in the inlet section of the test chamber is expanded through the nozzle to produce the supersonic free jet. The engine is mounted directly downstream of the nozzle outlet on a suitable stand. The pressure bulkhead seals the inlet from the test section of the chamber and thus permits the necessary pressure ratio to be maintained across the nozzle. The altitude pressure and the inlet air temperature may be controlled by simply varying the stagnation conditions in the inlet section of the test chamber. The flight Mach number is determined by the design of the nozzle, which may be either fixed or adjustable. If fixed, a separate nozzle is required for each flight Mach number. Angle-of-attack conditions are simulated by inclining the axis of the nozzle relative to the axis of the engine inlet as shown in the figure.

The pressure ratio required to operate a free jet facility is determined by the pressure drop across the supersonic nozzle and the pressure recovery experienced by the secondary or bypass flow that passes around the engine rather than by the pressure recovery of the engine itself. Additional equipment is, therefore, frequently used in order to increase the pressure recovery in the secondary flow and thus reduce the over-all

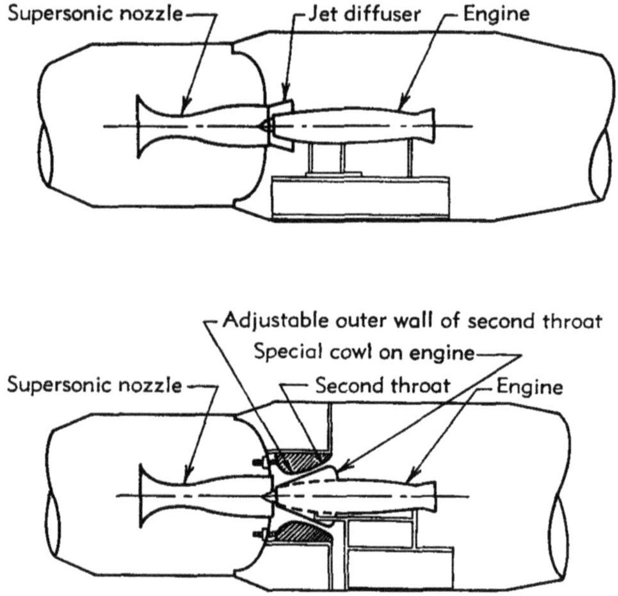

Fig. E,33b. Schematic diagram of free jet facility with jet diffuser and second throat for pressure recovery. Top, with jet diffuser at nozzle outlet; bottom, with second throat around engine cowl.

pressure ratio required. Two arrangements that have been used successfully are shown schematically in Fig. E,33b. The jet diffuser, shown in Fig. E,33b, top, consists of a divergent shroud attached to the outlet end of the jet nozzle. The annular passage thus formed around the cowl of the engine serves to position a system of shocks that provides a more complete pressure recovery than is obtained in a free flow field around the engine alone. As shown in the figure, the diffuser is a simple conical section about $1\frac{1}{2}$ engine-lip diameters long; the divergence angle is usually made equal to or slightly greater than the engine cowl angle. A photograph of this type of installation is presented in Plate E,33a. Visible in this view of the test section is the engine and its mounting stand, the shroud of the jet diffuser, and the pressure bulkhead. The engine inlet cowl is located in the region of the shadowgraph or schlieren port installed through the shroud.

The installation of a somewhat more elaborate second throat is schematically shown in Fig. E,33b, bottom. This secondary flow passage is essentially an axially symmetric convergent-divergent diffuser with a flow direction having both radial and axial components. The secondary air is thus decelerated into the throat, followed by a normal shock process at a relatively low Mach number in the divergent section. The inner wall of the diffuser is a conic frustum of about 20° half angle; translation of the outer wall permits adjustment of the second throat area.

The prime requirement in the design and operation of a free jet facility is that the stream flow into the engine inlet be free of shocks or expansions. Principal factors influencing these conditions are the position of the engine inlet relative to the nozzle outlet, the size or flow rate of the engine relative to that of the nozzle and the pressure ratio across the facility. Although each of these factors is treated separately in the following articles for orderly presentation, it should be recognized that they are closely interrelated. Following this discussion, various secondary factors such as inlet flow conditions into the jet nozzle, condensation shocks, and nozzle design procedures are briefly reviewed.

Compared with the other methods of ramjet testing discussed, the free jet method is relatively new. Consequently, the design of such facilities has not yet been subjected to systematic investigation or thorough analysis. In fact, most of the useful information available to date has been obtained from only three or four full scale engine installations. Most of these installations were, of course, intended primarily for engine research programs rather than facility design information. They were designed largely by intuition with sufficient conservatism to assure successful operation without expensive modification. A further difficulty is that the flow field downstream of the nozzle outlet and around the engine is extremely complex and is not amenable to simple theoretical treatment. As a result, the present discussion consists mainly of a review of operating experiences to date in a few facilities rather than a thorough or accurate analysis of design and performance requirements.

Position of engine. The engine inlet must be placed so that the diffuser cone and cowl are within the cones defined by the upstream and downstream Mach lines originating from the nozzle lip. Thus the stream entering the engine is not influenced by the chamber pressure at the jet boundary so long as this pressure is equal to or less than the jet pressure. The size of the jet required to test a given engine may therefore be reduced as the engine cowl is moved closer to the nozzle outlet. Placing the engine cowl close to the nozzle outlet, or even within the nozzle, has also been observed to have a beneficial effect in preventing shock disturbances from entering the engine cowl and in reducing the pressure ratio required for operation. In one installation, moving the engine cowl forward from a position about 0.75 cowl diameter downstream of the nozzle outlet (with a nozzle area 1.35 times the cowl area) to the plane of the nozzle outlet reduced the starting pressure ratio about 15 per cent and the operating pressure ratio, about 25 per cent. Engine operation during these tests was supercritical.

The above considerations indicate that the engine inlet or cowl should be placed at, or at least very near, the outlet of the jet nozzle. Such a position would, however, interfere with a free light path forward of the engine cowl for observation of the engine inlet shock systems. Most full scale engine installations have therefore provided a 4- to 6-inch space between the nozzle outlet and the engine cowl. An even larger space may be necessary if observation of a greater forward motion of the expelled shock under subcritical engine operating conditions is desired.

Facility mass flow ratio. The facility mass flow ratio is defined as the total flow rate through the nozzle divided by the flow rate through the engine under supercritical conditions. It is thus approximately equal to the nozzle outlet area divided by the projected area of the engine cowl. If the engine is operating subcritically, the actual mass flow ratio of a given installation would, of course, increase because of the reduction in engine air flow rate under such conditions. The facility mass flow ratio is, however, still defined as above because this definition provides a convenient description of a given installation that is independent of engine operating conditions.

Facility mass flow ratios used in installations in which only supercritical engine operation was desired have varied from about 2.0 to 2.6. These values were determined largely from consideration of the flow capacity of the central plant that was available, compared to the air flow rate required to test a particular engine. No attempt was made to minimize the flow ratios and they could undoubtedly have been somewhat smaller and still would provide satisfactory operation. If a second throat is used to increase the pressure recovery of the bypass air there is, however, some advantage in using a mass flow ratio that is larger than the minimum value for satisfactory engine inlet flow conditions. By increasing the mass flow ratio, the amount of high energy air passing through the second throat is increased, with attendant increased pressure recovery and reduced over-all facility pressure ratio. The amount that the mass flow ratio may be economically traded for pressure ratio depends, of course, on the characteristics of the plant equipment available and the desired range of test conditions. Unfortunately, only a very limited amount of data exist to guide this choice. It was observed in one series of tests that an increase in the mass flow ratio from 1.35 to 1.85 (for a Mach number of 2.98) reduced both the starting and operating pressure ratios by about 30 per cent.

If a jet diffuser or second throat is not used, the mass flow ratio probably need be only large enough to keep the engine inlet within the

Mach lines or oblique shocks emanating from the nozzle lip and within the jet boundary layer. The mass flow ratio of 1.35 mentioned previously provided satisfactory operation at Mach numbers of 2.18 and 2.98 and with the engine cowl located about 0.2 cowl diameter downstream of the nozzle outlet. This mass flow ratio was about the smallest feasible because the engine inlet was close to the edge of the jet boundary layer.

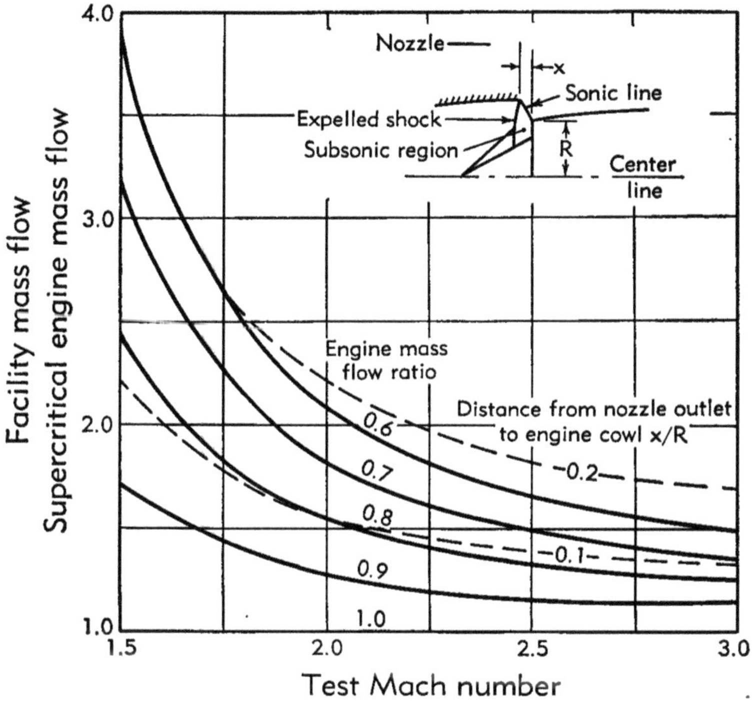

Fig. E,33c. Minimum theoretical facility mass flow ratios for subcritical engine operation in free jet. Intersection of expelled shock and sonic line at nozzle outlet. No allowance for nozzle boundary layer.

For subcritical engine operation, not only the engine inlet but also the subsonic region downstream of the expelled normal shock must be within the region of undisturbed nozzle flow. Higher mass flow ratios are, therefore, usually required for subcritical operation than for supercritical engine operation, particularly at low Mach numbers and large engine mass flow ratios (ratio of engine flow with subcritical spillage to critical air flow). Sufficient data to define these requirements are not available although the general trends of required facility mass flow ratio with test Mach number and engine mass flow ratio are presented in Fig. E,33c. The values presented in this figure are minimum theoretical requirements. They are based on the assumption that the intersection of the expelled shock and the sonic line is always at the nozzle outlet and no allowance is made for jet boundary layer. If the engine is located further downstream, allowance must be made to keep the subsonic region within any oblique shocks from the nozzle (the position of which depends upon the nozzle back pressure). The dashed lines in Fig. E,33c indicate the spacing between the nozzle outlet and engine cowl available for flow observation under the above assumptions.

Facility pressure ratio. As previously mentioned, the pressure ratio required to operate a simple "open-type" of installation, such as illustrated in Fig. E,33a, will be higher than those that include a diffuser or second throat for pressure recovery. This open-type facility is, however, of considerable practical interest because it is the only type in which a full scale, burning engine has been operated very far subcritically. Some range of subcritical operation has been achieved with a jet diffuser on the nozzle, but large facility mass flow ratios were used (over 3.0) and the range of operation was rather small. With a second-throat installation or with a jet diffuser at low engine mass flow ratios, either the nozzle flow broke down or an oblique shock from the nozzle intersected the expelled engine shock in such a manner that proper engine inlet flow was not simulated. On the other hand, satisfactory engine operation with an open facility has been obtained down to engine mass flow ratios of 0.6 or 0.7.

With regard to the pressure ratio required to operate such a facility, an oblique shock is produced at the nozzle outlet as the facility exhaust pressure is increased above the nozzle outlet static pressure. The engine must be so placed that this oblique shock does not fall upstream of the engine inlet. Furthermore, it can be inferred that, if this shock falls close to the lip of the cowl, the maximum strength it may have is that which permits a simple reflection at the engine cowl. Defining a maximum shock strength (or pressure recovery) in this manner thus defines the minimum facility pressure ratios required for various engine cowl angles. Minimum facility pressure ratios defined in this manner are plotted in Fig. E,33d for a range of test Mach numbers and engine cowl angles. The required pressure ratio is seen to increase rapidly as either the cowl angle or test Mach number is increased. For a cowl angle of 20°, for example, the facility pressure ratio increases from 7 to 18 as the test Mach number is increased from 2.0 to 3.0. For the Mach number of 1.92, the cowl angle was 20° and the oblique shock was computed to fall 0.09 cowl diameter (1.6 in.) downstream of the cowl lip. Reduction in facility pressure ratio below the value of 6.6 shown in the figure resulted in pressure variations at the cowl lip, and thus represented the limit of proper inlet flow simulation.

If the engine is placed closer to the nozzle outlet, the reflection of the oblique shock off the engine cowl will not influence the engine inlet flow and somewhat lower facility pressure ratios may be used. Data obtained under such conditions are plotted in Fig. E,33d at test Mach numbers of 2.35, 2.50, and 2.70. The pressure ratios noted are the minimum values that provided satisfactory engine inlet flow conditions as determined by pressure measurements at the engine cowl lip. The curve drawn through these data points is for a constant static pressure ratio of

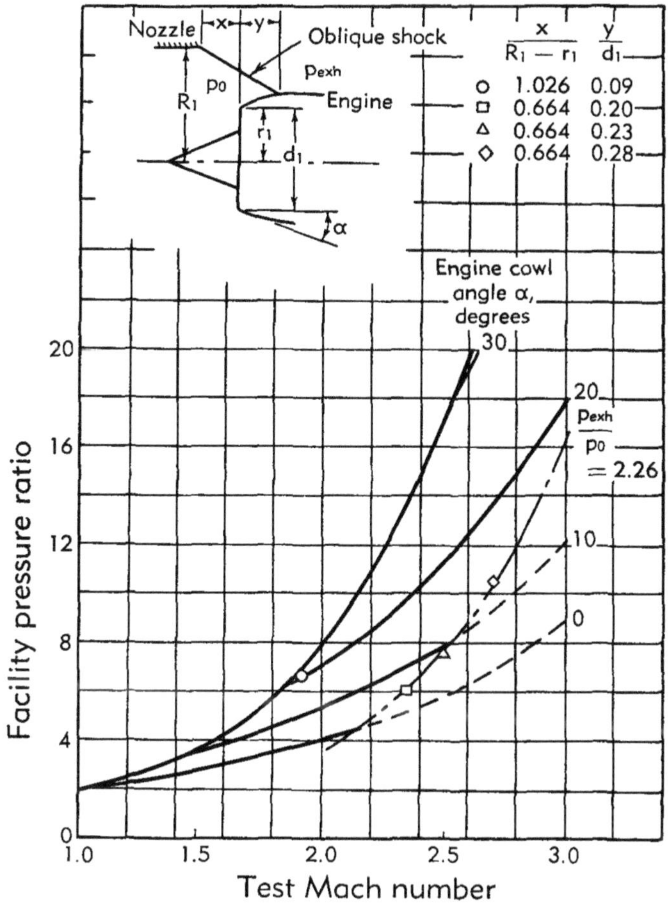

Fig. E,33d. Pressure ratio required to operate free *jet facility without jet diffuser or second throat.*

2.26 across the oblique shock (ratio of exhaust to nozzle outlet pressure). The oblique shock was computed to fall between 0.20 and 0.28 cowl diameters downstream of the cowl lip for these three points. The fundamental significance, if any, of the constant pressure recovery of 2.26 is not known; however, it was observed in a different facility (Mach number of 3.0) that the flow from the nozzle first broke down close to this shock pressure ratio.

It may thus be inferred that two limits exist for the minimum pressure ratio of open-type facilities. If the oblique shock falls near the engine inlet, its strength must be limited to the value that permits a simple reflection from the engine cowl. If the oblique shock falls further downstream on the engine, the presently available data define a limiting shock pressure ratio of 2.26. Shocks of greater strength have been observed to result in a flow breakdown from the nozzle.

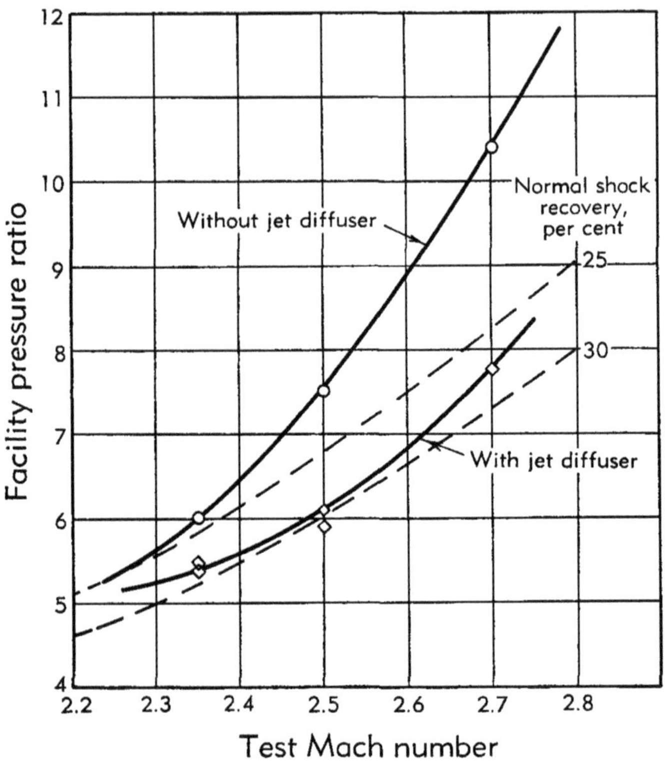

Fig. E,33e. Effect of jet diffuser on pressure ratio required to operate free jet facility.

The pressure ratios required to operate these same facilities when provided with a jet diffuser as illustrated in Fig. E,33b, top, and Plate E,33a are shown in Fig. E,33e. Reductions in pressure ratio of from 10 to 25 per cent were obtained by the diffuser, with the largest reduction occurring at the higher test Mach numbers. As illustrated by the dashed curves in the figure, the pressure recovery in the jet diffuser was between 25 and 30 per cent of the pressure recovery of a normal shock at the test Mach number.

If a second throat (Fig. E,33b, bottom) is used, the external flow of the facility is comparable to the behavior of a one-dimensional system of two successive throats. To establish or start the flow, the back pressure must be reduced sufficiently to cause the shock system to pass from between the throats to a point downstream of the second throat. After this is accomplished, the back pressure may be increased, which causes the shock system to move forward in the diffuser toward the throat. The limiting, or breakdown, pressure ratio is reached when the shock system is at the minimum area of the diffuser.

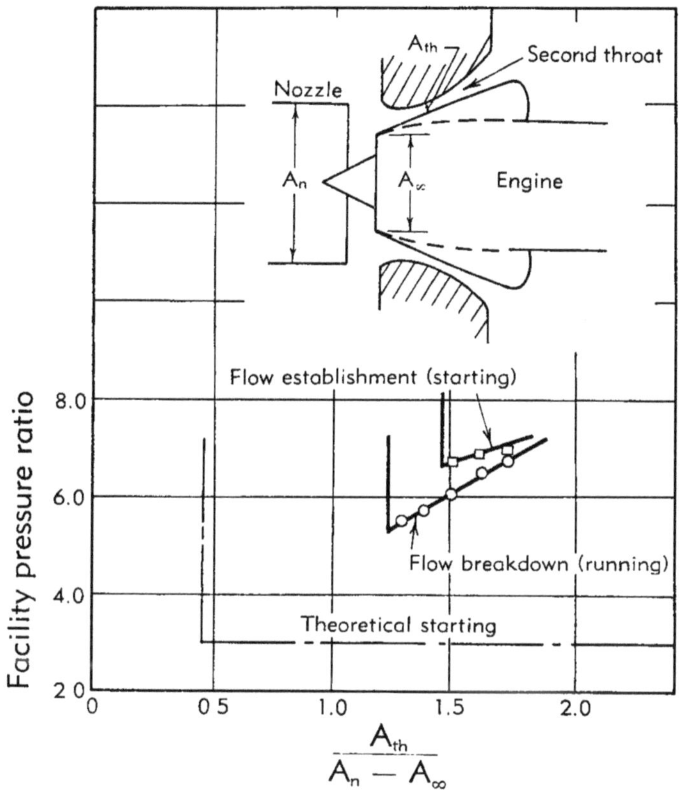

Fig. E,33f. Effect of second-throat area on starting and running pressure ratio of a free jet facility. Test Mach number 3.0.

The minimum pressure ratios required to start and to run the facility are thus dependent on the throat area of the diffuser. Some data for a range of second-throat areas for a facility with a Mach number of 3.0 are presented in Fig. E,33f. The abscissa of this figure is the ratio of the second-throat area A_{th} to the difference between the nozzle outlet area A_n and the free stream capture area of the engine A_∞. As the throat area is decreased, the throat Mach number is reduced and, hence, greater supersonic compression is obtained. Consequently, the facility pressure ratio may be reduced below the starting value before the flow breaks down. The lower limit to this process occurs when the throat Mach number is 1; further decrease in diffuser area causes the shock system to move to the nozzle, thus accounting for the vertical part of the running curve. Optimum running conditions are thus obtained in this case at a throat area ratio of about 1.25 and a facility pressure ratio of 5.40. The vertical part of the starting curve at an area ratio of 1.45 indicates that the diffuser throat is choked even though the entering flow is subsonic; hence, decreases in exhaust pressure cannot affect the shock system between

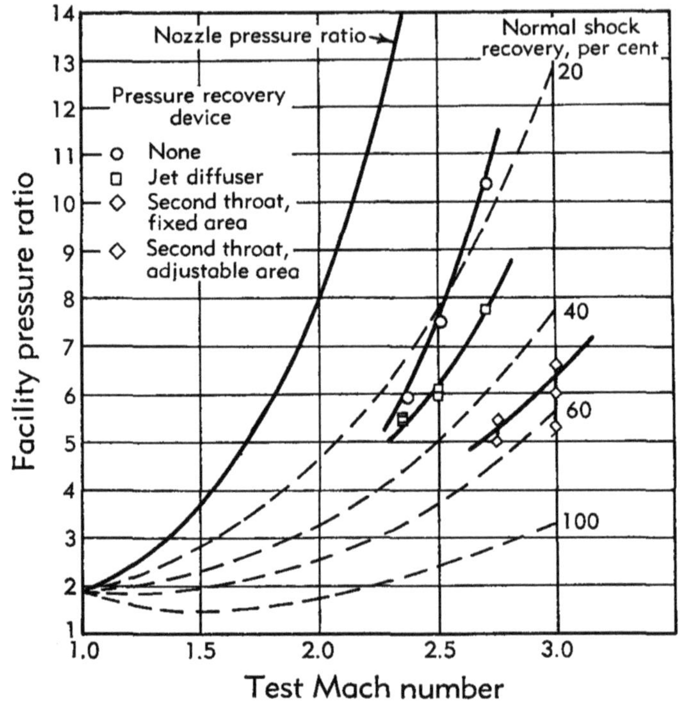

Fig. E,33g. Pressure ratios required to operate various types of free jet facility.

the throats and make starting impossible at any pressure ratio. The minimum starting pressure ratio is seen to be about 6.6 for this facility.

For an ideal system of two successive throats, the shock movement that results in starting the free jet flow occurs when the shock has attained the greatest strength possible for a position between the throats. For a facility with a Mach number of 3.0, this would occur at a pressure ratio of 3.05 (the pressure ratio across a normal shock at Mach 3.0). The fact that the actual pressure ratio is much larger (6.6) indicates that additional pressure losses have occurred in the system. These losses arise from several sources, among which are (1) nozzle boundary layer (approximately 10 per cent of the flow in the second throat is this low energy boundary layer), (2) engine diffuser conical shocks, (3) shock waves at junction of engine cowl and diffuser inner wall, (4) turbulent mixing in the free jet, (5) shock waves attending the initial contact of the jet with the diffuser wall, and (6) boundary shear in the jet diffuser.

These pressure ratios for a facility with a Mach number of 3.0, together with those of a similar second-throat facility (Mach number 2.75), are compared with the values required to run the other types of facility in Fig. E,33g. For both the open-type and the jet diffuser facility, the starting and breakdown pressure ratios were about the same. In the second-throat facilities, the different starting and running pressure ratios are shown by two symbols connected by a vertical line. The considerable

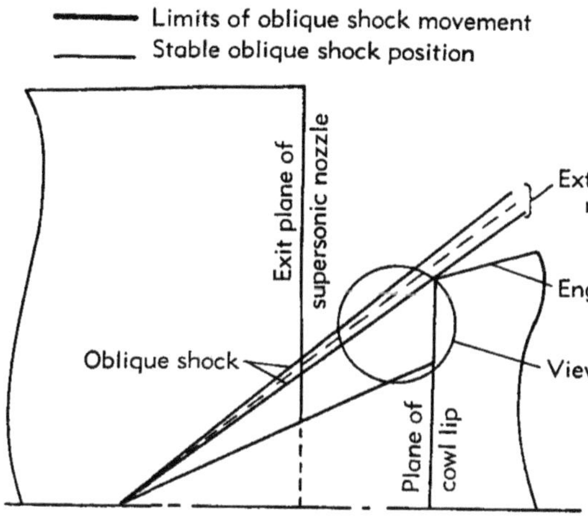

Fig. E,33h. Oscillations in oblique shock from engine inlet cone arising from uneven flow conditions at free jet nozzle inlet. Test Mach number 2.35; no jet diffuser or second throat.

reduction in facility pressure ratio afforded by the second throat is apparent in this figure. Whereas the simple jet diffuser provided a pressure recovery of only about 30 per cent of a normal shock, the pressure recovery in the second throat is at least 50 per cent of normal shock recovery; with the adjustable area throat, the pressure recovery was over 60 per cent of normal shock recovery.

Nozzle inlet flow conditions. If the flow entering the free jet nozzle is not uniform, unsatisfactory flow conditions result at the nozzle outlet. Such unsatisfactory flow conditions were obtained in one installation similar to that illustrated in Fig. E,33a. Although the installation in the test chamber was symmetrical, the air-supply pipe had a 180° turn immediately upstream of the test chamber. In this case, the nozzle was not calibrated in the facility, reliance being placed on previous calibrations in another facility. The first knowledge of unsatisfactory nozzle flow conditions was therefore obtained from visual observations of the engine inlet shock. Typical results are reproduced in Plate E,33b. The shadowgraph photographs presented in this plate illustrate the oscillatory nature of the oblique shock and the diagram of Fig. E,33h indicates the extent of shock movement. The inlet flow was therefore unsteady, as well as distorted. The flow variations were random in nature, with no regular periodicity in either amplitude or frequency of the oblique shock movement. The time interval between the photographs of Plate E,33b is

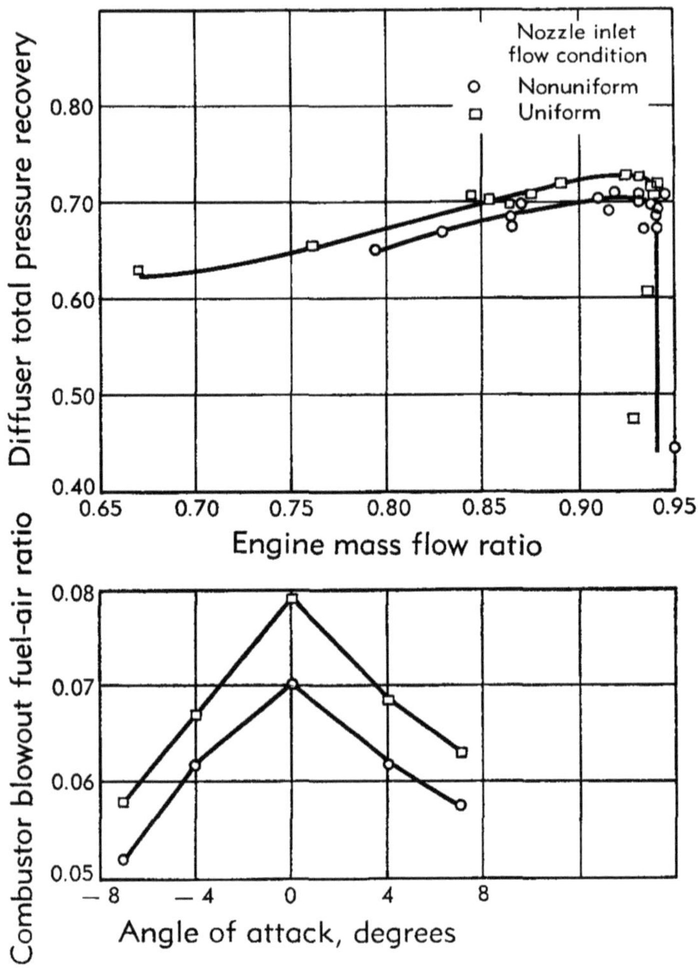

Fig. E,33i. Effect of flow condition at free jet nozzle inlet on engine performance. Top, engine diffuser pressure recovery; bottom, combustor blowout limits.

0.06 seconds. The extent of the oblique shock movement corresponds to a maximum variation in test Mach number (at least in the region under observation) of from 2.20 to 2.50.

Steady and uniform flow conditions at the nozzle outlet were restored by the installation of flow-straightening screens across both the inlet section of the test chamber and across the inlet bellmouth of the nozzle.

The effect of this unsteady inlet flow on the performance of an engine is indicated in Fig. E,33i. In Fig. E,33i, top, are plotted the pressure recovery characteristics of the inlet diffuser and in Fig. E,33i, bottom, are presented the maximum fuel-air ratio blowout limits of the combustor used. The nonuniform flow conditions resulted in a reduction of about 2.5 points in pressure recovery and of about 10 per cent in the maximum fuel-air ratio limit.

Condensation shocks in nozzle. If the moisture content of the air entering the nozzle is too high, condensation shocks occur during the expansion process in the nozzle. These condensation shocks act to decrease the jet Mach number and otherwise destroy proper flow conditions at the nozzle outlet. The inlet air to the facility must therefore be dried well below the moisture content normally existing in the atmosphere at ground level. A method of computing the maximum moisture content necessary to avoid the occurrence of condensation shocks in free jet nozzles is presented in [*11*]. This method includes, in addition to the usual equilibrium requirements among pressure, temperature, and moisture content, the effect of the time delay necessary for nuclei formation.

Fig. E,33j, which is reproduced from the results of [*11*], presents the maximum moisture content for condensation-free flow as a function of jet static temperature (simulated-altitude ambient temperature) and simulated altitude. The allowable moisture content decreases rapidly with both decreasing altitude and decreasing jet static temperature. The effect of temperature is particularly large. At an altitude of 60,000 feet, for example, a reduction in ambient temperature from −67 to −100°F would require a decrease in moisture from 29.5 to 5.5 grains per pound of dry air.

The conditions presented in Fig. E,33j are those for the first occurrence of condensation of moisture in the nozzle. If the moisture content is only slightly greater than the values indicated, the shocks encountered will be weak and the associated flow disturbance may not have any measurable effect on engine performance. Some relaxation of these requirements is therefore usually possible in practice. To help define these more practical limits, a series of tests were run in which the moisture content of the air was progressively increased and the conditions at which the nozzle-outlet static pressure and the engine air flow first began to depart from the condensation-free values were noted. Data defining these conditions for a range of inlet-air temperatures at three altitudes are presented in Fig. E,33k. Although these data are not adequate to completely define the moisture requirements for all possible conditions of pressure and temperature, the calculated curves from Fig. E,33j that are included provide some basis for comparison of the two criteria of maximum moisture content. Unfortunately, the amount the moisture content can be relaxed is smallest at the lower altitudes where they were most severe in the first place. The maximum moisture content required for proper flight simulation of engine performance would thus appear to be of the order of 4 or 5 grains per pound of dry air for altitudes of 40,000 or 50,000 feet, if ambient air temperatures much below standard are required.

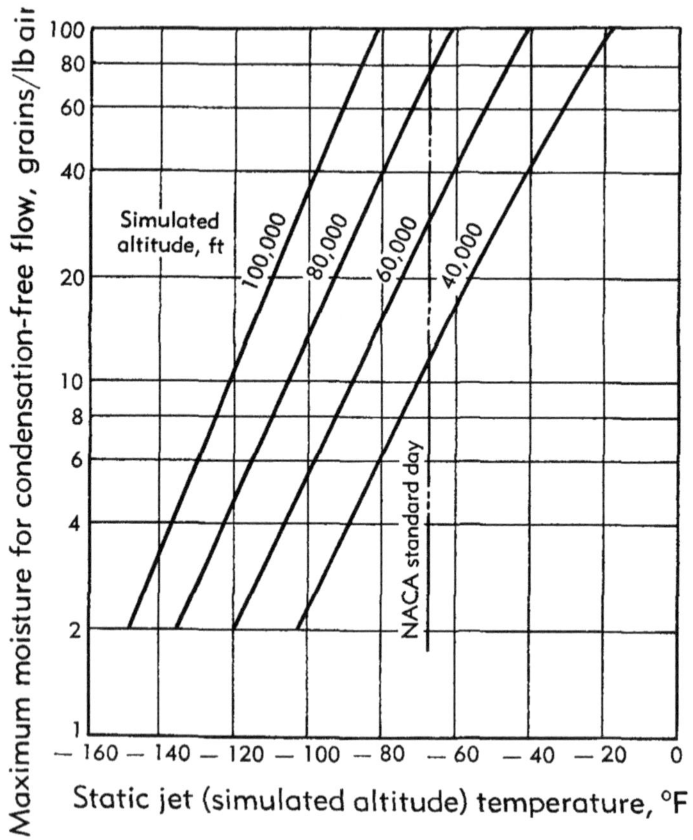

Fig. E,33j. Moisture content of inlet air required to avoid condensation shocks in inlet nozzles.

Free jet nozzle design. The free jet nozzle must be so designed that the flow at the nozzle outlet is uniform and parallel to the axis of the nozzle. The wall contour in the subsonic region upstream of the throat may be arbitrarily chosen; circular arc contours with a radius as small as 4.75 throat radii have provided satisfactory results. The flow solution immediately downstream of the throat may be determined by the series expansion of [*12*]. The remainder of the supersonic design, exclusive of boundary layer, is determined by a numerical method of characteristics for three-dimensional flows having axial symmetry. The equations of [*13*] may be used for this solution. Good accuracy is obtained if a vertical spacing of 0.10 throat radius is taken at the start of the characteristics net. After completion of this design, a correction to the nozzle contour for boundary layer effect must be made. In the absence of a rigorous solution for the flow of a three-dimensional turbulent boundary layer, the method of [*14*] for plane radial flow has been used. The nozzle flow field in a plane through the axis of symmetry is approximated by a radial

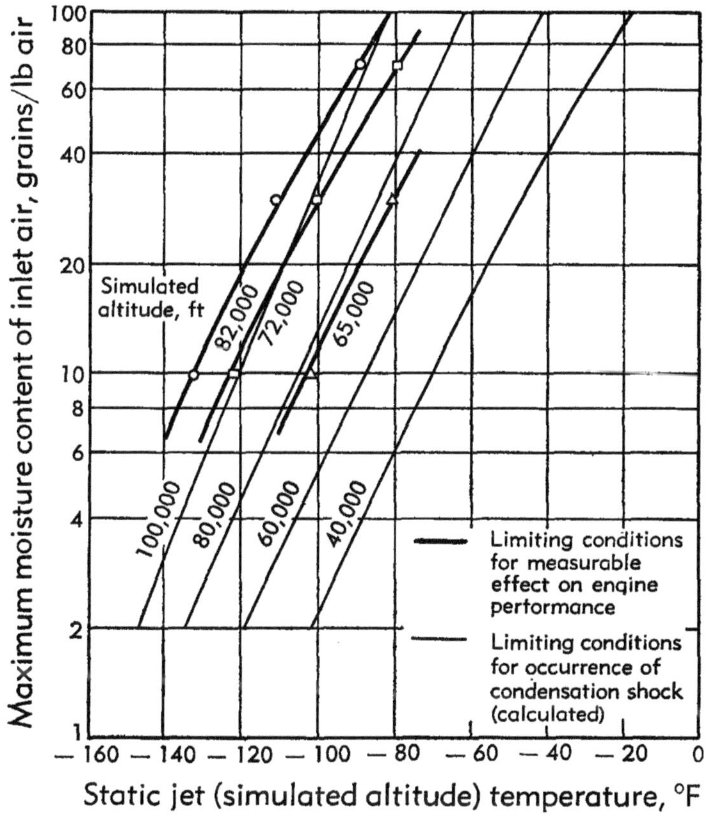

Fig. E,33k. Moisture content of inlet air required to avoid measureable effect of condensation shocks in free jet inlet nozzles on engine performance.

flow pattern with a varying source (or sink), but the pressure gradient used is that of a three-dimensional nonviscous flow field. The adjustment to the wall contour is made equal to that of the displacement thickness. A final step in the nozzle design is a smoothing of the contour by the method of item differences [*15*].

To conserve the facility air flow capacity, the shape of the nozzle outlet may have to correspond rather closely to the shape of the engine inlet. In this event, the axially symmetric nozzles provided by the foregoing method are satisfactory only for engine inlets of the circular, or nose, type. For semicircular inlets such as used on the sides of a fuselage, the nozzle outlet may have to be in the shape of only part, or a segment, of a complete circle. Semicircular nozzles have been formed by simply passing a plane through the axis of the nozzle and adding the necessary boundary layer correction. A photograph of the discharge end of one such nozzle, which has a radius of over 33 inches and a length of 19 feet, is presented in Plate E,33c.

Nozzles having an outlet cross-sectional area that is a circular segment other than that of a semicircle may also be formed by adding a properly contoured plate to a circular nozzle. The contour of the plate, including boundary layer adjustment, is laid out to follow the flow

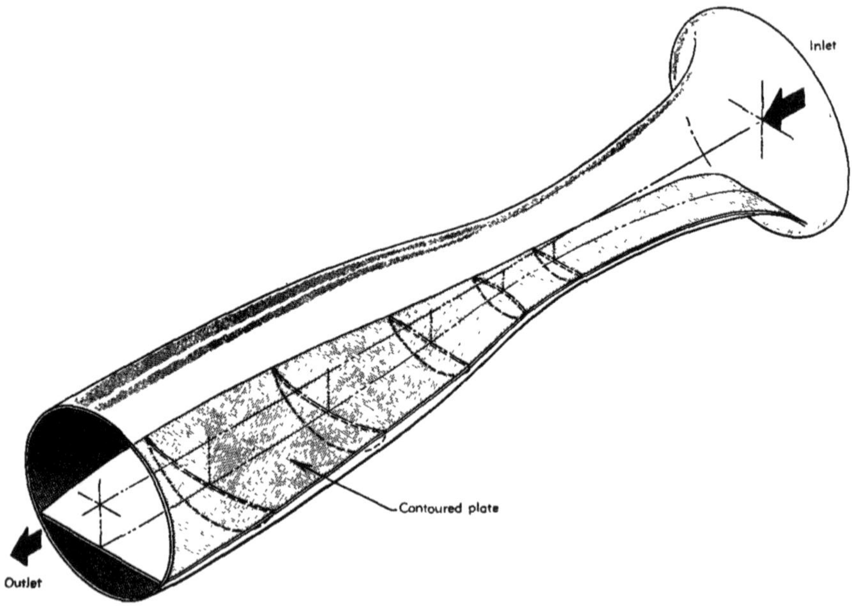

Fig. E,331. Free jet nozzle having a circular segment discharge area.

streamlines that terminate along the desired chord. The plate is, of course, flat at the throat section and within the parabolic area described by the intersection of the plate with the outlet Mach cone; at intermediate stations, the plate is curved to fit the streamlines. A three-dimensional sketch of a nozzle that has a circular-segment outlet area of 75 per cent of a full circle is presented in Fig. E,331.

It is, of course, usually desirable to keep the over-all length of the nozzle as short as possible; this is done by selecting a wall contour near the throat that expands the flow rapidly to the design Mach number. The over-all length of the nozzles that have been used for test Mach numbers between 2.0 and 3.0 has been from 3.1 to 3.5 times the outlet diameter.

The flow field at several axial stations at and near the nozzle outlet has been extensively surveyed for two nozzles. One of these nozzles was axially symmetric and designed for a Mach number of 3.0; the other was semicircular in cross section and designed for a Mach number of 2.75. With both nozzles, the maximum variation of Mach number within the flow field that would be captured by an engine was within ±1.5 per cent of the average value. The maximum deviation in flow angle was measured as ±1.7°; this flow angularity seems somewhat high and may be due to the inaccuracy of measurements obtained from the wedge rakes that were used. The area-weighted average Mach number of the Mach 2.75 asymmetric nozzle was 0.4 per cent higher than the design value; for the symmetric Mach 3.0 nozzle, it was about 1.0 per cent lower than the design value. In both nozzles, no measurable effect of pressure was observed on either the Mach number distribution or the average Mach number.

Engine performance measurements. Most of the engine performance variables may be determined according to the procedures previously discussed for the diffuser, combustor, and nozzle components. Brief mention of a method of conveniently measuring the engine air flow for supercritical operating conditions is, however, probably desirable. Because the amount of flow bypassed around the engine is not accurately known, the air flow through the engine cannot be determined from the total facility air flow rate. The air flow characteristics of the engine inlet are therefore calibrated under cold-flow conditions by an air flow calibrator. This calibrator consists of a length of straight pipe that replaces the combustor of the engine; it is provided with a valve near its upstream end to vary the engine back pressure, several flow-straightening screens or grids near its midpoint, and a metering nozzle of known flow characteristics at its outlet. If the nozzle is maintained choked, the air flow through it may be determined from

$$w_a = \frac{p_5^0 A_5 C_d}{\sqrt{T_5^0}} \sqrt{\frac{\gamma g}{\mathcal{R}}} \left(\frac{2}{\gamma + 1}\right)^{\frac{\gamma+1}{2(\gamma-1)}} \tag{33-1}$$

Because $T_\infty^0 = T_5^0$ for cold flow, Eq. 33-1 may, for a given engine, be replaced by

$$\frac{w_a \sqrt{T_\infty^0}}{p_\infty^0} = B\left(\frac{p_5^0}{p_\infty^0}\right) \tag{33-2}$$

For supercritical diffuser operation, the value of $w_a \sqrt{T_\infty^0}/p_\infty^0$ is a constant for a given angle of attack and inlet temperature. Determination of this constant for a range of angle of attack and inlet temperature thus provides an inlet calibration that may be conveniently applied during burning runs.

With the engine air flow determined, the outlet temperature may be evaluated from measurements of the total pressure in the nozzle according to Eq. 32-1.

The engine thrust, of course, cannot be directly measured. It is therefore computed from momentum considerations based on the measured pressures and gas flows at the exhaust nozzle and assumed or previously determined nozzle flow and thrust coefficients.

E,34. Supersonic Wind Tunnels. The outstanding characteristic of supersonic wind tunnels for ramjet engine testing is that they permit determination of external drag forces as well as internal propulsive forces. They have accordingly been used for two, frequently overlapping areas of research: (1) for inlet diffuser research, in which the external cowl drag is as important as internal performance, and (2) for complete engine performance investigations. Within their size and operating limits, supersonic wind tunnels therefore provide a more final or complete performance evaluation than is possible with any of the previously discussed facilities.

For complete engine investigations, information regarding the various components of external drag is usually desired, as well as many of the internal performance characteristics previously discussed. The various drag components of interest may include additive inlet drag, cowl pressure drag, skin friction drag, and boattail or base pressure drag. Principal parameters of diffuser performance are, of course, the pressure recovery and the mass flow ratio. In addition, information on combustor stability limits, the amplitude and frequency of pressure fluctuations during diffuser buzz, and diffuser-outlet-flow profiles is frequently required.

Engine size and tunnel operating limits. One limit of the size of the engine that may be tested in a wind tunnel is that of the flow area required at the test section (second throat) to permit passage of the starting shock. This well-known contraction ratio, which relates the permissible projected area of the model and support strut to the test-section cross-sectional area, is presented in Fig. E,34a.

Most supersonic wind tunnels in operation today cannot be operated as a closed circuit with a burning engine because of the increase in temperature and contamination of the air by the engine exhaust. With the necessary open-circuit arrangement, the pressure altitude cannot be varied but will be a unique function of the test Mach number. The particular value of altitude for each Mach number depends, of course, on the arrangement of the tunnel. The relation between the test Mach number and the pressure altitude for two of the large propulsion wind tunnels at the NACA Lewis Laboratory is illustrated in Fig. E,34b.

Engine support and force measurements. Two commonly employed methods of mounting the model in the test section are on a strut and on a sting. A photograph of a complete ramjet engine mounted on an overhead strut is presented in Plate E,34a. If an external balance system is used for force measurement, it is necessary to separate the drag of the support strut from the engine drag. This is usually accomplished by the use of an additional dummy strut on the opposite side of the engine that is identical to the support strut. By running with and without this additional support strut, and assuming that the strut and strut-engine

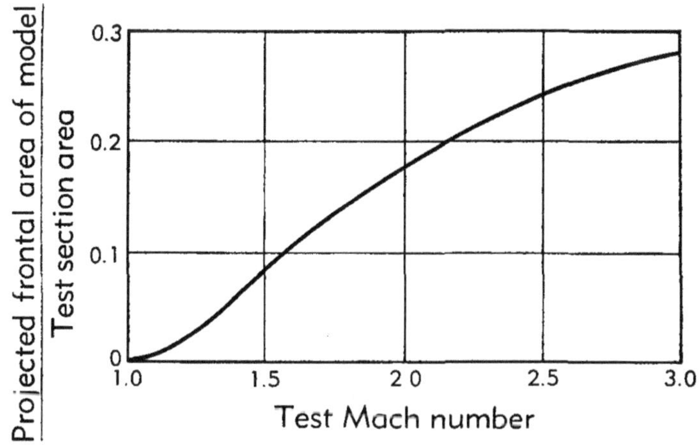

Fig. E,34a. Maximum contraction ratio for supersonic wind tunnels.

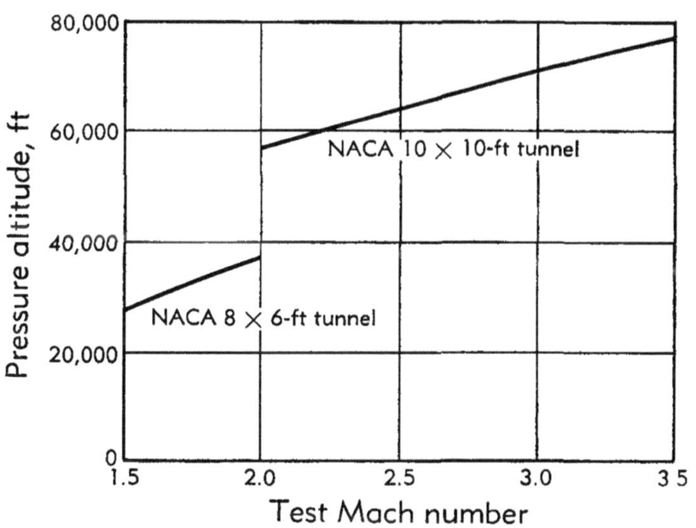

Fig. E,34b. Relation between pressure-altitude and test Mach number for two large propulsion wind tunnels.

interference drags are twice as great for two struts as for one, the drag of the engine alone may be obtained.

A photograph of a sting-mounted model is presented in Plate E,34b. This particular model was used for an investigation of side-inlet diffusers; hence the inclusion of the fuselage ahead of the inlets. Two inlets were used in order to determine the duct stability characteristics. Most sting-mounted models, as well as many strut-mounted models, have used internal straingauges instead of external balances for force measurements. The shroud visible in the downstream portion of the model in Plate E,34b was connected to the sting and used to cover the drive mechanisms and to form a continuation of the fuselage.

A schematic drawing of an internal straingauge balance between a sting support and a diffuser model is presented in Plate E,34c. Of course, similar balances may be used with strut-mounted arrangements. The straingauges G_1 to G_8 are mounted on flexural members of the link to provide indications of axial and normal forces and pitching moments. Interaction among the various forces of the system may be accounted for by calibration. Tare forces acting on the base of the model and within the sting balance chamber also must be determined and subtracted out of the axial force data. The forces thus determined are the combined internal duct forces and total external drag, but do not include the forces acting on the adjustable exit plug.

Instrumentation. The internal performance of the diffuser or complete engine is determined in much the same manner as previously discussed for the other facilities. Internal instrumentation, therefore, usually includes total and static pressure surveys at the diffuser inlet, at the diffuser outlet, and at the exit of the model. These measurements permit determination of such factors as the diffuser pressure recovery, the diffuser mass flow ratio, and the base pressure forces of a diffuser model or the internal thrust of a complete engine.

The various external drag forces acting on a model are determined by pressure surveys over the appropriate areas. Thus external cowl drag is evaluated by integration of cowl-static pressure distributions from two or more rows of static taps along the forward portion of the body. Integration of boundary layer profiles along the external shell in the vicinity of the base provides an evaluation of the friction drag. (A constant static pressure and total temperature throughout the boundary layer is usually assumed here.) Additive drag is determined either from a momentum balance between the free stream and the engine inlet or from analytical methods such as those presented in [*16*]. A comparison of the sum of the various component drag forces and the internal thrust with the over-all net thrust from the external balance system provides a check on the measurements.

Typical diffuser performance data. Some typical pressure recovery characteristics of a supersonic inlet diffuser are presented in Fig. E,34c, top. The pressure recovery, defined as the average diffuser-outlet total pressure p_3^0 divided by the free stream total pressure p_∞^0 is plotted in the conventional manner against the diffuser mass flow ratio. This mass flow ratio is the ratio of the air flow passing through the diffuser to the mass flow in a free stream tube of the same area as the inlet cowl. For diffuser tests, this mass flow ratio is varied by means of an adjustable outlet valve such as the plug shown in Plate E,34c. The vertical portion of these curves represents supercritical operation of the diffuser (with the normal shock swallowed) during which the mass flow is independent of the

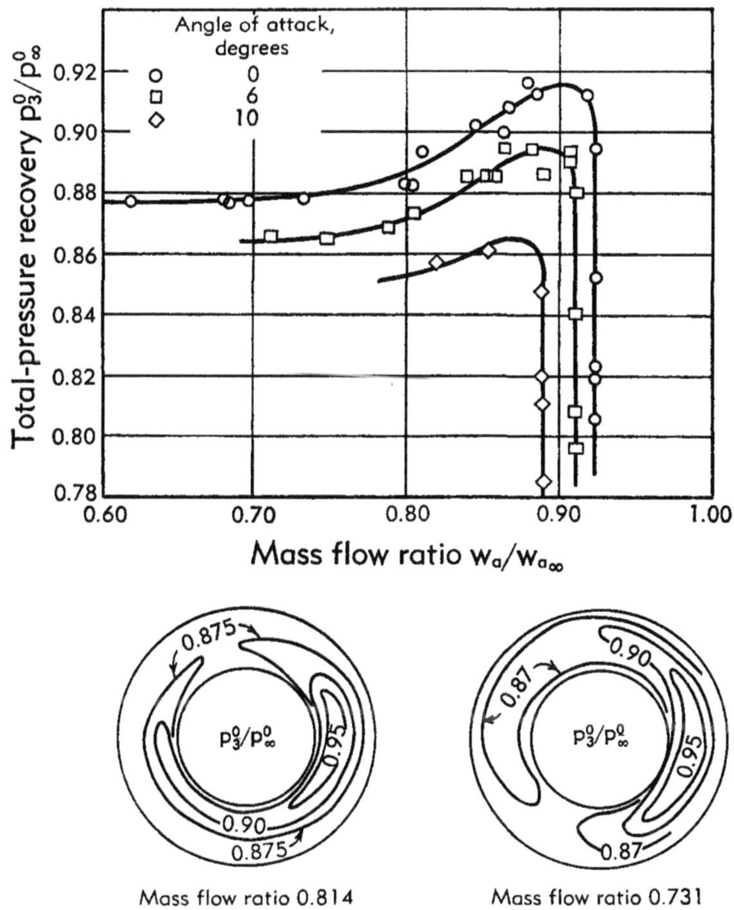

Fig. E,34c. Some representative inlet diffuser performance characteristics. Top, pressure recovery characteristics; bottom, contours of diffuser-exit total pressure recovery.

diffuser back pressure. When the back pressure is increased sufficiently to cause the normal shock to be expelled (or in front of the cowl lip), the mass flow through the diffuser starts to decrease and the operation is subcritical. Further reduction in the exit area results in a rapid decrease in the mass flow ratio, with an attendant moderate loss in pressure recovery.

Some diffuser-outlet total pressure profiles are shown in Fig. E,34c, bottom. The nature of these profiles varies greatly, of course, with both the type of diffuser and the operating conditions. The data shown are for the compressor-inlet station of a turbojet installation (or combustor inlet of a ramjet engine) which is fed by two side-fuselage inlets. The operating conditions shown are for a small angle of yaw, which produces the high pressure region on one side of the duct.

E,35. Flight Testing. Historically, the first supersonic operation of a ramjet engine was accomplished by means of flight testing. Although ground test facilities now contribute the major part of ramjet engine research and development, flight testing remains a very important phase in the over-all picture. No other method of testing provides a simultaneous variation of altitude, flight speed, angle of attack, and other related conditions as they occur in practice. Many phases of ramjet operation that are highly sensitive to environmental factors, such as burner ignition in the transient following booster separation, can therefore be adequately investigated only in flight. Flight testing is also the only available method of obtaining over-all performance data on an operating engine in the transonic speed range. Of perhaps even greater importance is the contribution of flight testing to systems research and development; typical of these problems are the reliability and performance of ignition, fuel metering, and control systems as well as engine structural integrity under conditions of acceleration loads, vibration, and widely varying environmental pressures and temperatures. If for no other purpose, there is also a strong psychological advantage to flight testing; no matter how extensive and thorough the theoretical analysis and ground testing, one can never be sure that all the problems are known and solved until a flight is performed.

Many different methods have been used for flight-testing ramjet engines. One of the very early methods was to mount the test engine on a conventional aircraft and operate it in flight as a sort of auxiliary power plant. Testing under these conditions was, of course, limited to the flight capabilities of the carrier aircraft. This limitation has led to the development of special, rocket-boosted test vehicles that are capable of carrying a test engine over considerable distances and at high speeds and altitudes. The test engine provides the propulsive thrust. These vehicles are thus essentially supersonic aircraft or missiles that are especially designed for test purposes. Their application to ramjet flight testing is very similar to the flight testing of actual prototype or operational missiles.

The test vehicles or missiles mentioned above are used for aerodynamic, structural, guidance, control, and many other phases of testing in addition to engine testing. An extensive amount of flight testing of engines without the complication and expense of a complete airframe has also been carried out in either air-launched or ground-launched missiles that follow a zero-lift trajectory.

Ground-launched vehicles. A photograph of a representative ground-launched test vehicle is presented in Plate E,35a. Two ramjet engines are mounted in nacelles on the tail of the vehicle. This arrangement not only simplifies ground testing of the engines but also leaves the fuselage free for storage and for access to instruments and controls. One of the primary design requirements of the vehicle is that it be small, light, and compact in order to minimize boosting and launching problems. A high thrust-to-weight ratio is desired so that the test vehicle can accelerate to high Mach numbers.

The model is usually launched at elevation angles of from 45 to 75° and follows a zero-lift trajectory in flight. Commercially available rockets are used for boost; two such rockets are shown in the photograph of Plate E,35a. The rocket boost period may last about 3 seconds and accelerations of the order of 30g have commonly been used. The engines are started in flight during the latter part of the boosting period. Ignition is accomplished by an electric squib or flare; the combustion chamber may be partially blocked by an aluminum or magnesium disk to facilitate starting.

Air-launched vehicle. The ramjet engine is launched from a carrier aircraft at high altitudes and accelerated to supersonic velocities by the engine thrust and the force of gravity. In some cases, a rocket booster is added to obtain higher speeds. A photograph of a typical flight-test engine mounted on a carrier aircraft is presented in Plate E,35b, top, and a cutaway view of an engine in free flight is shown in Plate E,35b, bottom. The fuel system and instrumentation is located in the center-body. The engine is fin-stabilized, has no guidance equipment, and follows a zero-lift trajectory. In addition to vehicles that are propelled by ramjet engines, rocket-propelled cold models have also been used for detailed drag evaluations at transonic Mach numbers (elimination of ramjet fuel providing greater space for instrumentation).

If a rocket booster is not used, the engine is ignited just prior to release from the carrier airplane. For models utilizing a rocket booster, the rockets are mounted within the combustion chamber on rollers and retained in the engine by a shear pin. This internal mounting of the booster is used to produce a smaller shift in the center of gravity than would be obtained with external mounting and to eliminate the need for large booster fins. Electric heating blankets and insulation are carried within the rocket-housing for temperature control of the rocket grain. The rockets are ignited by time-delay squibs that are fired by a trailing line as the engine leaves the airplane. The thrust of the rocket shears the retaining pin and is ejected by drag forces after burnout. A lanyard attached to the ejected rocket triggers the fuel release and closes the ignition switch.

Typical flight trajectories of both ground-launched and air-launched models are shown in Fig. E,35a. With the ground-launched method, the rocket booster accelerates the model to a fairly high speed at low altitudes

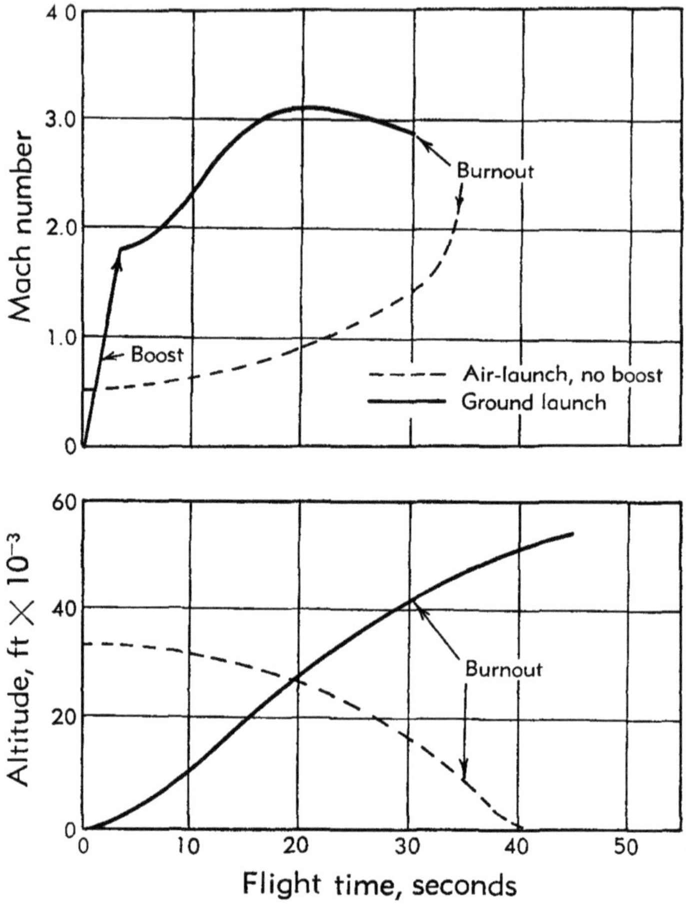

Fig. E,35a. Typical zero-lift trajectories for ground- and air-launched ramjet research vehicles.

before full ramjet power is obtained. A large portion of the flight is therefore under simultaneously increasing speeds and altitudes; high altitudes are accompanied by high flight speeds. The air-launched models, on the other hand, reach their highest speeds at low altitudes. This method therefore provides engine operation both in the transonic speed range and under more severe structural conditions.

Fuel systems. Because of the relatively short duration of flight, a pressurized fuel system is usually used. This is accomplished by using either a vapor fuel that is stored at high pressure or a liquid fuel that is forced by high pressure helium from the tank through the fuel nozzles. In the latter case, fuel tanks of either a collapsible bladder or a piston and cylinder type have been successfully used; a schematic diagram of each type of system is shown in Fig. E,35b. Although the two systems differ

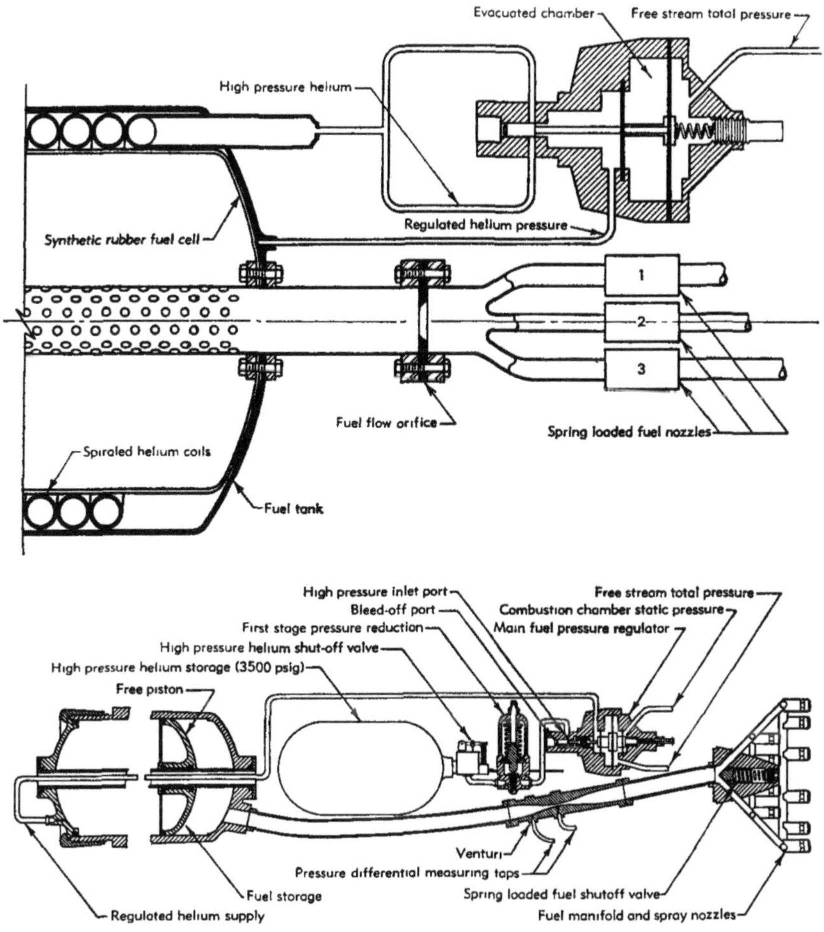

Fig. E,35b. Schematic diagrams of typical pressurized fuel systems for liquid-fueled, flight-test, ramjet engines. Top, collapsible-bladder fuel tank; bottom, piston and cylinder fuel tank.

in arrangement and detail, the principles of operation are the same for both. Helium at either the storage pressure or at some reduced pressure is passed through a pressure regulator that controls the fuel pressure, and hence the fuel flow. With the regulator shown in Fig. E,35b, top, the fuel pressure to the nozzles is proportional to the free stream total pressure; with the regulator of Fig. E,35b, bottom, the pressure differential across the nozzles is proportional to free stream total pressure. In either case, control is at approximately constant fuel-air ratio and various values may be obtained by presetting the regulator.

Instrumentation and measuring techniques. For the air-launched models, the position of the model in space is determined with a type SCR-584 radar-tracking unit with optical tracking facilities. Velocities of the winds aloft are obtained by radar-tracking a weather balloon that is released immediately after the drop. These velocities are applied to the computed space velocities of the model to obtain the velocity of the model relative to the air. The ambient pressure and temperature throughout the flight altitude range are determined from an atmospheric survey made by the descending carrier airplane. True altitude is determined by radar-tracking the airplane.

For ground-launched models, the most accurate measurement of velocity is obtained by a Doppler radar in conjunction with a tracking radar. The Doppler radar unit is usually mounted within 100 feet of the launcher and, because of this and the high elevation angles used (45 to 75°), only small corrections are required for flight path deviation. The tracking-radar unit provides flight path information for correcting Doppler radar velocities and in combination with a radiosonde provides atmospheric data as a function of altitude during the flight.

Performance data are transmitted from the model to ground receiving stations by an FM-AM frequency-division telemeter located in the nose section of the model. A typical eight-channel telemeter nose section is shown in Plate E,35c. Although space consideration usually limits the telemeter to 8 or 10 continuous channels, switching units have been added to cold-flow models that allow as many as 30 different measurements to be transmitted within a time period of 0.17 seconds.

With burning ramjet engines, typical instrumentation will consist of one or two accelerometers, an orifice or venturi in the fuel line, averaging total pressure probes in the diffuser and at the combustor inlet, and a static pressure or total pressure rake at the nozzle outlet. A Pitot static tube is also used to obtain measurement of free stream total and static pressures; this probe may also serve as the antenna for the telemeter unit.

With cold-flow models used for drag data, many additional pressure measurements must be obtained to evaluate the various components of drag. These measurements include a series of static pressure taps on the cowl surface, along the centerbody, along the diffuser cone, and at the base of the model as well as boundary layer total pressure profiles.

The free stream flight Mach number may be calculated from the free stream static and total pressure measured by the Pitot static tube, by differentiating the flight path from the tracking radar, or by integrating the total acceleration. The air flow through the engine and the outlet gas temperature is calculated from the measured internal pressures and the free stream total temperature, according to the equation of flow continuity previously discussed. The net thrust (or internal drag) is calculated as the difference between the momentum of the exhaust gases at the engine outlet and the momentum of the free stream air. External drag is determined as the difference between the net thrust and the product of the net acceleration times the instantaneous mass of the model.

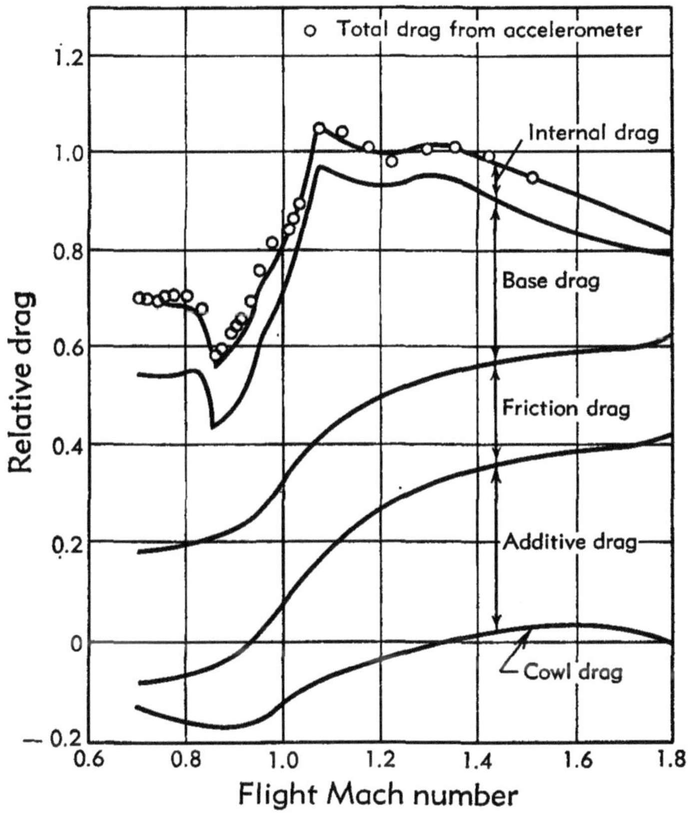

Fig. E,35c. Comparison of component drag from pressure measurements with total drag from accelerometer. Air-launched ramjet engine.

Various components of total drag may be evaluated in a manner similar to that for wind tunnel testing. Additive drag is determined by a force summation as the difference between the total momentum of the engine air flow at the inlet and in the free stream, plus the axial force on the diffuser cone, as calculated from cone-surface static pressures. Cowl drag is calculated by graphical integration of cowl static pressures and friction drag from the momentum decrement obtained from boundary layer surveys.

Typical test data. Engine performance data usually include diffuser pressure recovery, diffuser mass flow ratio, combustor temperature rise, combustion efficiency, propulsive thrust, over-all drag, engine efficiency, and specific fuel consumption. The principal difference between flight test data and data obtained in ground test facilities is that, in flight, they are obtained under simultaneously varying conditions of altitude and flight speed. Each flight, therefore, provides data over a wide range of operating conditions. Very good agreement between wind tunnel and flight test data has been obtained for several engines.

When the model or engine is instrumented for various drag components, a check on the measurement and calculations is available by comparing the summation of each drag component with the total drag from the accelerometer data. A sample of this comparison for one engine is presented in Fig. E,35c. Because these data were obtained from an air-launched model, the range of flight Mach numbers covered is mainly in the transonic region. In the case illustrated, the two independent methods of determining total drag agree very closely.

E,36. Cited References.

1. Keenan, J. H., and Kaye, J. *Gas Tables.* Wiley, 1948.
2. English, R. E., and Wachtl, W. W. Charts of thermodynamic properties of air and combustion products from 300° to 3500°R. *NACA Tech. Note 2071*, 1950.
3. Turner, L. R., and Lord, A. M. Thermodynamic charts for the computation of combustion and mixture temperatures at constant pressure. *NACA Tech. Note 1086*, 1946.
4. Hottel, H. C., and Williams, G. C. Charts of thermodynamic properties of fluids encountered in calculations of internal combustion engine cycles. *NACA Tech. Note 1026*, 1946.
5. Turner, L. R., and Bauer, E. A. Thermodynamic charts for internal combustion engine fluids. *NACA Tech. Note 1883*, 1949.
6. Henry, J. R., and Bennett, J. B. Method for calculation of ramjet performance. *NACA Tech. Note 2357*, 1951.
7. Pinkel, I., and Shames, H. Analysis of jet-propulsion-engine combustion-chamber pressure losses. *NACA Rept. 880*, 1947.
8. Pinkel, B., and Turner, L. R. Thermodynamic data for the computation of the performance of exhaust-gas turbines. *NACA Advance Restricted Rept. 4B25*, 1944.
9. Kantrowitz, A., and Donaldson, C. duP. Preliminary investigation of supersonic diffusers. *NACA Wartime Rept. 1-713*, 1945. (Supersedes *NACA Advance Confid. Rept. L5D20*).
10. Jack, J. R. Theoretical wave drags and pressure distributions for axially symmetric open-nose bodies. *NACA Tech. Note 2115*, 1950.
11. Burgess, W. C., Jr., and Seashore, F. L. Criterions for condensation-free flow in supersonic tunnels. *NACA Tech. Note 2518*, 1951. (Supersedes *NACA Research Mem. E9E02*.)
12. Hooker, S. G. The flow of a compressible fluid in the neighbourhood of the throat of a constriction in a circular wind channel. *Brit. Aeronaut. Research Council Repts. and Mem. 1429*, 1932.
13. Cronvich, L. L. A numerical-graphical method of characteristics for axially symmetric isentropic flow. *J. Aeronaut. Sci. 15*, 156–162 (1948).
14. Tucker, M. Approximate calculation of turbulent boundary-layer development in compressible flow. *NACA Tech. Note 2337*, 1951.
15. Worthing, A. G., and Geffner, J. *Treatment of Experimental Data.* Wiley, 1943.
16. Sibulkin, M. Theoretical and experimental investigation of additive drag. *NACA Research Mem. E51B13*, 1951.

SECTION F

INTERMITTENT JETS

JOSEPH V. FOA

CHAPTER 1. GENERAL PERFORMANCE EQUATIONS

F,1. Introduction. The first schemes of air-breathing jet engines and gas turbines were conceived in the wake of the successful development of the reciprocating internal combustion engine. The extent to which this development determined the thought of jet propulsion pioneers is clearly evinced in a scheme which was proposed by Lorin in

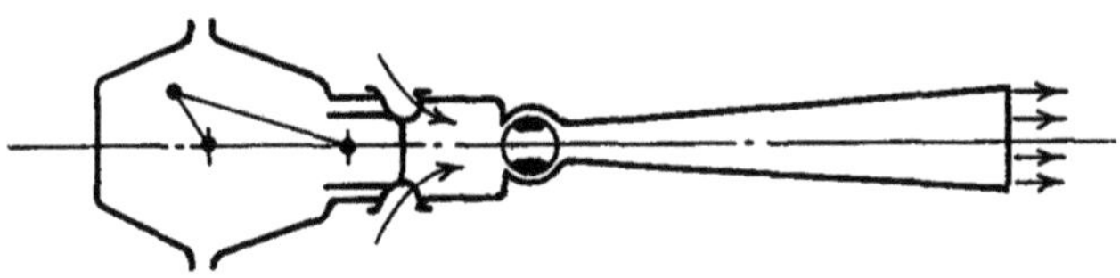

Fig. F,1a.

Fig. F,1b.

1908 [*1*], whereby the combustion products of a reciprocating engine would be expanded through nozzles for the direct generation of thrust, letting the shaft be supplied only with the power required to produce compression and to overcome losses (Fig. F,1a). It is also noteworthy that Marconnet [*2*], having invented in 1909 a jet engine which was substantially a ramjet (Fig. F,1b), failed to recognize it for what it was and described it in his patents as operating intermittently; and that it was not until about four years later that the idea of a steady flow, thermal jet engine was first conceived [*3*].

Some of the most significant contributions to the development of intermittent jet engines came from the field of gas turbine engineering. The possibility of utilizing inertia effects associated with the nonsteady motion of the combustion products for the purpose of scavenging the combustion chamber and of drawing in a fresh charge after each explosion was discovered by gas turbine engineers about the turn of the century. In this connection, it is interesting to quote the words of Amsler, who in 1900, in an article on gas turbines [*4*], wrote:

"From this it will be seen that the idea of driving a wheel by the explosion of gases bears about the same relation to the present gas-engine that the steam turbine bears to the steam engine. With so much inventive energy directed along the line of steam turbines, it is natural that a machine to use gas in a similar manner should have been thought of, and as it is not generally known that such a machine has been designed and constructed, numbers of inventors have entered this field under the impression that it is altogether new.

"A machine of this kind was designed and constructed under the supervision of a gentleman who was formerly with the Westinghouse Company, but who is now devoting his whole time to perfecting his invention, as he has sufficient financial backing for that purpose. This machine has two cylindrical explosion-chambers which communicate at either end with a common nozzle and a gas is to be exploded in each chamber alternately, and the discharge directed upon the vanes of a turbine by means of these nozzles . . .

"The fact that immediately after the explosion a reduction of pressure takes place inside the explosion chamber, has been developed by experiment with this machine. It was found that it was not necessary to scavenge the chamber by introducing the air and gas under pressure, as was at first thought necessary, but that there seemed to be a sort of injector action at the nozzle that drew the air and products of combustion out of the chamber if only a check valve were supplied through which the air and gas might enter. To whatever this effect is due, it made it possible to considerably simplify the construction of the machine, and made the method of the gas-turbine more promising."

The first successful heat engine based on wave processes was the gas turbine of Holtzwarth [*5*], but gas inertia effects were more fully utilized by Karavodine [*6*], who in 1908 succeeded in driving a turbine by means of the gases exhausting, through a long pipe, from a resonating explosion chamber (Fig. F,1c). The inertia of the gas contained in the pipe was here utilized not only for scavenging but also to permit the pressure to rise in the combustion chamber during combustion. The application of a similar principle to the direct generation of thrust was proposed a year later by Marconnet [*2*], whose now-famous "reacteur-pulsateur" (Fig. F,1d) was in every respect the precursor of the present-day pulsejet. Marconnet's idea was revived about twenty years later by Schmidt [*7*] with the additional suggestion that some of the pressure waves, which are generated in intermittent combustion, could be used to transfer energy from the combustion products to a "momentum medium" in a novel and highly efficient form of thrust augmentation (Art. 15).

The utilization of pressure waves as vehicles for the transfer of energy from the combustion products to the fresh charge for the purpose of precompression (a mechanism which has but recently been applied to jet

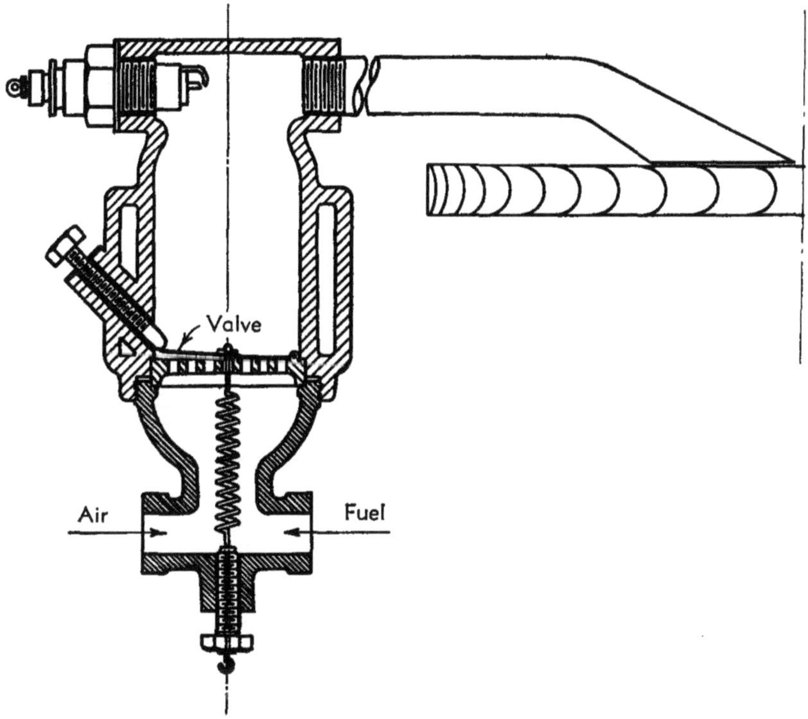

Fig. F,1c.

Fig. F,1d.

engines) was also attempted very early in the history of gas turbine development, as a means for reducing the power consumption of the compressor. Thus, in the Westinghouse and Bischof turbines of 1913 [*5*], the expanding gases for each combustion chamber, after having done work on the turbine wheel, were used to precompress a fresh charge in a neighboring chamber; and in the Esnault-Pelterie gas generator [*5*], also of 1913, the pressure wave generated by the explosion at one end of a long chamber was used to precompress the charge at the other end

(Fig. F,1e). Similar techniques of "wave precompression" have been used more recently in the pressure exchangers of Von Ohain [*8*] and Seippel [*9,10*] and in a large variety of internal combustion wave engines (Art. 17 and 18).

It is clear that the chronological precedence of the conception of the intermittent combustion jet engine over that of its steady combustion counterpart was not due to full discernment of the potential usefulness of certain features which are peculiar of nonstationary flow. In fact, the techniques of exploitation of these features—as, e.g. in wave precompression, inertial confinement of the burning gases and inertia scavenging

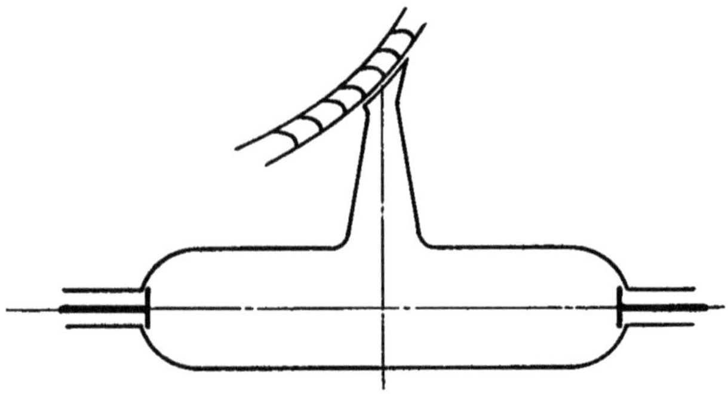

Fig. F,1e.

—cannot be said to have been adequately mastered even to date. Neither is the understanding of these mechanisms yet sufficient to permit a full appraisal of the relative potential merits of steady and intermittent jets.

F,2. Single-Flow and Multiple-Flow Engines. Nonsteady flow jet engines can be classified according to the number of flows which participate in the generation of thrust in each engine, but this classification must be based on thermodynamic rather than on topological considerations. It is indeed important to remember that, whereas in a steady flow engine all fluid particles which follow the same path may be assumed to be subjected to the same sequence of thermodynamic transformations, this assumption is not permissible when one deals with nonsteady flows. To be sure, there exist nonsteady flow jet engines in which the various flows are topologically as well as thermodynamically distinct, but it is more generally true that, in nonsteady flow engines, fluid aggregates which move along identical paths may undergo entirely different thermodynamic cycles. Thus, as a rule, each "flow" in such engines can be properly identified only with reference to the thermodynamic cycle to which it is subjected.

A single-flow jet engine will accordingly be defined as one in which all particles of the working medium undergo the same thermal cycle, and a multiple-flow engine as one in which various aggregates of the working medium undergo different cycles.

F,3. Thrust and Impulse of Single-Flow Engines. Consider a cylindrical control surface enclosing the engine, extending upstream and transversally to regions where the flow may be regarded as undisturbed and terminating at the engine exhaust station e (Fig. F,3a). The velocity V is also assumed to be parallel to the free stream velocity V_∞ at all points of station e, and the static pressure p is assumed to be equal to the free stream static pressure p_∞ at all points of this station except in

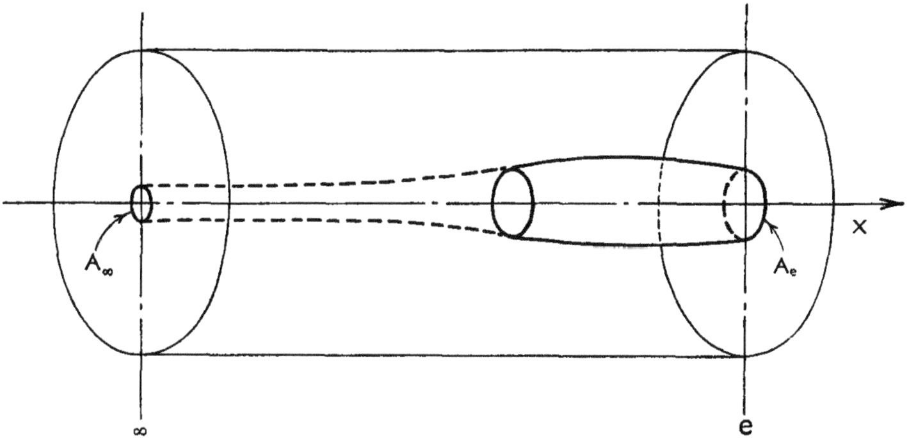

Fig. F,3a.

that portion which contains the exhaust jet (the area A_e), where the static pressure is p_e. Furthermore, the flow external to the engine is assumed to be isentropic and unaffected by the flow pulsations. This assumption implies the absence of frictional and wave drag and flow separation on the external surface of the engine. Then the instantaneous force acting in the positive x direction on the flow region bounded by the control surface is $(p_\infty - p_e)A_e + F_i$, where F_i is the instantaneous thrust produced by the engine. It also follows that the state and velocity of the external flow at all points of station e are the same as in the free stream. The momentum equation may therefore be written in the form

$$F_i - (p_e - p_\infty)A_e = \rho_e A_e V_e^2 - \rho_\infty A_\infty V_\infty^2 + \frac{\partial}{\partial t}\int_\tau \rho V d\tau \qquad (3\text{-}1)$$

where τ is the volume of the internal flow region bounded by stations ∞ and e and subscript $_e$ denotes conditions in the exhaust jet at station e, the state and velocity being assumed to be uniform at each instant across the area A_e.

The impulse I_c produced by the engine during one cycle of periodic operation at the frequency f is obtained by timewise integration of Eq. 3-1 over the period $1/f$. In this integration, the term containing the volume integral vanishes because the flow is periodic. Thus, letting

$$\int_{1/f} \rho_e A_e V_e \, dt = \frac{1}{f} \rho_\infty A_\infty V_\infty = m_c$$

(i.e. regarding the mass flow rate of fuel as negligible in comparison with the mass flow rate of air), one has

$$I_c = \int_{m_c} V_e \, dm - m_c V_\infty + \int_{1/f} (p_e - p_\infty) A_e dt \tag{3-2}$$

With the notation

$$\overline{V} = \frac{1}{m_c} \int_{m_c} V dm$$

$$F = f I_c \text{ (average thrust)}$$

$$m = f m_c \text{ (mass flow rate through the engine)}$$

one obtains

$$F = m(\overline{V}_e - V_\infty) + f \int_{1/f} (p_e - p_\infty) A_e dt \tag{3-3}$$

Now let

$$m V_e + (p_e - p_\infty) A_e = \varphi m V_f$$

where subscript $_f$ denotes the flow conditions that would be established at the exit if the exhaust were fully expanded. A quasi-steady isentropic expansion of the exhaust flow from $p = p_e$ to $p = p_\infty$ would produce

$$T_e + \frac{V_e^2}{2Jc_p} = T_e \left(\frac{p_\infty}{p_e}\right)^{\frac{\gamma-1}{\gamma}} + \frac{V_f^2}{2Jc_p}$$

where T denotes static temperature, or

$$\left(\frac{V_f}{V_e}\right)^2 = 1 + \frac{1 - \left(\frac{p_\infty}{p_e}\right)^{\frac{\gamma-1}{\gamma}}}{\frac{\gamma - 1}{2} M_e^2}$$

Therefore,

$$\varphi = \frac{m V_e + (p_e - p_\infty) A_e}{m V_f}$$

$$= \left[1 + \frac{1 - \left(\frac{p_\infty}{p_e}\right)^{\frac{\gamma-1}{\gamma}}}{\frac{\gamma - 1}{2} M_e^2} \right]^{-\frac{1}{2}} \left(1 + \frac{1 - \frac{p_\infty}{p_e}}{\gamma M_e^2} \right) \tag{3-4}$$

A plot of values of φ vs. underexpansion ratio for $M_e = 1.0$ and $M_e = 2.0$ and for $\gamma = 1.4$ and $\gamma = \frac{4}{3}$ is presented in Fig. F,3b. Since $\varphi \leqq 1.0$, the

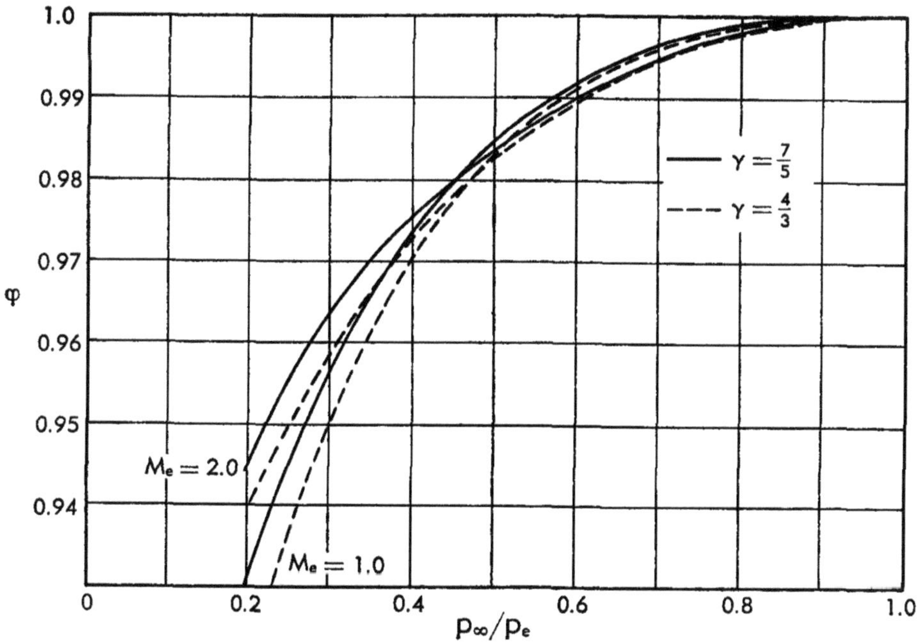

Fig. F,3b.

maximum thrust is obtained with complete expansion, but this condition can seldom be steadily maintained in a nonsteady exhaust. On the other hand, the value of φ is normally so close to 1.0 that the exhaust may be treated, in good approximation, as completely expanded in most cases of interest. With this assumption, Eq. 3-2 and 3-3 reduce to

$$I_c = m_c(\overline{V}_f - V_\infty) \tag{3-5}$$

$$F = m(\overline{V}_f - V_\infty) \tag{3-6}$$

F,4. Cycle Efficiency. The working fluid of each propulsive flow undergoes an open cycle within the engine. The assumption of complete expansion implies that the closure of the cycle in the free exhaust stream is an isobaric process.

The cycle efficiency of a propulsive flow is defined as

$$\eta_{cy} = \frac{\overline{V_f^2} - V_\infty^2}{2E} \tag{4-1}$$

where E represents the total energy (heat and mechanical work) added to each unit of mass of the working fluid in that flow and

$$\overline{V_f^2} = \frac{1}{m_c}\int_{m_c} V_f^2 dm$$

Timewise integration of the energy equation between stations ∞ and e (Fig. F,3a) over the period of one cycle yields

$$\int_{m_0}\left(\frac{V_f^2}{2} + J\bar{c}_{p_{\infty f}}T_f\right)dm - m_0\left(\frac{V_\infty^2}{2} + J\bar{c}_{p_{\infty f}}T_\infty\right) = Em_0 \qquad (4\text{-}2)$$

where

$$\bar{c}_{p_{\infty f}} = \frac{1}{T_f - T_\infty}\int_{T_\infty}^{T_f} c_p dT$$

All particles of a same flow emerge from the engine at the same entropy level and at the same pressure $p_f = p_\infty$, hence also at the same temperature:

$$T_f = T_\infty \exp\left(\frac{s_f - s_\infty}{\bar{\bar{c}}_{p_{\infty f}}}\right) \qquad (4\text{-}3)$$

where s represents entropy per unit mass and

$$\bar{\bar{c}}_{p_{\infty f}} = \frac{1}{\ln (T_f/T_\infty)}\int_{T_\infty}^{T_f} c_p \frac{dT}{T}$$

From Eq. 4-1, 4-2, and 4-3 one obtains

$$\eta_{cy} = 1 - \frac{J\bar{c}_{p_{\infty f}}T_\infty}{E}\left[\exp\left(\frac{s_f - s_\infty}{\bar{\bar{c}}_{p_{\infty f}}}\right) - 1\right] \qquad (4\text{-}4)$$

When energy is added to the gas only in the form of heat, the cycle efficiency reduces to the thermal efficiency. Eq. 4-4 permits the calculation of the cycle efficiency pertaining to each propulsive flow through computation of the entropy increments associated with all the processes to which the flow is subjected within the engine. Then

$$\overline{V_e^2} = 2\eta_{cy}E + V_\infty^2 \qquad (4\text{-}5)$$

F,5. Efficiency of Nonuniformity. Thrust and impulse can be calculated by means of Eq. 4-4, 4-5, 3-5, and 3-6, provided that a relation is established between $\overline{V}_f$ and $\overline{V_f^2}$.

The use of an "efficiency of nonuniformity," defined as $\eta_\nu = (\overline{V}_f)^2/\overline{V_f^2}$, has been proposed by Busemann [*11*]. If $V_f = \overline{V}_f + V_f'$, where V_f' is a periodic perturbation, then $\int_{m_0}V_f'dm = 0$ and $\overline{V_f^2} = (\overline{V}_f)^2 + (1/m_0)\int_{m_0}V_f'^2dm$, or

$$\eta_\nu = 1 - \frac{1}{m_0\overline{V_f^2}}\int_{m_0} V_f'^2dm$$

If one accepts the premise that the value of η_ν is rather insensitive to changes of shape of the function $V_f(m)$, it becomes permissible, in the computation of the performance of intermittent jets, to substitute an arbitrary function of convenient simplicity for the actual (and generally unknown) perturbation function $V_f'(m)$. The arbitrary function is merely required to satisfy the energy equation and to provide a "reasonably" approximate description of the velocity fluctuations. Thus, e.g. in Busemann's analysis of the pulsejet, the perturbation function was approximated by a sinusoidal fluctuation. Substantially similar procedures have been used by Rudinger [*12*], Brumfield and Berggren [*13*], and Kahane [*14*].

It should be noted, however, that the value of the thrust is quite sensitive to changes of η_ν under certain conditions.

Indeed, from Eq. 3-6 and 4-5 one obtains

$$F = m\{[\eta_\nu(2\eta_{cy}E + V_\infty^2)]^{\frac{1}{2}} - V_\infty\} \tag{5-1}$$

It is clear that the percentage error involved in this type of approximation may be very appreciable when the flight speed is high. It becomes necessary therefore to test the validity of the premise on which this type of procedure is based. In periodic operation, V_f is a periodic function of the amount of fluid which has issued through the exhaust, beginning at any arbitrary instant. Let

$$V_f = \overline{V}_f + \sum_{k=1}^{\infty}\left(a_k \cos\frac{2\pi km}{m_c} + b_k \sin\frac{2\pi km}{m_c}\right)$$

Then

$$\frac{2}{m_c}\int_0^{m_c} V_f^2 dm = 2(\overline{V}_f)^2 + \sum_{k=1}^{\infty}(a_k^2 + b_k^2)$$

Therefore

$$\eta_\nu = 1 - \frac{1}{2\overline{V}_f^2}\sum_{k=1}^{\infty}(a_k^2 + b_k^2) \tag{5-2}$$

and Eq. 5-1 becomes

$$F = m\left\{\left[2\eta_{cy}E + V_\infty^2 - \frac{1}{2}\sum_{k=1}^{\infty}(a_k^2 + b_k^2)\right]^{\frac{1}{2}} - V_\infty\right\} \tag{5-3}$$

The value of η_ν depends in equal measure on the amplitude of all the harmonic components of the function $V_f(m)$. Therefore caution must be exercised in the selection of the approximating function. The possibility of large variations of η_ν has also been recognized by Kahane [*14*]. In the absence of information on the exhaust velocity function, the usefulness of this approximation appears to be limited to the investigation of cases in which the amplitude of the exhaust velocity fluctuations is small or to the performance estimate of intermittent jets operating at low speeds or in low velocity streams (e.g. ducted power plants).

Eq. 5-2 also reveals that the maximum value of the efficiency of nonuniformity is obtained when all the Fourier constants are reduced to zero, i.e. when $V_f(m)$ = constant. This result indicates that, for maximum propulsive work, all the gas should leave the engine at the same velocity. It should be noted that the condition $V_f(m)$ = const may be interpreted to represent either a steady flow at the exhaust or an exhaust velocity which is a square-wave function of time, the gas being discharged all at the same velocity but intermittently with respect to time.

F,6. Multiple-Flow Jet Engines. In dealing with a jet engine which comprises two or more thrust-generating flows, the equations developed above must be applied to each individual flow. The total thrust is then the sum of the thrusts produced by the various flows. Thus, if m_1, m_2, . . . , m_n are the individual mass flow rates, the total thrust, with $\varphi = 1.0$, is

$$F = \sum_{k=1}^{n} m_k \overline{V}_{f_k} - V_\infty \sum_{k=1}^{n} m_k$$

The air specific impulse (impulse per pound of air) is, for each flow,

$$I_{a_k} = \frac{F_k}{m_k g} = \frac{\overline{V}_{f_k} - V_\infty}{g}$$

and for a multiple-flow engine

$$I_a = \frac{\sum_{k=1}^{n} F_k}{g \sum_{k=1}^{n} m_k} = \frac{\sum_{k=1}^{n} m_k I_{ak}}{\sum_{k=1}^{n} m_k}$$

Thus the air specific impulse of a multiple-flow engine is the weighted average of the air specific impulses of its individual flows. Similarly, the fuel specific impulse I_t (impulse per pound of fuel) of a multiple-flow engine is the weighted average of the fuel specific impulses of its individual flows.

CHAPTER 2. PROPULSIVE CYCLES

F,7. Entropy Increments. In order to apply the equations developed in the preceding chapter to the performance analysis of a jet engine, it is in the first place necessary to calculate the value of the entropy term of Eq. 4-4. Each flow may undergo any one of the great variety of cycles but, with regard to the computation of the entropy increments, the various processes which may generically make up a cycle can be conveniently classed as follows:

1. isoenergetic processes
2. flow through a fan or compressor
3. flow through a turbine
4. combustion
5. pressure exchange

In double-flow or multiple-flow engines, an additional entropy increment to be considered in some cases is that which is associated with the mixing of two or more flows.

These various increments will now be considered in some detail.

Isoenergetic processes. Entropy-producing energy transformations of this class are those which take place through stationary shocks or those resulting from wall friction and flow separation in diffusers, sharp changes of cross section, narrow passages and bends, or around such obstacles as fuel injectors, flame holders, etc. All these transformations are characterized by changes of stagnation pressure without change of stagnation temperature. Therefore the entropy increment produced by all such transformations in a flow through the engine is

$$\Delta s_\lambda = -\mathcal{R} \ln \lambda \tag{7-1}$$

where λ is the product of all the downstream-to-upstream stagnation pressure ratios associated with such losses within the engine.

Flow through a fan or compressor. From the definition of polytropic or small stage compressor efficiency η_c,

$$T^0_{rc} = (p^0_{rc})^{\frac{\gamma-1}{\eta_c\gamma}}$$

where T^0_{rc} is the exit-to-inlet stagnation temperature ratio across the fan or compressor and p^0_{rc} is the compression ratio, it follows that the entropy increase produced by a fan or compressor is

$$\Delta s_c = \frac{1-\eta_c}{\eta_c} \mathcal{R} \ln p^0_{rc} \tag{7-2}$$

Flow through a turbine. From the definition of polytropic turbine efficiency, η_t,

$$T^0_{rt} = (p^0_{rt})^{\eta_t\frac{\gamma-1}{\gamma}}$$

where T^0_{rt} and p^0_{rt} are the exit-to-inlet stagnation temperature ratio and stagnation pressure ratio, respectively, it follows that the entropy increase through a turbine is

$$\Delta s_t = -(1 - \eta_t)\mathcal{R} \ln p_{rt}^0 \tag{7-3}$$

Combustion. The most convenient representations of the process of entropy production by combustion in intermittent jet engines are those whereby combustion is assumed to take place simultaneously at all points of a certain region of space or of a certain portion of the gas ("volume mode"), or through a thin, plane propagating surface ("front mode"). Neither the "front mode" nor the "volume mode" can provide rigorous descriptions of what happens in the combustion chambers of jet engines, but one of the two is often found to represent a plausible approximation to the actual process. For example, although the combustion reaction is propagated through a flame front, this front is often distorted and broken up into a number of distributed surfaces, so that the combustion region becomes filled with interspersed "islands" of burned gas, in random distribution. Despite the great complexity of the details of this process, its macroscopic aspects are amenable to one-dimensional treatment through the "volume mode" approximation. We are dealing here, however, with an entire class of possible modes, rather than with a specific mechanism. The choice of a specific mode to fit any particular situation is generally dictated by considerations of analytical convenience as well as of plausibility in the light of experimental observations.

There is actually a much broader variety of possible transformation paths for nonsteady volume modes of combustion than there is for the steady modes, which are confined to the weak-deflagration and strong-detonation domains [*15*].

In steady combustion, the distinction between front and volume mode is relevant, from the thermodynamic standpoint, only with regard to the physical possibility of transformation paths in the weak-detonation domain. Indeed, an extended "homogeneous" reaction zone may be treated, in steady flow, as a front which is capable of propagating at a very high speed. In nonsteady combustion, on the other hand, the distinction is of greater significance, in that it pertains to the nature of the transformation path as well as to the domain in which this path lies.

Nonsteady volume modes whose transformation path lines are represented by the linear pressure-volume relations of the type

$$\frac{p}{p_1} - 1 = -\nu\left(\frac{v}{v_1} - 1\right)$$

where subscript $_1$ denotes initial conditions and ν is a constant, are called "linear explosions." Pangburn and Foa [*16*] have shown that the linear explosion domain covers, in the p, v plane, the entire area which lies above the locus of the Chapman-Jouguet states (region c in Fig. F,7). This area overlaps the weak-deflagration domain, where a linear transformation path line may represent either an explosion or a steady deflagration process. In the latter case, $\nu = \gamma M_1^2$, where M_1 is the propagation Mach number of the reaction front relative to the unburned gas. It is noteworthy that, for $-\infty < \nu < 0$, the path line lies in the positive quadrant, an area which is "forbidden" to Rayleigh processes. Thus the

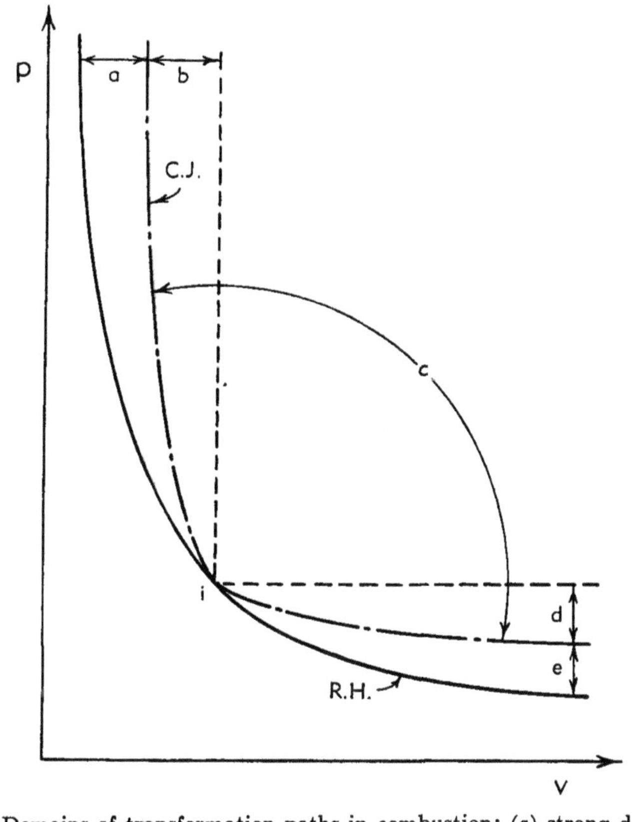

Fig. F,7. Domains of transformation paths in combustion: (a) strong detonations, (b) weak detonations, (c) explosions, (d) weak deflagrations, and (e) strong deflagrations. *C.J.* is the locus of the Chapman-Jouguet states, *R.H.* the Rankine-Hugoniot curve.

condition $\nu < 0$ is a sufficient condition for a linear combustion process to be unsteady.

Whereas any departure from the linearity of the pressure-volume relation would be incompatible with the conservation equations pertaining to the Rayleigh process, no such incompatibility is to be found in the case of nonsteady volume modes. Indeed several pressure-volume relations may suggest themselves as providing convenient and physically significant descriptions of transformation paths in nonlinear explosions. Of these the most plausible are those in which pressure and volume vary monotonically with one another ("monotonic explosions"). All monotonic explosions have at least two linear transformation paths in common, viz. the isobaric and the isochoric. Therefore the corresponding Hugoniot curves (loci of transformation end points for any given amount of heat liberated in combustion) have at least two points in common. Between these two points, i.e. within the "forbidden quadrant," the shape of the Hugoniot curve is found to be relatively insensitive to moderate changes of the transformation paths, thus indicating that any point in this quadrant may be regarded as the terminal point of an infinite variety of monotonic explosion paths, all involving the same amount of heat release. It is therefore permissible, as far as the computation of the entropy rise in combustion is concerned, to select the transformation function merely on the basis of analytical convenience. In this respect, the choice of the polytropic $pv^n = \text{const}$ as a description of the explosion path has proved particularly useful in generalized studies of jet engine performance [*14,17,18*].

Polytropic explosions have three transformation paths in common with the linear explosions. When $n = 0, -1$, or $\pm\infty$, the polytropic process reduces to a linear process with $\nu = 0, -1$, or $\pm\infty$, respectively. It is therefore not surprising to find [*16*] that the Hugoniot curves for polytropic processes are practically coincident with the corresponding Hugoniot curves for linear processes over the entire region bounded by the Chapman-Jouguet states. The line of demarcation between possible and impossible steady polytropic combustion also follows closely (although not exactly) the boundaries of the forbidden quadrant of Rayleigh processes [*17*], and for all practical purposes it may be stated that only nonsteady flow processes can follow polytropic paths with $n < 0$ or $n > \gamma$.

For a polytropic change, the first law of thermodynamics yields

$$\frac{T}{T_1} = 1 + \frac{n-1}{n-\gamma}\frac{q}{c_v T_1} \tag{7-4}$$

where q is the heat added per unit mass, from the beginning of the transformation. Therefore, when the total amount of heat added is q_a, the final temperature and pressure are given by

$$\frac{T_2}{T_1} = 1 + \frac{n-1}{n-\gamma_h}\frac{q_a}{c_{v_h} T_1} \tag{7-5}$$

$$\frac{p_2}{p_1} = \left(1 + \frac{n-1}{n-\gamma_h}\frac{q_a}{c_{v_h} T_1}\right)^{\frac{n}{n-1}} \tag{7-6}$$

where the subscript h denotes average values in the heating process.

Thus, when $n < 0$ or $n > \gamma$, the static pressure of the gas increases as heat is added. The amplitude of the pressure fluctuations increases with the absolute value of n and with the total amount of heat that is added to the flow. Conversely, one may also conclude that if the static pressure increases during the heating process, then n is either negative or greater than γ. In other words, just as the heating mode may be responsible for the generation of flow fluctuations, so may flow fluctuations which are otherwise excited be responsible for modifications of the heating mode.

From Eq. 7-5 and 7-6 one obtains for the entropy increase due to heat addition:

$$\Delta s_h = c_{v_h} \frac{n - \gamma_h}{n - 1} \ln \left(1 + \frac{n - 1}{n - \gamma_h} \frac{q_a}{c_{v_h} T_1}\right) \tag{7-7}$$

or

$$\Delta s_h = -c_{v_h} \frac{n - \gamma_h}{n - 1} \ln \left(1 - \frac{n - 1}{n - \gamma_h} \frac{q_a}{c_{v_h} T_2}\right) \tag{7-8}$$

While the extreme idealization of combustion as a spatially uniform process (volume mode) is justified in the performance analysis of many jet engines, there are cases (e.g. some of the internal combustion wave engines) in which the other extreme, the front mode representation, is believed to provide a more plausible picture of the actual process. In this idealization, the combustion reaction is assumed to take place across a plane surface of discontinuity, moving at the flame propagation velocity relative to the unburned gas. The transformation is thus considered to be, for each particle, an instantaneous Rayleigh process. If the flame propagation Mach number does not exceed about 0.1, the slope $(-\gamma M_1^2)$ of the transformation path line in the p, v plane is negligible and the process can be assumed to be isobaric ($n = 0$). Transformations pertaining to higher flame velocities can usually be approximated by polytropics with small positive exponents (in the range $0 < n < 0.3$).

Pressure exchange. Pressure exchange is commonly defined as a process in which a flow, or a region of a flow, is compressed by the piston-like action of another flow (or another region of the same flow) expanding between the same pressure limits. In dealing with nonsteady flow machines, this same designation is applied here to all energy transfer processes that are carried out, from one flow to another or from one region to another of the same flow, by pressure waves. Pressure exchange plays an important role in the operation of wave engines (Art. 17 and 18), where pressure waves are utilized to transfer energy from the combustion products to the fresh charge for the purpose of precompression beyond the ram level (the function which, in conventional steady flow engines, is performed by the turbine and the compressor), and also in some forms of thrust augmentation (Art. 15). Since the only isoenergetic shock processes are those which take place across stationary shocks, the entropy increment of pressure exchange is the sum of all increments produced on any given particle by moving shocks within the engine. This entropy increment is negligible when the pressure exchange process involves only weak shocks or pressure waves which are too short-lived to form shocks. On the other hand, the pistonlike action of the expanding hot gases is, in some engines, capable of producing rather powerful shocks, and there are cases (Art. 18) in which the shock that is generated by each explosion is forced to reverberate through the engine so that it may be utilized to precompress the charge for the next or a later explosion. Under such conditions, each particle of the working fluid may be subjected to several shock wave passes before it finally undergoes combustion, and the entropy increment associated with pressure exchange may become very appreciable. This increment, which will be denoted by Δs_σ, can only be calculated on the basis of a detailed study of the pressure exchange process in each individual case.

Mixing. When the amplitude of the flow fluctuations in the merging streams is very small, the entropy rise produced by mixing in each flow may be calculated by means of the standard mixing relations for steady flow. When, on the other hand, the amplitude of the flow fluctuations is large, pressure waves can be expected to play a dominant role in the early phases of the energy exchange process. Very little is known about this combined mechanism of pressure exchange and mixing, except for such theoretical information as that which has been acquired through quasi-one-dimensional treatment [*19*]. Therefore the calculation of the entropy increments Δs_m produced by mixing is still largely a matter of conjecture. In most cases of interest, pressure exchange is believed to be accomplished over a considerably shorter distance than mixing, because the propagation rate of transverse pressure waves is much faster than the transverse transport rates which determine the extent of the mixing region. In such cases it seems permissible to treat the transfer of energy as taking place in two successive steps and to assume the pressure exchange process to be completed when mixing begins. Then mixing may again, in fair approximation, be analyzed as a steady flow process.

F,8. Factors Affecting Cycle Efficiency. The effect of precompression is to increase the initial temperature T_1 of the combustion process, hence to decrease the entropy rise associated with combustion, in accordance with Eq. 7-7. Therefore, pressure exchange may be a very effective means of increasing the cycle efficiency. However, when account is taken of the entropy increment produced by the shock waves in the pressure exchange process, the best cycle efficiency is found to be obtained with a relatively moderate precompression ratio, in the same manner as moderate compression ratios are found to maximize the efficiency of gas turbine power plants, and for the same reason. The "optimum" pressure exchange process for engines utilizing front modes of combustion has been calculated by Pangburn [*20*]. This process is discussed in Art. 18.

With regard to the combustion process itself, the effect of changes of n on the cycle efficiency, with fixed values of T_1 and q_a, can be evaluated by means of Eq. 7-7; and it can be verified that, under such conditions, the entropy increment Δs_h decreases monotonically as n approaches $-\infty$, indicating that, as far as combustion alone is concerned, the highest cycle efficiency would be obtained with an isochoric process. On the other hand, the opposite trend is revealed by Eq. 7-8, when $n < 0$, for the condition in which T_2 and q_a are fixed. In other words, when the final rather than the initial temperature in the heating process is prescribed, the cycle efficiency is found to decrease monotonically as the heat addition process approaches the isochoric mode. The latter result would seem to put steady flow engines at an advantage over their nonsteady flow counterparts in those cases in which material properties impose a limitation on the peak temperature of the cycle. It should be noted, however, that the use of nonsteady processes offers the possibility of avoiding the continuous exposure of any part of the engine to very hot gases. Therefore, the maximum allowable gas temperature is often considerably higher in nonsteady than in steady flow engines. For this reason, a comparison of cycle efficiencies of the two classes of engines for equal peak cyclic temperatures would be of doubtful significance.

The cycle efficiency of a jet engine may be adversely affected by flow fluctuations occurring in certain regions of the engine. For example, flow fluctuations in a supersonic intake diffuser can generally be expected to reduce the pressure recovery and hence the cycle efficiency of the engine. Indeed, it is often possible to establish a shock pattern that will minimize the diffuser shock losses for a given operating condition, but it is obvious that this pattern can be steadily maintained only if the flow itself is steady. Fortunately, however, the amplitude of shock motion in any given diffuser in the presence of fluctuating back pressure decreases considerably when the frequency of these pressure pulsations increases, their amplitude remaining constant [*21,22*], because, compression and expansion waves being generated in closer succession, their neutralizing interaction takes place over a shorter distance. Therefore the diffuser length required to prevent the back pressure disturbances from reaching the intake is inversely proportional to the disturbance frequency. For example, with a sinusoidal disturbing wave at the back end of a Kantrowitz-Donaldson diffuser designed for and operating at $M_\infty = 1.5$, the length required to constrain the shock to oscillate downstream of the throat about the station where the steady flow Mach number is 1.5 can be estimated from the following table [*14*]:

Disturbance pressure ratio	(Length of diverging portion) × frequency (feet × cps)
1.5	360
2	560
3	840
4	1110

It appears, then, that the losses associated with the oscillations of diffuser shocks may be greatly reduced by increasing the diffuser length or the operating frequency. In a diffuser conveying the air to a sufficiently large number of intermittent flow chambers operating out of phase, it should be possible to achieve a diffuser efficiency comparable to those of steady flow diffusers. Shock losses in pulsating-flow diffusers may also be reduced by the use of surge chambers, connected to the diffuser passage through slits in the walls [*23,24*]. In any case, the effect of diffuser shock fluctuations may conveniently be accounted for by an "equivalent" reduction of λ in Eq. 7-1.

CHAPTER 3. ANALYSIS OF FLOW PHENOMENA

F,9. Introductory Remarks. The gas dynamic treatment of the operation of intermittent jet engines is generally based on the assumption that the cross-sectional averages of state and flow parameters can be closely approximated by the use of quasi-one-dimensional models. Even with this assumption, however, the solution of the problem presents great difficulties, because the equations of motion for nonsteady quasi-one-dimensional flow (III,C) cannot be solved analytically except for particularly simple transient phenomena which may involve limited regions of the flow (Art. 10) or by the use of processes of linearization which are of doubtful validity in the treatment of essentially nonlinear phenomena. The most rigorous method of solution of these equations is still the numerical-graphical method of characteristics (III,C): When the duct configuration and the initial and boundary conditions are given and the rates of entropy production are known or assumed, the equations of motion permit the construction, in the time-distance plane, of a network of characteristics, representing the lines along which changes of conditions are propagated. The network yields a detailed history of the changes of flow and state parameters along the duct. The usefulness of the method of characteristics is often limited by its extreme laboriousness and by the accumulation of errors which inevitably occurs as the construction of the network is carried further and further from the initial conditions. Such is typically the case when attempts are made to investigate periodic phenomena, either through an iterative application of the method (which would in any case leave serious doubts as to the uniqueness of the solution) or through the construction of extended networks to follow the gradual approach to periodicity, through a large number of cycles, from given initial conditions. A further drawback of the method of characteristics is that it cannot provide general information on the separate influence of the many parameters which affect the operation of intermittent jet engines, such information being obtainable only through systematic application of the method to a large number of cases representing properly selected combinations of the various parameters. Nevertheless, the method of characteristics is still the only available tool for the detailed analysis of the operation of intermittent jet engines, for the prediction of the effects of changes of duct configuration, combustion mode, and boundary conditions; and for a reliable estimate of pressure exchange losses, efficiency of nonuniformity, and mass flow rate through any given engine.

F,10. Analytic Solutions. There are flow situations which are simple enough—or can be simplified enough, in fair approximation—to lend themselves to explicit analytic solution. The most notable simplifications of this sort are those which are introduced in the treatment of nonsteady combustion processes. These simplifications may appear to be extreme in some cases, but they are well justified when it is considered that the use of quite arbitrary and far-reaching approximations is at any rate inevitable in the description of what goes on in the combustion chamber of an intermittent jet engine. Indeed, so little is known as yet about these processes that the only criterion for preference in the selection of an approximation would seem to be that of analytical convenience, so long as the underlying assumptions appear to be tenable in the light of current experience.

The stipulation that the state be spatially uniform in the combustion region is the first simplifying assumption that suggests itself in the analysis of a volume mode of combustion involving high rates of entropy production, particularly when the initial conditions are substantially uniform.

Consider a volume-mode combustion process taking place within a body of gas in a frictionless flow in a duct of constant cross section. If the initial conditions are substantially uniform throughout, one may assume, within the combustion region, $\partial s/\partial x = 0$ and $\partial a/\partial x = 0$, where a denotes the local velocity of sound. Then the continuity and momentum equations (III,C) become

$$\frac{2}{\gamma - 1}\frac{\partial \ln a}{\partial t} + \frac{\partial V}{\partial x} - \frac{1}{\mathcal{R}}\frac{\partial s}{\partial t} = 0 \tag{10-1}$$

and

$$\frac{\partial V}{\partial t} + V\frac{\partial V}{\partial x} = 0 \tag{10-2}$$

Eq. 10-2 indicates that each particle maintains a constant velocity throughout the process.

From Eq. 10-1 one obtains $V = gx + h$, where g and h are functions of time only. Using a coordinate system with respect to which the gas is at rest at the time at which the process starts ($t = 0$), and with the origin at the midsection of the combustible region, h vanishes and Eq. 10-2 yields

$$g = \frac{1}{t + \beta}$$

where β is a constant. Therefore

$$V = \frac{x}{t + \beta} \tag{10-3}$$

During the time t, the layer which is initially at $x = x_1$ moves, at the constant velocity x_1/β, to the position $x = x_1 + (x_1/\beta)t$. Therefore

$$\frac{v}{v_1} = 1 + \frac{t}{\beta} \tag{10-4}$$

where subscript $_1$ denotes the initial condition.

For a polytropic transformation, it follows from Eq. 10-4 that

$$\frac{p}{p_1} = \left(1 + \frac{t}{\beta}\right)^{-n} \tag{10-5}$$

$$\frac{a}{a_1} = \left(1 + \frac{t}{\beta}\right)^{\frac{1-n}{2}} \tag{10-6}$$

If, during the total combustion time t_c, the volume of the burning gas increases to k times the initial volume, one has from Eq. 10-4,

$$\beta = \frac{t_c}{(k - 1)}$$

From the first law of thermodynamics and the above relations, one obtains for the rate of heat addition in a polytropic process,

$$\frac{dq}{dt} = \frac{\gamma - n}{\gamma - 1}\, p\, \frac{dv}{dt}$$

$$= \frac{k - 1}{t_c}(\gamma - n)c_v T_1 \left[1 + (k - 1)\frac{t}{t_c}\right]^{-n} \tag{10-7}$$

whence

$$\frac{q_a}{c_v T_1} = \frac{\gamma - n}{1 - n} k^{1-n} \tag{10-8}$$

Eq. 10-7 indicates that, in a polytropic explosion with $n < 0$, the time rate of heat addition increases with time during the process.

Eq. 10-8 reveals that, in a polytropic explosion with $n < 0$ and with any given value of $q_a/c_v T_1$, the absolute value of n may be increased by decreasing k. This can be accomplished by means of mechanical valves or by the use of "rectifiers," enlarged combustion chambers or constrictions in the passages leading out of the combustion chamber. Effective confinement is also provided by the inertia of the gas columns adjacent

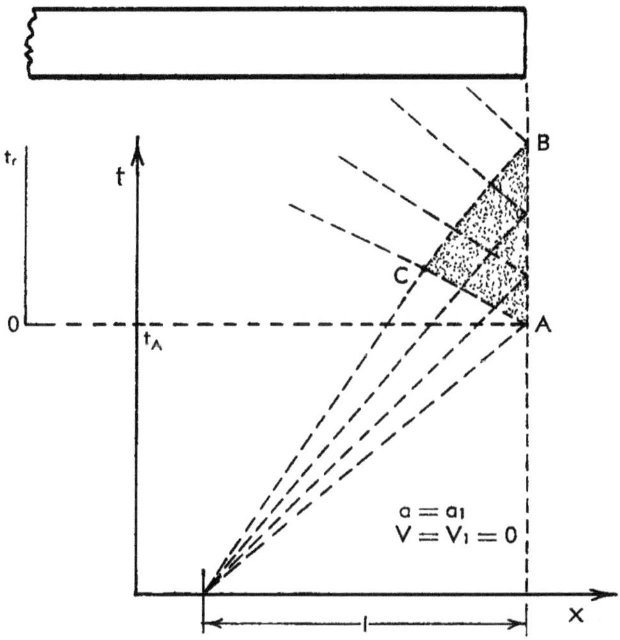

Fig. F,10.

to the combustion region [*17*] when the rate of heat release is very high or when the density of the gas in these columns is increased through cooling or dilution with cold air.

The assumption of spatial uniformity of state has also been found to be plausible in another nonsteady flow situation of considerable interest, not involving the production of entropy.

Poggi [*25,26*] has observed that, when a simple centered wave is reflected at the closed end of a duct of constant cross-sectional area, as in Fig. F,10, the state of the gas can, in good approximation, be assumed to be spatially uniform throughout the "reflection region" *ABC* of the x, t plane. With this assumption, the history of the flow changes in the reflection region is again provided by Eq. 10-5 and 10-6, but in this case $n = \gamma$. Thus Eq. 10-6 yields

$$a = a_1 \left(1 + \frac{t_r}{\beta}\right)^{\frac{1-\gamma}{2}} \qquad \text{(10-9)}$$

where t_r is the time measured from the instant t_A at which reflection starts ($t_r = t - t_A$). The state of the gas at A is still the initial state ($a_A = a_1$), and the value of the constant β can be determined as follows. From consideration of the time required by the various characteristics of the incident wave to cover the distance l, and letting $P = [2a/(\gamma - 1)] + V$, one obtains

$$\left(\frac{dt}{dP}\right)_A = -\frac{\gamma + 1}{4}\frac{l}{a_1^2}$$

At the closed end, $V = 0$, hence $a = P(\gamma - 1)/2$. It follows that

$$\left(\frac{da}{dt}\right)_A = -\frac{2(\gamma - 1)}{\gamma + 1}\frac{a_1^2}{l} \tag{10-10}$$

On the other hand, Eq. 10-9 yields

$$\left(\frac{da}{dt}\right)_A = \left(\frac{da}{dt_r}\right)_{t_r=0} = -\frac{\gamma - 1}{2}\frac{a_1}{\beta} \tag{10-11}$$

From Eq. 10-10 and 10-11 one obtains

$$\beta = \frac{\gamma + 1}{4}\frac{l}{a_1}$$

Therefore

$$a = a_1\left(1 + \frac{4}{\gamma + 1}\frac{a_1 t_r}{l}\right)^{\frac{1-\gamma}{2}}$$

The pressure at the closed end remains constant ($= p_1$) up to $t = t_A$. At any successive instant $t = t_r + t_A$ in the interval AB, the pressure at the closed end is

$$p = p_1\left(\frac{a}{a_1}\right)^{\frac{2\gamma}{\gamma-1}} = p_1\left(1 + \frac{4}{\gamma + 1}\frac{a_1 t_r}{l}\right)^{-\gamma} \tag{10-12}$$

This method of analysis has found useful application [*25*] in the study of thrust generation in pulse jets of simple shapes.

CHAPTER 4. THE PULSE JET

F,11. Introduction. The pulse jet may be defined as a compressorless, nonsteady flow, jet engine without wave precompression of the combustible charge. Actually, a clear line of demarcation between pulse jets and wave engines cannot be drawn, the pulse jet being in effect a wave engine with a wave precompression ratio in the neighborhood of 1.0. The distinction is widely accepted, however, primarily because of the historical precedence of the pulse jet, to the development of which much separate effort has been applied even in recent years.

In its standard form, which will be discussed first, the pulse jet consists of a shaped tube fitted with flow-check valves at the front end (Fig. F,11). The air flowing into the engine through the valves is mixed with a continuously sprayed fuel. The mixture is then fired. As a result of the pressure rise which accompanies the explosion, the inlet valves close and the exhaust gases are forced out through the tailpipe. Expansion waves which are generated in the discharge reduce the pressure behind the check valves until they open again to admit a fresh charge of air, and the cycle is repeated. A spark is needed only for starting; once a regular cycle is established, each fresh charge of mixture is ignited

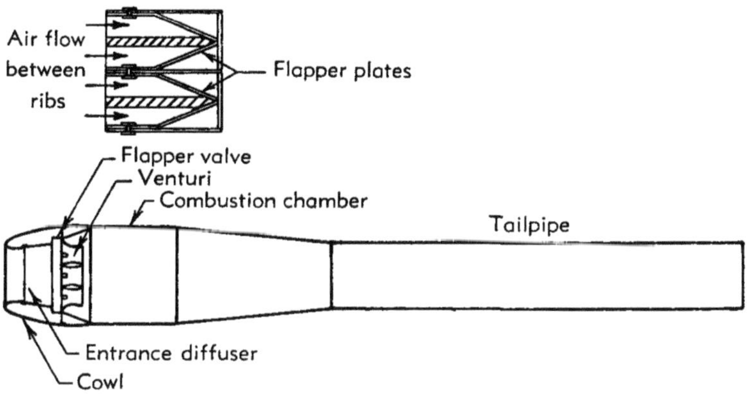

Fig. F,11.

by the hot gases from the preceding explosion and operation proceeds without further use of the spark plug.

F,12. Analysis by Method of Characteristics. Despite the mechanical simplicity of the pulse jet, a great deal of information is still lacking about such important details of its operating cycle as the flow behavior through the intake valves and at the discharge end, the timing and mechanism of ignition, and, above all, the process of combustion. This lack of basic information has necessitated the use of quite arbitrary assumptions in all theoretical investigations of the pulse jet cycle. Thus combustion has been assumed to take place at constant volume [*25*] or in the mode of a plane-propagating flame layer [*27*], or it has been approximated by an exponential change of temperature with time in the combustion chamber [*28*], by an isentropic pressure rise occurring simultaneously and instantaneously (or in a prescribed time) over the whole column of fresh charge [*29*], or by a polytropic change of state of the heated gases [*17*], and so forth.

The first extensive analysis of pulse jet operation was carried out by Schultz-Grunow [*29*], using the method of characteristics. The results of this analysis are significant in that they help to explain the experimentally observed effects of some design parameters and operating conditions on the performance of the pulse jet. In addition to the abovementioned representation of combustion as an isentropic and uniform pressure rise, the following simplifying assumptions are made in the Schultz-Grunow analysis: (1) all changes of state are assumed to be isentropic, (2) the spring action of the valves is disregarded, and it is

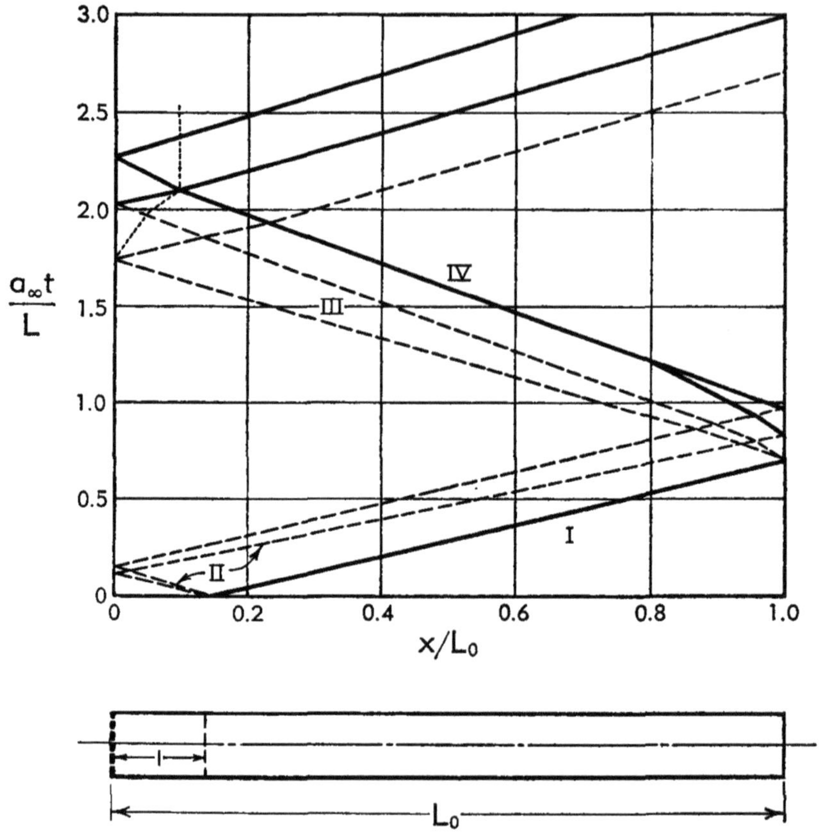

Fig. F,12a. Starting cycle of pulse jet in static operation [*29*].

assumed that the valves are closed when there is an excess pressure against their inner sides and fully open when the inside pressure is less than that outside, and (3) it is assumed that there is no pressure recovery from the minimum cross section of the open valve passage to the combustion chamber. The wave diagram of Fig. F,12a in which compression and expansion waves are represented by solid and dashed lines, respectively, and the dotted line denotes the boundary of the inflowing charge, illustrates the idealized starting-cycle operation of the pulse jet in its simplest form.

The establishment of design criteria for the improvement of pulse jet performance will require a more adequate account of periodic phenomena than the method of characteristics is capable of providing, but a good deal of qualitative information concerning the gross effects of some factors can be obtained from inspection of wave diagrams for the starting cycles of pulse jets of various configurations [*29*]. For example, the wave diagram of Fig. F,12a suggests the following: (1) increasing the combustion chamber length l without changing the total length L_0 of the engine will increase the path of expansion wave II, hence delay the formation of shock IV and increase the duration of the intake period and of the

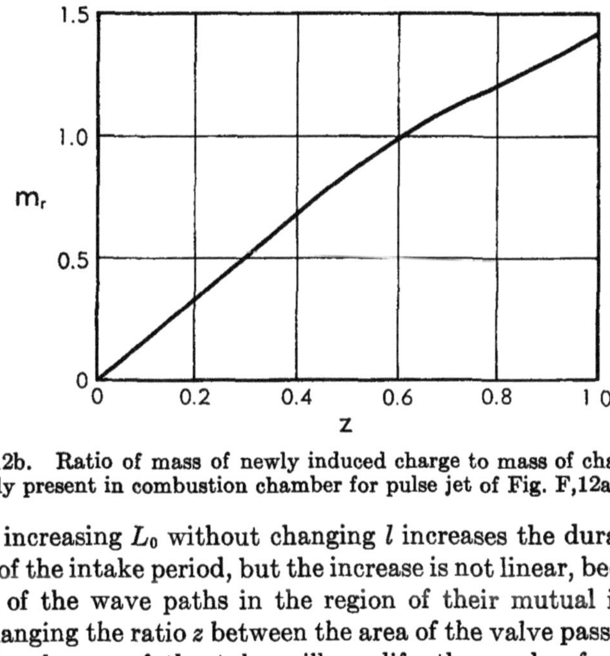

Fig. F,12b. Ratio of mass of newly induced charge to mass of charge originally present in combustion chamber for pulse jet of Fig. F,12a [*29*].

cycle; (2) increasing L_0 without changing l increases the duration of the cycle and of the intake period, but the increase is not linear, because of the deflection of the wave paths in the region of their mutual interaction; and (3) changing the ratio z between the area of the valve passage and the cross-sectional area of the tube will modify the mode of reflection of expansion wave III and therefore change the ratio m_r between the mass of the charge induced at the end of the cycle and that of the charge originally present in the tube (Fig. F,12b).

With proper assumptions concerning the ignition of the new charge, the construction of the wave diagram can be carried on to the following cycle (which will of course differ from the first because of the changed initial conditions), and so on. This extension of the analysis may bring to light important effects which only make their first appearance in the second or in later cycles. Thus, e.g. since the mode of reflection of waves III and IV depends on the valve-opening ratio, the tailpipe flow in all cycles but the first depends rather critically on this ratio. When z is increased beyond a certain value, the reflection of waves III and IV is predominantly a compression wave, which is in turn reflected at the tail end as an expansion wave, thereby producing a premature outflow which weakens the effect of the next explosion. This argument explains the experimentally observed fact that, when the valve-opening ratio is increased beyond a certain value, the amplitude of the flow pulsations decreases and "the limiting case of combustion at constant pressure is approached" [*29*].

The effect of an increase of pressure ahead of the valves is similar to that of an increase of valve-opening ratio, unless the ambient static pressure at the exit is also increased. Therefore, as the flight speed increases, the standard pulse jet can be expected to approach the mode of operation of the ramjet. This result is in qualitative agreement with the observed behavior of pulse jets at high speeds. It has also been suggested by Schultz-Grunow [*29*] that the operating speed range could be increased either by reducing the valve-opening ratio as the speed increases or by shielding the front of the valve bank by means of a cap (Fig. F,12c), to reduce the pressure in that region.[1]

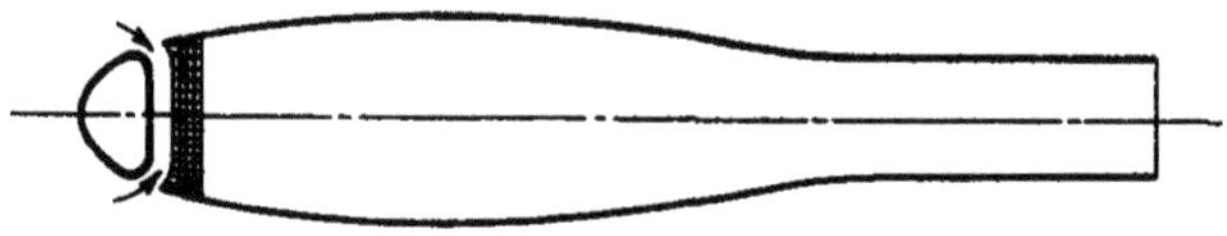

Fig. F,12c.

At low flight speeds and with moderate valve-opening ratios, waves III and IV form, after two reflections, a forward-facing compression wave which produces a reversal of the flow in the tailpipe, or "backflow" (Art. 15). This result is also in qualitative agreement with experimental information, e.g. [*30*].

Schultz-Grunow's analysis [*29*] predicts similar trends of behavior for constant-area pulse jets with noninstantaneous combustion and also provides useful information on the effects of specific changes of duct configuration, but the complexity of the wave diagrams makes it difficult to discern any general trend for the latter effects. On the other hand, systematic analyses by the method of characteristics have so far indicated that, in the discharge of a given mass of gas from given initial reservoir conditions to a given exhaust pressure, changes of shape of the reservoir and of the discharge duct have important effects on the duration and detailed history of the blowdown discharge, but hardly any effect on the total impulse which is generated thereby. This observation

[1] The opposite effect would be produced by a long intake duct ahead of the valves. Within such a duct, the closing of the valves would produce an air hammer effect, whereby the pressure against which the valves have to close would be increased above the stagnation pressure level. Thus the presence of a long intake duct would have the same effect on pulse jet performance as an increase of flight speed. The purpose of establishing a substantially uniform pressure on the upstream side of the valve bank can adequately be served by a very short intake diffuser.

may provide the basis for some speculation on the effect of shape on the performance characteristics of the pulse jet (Art. 13).

F,13. Performance Analysis. In the pulse jet, wave precompression of the fresh charge is, by definition, absent or negligible. Pressure exchange processes do, however, take place in the tailpipe and may under certain conditions introduce appreciable entropy increments. In general, $s_f - s_\infty = \Delta s_\lambda + \Delta s_h + \Delta s_\sigma$.

The flow velocity in the combustion chamber at the time of ignition is usually very low, and because of the absence of a compressor and of any appreciable pressure exchange before combustion, the temperature of the charge just before ignition can be assumed to be equal to the free stream stagnation temperature T^0_∞. Furthermore, energy being added to the flow only in the form of heat, one has $E = Jq_a$, where J is the mechanical equivalent of heat. Therefore, assuming constant specific heats, Eq. 4-4, 7-1, 7-7, and 5-1 yield

$$\eta_{cy} = 1 - \frac{c_p T_\infty}{q_a}\left[\lambda^{-\frac{\gamma-1}{\gamma}}\left(1 + \frac{n-1}{n-\gamma}\frac{q_a}{c_v T^0_\infty}\right)^{\frac{n-\gamma}{\gamma(n-1)}} \exp\frac{\Delta s_\sigma}{c_p} - 1\right] \quad (13\text{-}1)$$

$$I_a = \frac{F}{mg} = \frac{1}{g}\{[\eta_\nu(2Jq_a\eta_{cy} + V^2_\infty)]^{\frac{1}{2}} - V_\infty\} \quad (13\text{-}2)$$

The ideal pulse jet is one in which ram precompression of the charge is fully utilized, all flow processes except combustion are isentropic, and the exhaust velocity is a square-wave function of time. The performance of this ideal pulse jet can be calculated by means of Eq. 13-1 and 13-2 with $\lambda = \exp(\Delta s_\sigma/c_p) = \eta_\nu = 1.0$.

When $n = 0$ (isobaric combustion) the amplitude of the pressure fluctuations becomes zero and the ideal pulse jet becomes the ideal ramjet. More generally (with any value of n), as M_∞ increases, T^0_∞ also increases and the cycle efficiency of the ideal pulse jet approaches the value

$$\eta_{cy_i} = 1 - \frac{T_\infty}{T^0_\infty}$$

which is the ideal cycle efficiency of the ramjet. Thus the air specific impulse of the ideal pulse jet approaches asymptotically that of the ideal ramjet as the flight Mach number increases indefinitely. A comparison of ideal ramjet and pulse jet performance is presented in Fig. F,13a and F,13b (curves a and d in both figures) for the following conditions: $T_\infty = 500°R$; $n = -1$; heat liberated in combustion (heating value of fuel × combustion efficiency) = 18,000 BTU/lb of fuel; air/fuel ratio = 30; and $\gamma = 1.4$ throughout. The value of the polytropic exponent $n = -1$ corresponds to explosion pressure ratios in the neighborhood of

2.5 (according to Eq. 7-6) under the stipulated conditions and is believed to lead to a plausible description of the explosion process in current pulse jets.

In the standard pulse jet, the most important departure from the ideal behavior is that resulting from the inability of the engine to sustain ram pressure in the combustion chamber during the charging phase of the cycle. Indeed, the wave diagrams reveal that the charge pressure at the

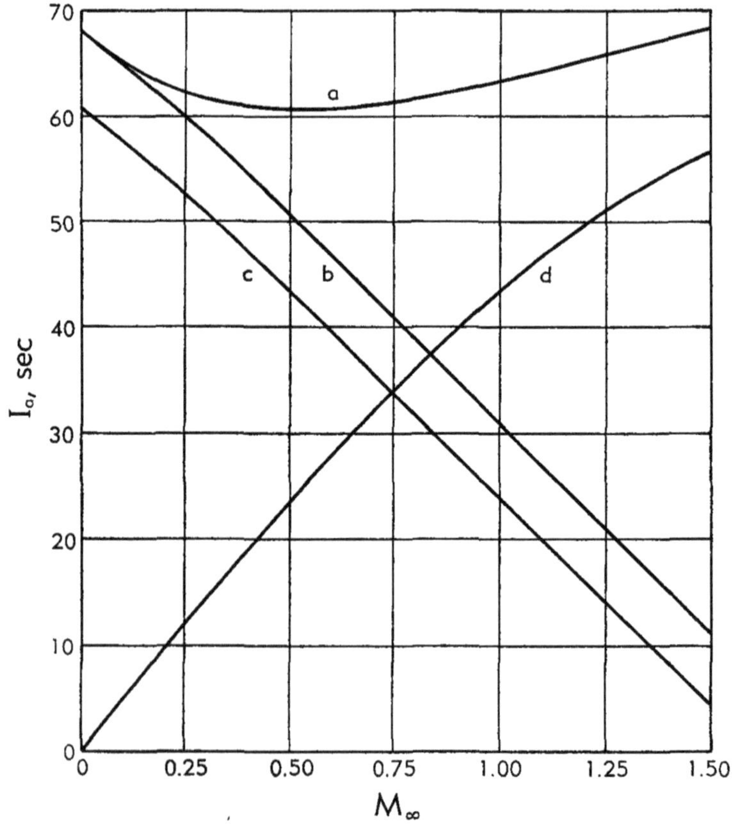

Fig. F,13a. Air specific impulse of pulse jets without backflow, under conditions specified in text: (a) ideal pulse jet, (b) standard pulse jet with $\eta_p = 1.0$, (c) standard pulse jet with $\eta_p = 0.8$, and (d) ideal ramjet.

time of ignition, being primarily determined by the exit boundary conditions, is generally very close to the free stream static pressure. As a matter of fact, the charge pressure level is believed to be depressed further as the flight speed increases, because, during those phases of the cycle in which the exhaust velocity is lower than that of the surrounding stream, the exhaust pressure approaches a base-pressure level. This difficulty can be remedied, however, by means of a tailpipe constriction. Therefore, the charge pressure level will be assumed to be equal to the

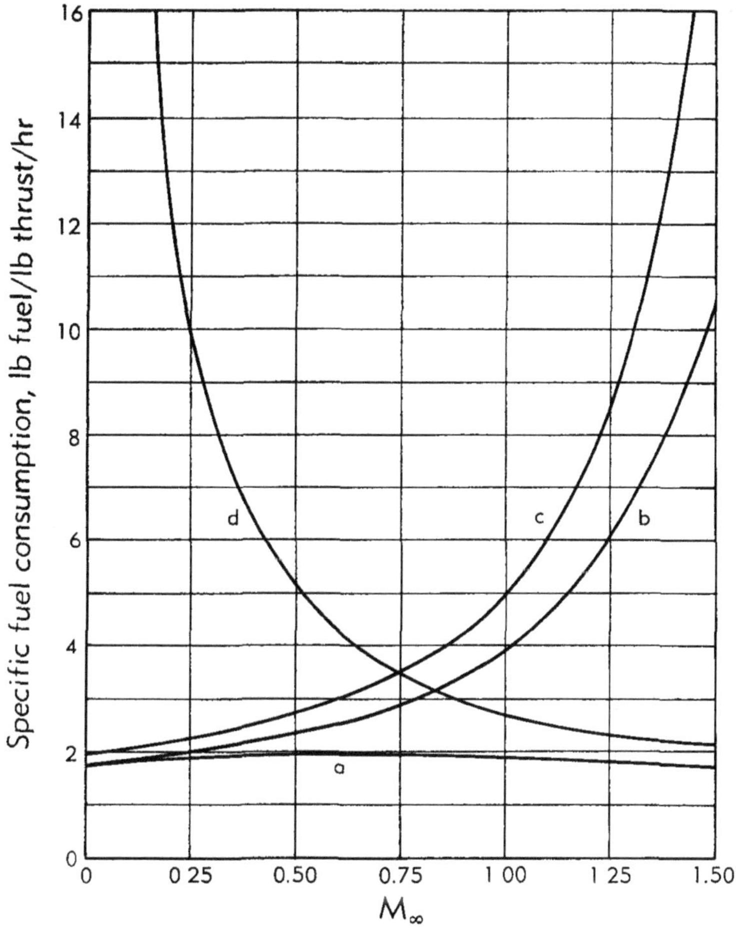

Fig. F,13b. Specific fuel consumption of pulse jets without backflow, under conditions specified in text: (a) ideal pulse jet, (b) standard pulse jet with $\eta_\nu = 1.0$, (c) standard pulse jet with $\eta_\nu = 0.8$, and (d) ideal ramjet.

free stream static pressure at all speeds. With this assumption, the entropy rise in the inflow process is

$$\begin{aligned}\Delta s_\lambda &= -\mathcal{R} \ln \frac{p_\infty}{p_\infty^0} \\ &= c_p \ln \frac{T_\infty^0}{T_\infty}\end{aligned}$$

and the cycle efficiency becomes

$$\eta_{cy} = 1 - \frac{c_p T_\infty}{q_a} \left[\frac{T_\infty^0}{T_\infty} \left(1 + \frac{n-1}{n-\gamma} \frac{q_a}{c_v T_\infty^0} \right)^{\frac{n-\gamma}{\gamma(n-1)}} - 1 \right] \qquad (13\text{-}3)$$

As for the entropy rise associated with pressure exchange, it should be borne in mind that this increment must be considered only insofar as it results from wave phenomena occurring outside of the burning region, because the effect of pressure waves within this region is already accounted for in the description of the heating mode. Therefore, consideration must be given to this entropy increment only if the heating mode is such as to produce pressure waves that are steep enough to form early shocks and thus to affect a significant portion of the tailpipe gases. In general, the entropy rise caused by these shocks is found to be insignificant unless

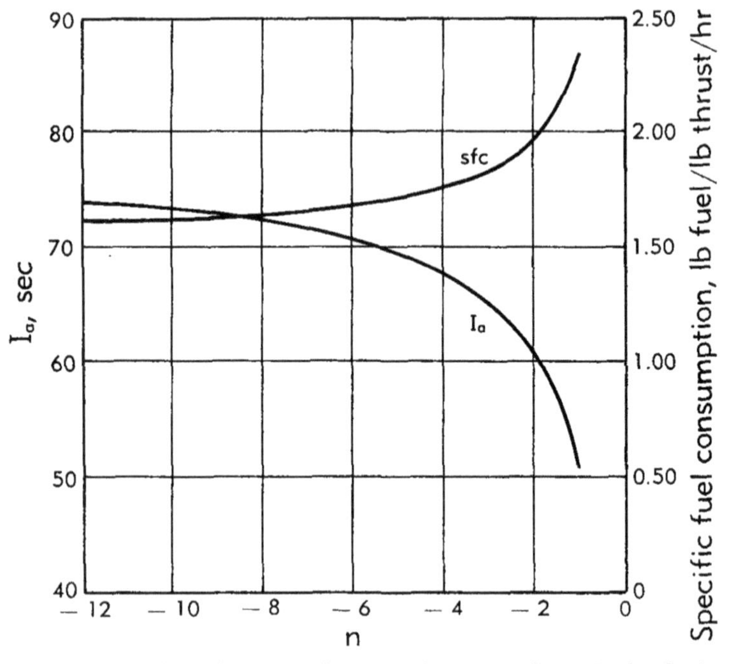

Fig. F,13c. Effect of change of n on performance of standard pulse jet with $\eta_p = 1.0$ at $M_\infty = 0.5$ under conditions specified for Fig. F,13b.

the rate of heat release is very high, i.e. unless the absolute value of n is very large, as in isochoric combustion. In the latter case, the shock losses of pressure exchange may under certain conditions be high enough to offset much of the entropy advantage that would otherwise be associated with a high value of $|n|$ [*17,18*]. On the other hand, Fig. F,13c indicates that, even in the absence of pressure exchange losses, it is only in the lower range of $|n|$ that changes of heating mode have a pronounced effect on performance. Since it has been shown that $|n|$ increases with the time rate of heat release (Art. 10), it follows that it would not be profitable to seek performance improvements by increasing the rate of heat release to the point at which pressure exchange losses become appreciable. Therefore Eq. 13-3 may be used to calculate the cycle efficiency of the standard pulse jet in all operating conditions that may be of practical interest. The performance of the standard pulse jet can then be calculated by means of Eq. 13-3 and 13-2. An accurate estimate of the efficiency of nonuniformity, which appears in the latter equation, can only be obtained through inspection of pertinent wave diagrams or experimental data (Art. 5). Values of η_ν in excess of 0.8 are believed to be easily achievable in most pulse jets.

The calculated performance of the standard pulse jet, for the above stipulated conditions, is also plotted in Fig. F,13a and F,13b, where curves *b* and *c* pertain to $\eta_\nu = 1.0$ and $\eta_\nu = 0.8$, respectively.

A more careful analysis would of course account for the changes of specific heats, the effect of which may be appreciable, as indicated by Eq. 4-4 and 7-7.

Major discrepancies between calculated and measured specific fuel consumptions are generally attributable to errors in the estimate of the combustion efficiency.

A qualitative evaluation of the effect of shape on pulse jet performance may be made on the basis of the observation that the impulse generated in each cycle is substantially independent of the shape of the duct (Art. 12). Changes of shape will primarily affect the blow-down discharge time—and therefore the frequency of the pulsations—and the mode of combustion (Art. 10). A reduction of tailpipe cross-sectional area can be expected to produce a decrease of frequency and an increase of $|n|$. The former effect alone would be reflected in a decrease of thrust without change of specific impulse, while the increase of $|n|$ has a beneficial effect on both thrust and specific impulse. Therefore a reduction of tailpipe area is likely to result in a decrease of specific fuel consumption and may produce a decrease or an increase of thrust, depending on the relative importance of the effects of shape on the frequency and on the mode of combustion. There may exist, for some operating conditions, a tailpipe contraction ratio that maximizes the thrust, but the value of this "optimum" ratio is best determined through systematic experimentation in each specific case.

F,14. Valveless Pulse Jets. The function of the intake valves in the conventional pulse jet is to prevent reversal of the flow at the inlet, and therefore a negative momentum transport, when the combustion chamber pressure becomes higher than the free stream stagnation pressure. The same purpose can be achieved without valves, by proper utilization of wave phenomena in an inlet duct of adequate length. Consider, e.g., the configuration shown in Fig. F,14, top. The combustion chamber pressure required to stop the flow in the inlet duct is the hammer pressure corresponding to the initial conditions of this flow. This hammer pressure is generally higher than the free stream stagnation pressure, but may still be exceeded by the peak explosion pressure. Therefore, the pressure waves that are generated by each explosion may still produce a reversal of the flow in the inlet duct. The occurrence of flow reversal can, however, be kept from extending all the way to the intake, by making the inlet duct long enough to permit each pressure wave to be weakened to hammer strength by the expansion wave which follows and overtakes it. A pulse jet working on this principle was successfully developed by Schubert in 1944 [*31*].

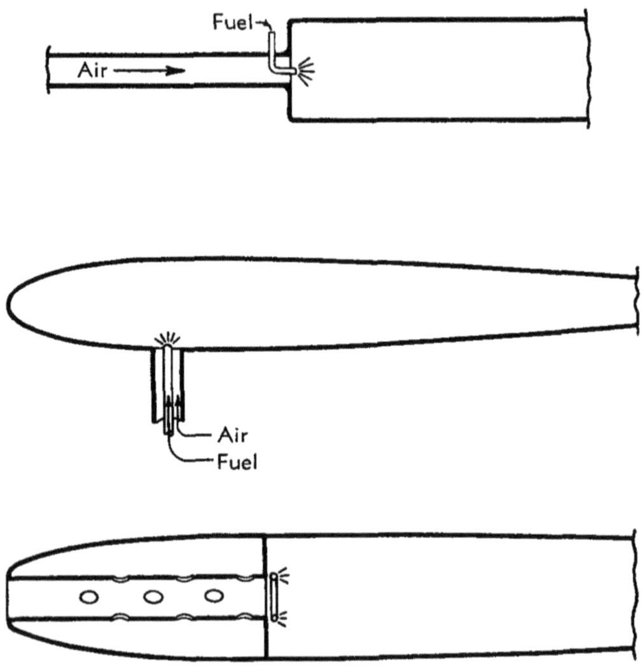

Fig. F,14. Valveless pulse jets: (top) nose inlet, (middle) side inlet, (bottom) surge-chamber inlet.

It is noteworthy that the only feature which distinguishes the configuration of this power plant from that of the ramjet is the presence in the former of a sharp change of cross section at the combustion chamber inlet. Apparently, the sudden change of cross section causes the collection of residual hot gases in the annular dead-water region behind the shoulder and the turbulent interpenetration of these gases with the inflowing fresh charge. The area of the contact surface, and with it the instantaneous rate of heat release, are thus increased to the extent that n for the equivalent polytropic process becomes appreciably lower than zero and combustion takes place intermittently, even though the fuel is injected continuously. The distinction between ramjet and valveless pulse jet is actually debatable, because severe burner pressure fluctuations are known to be associated with the spontaneous occurrence of flow oscillations in ramjets under certain conditions. These flow oscillations have an adverse effect on ram precompression and on the efficiency of nonuniformity at the exhaust. On the other hand, the burner pressure fluctuations may represent beneficial modifications of the heating mode (Art. 7), and there are indeed subsonic flight conditions under which the performance of ramjets is favorably affected by such fluctuations [*17*]. Ramjets which operate in this manner are also sometimes referred to as valveless pulse jets, despite their inability to operate statically like the pulse jet proper.

The mechanism of operation of Logan's side-intake pulse jet [*32*], shown in Fig. F,14, middle, is substantially the same as that of the Schubert engine. The side intake arrangement appears to be particularly effective in increasing the instantaneous rate of heat release, and therefore the thermal efficiency of the engine, by the process just described.

The decay of the upstream-propagating pressure waves in the inlet duct can be hastened by the use of surge chambers (Art. 8), as illustrated in Fig. F,14, bottom. Surge-chamber inlets were successfully used by Schmidt in 1944 [*23*].

The use of "flow rectifiers," i.e. of passages offering lesser resistance to the flow in one direction than in the opposite direction has been attempted by various workers for the purpose of inhibiting the reversal of the flow in the intake duct [*23*; *33*; *34*, pp. 490–505], but none of these rectifiers has so far proved more effective than a simple intake duct of adequate length. With low frequency pulse jets the inlet-duct length required to prevent spillage at the intake may be excessive. In such cases, it is profitable to accept some measure of spillage. The momentum of the reverse flow may be conserved for thrust by turning the intake duct back 180°, as in the SNECMA "Escopette" and "Ecrevisse."

The potential performance of the valveless pulse jet is the same as that of the pulse jet with valves. The main advantages of the valveless pulse jet are, of course, its mechanical simplicity and long service life. In addition, throttling and wave energy losses can be made lower in the valveless than in the valved pulse jet, and the valveless configuration lends itself better to external streamlining.

F,15. Thrust Augmentation. It is often possible to increase the thrust of a pulse jet very appreciably without increasing the flow rate of fuel, by transferring some energy from the flow of the hot combustion products (primary flow) to other propulsive flows (secondary flows). This "augmentation" of thrust may be accomplished in several different ways, and one of its forms (backflow augmentation) is known to occur spontaneously in most standard pulse jets at low flight speeds. In every case the process involves, by definition, a double-flow or even a multiple-flow system, the primary component of which is the combustion chamber flow in the pulse jet proper.

"Thrust augmentation ratio" is defined as the ratio between the thrust of such a multiple-flow system and the thrust which would be produced by the primary component alone under the same conditions. The words "under the same conditions" specify that, in the computation of the augmentation ratio, the early phases of the primary cycle (up to the energy transfer process) must be assumed to be identical in the two systems.

In the following, subscripts $_{\text{I}}$ and $_{\text{II}}$ denote quantities pertaining to the primary flow and to the secondary flows, respectively, in the "augmented" system, and subscript $_{\text{I}'}$ refers to the primary component operating alone under the conditions just stipulated. The ratio $m_{o\text{II}}/m_{o\text{I}} = m_{o\text{II}}/m_{o\text{I}'}$ is denoted by μ.

Only two forms of thrust augmentation are discussed here: (1) augmentation by backflow and (2) augmentation by shrouding. It is noteworthy that thrust augmentation is accompanied in both instances by modifications of the primary cycle, which in turn produce a further increase of thrust. Thus shrouding makes possible the utilization of ram precompression in the primary engine and backflow is believed to produce a beneficial modification of the heating mode. However, in accordance with the definition given above, the augmentation ratio will be computed without reference to these modifications of the primary cycle.

Backflow. At low flight speeds the standard pulse jet is a double-flow jet engine, because of the occurrence of backflow (Art. 12). Indeed, the backflow air does not undergo the same thermal cycle as the air which flows through the combustion chamber. The only energy received by the backflow air is that which is transferred to it from the hot combustion products in a substantially isentropic pressure exchange process which takes place in the tailpipe. Thus $s_{f\text{I}} = s_{f\text{I}'}$, and assuming complete expansion in all cases, $T_{f\text{I}} = T_{f\text{I}'}$. Therefore the energy transferred from the primary to the secondary flow in each cycle is $(m_{o\text{I}}/2)(\overline{V_{f\text{I}'}^2} - \overline{V_{f\text{I}}^2})$.

In the pressure exchange process, the velocities on the two sides of the interface separating the two flows become equal. Therefore, neglecting the waves which are generated, under certain conditions, by the arrival of entropy discontinuities at the duct exit, and assuming $\eta_{v\text{I}'} = \eta_{v\text{I}} = \eta_{v\text{II}} = 1.0$, one has $V_{f\text{I}} = V_{f\text{II}} = V_f$ and

$$E_{\text{II}} = \frac{V_{f\text{I}'}^2 - V_f^2}{2\mu} \tag{15-1}$$

$$\eta_{oy\text{II}} = \frac{V_f^2 - V_\infty^2}{2E_{\text{II}}} \tag{15-2}$$

The backflow air loses its entire free stream kinetic energy without

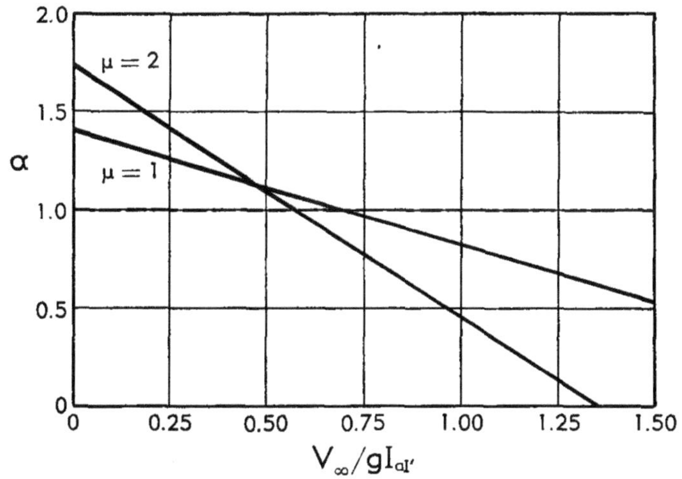

Fig. F,15a. Thrust augmentation by backflow.

pressure recovery as it is brought to rest within the engine. Therefore,

$$\Delta s_{\text{II}} = c_p \ln \frac{T^0_\infty}{T_\infty}$$

and

$$\eta_{\text{cyII}} = 1 - \frac{Jc_pT_\infty}{E_{\text{II}}}\left(\frac{T^0_\infty}{T_\infty} - 1\right)$$

$$= 1 - \frac{V^2_\infty}{2E_{\text{II}}} \qquad (15\text{-}3)$$

From Eq. 15-1, 15-2, and 15-3 one obtains

$$V_f = \frac{V_{f\text{I}'}}{\sqrt{1 + \mu}} \qquad (15\text{-}4)$$

The augmentation ratio α is therefore

$$\alpha = \frac{(1 + \mu)(V_f - V_\infty)}{V_{f\text{I}'} - V_\infty}$$

$$= \sqrt{1 + \mu}\left[1 - (\sqrt{1 + \mu} - 1)\frac{V_\infty}{gI_{\text{aI}'}}\right] \qquad (15\text{-}5)$$

This relation is plotted in Fig. F,15a for two values of μ. With any fixed value of μ, the thrust augmentation ratio decreases as the flight speed increases. The thrust-augmenting effect of backflow vanishes when α becomes 1.0, i.e. when

$$\frac{V_\infty}{gI_{\text{aI}'}} = \frac{\sqrt{1 + \mu} - 1}{1 + \mu - \sqrt{1 + \mu}}$$

The value of μ which maximizes α for any value of $V_\infty/gI_{aI'}$ is obtained from Eq. 15-5 as

$$\mu_{opt} = \frac{1}{4}\left(\frac{gI_{aI'}}{V_\infty} + 1\right)^2 - 1$$

and the corresponding augmentation ratio is

$$\alpha_{max} = \frac{1}{4}\left(\frac{gI_{aI'}}{V_\infty} + 1\right)^2 \frac{V_\infty}{gI_{aI'}}$$

α_{max} becomes 1.0 when $V_\infty/gI_{aI'} = 1.0$. Therefore no augmentation can be derived from the utilization of backflow, regardless of the value of μ, when $V_\infty > gI_{aI'}$, i.e. when $V_\infty > V_{fI'}/2$. These results are plotted in Fig. F,15b.

In the standard pulse jet, the backflow mass ratio decreases very rapidly as the speed increases, from a maximum which may be higher than 2 in static operation to zero at a flight speed which is seldom believed to exceed about 400 ft/sec. Therefore, backflow normally ceases

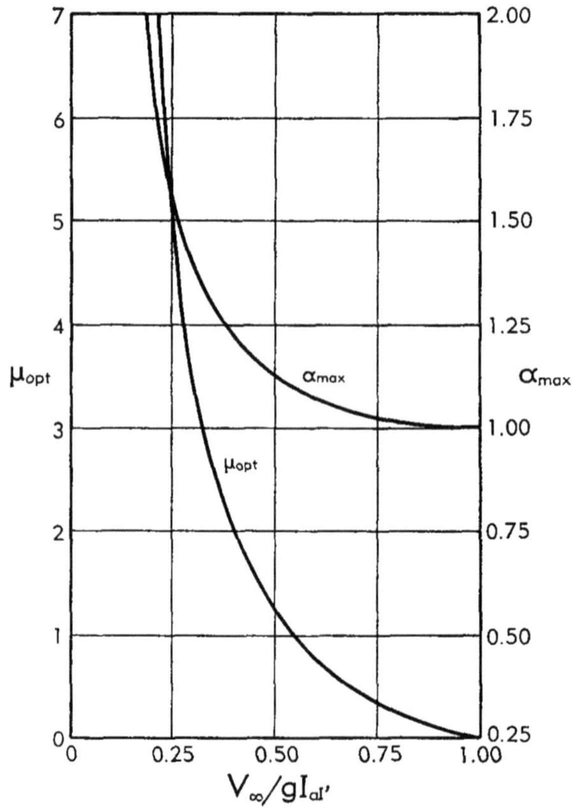

Fig. F,15b. Optimum backflow mass ratio and maximum thrust augmentation by backflow.

to take place well below the flight speeds at which its occurrence would be detrimental. At low flight speeds, it is possible to increase the backflow mass ratio by flaring out the end of the tailpipe. Such flared-out exits are sometimes called "augmenters," because their effect on backflow has been found sufficient to increase the static thrust by 20 per cent or more. As the flight speed increases, these "augmenters" become less and less effective until their presence may actually result in a net loss of thrust. Introduction of cold air into the tailpipe through side ports, as in the Schmidt designs of 1938 [*33*], may make it possible to create there, even at relatively high speeds, a condition similar to that which is produced by backflow at the lower speeds. It is clear, however, from the preceding discussion, that such provisions may have a detrimental effect on performance at high speeds.

It can be verified that the effect of redistributing the energy released in combustion over a larger mass of gas by pressure exchange in the manner just described is not equivalent to that of releasing the same total amount of heat from the start over this larger mass of gas in a single-flow system. The latter conditions can be produced by increasing the size of the pulse jet, without changing the flow rate of fuel, so that the air-fuel ratio will be increased in the proportion $(1 + \mu)$. This is not "thrust augmentation" in accordance with the definition given at the beginning of this article. For the purpose of comparison, consider a standard pulse jet under the conditions stipulated for the performance calculation of Art. 13. For static operation without backflow and with $\eta_v = 1.0$, Fig. F,13a yields $I_a = 68$ sec. Doubling the air-fuel ratio in this single-flow system would, under the same conditions, yield $I_a = 40.2$ sec. The thrust would therefore increase in the ratio $(2 \times 40.2)/68 = 1.18$. In contrast, Eq. 15-5 indicates that introducing the same additional amount of air as a separate flow (backflow) and transferring energy to it by pressure exchange would produce a thrust augmentation ratio of 1.414.

It should also be noted that backflow has a beneficial effect on the heating mode, in that the lowered temperature of the gas column in the tailpipe increases the inertial confinement of the burning region, thereby increasing the value of $|n|$ for the equivalent polytropic process [*17*]. This effect is believed to be appreciable only in static operation and at very low flight speeds.

Ducted pulse jet. Shrouding can be used to improve the performance of the pulse jet in two important ways: one through thrust augmentation, the other through the utilization of ram precompression at high flight speeds. Thrust augmentation results from the fact that shrouding brings an additional flow into play, thereby reducing the pulse jet to a component of a multiple-flow system. As for the utilization of ram precompression, this purpose can obviously be accomplished by full shrouding, i.e. by submerging the entire pulse jet in a flow which has been almost isentropically decelerated from the free stream velocity (Fig. F,15c). It is also clear that substantially the same result can be achieved by shrouding the exit region alone, so as to raise the exhaust static pressure almost to the ram level, as was proposed by Diedrich in 1943 (Fig. F,15d). The latter solution is actually preferable, because it eliminates most of the secondary flow losses in the forward portion of the engine. In either case, the cycle of the primary flow is brought close to the ideal and $I_{aI'}$ can be calculated, in fair approximation, as the air specific impulse of an ideal pulse jet (Art. 13).

Strictly speaking, a ducted pulse jet comprises three flows, the direct flow and the backflow in the primary engine and the flow in the annular

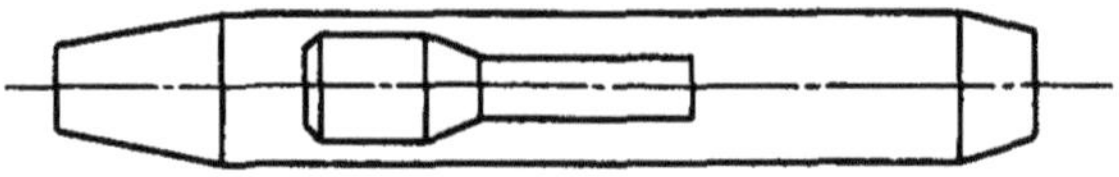

Fig. F,15c.

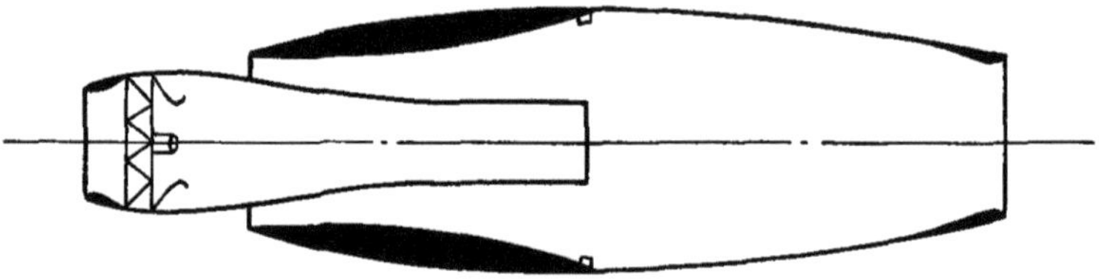

Fig. F,15d.

space surrounding it, but backflow is believed to be negligible in most configurations of practical interest. Indeed, considerations of shroud size, weight, and drag preclude the establishment of such low flow velocities around the primary exhaust as would normally be expected to produce an appreciable backflow. Therefore it seems permissible to neglect the backflow and to treat the ducted pulse jet as a double-flow engine. The designation "secondary flow" (flow II) is accordingly applied only to the flow through the annular space surrounding the primary component. It is also stipulated, at first, that the only energy received by the secondary flow be that which is transferred to it from the primary flow.

The transfer of energy takes place both by pressure exchange and by mixing. These two processes may be assumed to take place in succession, the former being completed when the latter begins (Art. 7). The process as a whole is believed to be greatly complicated by the large vortex rings which are generated at the primary exhaust in each discharge, but it seems reasonable to postulate that the initial effect of such rings will generally be to promote rather than to inhibit the transfer of energy by pressure exchange. It is also known that, in the mixing of steady parallel streams differing greatly in their initial static temperatures, the dominant component of the entropy rise is that which is associated with the transfer of heat between the two streams. On the basis of this observation, and in consideration of the large difference between the temperatures of the two flows at their merging station in the ducted pulse jet, it appears permissible to neglect the entropy rise produced by the viscous dissipation of the kinetic energy associated with any residual nonuniformity of the velocity at the end of the pressure exchange process. According to this view, pressures and velocities of the two flows are assumed to be substantially equalized through pressure exchange, and mixing is essentially reduced to a heat transfer process.

This leads to consideration of two distinct cases of special interest: (1) if the distance l to which the shroud extends downstream of the primary exit is very short, the only energy transfer process is that of pressure exchange; (2) if l is sufficiently long to permit mixing to be completed, both processes take place, in the order indicated above. The analysis of intermediate situations is not attempted here.

In both cases, the efficiency of nonuniformity is assumed to be 1.0 both at the primary exhaust and at exhaust from the ducted system.

In the first case (l short), the energy transfer process is the same as with backflow, except for the fact that the secondary inflow losses are now so low that η_{cyII} can be assumed to be 1.0, in very good approximation. Therefore, Eq. 15-1 and 15-2 yield

$$V_{f\text{I}'}^2 - V_f^2 = \mu(V_f^2 - V_\infty^2) \tag{15-6}$$

or

$$gI_{\text{a}} = \left[\frac{(gI_{\text{aI}'} + V_\infty)^2 + \mu V_\infty^2}{1 + \mu}\right]^{\frac{1}{2}} - V_\infty$$

Thus the augmentation ratio is

$$\alpha = (1 + \mu)\frac{I_{\text{a}}}{I_{\text{aI}'}} = \sqrt{1 + \mu}\left[\left(1 + \frac{V_\infty}{gI_{\text{aI}'}}\right)^2 + \mu\left(\frac{V_\infty}{gI_{\text{aI}'}}\right)^2\right]^{\frac{1}{2}} - (1 + \mu)\frac{V_\infty}{gI_{\text{aI}'}} \tag{15-7}$$

This relation is plotted in Fig. F,15e for two values of μ. The augmentation ratio is again found to decrease, with any fixed value of μ, as the flight speed increases, but at a slower rate than the augmentation due to backflow. It can indeed be verified from Eq. 15-7 that the augmentation ratio produced by shrouding decreases asymptotically to 1.0 as the flight speed increases indefinitely.

In every case the maximum augmentation ratio is obtained in static operation, this maximum value being $\alpha_0 = \sqrt{1 + \mu}$, as in the case of backflow.

When l is long enough to permit complete mixing, the two flows emerge from the ducted engine as a single homogeneous stream. Therefore, not only is $V_{fI} = V_{fII} = V_f$, but also $s_{fI} - s_\infty = s_{fII} - s_\infty = \Delta s$. This entropy rise can be calculated as follows.

Let T_1 and T_2 denote the static temperatures of flows I and II, respectively, at the beginning of mixing, i.e. when the pressures and velocities

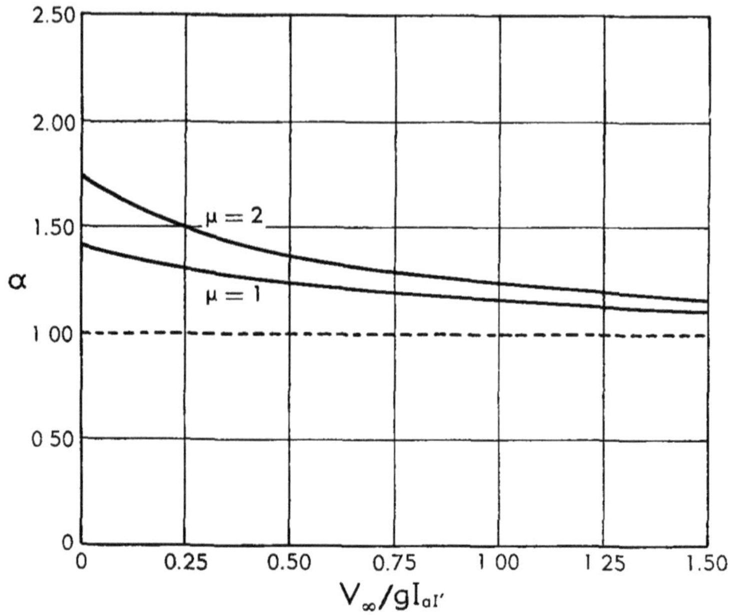

Fig. F,15e. Thrust augmentation by ducting.

of the two streams have become equal. Since pressure exchange is substantially isentropic, one has

$$\frac{T_1}{T_2} = \exp \frac{\Delta s_{I'}}{c_p}$$

"Mixing" is now represented by an isobaric heat transfer process, at the end of which the uniform temperature

$$T = \frac{T_1 + \mu T_2}{1 + \mu}$$

is established across the duct. Therefore,

$$\Delta s = \Delta s_{II} = c_p \ln \frac{T_1 + \mu T_2}{(1 + \mu)T_2}$$

or

$$\exp \frac{\Delta s}{c_p} = \frac{1}{1 + \mu}\left(\exp \frac{\Delta s_{I'}}{c_p} + \mu\right)$$

Since the entropy rise is the same for all particles, the ducted pulse jet may now be treated as a single-flow jet engine in which the energy added per unit mass of the fluid is $Jq_a/(1 + \mu)$ and through which the specific entropy rise is the calculated increment Δs. The cycle efficiency of this engine is then

$$\eta_{cy} = 1 - \frac{(1 + \mu)c_pT_\infty}{q_a}\left(\frac{\exp\frac{\Delta s_{I'}}{c_p} + \mu}{1 + \mu} - 1\right)$$

$$= 1 - \frac{c_pT_\infty}{q_a}\left(\exp\frac{\Delta s_{I'}}{c_p} - 1\right)$$

$$= \eta_{cyI'}$$

Thus, *the cycle efficiency of the ducted pulse jet is equal to the cycle efficiency of its primary component operating alone with the same utilization of ram precompression*. Therefore,

$$\frac{V_{fI'}^2 - V_\infty^2}{2Jq_a} = \frac{V_f^2 - V_\infty^2}{2Jq_a}(1 + \mu)$$

or

$$V_{fI'}^2 - V_f^2 = \mu(V_f^2 - V_\infty^2)$$

as in the ducted pulse jet without mixing (Eq. 15-6). It follows that the augmentation ratio is given by Eq. 15-7 in this case also.

On the basis of this argument, the weight and drag penalties of a long shroud extension downstream of the pulse jet exit would seem unwarranted. On the other hand, a distinct benefit which can be derived from mixing is the utilization of some of the chemical energy which may still be available in the primary exhaust. For this reason, the fuel specific impulse of most ducted power plants with mixing is relatively insensitive, particularly at high flight speeds, to changes of combustion efficiency in the primary component. Utilization of the mixing region for afterburning of fuel injected in the secondary flow has also been proposed by Diedrich [*33*].

It should also be noted that the augmentation ratio applies to the thrust obtainable with a given fuel flow, hence also to the fuel specific impulse, but not to the air specific impulse. Indeed, since $(1 + \mu)I_a = \alpha I_{aI'}$ and $\alpha \leqq \sqrt{1 + \mu}$, it follows that $I_a \leqq I_{aI'}/\sqrt{1 + \mu}$. Therefore the air specific impulse of the ducted pulse jet—with or without mixing—is lower than that of the unducted pulse jet with ram utilization. Although this observation may also be made with regard to the augmenting effect of backflow, its significance is far greater in the case of the ducted pulse jet where μ and α, with any given primary engine, depend on the over-all size of the power plant.

As the flight Mach number increases, the thrust of the ducted pulse jet approaches that of the ideal ramjet with the same fuel flow rate. This is so because α tends to 1.0 and the thrust of the primary component approaches that of the ideal ramjet (Art. 13) which, at very high speeds, becomes only a function of the total rate of heat release and not of the mass flow rate of air. It follows that, as the flight Mach number increases, the fuel specific impulse of the ducted pulse jet approaches that of the ramjet. This is, of course, also true when additional

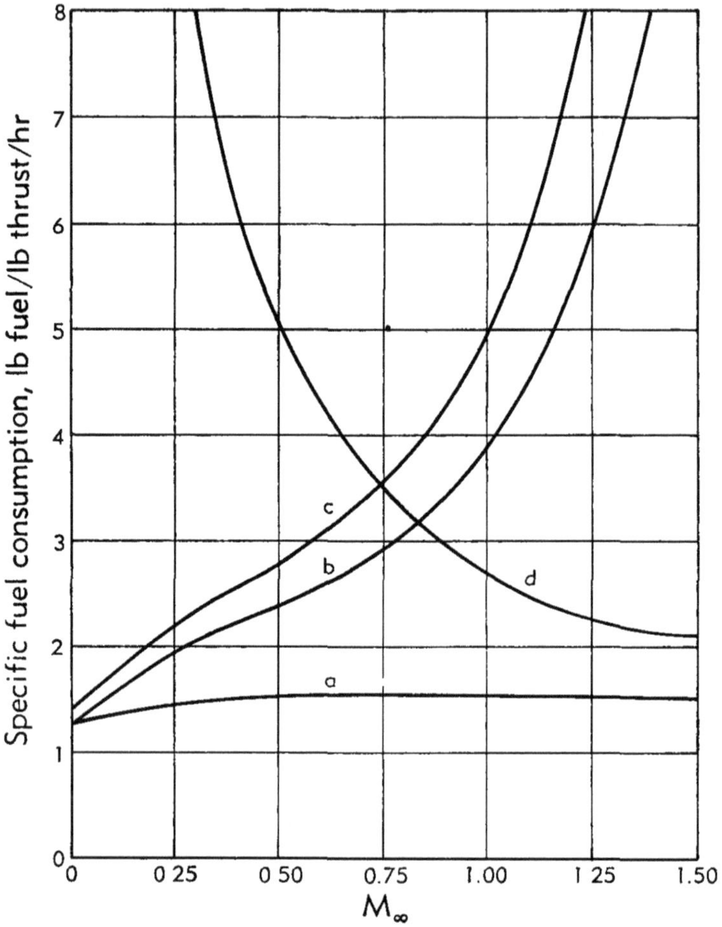

Fig. F,15f. Specific fuel consumption, under conditions specified in text: (a) ducted pulse jet, (b) bare pulse jet with $\eta_v = 1.0$, (c) bare pulse jet with $\eta_v = 0.8$, and (d) ideal ramjet.

fuel is injected in the secondary flow of the ducted pulse jet for afterburning in the mixing duct.

A comparison of specific fuel consumptions of bare and ducted pulse jets is presented in Fig. F,15f. The stipulated conditions are again those of the numerical illustration of Art. 13, but the effect of a backflow mass ratio decreasing from $\mu = 1.0$ at $M_\infty = 0$ to $\mu = 0$ at $M_\infty = 0.4$ is accounted for in the case of the bare pulse jet. The value of μ for the ducted pulse jet is assumed to be 1.0 at all speeds. Two values of the efficiency of nonuniformity ($\eta_\nu = 1.0$ and $\eta_\nu = 0.8$) are considered in the case of the pulse jet alone. The value of the efficiency of nonuniformity in the ducted pulse jet is not known but is believed to be closer to 1.0 than in the pulse jet alone, because of the decay of the flow fluctuations in the energy transfer process. Therefore only the ducted pulse jet curve for $\eta_\nu = 1.0$ is presented in Fig. F,15f.

Of course it should be remembered that external drag effects have not been taken into consideration in this analysis. When account is taken of these effects, the margin of superiority of the ducted pulse jet may be expected to be narrowed down very considerably.

It should also be noted that the unducted pulse jet with utilization of backflow augmentation at low flight speeds and of ram precompression in the high speed range would have a considerably higher air specific impulse (with approximately the same specific fuel consumption) than the best ducted pulse jet for the same thrust. However, at the writing of this manuscript, no way for achieving the utilization of ram precompression in pulse jets without shrouding has yet been successfully demonstrated.

CHAPTER 5. WAVE ENGINES

F,16. Introduction. Although all intermittent jet engines operate on the basis of wave processes, the designation "wave engine" is now customarily applied only to compressorless jet engines in which a pressure exchange process is used to precompress the fresh charge.

The utilization of pressure exchange in gas generators or "topping units" for gas turbines has received considerable attention in recent years [*8,9,10*]. In these applications, the excess mechanical energy available in the combustion products is used to drive a turbine, the work output of which may be partly utilized to produce additional precompression through a conventional compressor. Pressure exchangers can find similar applications in turbojets, in which the entire work output of the turbine is utilized to drive the compressor and the final stage of expansion of the combustion products is used to produce a jet for the direct generation of thrust. Finally, in the wave engine, turbine and compressor are eliminated and all the excess mechanical energy available in the combustion products after pressure exchange is utilized to produce a propulsive jet.

The manner in which pressure exchange is effected and utilized in wave engines varies considerably from type to type, and a detailed description of the various schemes which have been proposed to date would exceed the scope of this section. The following discussion considers only the salient features of the two classes into which wave engines are commonly grouped: the "external combustion wave engines," in which pressure exchange and combustion take place in separate ducts or chambers, and the "internal combustion wave engines" in which the two processes take place in the same duct.

F,17. External Combustion Wave Engines. Seippel [*9,10*] was the first to conceive (1940) and to develop, in the Brown-Boveri "Comprex," a successful pressure exchanger for external combustion applications.

The Comprex consists of a cylinder which carries on its periphery a large number of axial ducts or "cells." The cylinder rotates between two fixed end shields with two ports on each side, through which fresh air and hot combustion products from a high pressure combustion chamber are alternately supplied to and exhausted from each cell. The pressure exchange process (Fig. F,17) is the same for all cells and involves two stages of compression of the fresh charge. The first stage is provided by the hammer wave, which is generated when the flow of the fresh charge through the cell is suddenly interrupted by a closed portion of the shield at the downstream end. The second stage is provided by the shock resulting from the pistonlike action of the combustion products, which suddenly enter the cell when its front end is brought to the high pressure gas port. Upon completion of the second stage of compression, the charge is led through the high pressure air port to the combustion chamber.

Since pressures and velocities are the same on the two sides of the interface separating the hot gas from the fresh charge, the stagnation pressure of the latter is higher than that of the former. Therefore the pressure in the combustion chamber can spontaneously build up to a high level.

The volume rates of flow of hot and cold gases through each cell are very nearly the same. Therefore, in the rotor passages, the mass flow rate of cold air greatly exceeds that of the hot gas. It follows that a considerable portion of the high pressure gases may be expanded directly to form a propulsive jet. The remainder returns to the rotor, where it is deenergized in the pressure exchange process and is finally discharged through the low pressure gas port to form another propulsive jet.

Several important variations of this scheme have been suggested and tried, but the mechanism of operation of most pressure exchangers for external combustion is still basically the same as that of the Comprex.

Although the amplitude of the flow fluctuations in each cell is very large (and with some arrangements the flow is even fully reversed twice in each cycle), each port finds identical flow situations in all cells that come to face it. Therefore the flow through the inlet and exhaust ports of these pressure exchangers is substantially steady. The small pulsations which are caused by the passing of the cells decay very rapidly because of their high frequency (5000 cps for a 30-cell rotor turning at 10,000 rpm). In particular, combustion is essentially a steady isobaric process and the intake and exhaust flows of the engine are practically steady.

An accurate calculation of the performance of this engine requires the construction of a characteristics diagram and a careful estimate of

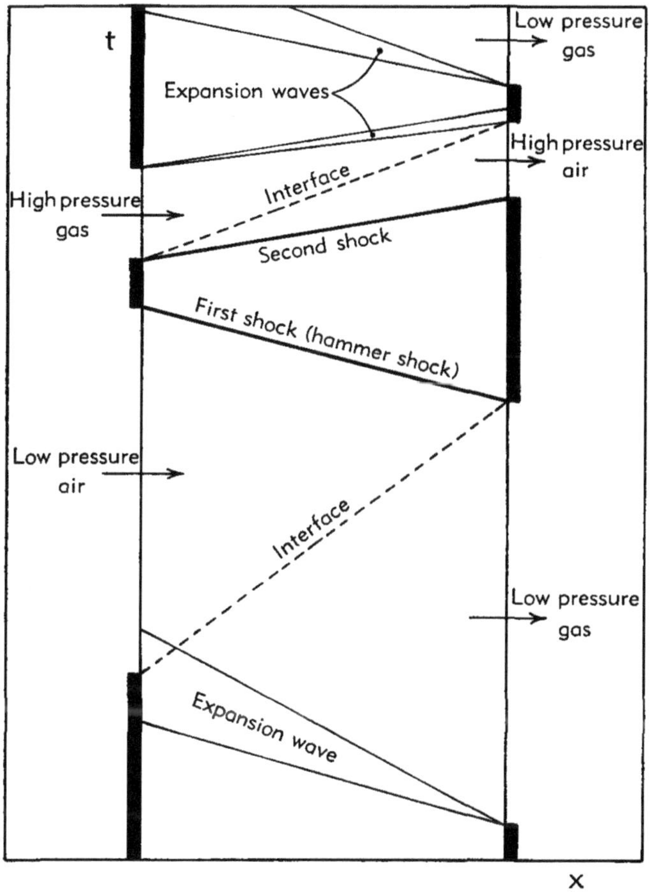

Fig. F,17. Pressure exchange process in the Comprex.

throttling, leakage, and other losses through the engine. A crude evaluation of this performance can be made, however, on the basis of the following argument.

According to the definition of Art. 2, the external combustion wave engine is a double-flow thrust generator. Now let subscript $_{\mathrm{I}}$ refer to that portion of the gas which reenters the rotor passages after combustion to participate in the pressure exchange process and subscript $_{\mathrm{II}}$ to the remaining portion, which is here assumed to be directly discharged from the combustion chamber as a jet, without dilution with bleed-off high pressure air; and let $m_{\text{II}}/m_{\text{I}} = \mu$.

The unit mass of each gas receives the amount of energy w in the pressure exchanger before entering the burner and the amount of energy Jq_a in the burner, but then the unit mass of gas I, in its second pass through the pressure exchanger, relinquishes an amount of energy $w(1 + \mu)$ to the new charge. Therefore

$$E_{\text{I}} = Jq_a - \mu w$$
$$E_{\text{II}} = Jq_a + w$$

The specific entropy rise through the engine, Δs, is substantially the same for both flows, the major increments being those associated with combustion and with the pressure exchange shocks, to which both gases are subjected. Since $E_{\text{I}} \neq E_{\text{II}}$, it follows (Art. 4) that the velocities $V_{f\text{I}}$ and $V_{f\text{II}}$ of the two fully expanded jets are different. The air specific impulse of the engine is (Art. 6)

$$I_a = \frac{V_{f\text{I}} - V_\infty + \mu(V_{f\text{II}} - V_\infty)}{g(1 + \mu)}$$
$$= \frac{1}{g}\left(\frac{V_{f\text{I}} + \mu V_{f\text{II}}}{1 + \mu} - V_\infty\right)$$

and the fuel specific impulse is obtained by multiplying this quantity by the air-fuel ratio of the combustible mixture.

Air and fuel specific impulses are unmodified if the two *fully expanded* flows are completely mixed at constant pressure before emerging from the engine, because the momentum outflow is unmodified in this process. Letting primes refer to this new situation,

$$V_f' = \frac{V_{f\text{I}} + \mu V_{f\text{II}}}{1 + \mu}$$

Hence $I_a' = I_a$.

Now, if the exhaust flow is uniform, the wave engine may be treated as a single-flow system, all particles of the emerging flow having been subjected to the same entropy increment and having received the same amount of energy Jq_a per unit mass. Thus the performance of an external combustion wave engine may be computed as that of a turbojet operating with the same compression ratio and with the same air-fuel ratio and combustion efficiency, provided that the entropy rise through turbine and compressor be equal to the sum of the entropy increments of the wave engine in pressure exchange *and in the final mixing process*. It is noteworthy that this result applies regardless of whether or not the final mixing process is actually made to occur, i.e. regardless of whether the two wave-engine flows are completely mixed after full expansion or are made to emerge as independent jets. This means that, under the specified conditions, the external combustion wave engine cannot excel the turbojet unless the over-all efficiency of the pressure exchanger is appreciably higher than the combined efficiency of the turbine-compressor system. It should be noted, however, that the specified conditions (mixing after full expansion or no mixing at all) are not the best ones for the wave engine. Indeed, the performance of this engine is improved considerably when the two flows are mixed *before* they are fully expanded, i.e. when their initial velocity relative to one another is low. This situation is easily obtainable at high flight speeds. As the relative velocity of the two flows at their merging station (where their pressures and temperatures become substantially equal) is decreased, the entropy rise in mixing approaches zero and the final exhaust velocity approaches the value

$$V_f'' = \left(\frac{V_{f\text{I}}^2 + \mu V_{f\text{II}}^2}{1 + \mu}\right)^{\frac{1}{2}}$$

which can be appreciably higher than V_f'.

When consideration is given to this possibility of decreasing the entropy increment of mixing, it must be concluded that the potential performance of the external combustion wave engine at high flight speeds is better with than without mixing of the exhaust flows. At low flight speeds, mixing can serve the useful purpose of providing an ejector action whereby the high energy flow II aids in scavenging the low energy gas I.

In view of these possible improvements and of the high pressure exchange efficiencies which have already been attained, there is reason to believe that the entropy rise through pressure exchange and mixing may be reduced (at least for moderate pressure ratios) to a value comparable to that associated with the flow through compressor and turbine in a turbojet. It should also be remembered that the permissible peak cycle temperature is considerably higher in the wave engine than in the turbojet, because all moving parts in the wave engine are alternately exposed to hot and cold gases. The frequency of these temperature fluctuations is so high that the rotor material is almost steadily maintained at a temperature which is the average of the averages of the entrance and exit temperatures of the hot and cold gases [*10*].

Thus the potential performance of the external combustion wave engine can be calculated by means of the performance equations of the turbojet, with current values of the compressor and turbine efficiencies, but with much higher peak cycle temperatures than are currently permissible in turbojets.

F,18. Internal Combustion Wave Engines. The nature of pressure exchange in internal combustion wave engines is determined primarily by the mode of combustion and by the boundary conditions.

A simple form of pressure exchange is that which is associated with nonsteady front modes of combustion.

Consider a frictionless and initially steady flow of air in a constant area tube of infinite length. Injection of fuel into the stream is suddenly started at some point of the tube, forming a combustible air-fuel mixture, the downstream end of which is suddenly ignited when it reaches a certain section of the tube. Combustion then proceeds upstream through the

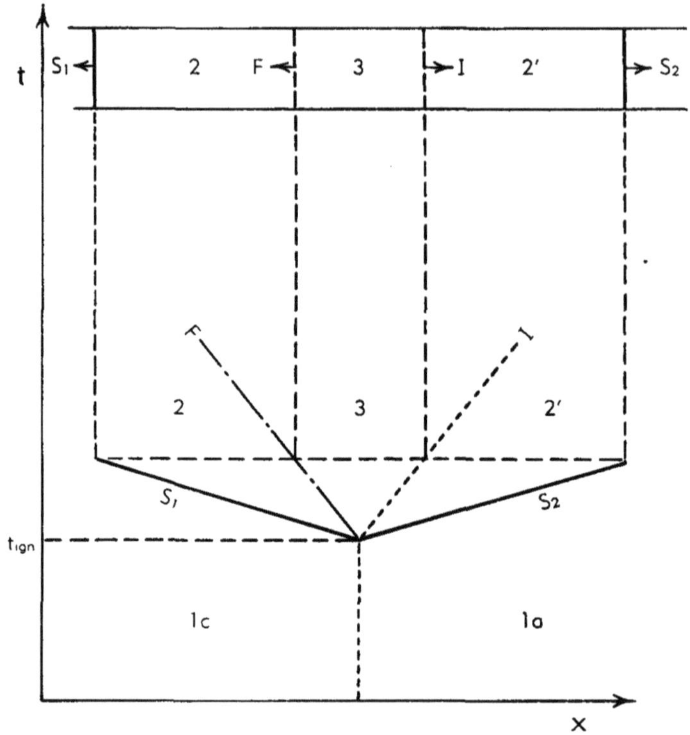

Fig. F,18a.

mixture in a plane normal flame front F, propagating at velocity V_F relative to the unburned gas.

Except for the dependence of the ignition time on the position of the ignition section, no part of the process is in any way "fixed" to the walls of the duct. Therefore the gas may be assumed to be at rest, relative to the walls, before ignition, and separated into a fresh air region a and a combustible region c by an interface I, at which ignition suddenly occurs at $t = t_{ign}$ (Fig. F,18a).

For the sake of simplicity, the temperature is assumed to be uniform throughout the duct before ignition, and molecular weight and specific heats are assumed to be uniform throughout the duct and constant throughout the process.

Ignition produces a disturbance S_1, propagating into the combustible mixture and a disturbance S_2 propagating into the fresh air. These two disturbances are found to be weak shock waves. Since the change of state across the flame can be assumed to be isobaric, the pressure is constant between the two shocks and therefore S_1 and S_2 are shocks of equal strength. It follows that $V_2 = -V_2'$. Furthermore, since the velocity must be the same on the two sides of the interface between burned gas and fresh air, we have $V_3 = V_2' = -V_2$. Thus, in this frame of reference, the flow velocities on the two sides of the flame are equal but of opposite signs. Then the continuity equation across the flame becomes

$$\rho_2 V_F = \rho_3(V_F - 2V_2)$$

Since the shocks are weak, it is permissible (III,C) to assume $s_2 = s_1$, hence

$$V_2 = \frac{2}{\gamma - 1}(a_1 - a_2)$$

and, since $p_2 = p_3$,

$$\frac{\rho_2}{\rho_3} = \frac{T_3}{T_2} = 1 + \frac{q_a}{c_p T_2}$$

Letting $V_F/a_2 = M_F$, one obtains

$$\frac{a_1}{a_2} = 1 + \frac{\gamma - 1}{4} M_F \frac{q_a}{c_p T_2}$$

i.e.

$$\frac{T_1}{T_2} = \left(1 - \frac{\gamma - 1}{4} M_F \frac{q_a}{c_p T_2}\right)^2$$

whence, considering that $M_F \ll 1.0$,

$$\frac{T_2}{T_1} \cong 1 + \frac{\gamma - 1}{2} M_F \frac{q_a}{c_p T_1} \tag{18-1}$$

Therefore

$$\frac{p_2}{p_1} \cong \left(1 + \frac{\gamma - 1}{2} M_F \frac{q_a}{c_p T_1}\right)^{\frac{\gamma}{\gamma - 1}} \tag{18-2}$$

and it can be verified that this pressure ratio is generally small enough to justify the premise that the shocks be weak and substantially isentropic.

Shock S_1 causes the combustible charge to undergo compression before combustion, and p_2/p_1 is the precompression ratio. An attempt to utilize pressure exchange processes of this sort was made by Kahane in the "intermittent ramjet" [*14*].

Changes of boundary conditions may have an appreciable effect on the pressure exchange process with this mode of combustion. Thus, if the ignition plane is located at a closed end of a tube, the boundary condition $V_3 = 0$ reduces the continuity equation to

$$\rho_2 V_F = \rho_3(V_F + V_2)$$

whence

$$V_2 = \frac{q_a V_F}{c_p T_2}$$

Therefore, with the simplifying assumptions used in the preceding illustration, one obtains

$$\frac{p_2}{p_1} \cong \left[1 + (\gamma - 1)M_F \frac{q_a}{c_p T_1}\right]^{\frac{\gamma}{\gamma - 1}} \tag{18-3}$$

The pressure rise produced by shock S_1 is appreciably higher in the latter than in the former case.

Further increases in precompression are obtained when the waves which are produced in the combustion of each charge are made to participate in the precompression of another charge, either in another chamber, as in the Esnault-Pelterie device (Fig. F,1e), or in the same chamber in a later cycle. This extended form of pressure exchange is used in most of the present-day internal combustion wave engines. The extension is often indirect, as when the waves resulting from each deflagration are used to create the flow conditions which permit the generation of a precompressing hammer wave in the next cycle. In any case, when the extended pressure exchange is properly carried out, the effect of each deflagration is not so much to produce a new pressure wave as it is to provide new energy to strengthen a carry-over precompressing wave from the previous cycle. Under these conditions, the combustion-energized shocks may build up to great strength, and the entropy rise which is produced in their repeated passes may become very appreciable (Art. 7).

The first internal combustion wave engine to utilize extended pressure exchange was the Hertzberg-Logan wave engine of 1949 [*35*], the cyclic operation of which is described by the wave diagram of Fig. F,18b. This engine consists of a cluster of parallel tubes, with two controlled valves which periodically open and close the inlet and exit of each tube. Fuel is injected in each tube to form a combustible mixture in the tube region adjacent to the exit valve. Sudden closure of this valve produces an upstream-propagating hammer wave which precompresses the combustible charge. When this hammer wave, strengthened by the deflagration which immediately follows precompression, reaches the front end, the inlet valve closes and the wave is reflected back in the duct. Meanwhile, the exit valve has been opened to permit the combustion products to expand in a propulsive jet, and an upstream-propagating expansion wave has been generated in this process. The inlet valve opens after this expansion wave has reduced the pressure on its inner side enough to permit the inflow of a new charge of air. When the returning pressure wave arrives at the downstream end of the tube, the exit valve closes again and the cycle is repeated.

The same mechanism of extended pressure exchange is used in the RPI internal combustion wave engine [*36*], which differs from the Hertzberg-Logan engine in that it operates without controlled mechanical

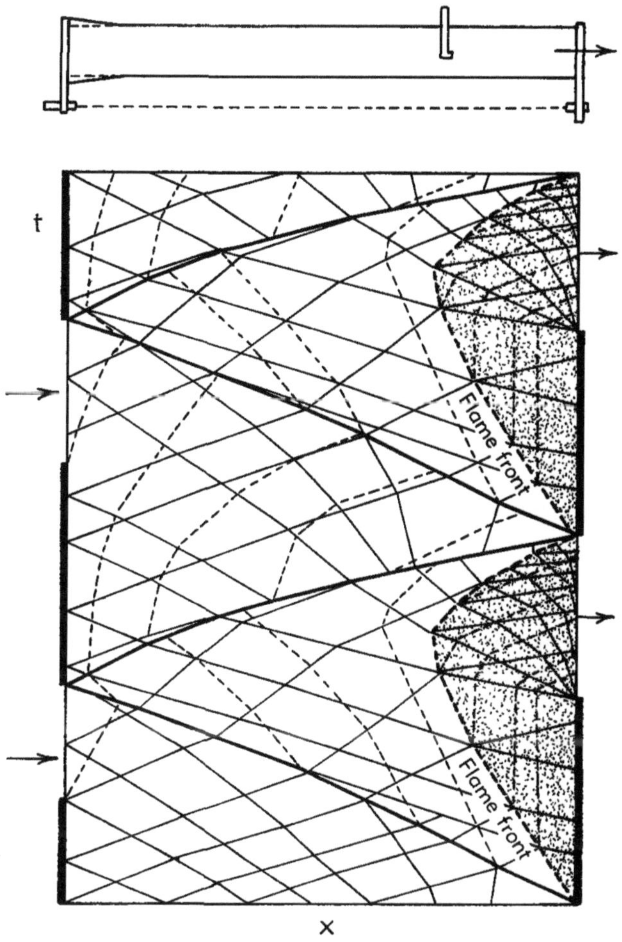

Fig. F,18b. Cyclic operation of Hertzberg-Logan wave engine of 1949.

valves. The RPI wave engine consists of a duct with an inlet flapper valve, a fuel injector, and a combustion chamber at the downstream end, terminating in a converging (or converging-diverging) exhaust nozzle. The flapper valve automatically fulfills the timing requirements at the intake end, just as it would in the Hertzberg-Logan engine. At the exhaust end, the valving action is provided by the arrival of interfaces at the converging nozzle passage. When an interface reaches a constricted passage, an upstream-propagating compression wave is generated at the constriction if the interface is a cold front, i.e. if the upstream region is occupied by the colder gas. Conversely, an upstream-propagating expansion wave is generated at the constriction if the interface is a hot front. If the entropy increment across the interface is very large, as at the contact surface between burned gas and fresh charge in a jet engine, the strength of these waves may be comparable to that of the waves which would be produced by the action of mechanical valving. For example, in the RPI engine, the upstream-propagating compression wave which is generated upon completion of scavenging (i.e. with the arrival of a cold front at the exit) is strong enough to bring the fresh charge in the combustion chamber almost completely to rest. This wave is called a "quasi-hammer" wave, because of the similarity between its action and the effect of sudden mechanical closing of the downstream end of the duct.

The actual mode of combustion in most internal combustion wave engines is believed to be intermediate between the one just discussed and that of a simultaneous change of state through the entire mixture (volume mode). The flame will usually be distorted or broken up enough to create a condition approaching that of volume combustion in the burning region, but this region will seldom—if ever—extend simultaneously over the entire length of the combustion chamber. As the extreme condition of volume combustion is approached, the strength of the pressure waves generated in the burning region increases but a smaller and smaller portion of the unburned mixture is left to benefit from pressure exchange within the cycle. However, in the same manner as with the front mode of combustion, the explosion wave energy can be partially retained within the engine for delayed utilization. In the extreme case of a true volume mode, precompression is obtained entirely from the extended phase of pressure exchange.

This mechanism is illustrated by the characteristics diagram of Fig. F,18c, which describes the cyclic operation, at $M_\infty = 1.0$ of an RPI wave engine with volume-mode combustion. The explosion, which is assumed to take place instantaneously (hence at constant volume), is followed by the generation of two expansion waves and by the reinforcement of a carry-over shock from the preceding cycle (not shown in the diagram). This shock reaches the front end, closes the valve, and is reflected back. A little later the expansion waves arrive at the valve and are also reflected there. The pressure behind the valve is decreased until the valve opens and a new charge is drawn in. When scavenging is completed and the cold front reaches the constricted exit, an upstream-propagating quasi-hammer wave is generated. Almost simultaneously with the arrival of the interface, the reflected shock also reaches the constricted exit and is reflected there as a compression wave. The two upstream-propagating compression waves coalesce to form a shock as they precompress the new charge and slow it down so that combustion can again take place.

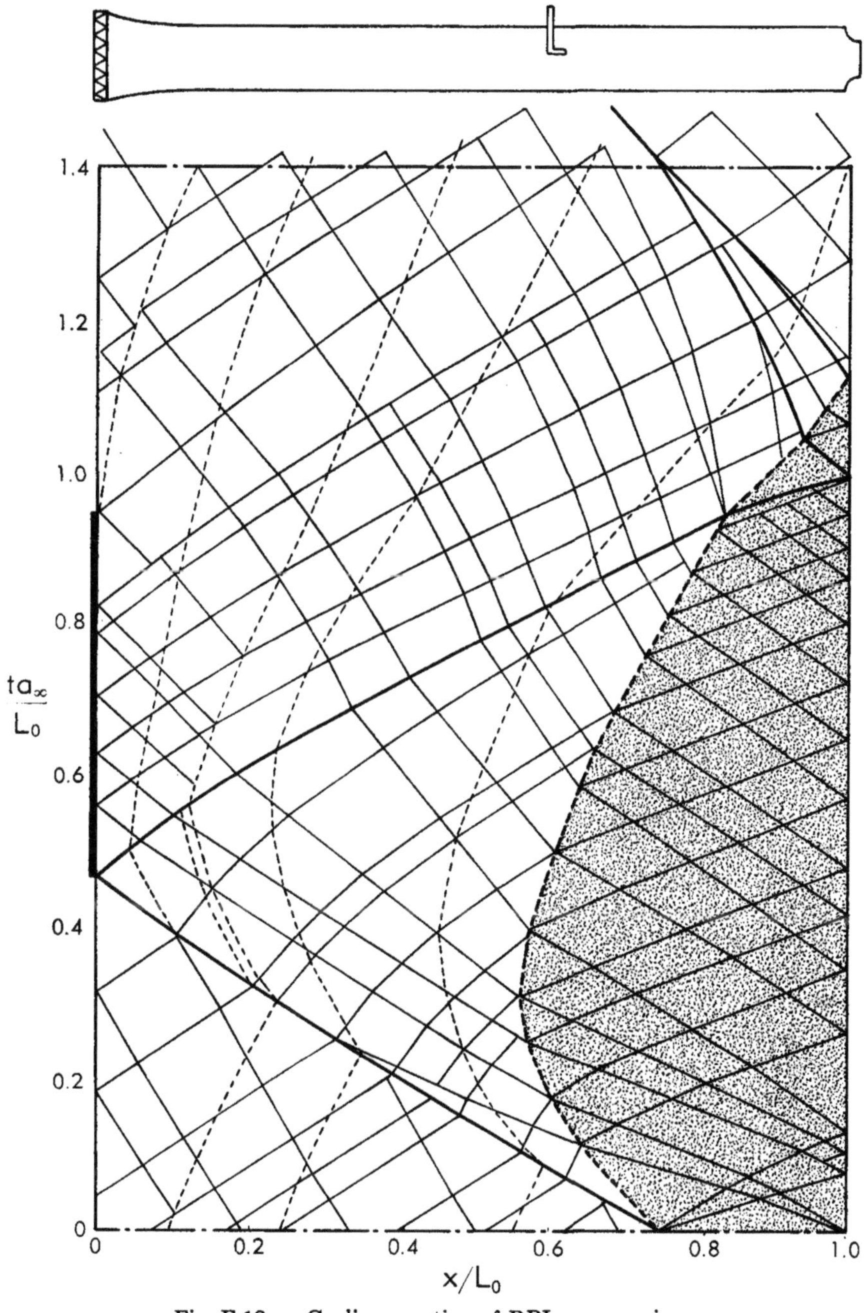

Fig. F,18c. Cyclic operation of RPI wave engine.

A fair measure of control over the combustion mode appears to be possible in wave engines. Thus, although both the front and the volume mode represent, as already noted, idealizations which can be approached but not achieved in practice, it seems likely that either mode can be approached more closely in wave engines than in other intermittent jet engines. Systematic analyses of wave engine operation with various modes of combustion are therefore of considerable practical significance, particularly when it is considered that a change of mode may in some cases (e.g. in the RPI wave engine) produce important modifications of the entire mechanism of operation of the engine.

Examination of the particle path lines in the characteristics diagram for most internal combustion wave engines reveals that each particle of the working fluid remains within the engine during several cycles and is therefore subjected to repeated pressure-exchange shock passes before it undergoes combustion. The greatest portion of the entropy rise produced by pressure exchange is due to the K passes of the upstream-propagating, flame-energized shock. In fact, the returning shock is generally weak enough so that the entropy rise associated with it can be neglected. In each type of engine, the number Ω of cycles during which each particle remains in the engine depends primarily on the ratio between the volume of the entire duct and the volume of the combustion chamber. The wave diagrams also reveal that, with front modes of combustion and ignition at the downstream end, $K = \Omega$; while in the case of a pure volume mode, $K = \Omega - 1$. This is, of course, in accord with the preceding remarks concerning the different manner in which pressure exchange is carried out with these two modes of combustion.

In the engine illustrated in Fig. F,18c, the cross-sectional area of the duct is constant, except for an exit contraction (ratio of nozzle throat area to duct area) of 0.5, and the combustion chamber length is one fourth of the total length L_0 of the engine. The particle path lines (dashed lines in the diagram) indicate that $\Omega = 4$ and $K = 3$. The entropy distribution along the duct just before combustion is shown, for the same operating conditions, in Fig. F,18d. The stepped shape of this distribution is due to the fact that successive charges have been subjected to an increasing number of passes of shocks in the forward portion of the tube. The entropy rise associated with the extended pressure exchange process is seen to be considerable.

The flow conditions in the combustion chamber just before the arrival of the quasi-hammer wave are of particular interest. The static pressure, as indicated by the pressure distribution curve of Fig. F,18d, is substantially equal to the free stream static pressure, but the flow velocity is high and, consequently, the pressure exchange process is vigorous enough to produce precompression in excess of the ram level. Thus, ram precompression is absent, but this absence does not necessarily represent a loss as it does in the case of the standard pulse jet (Art. 13).[2]

Observations of this type suggest a set of plausible assumptions which, through application of the procedures discussed in Chap. 1 and 2, permit the calculation of useful extrema and limiting values in the potential performance of internal combustion wave engines [*20*]. Only

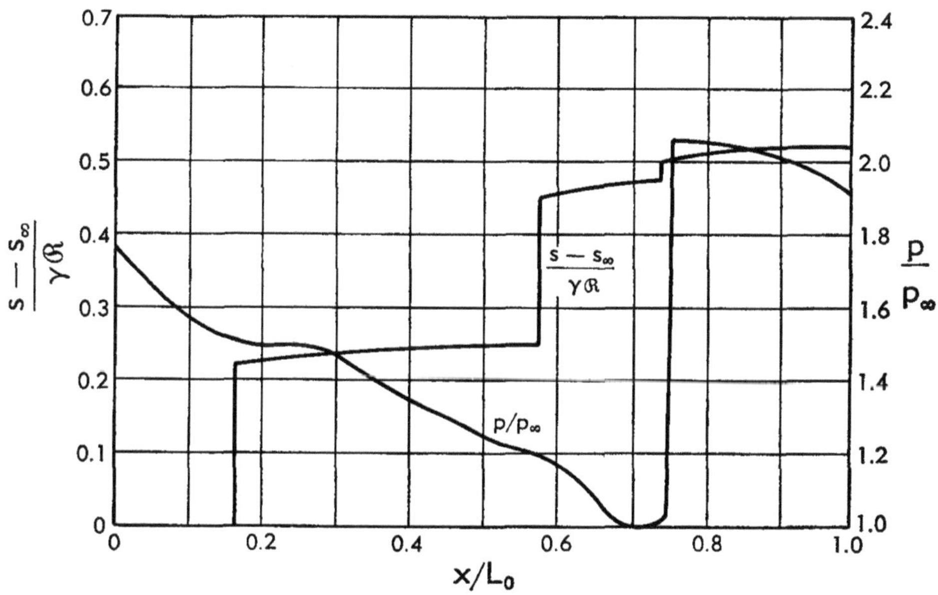

Fig. F,18d. Specific entropy and static pressure distribution at ignition time in RPI wave engine.

the treatment of front modes is reviewed here. This analysis is strictly valid only for front modes involving flame propagation Mach numbers that are low enough to permit the assumption that the static pressure be constant across the flame. On the other hand, it can be verified that, within the narrow domain of the deflagration processes which are utilized in jet engines, the entropy rise associated with combustion is quite insensitive to changes of propagation Mach number. Therefore the results of this analysis, which is concerned only with entropy increments, may be applied in good approximation to most front-mode situations of practical interest.

Since, in periodic operation, the K strong shocks to which each particle is subjected are all of equal strength, one has, assuming constant

[2] In a broad sense, ram precompression may itself be regarded as a form of extended pressure exchange, in that the energy required to produce ram compression of the fresh charge is furnished by the expanding hot gases. In the standard pulse jet, the work of ram precompression is completely dissipated in the throttled inflow. In contrast, in the wave engine, ram precompression is replaced by wave precompression (pressure exchange in the stricter sense).

specific heats,

$$\begin{aligned}\Delta s_\lambda + \Delta s_\sigma &= \Delta s_\lambda + K\Delta s_s \\ &= c_p \ln \frac{T_i}{T_\infty} - R \ln \frac{p_i}{p_\infty} \qquad (18\text{-}4)\end{aligned}$$

where Δs_s represents the entropy rise through each of the K shocks and subscript i denotes conditions immediately before combustion (i.e. after the last shock pass). If ram precompression is assumed to be completely absent, the precompression ratio p_i/p_∞ is merely the static pressure ratio

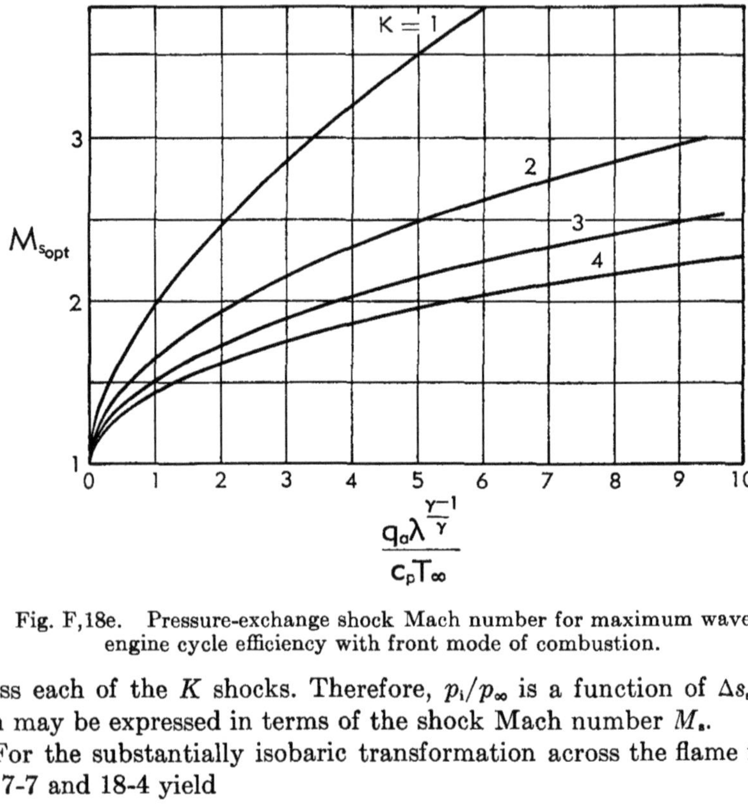

Fig. F,18e. Pressure-exchange shock Mach number for maximum wave engine cycle efficiency with front mode of combustion.

across each of the K shocks. Therefore, p_i/p_∞ is a function of Δs_s, and both may be expressed in terms of the shock Mach number M_s.

For the substantially isobaric transformation across the flame front, Eq. 7-7 and 18-4 yield

$$\begin{aligned}\Delta s_h &= c_p \ln \left(1 + \frac{q_a}{c_p T_i}\right) \\ &= c_p \ln \left[1 + \frac{q_a \lambda^{\frac{\gamma-1}{\gamma}}}{c_p T_\infty} \left(\frac{p_\infty}{p_i}\right)^{\frac{\gamma-1}{\gamma}} \exp\left(-\frac{K\Delta s_s}{c_p}\right)\right]\end{aligned}$$

The specific entropy rise through the engine is $\Delta s = \Delta s_\lambda + \Delta s_\sigma + \Delta s_h$. Therefore,

$$\exp \frac{\Delta s}{c_p} = \lambda^{-\frac{\gamma-1}{\gamma}} \exp \frac{K\Delta s_s}{c_p} + \frac{q_a}{c_p T_\infty} \left(\frac{p_\infty}{p_i}\right)^{\frac{\gamma-1}{\gamma}} \qquad (18\text{-}5)$$

With fixed values of q_a and T_∞, the maximum cycle efficiency and air specific impulse and the minimum sfc are obtained when $\exp(\Delta s/c_p)$ is minimum. The vanishing of the first derivative of the right-hand member of Eq. 18-5 with respect to M_s provides, for any value of K, the optimum Mach number, $M_{s_{opt}}$, of the pressure exchange shock. The

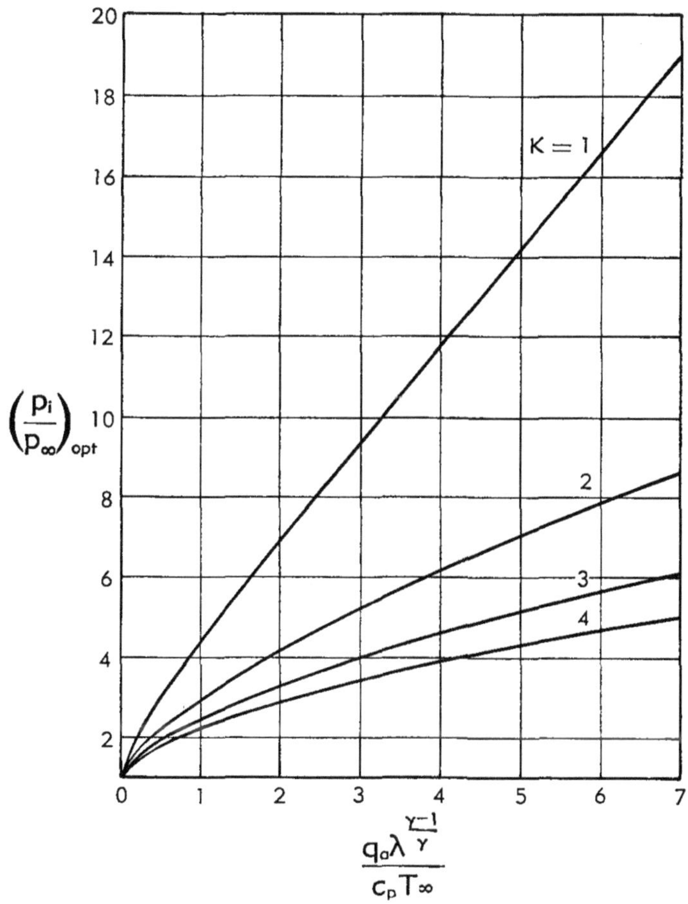

Fig. F,18f. Compression ratio for maximum wave engine cycle efficiency with front mode of combustion.

results of this calculation for $\gamma = 1.4$ are plotted in Fig. F,18e. The corresponding values of the compression ratio and of the entropy exponential pertaining to this "optimum" pressure exchange process are plotted, again for $\gamma = 1.4$, in Fig. F,18f and F,18g. Finally, the performance of the internal combustion wave engine with the optimum pressure exchange is presented in Fig. F,18h for two values of K and for the following conditions: $T_\infty = 500°R$, heat liberated in combustion = 18,000 BTU/lb of fuel, air-fuel ratio = 30, $\lambda = 0.95$, $\eta_v = 0.95$, and $\gamma = 1.4$ throughout.

Departures from the optimum mode of pressure exchange are often found to produce a sharp deterioration of wave engine performance. On the other hand, the achievement of this optimum is very difficult in practice, because the strength of the pressure exchange shock is determined not only by the boundary conditions, which are easily controlled, but also by the flame propagation velocity for which no effective method of control has yet been found.

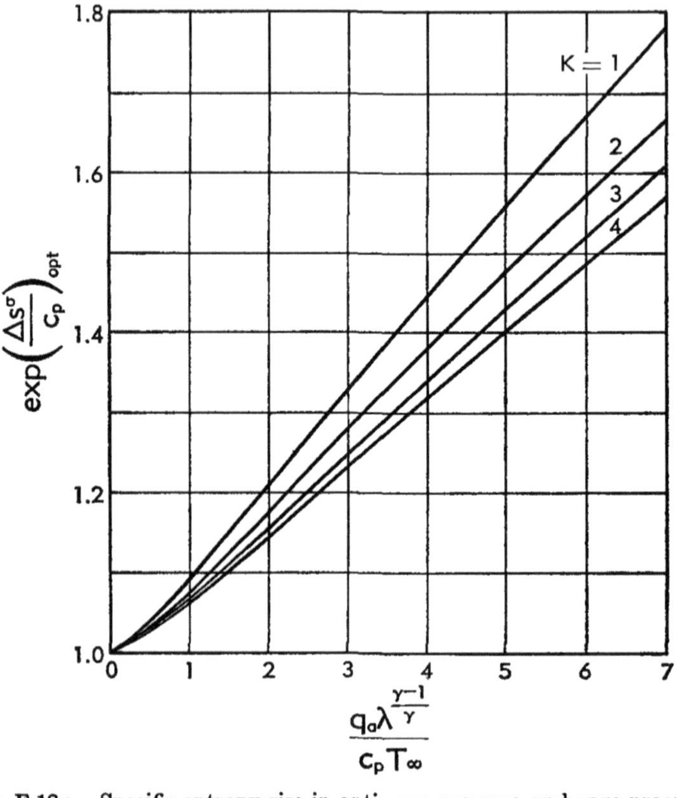

Fig. F,18g. Specific entropy rise in optimum pressure exchange process.

An idea of the importance of the losses associated with the extended phase of pressure exchange may be obtained by comparing the performance values of Fig. F,18h with the overoptimistic results of previous analyses [*17,18*] in which pressure exchange was represented by a stepped but monotonic process.

The treatment of volume modes is essentially similar to that of the front mode, but there are cases in which a change of combustion process from the front to the volume mode has the effect of increasing the number of flows through the engine. An illustration of this effect is provided by the RPI wave engine, which operates as a single-flow or as a double-flow engine, depending on whether it utilizes a front or a volume mode of combustion.

The operation described by the wave diagram of Fig. F,18c is that of a double-flow engine, because the time delay between the end of scavenging and reignition results in the discharge of a considerable amount

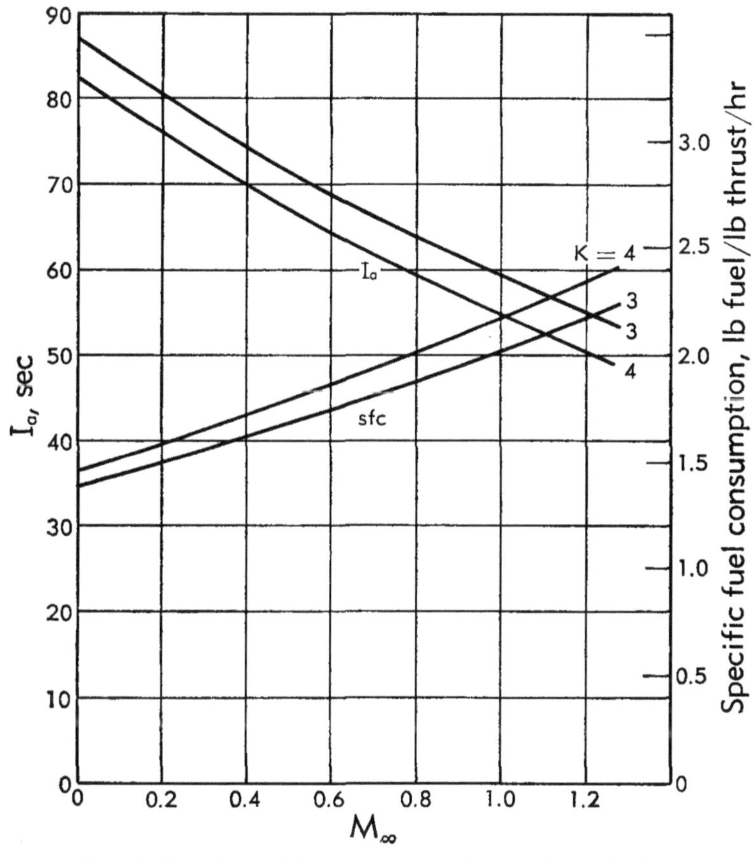

Fig. F,18h. Potential performance of internal combustion wave engine, under conditions specified in text.

of air which has been energized by the pressure exchange waves but has not undergone combustion. Let this flow be denoted by subscript $_{\text{II}}$ and the combustion air by subscript $_{\text{I}}$. Precompression brings both flows almost to rest and to the same temperature T_i. The exhaust stagnation temperature of flow II is substantially constant and equal to T_i. It follows that

$$
\begin{aligned}
E_{\text{II}} &= Jc_p(T_i - T_\infty^0) \\
&= Jc_p\left[T_\infty\left(\frac{p_i}{p_\infty}\right)^{\frac{\gamma-1}{\gamma}} \exp\frac{\Delta s_\sigma}{c_p} - T_\infty^0\right]
\end{aligned}
$$

The unit mass of gas I receives an amount of energy equal to E_{II} in precompression, and the additional energy Jq_a in combustion, but then relinquishes the energy $(1 + \mu)E_{II}$ in pressure exchange. Therefore,

$$E_I = Jq_a - \mu E_{II}$$

The specific entropy rise due to flow losses and pressure exchange is the same for the two flows, but flow I also suffers the increment Δs_h associated with constant volume combustion. Thus

$$\exp \frac{\Delta s_{II}}{c_p} = \lambda^{-\frac{\gamma-1}{\gamma}} \exp \frac{\Delta s_\sigma}{c_p}$$

$$\exp \frac{\Delta s_I}{c_p} = \lambda^{-\frac{\gamma-1}{\gamma}} \left(1 + \frac{q_a}{c_v T_1}\right)^{\frac{1}{\gamma}} \exp \frac{\Delta s_\sigma}{c_p}$$

$$= \lambda^{-\frac{\gamma-1}{\gamma}} \left[1 + \frac{q_a}{c_v T_\infty} \left(\frac{p_\infty}{p_1}\right)^{\frac{\gamma-1}{\gamma}} \exp\left(-\frac{\Delta s_\sigma}{c_p}\right)\right]^{\frac{1}{\gamma}} \exp \frac{\Delta s_\sigma}{c_p}$$

The exhaust velocity is, for both flows, very nearly a square-wave function of time.

Thus the cycle efficiencies of the two flows and the over-all air specific impulse can be calculated in the usual manner. A generalized analytical search for extrema would be of no practical significance in this case, because the relation between Δs_σ, p_1/p_∞, and μ, in engines operating in this manner, can only be established through a systematic series of calculations by the method of characteristics for each individual case.

The augmenting effect of the secondary flow in this type of engine can be appreciably greater than that of backflow in the pulse jet and can also extend over a wider range of speeds, but its utilization requires the use of intermittent fuel injection to prevent the discharge of unburned fuel with the cold air. In view of the mechanical complication and timing difficulties of such an arrangement, this type of double-flow operation is at present considered to be only of academic interest.

A special type of internal combustion wave engine is the "hammerjet," in which pressure exchange is reduced to a single sweep of a hammer wave. In its simplest form, the hammerjet consists of a tube with controlled valves at both ends. Combustion takes place in the entire tube, with both valves closed, hence at constant volume. The exhaust valve is opened when combustion is completed, thereby letting the hot gases expand in a propulsive jet. When the pressure on the inner side of the front valve becomes lower than the external ram pressure, this valve is opened and the entering fresh charge forces out the residual hot gases. When scavenging is completed, sudden closing of the exhaust valve produces the precompressing hammer wave. The inlet valve is closed when this wave reaches the front end. At this point the precompressed charge, which now fills the entire tube, is ignited; and the cycle is repeated. The main advantages of the hammerjet are its excellent mode of combustion and its low pressure exchange losses. So far, however, these advantages have not been found to be important enough to offset the disadvantages which derive from the sluggishness of the hammerjet's scavenging process, viz. low efficiency of nonuniformity and low frequency of operation.

F,19. Cited References.

1. Lorin, L. *L'Aerophile 16*, 332 (1908).
2. Marconnet, G. *French Patent 412,478*, 1910.
3. Lorin, L. *L'Aerophile 21*, 229 (1913).
4. Amsler, W. O. *The Sibley J. Eng. 14*, 333 (1900).
5. Stodola, A. *Steam and Gas Turbines.* Peter Smith, New York, 1945.
6. Barbezat, A. Praktische Versuche an Gasturbinen. *Die Turbine 5*, 305–312 (1909). Transl. by G. Rudinger, *Cornell Aeronaut. Lab. Rept. CAL-38*, Mar. 1951.
7. Schmidt, P. *German Patent 523,655*, 1930; *567,586*, 1931.
8. Von Ohain, H. J. P. Report on a special gas turbine principle. *Wright-Patterson Air Force Base Tech. Rept. F-TR-2186-ND, Air Tech. Index 31,133*, Mar. 1948.
9. Seippel, C. *U.S. Patent 2,399,394*, 1946.
10. Meyer, A. Recent developments in gas turbines. *Mech. Eng. 69*, 273–277 (1947).
11. Busemann, A. Bericht über den Paul Schmidt'schen Strahlrohrantrieb. *Deut. Versuchsanstalt für Luftfahrtforschung FB530*, 1936. Transl. *Air Tech. Index 36656.*
12. Rudinger, G. On the performance analysis of the ducted pulsejet. *Cornell Aeronaut. Lab. Project Squid Tech. Mem. CAL-36*, 1951.
13. Brumfield, R. C., and Berggren, W. P. An application of the "Universal Thrust Equation" to the calculation of the aeropulse engine efficiency. *Aerojet Eng. Corp.*, Jan. 1946.
14. Kahane, A., Marino, A. A., Messinger, C. W., and Shafer, H. J. A theoretical and experimental investigation of the feasibility of the intermittent ram-jet engine. *Princeton Univ. Project Squid Tech. Rept. 35*, 1951.
15. Courant, R., and Friedrichs, K. O. *Supersonic Flow and Shock Waves.* Interscience, 1948.
16. Pangburn, D. F., and Foa, J. V. An extension of the Chapman-Jouguet gasdynamics of combustion. *Rensselaer Polytech. Inst. Tech. Rept. TR-AE5402*, 1954.
17. Foa, J. V. Performance of single-flow jet engines. *Cornell Aeronaut. Lab. Project Squid Tech. Rept. 20*, 1950.
18. Foa, J. V. Single-flow jet engines—A generalized treatment. *J. Am. Rocket Soc. 21*, 115–126 (1951).
19. Rudinger, G. A problem arising in the performance analysis of the ducted pulsejet. *Proc. First Natl. Congress Appl. Mech.*, Chicago, 1951, *ASME*, New York, 935–939, 1952.
20. Pangburn, D. F. A comparison of combustion modes in intermittent jet engines. *Rensselaer Polytech. Inst. Tech. Rept. TR-AE5504*, 1955.
21. Kantrowitz, A. R., and Perry, R. The response of a normal shock in a channel flow to small disturbances coming from the rear of the channel. *Cornell Aeronaut. Lab. Project Squid Tech. Mem. 28*, 1949.
22. Logan, J. G. Forced shock oscillations in diffusers. *Cornell Aeronaut. Lab. Project Squid Tech. Mem. CAL-12*, 1950.
23. Lembcke, H. Das Schmidtrohr. *Z. Ver. deut. Ing. 94*, 1005–1008 (1952).
24. Kantrowitz, A. R., and Neice, S. E. *J. Aeronaut. Sci. 17*, 316 (1950).
25. Poggi, L. *Aerotecnica 29*, 288–296, 356–363 (1949).
26. Poggi, L. *J. Aeronaut. Sci. 17*, 813 (1950).
27. MacDonald, J. K. L. A gas-dynamical formulation for waves and combustion in pulsejets. *Appl. Math. Group, New York Univ. Mem. 151*, 1946.

28. Bailey, N. P., and Wilson, H. A. The intermittent jet engine. *Office Sci. Research and Develop. Jet Propelled Missiles Panel*, 1945.
29. Schultz-Grunow, F. Gasdynamische Untersuchungen am Verpuffungstrahlrohr. *Inst. Mech., Tech. Hochschule Aachen (ZWB) FB 2015/1 u.2*, 1943, 1944. Transl. *NACA Tech. Mem. 1131*, 1947.
30. Sutton, G. P. Backflow determination on aeroresonator. *Aerojet Eng. Corp. RTM-16*, 1945. See also: Berggren, W. P., Coxe, W. K., and McRoberts, G. Results of certain experiments on conventional pulsejets and related devices. *Aerojet Eng. Corp. Rept. 67*, 1946.
31. Schubert, W. Design, construction and testing of a 6-inch valveless resojet. *Nav. Eng. Expt. Sta. EES-B-5350-As (b)*, 1944.
32. Logan, J. G., Jr. Summary report on valveless pulsejet investigation. *Cornell Aeronaut. Lab. Project Squid Tech. Mem. CAL-42*, 1951.
33. Diedrich, G. The aero-resonator power plant of the V-1 flying bomb Trans. by A. Kahane, *Princeton Univ. Project Squid Tech. Mem. Pr-4*, 1948.
34. *AGARD Combustion Colloquium*, Cambridge Univ., 1953. Butterworths, London, 1954.
35. *Project Squid Ann. Progress Rept.*, 7–10 (1950).
36. Foa, J. V. The RPI wave engine. *Rensselaer Polytech. Inst. Tech. Rept. TR-AE5409*, 1954.

SECTION G

THE LIQUID PROPELLANT ROCKET ENGINE

MARTIN SUMMERFIELD

G,1. Introduction. The scientific development of the modern rocket engine has been made possible to a great extent by the results of research in the fields of chemical physics, high temperature combustion, high intensity heat transfer, gas dynamics, and heat-resistant materials. Much of this work has taken place in recent years and has not been incorporated heretofore in the standard textbook literature in a form suitable for systematic study. For an up-to-date comprehensive treatment of the liquid propellant rocket engine, therefore, it would have been necessary to introduce in this section all of the new developments mentioned. However, in the planning of this series, many of these topics were placed in other sections to allow a more logical arrangement of the entire subject matter. Therefore, in this section, the pertinent results or conclusions derived from these modern investigations are merely stated in their simplest form, and the reader is directed by cross references to the other volumes where the complete treatments can be found.

In rocketry, as in any field that has grown rapidly, the terminology has not yet been fully standardized and accepted. Therefore it is appropriate to start with a few definitions.

Rocket propulsion is a system of propulsion that depends on forward thrust created by rearward ejection of a fluid jet through a nozzle mounted in the vehicle, with the special condition that the fluid in the jet originate entirely from tanks within the vehicle. It is this special condition that distinguishes the rocket from other classes of jet engines that ingest the surrounding medium (air or water) to form the driving jet. Therefore the rocket engine is able to function not only under the usual conditions of flight through the atmosphere (or under water), but in the vacuum outside the atmosphere.

A *rocket propellant* is the fluid substance that forms the driving jet, although the term is used most frequently to refer to the driving fluid in its chemical state before combustion. The term is used also to denote *one* of the reactants in a multicomponent propellant system. A *fuel* is any propellant that can burn in the presence of oxygen, and the term includes not only hydrocarbons but other substances (e.g. ammonia, powdered aluminum) as well. An *oxidizer* is a propellant that can support the combustion of a fuel, and is applicable to substances that may not contain oxygen (e.g. fluorine) as well as those that do. A *bipropellant system* is one that consists of two reactants, usually a fuel and an oxidizer. A *monopropellant* is a single substance that can be caused to react in the combustion chamber to generate hot gas to form the driving jet. This term applies strictly only to a single compound (e.g. ethylene oxide) that undergoes a decomposition reaction, but it has been applied also to a propellant mixture that is stored in a single propellant tank (e.g. a mixture of methyl nitrate and methyl alcohol).

The *rocket motor* is usually understood to be the part of the engine in which the propellants are burned and the jet is formed, while the term *rocket engine* usually refers to the entire propulsion system, including the tanks if they are constructed integrally with the engine. The term *thrust cylinder* has been used in place of rocket motor in some writings, but the latter term is historically the oldest and the most widely preferred. In the conventional solid propellant rocket, the engine and the motor happen to be the same piece of apparatus because the propellant is stored in the combustion chamber of the rocket motor; this is not so in the liquid propellant rocket.

G,2. Performance Analysis of the Ideal Rocket Motor. The performance analysis of a rocket motor comprises calculations of the thrust F, the effective exhaust velocity c, the adiabatic combustion temperature in the chamber T_c, the thrust coefficient C_F, the characteristic velocity c^*, and certain efficiencies η. Performance parameters derived from c are the specific impulse I_{sp} and the specific propellant consumption w_{sp}.

The thrust equation is the fundamental starting point. In general, the thrust exerted on a duct of arbitrary shape can be calculated from the momentum equation written in integrated form appropriate for one-dimensional flow problems. (See III,B.) Let $\dot{m}_i$ be the rate of mass flow into the inlet, p_i the static pressure at the inlet, V_i the stream velocity at the inlet, and A_i the area of the inlet, and let the corresponding quantities at the exit be indicated by the subscript $_e$. Then,

$$\text{The stream thrust at the inlet} = (\dot{m}_i V_i + p_i A_i)$$

$$\text{The stream thrust at the exit} = (\dot{m}_e V_e + p_e A_e)$$

$$\text{The total force on the external duct surface} = F + p_\infty(A_e - A_i)$$

$$F + p_\infty(A_e - A_i) = (\dot{m}_e V_e + p_e A_e) - (\dot{m}_i w_i + p_i A_i)$$

The external force on the duct is expressed here as if the pressure on the external surface were identical with the ambient pressure p_∞ of the atmosphere, although in actual flight this is not so. Therefore, for a duct in flight, this equation implies a certain arbitrary separation between the thrust F and the aerodynamic drag D. Separation in this manner is justified by its convenience, because the thrust measured in a ground test of the propulsion system is closely equal to the thrust F thus calculated.

In the particular case of a rocket, $\dot{m}_i$ and A_i may be set equal to zero. Then,

$$F = \dot{m}V_e + (p_e - p_\infty)A_e \tag{2-1}$$

With reasonably well-designed exhaust nozzles, the exit pressure p_e is nearly or exactly equal to the ambient pressure p_∞, so that the second term is in the nature of a small correction to the thrust. This makes it convenient to define an *effective exhaust velocity* c such that

$$F = \dot{m}c \tag{2-2}$$

$$c = V_e + \frac{(p_e - p_\infty)A_e}{\dot{m}} \tag{2-3}$$

Clearly, c equals V_e if the nozzle is designed properly, that is, if the p_e equals p_∞. For any given values of chamber pressure p_c and of p_∞ and $\dot{m}$, both F and c reach their maximum values when the exit area A_e of the nozzle is chosen to produce a static pressure p_e at the exit exactly equal to p_∞. (This will be proved later.) The exit velocity and the correction term $(p_e - p_\infty)A_e/\dot{m}$ both vary strongly with A_e, but in opposite directions so that the sum is quite insensitive to A_e, that is, the maximum is very flat. As a result, the effective exhaust velocity measured by the ratio of F to $\dot{m}$ can be taken to be the value of $(V_e)_{\text{opt}}$, even if the actual nozzle used in the test is somewhat off-design. Herein lies the practical significance of the concept of the effective exhaust velocity (see Fig. G,2a.)

The specific propellant consumption w_{sp}, defined as the weight rate of consumption of propellant $\dot{w}$ per unit thrust, is another useful index of rocket engine performance. Let g_0 denote the standard acceleration due to gravity.

$$w_{sp} = \frac{\dot{w}}{F} = \frac{\dot{m}g_0}{F} = \frac{g_0}{c} \tag{2-4}$$

The specific impulse I_{sp} (called specific thrust in some writings) is defined as the propulsive impulse delivered by the engine per unit weight of propellant.

$$I_{sp} = \frac{Fdt}{dw} = \frac{F}{\dot{w}} = \frac{1}{w_{sp}} = \frac{c}{g_0} \tag{2-5}$$

A figure of merit that is sometimes quoted is the impulse-weight ratio, I/W, of a loaded rocket propulsion system or of a loaded rocket vehicle.

If the firing program calls for constant pressure and hence constant specific impulse,

$$\frac{I}{W} = I_{sp}\nu \tag{2-6}$$

where ν is the *propellant loading fraction*, that is, the ratio of the initial mass of propellant when the rocket is fully loaded to the gross mass of the loaded rocket. Obviously, ν approaches unity as the structural effectiveness is improved. Therefore, I/W measures both the performance of the engine and the effectiveness of the structure of the rocket.

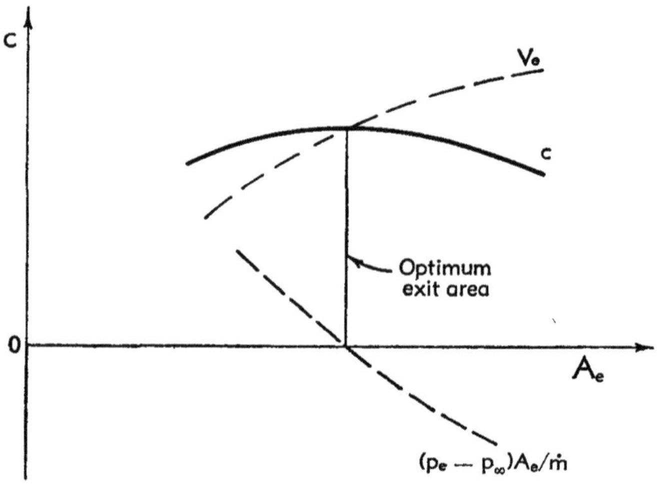

Fig. G,2a. Variation of effective exhaust velocity with exit area.

The matter of units of specific impulse deserves comment. The definition involves the mass rate of flow of propellant expressed in weight units, so that in the fps system I_{sp} should be lb sec/lb. It is common practice to denote this ratio simply as sec, canceling the lb in numerator and denominator, disregarding the fact that one lb refers to a force and the other to a mass. That this is an error becomes apparent when it is realized that the specific impulse has the dimensions of a velocity.

In order to develop several additional performance parameters, it is necessary to describe in detail the thermodynamic and gas dynamic processes in the rocket motor. As a starting point, it is a great simplification to deal with the so-called *ideal rocket motor*. From a practical standpoint, the ideal rocket motor is a useful concept because it leads to simple theoretical formulas for F, c, T_0, C_F, and c^*, which otherwise would have to be presented in tables or in graphs.

The ideal rocket motor analysis rests on the following simplifications: (1) the propellant gas obeys the perfect gas law; (2) its specific heat is constant, independent of temperature; (3) the flow is parallel to the axis of the motor and uniform in every plane normal to the axis, thus constituting a one-dimensional problem; (4) there is no frictional dissipation in the chamber or nozzle; (5) there is no heat transfer to the motor walls; (6) the flow velocity in the chamber before the nozzle entrance is zero; (7) combustion or heat addition is completed in the chamber at constant pressure and does not occur in the nozzle; and (8) the process is steady in time.

The thermodynamic process is indicated in Fig. G,2b, both in the p, V diagram and in the h, s diagram. Combustion at constant pressure

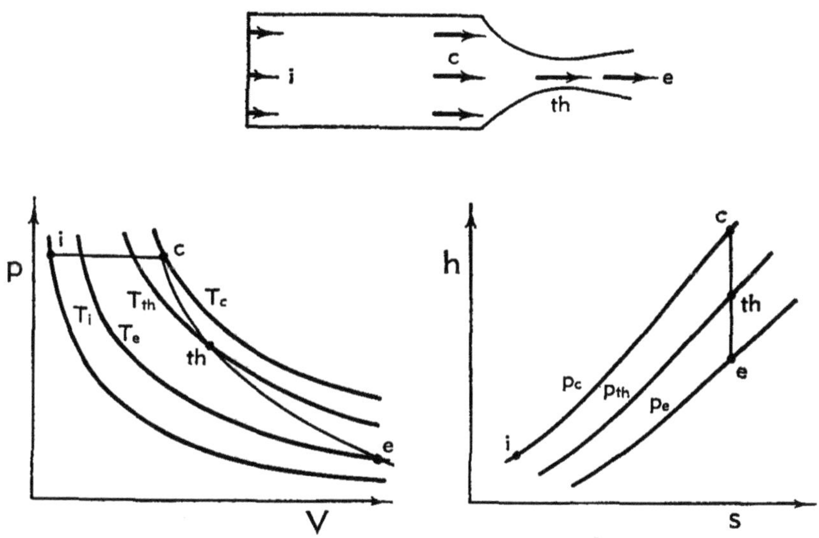

Fig. G,2b. Ideal thermodynamic processes in the combustion chamber and nozzle of a rocket motor.

moves the state point from i to c. The highest temperature occurs at c. Then, since the frictionless and adiabatic conditions assumed for the ideal rocket motor imply an isentropic expansion in the nozzle (see III,B for proof; also I,A), the state point moves along the constant entropy line from c to e during the expansion process. As indicated in the motor sketch of Fig. G,2b, the subscripts i, c, th, and e employed in the following analysis refer respectively to the unburned state at the injector, the all-burned state in the chamber just before expansion, the state at the throat of the de Laval nozzle, and the state at the exit of the nozzle.

The combustion temperature (or adiabatic flame temperature) T_c is determined by the heat of combustion at constant pressure per unit mass Δh_c.

$$\Delta h_c = c_p(T_c - T_i) \tag{2-7}$$

At any station in the nozzle, the entropy, pressure, temperature, velocity, and Mach number are given by the following relations (III,B):

$$p = \rho \frac{R}{\mathfrak{M}} T; \quad c_p - c_v = \frac{R}{\mathfrak{M}}; \quad c_p = \frac{\gamma}{\gamma - 1} \frac{R}{\mathfrak{M}} \tag{2-8}$$

$$s = s_c; \quad \frac{T}{T_c} = \left(\frac{p}{p_c}\right)^{\frac{\gamma-1}{\gamma}} \tag{2-9}$$

$$\left.\begin{aligned} \tfrac{1}{2}V^2 &= c_p(T_c - T) = \frac{\gamma}{\gamma - 1} \frac{RT_c}{\mathfrak{M}} \left[1 - \left(\frac{p}{p_c}\right)^{\frac{\gamma-1}{\gamma}}\right] \\ \frac{A}{A_{th}} &= \frac{1}{M} \left[\frac{1 + \frac{\gamma - 1}{2} M^2}{\frac{\gamma + 1}{2}}\right]^{\frac{\gamma+1}{2(\gamma-1)}} \end{aligned}\right\} \tag{2-10}$$

Since the over-all pressure ratio p_c/p_∞ is always sufficiently large in rockets to establish sonic flow at the throat, then

$$M_{th} = \frac{V_{th}}{a_{th}} = 1; \quad V_{th} = \left(\gamma \frac{R}{\mathfrak{M}} T_{th}\right)^{\frac{1}{2}} \tag{2-11}$$

$$\frac{p_{th}}{p_c} = \left(\frac{2}{\gamma + 1}\right)^{\frac{\gamma}{\gamma-1}}; \quad \frac{T_{th}}{T_c} = \frac{2}{\gamma + 1} \tag{2-12}$$

The specific heat ratios of typical rocket jet gases range between 1.1 and 1.3. The first figure corresponds to mixtures at very high temperatures with large concentrations of water vapor and large effective specific heats due to strong dissociation; the latter figure applies to moderate temperature gases with moderate concentrations of H_2O and CO_2. With $\gamma = 1.2$, it can be seen that the drop in pressure from the chamber to the throat is nearly half the chamber pressure, while the drop in temperature is only about one tenth the chamber temperature.

The mass flow through the nozzle can be expressed in terms of the flow conditions at any station. Let A be the cross-sectional area at any station:

$$\dot{m} = \rho V A = p_c A \left\{\frac{2\gamma}{\gamma - 1} \frac{\mathfrak{M}}{RT_c} \left(\frac{p}{p_c}\right)^{\frac{2}{\gamma}} \left[1 - \left(\frac{p}{p_c}\right)^{\frac{\gamma-1}{\gamma}}\right]\right\}^{\frac{1}{2}} \tag{2-13}$$

A plot of mass flow per unit area $\dot{m}/A$ against static pressure ratio p/p_c, shown in Fig. G,2c, exhibits a maximum at the throat of the de Laval nozzle, just as expected.

$$\frac{\dot{m}}{A_{th}} = p_c \left\{\gamma \left(\frac{2}{\gamma + 1}\right)^{\frac{\gamma+1}{\gamma-1}} \frac{\mathfrak{M}}{RT_c}\right\}^{\frac{1}{2}} \tag{2-14}$$

By equating the two expressions, Eq. 2-13 and 2-14, for $\dot{m}$, an expression for A/A_{th} is obtained.

$$\frac{A}{A_{th}} = \frac{\left(\frac{\gamma - 1}{2}\right)^{\frac{1}{2}} \left(\frac{2}{\gamma + 1}\right)^{\frac{\gamma+1}{2(\gamma-1)}}}{\left(\frac{p}{p_c}\right)^{\frac{1}{\gamma}} \left[1 - \left(\frac{p}{p_c}\right)^{\frac{\gamma-1}{\gamma}}\right]^{\frac{1}{2}}} \tag{2-15}$$

By inserting A_e and p_e for A and p in Eq. 2-15, the nozzle area ratio ϵ ($\equiv A_e/A_{th}$) can be expressed as a function of p_c/p_e. This relation is plotted in Fig. G,2d for several values of γ.

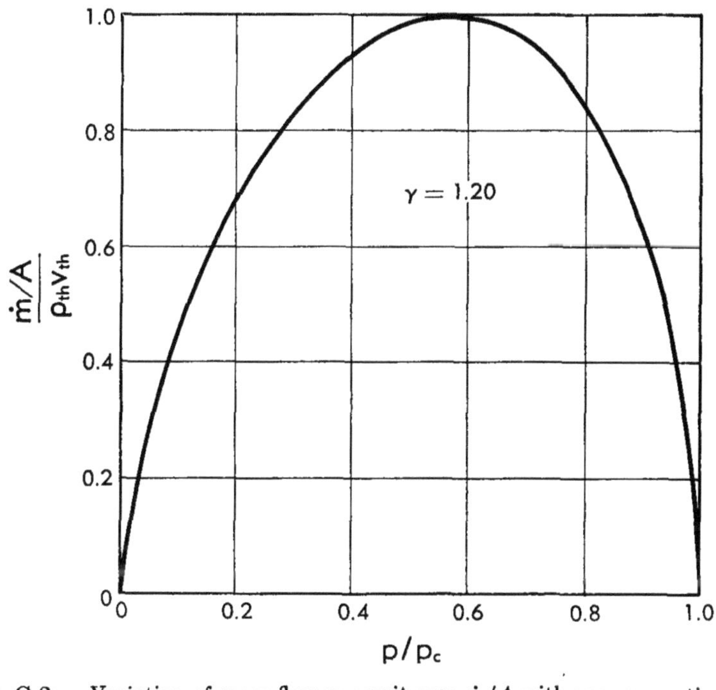

Fig. G,2c. Variation of mass flow per unit area $\dot{m}/A$ with pressure ratio p/p_c.

The maximum exit velocity $(V_e)_{max}$, obtained by setting p_e/p_c in Eq. 2-10 equal to zero, is

$$(V_e)_{max} = \sqrt{\frac{2\gamma}{\gamma - 1} \frac{RT_c}{\mathfrak{M}}} \tag{2-16}$$

It is interesting that the maximum exit velocity, obtained by expansion to zero pressure, is greater than the root-mean-square molecular velocity in the chamber by the factor $[2\gamma/3(\gamma - 1)]^{\frac{1}{2}}$, or about 2 for $\gamma = 1.2$. This result follows directly, of course, from molecular energy considerations.

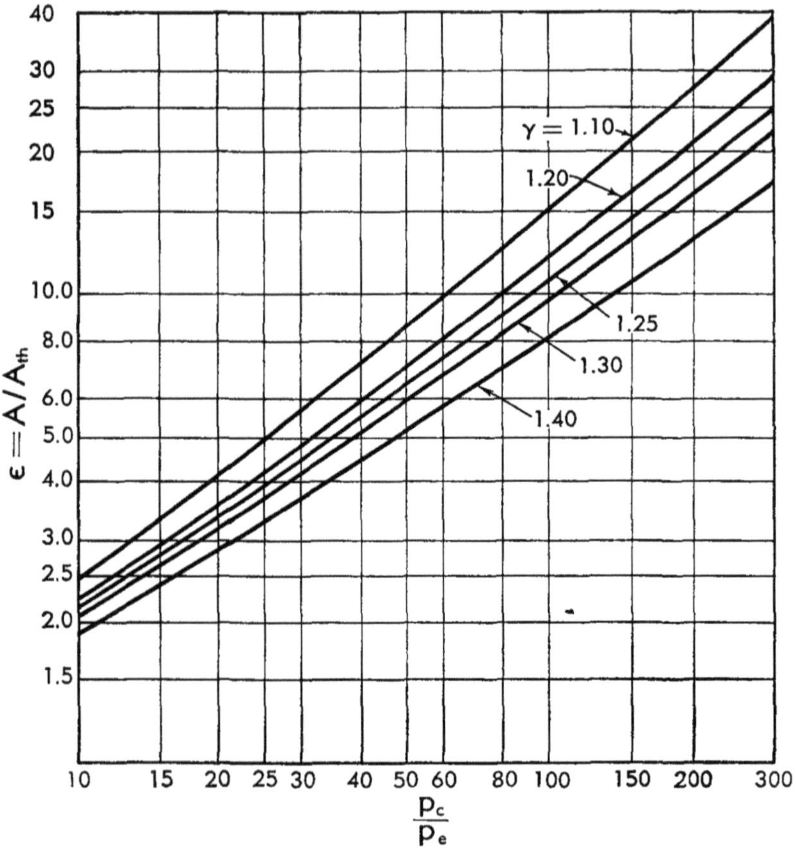

Fig. G,2d. Variation of nozzle area ratio with pressure ratio. After Sutton, G.P., *Rocket Propulsion Elements*, Wiley, 1956, by permission.

The thrust formula (Eq. 2-1) can now be expressed in terms of the pressures by substituting Eq. 2-10 for V_e and Eq. 2-14 for $\dot{m}$.

$$F = p_c A_{th} \left\{ \frac{2\gamma^2}{\gamma - 1} \left(\frac{2}{\gamma + 1} \right)^{\frac{\gamma+1}{\gamma-1}} \left[1 - \left(\frac{p_e}{p_c} \right)^{\frac{\gamma-1}{\gamma}} \right] \right\}^{\frac{1}{2}} + (p_e - p_\infty) A_e \quad (2\text{-}17)$$

From this formula, it is clear that the thrust does not depend at all on the combustion temperature T_c, but depends mainly on the dimensions of the nozzle A_e and A_{th}, and on the chamber pressure p_c.

An important performance parameter, the rocket thrust coefficient C_F, can now be deduced. The defining equation is

$$C_F \equiv \frac{F}{p_c A_{th}} \quad (2\text{-}18)$$

From Eq. 2-17 C_F can be evaluated:

$$C_F = \left\{ \frac{2\gamma^2}{\gamma - 1} \left(\frac{2}{\gamma + 1} \right)^{\frac{\gamma+1}{\gamma-1}} \left[1 - \left(\frac{p_e}{p_c} \right)^{\frac{\gamma-1}{\gamma}} \right] \right\}^{\frac{1}{2}} + \left(\frac{p_e}{p_c} - \frac{p_\infty}{p_c} \right) \epsilon \quad (2\text{-}19)$$

Since p_e/p_c is a function of ϵ according to Eq. 2-15, C_F depends only on the three independent variables γ, p_c/p_∞, and ϵ. Graphs of this function are presented in Fig. G,2e and G,2f for $\gamma = 1.2$ and $\gamma = 1.3$, respectively.

There are several features of these curves that deserve attention. First, each curve shows a maximum value of C_F at a certain area ratio which may, for this purpose, be called ϵ_{opt}. It can be shown analytically, by differentiating Eq. 2-19 and setting the derivative equal to zero, that the peak value occurs for a value of ϵ such that $p_e = p_\infty$, that is, for a

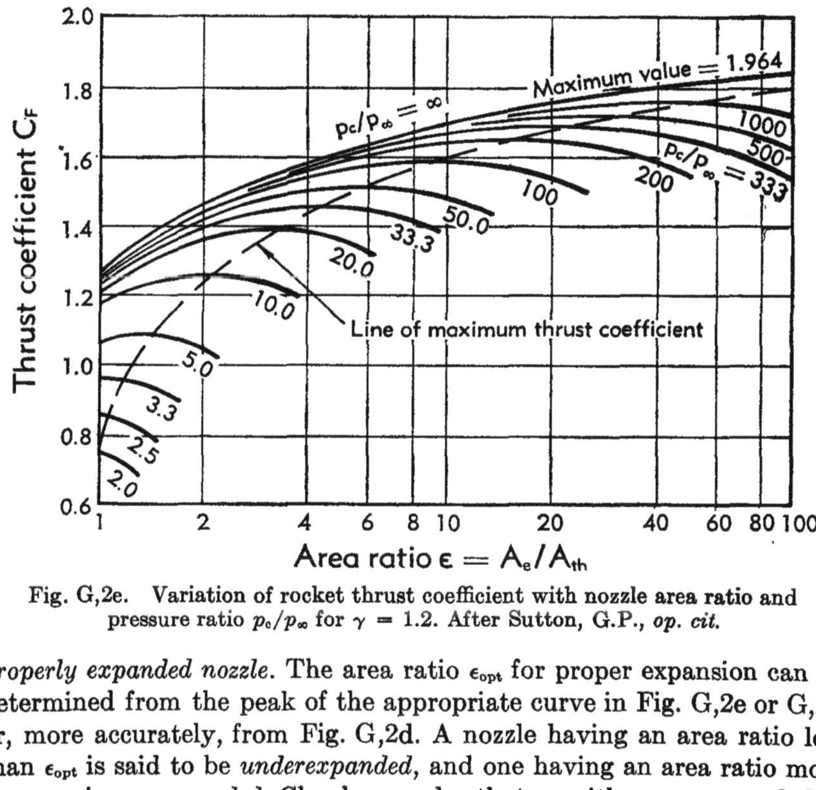

Fig. G,2e. Variation of rocket thrust coefficient with nozzle area ratio and pressure ratio p_c/p_∞ for $\gamma = 1.2$. After Sutton, G.P., *op. cit.*

properly expanded nozzle. The area ratio ϵ_{opt} for proper expansion can be determined from the peak of the appropriate curve in Fig. G,2e or G,2f, or, more accurately, from Fig. G,2d. A nozzle having an area ratio less than ϵ_{opt} is said to be *underexpanded*, and one having an area ratio more than ϵ_{opt} is *overexpanded*. Clearly, nozzles that are either overexpanded or underexpanded produce less thrust than a properly expanded nozzle. This conclusion can be proved in another way, by considering the distribution of pressure on the inner and outer surfaces of the rocket motor, as shown in Fig. G,2g. Downstream of the section indicated as ϵ_{opt}, the internal pressure is less than the external pressure, so that this portion of the cone acts to produce a force opposed to the thrust of the rocket motor as a whole. Therefore it is better to dispense with it and to terminate the nozzle at ϵ_{opt}. Consequently the highest thrust is produced with a properly expanded nozzle.

Examination of the pressure distribution pictured in Fig. G,2f shows that C_F must be somewhat greater than unity, except for the unusual case of low chamber pressure ($p_c/p_0 \cong 1$) and for possibly greatly overexpanded nozzles. Viewed in the simplest way, the rocket motor is a pressurized vessel with a hole of area A_{th} in the aft wall, and so it would be

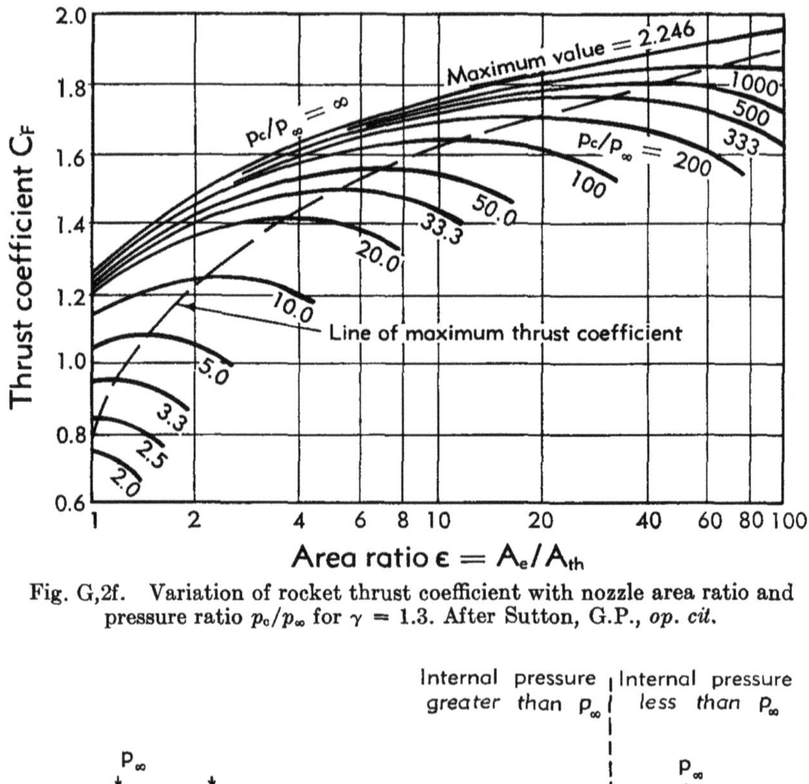

Fig. G,2f. Variation of rocket thrust coefficient with nozzle area ratio and pressure ratio p_c/p_∞ for $\gamma = 1.3$. After Sutton, G.P., *op. cit.*

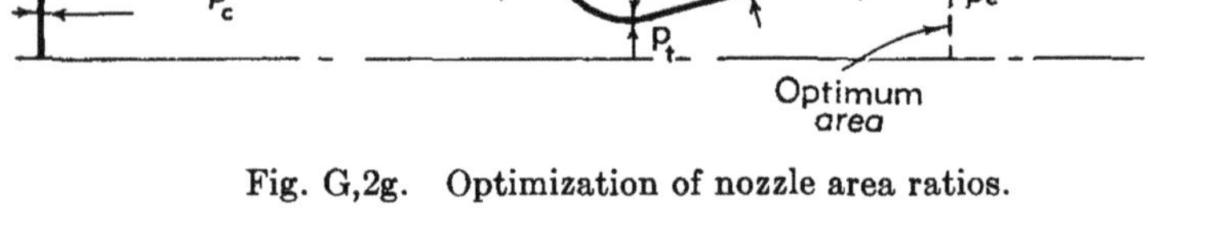

Fig. G,2g. Optimization of nozzle area ratios.

acted upon by an unbalanced force $(p_c - p_0)A_{th}$. To this must be added the additional unbalance due to the pressure reduction in the entrance cone of the nozzle. Then, if $p_0 \ll p_c$, the ratio F/p_cA_{th} must be somewhat larger than 1.

The maximum values of C_F on the families of curves in Fig. G,2e and G,2f can be connected by a smooth curve obtained by setting $p_e = p_\infty$

in Eq. 2-19.

$$(C_F)_{\max} = \left\{\frac{2\gamma^2}{\gamma - 1}\left(\frac{2}{\gamma + 1}\right)^{\frac{\gamma+1}{\gamma-1}}\left[1 - \left(\frac{p_\infty}{p_c}\right)^{\frac{\gamma-1}{\gamma}}\right]\right\}^{\frac{1}{2}} \tag{2-20}$$

The dependence on (p_∞/p_c) can be replaced by dependence on ϵ_{opt} through the relation (Eq. 2-15). The curve $(C_F)_{\max}$ vs. ϵ_{opt} itself reaches an ultimate value for infinite expansion. Thus, if (p_∞/p_c) is set equal to zero in Eq. 2-20,

$$(C_F)_{ult} = \left[\frac{2\gamma^2}{\gamma - 1}\left(\frac{2}{\gamma + 1}\right)^{\frac{\gamma+1}{\gamma-1}}\right]^{\frac{1}{2}} \tag{2-21}$$

For example, $(C_F)_{ult} = 2.246$ for $\gamma = 1.2$.

Inspection of Fig. G,2e and G,2f discloses that, for a prescribed nozzle area ratio ϵ, the thrust coefficient increases monotonically as p_c/p_∞ increases. This becomes clear when Eq. 2-19 is written in terms of the thrust coefficient for a given nozzle when it is operating in a vacuum:

$$C_F = (C_F)_{vac} - \epsilon\frac{p_\infty}{p_c} \tag{2-22}$$

Thus, for given ϵ and p_c, the increase in thrust with increasing p_c/p_∞ stems entirely from the reduction of the pressure acting on the external surfaces of the rocket motor.

It is significant that C_F is completely independent of combustion temperature T_c and of molecular weight $\mathfrak{M}$. Consequently, as a figure of merit, it is insensitive to the efficiency of combustion, but it is sensitive to the quality of the exhaust nozzle. In practice, the test engineer compares the measured C_F, computed from p_c, A_{th}, and F by means of Eq. 2-18, with the theoretical C_F computed from Eq. 2-19 to determine whether the nozzle is functioning properly, and in this way he can localize to some extent the cause of an unexpected defect in specific impulse. The other possible area for loss is in the combustion process. To detect combustion inefficiency, the performance parameter c^* is useful.

The characteristic velocity c^* is defined as follows:

$$c^* \equiv \frac{p_c A_{th}}{\dot{m}} \tag{2-23}$$

It follows immediately from Eq. 2-2 and 2-18 that

$$c = C_F c^* \tag{2-23a}$$

A theoretical expression for c^* is obtainable from Eq. 2-14.

$$c^* = \left[\frac{1}{\gamma}\left(\frac{\gamma + 1}{2}\right)^{\frac{\gamma+1}{\gamma-1}}\frac{RT_c}{\mathfrak{M}}\right]^{\frac{1}{2}} \tag{2-24}$$

From this formula it appears that c^* depends mainly on conditions in the combustion chamber, that is, on flame temperature and combustion product composition. Consequently, just as C_F is used as an index of the quality of the exhaust nozzle, so c^* is used in practice as an index of the efficiency of combustion. The test engineer determines c^* from measured values of p_c, A_{th}, $\dot{m}$, and compares it with the theoretical value (Eq. 2-24). In this way, a defect in specific impulse can be traced to a possible loss in the combustion process. Although the performance of a rocket motor is adequately described by the exhaust velocity c, which requires only the measurement of F and $\dot{m}$, it is the usual practice to measure at the same time p_c and A_{th} in order to compute C_F and c^* for diagnostic purposes.

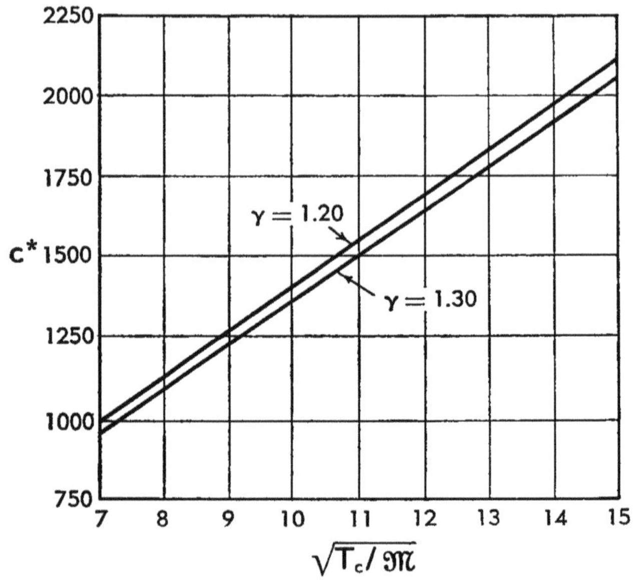

Fig. G,2h. Variation of characteristic velocity c^* with $\sqrt{T_c/\mathfrak{M}}$.

(Careful consideration of the flow process reveals that there exists some slight cross-dependence of C_F and c^*, that is, the former is slightly sensitive to the combustion process and the latter is somewhat affected by the flow conditions in the nozzle, but this is usually ignored.)

Curves of c^* vs. $(T_c/\mathfrak{M})^{\frac{1}{2}}$ for $\gamma = 1.2$ and $\gamma = 1.3$ are plotted in Fig. G,2h. It is significant that c^*, and therefore the specific impulse, depends as much on molecular weight $\mathfrak{M}$ as on flame temperature. Thus, it is just as important for the product gas to have a low mean molecular weight as a high temperature. This point will arise later when actual propellants are considered. It will be pointed out then that the optimum fuel-oxidizer mixture ratio is not necessarily the one that produces the highest flame temperature, and that a particular propellant combination may be very hot but no better than another much cooler one from the standpoint of specific impulse.

The magnitude of c^* is of interest. Because of its dependence on $(RT_c/\mathfrak{M})^{\frac{1}{2}}$, it can be compared directly with the velocity at the nozzle throat. Thus, for $\gamma = 1.2$, $c^* = 1.5V_{th}$. In general, for a properly designed nozzle, $V_{th} < c^* < V_e$, so that c^* equals the gas velocity at some station in the divergent part of the exhaust nozzle.

The efficiency of the rocket engine can be discussed in terms of the concepts of the ideal rocket motor. Five efficiencies deserve discussion here: combustion efficiency, expansion or "cycle" efficiency, nozzle efficiency, thermal efficiency, and total efficiency. In addition, the attempts to define a so-called propulsive efficiency will be examined.

The *combustion efficiency* η_c is defined as the ratio of the actual enthalpy released by combustion to the ideal enthalpy that would be released if the reaction were to go to completion.

$$\eta_c = \frac{(\Delta h_c + h_i)_{actual}}{(\Delta h_c + h_i)_{ideal}} = \frac{(T_c)_{actual}}{(T_c)_{ideal}} \tag{2-25}$$

The *ideal expansion or cycle efficiency* η_i expresses the fraction of the enthalpy available in the combustion chamber that can ideally be converted to kinetic energy in the exhaust jet. Let h_i represent the enthalpy of the product gas at the injection temperature with reference to 0°K.

$$\eta_i = \frac{(\frac{1}{2}V_e^2)_{ideal}}{\Delta h_c + h_i} = \frac{(\frac{1}{2}V_e^2)_{ideal}}{c_pT_c} \tag{2-26}$$

From Eq. 2-10,

$$\eta_i = 1 - \left(\frac{p_\infty}{p_c}\right)^{\frac{\gamma-1}{\gamma}} \tag{2-27}$$

The *nozzle efficiency* η_n is defined as the ratio of the actual kinetic energy in the exhaust jet to that which could be produced ideally at the specified pressure ratio. (Compare with diffuser efficiency in ramjets, Sec. E.)

$$\eta_n = \frac{(\frac{1}{2}V_e^2)_{actual}}{(\frac{1}{2}V_e^2)_{ideal}} = \frac{(T_c - T_e)_{actual}}{(T_c - T_e)_{ideal}} \tag{2-28}$$

The *thermal efficiency* η_{th} is the ratio of actual kinetic energy in the exhaust jet to the total enthalpy that could ideally be produced by the combustion reaction.

$$\eta_{th} = \frac{(\frac{1}{2}V_e^2)_{actual}}{(\Delta h_c)_{ideal} + h_i} \tag{2-29}$$

Conversion losses in the combustion process and in the expansion process, and the enthalpy discarded in the hot exhaust jet, are all represented in the thermal efficiency.

$$\eta_{th} = \eta_c\eta_n\eta_i = \frac{(V_e^2)_{actual}}{(V_e^2)_{max}} \tag{2-30}$$

The four efficiencies mentioned so far refer to the rocket motor simply as a heat engine and do not involve the energy quantities connected with flight. When flight is considered, it is possible to define an *over-all* or a *total efficiency* η_0 as the ratio of propulsive power (output) to the rate of consumption of energy in the rocket motor (input). The output is simply $V_\infty V_e$ per unit mass flow of propellant, where V_∞ is the flight speed. The input is the sum of $(\Delta h_c + h_i)$ plus the kinetic energy $\frac{1}{2}V_\infty^2$ of the propellant. The latter quantity results from the recognition that a quantity of propellant possesses more total energy when it is in motion than when at rest.

$$\begin{aligned}\eta_0 &= \frac{V_\infty V_e}{(\Delta h_c + h_i)_{\text{ideal}} + \frac{1}{2}V_\infty^2} \\ &= \sqrt{\eta_{\text{th}}}\left[\frac{2V_\infty (V_e)_{\max}}{(V_e^2)_{\max} + V_\infty^2}\right]\end{aligned} \tag{2-31}$$

The maximum value of η_0 is $\sqrt{\eta_{\text{th}}}$, and this value is reached when the flight velocity equals the maximum theoretical exhaust velocity. (This result has led to statements in the early literature that a rocket's flight speed could not exceed the maximum theoretical exhaust velocity. Clearly, this conclusion is not justified.)

There have been several attempts in the past to define a suitable "propulsive efficiency" by analogy with the corresponding case of a propeller-driven airplane, but these attempts have always failed because of the lack of a logical definition of mechanical input. If the analogy is to be carried through, the so-called propulsive efficiency will have to satisfy three requirements: (1) it should be a ratio of a mechanical output to a mechanical input; (2) the product of the propulsive efficiency and the thermal efficiency η_{th} should be equal to the total efficiency η_0; and (3) it must under no flight condition exceed unity.

The definition of propulsive efficiency that has been most prominent in the literature is

$$\eta_p = \frac{\text{output}}{\text{input}} = \frac{\text{output}}{\text{output} + \text{loss}} = \frac{V_\infty V_e}{V_\infty V_e + \frac{1}{2}(V_e - V_\infty)^2} = \frac{2V_\infty V_e}{(V_\infty^2 + V_e^2)}$$

The denominator is supposed to represent the sum of the propulsive work and the absolute kinetic energy of the jet (a loss). The present author objects to this propulsive efficiency on the grounds that it offers no clear definition of mechanical input and that, even if this particular definition were allowed, it fails to meet the requirement that $\eta_p \eta_{\text{th}} = \eta_0$.[1]

As a conclusion to this discussion of efficiency, it may be remarked that the concept of efficiency has not been at all useful in the field of rocket engines. Rockets are always compared on the basis of I_{sp} or c^* values, or on specific propellant consumption, and rarely is efficiency

[1] Editor's note: The author would find the same difficulty with any engine whenever the kinetic energy of the fuel is included.

mentioned. This is logical since flight performance depends directly on I_{sp}, and only indirectly on η.

Consideration of the thermal efficiency leads to the idea of a heat balance for the rocket motor. The chemical energy introduced with the injected propellants is distributed in four directions: (1) a small part (up to 15 per cent) remains unconverted due to incomplete reaction; (2) about 1 per cent, more or less, of the heat reaction is transferred to the motor walls where it may be lost unless the motor is cooled by the liquid propellant (regenerative cooling); (3) the kinetic energy of the jet comprises a large part, from one third to two thirds, of the heat of reaction; and (4) finally, the remainder of the reaction heat is carried away in the hot jet as thermal energy.

Typical values of the performance parameters of rocket motors in the moderate and high performance classes, respectively, are indicated in Table G,2. It is notable that the flame temperatures range from 2000

Table G,2. Typical performance characteristics.

Performance parameter	Ordinary range	High range
T_c	2000–3000°K	3000–5000°K
c^*	4000–5500 ft/sec	5000–8000 ft/sec
C_F	1.3–1.5	1.5–1.6
I_{sp}	200–270 lb-sec/lb	270–400 lb-sec/lb
$\overline{\mathfrak{M}}$	20–25	8–20
$\bar{\gamma}$	1.15–1.25	1.15–1.20

to 5000°K, and that dissociation is prominent in most of the range. The effects of dissociation constitute the greatest source of inaccuracy in the application of the ideal analysis just presented to actual rocket processes. The modification of the ideal analysis to handle dissociation will be discussed in Art. 4. Modifications due to other departures from the ideal conditions of this analysis are discussed in the following article.

G,3. Departures from Ideal Performance. The preceding ideal performance analysis requires correction to take care of the following actual conditions:

a. Conical divergence of the exhaust jet.
b. Surface friction and flow disturbances in the exhaust nozzle.
c. Constriction of the exit area due to boundary layer build-up.
d. Jet detachment.
e. Heat loss from the hot gas to the cold motor walls.
f. Suspended liquid or solid particles in the exhaust jet.
g. Pressure drop in the combustion chamber due to heat release.

Consideration of the effects of temperature dependence of the specific heats and of chemical reaction during the expansion in the nozzle will be postponed until Art. 4.

The effect of exhaust nozzle divergence. A loss in thrust occurs as a result of the divergence of the exhaust jet as it leaves the nozzle, in comparison with the ideal case of parallel flow. The loss may be viewed as the decrease in the axial component of momentum due to the outward inclination of the streamlines. Inasmuch as the loss is usually only a few per cent, a fairly approximate theory is adequate to determine its magnitude.

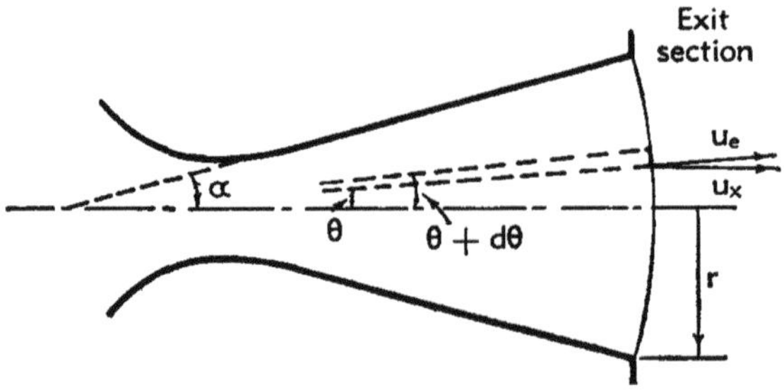

Fig. G,3a. Radical flow in exhaust nozzle.

If the exit-to-throat area ratio ϵ is sufficiently large, say at least 3, and if the half angle α of the exit cone is sufficiently small, say not more than 20°, it is reasonable to assume a radial flow pattern in the exhaust cone with concentric spherical caps for isobaric surfaces. The exit cone defines the location of the effective center for this radial flow pattern. The axis of the nozzle corresponds to $\theta = 0$, and the surface of the exit cone corresponds to $\theta = \alpha$ (Fig. G,3a).

To calculate the thrust, divide the exit flow area into differential annular areas between θ and $\theta + d\theta$. The axial component of the thrust of this element of flow is

$$dF = \rho_e V_e(2\pi r \sin\theta \cdot r d\theta)(V_e \cos\theta) + (p_e - p_\infty)(2\pi r \sin\theta \cdot r d\theta) \cos\theta$$

Integrating with respect to θ from 0 to α, F becomes

$$F_\alpha = \left(\frac{1 + \cos\alpha}{2}\right)[\dot{m}V_e + (p_e - p_\infty)A'_e] \tag{3-1}$$

In this formula, V_e, p_e, and A'_e come out in the analysis to be the velocity, pressure, and area of the *spherical isobaric cap* at the exit, and not the planar area at the exit. The quantity in brackets is the thrust of a zero divergence motor whose exit area equals A'_e. However, for small angle α, the planar exit area A_e is so close to A'_e that the thrust of the zero divergence motor would be almost unaffected if A'_e were replaced by A_e, and V_e and p_e were interpreted to apply to A_e instead of

A'_e. Therefore, if F_0 is the thrust of a zero divergence motor having an exit area ratio ϵ equal to the *planar* exit area ratio of the α divergence motor,

$$F_\alpha = \lambda F_0, \quad \text{where} \quad \lambda = \tfrac{1}{2}(1 + \cos \alpha) \tag{3-2}$$

For $\alpha = 10°$, $\lambda = 0.992$, and for $\alpha = 20°$, $\lambda = 0.970$. Consequently, the correction is a small one, and the preceding simplified model for the flow field is justified. To measure the loss accurately in order to check the theory requires experiments of high precision. Some experimental results have been reported, which show that the formula (Eq. 3-2) holds reasonably well from $\alpha = 10°$ to $\alpha = 30°$. For smaller angles, the nozzles are so long that frictional losses dominate the results, and for angles in excess of 40° there is clear evidence of flow separation.

The most commonly employed divergence angle in practice is about 15°. This produces a loss of only 1.7 per cent, yet the nozzle is not so long as to be excessive in either weight or length. It is possible to design an exit section with a curved contour so as to produce a parallel jet and thereby recover this divergence loss, but the penalties of complexity and cost must then be accepted. For a design to produce the utmost in performance, e.g., a satellite launcher, this is worthwhile.

The effects of fluid friction and flow losses. It is not possible to calculate reliably the thrust loss due to surface friction, because so little is known of the nature of the boundary layer under the conditions of pressure gradient and curvature that exist in the nozzle. Also, the flow field in a practical nozzle is likely to exhibit strong pressure waves originating in the sharply curved region near the throat. The rocket designer ordinarily cannot eliminate such sharp curvature, as can the supersonic wind tunnel designer, because rocket nozzles must be made cheaply (hence simply) and must be compact. Consequently, the flow disturbances are accepted as long as the loss in thrust is small. Fortunately, it seems that drastically poor design is necessary to cause a thrust loss greater than, say 10 per cent, due to bad flow, and if the designer merely uses his French curves with good technique, the loss can be cut to about 1 or 2 per cent.

For example, if the nozzle contour consists of a converging cone of 30° half angle, a diverging cone of 15° half angle, and a throat section with a longitudinal radius of curvature equal to twice the radius of curvature of the throat section, the loss will ordinarily not exceed 3 per cent and will probably be nearer to 1 per cent if there are no sharp corners.

In test results, this type of loss is reported in the form of a discharge coefficient C_d that normally lies between 0.97 and 0.99.

$$(C_F)_{\text{actual}} = C_d \lambda [(C_F)_{\alpha=0}]_{\text{ideal}} \tag{3-3}$$

Constriction of the exit area by the boundary layer. The effect of the boundary layer in constricting the flow passage of the nozzle can be considered in two places, at the throat and at the exit. At the throat, a boundary layer would reduce the mass flow $\dot{m}$; at the exit, it would reduce the effective expansion by reducing ϵ. It turns out that the first effect is negligible and the second one is small.

Observations of boundary layer growth in supersonic wind tunnels indicate that, in the Reynolds number range from 5×10^5 to 5×10^6, and in the Mach number range from 1.5 to 2.5, the boundary layer is turbulent and the displacement thickness grows in proportion to the distance from the throat. This Reynolds number is based on the velocity and kinematic viscosity at the particular station and the distance from the throat to this station. Although the strong cooling of the boundary layer and the much larger pressure gradient in the rocket nozzle would suggest caution in applying wind tunnel results to rocket nozzles, there are some observations in cold flow through rocket-type nozzles which indicate similar behavior. (This conclusion will be utilized in Art. 5 in the discussion of convective heat transfer in rocket nozzles.)

$$\delta^* \text{ (displacement thickness)} \cong 0.004 l_n \text{ (distance from throat)} \qquad (3\text{-}4)$$

With this relation it turns out that the exit area constriction for a 15° half angle nozzle is only 3 per cent. Inspection of Fig. G,2e and G,2f shows that the effect of a 3 per cent area reduction can be neglected if the nozzle area ratio is near the optimum. For narrower exit cones and for off-design nozzles, it may be necessary to allow for this effect.

This is another instance of the remarkable insensitivity of rocket nozzle performance to flow disturbances that are of great importance in wind tunnel nozzle design.

Jet detachment or flow separation. It is not always possible to provide a nozzle with the optimum area ratio for each condition of ambient pressure. In particular, the exhaust nozzle of a rocket airplane that must fly at altitudes from sea level to perhaps 60,000 feet, or of a guided missile that starts at sea level and reaches 100,000 at burnout, generally operates in a very much overexpanded condition at the lower altitudes. The question arises whether the flow in the divergent section remains isentropic.

The possibility of a stationary shock wave perpendicular to the flow axis has been recognized for a long time, and the early literature on rocket performance indicates that this is to be expected. However, although such normal shocks have been observed in supersonic flow channels with small angles of divergence, normal shocks have not been found at all in rocket nozzles. Instead, it is found that when the back pressure (pressure of the surrounding atmosphere) exceeds the isentropic exit pressure by a large enough amount, the flow separates symmetrically from the nozzle wall, and a complicated pattern of oblique shock waves appears in the stream beyond the point of separation (Plate G,3).

The fundamental fluid mechanics of this type of separation phenomenon is discussed in IV,B. In brief, the region of importance is the place near the wall where the shock wave penetrates the boundary layer. Here the sharp adverse pressure gradient connected with the shock thickens the boundary layer and leads to deflection of the flow, while the pressure rise across the shock is supported by the momentum transfer from the

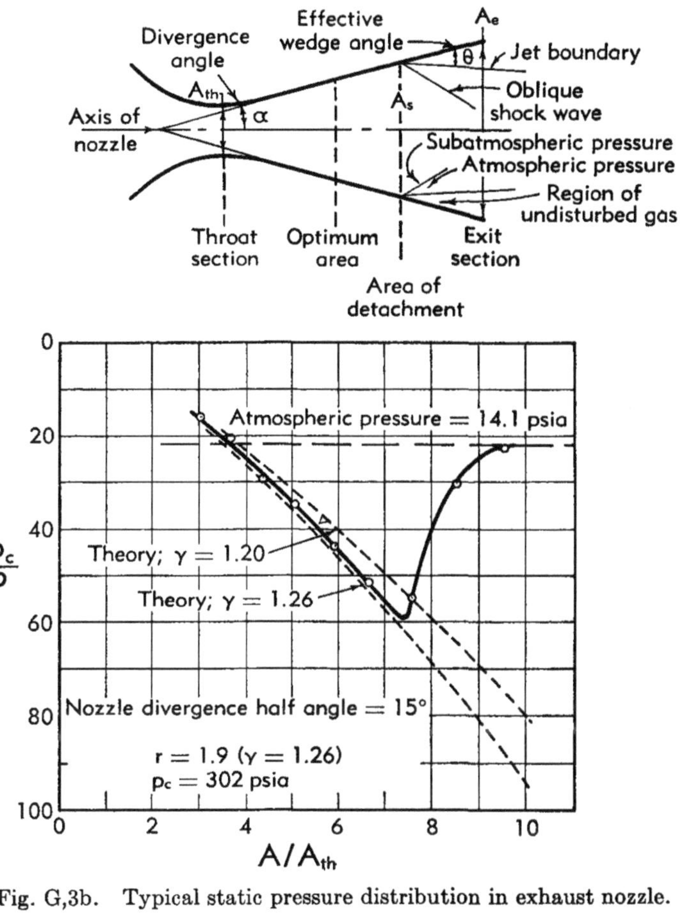

Fig. G,3b. Typical static pressure distribution in exhaust nozzle.

supersonic stream to the "stagnant" fluid in the flow corner. Consequently the strength of the shock and the separation location depend on the state of the boundary layer (laminar or turbulent) and particularly on its development along the nozzle wall prior to separation. For the purpose of the reader interested in rockets, the subject can be presented most usefully in terms of the experimental results and the empirical correlations.

The static pressure distribution in the nozzle is shown in Fig. G,3b. The pressure follows the isentropic expansion curve until the separation point is reached. At this point, the pressure rises steeply to almost the level of the ambient pressure at the exit, indicating a rather clean breakaway of the flow. From the standpoint of thrust, the nozzle may well be cut off at the separation station, since the internal and external pressures are very nearly in balance beyond this point. Experimentally, it makes little difference if a longer nozzle is used, once separation has occurred. The extra length of nozzle does not affect the location of separation and has no appreciable effect on the thrust. (There is a small reduction in thrust due to the extra length of cone, because the internal pressure is slightly less than the external pressure, but this is usually neglected in performance estimates.)

It is obvious that the complicated flow structure downstream of separation is of no consequence in the computation of thrust. The thrust of a rocket motor with separation in the nozzle is calculated on the assumption that the nozzle area ratio is not that corresponding to the actual exit but rather that of the separation station. Therefore it is necessary to be able to predict the location of separation.

In tests with nitric acid-aniline rocket motors employing nozzles with half angles α of 10°, 15°, 20°, and 30°, with actual area ratios ϵ of 10 and 20, and with pressure ratios p_c/p_∞ from 11 to 25, it was found that

$$\frac{p_s}{p_\infty} \cong 0.36 \pm 0.02 \tag{3-5}$$

Almost identical results were obtained at all mixture ratios, and in fact, the results obtained with the nitrogen flow channel previously mentioned were also nearly the same. Consequently, as a general rule for estimating the performance of overexpanded nozzles in the absence of specific test data, the relation (Eq. 3-5) may be used.

Heat loss from the combustion gas. The transfer of heat from the combustion gas to the relatively cold motor walls, by convection and by radiation, affects the performance of the motor. Two cases of heat transfer are of practical interest: (a) the heat transferred is completely removed from the thermodynamic process, as in a water-cooled motor; and (b) the heat is reintroduced into the process as in a motor cooled by the propellants (regenerative cooling). These two processes, together with the adiabatic one, are shown in the h, s diagrams of Fig. G,3c.

In diagram 1, the case of no heat loss, the working fluid (product gas) is injected into the chamber cold at A, is heated adiabatically at constant pressure by combustion to point B, and is expanded isentropically to ambient pressure at C. The enthalpy drop from B to C is transformed to the kinetic energy of the jet (see Eq. 2-10). In diagram 2, the case of heat lost to the cooling water, it is assumed in order to simplify the calculations that the heat removal is concentrated at two stations, the hot end of the chamber and the throat of the exhaust nozzle. (This is reasonable on the basis of experimental heat transfer distributions.) The process is then as follows: adiabatic combustion from A to B, constant pressure heat loss from B to D, isentropic expansion from D to E, constant pressure heat loss from E to F, and isentropic expansion from F to G. The regenerative process, diagram 3, traces the following path: constant pressure heating from A to J, adiabatic combustion from J to K, heat loss from K to L, expansion from L to M, heat loss from M to N, and expansion from N to O.

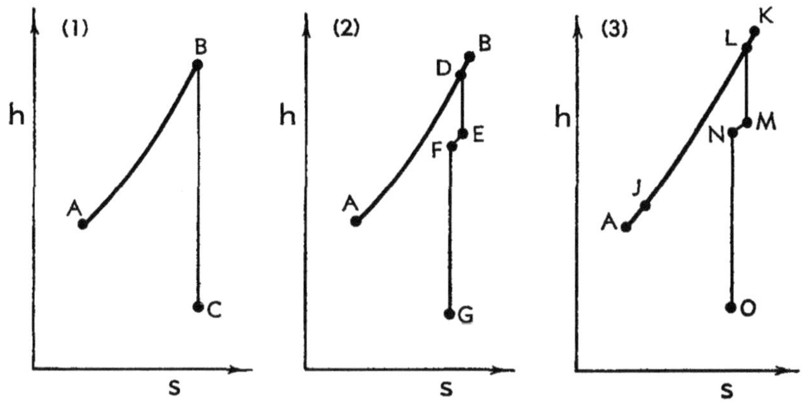

Fig. G,3c. Thermodynamic processes for (1) adiabatic, (2) water-cooled, and (3) regeneratively cooled rocket motors.

The exhaust velocities for the three diagrams are:

$$\left.\begin{aligned} \tfrac{1}{2}V_{e(1)}^2 &= h_B - h_C \\ \tfrac{1}{2}V_{e(2)}^2 &= (h_D - h_E) + (h_F - h_G) \\ \tfrac{1}{2}V_{e(3)}^2 &= (h_L - h_M) + (h_N - h_O) \end{aligned}\right\} \tag{3-6}$$

The heat of combustion: $\Delta h_C = h_B - h_A = h_K - h_J$ (3-7)

In cases 2 and 3, the heat transferred per unit mass is given by:

$$\left.\begin{aligned} \Delta q_2 &= (h_B - h_D) + (h_E - h_F) \\ \Delta q_3 &= (h_K - h_L) + (h_M - h_N) = (h_J - h_A) \end{aligned}\right\} \tag{3-8}$$

For a perfect-gas working fluid with constant specific heat, the ratio of the exhaust velocity of the cooled motor to that of the ideal motor can be given explicit form, in each case, in terms of the heat of combustion and the pressure ratio. In general, to calculate the correction with precision, it is necessary to refer to actual h,s computations on diagrams that incorporate the temperature-dependent specific heats and the effects of dissociation. It is immediately evident from the diagrams that the water-cooled motor produces a lower specific impulse than the ideal one. In practice, the approximate formula: $V_{e(2)} \cong V_{e(1)}[(\Delta h_c - \Delta q_2)/\Delta h_c]^{\frac{1}{2}}$ is applied to relate the two results. For small test motors, this can amount to as much as 5 per cent. It is also evident that the specific impulse of a regeneratively cooled motor is never less than that of the ideal one, and can *exceed* the ideal performance if the heat transfer $h_M - h_N$ is large in comparison with $h_K - h_L$. In practice, this effect is too small to be measured.

Suspended liquid or solid particles in the exhaust jet. Condensed phases can appear in the exhaust jet if elements are present in the reactants that form refractory products. For example, fuel compounds based on the light metals Li, Be, Al, and Mg produce oxides Li_2O, BeO, Al_2O_3, and MgO when reacted with oxygen. These oxides have normal vaporization temperatures of 2250, 3900, >1700, and >2800°C, respectively, and therefore these vapors would condense to droplets or solid particles whenever the temperature drops below these levels, either in the combustion or in the expansion process. Similarly, carbon (sublimation point >4200°C) which can form in large amounts in the decomposition of acetylenic monopropellants, and KCl (sublimation point, 1500°C) which forms in solid propellant combustion where potassium perchlorate is used as the oxidizer, present the problem of flow with condensed phases.

The occurrence of condensation leads to a reduction in specific impulse for two reasons. The portion of the working fluid that condenses cannot perform any expansion work and therefore cannot contribute to the acceleration of the jet; also, the heat in the condensed phase is partly ejected with the jet and not transformed to kinetic energy, because of the low rate of heat transfer from the hot particles to the surrounding gas. The rate of heat transfer depends on the particle size, which in turn depends on whether condensation occurs in the chamber or at some position in the nozzle. Average particle sizes have been observed in the range from 10μ to 50μ.

To analyze the problem in the most exact way, it is necessary to know at what place in the expansion process nucleation occurs, the rate of growth of the condensed nuclei, the rate of heat transfer from the hot particles to the cooler gas stream, and the velocity lag between the suspended particles and the gas stream. For practical purposes, it is usually adequate to simplify the analysis by assuming that the velocity lag is zero. Concerning heat transfer, two extreme assumptions are possible: (1) the temperature of the condensed phase remains the same throughout the expansion process as that of the gas where condensation took place, that is, zero heat transfer; and (2) the condensed phase temperature follows exactly the temperature history of the gas, that is, perfect heat exchange. The first assumption would apply to very large particles; the second assumption would apply to very small particles.

The equation of state of the mixture of gas and suspended particles, and the density of the mixture ρ, are definable in terms of the mean molecular weight $\mathfrak{M}_g$ of the gas, the molecular weight of the solid $\mathfrak{M}_s$, the mean molecular weight of the mixture $\overline{\mathfrak{M}}$, and the respective numbers of moles of gas and solid, n_g and n_s, in a standard molar volume $\mathcal{V}_0$.

$$\left.\begin{aligned} p &= \frac{n_g}{\mathcal{V}_0} RT_g = p_g \\ \rho &= \frac{n_s\mathfrak{M}_s + n_g\mathfrak{M}_g}{\mathcal{V}_0} \end{aligned}\right\} \tag{3-9}$$

For adiabatic, frictionless flow, $dh = (1/\rho)dp$. For the mixture,

$$\frac{n_g C_{p_g} dT_g + n_s C_{p_s} dT_s}{n_g\mathfrak{M}_g + n_s\mathfrak{M}_s} = \frac{n_g RT_g}{n_g\mathfrak{M}_g + n_s\mathfrak{M}_s} \frac{dp_g}{p_g} \tag{3-10}$$

Introducing mole fractions X_g and X_s, this equation becomes

$$X_g C_{p_g} \frac{dT_g}{T_g} + X_s C_{p_s} \frac{dT_s}{T_s} = X_g R \frac{dp_g}{p_g} \tag{3-11}$$

For adiabatic flow, $VdV + dh = 0$. For the mixture, with zero velocity lag,

$$\tfrac{1}{2}(X_g\mathfrak{M}_g + X_s\mathfrak{M}_s)VdV + X_g C_{p_g} dT_g + X_s C_{p_s} dT_s = 0 \tag{3-12}$$

Eq. 3-11 and 3-12 are applicable upstream of the condensation location, if X_g is set equal to unity, and X_s equal to zero. Downstream of this location, these equations can be applied if it is assumed that the mole fractions X_g and X_s are fixed. This is a reasonable assumption since most of the condensation takes place in a very short region.

A satisfactory solution of the problem is to obtain V_e in terms of the pressure ratio p_e/p_c. This can be obtained by integrating Eq. 3-11 and 3-12 if the ratio dT_s/dT_g can be specified. For case 1 mentioned above, the ratio is zero; for case 2, the ratio is unity; and other cases can be treated. The most extreme case is to assume that dT_s/dT_g is zero and that condensation occurs in the combustion chamber, so that the particles are at temperature T_c at the exit. This leads to:

$$V_e = \left\{\frac{2X_g C_{p_g} T_c}{\overline{\mathfrak{M}}}\left[1 - \left(\frac{p_e}{p_c}\right)^{\frac{\gamma-1}{\gamma}}\right]\right\}^{\frac{1}{2}} \tag{3-13}$$

$$\overline{\mathfrak{M}} = X_g\mathfrak{M}_g + X_s\mathfrak{M}_s$$

$$\gamma = C_{p_g}/C_{v_g}$$

For a typical propellant gas composition, based on hydrocarbon combustion, the condensation of 20 per cent of the mass in the combustion chamber leads to about 10 per cent loss in I_{sp}, by formula (Eq. 3-13). The effect can therefore be of considerable importance for some propellant systems.

Pressure drop in the combustor due to heat release. The analysis of the ideal rocket motor is based on the postulate that combustion takes place at constant pressure. In practice, the release of heat in the gas stream in the combustor is accompanied by a decrease of both static and stagnation pressures. The decrease is dependent on the Mach number at the hot end of the combustion chamber, and becomes larger as the ratio of combustor cross-sectional area A_c to throat area A_{th} approaches unity. The loss in stagnation pressure represents an undesirable increase in entropy, a decrease in the effective pressure ratio for expansion, and hence a loss in specific impulse. The following analysis deals with these effects.

As before, the subscripts i, c, th, and *e* refer to states at the injector end of the combustor, the hot end of the combustor, the nozzle throat, and the nozzle exit. Except for the nonzero velocity in the combustor, the conditions of the analysis are the same as for the ideal rocket motor.

Since the expansion in the nozzle is isentropic (heat release completed in the chamber) the Mach number at c is obtainable from Eq. 2-10:

$$\frac{A_c}{A_{th}} = \frac{1}{M_c}\left(\frac{1 + \frac{\gamma - 1}{2} M_c^2}{\frac{\gamma + 1}{2}}\right)^{\frac{\gamma+1}{2(\gamma-1)}} \tag{3-14}$$

The integrated momentum equation, for frictionless flow in a constant area duct, provides an expression for the decrease in static pressure in the chamber. It is assumed that $M_i \cong 0$.

$$p_i = p_c(1 + \gamma M_c^2) \tag{3-15}$$

The simplest way to treat the performance for the case of nonzero M_c is first to calculate the equivalent plenum chamber pressure and temperature corresponding to the conditions at c, and then to calculate the flow properties in the nozzle on the assumption that the gas flow originates from this plenum chamber. The temperature T_c^0 and pressure p_c^0 of the equivalent plenum chamber are really the stagnation temperature and pressure at station c. (See III,A for discussion of stagnation quantities.)

$$p_c^0 = p_c\left(1 + \frac{\gamma - 1}{2} M_c^2\right)^{\frac{\gamma}{\gamma-1}} \tag{3-16}$$

$$T_c^0 = T_c\left(1 + \frac{\gamma - 1}{2} M_c^2\right) \tag{3-17}$$

$$\Delta h_c = c_p(T_c^0 - T_i) \tag{3-18}$$

Eq. 3-18 is simply a statement of the conservation of energy, with the kinetic energy term $\frac{1}{2}V_c^2$ absorbed in the stagnation enthalpy $c_pT_c^0$.

Therefore, T_c^0 is the same as T_c in the ideal rocket motor of Eq. 2-7, that is, the so-called adiabatic flame temperature, or simply the chamber temperature. The static temperature T_c is less than the adiabatic flame temperature, by as much as 10 or 15 per cent for a throatless motor, but this is of no consequence for the exhaust velocity. Only the stagnation temperature matters, and this depends only on the thermochemical heat of reaction Δh_c.

Fig. G,3d compares the conditions of the equivalent plenum chamber with those of the actual chamber by presenting the ratios p_c/p_i, p_c^0/p_i,

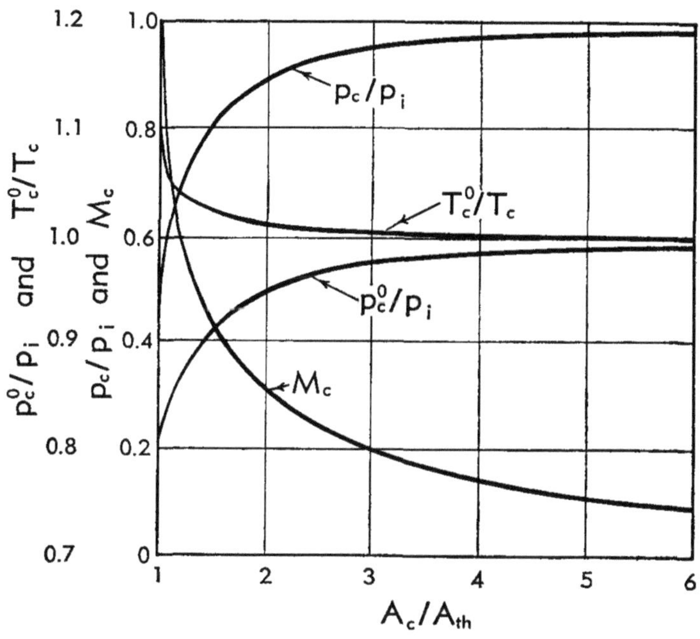

Fig. G,3d. Comparison of conditions in equivalent plenum chamber with those in actual combustion chamber.

M_c, and T_c/T_c^0 as functions of A_c/A_{th}. This figure is useful as a guide in carrying out the following computational procedure recommended for the performance analysis of a rocket motor.

In general, the particular propellant combination, the feed pressure, the nozzle throat area, the external pressure, and the chamber dimensions are specified in advance. The problem is to compute the effective exhaust velocity c, taking into account the effect of the pressure drop in the combustion chamber.

1. From the specified A_c/A_{th} and γ, compute M_c from Eq. 3-14.
2. With this M_c and the specified p_i, calculate p_c and then p_c^0 from Eq. 3-15 and 3-16.
3. From the specified ϵ of the nozzle, and the effective pressure ratio

p_c^0/p_∞, determine C_F^0 for the equivalent plenum chamber, from Fig. G,2d and G,2e.

4. The thrust is then calculated as follows:

$$F = C_F^0 p_c^0 A_{th} \tag{3-19}$$

5. The characteristic velocity c^* depends on the stagnation temperature, which is exactly equal to the adiabatic flame temperature. Therefore, $(c^*)^0$ is known, either by thermochemical calculation or by experiments with plenum-type combustors.
6. The mass flow is then calculated in terms of $(c^*)^0$:

$$\dot{m} = \frac{p_c^0 A_{th}}{(c^*)^0} \tag{3-20}$$

7. Finally, the effective exhaust velocity is calculated:

$$c = C_F^0 (c^*)^0 \tag{3-21}$$

In rocket testing, the chamber pressure is frequently measured at the forward end of the chamber, through a pressure tap drilled into the injector face. Clearly, this is p_i and not p_c^0. Unless the measured p_i is converted by Eq. 3-15 and 3-16 to p_c^0, the reported value of c^* will be too high by as much as 25 per cent (Fig. G,3b) for a throatless motor, and the reported value of C_F (experimental)/C_F (theoretical) will be low by a comparable error. These effects are significant for a rocket motor with a throat diameter equal to $\frac{1}{2}$ or more of the chamber diameter. If the combustor is not cylindrical, the preceding analysis must be modified. In particular, Eq. 3-15 holds only for cylindrical combustors; for other shapes, it is necessary to know the axial distribution of the heat release in order to integrate the momentum equation.

It may be noted that the same theory can be used to determine the axial distribution of heat release, i.e. the over-all kinetics of combustion. Thus, in a cylindrical chamber, the variation of M with axial position x is obtainable from the pressure distribution by means of Eq. 3-15. Then, from the equation of continuity (Eq. 3-22), the point-by-point static temperature can be computed, and from this, the stagnation temperature and the heat release.

$$\dot{m} = \frac{pA_c\mathfrak{M}}{[RT/\gamma M]^{\frac{1}{2}}} \tag{3-22}$$

G,4. Theoretical Specific Impulse Calculations. The calculation of specific impulse in the theory of the ideal rocket motor in Art. 2 is deficient in two very important respects: the combustion gas mixture is chemically reactive, and the specific heats vary strongly with temperature. Both these factors were ignored there. It is the purpose of this article to develop the complete theory, with both effects present. Inasmuch as the thermodynamics of combustion and the theory of expansion of reactive mixtures are covered in detail in II,A, B, and C, the theory is given here only in condensed form. The reader is referred to Vol. II for a more complete discussion.

The problem is the following one: for a given set of reactants (propellants) and a specified constant combustion pressure, calculate the adiabatic flame temperature and the composition of the product mixture, and then, for expansion to a specified exit pressure, calculate the specific impulse and the proper nozzle area ratio.

The analysis is carried out most conveniently in terms of the *total enthalpy* H of the reacting mixture and the *total entropy* S, both quantities being computed for a definite amount of the mixture.

$$H = \sum_i n_i[(H_T^0 - H_{T_0}^0) + \Delta H_{fT_0}^0]_i \qquad (4\text{-}1)$$

$$S = \sum_i n_i(S_{T,P}^0)_i = \sum_j n_j[S_{T,1}^0 - R \ln p_j]_j + \sum_k n_k(S_{T,1}^0)_k \qquad (4\text{-}2)$$

where $\Delta H_{fT_0}^0$ = molar enthalpy of formation of component i at reference temperature T_0 (usually 0°K or 298°K) in the ideal state.

$H_T^0 - H_{T_0}^0$ = molar thermal enthalpy or heat content at temperature T with respect to temperature T_0, also in the ideal state. Equivalent to

$$\int_{T_0}^{T} C_p^0 dT + \sum \Delta H_T^0$$

The latter sum refers to all the heats of transition between T_0 and T.

$S_{T,P}^0$ = absolute molar entropy (exclusive of nuclear spin) of component i at temperature T and pressure p, in the ideal state.

$S_{T,1}^0$ = value of $S_{T,P}$ at the standard pressure of one atmosphere.

p_j = partial pressure (in atmospheres) of gaseous component j, considered to be an ideal gas. $[p_j = \rho_j RT/\mathfrak{M}_j]$

n_j = number of moles of gaseous component j in the specified mixture.

n_k = number of moles of condensed component k (liquid or solid) in the mixture.

n_i = number of moles of any component (gas, liquid, or solid); $\sum n_i = \sum n_j + \sum n_k$.

The following physical conditions are implicit in Eq. 4-1 and 4-2. Each substance is in its ideal state corresponding to the particular temperature and partial pressure that prevails. This is indicated by the superscript 0. Thus, each gaseous component obeys the perfect gas law, and in particular, the specific heats are dependent only on temperature and not on pressure. For the condensed phases, the specific heat also depends only on the temperature. In ideal mixtures, there is no energy of mixing, so that the total enthalpy is simply the sum of the component enthalpies, as given by Eq. 4-1.

Similarly, the total entropy is simply the sum of the component entropies. The gaseous components, being perfect gases, contribute logarithmic pressure terms, as in Eq. 4-2. The liquid and solid components in their ideal states have entropies that are independent of the pressure, and therefore the terms appear in a separate sum in Eq. 4-2.

For the hot combustion products encountered in rockets, the assumption of the ideal state and ideal mixing is quite accurate up to pressures of 200 atmospheres. Above that, the gas density is 10 per cent or more of the average critical point density of the gas mixture, and the intermolecular forces begin to affect seriously the equations of state and the enthalpies. Therefore, for pressures of 200 atmospheres or more, the analysis described here needs modification. Details are given in II,A. For the condensed products, the ideal state assumption is generally good to even higher pressures. On the reactant side, the liquid propellants may be treated as ideal substances with entropy and enthalpy both independent of pressure. This is acceptable because the pressure effect on the enthalpy is usually quite small, while the effect on the entropy does not matter in the analysis.

The notations and units employed in this article for the treatment of combustion are those normally familiar to physical chemists rather than to mechanical engineers. The reason is simply one of convenience: the standard published collections of thermochemical data make use of the c.g.s. system of units, the Kelvin temperature scale, molar quantities, and the same symbols as those in Eq. 4-1 and 4-2. Another important difference from the mechanical engineering literature pertains to the zeros of the enthalpy and entropy scales. For example, in combustion engineering, there exists the convention of assigning zero to the enthalpy of formation of CO_2, H_2O, O_2, and N_2. In this article, the physical chemists' assignment of zero to all the elements in their standard states is used. Thus, in the combustion engineer's system, H_2 gas has a positive heat of formation, but zero in the system in use here. Finally, when in the treatment of the expansion process the mass flow and the jet velocity are to be calculated, the fps system of units is employed in this article in order to conform to current rocket practice.

As a first step, the combustion reaction is written down with all the products that are likely to be present in appreciable amounts. For example:

$$\underset{\text{(reactants)}}{n_{H_2}H_2 + n_{O_2}O_2} \rightarrow \underset{\text{(products)}}{n_{H_2O}H_2O + n_{OH}OH + n_{H_2}H_2 + n_HH + n_{O_2}O_2 + n_OO}$$

The weight ratio r of oxidizer to fuel is $n_{O_2}\mathfrak{M}_{O_2}/n_{H_2}\mathfrak{M}_{H_2}$.

The energy equation is

$$H_c \text{ (products)} = H_i \text{ (reactants)} - \Delta Q_{\text{loss}}$$

$$\sum_{\text{prod}} n_k[(H^0_{T_c} - H^0_{T_0}) + \Delta H^0_{fT_0}]_k = \sum_{\text{react}} n_j[(H^0_{T_i} - H^0_{T_0}) + \Delta H^0_{fT_0}]_j - \Delta Q_{\text{loss}} \tag{4-3}$$

The flame temperature T_c and the composition n_k's are unknown. The injection temperatures T_{ij} are specified, and so the entire right member of Eq. 4-3 can be computed. ΔQ_{loss} is a measured heat loss, if any.

Additional equations that make possible the determination of the n_k's and T_c are those governing the chemical equilibria at a specified pressure p_c:

(a) Mass conservation, one for each element in the system (linear equations). For example: $2n_{H_2} = 2n_{H_2O} + n_{OH} + 2n_{H_2} + n_H$

(b) Chemical equilibrium, one for each *independent* reaction (nonlinear equations). For example: $\dfrac{n_H n_{OH}}{n_{H_2O}} = K_p \dfrac{\sum n \text{ (products)}}{P_c}$

All together, there are enough equations to solve for the unknown composition and flame temperature. Because the chemical equilibria are nonlinear and because the energy equation is not explicit in T_c, numerical or graphical methods of solution are necessary. Specific procedures are discussed in II,A and C.

The next step is the analysis of the expansion process. If the expansion is adiabatic and nondissipative, then it is isentropic:

$$S_e \text{ (exhaust products)} = S_c \text{ (combustor products)}$$

Concerning the physical nature of the expansion process, several alternative assumptions are possible. First, if condensed phases are produced either in the combustor or in the nozzle, special analysis is required (see Art. 3). If there is no condensation, then it is necessary to decide whether chemical equilibrium is maintained during the entire expansion. If not, a satisfactory approximation would be to postulate full chemical equilibrium down to a certain temperature level and then frozen composition below that level. A more exact analysis would involve the explicit introduction of the chemical kinetic relationships in place of the chemical equilibrium equations. This problem is discussed in III,B. Finally, it is necessary to decide whether the internal degrees of freedom of the molecules can adjust their energies to the rapid expansion, that is, whether there is an appreciable vibrational energy lag. This is also discussed in III,B.

Both the chemical and vibrational lags are dependent on the expansion time. For motors of less than 50-lb thrust, nozzles less than 2 inches long, expansion times of 10^{-4} sec or less, the chemical lags might be appreciable, but for motors of 1000-lb thrust or more, it is likely that these effects are unimportant. The vibrational lags do not seem to occur at all, probably because of the high temperature level and the consequent rapid rates of adjustment. In general, the chemical lag is difficult to detect by its effect on specific impulse because the difference in I_{sp} is rarely more than a few per cent, even for very hot systems with considerable dissociation.

The most effective ways to detect a lag in radical recombination would be by the exit temperature or by the pressure distribution in the exhaust nozzle. However, both types of experiments have been tried and both leave much to be desired in the interpretation of the results. It seems the best practice, therefore, to compute both the constant composition performance and the equilibrium composition performance, and to regard them as lower and upper limits, respectively. Another possible method is to determine experimentally the area ratio that expands the gas to an exit pressure of exactly one atmosphere, this condition being observable from the exhaust shock wave pattern.

Expansion with complete chemical and vibrational equilibrium, and no condensation of solid or liquid substances, is governed by the following entropy equation and a set of equations governing the chemical equilibrium:

$$\sum_{\text{exhaust}} n_l \left[S^0_{T_e,1} - R \ln \left(\frac{p_e n_l}{\sum n_l} \right) \right]_l = \sum_{\text{combustor}} n_k \left[S^0_{T_c,1} - R \ln \left(\frac{p_c n_k}{\sum n_k} \right) \right]_k \tag{4-4}$$

The exhaust temperature T_e and the exhaust composition n_l's are unknown. The entire right member is known from the solution of the first part of the problem. The additional equations are, as before,

(a) Mass conservation, one for each element in the system (linear equations).
(b) Chemical equilibrium, one for each *independent* reaction (nonlinear equations).

As before, numerical or graphical methods are required to solve this system of equations for the n_l's and T_e (see III,B for details).

The exhaust jet velocity V_e is obtainable from the energy relation for the expansion, after evaluating H_e from the now-known exhaust composition and temperature.

$$\tfrac{1}{2}V_e^2 \sum_{\text{react}} n_j \mathfrak{M}_j = H_c\ (\text{combustor}) - H_e\ (\text{exit}) \tag{4-5}$$

The mean molecular weight, the gas density, and the area per unit mass flow, all at the exit, are given by

$$\overline{\mathfrak{M}}_e = \frac{\sum n_l \mathfrak{M}_l}{\sum n_l} \tag{4-6}$$

$$\rho_e = \frac{\overline{\mathfrak{M}}_e p_e}{RT_e} \tag{4-7}$$

$$\frac{A_e}{\dot{m}} = \frac{1}{\rho_e V_e} \tag{4-8}$$

To determine the proper nozzle area ratio, the method just outlined can be utilized. A series of pressures p is constructed, ranging from p_c down to p_e, and then Eq. 4-4, 4-5, 4-6, 4-7, and 4-8 are solved for each pressure. A curve is then plotted of $A/\dot{m}$ against p. The minimum of this curve corresponds to the throat of the de Laval nozzle and is designated as $A_{th}/\dot{m}$. The ratio of $A_e/\dot{m}$ determined in Eq. 4-8 to $A_{th}/\dot{m}$ constitutes the desired nozzle area ratio ϵ_{opt}.

This procedure produces, of course, more information than merely ϵ_{opt}. The theoretical value of c^* is obtained as the product $p_c(A_{th}/\dot{m})$, and the theoretical value of C_F is obtained as V_e/c^*. Also, the axial distribution of pressure along the nozzle is given by the curve just plotted.

The general scheme of the performance analysis can be summarized by tracing the process on a suitable Mollier diagram. To construct this diagram, a definite atomic composition is specified, that is, a definite number of gram-atoms of C, H, O, N, etc., respectively. These atoms are assumed to form components of the product mixture that are in chemical equilibrium at every point of the diagram. In principle, lines of constant pressure can be mapped in the following way: Specify a total enthalpy H as defined in Eq. 4-1 and a total pressure p, then solve for the total entropy S. This involves, of course, solving for the equilibrium composition and the temperature. Because the equilibrium constants depend on the temperature in a complicated way, it is easier to locate the isobars by starting with a given T and a given p, and then solving for H and S. It is to be noted that this Mollier diagram differs from that shown in Fig. G,3a for the analysis of the ideal rocket motor. In the latter diagram, the enthalpy included only the thermal enthalpy and not the chemical enthalpy, that is, the gas mixture was assumed to be nonreactive and the mixture composition was the same everywhere on the diagram. In Art. 3, the enthalpy of the propellant gas was raised from h_A to h_B by the heat of combustion (see Eq. 3-7); in the present analysis, because the chemical enthalpy is included, the total enthalpy of the reactants equals the total enthalpy of the products, except for heat transfer.

Once the Mollier diagram is constructed, the analysis proceeds as follows. By means of Eq. 4-1, the total enthalpy of the propellant combination is computed, and from this is subtracted the heat loss, if any. The resulting enthalpy is marked on the H scale. A horizontal line from this starting point (i) to the intersection (c) with the proper isobar p_c can be considered to represent the process of combustion. Then, a vertical line from state c to the intersection (e) with the proper isobar p_e represents the isentropic expansion. The enthalpy drop from c to e is the enthalpy converted to jet kinetic energy, according to Eq. 4-5. It is clear that the same Mollier diagram can be used for any propellant combination having the proper atomic constitution, the only difference being in the location of the injection state (i) on the H scale. It should be noted that, as a practical matter, it is not necessary to construct the entire Mollier diagram unless it is to be used ultimately for a large number of performance analyses for widely varying propellant systems and combustion pressures. Otherwise, it is sufficient to compute merely the states c and e, and such other specific states as may be of interest.

It is sometimes convenient to report the results of a performance analysis in terms not only of V_e, but also of an effective value of the specific heat ratio $\bar{\gamma}$ so as to fit a formula like Eq. 2-24:

$$c^* = \left[\frac{1}{\bar{\gamma}}\left(\frac{\bar{\gamma}+1}{2}\right)^{\frac{\bar{\gamma}+1}{\bar{\gamma}-1}}\frac{RT_c}{\bar{\mathfrak{M}}}\right]^{\frac{1}{2}} \tag{4-9}$$

A definition of $\bar{\gamma}$ that meets this requirement is the following:

$$\bar{\gamma} = \frac{\bar{C}_p}{\bar{C}_p - R}, \quad \bar{C}_p = \frac{(H_c - H_e)}{(T_c - T_e)(\overline{\sum n_i})}, \quad \overline{\sum n_i} = \frac{1}{2}\left(\sum_c n_i + \sum_e n_i\right) \tag{4-10}$$

The sum of the number of moles in the specified amount of gas in the combustion chamber is $\sum_c n_i$. Due to radical recombination during expansion, the similar sum $\sum_e n_i$ at the exit is somewhat less. $\overline{\sum n_i}$ is the arithmetic mean of the two and $\bar{\mathfrak{M}}$ is based on this sum. It is obvious that $\bar{C}_p$ includes the heat of recombination and is therefore considerably larger than the specific heat for the case of no reaction. Correspondingly, the value of $\bar{\gamma}$ is lower than the no-reaction γ. The exit temperature T_e is consistent with this value of $\bar{\gamma}$:

$$\frac{T_e}{T_c} = \left(\frac{p_e}{p_c}\right)^{\frac{(\bar{\gamma}-1)}{\bar{\gamma}}} \tag{4-11}$$

Another way to compute the state of the gas mixture at e is to make use of Eq. 4-10 and 4-11. The procedure starts by taking a trial value of $\bar{\gamma}$, and then computing in succession T_e, n_i at e, $\overline{\sum n_i}$, $(H_c - H_e)$, $\bar{C}_p$, and again $\bar{\gamma}$. If the last value of $\bar{\gamma}$ does not agree with the first value, a new trial value is selected until consistency is obtained. In much of the literature, a performance analysis is reported not only in terms of I_{sp} and T_c, but with $\bar{\gamma}$ and $\bar{\mathfrak{M}}$ as well.

It was pointed out above that the alternate assumption of constant composition expansion is also possible. The analysis of this case is greatly simplified because the composition of the end state is known. The same equation (Eq. 4-4) is used to evaluate the temperature and the enthalpy at the exit, or Eq. 4-10 and 4-11 can be used for this purpose. In either case, the general result is that the exit temperature T_e for a constant-composition expansion is lower than that for an equilibrium-composition expansion, particularly if there is appreciable dissociation in the state at T_c. However, the enthalpy difference is always slightly *less* for the constant-composition expansion, and therefore, by Eq. 4-5, the constant-composition case produces a slightly lower specific impulse.

The difference is usually only a few per cent. As a qualitative explanation, the larger I_{sp} of the equilibrium composition case is due to the heat released during the expansion by the recombination of free atoms and radicals, but although the recombination heat may be large, the effect on I_{sp} is weakened by the low pressure level at which it is released, that is, the thermodynamic conversion of the recombination heat is not as favorable as that of the main heat of combustion in the chamber. Of course, for a rocket operating with very low exhaust pressure, such as the upper stage of a satellite launcher, the large pressure ratio permits a favorable conversion to kinetic energy of the recombination heat released in the nozzle. In this case, the difference in I_{sp} may be significant.

Table G,4a. Calculated performance characteristics of several monopropellants and bipropellant combinations.

Propellant	I_{sp}	T_c	$\bar{\gamma}$	r	$\mathfrak{M}$
Monopropellants:					
Ethylene oxide	160	1860	1.17	—	44
Hydrazine	170	1125	1.37	—	32.1
Nitromethane	218	3950	1.25	—	61
Bipropellants:					
Liquid oxygen + 75% ethyl alcohol + 25% water	227	4875	1.19	1.38	23.8
Liquid oxygen + hydrazine	265	5200	1.23	0.68	16.6
Liquid oxygen + liquid hydrogen	346	4150	1.28	3.30	8.6
Liquid oxygen + gasoline	243	5450	1.22	2.50	22.7
RFNA + 15% NO_2 + aniline	222	5050	1.23	3.00	24.5
RFNA + 13% NO_2 ethyl alcohol	218	4750	1.21	2.80	24.4

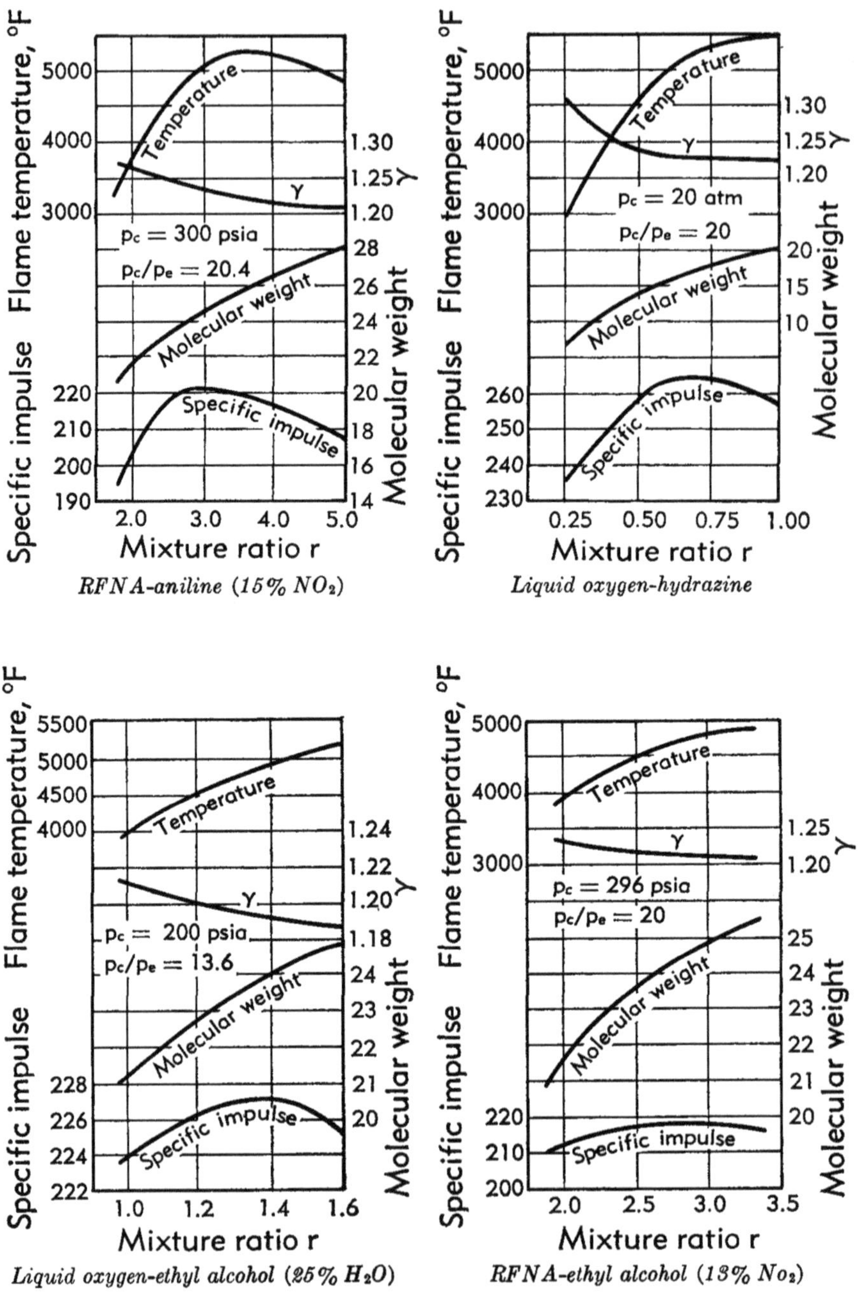

Fig. G,4. Variation of performance characteristics with mixture ratio for several bipropellant combinations. After Sutton, G.P., *op. cit.*

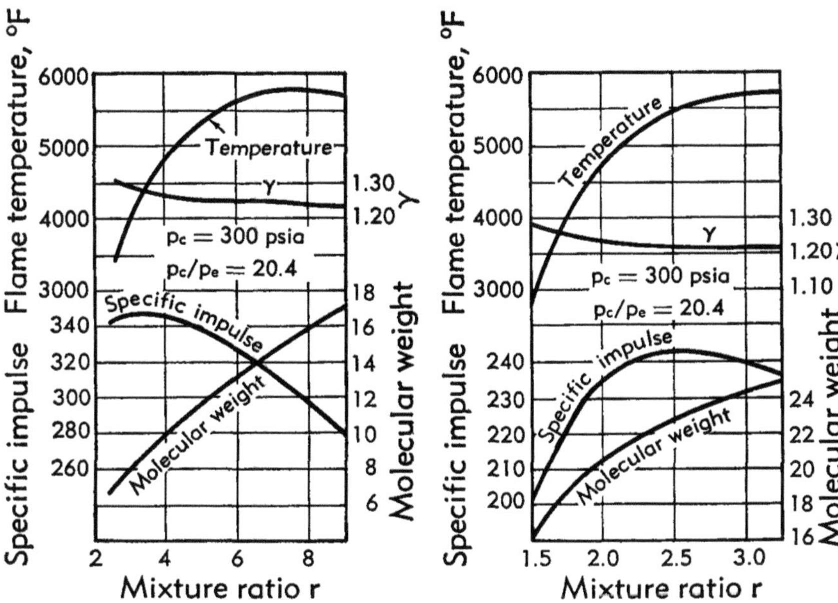

Liquid oxygen-liquid hydrogen *Liquid oxygen-gasoline*

Fig. G,4 (continued)

This concludes the general theory of specific impulse calculations. The remainder of this article is devoted to a presentation of some typical results and a discussion of the influence of certain factors on the specific impulse and the flame temperature.

Calculated performance characteristics of a series of monopropellants and bipropellant combinations are listed in Table G,4a. The significant characteristics are: I_{sp}, T_c, $\bar{\gamma}$, r, and $\overline{\mathfrak{M}}$. Unless otherwise noted, the reported

Table G,4b. Equilibrium gas composition in the combustion chamber of several monopropellants.

Propellant	Ethylene oxide	Hydrazine	Nitromethane
	C_2H_4O	N_2H_4	CH_3NO_2
Theoretical decomposition products in mole fractions	0.481 CO	0.656 H_2	0.057 CO_2
	0.404 CH_4	0.331 N_2	0.277 CO
	0.077 H_2	0.013 NH_3	0.277 H_2O
	0.038 C_2H_2		0.223 H_2
			0.166 N_2

values correspond to expansion with complete equilibrium from 300 lb/in.^{2} abs chamber pressure to 14.7 lb/in.^{2} abs exhaust pressure, at the optimum mixture ratio. The stoichiometric ratio is shown in the table for comparison.

Table G,4b lists the equilibrium gas composition in the combustion chamber for the monopropellants C_2H_4O, CH_3NO_2, and N_2H_4.

Curves are presented in Fig. G,4 of the variation of I_{sp}, T_c, $\bar{\gamma}$, and $\overline{\mathfrak{M}}$ with mixture ratio r for the bipropellant combinations listed in Table G,4a.

Table G,4c. Equilibrium gas composition in the combustion chamber of several bipropellant combinations at mixture ratios for optimum impulse.

Propellant	Gasoline-oxygen 300 psi	Hydrogen-oxygen 300 psi	RFNA-aniline 300 psi	Fluorine-ammonia 500 psi
Products of reaction in mole fractions	CO 0.322	H_2O 0.312	CO_2 0.159	HF 0.676
	CO_2 0.137	H_2 0.687	CO 0.251	H_2 0.041
	H_2 0.105	H 0.001	H_2O 0.283	N_2 0.137
	H_2O 0.330		H_2 0.063	H 0.085
	H 0.028		N_2 0.192	F 0.054
	OH 0.061		H 0.011	N 0.007
	O_2 0.008		OH 0.032	
	O 0.006			

Finally, equilibrium gas compositions are given in Table G,4c for oxygen-gasoline, oxygen-hydrogen, RFNA-aniline, and fluorine ammonia at mixture ratio for optimum impulse.

Several conclusions of general applicability can be deduced from an examination of these performance characteristics and the preceding discussion of the combustion and expansion processes:

1. The peak flame temperature occurs close to the stoichiometric mixture ratio in C-H-O-N systems, although some deviations can be observed. Thus, because of the relatively strong stability of nitric acid compared with aniline, this system shows a peak temperature on the fuel-rich side of stoichiometric, whereas no appreciable deviation is observed in oxygen-gasoline. This is significant for the cooling problem.
2. The peak specific impulse in C-H-O-N systems always occurs far to the fuel-rich side of stoichiometric. Although the temperature is less than the stoichiometric value, the mean molecular weight $\overline{\mathfrak{M}}$ is even less proportionately, due to the higher concentrations of H_2 and CO on the rich side instead of CO_2 and O_2 on the lean side. The strong effect

of a reduction in $\overline{\mathfrak{M}}$ in raising the specific impulse despite a reduction in temperature is seen in the decomposition of hydrazine monopropellant. Water addition to an alcohol-oxygen propellant to depress the temperature as an aid to cooling results in only a modest reduction of specific impulse because of the low molecular weight of H_2O and its dissociation products.

3. The pronounced effect of low $\overline{\mathfrak{M}}$ on specific impulse suggests the use of noncarbonaceous fuels such as NH_3, N_2H_4, and H_2 in order to avoid the formation of heavy CO and CO_2 molecules. The calculated performance data bear out this prediction.
4. The difference in specific impulse between equilibrium composition flow and constant composition flow is at most a few per cent, in the expansion from 300 lb/in.^{2} abs to 14.7 lb/in.^{2} abs, but exhausting to a lower pressure magnifies the difference because of the more favorable conversion of the recombination energy into kinetic energy. Raising the combustion pressure tends to produce a similar effect, but a large increase in pressure narrows the difference again because the dissociation is suppressed.
5. In highly dissociated mixtures, c^* increases strongly with increasing pressure if equilibrium flow is assumed. In nondissociated systems, c^* is practically insensitive to pressure. Therefore, the shape of the c^* versus r curve can change considerably with increasing pressure.
6. At 300 lb/in.^{2} abs pressure, flame temperatures in the C-H-O-N system cannot get much above 3500°K because of the strong increase in CO_2 and H_2O dissociation. This is true regardless of the fuel or oxidizer employed. Higher temperatures are possible at higher pressures where dissociation is suppressed.
7. Since flame temperatures in the C-H-O-N system at 300 lb/in.^{2} abs are limited, and since no reaction is known that will produce lighter products than the hydrogen-oxygen system, there exists an upper limit of $(T_c/\overline{\mathfrak{M}})$ and therefore of I_{sp} at 300 lb/in.^{2} abs. This is about 340 lb sec/lb.
8. The limits on T_c and I_{sp} for C-H-O-N systems can perhaps be evaded by not using compositions that will form CO_2 or H_2O. Thus, the light metals, Li, Be, and B, have become interesting as high energy fuels. Unfortunately, some of the products are not gases, e.g. B_2O_3, and in their most convenient form as liquids they contain hydrogen, e.g. B_2H_6. Switching to fluorine as the oxidizer allows the use of these high energy fuels with very little dissociation. Nevertheless, no system is yet known that exceeds the performance of fuel-rich, hydrogen-fluorine (350 lb sec/lb).

G,5. Combustor Design Principles. With regard to the individual processes that take place in a rocket motor, the main components of the motor are the injector, the combustion chamber, the exhaust nozzle, the cooling system, the igniter, and the starting device. The problem of exhaust nozzle design is treated in Art. 2 and 3, and the cooling problem is discussed in Art. 6. The starting system is included as part of the total propellant control and feed system in Art. 7, but the combustion aspects are considered here. This article is devoted, therefore, mainly to the problems of injection, combustion, and ignition.

Injection. The injector usually has the form of a single plate or shell containing many orifices or slots through which the propellants flow into the combustion chamber. As a rule, to provide adequate residence time for complete burning of all particles of propellant, it is located at the forward end or head of the chamber. To prevent burnout, it is cooled regeneratively or otherwise protected by the methods described in Art. 6.

The functions of the injector are: (1) to introduce the propellants in the desired ratio into the combustion chamber and into the combustion space, (2) to cause the necessary degree of liquid mixing, or interpenetration plus atomization, or both, in order to achieve rapid and complete reaction, (3) to distribute the propellant spray uniformly in the provided combustion space to make the combustor as compact as possible, and (4) to provide the proper pressure drop to fix the flow ratio and to dampen flow disturbances.

Experience with many injectors leads to the following understanding of the combustion process, according to whether the bipropellant combination is self-igniting or nonself-igniting, or whether a monopropellant is to be used.

Considering the self-igniting combination of nitric acid and UDMH (unsymmetrical dimethyl hydrazine), the rapidity with which flame occurs upon contact, suggests that the first step should be liquid-phase mixing, that is, the generation of as much interfacial contact in the liquid phase as possible. This is borne out by the better efficiency of the injectors with multiple pairs of unlike impinging streams as compared with the atomizing-type spray injection. Of course, the complete reaction cannot take place entirely in the liquid phase: heat generation immediately vaporizes the liquids so that the largest part of the reaction must take place in the gas phase. The unlike impinging stream injector was the first to be used successfully in this country with nitric acid propellants.

A nonself-igniting propellant combination such as liquid oxygen plus kerosene (or JP-4) can burn only in the mixed vapor phase. Consequently, the best injector for this combination is one that vaporizes the liquids as rapidly as possible and disperses them without regard to liquid-phase mixing. Suitable for this purpose is the showerhead-type injector that subdivides the incoming propellant into a large number of very thin streams, or the swirl-nozzle type that atomizes the injected streams, as well as the impinging-stream type. Burning probably takes place in a diffusion flame surrounding each droplet, with the last stages of reaction occurring in the hot gas phase.

With a monopropellant such as nitromethane, mixing is obviously not important, but atomization and distribution of the spray are. To date, some success has been achieved with atomizing swirl injectors.

The importance of proper distribution of the spray in the combustion chamber was mentioned. Poor distribution may result in ineffective utilization of the chamber volume so that an oversize chamber may be needed to attain full efficiency. This may create not merely a size and weight problem, but more seriously, a cooling problem.

The pressure drop across the injector under operating conditions is an important design characteristic. A typical magnitude is about 100 lb/in.^2 A high pressure differential has the advantage of producing a high velocity stream with good mixing and atomization, but the disadvantage of imposing a higher feed pressure requirement on the pumping system. Too low a differential runs the danger of allowing the flow rates or the mixture ratio to wander from the design point, since the injector pressure drop is an essential part of the hydraulic balance of the feed system, as explained in Art. 7. Too low a differential may also invite combustion instability of the low frequency type. (See below.)

The pressure drop Δp can be calculated if the empirical pressure loss coefficient K is known. For the fuel side,

$$\Delta p_f = (p_f - p_c) = K_f \tfrac{1}{2} \rho_f V_f^2 \tag{5-1}$$

$$V_f = \frac{\dot{m}_f}{\rho_f A_f} \tag{5-2}$$

The stream velocity V_f is determined by the total mass rate of flow of fuel $\dot{m}_f$ divided by the sum of the fuel orifice areas A_f, on the acceptable assumption that there is no appreciable contraction of the stream as it emerges from the orifice. The pressure loss coefficient is always greater than unity since even the smoothest orifices cause some momentum loss. In the hydraulics literature, the discharge coefficient C_d is frequently used:

$$V_f = C_{d_f} \left(\frac{2 \Delta p_f}{\rho_f} \right)^{\frac{1}{2}} \tag{5-3}$$

$$C_d = \frac{1}{\sqrt{K}} \tag{5-4}$$

Typical orifice sizes for most injectors range between $\frac{1}{16}$ to $\frac{1}{8}$ inch. Holes that are too small may easily become plugged during operation by foreign matter; holes that are too large produce a poorly atomized or mixed spray. The quality of the streams issuing from the orifices is always affected by the contour of the orifice approach and the shape of the connecting liquid manifolds.

A fixed area injector limits the possibility of controlling the thrust of a liquid propellant rocket motor. The quadratic pressure loss relation (Eq. 5-1) shows that a reduction of the flow rate by 50 per cent would reduce the differential to 25 lb/in.^{2}, if it were 100 lb/in.^{2} at rated flow. Trouble would be encountered at this low level with mixture ratio deviations and combustion instability. The solution is obviously a mechanically variable-area injector or a multiple injection system, and both have been used.

An important step in the development of an injector is the water test. By forcing water through the injector at about the rated flow, the discharge coefficient can be measured, and mixing atomization and distribution characteristics can be observed. Visual determination of the quality of the spray is aided by coloring the fuel and oxidizer water supplies with contrasting dyes, e.g. yellow and purple. It may be necessary to discharge into a simulated pressurized chamber in order to prevent misleading cavitation in the orifices at atmospheric pressure.

Another useful test method employed in injector development is the combustion chamber traverse with cooled temperature and composition probes. Such tests reveal the uniformity of mixing and the origin of any unusual hot spots on the chamber or the nozzle wall. Transparent two-dimensional motors have also provided valuable information about injector operation.

Combustion. From the standpoint of combustion efficiency, the chamber volume must be made large enough to allow the necessary residence time for the reaction to reach completion, and the length from the injector to the exhaust nozzle must be greater than the ballistic path of the largest droplets before they are consumed. Thus, for each propellant combination and each injector, at a given pressure, the fundamental design specifications are minimum chamber volume and minimum chamber length. A volume larger than the minimum leads to weight, size, and cooling penalties; a length greater than the minimum, for a given volume, means that the convective heat flux in the chamber is unnecessarily high because of the small cross-sectional area.

The residence time in a chamber of volume $\mathcal{V}_c$ is:

$$\tau_c = \frac{\mathcal{V}_c}{\dot{m}/\rho_{av}} \cong \left(\frac{\rho_{av}}{\rho_{min}}\right)\frac{L^*}{c^*}\frac{1}{\gamma}\left(\frac{\gamma+1}{2}\right)^{\frac{\gamma+1}{\gamma-1}} \cong 5\,\frac{L^*}{c^*} \tag{5-5}$$

The following relations were used in this derivation:
Characteristic velocity (see Art. 2):

$$c^* = \frac{p_c A_{th}}{\dot{m}} \tag{5-6}$$

Theoretical c^* (ideal motor theory):

$$c^* = \left(\frac{RT_c}{\mathfrak{M}}\right)^{\frac{1}{2}} \gamma^{-\frac{1}{2}} \left[\frac{(\gamma + 1)}{2}\right]^{\frac{(\gamma+1)}{2(\gamma-1)}} \tag{5-7}$$

Perfect gas law

$$\rho_{\min} = \frac{p_c\mathfrak{M}}{RT_c} \tag{5-8}$$

Definition of "characteristic chamber length"

$$L^* = \frac{\mathcal{V}_c}{A_t} \tag{5-9}$$

The numerical factor at the right side of Eq. 5-5 is based on the values $\gamma \cong 1.20$ and $(\rho_{av}/\rho_{\min}) \cong 2.0$. For a typical rocket motor using oxygen-gasoline, with $L^* \cong 60$ inches and $c^* \cong 5000$ ft/sec, at 300 lb/in.^{2} chamber pressure, the residence time calculated according to Eq. 5-5 is about 5 milliseconds. This time is to be regarded properly as the gas phase part of the total residence time, since the propellant spends a few milliseconds in the liquid phase as droplets, and this would not show up in Eq. 5-5 unless the time-average density ρ_{av} were made very much larger than $2\rho_{\min}$ to reflect the liquid-phase history.

It is the practice to determine $L^*_{\min}$ by measuring c^* for a series of chambers of progressively smaller volume, all at one pressure, and then plotting c^* as a function of L^*. For liquid oxygen-liquid hydrogen at 300 lb/in.^{2}, the minimum is about 10 inches; for nitric acid-gasoline, 50 inches; for nitric acid-aniline, 40 inches; for oxygen-alcohol, 50 inches; and for nitromethane, about 300 inches. Inasmuch as these minima were established by trying various kinds of injectors at different laboratories with different chamber configurations, it appears that they correspond to fundamental chemical reaction times. A calculation of the reaction time for NO decomposition indicates that this may be the step that determines the minimum L^* for nitric acid motors.

This interpretation of L^* suggests that, in scaling up a motor from, say, a 1000-lb thrust laboratory unit to a larger practical one, the proper rule is to increase the volume V_c in proportion to $\dot{m}$, at the same pressure. It has been found that this rule is a good guide. Furthermore, if the over-all reaction kinetics can be assumed to be of second order, the time to completion would vary inversely as the pressure. (See II,D.) According to Eq. 5-5, this means $L^*_{\min}$ can be reduced in the extrapolation of test data to higher pressures. This trend is also confirmed experimentally.

Although the L^* concept determines the volume of the chamber, its proportions have to be decided according to other considerations. It is obvious that the length cannot be shorter than the ballistic path of the larger droplets, that is, the penetration distance. At the present time, experiments to determine the minimum length for any particular injector are a more reliable guide than theory. The longest length for a given L^* is achieved by making the chamber diameter equal to the nozzle throat diameter, but this results in a prohibitive heat flux by convection. The usual practice is to make the chamber diameter about 1.5 to 3 times the nozzle throat diameter. If high frequency instability occurs, modification of the length, either larger or smaller, can stabilize the motor. (See below.)

Ignition. A nonself-igniting propellant combination such as alcohol plus liquid oxygen, or a monopropellant such as ethylene oxide, requires an ignition energy source to get started. There are a number of practical considerations that govern the selection of an igniter: (a) whether one-shot or repeated starts are desired; (b) availability of the energy or substance, for example, electrical power; (c) the necessary degree of reliability (with nitric-acid-gasoline, for example, a misfire can lead to violent explosion); (d) the complexity of the system; (e) the maximum allowable waiting time from the "fire" signal to the moment of ignition, and the desired rate of build-up. It is usually impossible to meet all these specifications to the highest degree simultaneously.

Some practical ignition devices are:

1. Pyrotechnic squibs, which may be mounted in the injector head, the chamber wall, or on an expendable rod pointing forward through the exhaust nozzle.
2. High energy spark plugs, the fastest to start but the most prone to fouling by deposits or short-circuiting by acid spray.
3. Self-igniting chemical starting fuel, for example, xylidene injected just before admission of gasoline, in a nitric acid-gasoline motor.
4. Catalytic surfaces in the chamber, such as ceramic elements saturated with potassium permanganate, for starting hydrogen peroxide decomposition.
5. Auxiliary oxygen, for starting the combustion of JP-4 before admission of the main oxidizer, nitric acid, or for start-up of ethylene oxide.
6. Electrically heated glow plugs, to provide a hot surface that can ignite the initial spray of the mixture (convenient, but slow to start).

There are several special problems connected with ignition that deserve mention. One is the starting of very large motors, 10,000-lb thrust and up. In order to provide a spray that can be ignited, the initial flow rate through a fixed area injector has to be at least about $\frac{1}{4}$ of the rated flow, taking an approximate lower limit. For a larger motor, even this flow may drown the igniter of usual size. A practical solution is to have a starting injector recessed in a small ignition chamber which is connected to the main chamber through a critical flow orifice; the ignition can take place in the small chamber just as in a small motor, and the high intensity flame shooting into the large chamber then ignites the main flow. A pressure-actuated relay makes sure that the main flow is not started until the starting chamber is up to rated pressure.

The start-up of a large motor must be programmed carefully, the usual scheme being to provide for two stages of flow, a warm-up stage of a few seconds, and then rapid rise to rated flow. Without adequate warm-up, there is danger of flame-out by flooding the fire.

A "hard start," or initial pressure overshoot in the chamber, is almost inevitably the result of a delay in ignition, particularly of propellants such as nitric acid-gasoline that can burn explosively or detonate when premixed. Nitric acid reacts immediately upon contact with hydrocarbons to form nitration products that are explosive. If the igniter fails to initiate combustion immediately after the propellants enter the chamber, the sensitive mixture accumulates and explodes almost instantaneously when ignition does take hold. The resulting overshoot in chamber pressure may be sufficient to damage the rocket motor or upset the guidance system or instrumentation in the vehicle.

For the case of an accumulated mixture that is nondetonable, a rough estimate of the maximum allowable delay can be deduced as follows. If the exploded propellant accumulation produces a peak temperature about twice the space average flame temperature for normal constant pressure burning, and if no gas escapes from the nozzle during the explosion, then the mass of gas that would produce a transient explosion pressure equal to the rated pressure is given by the normal flow rate $\dot{m}$ multiplied by one half the normal gas-phase residence time τ_c. Then, using Eq. 5-5,

$$\Delta t_{\text{delay}} \cong 2.5 \frac{L^*}{c^*} \left(\frac{\dot{m}\ (\text{rated})}{\dot{m}\ (\text{start})}\right) \left(\frac{p_{c(\text{trans})}}{p_{c(\text{rated})}}\right) \tag{5-10}$$

If we take $L^* = 60$ in., $c^* = 5000$ ft/sec, $\dot{m}_{(\text{start})} = \frac{1}{4}\dot{m}_{(\text{rated})}$, and $p_{c(\text{trans})} = p_{c(\text{rated})}$, then $\Delta t_{\max} \cong 10$ milliseconds.

It is clear from this result that prompt ignition is essential. Alternatively, as a safety measure, a means must be provided to shut off the propellant and purge the chamber promptly of the accumulated propellant in the event of excessive ignition delay.

The problem of ignition is greatly simplified, particularly for repeated starting, in the case of self-igniting propellants. The most common examples of self-igniting combinations are hydrocarbon amines with concentrated nitric acid. Fuels of this class are aniline (phenylamine), xylidine, hydrazine, unsymmetrical dimethyl hydrazine (UDMH), etc. Furfuryl alcohol, which is not an amine, reacts in a similar fashion, however, with nitric acid. Spontaneous ignition occurs also between fluorine and all of the common fuels. Oxygen ignites upon contact with diborane. Hydrazine and hydrogen peroxide are also self-igniting. It should be observed, of course, that a combination that is spontaneous at room temperature may not ignite at very low temperature, and that a combination that is not spontaneous at room temperature may be self-igniting at elevated temperature. Consequently, the distinction between self-igniting and nonself-igniting propellants is largely one of the degree of reactivity in the liquid phase at room temperature.

From the standpoint of rocket ignition, the appropriate measure of reactivity is the *ignition lag*, that is, the measured time interval between the moment of contact of the two liquids and the subsequent appearance of flame. Such ignition lags lie generally in the range from 5 to 100 milliseconds, but experience has shown that the method of testing can greatly influence the absolute value. Various methods of measuring ignition lag have been used: (1) the open cup test, (2) the closed chamber with forced injection, (3) the falling drop test, and (4) small rocket motors. The values for the identical propellants at the same temperature differ by as much as a factor of ten, depending on the type of test. Nevertheless, the different methods generally agree as to the *order* of reactivity of various propellants and as to the effects of temperature, diluents, and catalysts. Typical ignition lag measurements are given in Table G,5.

Table G,5. Typical ignition delay times for several propellant combinations.

Propellants	*Ignition delay, msec*
Triethylamine + WFNA	30
Allylamine + WFNA	50
Cyclohexene + WFNA	90
80% Furfuryl alcohol + 20% aniline + WFNA	20
Furfuryl alcohol + WFNA	40

One of the interesting aspects of ignition lag testing is the evidence that the lag depends as strongly on the intimacy of mixing and on the heat insulation as on the chemical kinetics. Thus, forced penetration of nitric acid and aniline by injection results in a much smaller lag. Tests involving large quantities of propellant as compared with small quantities show shorter lags, undoubtedly the result of better heat retention during the first stage of reaction. The addition of "wetting agents" to reduce the interfacial tension between the two liquids and thus promote surface contact has been found very effective: 2 per cent of oleate of polyoxyethylene glycol added to furfuryl alcohol reduced the lag from 25 to 12 millisec in the reaction with nitric acid; certain other substances were similarly effective.

The ignition lag in the rocket motor is an important characteristic of the propellant combination in connection with the "hard start" pressure evaluated in Eq. 5-10. A more complete discussion of the ignition problem from the standpoint of combustion mechanisms can be found in II,L.

Combustion instability. One of the difficult problems of combustor design is that of preventing unstable combustion during the so-called steady thrust period of firing. The phenomenon is observed in all types of liquid propellant rocket motors, but the consequences are more serious the larger the size, and so it has only been in recent years that the problem has become an urgent one. Also, the phenomenon became quite prominent only when part of the liquid rocket engine industry in the U.S. switched strongly over to nitric acid-kerosene-type propellants from the earlier nitric acid-aniline types; it has not been as troublesome in oxygen systems. The reasons for these differences are discussed below.

Instability manifests itself as a strong oscillation in combustion pressure, with amplitudes as large as 50 per cent of the mean pressure, which occurs in the absence of any external stimulation while the control valve positions and feed pressures are absolutely steady. Observed frequencies range from as low as 10 cps to several thousand per second. In the low frequency range, the propellant flow rate oscillates along with the combustion pressure, due to dynamic coupling, but the high frequencies appear only in the combustion chamber because the inertia of the liquid in the lines is too great to follow the rapid pressure changes. The practical consequences of combustion instability are, in the low frequency or "chugging" range, severe vibration of the vehicle and its delicate instrumentation, a reduction in specific impulse, and quite often, rupture of the rocket motor. In the high frequency range, usually called "screaming," the most serious consequence is a large increase in heat flux and usually a burnout of the motor.

Although superficial observations of instability have resulted in a two-fold classification, either "chugging" or "screaming," more refined observations have disclosed at least three basic types of instability, according to the driving mechanism: low frequency, high frequency, and intermediate frequency. The *low frequency type*, below 100 cycles in practical motors, is characterized by a uniform pressure in the chamber which oscillates with time, coupled to a similar oscillation in the propellant flow rate. The basic situation is that of a servomechanism with excessive time delay in the feedback loop, which is therefore unstable. In the case of the rocket combustor, the feedback is either the effect of a p_c fluctuation on the instantaneous $\dot{m}$ (that is, hydraulic or structural coupling with the feed system), or the effect of a similar fluctuation on the combustion reaction rate (chemical kinetics coupling). Both feedback effects may exist simultaneously.

The simplest theory was evolved in 1941 after it was observed that nitric acid motors, which could not be made to burn smoothly on gasoline, did burn smoothly when the tank was filled with aniline. The fact that the former combination was nonself-igniting and not easily ignited, whereas the latter was self-igniting, suggested the idea that "chugging" of the former was due to a slow reaction rate. This concept has been introduced into the theory of instability as the "combustion time lag." In its initial form, the time lag was regarded as a property of the propellant and the conditions of injection, but invariant during the fluctuation of pressure. Therefore, it was not time-dependent. Subsequently it was shown theoretically that important types of instability can occur if the time lag is sensitive to the local instantaneous conditions of pressure and temperature. Since the local temperature during an oscillation is coupled to the pressure in a unique fashion, it is sufficient mathematically to describe this as simply a pressure-sensitive time lag.

The pressure-insensitive time lag is defined by the observation that the rate of evolution of gas at any instant is equal to the rate of injection of propellant at a time τ millisec earlier:

$$\dot{m}_b(t) = \dot{m}_i(t - \tau) \tag{5-11}$$

The interval τ is the time required for the physical processes of vaporization and mixing and the chemical processes of reaction. When this time lag is inserted in the theory of small oscillations of a liquid monopropellant engine with a compressed gas feed system, a simple approximate condition for stability emerges:

$$\frac{l_p \dot{m}}{p_c A_p} + \frac{4.8 L^* \Delta p_p}{c^* p_c} > \tau \tag{5-12}$$

According to this equation, reduction of the time lag τ, either by better mixing and atomization or by substitution of a more reactive propellant for a sluggish one, will tend to stabilize an otherwise chugging engine. (This corresponds to the experience noted above of switching from gasoline to aniline.) Increasing the length of the feed line l_p or the velocity of flow $\dot{m}/A_p$ will also stabilize the system, by increasing the momentum in the feed lines. An increase in L^* produces stability by the capacitance effect of the chamber, and an increase in the injection pressure drop tends to stiffen the flow rate against perturbations. There are experimental results tending to corroborate these predictions. As to the magnitudes of the various terms, the left side is generally of the order of 5 to 10 milliseconds, and time lags have been measured in the range of 1 to 5 milliseconds, indicating the possibility of instability.

The introduction of pressure sensitivity in the time lag (or in part of it) makes instability possible even with a perfectly "stiff" feed system, that is, a system having an absolutely constant flow rate. In this case, the oscillating pressure arises in fact from an *oscillating burning rate*, the coupling being affected by the sensitivity of the time lag. This type of "*intrinsic*" *low frequency instability* can occur only if the pressure-sensitive time lag is long enough and if the sensitivity is sufficiently strong. The sensitivity to the pressure is described in terms of an interaction index n, which can be defined by saying that the time lag required for the pressure-sensitive process to be completed, if carried out at constant pressure, varies as p^{-n}. The critical value of the time lag for any index n is shown to be:

$$\tau = \left(\frac{L^*c^*}{RT_c}\right) \frac{\left[\pi - \cos^{-1}\frac{1-n}{n}\right]}{[2n-1]^{-\frac{1}{2}}} \tag{5-13}$$

This curve is plotted in Fig. G,5a. It shows, first, that no such instability is possible if n is less than $\frac{1}{2}$. Secondly, if the interaction is strong (large n), the motor will easily become unstable. The pressure-sensitive

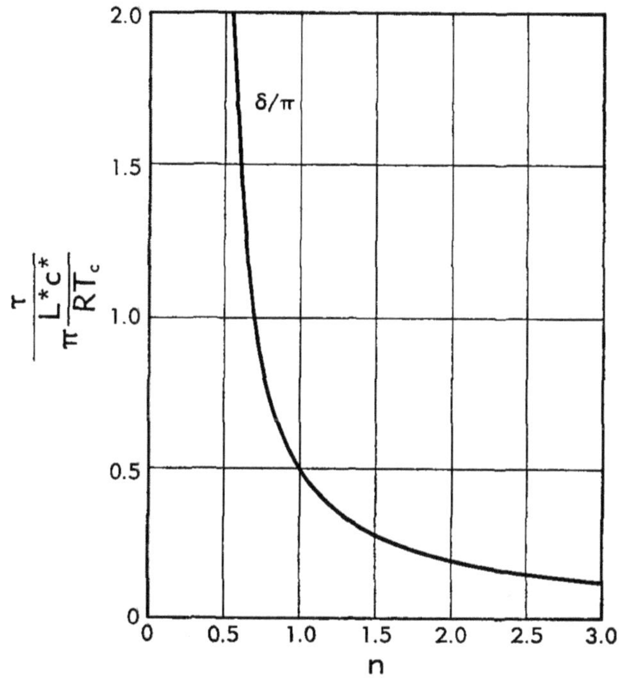

Fig. G,5a. Variation of critical time lag with index n.

time lag and the interaction index have been measured experimentally, by imposing forced flow oscillations on the pressurized feed system of an oxygen-alcohol rocket engine in stable operation and by observing the phase and amplitude of the chamber pressure as a function of the forcing frequency. Typical values are (at 300 lb/in.^{2} pressure, 1.34 oxygen-alcohol ratio):

$$\begin{aligned} \text{total } \tau &= 0.25 \text{ millisec} \\ \text{sensitive } \tau &= 0.08 \text{ millisec} \\ \text{index } n &= 1.0 \end{aligned}$$

The interaction of the combustion process with the pressure, expressed through the index n, serves to explain also the occurrence of *high frequency oscillations*. The explanation rests on the observed fact that the frequencies are very close to the natural frequencies of the combustion chamber treated as a closed, long cylindrical column of stationary gas. Thus the dominant frequency is usually that of the fundamental axial mode of the "organ pipe." (A two-foot-long chamber filled with typical combustion gas at 3000°K would show a fundamental of about 1000 cycles.) Higher frequencies can be identified with longitudinal overtones, radial modes, and angular modes, all of which can occur

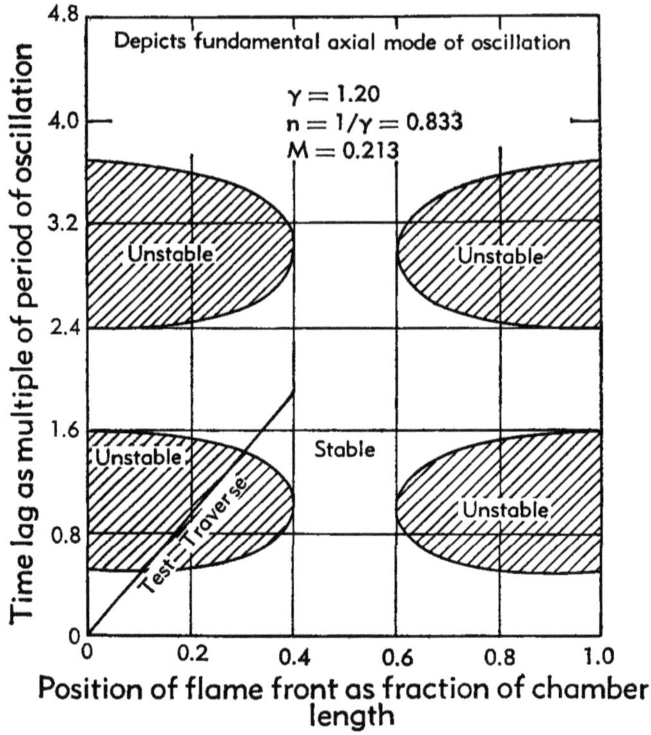

Fig. G,5b. Stability diagram for longitudinal oscillations.

simultaneously. These wave motions would ordinarily decay as a result of viscous dissipation and loss of wave energy through the exhaust nozzle. However, the local speed-up of the exothermic reaction when a compression wave passes through an element of the combustion gas (assuming that $n > 0$) can serve as the mechanism for injecting energy into the wave to compensate for losses. It can be shown that, in order to produce instability of a particular mode, this energy release must take place in regions of the chamber near the pressure antinodes of the stationary wave and it must occur with the right phase lag, about one half the period. These conditions can be expressed, in the case of longitudinal oscillations, by a stability diagram such as Fig. G,5b, in which the unstable values of *space lag* and sensitive *time lag* are defined, for a specified interaction index n. The diagram shows, for example, that the fundamental mode is unstable only if the sensitive part of the combustion process is concentrated either near the injector or near the end of the chamber, with the simultaneous condition that the time lag must have a particular value or a certain larger value (so as to straddle an odd multiple of half-periods). Similar regions of instability are shown for the first overtone.

It is clear that instability can be avoided by choosing an injector type and a chamber length so as to place the sensitive combustion reaction in one of the stable areas of the diagram. Inasmuch as the fundamental is the most powerful mode, the designer must start by ignoring the unstable regions connected with the higher modes. Recent tests of an oxygen-alcohol motor, conducted by varying the length of the combustion chamber while keeping all other conditions the same, have demonstrated the actual existence of such zones of instability for the fundamental longitudinal mode. This series of tests were carried out along a traverse such as the inclined straight line in Fig. G,5b.

In modern large rocket motors, the most important unstable mode seems to be the transverse "sloshing" mode instead of the longitudinal one. The stability theory for this case has not yet been published, but the principles for avoiding instability are undoubtedly the same.

An important instability with frequencies intermediate between the chugging frequency and the organ-pipe frequencies has been traced to an interaction between pressure waves in the chamber and the mixing or injection process near the injector head. A pressure pulse that starts at the exhaust nozzle moves upstream with the velocity of sound, strikes the injector where it causes unequal fluctuations in the two propellant injection rates or it disturbs the mixing process; the off-mixture layer of propellant is swept downstream toward the nozzle, burning on the way; a new pressure pulse is reflected from the nozzle when this layer (called an "entropy wave") enters the nozzle; in this way, the cycle is repeated. Although an exact theory has yet to be developed for the stability condition, it is evident that the period of oscillation is the sum of the wave travel time and the mean chamber residence time, and that the frequency is therefore intermediate between the organ-pipe fundamental and the chugging frequency.

Other hypothetical instability mechanisms have been suggested but are as yet unverified experimentally: (1) wave propagation in the liquid feed system, supported by oscillations of combustion pressure, (2) oscillatory injection of the propellant, stimulated by elastic deformations of the cooling jacket and the injector manifolds due to the variations in chamber pressure, (3) hydraulic instabilities in injection orifices, (4) spray instabilities due to hydrodynamic forces, (5) self-excited oscillations of the chemical kinetics of consecutive reactions with feedback; and (6) coupling of the rocket engine with a natural vibration of the supporting structure.

Some of these mechanisms are plausible, but it remains for the future to clarify them.

Although the preferred way for a designer to cure a case of combustion instability is to modify the combustor or the operating conditions according to the above theories, some attention has been given to the use of feedback servos to provide automatic stability. The idea is to use the combustion pressure signal to actuate propellant flow controls to modify the injection rate to overcome at least low frequency instability. No such device has yet been put into practice.

Scaling of rocket motors. Scaling is the problem of extrapolating the performance characteristics of one rocket motor, acquired by firing tests, to another rocket motor as yet untested. A narrower version of the problem is to design a large rocket motor that will have the same characteristics as the smaller one that has been successfully developed. Among the characteristics to be preserved or predicted are the combustion efficiency (c^*), the nozzle effectiveness (C_F), the heat flux distribution (q), the margin of combustion stability at both low and high frequencies, the starting chamber pressure pulse, and the working stresses in the important chamber structure.

The present approach of the designer is the obvious one of dealing with each problem separately on the basis of either experimental correlations or theory. The simplest case is that of relating a large motor to a small one, keeping the propellants, their mixture ratio, and the chamber pressure the same. Thus the exhaust nozzle is scaled according to the theory and data in Art. 2 and 3; the heat flux, both convective and radiative, according to Art. 6; the chamber volume according to the L^* concept in Art. 5; the stability margin according to the theories just presented; etc. Incompatible requirements frequently arise: for example, the choice of chamber length is dictated on the one hand by the volume required for combustion together with the cross-sectional area required by convective heat flux considerations, and on the other hand by the condition for stability in the fundamental longitudinal mode. It is not possible at this time to write a universal prescription for such difficulties except to say that the designer must decide to relax one specification or the other, or strike a compromise. The more difficult problem is that of predicting the proper design at a different pressure level as well as a larger size. For this purpose, experimental data are needed on such empirical parameters as c^*, nozzle C_d, time lag τ, interaction index n, radiative emissivity ϵ, etc., all as functions of p_c.

An interesting theoretical approach has been proposed recently, to derive dimensionless similarity parameters from the basic physical equations governing rocket motor performance, just as the Reynolds number emerges from the Navier-Stokes equations for viscous flow. In the case of the rocket motor, regarded as a chemical reactor with heterogeneous chemical kinetics, convective, conductive and radiative heat transfer, steady and unsteady gas dynamics, wall stresses, etc., the problem is a complicated one. This approach is not yet fruitful but is receiving active attention.

Mechanical design and construction. The principles of rocket combustor design, having emerged as a result of experience over the years, can be appreciated only by careful study of a number of actual rocket motors, the early types as well as the modern ones. Many are described in the engineering literature.

Several of the design principles deserve explicit mention:

1. The chamber and nozzle wall and the injector face plate must be made of metals selected for high strength at elevated temperature coupled with good thermal conductivity, resistance to high temperature oxidation, chemical inertness on the coolant side, and suitability for the fabrication method to be employed. In the case of certain monopropellants, the metal must not catalyze the decomposition. Although aluminum and copper alloys have been used successfully for combustion chambers and nozzles, stainless steels and carbon steels are in widest use today.
2. Large scale motors today are fabricated by sheet-forming and welding techniques. Experimental small motors are usually made by machining with the lathe and milling machine, and then assembled with bolts.
3. The chamber wall must be carefully supported to stiffen it against the compressive load of the pressure in the cooling jacket, both during normal operation and after shut-down if the propellant valve is between the cooling jacket and the injector. Similarly, the injector face plate must be supported by stiffening ribs to prevent collapse under pressure.
4. Thermal expansion of the chamber wall must be accommodated during firing, usually by providing an expansion joint on the outer casing. Allowance must also be made for thermal expansion of the injector plate, usually by elastic deformation.
5. Dimensional accuracy of the cooling passages must be assured, during firing as well as in assembly, in order to guarantee cooling performance. Definite spacers must be provided in large, coaxial shell-type combustors. Longitudinal cooling passages are usually used for large motors, but spiral passages are needed for small motors to preserve a reasonable duct width that is not too narrow to guarantee. The ducts are not of uniform width, being usually narrower at the throat than at the chamber. The method of fabrication should assure smooth surfaces in the cooling ducts to minimize the pressure drop in the feed system.
6. The manifold passages in the injector head should be designed to serve also as regenerative cooling ducts with prescribed coolant velocities.

7. The coolant inlet should be at the exhaust nozzle end to take advantage of the lowest liquid temperature at the most severely heated section of the motor.
8. Provision must be made for close alignment of the nozzle to produce a thrust vector accurately perpendicular to the mounting ring.

Finally, it may be remarked that the problems of design are all made more difficult by the need to keep the weight at a minimum. Typical weight figures for modern rocket motors center around 0.01 lb per lb thrust, the ratio being somewhat less for large missile motors and greater for small aircraft motors.

G,6. Cooling of Rocket Motors.

Heat transfer processes in rocket motors. The problem of designing a cooling system for a liquid propellant rocket motor can be divided into two parts. The first is an examination of the expected magnitude of the heat flux into the walls, based on conditions inside the combustor; the second is an analysis of cooling schemes capable of keeping the wall temperature below the structural failure point.

Two processes account for the heat flow from the combustion gas to the wall, convection and radiation, and the heat flux is the sum of these two contributions. Each depends in a complicated way on the flow velocity distribution, the composition distribution, and the temperature distribution inside the combustor and the nozzle, and these distributions are never known or predictable except in a very rough way. Consequently, precise heat flux predictions are not feasible. The most the designer can hope to calculate is a probable upper limit of the heat flux, or if this is known at one pressure for a given rocket motor, he can make an estimate of the variation of heat flux with chamber pressure or mixture ratio or size.

The cooling of rocket motors can be accomplished in a variety of ways. In this discussion, the meaning of "cooling system" is generalized to include any scheme designed to limit the wall temperature, even if no fluid coolant is employed. Thus the most direct way to limit the internal surface temperature is to provide a sufficiently thick chamber or nozzle wall with the necessary heat capacity to soak up the heat transferred during the prescribed firing duration. This method has the advantages of simplicity and cheapness of manufacture, but for durations greater than 10 to 20 seconds it results in a weight penalty. For longer durations, a better solution is to use one or both of the propellants as coolants ("regenerative" cooling) with lightweight, thin metal, combustion chambers and nozzles. Either the heat capacity motor or the regeneratively cooled motor can be aided by internal protective techniques, such as: (1) refractory insulating nonmetallic liners or inserts to reduce the heat flux, (2) internal coatings applied either as a paint before firing or by continuous deposition from the combustion gas to the wall, (3) liquid or gaseous films on the inner surface supplied by continuous injection through either a porous wall or small orifices in the wall, and (4) low temperature gas or even nonburning propellant in the region near the wall, supplied by appropriately modified injector orifices.

For the practical designer, the following subarticles on heat transfer from the hot combustion gas to the wall by convection and radiation may be less useful than that dealing with cooling methods, since the usual practice is to test a water-cooled experimental motor of the proposed configuration first, and to accept the measured heat flux distribution as the basis of the cooling system design. The study of the high temperature side is useful only for making predictions of heat flux in motors not yet tested and still being designed.

It is not possible in the space of this article to develop all of these processes of heat transfer and cooling in full detail. Instead, only the main principles are presented, and references are given to more complete treatments. The observed heat transfer distributions are discussed first, since they serve to illuminate the more theoretical material to follow.

Observations of heat transfer in experimental rocket motors. The heat transfer characteristics of a rocket motor at a particular operating condition are most conveniently presented in terms of longitudinal and circumferential distributions of the heat flux per unit area. This presentation is useful only because the flow of heat is practically independent of the manner of cooling the chamber. This is so because the inner wall temperature of a cooled motor must be kept so much lower than the flame temperature that differences in wall temperature due to different cooling methods cause only small differences in heat flux. If the heat flux distribution were sensitive to the particular conditions of the cooling system, it would be necessary to present the results in terms of conductance coefficients, as in conventional heat exchanges.

Measurements of heat flux distribution are made in several ways, but the most convenient method makes use of a sectional, water-cooled rocket motor. The internal geometries of the combustion space, the injector, and the nozzle are made identical with the practical motor. In the test motor, each section is cooled by water, and the *average* heat flux in the section is calculated from the increase in water temperature from inlet to outlet. If measurement of *local* heat flux is desired, it is possible to position two thermocouples at different radial distances from the inner surface, and on the assumption of purely radial heat flow, the heat flux is calculable from the temperature difference and the thermal conductivity of the wall material. One of the simplest methods is to use a "heat capacity" motor, for example, a rocket motor made of sections of solid copper, one insulated from the other; the average heat flux can be calculated from the transient rate of temperature rise of each section. Unfortunately, the methods based on the use of sectional motors result in circumferential averages of the heat flux and fail to indicate variations in that direction. Such circumferential variations may be important if the injector is not exactly axially symmetric, and most are not. The most obvious experimental clues to the existence of heat flux peaks in the circumferential direction are longitudinal discolorations (streaks) of the inner surface of the chamber after a firing. Such hot zones are frequently visible in the exhaust jet also.

In the evaluation of a measured heat flux distribution, it is important to realize that two motors operating on the same propellants, at the same mixture ratio and at the same pressure, can exhibit different heat fluxes by a factor of two, because of various unpredictable effects. Thus, in a nitric acid motor at 300 lb/in.^{2} pressure, it was found that an annular spray-type injector produced 36 per cent higher average heat flux (averaged over the entire inner surface) than an impinging jet injector, although the specific impulse was increased only 5 per cent. A similar sensitivity to injector design is shown by the fact that a slight change of about 5° outward in the direction of the mixed streams produced by pairs of impinging jets can result in the burnout of an otherwise satisfactory motor. As a practical technique, it is possible to design a large injector plate so as to produce a relatively cool layer of combustion gas at the wall, while the bulk of the injected propellant burns at high temperature, and thereby reduces the heat transfer. Rough or unsteady combustion generally results in higher heat fluxes in the chamber, and under the extreme condition of "screaming" combustion, the heat flux in the same motor has been found to increase by more than 100 per cent. Another point well known to rocket designers is that any break in the smooth contour of the inner surface can raise the heat flux locally; such breaks are made by a pressure measurement orifice, a sharp nozzle throat, or a turbulence promotor or flame holder in the combustion chamber. Finally, it must be observed that solid deposits formed on the walls, such as carbon, iron oxides (from iron dissolved in nitric acid in steel tanks), boron oxides, etc. tend to reduce the heat flux through their insulating effect. This phenomenon has been used intentionally to cool a motor, by adding a few per cent of a silicone oil to the fuel, to produce a protective deposit of silica.

If one considers the longitudinal distribution of the circumferentially averaged heat flux in a typical nitric acid rocket motor, some noteworthy features are: (1) the order of magnitude of heat flux, 1 BTU/in.^{2}-sec, about 10 to 20 times that of a high-output, aircraft-type combustion heater, (2) the high flux in the vicinity of the throat, due mainly to strong convective heat transfer, (3) the sharp fall-off after the throat, due to the expansion cooling and the build-up of the boundary layer, and (4) the slow rise along the combustion chamber wall corresponding to the axial progress of combustion.

Considering the effect of mixture ratio on the heat flux in a rocket motor, one finds that the peak values of heat flux occur near the peak flame temperature. For hydrocarbon fuels, this occurs on the fuel-lean side of the ratio for peak specific impulse, as explained in Art. 4.

The effect of pressure on the heat flux in a nozzle is in general an increase in the heat flux with increasing pressure due to a progressive increase in boundary layer conductance with increasing gas density.

Finally, it should be observed that a typical value of the integrated heat flux over the entire surface, for a well-designed, 1000-lb thrust motor operating at 300 $lb/in.^2$ on either nitric acid-hydrocarbon or oxygen-hydrocarbon, is about 2 per cent of the heat of combustion. This percentage is considerably smaller for larger thrust motors, because the combustor volume increases approximately in proportion to the mass flow, while the surface of the combustor increases only as the two-thirds power, for similar shapes. Thus the heat transfer in the V-2 motor represents about 0.7 per cent of the heat of combustion.

The practicability of running a rocket motor continuously with only regenerative cooling, using either one or both propellants, is directly the result of this low percentage of heat transfer. None of the common propellants, such as JP-4 or alcohol, can absorb more than a few per cent of the heat of combustion without vaporizing or decomposing and thus becoming unsuitable as coolants. This observation can be presented another way, that for each propellant combination and combustion pressure, there exists a certain minimum thrust rocket motor below which simple regenerative cooling is inadequate. Fortunately, for the common propellant combinations based on nitric acid or oxygen, this is well below 1000-lb thrust. For propellants that burn slowly and require large residence times, such as nitromethane, a simple regenerative motor may not really be practical except for a thrust in excess of, say, 10,000 lbs.

Convective heat transfer from the combustion gases to the wall. Convection generally accounts for the main part of the heat flow from the hot combustion gases to the walls of the rocket motor, usually two-thirds or more, depending on the particular location of the surface element in question and the temperature and composition of the gas mixture. Unfortunately, there is no firm basis for theoretical prediction of the heat flux, although several useful empirical correlations of restricted applicability have been developed. The main difficulty is that the temperature and gas flow patterns inside the rocket motor seem to vary strongly in an unpredictable manner from one injector to another, even in some instances where the configurations are similar and the specific impulses are nearly the same. Thus, the practical engineer refers to "hot" injectors and "cool" injectors, the latter class being preferred. Another obstacle to theoretical treatment is that there is not yet sufficient information on the boundary layers involved, e.g. at the throat of a cooled de Laval nozzle. Finally, the transport properties of the gas mixture, viscosity, and thermal conductivity are not well known at the elevated temperatures involved.

Despite these uncertainties, empirical correlations have emerged which can be applied to determine within about ±20 per cent the convective heat transfer, although the "hot spots" produced characteristically by "hot" injectors are not predicted by these methods. The discussion can be divided into three parts dealing, respectively, with the subsonic cylindrical combustion chamber, the supersonic portion of the de Laval nozzle, and the throat of the nozzle.

In the chamber, strongly nonsteady flow conditions prevail, and so the boundary layer is undoubtedly turbulent. It has been found that the heat transfer at the hot end of the chamber can be correlated by an equation of the Colburn type in which the pertinent length is the chamber diameter d_c:

$$Nu_d = aRe_d^{0.8}Pr^{0.33} \tag{6-1}$$

where Nu_d is the Nusselt number, Re_d is the Reynolds number, and Pr is the Prandtl number defined as follows:

$$Nu_d = \frac{hd_c}{k}$$
$$Re_d = \frac{\rho V d_c}{\mu} \tag{6-2}$$
$$Pr = \frac{\mu c_p}{k}$$

The numerical coefficient a has the value 0.023 in fully developed turbulent pipe flow, but there is no a priori reason to apply to the short-tube rocket combustor the same numerical coefficient. Nevertheless, test results in rocket chambers indicate that Eq. 6-1 gives a reasonable estimate of the convective heat transfer coefficient h, with about the same numerical coefficient. This applies particularly to the downstream end of the cylindrical combustion chamber.

The transport properties μ and k are sensitive to gas temperature, and so it is necessary to specify the reference temperature at which they are to be evaluated for use in Eq. 6-2. The best correlation, at least in the case of pipe flow, is obtained by using a reference "film" temperature

$$T_f = T_w + 0.5(T_g - T_w) \tag{6-3}$$

where T_w is the wall temperature and T_g is the local gas temperature. Use of this "average" temperature seems justified also for the rocket motor. Fortunately, because the Prandtl number is nearly independent of temperature and because h depends only on the 0.2 power of μ, the sensitivity of the result to the reference temperature is not very great.

(For a general discussion of the dependence of transport properties on gas temperature, consult I,D.) Finally, it is general practice to evaluate ρ and c_p at this reference temperature, too.

The heat flux from the hot gas to the wall becomes:

$$q = h(T_g - T_w) \tag{6-4}$$

In the supersonic portion of the deLaval nozzle, as in the combustion chamber, the flow conditions are not like those of the classical flat plate or straight cone which have been the objects of many researches. Therefore it is necessary to rely heavily on the few systematic nozzle tests that have been performed. It appears that, in rocket nozzles of practical size, the boundary layer in the major portion of the diverging section is turbulent, and that the proper Reynolds number for the situation is based on the distance L downstream of the throat. The boundary layer at the throat is very thin, as a result of the steep negative pressure gradient, and so the boundary layer downstream may be considered to have started at the throat as a leading edge. By analogy with the flat plate in a uniform supersonic stream, a correlation of the following form may be expected for the local heat transfer coefficient:

$$Nu_L = aRe_L^{0.8}Pr^{0.33} \tag{6-5}$$

The numerical factor a has the value 0.0296 for the case of a turbulent boundary layer on a flat plate with zero pressure gradient and small temperature difference. For the divergent cone of a Laval nozzle, experiments indicate values of a ranging from 0.025 to 0.028. Like the case of the flat plate in supersonic flow, the heat transfer coefficient h that enters into Nu is defined in terms of the adiabatic wall or recovery temperature T_r, thus

$$q = h(T_r - T_w) \tag{6-6}$$

This relation serves to define also T_r, namely, the local temperature that would be created on the surface of a wall made of nonconducting material. In low speed incompressible flow, T_r would equal the static temperature of the gas T_g but, in high speed compressible flow, T_r is more nearly equal to the stagnation temperature of the gas T_g^0. This effect, sometimes referred to as frictional heating, is expressed in terms of a temperature recovery factor r:

$$r = \frac{T_r - T_g}{T_g^0 - T_g} \tag{6-7}$$

In steady, adiabatic flow it is impossible for a gas with Prandtl number less than unity to develop a temperature anywhere that exceeds stagnation temperature, so that r must be less than unity. For compressible, turbulent boundary layers along flat plates, wedges, and cones, with zero or near-zero pressure gradient, experiments in the supersonic range up to $M = 4$ indicate that:

$$r \cong Pr^{\frac{1}{3}} = 0.90 \text{ (for air) and } 0.93 \text{ (for } CO_2) \tag{6-8}$$

A mixture of diatomic and polyatomic gases, such as that which occurs in rocket combustion, would exhibit a recovery factor of about 0.91.

The gas properties, Pr, c_p, μ, k, and ρ vary with temperature, particularly the transport coefficients. It is therefore necessary to specify a reference temperature at which they are to be evaluated. The best correlation of the experimental results has been achieved with the following reference temperature:

$$T_f = T_w + 0.23(T_g - T_w) + 0.19(T_r - T_w) \tag{6-9}$$

This relation converges approximately to Eq. 6-3 as M goes to zero.

For the laminar compressible boundary layer, the defining equations (Eq. 6-6 and 6-7) are still employed, but the following relations for Nu and r replace Eq. 6-5 and 6-8:

$$Nu_L = 0.33 Re_L^{0.5} Pr^{0.33} \tag{6-10}$$

$$r \cong Pr^{\frac{1}{2}} = 0.86 \text{ (for air) and } 0.90 \text{ (for } CO_2) \tag{6-11}$$

Eq. 6-10 is the same correlation as that for flat plates in incompressible flow with zero pressure gradient. The effects of compressibility enter in the use of T_r instead of T_g in Eq. 6-6, the use of r to obtain T_r, and the use of Eq. 6-9 for T_f instead of Eq. 6-3. It is possible that the correlations for a laminar boundary layer should be applied to the region of the exit cone near the throat, but there is not sufficient experimental information to support this idea. It is current practice to apply the turbulent correlations throughout the exit cone.

A remarkable fact, however, is that the heat transfer at the nozzle throat is *also* predicted with fair precision by an equation like Eq. 6-5 in which the significant dimension seems to be the diameter d_{th} of the throat rather than some length along the surface. This equation has been found to hold for a throat having a contour radius approximately equal to the throat diameter; it would not apply to "sharp" throats:

$$Nu_d = 0.023 Re_d^{0.8} Pr^{0.33} \tag{6-12}$$

This is exactly the same as Eq. 6-1, the formula for incompressible flow in tubes. However, because of compressibility, the gas properties are evaluated at T_f, as in Eq. 6-9, and T_r is introduced in the heat flux equation, as in Eq. 6-6. The recovery factor can be taken equal to unity, with small error.

In summary, there is some experimental support for the utilization of the correlations just presented to predict the convective heat transfer at the downstream end of the cylindrical chamber, at the nozzle throat, and in the exit cone. From the viewpoint of boundary layer theory, there is considerable basis for misgiving when correlations that have been established for the idealized cases of the flat plate and the infinite tube are used for the much less ideal rocket motor. All that can be said in defense of such practice at the present time is that it gives reasonable results. Future investigations may very well disclose the existence of some coincidences that enable a doubtful treatment to produce acceptable results.

In the application of the preceding correlations, it is necessary to evaluate V, ρ, μ, k, c_p, T_g, T_g^0, T_r, T_w, and T_f. On the basis of theoretical combustion calculations (Art. 4), V, T_g, and T_g^0 are known. Some accuracy can be gained by adjusting T_g^0 by means of the factor $(c^*_{\text{exptl}}/c^*_{\text{theo}})^2$ to allow for combustion inefficiency and a consequent reduction in the actual flame temperature. Correspondingly, V and T_g can be adjusted. The surface temperature T_w is either prescribed in the structural design analysis or determined by the over-all heat transfer analysis to be described later. With very little error, the recovery factor r can be regarded as independent of temperature and assigned a fixed value, and thus T_r is determined. With T_g^0, T_g, T_w, and T_r known, T_f can be calculated. Two questions arise in the evaluation of the proper physical properties: how to deal with a mixture of gases and how to allow for the effect of high temperature.

Inasmuch as data are generally lacking on the transport properties of mixtures, it is common practice to treat each of the properties as a molar weighted average of the pure gas properties. Thus,

$$k = \frac{\sum n_i k_i}{\sum n_i} \tag{6-13}$$

It is known that serious exceptions to this rule exist, but when specific data are lacking, this formula is probably the best to adopt. A similar average is recommended for μ, the viscosity of the gas mixture. (These problems are discussed in I,D.) The specific heat c_p is a molar average, quite rigorously

$$c_p = \frac{\sum n_i C_{pi}}{\sum n_i m_i} = \frac{\overline{C}_p}{\overline{m}} \tag{6-14}$$

Similarly, the mixture density is a molar average of the component densities at the same temperature and total pressure:

$$\rho = \frac{\sum n_i \rho_i}{\sum n_i} \tag{6-15}$$

The preceding treatment is appropriate for heat transfer without dissociation in the main flow outside the boundary layer, that is, for the case in which the composition everywhere in the boundary layer is the same as the main stream.

The effects of elevated temperature on ρ and c_p are well known and require no discussion here. The viscosity μ increases with the absolute temperature somewhat more rapidly than the $\frac{1}{2}$ power predicted by gas kinetic theory for hard-sphere nonattracting molecules. For a combustion gas mixture, the viscosity is nearly proportional to the first power of the temperature. Less is known about the effect of temperature on thermal conductivity, but the Eucken relation which is based on the theory of heat conduction indicates that the Prandtl number should be nearly independent of T:

$$\left.\begin{aligned} &\text{Eucken (1913):} \qquad k = \tfrac{1}{4}(9\gamma - 5)\mu c_v \\ &\text{equivalent to:} \qquad Pr = \frac{4\gamma}{(9\gamma - 5)} \end{aligned}\right\} \qquad (6\text{-}16)$$

From the variation of μ and c_p with T, it is possible then to deduce the variation of k.

With regard to the effect of dissociation on the convective heat transfer, theoretical analysis indicates that, if the Lewis number Le for the gas mixture is near unity, that is, the molecular diffusivity equals the thermal diffusivity, and if the dissociative composition corresponds to the equilibrium value at each point in the cooled boundary layer, that is, recombination takes place fast enough to keep up with the diffusion of the gas toward the wall, then the heat transfer can be correctly calculated by means of the conventional formulas for nonreactive air, provided the specific heat is modified to include the energy of dissociation. In other words, this statement is that the driving enthalpy difference $c_p(T_g - T_w)$ is to be replaced by $(h_g - h_w)$ so as to include the dissociation enthalpy. Surprisingly, it makes practically no difference if it is assumed that the recombination process is too slow to be effective, that is, if the boundary layer composition is frozen, provided that total recombination occurs at the wall. This is so because the transport of energy by free atom diffusion is just as effective as transport by thermal diffusion, if Le is unity. But, if the composition in the boundary layer is frozen, and if the wall is totally noncatalytic so that no recombination occurs anywhere, the heat transfer rate is reduced. In this case, the correct formula makes use of the average specific heat of the frozen gas mixture, and the driving enthalpy difference is simply $c_p(T_g - T_w)$.

It is general practice to ignore the contribution due to recombination in the boundary layer or at the wall, since the additional heat transfer is well within the range of uncertainty of these calculations, for the common propellants at optimum specific impulse mixture ratios. It should probably be included in such high performance, highly dissociated propellants as oxygen-hydrogen.

Radiative heat transfer in rocket combustion chambers. The high temperature combustion gas mixture in the chamber can radiate strongly to the internal walls of the combustor, and there are some experimental results that indicate that such heat transfer can amount to as much as 25 per cent or more of the total heat flux at the wall. In general, the radiative heat flux increases with increasing temperature, with increasing pressure, and with increasing chamber diameter. Polyatomic gases radiate more strongly than diatomic gases, so that, for example, the products of hydrocarbon combustion (CO_2, H_2O) radiate more energy than the decomposition products of hydrazine monopropellant (N_2, H_2). Incandescent particles such as carbonaceous soot and light metal oxides produced in the combustion process may easily contribute the largest fraction of the radiative heat flux. Chemiluminescent radiation from the active reaction zone, observable mainly in the visible and near ultraviolet, seems to be a negligible part of the heat flux, in comparison with the thermal radiation emitted by the gaseous and solid reaction products.

It is necessary at the outset to define precisely the several types of radiation that may be encountered in a combustion chamber, and for this purpose we rely heavily on the extensive theoretical and experimental research carried out by physicists since the latter part of the nineteenth century. Thermal radiation is emitted by a hot body (solid, liquid, or gaseous) merely by virtue of its temperature. Radiation can be stimulated by nonthermal processes as well, such as electron bombardment, X-ray absorption, etc. In a combustor, an important nonthermal stimulus for radiation is chemical reaction; the radiation in this case is called chemiluminescence. With regard to thermal radiation, it is usually necessary for calculation purposes to assume that equilibrium conditions prevail, that is, that the temperature of the gases, of the enclosure walls, and of the contained radiation are uniform and equal. In the literature, the term "thermal radiation" usually is intended to imply the existence of such equilibrium. In a practical combustor, however, radiative equilibrium cannot exist, because of the relatively cold walls; but because the radiated energy is only a small fraction of the thermal energy of the gas, the thermodynamic equilibrium in the gas is not appreciably upset, and equilibrium theory may be used for calculating the radiation intensity emitted by the gas. A similar justification allows the use of Kirchhoff's law in practical heat transfer calculations to equate the spectral absorptivity of a combustion gas to its emissivity. In this case, however, it is more correct to evaluate the total absorptivity at the temperature characteristic of the incident radiation rather than at the temperature of the gas.

Consideration of thermal radiation starts most conveniently with so-called black body radiation. In a strict sense, black body radiation can be produced only inside a sealed cavity with uniformly heated containing surfaces. For experimental measurement of the characteristics of the black body radiation, a heated cavity is used as a source, except that a very small aperture allows a beam to reach the spectrophotometer. The spectral energy distribution as a function of temperature is described by Planck's equation, and the integrated energy flux is given by the Stefan-Boltzmann equation. The practical significance of the concept of black body radiation is that the emissive power of a black body (the aperture in the heated cavity) cannot be exceeded by that of any material or surface in any spectral interval if the excitation is purely thermal. This means that the spectral emissivity of any heated body is less than unity, and so is the total emissivity.

A "grey body" or a "grey" emitter is one that radiates, when heated, an amount of energy in each spectral interval that is a uniform fraction of that of a black body of the same temperature, that is, a grey emitter has a uniform spectral emissivity, which is therefore equal to its total emissivity. No substance is known that is strictly grey, but many opaque solids display nearly constant emissivity over a wide spectral range.

Of particular interest for this discussion of the rocket combustion chamber is the thermal radiation emitted by a cloud of carbon particles or soot that may be formed by the combustion of a carbonaceous fuel, especially with a fuel-rich mixture. Such flames have been called "luminous flames" to distinguish them from the nonsooty flames that radiate very little in the visible region of the spectrum. The emissivity of a luminous flame depends on the type of particles, the size distribution, and particularly on the optical thickness, that is, the number density of the particles multiplied by the path length in the line of sight. Luminous flames with small numbers of particles are generally nongrey, that is, their spectral emissivities are usually greater in the blue than in the red end of the spectrum, and this must be taken into account in the evaluation of the total emissivity of the flame. However, if the particle density is sufficient to produce a total emissivity of 0.2 or more, the contribution to the emission spectrum due to the luminous soot or ash can be regarded as "grey." This contribution is, of course, superimposed on the nonluminous emission of the flame. The luminous emissivity can easily attain values of 0.7 to 0.9 in fuel-rich rocket combustors, and probably also in chambers fed with such fuels as the borohydrides that would produce a smoky ash. In the range in which grey emission holds, the emissivity increases with increasing optical thickness, asymptotically approaching unity by an exponential formula (Beer's Law).

The nonluminous radiation emitted by the hot combustion gas consists mainly of the characteristic vibration-rotation and pure rotation bands of the various polyatomic and heteropolar diatomic molecules. These gases include, in the approximate descending order of their emissive powers: CO_2, H_2O, CO, NO, OH, and HF. The bands all appear in the infrared region of the spectrum ($\lambda > 1\mu$) and are therefore not visible; that is, the gases are transparent and colorless. At temperatures above 5000°K, the ultraviolet electronic bands of these and other gases would probably make significant contributions to the total emissivity, but there are no experimental data for such temperatures. Homopolar diatomic gases, such as N_2, O_2, H_2, and F_2, emit no radiation in the infrared or visible, and therefore contribute nothing to the total emissivity of the gas mixture at temperatures less than 5000°K.

Unlike the radiation emitted by a "grey" sooty flame, the radiation from a nonluminous flame is concentrated in relatively narrow spectral bands, which in turn consist of groups of closely spaced spectral lines. At very high pressures, above 100 atmospheres, the bands tend to broaden, but at the combustion pressures of liquid rockets, the band envelope is substantially the same as at atmospheric pressure. While the total emissivity of a sooty flame does not vary strongly with either pressure or temperature if the same optical thickness (particle density times path length) is considered, the pressure and temperature effects may be quite substantial in the case of banded spectra. The effect of temperature on the total emissivity at constant optical thickness is a consequence of three distinct physical processes: (1) the envelope of an emission band broadens as the temperature is raised because the molecular system is excited statistically to higher energy levels; (2) the maximum of the Planck curve shifts to shorter wavelengths (Wien's displacement law) as the temperature is raised, while the spectral regions of band emission occupy correspondingly weaker sections of the long wavelength end of the Planck curve; (3) the individual rotational lines, broadened mainly by intermolecular collisions, become narrower as the temperature is raised while the pressure is held constant. The first process tends to raise the emissivity of a gas, while the last two reduce it, as the temperature is raised. The second process is usually the dominant one above 1500°K.

The effect of pressure on the emissivity at pressures below 100 atm results from the collision broadening of the rotational lines: it increases with increasing pressure, but reaches a saturation limit asymptotically when the lines have become smeared out so as to overlap each other. For polyatomic molecules such as CO_2 and H_2O, the saturation occurs near atmospheric pressure; for CO and OH at somewhat higher pressures; but the effect is probably "saturated" at rocket combustion pressures.

Other complications that must be considered in the calculation of hot gas emissivities are: (1) the departure of the total emissivity from exact exponential dependence on optical thickness, a consequence of the line character of band emission, (2) overlapping of bands in the spectra of the various gases in a mixture, that is, nonadditivity of the component emissivities, and (3) similar nonadditivity of the thermal emissivities of the nonluminous gases and the luminous soot or ash.

Finally, chemiluminescent radiation should be considered. As stated earlier, the radiation in this category is produced mainly in the active reaction zone, and arises from excited intermediates such as CH, C_2, HCO, and CO_2. Quantitative measurements of the energy emitted by these species in typical hydrocarbon-air flames indicate that this can be ignored as a source of heat transfer; the total radiated energy in the visible and ultraviolet bands is of the order of 10^{-7} of the heat of combustion, as observed at 1 atm pressure. Because of the nonthermal origin, it is not surprising to find experimentally that such bands can exhibit spectral emissivities greater than unity and rotational or vibrational temperatures higher than the gas temperature. Such experiments indicate that the relative intensities of these bands drop as the pressure is raised. There do not seem to be any definitive measurements on the intensity of chemically excited radiation that might be emitted in the infrared, particularly by the much more abundant stable products such as CO_2, CO, and H_2O; it is the usual practice to ignore such radiative heat transfer on the assumption that it is small.

The preceding paragraphs represent a brief survey of the nature of the radiation that is emitted by the combustion gas in a rocket chamber. For engineering purposes, it is desirable to express the complex radiative character of the gas in terms of an emissivity factor which is then to be inserted in the heat transfer equation below to calculate the heat flux. It is evident from the foregoing discussion that there are still many unknown factors, so that experimental heat transfer measurements must be used as a guide in calculating emissivities. Additional complications that require consideration are: (1) the unknown absorption coefficient of the internal combustor wall for the particular radiation emitted by the gas, (2) the uncertain temperature and composition distribution within the chamber and the difficulty of handling theoretically the radiation from a nonuniform gaseous emitter, (3) the unknown particle density of carbon soot or ash, and (4) the complicated geometrical shape factors for the rocket combustor.

Under the circumstances, it has been the accepted practice to calculate the radiative heat transfer by making a number of broad simplifying assumptions. Surprisingly, the few experiments that have been made on the radiated inside rocket chambers have not disagreed seriously with the calculated results. For rockets operating on C-H-O-N propellants at pressures not over 50 atmospheres, these assumptions are:

1. The only important emitters are CO_2, H_2O, and CO, and these radiate thermally.
2. The gas temperature is uniform and equal to $(c^*_{\text{exptl}}/c^*_{\text{theo}})^2 T_{c(\text{theo})}$.
3. The gas composition is uniform and corresponds to equilibrium at the theoretical temperature.
4. The beam length for a cylindrical combustor is 0.9 times the diameter.

5. The component emissivities of CO_2 and H_2O are the same as those at 1 atm pressure, for the same temperature and optical density.
6. The emissivity of CO for a given optical path length is 0.5 of that of CO_2.
7. The correction for overlapping of the H_2O and CO_2 spectra is the same as that at 1 atm.
8. The absorptivity of the chamber wall is unity, that is, it is a black receiver.
9. The wall temperature is so low that back radiation is unimportant.
10. The intensity of the chemiluminescent radiation is negligible for heat transfer.
11. If soot is not observable in the rocket exhaust, the luminous component of the emissivity may be neglected.

With these assumptions, it is possible to make use of the emissivity data that have been compiled by Hottel and his collaborators. The necessary data for calculating the emissivity of a hot gas mixture of CO_2, H_2O, and CO are available in V,I. The heat flux per unit area into the chamber wall is then given by

$$q = \epsilon_g \sigma T_g^4 \tag{6-17}$$

Cooling methods. At the beginning of this article, a number of approaches to the design problem of cooling a rocket motor are mentioned, which need to be analyzed in greater detail. It is necessary to decide at the outset whether the simple but limited duration heat capacity motor or the more complex, unlimited duration, regeneratively cooled motor is most suitable for the intended application. For liquid propellant boosters, for short-duration aircraft JATOS (especially if jettisonable), and for anti-aircraft guided missiles such as NIKE, the heat capacity type is usually adequate; for auxiliary aircraft power plants and for missile sustainer motors, the regenerative type is best. In either case it is possible to employ one or more of the internal protective schemes mentioned earlier, such as refractory liners, refractory deposits, film cooling, sweat cooling, or peripheral low temperature combustion gases, as a supplement to the basic cooling scheme.

The only scheme to be discussed in detail in this article is straight regenerative cooling. References are given at the end to the published literature where full treatments can be found for the cooling methods not discussed here.

The basic idea of regenerative cooling is that one of the propellants, the fuel, is used as a coolant. Since the most intensely heated region is the exhaust nozzle throat, it is best to have the cooling jacket inlet at the nozzle end rather than at the chamber end. For a coolant such as JP-4 or kerosene, it may be assumed that there is a maximum allowable temperature, about 350°F, above which decomposition or coking may take place in the fuel jacket, with the resulting deposition of a harmful insulating carbonaceous layer on the surface to be cooled. It is necessary to avoid raising the fuel above this temperature anywhere in the cooling duct. For any coolant, there is a maximum allowable temperature: for hydrazine or nitromethane, it would be the explosion temperature; for liquid oxygen or nitric acid, or for any coolant, the boiling temperature might be the maximum. To determine whether the proposed design will allow the danger temperature to be reached, it is necessary to make temperature calculations for at least three points in the cooling duct: (1) the mixed stream at the outlet of the duct, (2) the wall at the nozzle throat, and (3) the wall at the injector end of the chamber. Other suspicious places may be indicated by the results of these three calculations.

The temperature of the fuel stream at the outlet is determined by the rate of fuel flow and the total heat flux integrated over the entire cooled area of the combustion chamber and nozzle. (For this discussion, the cooled injector head is included as part of the cooled chamber.)

$$T_{\mathrm{liq}2} = T_{\mathrm{liq}1} + \frac{\int_1^2 q\,dA}{\dot{m}_f c_f} \tag{6-18}$$

If $T_{\mathrm{liq}2}$ exceeds the boiling point of the fuel *at the pressure in the combustion chamber* (p_c), trouble will occur in the form of unsteady propellant flow due to bubbling in the fuel stream in the injector orifices. If $T_{\mathrm{liq}2}$ is high enough to exceed the boiling point *at the pressure at the outlet of the cooling jacket*, momentary "vapor-lock" of the fuel system will occur and result in either burnout or severe chugging of the motor. Inasmuch as the integrated heat flux is practically insensitive to the degree of cooling of the wall, as mentioned earlier, the value of $T_{\mathrm{liq}2}$ cannot be reduced by modifying the cooling duct but only by a change in the combustor or by use of one of the internal protective schemes, such as film cooling.

The calculation of the wall temperature on the liquid side at any station, such as the throat th, can be discussed by means of Eq. 6-18. With the subscript $_2$ replaced by subscript $_{\mathrm{th}}$, it is possible to calculate the bulk temperature in the cooling duct at station th; the integration of q between the two limits is then performed. Then, with the heat flux at th given by experiment, and with the heat transfer coefficient h_{liq} calculated by a formula similar to Eq. 6-1, the wall temperature on the liquid side $T_{w,\mathrm{liq}}$ can be calculated:

$$T_{w,\mathrm{liq}} = T_{\mathrm{liq},\mathrm{th}} + \frac{q_{\mathrm{liq},\mathrm{th}}}{h_{\mathrm{liq},\mathrm{th}}} \tag{6-19}$$

The evaluation of q_{th} and h_{liq} is not simple to carry out. The magnitude of q_{th} in Eq. 6-19 is usually less than the value given in the case when

q is the heat flux per unit area *on the gas side surface*, while the value of q_{liq} is based *on the effective area being cooled on the liquid side*. The two values of q are in the inverse ratio of the heated area to the effective cooled area. The heated area is usually easy to determine; the effective cooled area depends on the design of the cooling duct. The only simple case of duct design is that of the annular duct with longitudinal or spiral flow. Designs incorporating spiral tubes or fins are more complicated, but solutions of such problems can be found in the usual heat transfer literature.

The film conductance h_{liq} can be calculated for a formula such as Eq. 6-1 with the reference temperature given by Eq. 6-3. In all practical cases, this formula is justified because the flow is in the fully developed turbulent range. However, the coefficient a has been evaluated experimentally only for simple shapes such as flat plates, straight round tubes, and two-dimensional ducts. Coiled ducts, especially of noncircular cross section, spiral passages, finned passages, etc., all of them useful in rocket-cooling jacket design, have not been fully investigated, and probably never will, particularly under a special condition of heat flow which is not uniformly distributed around the surface of the passage. Although it is possible, in principle, to analyze such problems or to perform systematic model tests, the designer usually resorts to his intuition and to his experience with previous similar designs.

The values of $T_{w,liq}$ at the throat and at the chamber head, computed by Eq. 6-19 are then compared with the previously determined maximum allowable temperature. If either value is too high or too close to the limit, the remedy might be to reduce q_{liq} or to raise h_{liq} or both. For a given q on the gas side, q_{liq} can be reduced by the use of a finlike structure or even round channels on the cooled side of the chamber wall. The conductance h_{liq} can be raised by making the channel width smaller and increasing the flow velocity, as revealed by Eq. 6-1 with the condition that the volumetric flow rate is kept the same. Because the heat flux is always more intense in the throat region, the duct is usually designed to provide the highest velocity at that section. Thus, the channel width is rarely uniform over the entire motor. There are practical limits to how far h_{liq} can be raised in this way: (1) an increase in flow velocity and a reduction in the channel width results in an increase in pressure loss, which is approximately proportional to the minus 3 to minus 6 power of the width, depending on the type of duct, (2) a reduction in channel width makes the gap tolerances more severe and increases the production cost, and (3) a reduction in channel width makes the width more sensitive to small dimensional changes caused by heating during firing. Of these effects, the pressure loss is usually the most serious, since this imposes a penalty on the propellant feed system. Various design tricks have been used, however, to minimize the pressure loss penalty, but these are too specialized for treatment in this section. For the calculation of pressure loss, the reader is referred to standard works on hydraulics.

Once a satisfactory cooling duct has been designed, the next problem is to decide on the structure of the chamber wall. For the case of a simple, thin shell wall, the gas side wall surface temperature at each section is determined by the given heat flux q and the temperature $T_{w,liq}$ just calculated in the cooling duct design:

$$T_{w,g} = T_{w,liq} + q\frac{d_w}{k_w} \tag{6-20}$$

The designer must choose the wall thickness d_w and the wall metal (conductivity k_w). The choice is usually a complicated problem, involving several questions: (1) the strength of the chamber wall and nozzle against the collapsing pressure in the cooling jacket, while at elevated temperature (a short-time creep problem), (2) the chemical resistance of the metal to the hot coolant on one side (corrosion) and to the nonuniform, high velocity hot gas on the other (corrosion and erosion), and (3) the method of fabrication, whether by sheet metal forming and welding or by machining, casting, etc. Several typical combustor designs are discussed in Art. 5. It is obvious, however, that materials of high conductivity and high strength at elevated temperatures are necessary. Alloy steels of the type developed for turbojet blades and disks are usually acceptable, but wall thicknesses in excess of $\frac{1}{8}$ inch are rarely tolerable. In large motors, such thin shells of diameter 1 ft or greater cannot withstand the collapsing pressure, and so devices such as stiffening ribs, support attachments to the outer wall, etc., are employed. Then, these ribs usually do double duty as heat transfer fins and thus reduce the wall temperatures $T_{w,g}$ and $T_{w,liq}$.

This concludes the discussion of the pure regeneratively cooled motor. The following paragraphs present briefly the main ideas of other cooling schemes as well as references to the literature for more complete analyses.

Uncooled heat capacity motor. Except for the injector head which usually incorporates regenerative cooling, the chamber and nozzle are treated as a hollow circular cylinder, using the Fourier equation in cylindrical coordinates at each section, for simplicity, and boundary conditions corresponding to convective heating on the inner surface and total insulation on the outer surface. Solutions of this problem in the following form are obtained:

$$\left(\frac{T_g - T_r}{T_g - T_o}\right) = F\left[\left(\frac{\kappa_w t}{r_o^2}\right), \left(\frac{r}{d_w}\right), \left(\frac{r_o}{r_i}\right), \left(\frac{k_w}{h_g d_w}\right)\right] \tag{6-21}$$

The temperature of the metal T_r at the radial distance r, for a cylinder with outside and inside radii r_o and r_i, is given as a function of time t from the start of exposure to the gas of temperature T_g. The metal conductivity k_w, its thermal diffusivity κ_w, and the film conductance in the hot gas h_g enter into the equation. More details of such analyses can be found in V,D.

An approximation frequently employed for the case of thin-walled chambers for such short-duration applications as liquid propellant boosters or anti-aircraft rocket is:

$$\left.\begin{aligned} \rho_w c_w d_w \frac{dT_{av}}{dt} &= h_g(T_g - T_{w_i}) \\ T_{av} &\cong \tfrac{1}{2}(T_{w_i} + T_{w_o}) \\ h_g(T_g - T_{w_i}) &\cong \frac{2k_w(T_{w_i} - T_{w_o})}{d_w} \end{aligned}\right\} \tag{6-22}$$

The calculation is then carried out by numerical iteration. The value of h_g can be determined with a sectional, water-cooled motor.

Uncooled motor with refractory insulator. An example of this type of design is the NIKE anti-aircraft guided missile rocket motor. Use of an insulating, high temperature ceramic such as SiC, Al_2O_3, B_4C, or even graphite C usually reduces the heat flux to the outer metal wall to such an extent as to permit a reduction in thickness without loss of strength. The result is an over-all saving of weight. The wall heating problem can be analyzed in a manner similar to that of the simple, unprotected, heat capacity motor, but the final design is usually selected by firing tests. The analytical approach is frustrated by inadequate knowledge of the heat transfer conditions at the rough ceramic surface, of the thermal properties of the ceramic body, of the thermal resistance at the interface between the ceramic and the metal wall, etc.

The selection of a ceramic material is determined largely by considerations other than mere insulating effect: (1) high melting or sublimation temperature of the basic substance, (2) absence of low melting point binders in the ceramic body, (3) chemical resistance to oxidizing or reducing action, depending on the oxygen balance in the combustion gas, (4) low specific gravity, or more generally, low product of specific gravity and thermal conductivity, (5) resistance to thermal shock or "spalling," which in itself requires a high ultimate strength and a low thermal expansivity, and (6) absence of phase transitions with severe density changes within the expected operating range. The usual considerations of availability and cost apply here, particularly since the highest temperature refractories such as B_4C and BeC are available only in small sizes at relatively great expense.

Convective cooling with surface boiling. In the design of the regeneratively cooled motor, the liquid at the heated side of the cooling duct can be prevented from boiling only if the flow velocity is made sufficiently high. As already indicated, this may require an excessive pressure differential or a mechanically unacceptable channel width. It has been found, however, that a slower velocity is allowable, down to a certain lower limit, and in this range boiling occurs on the metal surface (not in the bulk liquid which may have a temperature far below the boiling point). This is the phenomenon of forced convection with surface or "nucleate" boiling discussed in V,E.

When the temperature of the wall adjacent to the liquid is allowed to reach the boiling point at the pressure in the duct, by reducing the velocity, bubbling is seen to begin at the wall. Further reduction of the velocity results in more violent bubbling, but the wall temperature climbs only a few degrees. In effect, the film conductance has been increased by the agitation in the boundary layer. Ultimately, when the velocity has been reduced by 50 per cent or more below the value required to prevent boiling, the vapor formation at the surface has become so rapid that a large part of the surface is blanketed with vapor and the cooling process is disrupted. At this stage, the surface rapidly overheats because of the continuous generation of heat on the hot side of the wall, and an immediate burnout occurs. The burnout heat flux depends on the nature of the liquid, the flow velocity, the degree of subcooling (boiling point minus bulk temperature), the pressure, and the presence of dissolved gases or low boiling point solvents.

The designer usually allows for surface boiling in his design in order to economize on pressure drop, but he is aware of the danger of allowing the velocity to approach too close to the burnout limit.

Cooling with liquefied gases. Rocket motors operating with such propellant combinations as liquid oxygen plus liquid hydrogen, liquid oxygen plus liquid ammonia, liquid fluorine plus liquid diborane, etc., will have to be cooled by a liquefied gas if regenerative cooling is to be employed. In principle, the above discussions concerning convective cooling, with or without surface boiling, apply just as well to this case. Several special problems exist, however, which can be discussed in terms of liquid oxygen as an example.

The atmospheric boiling temperature is 162°R; the critical temperature and pressure are 278°R and 730 lb/in.^{2} abs. If the injection pressure is substantially less than 730 lb/in.^{2} abs, and if the oxygen coolant is admitted to the jacket at 162°R and emerges above the boiling point, the fluid entering the injection orifice will be an uncontrolled mixture of liquid and vapor bubbles, the injection rate will be unsteady, and combustion will be very rough. It is therefore desirable to keep the injection pressure well above 730 psia to avoid bubbling.

There are difficulties in proportioning the dimensions of the cooling duct due to the severe density changes resulting from the progressive warm-up of the supercritical oxygen as it flows along the duct. For example, oxygen at 1000 lb/in.^{2} abs has a density of 71 lb/ft^3 at 162°R (at the inlet), 23 lb/ft^3 at 300°R, and 7.2 lb/ft^3 at 450°R. An additional difficulty revealed by laboratory tests of forced convection with liquid oxygen is that the film conductance goes through a severe minimum in the vicinity of the critical temperature, 278°R. The effect is unexplained.

As yet, there is very little practical experience with liquefied gas cooling of rocket motors.

Internal film cooling. This method has been used in the past mainly as an aid to regenerative cooling, as in the V-2 motor, where it was necessary to incorporate film cooling to rescue a design that was found in the test firings to be inadequately cooled. Liquid, usually the fuel, is injected at low velocity, tangential to the wall if possible, through a ring of injection orifices drilled in the chamber wall just upstream of the overheated zone that needs protection. Ideally, the streams issuing from the orifices collectively form an annular sheet of liquid that flows downstream along the wall. The stream gets progressively thinner as a result of evaporation, until it becomes so thin that surface tension and gas flow disturbances cause break-up of the sheet; this is the end of the protected area, although some protection persists downstream because the boundary layer gas is relatively cool.

It is obvious that the metal wall temperature is no higher than the boiling temperature of the fuel at the pressure in the chamber. As a cooling method, it is very effective. However, there is a specific impulse penalty associated with the imperfect combustion of the fuel that is evaporated into the cool boundary layer, and the fuel stream used to protect the nozzle has no chance to enter the combustion process at all.

A typical magnitude of the flow rate required to protect the inner surface of a nitric acid—JP-4 chamber is 0.005 lb/sec-in.^{2} and about 3 times this rate at the throat. (The fundamental theory to calculate the required flow under various heat transfer conditions, together with the results of experimental researches on film cooling, are described in V,G.) On the basis of this figure, it would take about 5 per cent of the total propellant flow (about 20 per cent of the fuel flow) to fully protect a 10,000-lb-thrust motor in order to dispense entirely with regenerative cooling. The loss in specific impulse would probably be about 3 or 4 per cent.

In general, therefore, film cooling is useful as an auxiliary to, not as a replacement for, regenerative cooling. However, for very large motors, over 100,000-lb thrust, the loss in specific impulse might be less and therefore tolerable, as a result of the lower ratio of surface area to the main propellant flow. Also, for a propellant system such as hydrogen-fluorine with neither substance being suitable as a conventional regenerative coolant, an internal film of hydrogen may be the most practical solution.

Sweat cooling. Some of the practical objections to film cooling using orifice or slot injectors are that perfect surface coverage is difficult to achieve, a large fraction of the injected coolant is ineffective because it fails to evaporate uniformly, and the method is suitable for liquid coolants not for gas coolants. Sweat cooling, that is, admission of the coolant through a porous wall, represents an attempt to get around these objections. In principle, it achieves more uniform coverage and is less wasteful. The theory that permits the calculation of the flow rate required to maintain a desired wall temperature is covered in V,G.

The foremost obstacle to practical use of sweat cooling is the great difficulty of making combustion chambers with the required distribution of permeability, and with the required strength and weight. The original technique was to make the chambers of sintered powdered metal compacts, but this never worked satisfactorily for any but the smallest specimens. A more recent development which has yet to be accepted is spiral winding of fine wires, and then sintering. Another great difficulty is that a porous wall is also an excellent filter for suspended dirt in the fuel; the smallest traces cause local clogging and burnout. Development work in this field is still continuing.

In conclusion, it should be noted that the design of the cooling system for a rocket motor cannot be completely separated from the other aspects of motor design. Thus, a cooling difficulty may sometimes be due to an excessively "hot" injector, and it may be better to try another injector design than to accept extreme difficulties in the cooling system design. Redesign of the chamber shape or exhaust nozzle contour can sometimes alleviate the cooling problem. Also, a small change in the mixture ratio to increase the proportion of the fluid used as the coolant and perhaps to lower the flame temperature may be the best way out of a difficulty. A reduction in combustion pressure may be required if the optimum mixture ratio is to be used. In other words, the design of the motor must be treated as an integrated task.

G,7. Liquid Rocket Systems. The liquid rocket system is the entire complex of devices in the rocket-propelled vehicle that stores, feeds, and burns the liquid propellant. Usually, it is the assignment of the power plant manufacturer to deliver the complete system minus the tanks, instead of merely the rocket motor, because of the necessity for integrated functioning of the entire system. (The power plant manufacturer frequently is concerned with the design of the tanks, too, even if the airframe manufacturer makes them.) Similarly, the designer cannot set the specifications for the rocket motor without regard for the system as a whole.

The components of a liquid rocket system can be classified into four groups according to their functions: the rocket motor assembly, the propellant tanks, the pressurizing equipment, and the engine controls. Many different types of components have been used in liquid rocket systems, to suit the intended purpose. At the end of this article, there is a discussion of the general principles that govern the selection of the optimum system for a particular application.

Compressed gas system. The principal feature that distinguishes a liquid rocket system is the method of pressurizing the propellants for injection into the motor. Although it is a straightforward task to provide the necessary pressure and flow rate in a rocket motor test cell where weight is unimportant and electric power or engine power are unlimited, the problem is much more difficult for a system that must fly. The simplest pressurizing scheme is the regulated compressed gas type. Having no moving machinery other than valves, it has the advantage of great reliability. Its disadvantages can be serious, however, for certain applications. First, its weight increases rapidly with firing duration in a linear manner, due to the proportional increase in the volume and weight of the propellant tanks and the gas tank. This weight is prohibitive for durations longer than about 30 to 60 seconds. Second, the tanks must be spherical or cylindrical to withstand the high pressures, which sometimes complicates their placement in an airplane or other vehicle, although cylindrical tanks can be accommodated nicely in guided missiles or high altitude research rockets. Third, under some circumstances, the field problem of supplying high pressure gas is troublesome.

The principal components are the gas pressure tank, the propellant tanks, the pressure regulator(s), the gas starting valve, and the propellant flow control valves.

The pressure regulator is a servo-operated metering valve that senses the delivery pressure, compares it with the predetermined pressure setting, and so throttles the outflow of gas as to keep the difference nearly zero. The design of high volume regulators is a specialized task, but several of its performance characteristics deserve mention. A regulator, like any servomechanism with feedback, can hunt or "chatter"; introduction of damping or modification of the orifice and pintle shapes may be required. Condensation or freezing of moisture or oil, if present in the gas supply, or even of the gas itself, may occur on the seat as a result of expansive cooling. The delivery pressure may climb or fall appreciably, as the inlet pressure diminishes during the course of a firing, because of inexact balance of pressures on the throttling pintle. Finally, if the propellant valves are shut off in the middle of a firing, the regulated pressure may climb considerably above that delivered during normal operation, due to the large actuating force required to close the valve tightly enough to stop the flow.

The gas starting valve located upstream of the regulator provides a secure means of retaining the high pressure gas until the start of firing, and serves to shut off the gas flow when necessary to avoid any dangerously high "lock-up" pressure delivered by the regulator.

The propellant flow control valves are of various types and serve several functions. A positive seal before the start of firing is provided by frangible diaphragm units, designed to burst open at some fraction of the starting feed pressure. Throttling valves are provided to gradually increase the flow rates according to a definite program during the start-up transient. Shut-off valves are provided in one or both lines to terminate the thrust promptly if desired. Many additional valves appear in even the simplest rocket system that have to do with filling the propellant tanks, charging the gas tank, venting the tanks, draining the system at various points, checking reverse flow, relieving excessive pressure, etc.

Several considerations determine the volume of the compressed gas tank. First is the charging pressure: the higher the pressure, the more compact the tank, but at the same time the more difficult the logistic supply problem. Pressures between 2000 and 5000 $lb/in.^2$ have been used, with the current trend toward the higher pressures as suitable sources become available. Second is the type of gas: it must be available at the required pressure at the launching site; it must not react with either propellant or dissolve too easily in them; it must not condense in the throttle of the regulator; it should have the lowest possible density. These considerations rule out, for example, hydrogen or carbon dioxide, but favor helium, nitrogen, or air.

The ideal throttling process is isothermal, for a perfect gas like helium, and is nearly so except for a small Joule-Thompson drop for nitrogen and air. The expansive cooling mentioned above is localized in the high velocity gas region of the throttle, where the process is nearly isentropic, but the temperature at the outlet is nearly equal to the inlet temperature. The inlet temperature falls steadily from the start of firing, if the system is adiabatic, as a consequence of the isentropic expansion in the gas tank. On the basis of this description, it is possible to integrate the gas flow rate to determine the mass of gas m_g needed in a gas tank of volume $\mathcal{V}_g$ to displace the propellants and fill the propellant tanks of volume $\mathcal{V}_p$ with gas at regulated pressure p_p, when the gas tank is charged at p_1 and runs down to pressure p_2:

$$m_g = \frac{p_p \mathcal{V}_p \mathfrak{M}}{RT_1}\left(\frac{\gamma}{1 - p_2/p_1}\right) = \frac{p_1 \mathcal{V}_g \mathfrak{M}}{RT_1} \tag{7-1}$$

If the over-all expansion were isothermal in the tank as well as in the regulator, and if all the gas in the gas tank were used in the propellant tanks, the quantity in brackets would be unity, and the remaining factor is easily recognized as the result of Boyle's law. In practice, the actual m_g can be somewhat less than Eq. 7-1 since some heat is actually transferred to the gas during the expansion, and since it is quite feasible to permit p_2 and p_p to fall below the nominal design levels in the last part of the firing period. The above formula makes no allowance, of course, for gas that dissolves in the propellants.

A typical figure for the mass of air required for a rocket like the Aerobee is 5 per cent of the propellant mass; for helium, it would be about 1 per cent. The switch from air to helium is obviously desirable for high performance missiles and research rockets, and especially so because of the solubility problem in the case of liquid oxygen systems. Additional savings in the inert gas weight can be effected by warming the gas before it is admitted to the propellant tank, by allowing it to absorb some of the heat of the rocket motor in a suitable heat exchanger. An important saving in the weight of the gas tank is possible in principle, but complicated in practice, by charging with liquefied gas into a well-insulated gas tank and gasifying it with an auxiliary heater when the gas is to flow into the propellant tanks.

Turbopump system. The weight of the gas tank, the pressure-stressed propellant tanks, and the gas in the compressed gas system usually amounts to about 15 to 20 per cent of the propellant weight. For some applications this weight penalty can be reduced sharply by using a pumping system with lightweight, thin shell, propellant tanks. The most common power source to drive the propellant pumps is a chemical gas-driven turbine, the combination comprising a turbopump. The following considerations determine the design of a turbopump.

The turbine shaft power required to drive the pumps to supply the rocket motors can be calculated as follows:

$$P_t = P_f + P_o = \frac{\Delta p_f \dot{m}_f / \rho_f}{\eta_f} + \frac{\Delta p_o \dot{m}_o / \rho_o}{\eta_o} \tag{7-2}$$

The pump efficiencies η_f and η_o are defined as the ratio of hydraulic output power to shaft input power for the fuel and oxidizer flows respectively, as implied in Eq. 7-2. The efficiency of a pump can be determined in advance in a water test flow system; it varies over a wide range according to the type of design and the particular operating point. For rocket pumps, with their particular limitations, an efficiency of 70 per cent is quite good, and 40 per cent may be acceptable.

The driving turbine is usually a simple-, single-, or two-stage, impulse-type unit, driven by gases generated at a moderate temperature by decomposition or combustion of a propellant. The mass flow $\dot{m}_t$ consumed by the turbine to deliver a shaft power P_t can be calculated according to a measured or computed efficiency η_t:

$$P_t = \dot{m}_t(h_1 - h_2)_a = \eta_t \dot{m}_t (h_1 - h_2)_i \cong \eta_t \dot{m}_t \bar{C}_p T_1 \left[1 - \left(\frac{p_2}{p_1}\right)^{\frac{\gamma-1}{\gamma}}\right] \tag{7-3}$$

In this equation, the ideal enthalpy drop $(h_1 - h_2)_i$ is replaced by the equivalent expression for a near-perfect gas that enters at pressure p_1

and temperature T_1 and leaves at pressure p_2 after an isentropic work process. The departure from isentropicity is measured by the efficiency η_t, which indicates the fraction of the ideal enthalpy drop that actually occurs in the turbine, $(h_1 - h_2)_s$. To carry out precise turbine performance calculations, a Mollier diagram is required. Typical values of η_t in existing rocket engines range from 30 to 40 per cent. These relatively low values are the result of adopting simple one- or two-stage turbines to minimize the fixed installation weight; the penalty is a higher rate of turbine propellant consumption and a larger gas generator. Some designers have kept the weight down also by using aluminum alloy turbines with aluminum blades, even though this necessitated a limit on the temperature of the driving gas of about 725°F.

From Eq. 7-2 and 7-3, it is clear that the mass flow of turbine propellant varies inversely as the products $\eta_f\eta_t$ and $\eta_o\eta_t$ and directly as the feed pressure. Typical values of turbine flow are as low as 1.5 per cent of the rocket engine flow for large missiles, and as much as 5 per cent for engines of less than 10,000-lb thrust. A large part of the difference results from the wide variations in pump and turbine efficiencies, which depend on optimization studies that seek to balance the larger dead weight of an efficient turbopump against the higher consumption rate of an inefficient one. Typical turbine powers are about 0.01 to 0.02 hp/lb thrust. The V-2 turbopumps delivered 465 hp at rated operating conditions.

Inasmuch as the over-all specific impulse of a rocket engine must be based on the total propellant flow which includes the turbine flow, it is clear that turbopump performance analysis is just as important as the analysis given in Art. 3 of the corrections to ideal rocket motor theory. For this purpose, the reader is referred to X,F for the principles of axial flow turbines and to X,J for the principles of centrifugal compressors. Certain special design features of rocket turbopumps deserve mention here, however.

Light weight is achieved by the use of a single shaft, without speed changer, to drive the pump impellers. The high rotary speed demanded by the single-stage turbine usually means that a single centrifugal impeller is adequate to develop the full pressure required by the rocket engine. Unfortunately, the development of high pressure in a single high speed impeller of small diameter means a high relative inlet velocity at the impeller eye and a strong tendency toward cavitation. Cavitation, the boiling of the liquid on the pump vanes and in the inlet passage due to the reduced static pressure, results in a reduction in fluid delivery and, if it is extensive, in damaging vibration. To suppress cavitation, it is necessary to boost the pressure at the pump inlet. Booster pumps, if they are not to cavitate themselves, must be slow speed, large, and therefore heavy devices, and need a driving power source. The alternative frequently employed is gas pressurization of the tanks to about 5 or 10 per cent of the main pump pressure. The choice between booster pumps and gas pressurization depends largely on the necessary tank configuration and is decided by a weight optimization analysis.

The hot gas supply to drive the turbine is created by a combustion or decomposition reaction in a gas generator or "combustion pot." The most commonly used gas generation propellant is concentrated hydrogen peroxide. The range of turbine inlet temperatures that can be obtained with various water concentrations in the peroxide is as follows: For aluminum alloy turbines, the maximum allowable temperature is about 1000°F; for steel alloys, about 2000°F. (These are somewhat higher than the temperatures allowed for relatively long-life turbomachines.) The gas generator system is quite similar to a rocket system, involving a propellant tank, a pressurizing device, flow controls, as well as a combustion chamber. Propellants other than concentrated hydrogen peroxide have been used or proposed as gas generants.

The hot gas that leaves the driving turbine can either be discharged rearward with moderate back pressure to add a small amount of thrust to the engine, or it can be used as a heat source to vaporize, for example, a liquefied pressurizing gas, or it can be used to actuate some devices in the flight control or engine control system. The back pressure can usually be accommodated with a small loss of turbine efficiency.

Auxiliary devices are sometimes connected to the driving shaft of the turbine, for example, an electric generator for large electric power requirements or a hydraulic pump to actuate various controls. The extra propellant consumption to provide this power must be included in the over-all specific impulse computation.

The first successful U. S. turbopump-type rocket engine, developed in 1945, operated on nitric acid with an aniline-furfuryl alcohol mixture as the fuel. The driving gas was generated by burning the same propellants together with a large diluent flow of water. The German ME-163 rocket interceptor plane was propelled by a turbopump-type engine in which the driving gas was produced by hydrogen peroxide decomposition and the propellants were peroxide and a hydrazine-alcohol-water mixture.

Other systems. It was pointed out above that the main distinguishing feature of a liquid rocket system is the propellant pressurizing scheme. In addition to the two principal methods based on pressurized gas and turbopumps, other methods have been proposed or actually used in one or more rockets:

1. A solid propellant cartridge can be used instead of the high pressure gas source, the propellants being pressurized by the combustion products. A system of this type was employed in the German antiaircraft "Taifun."
2. Controlled combustion can be allowed to take place in the fuel tank by

injecting a small spray of oxidizer, and vice versa, thereby creating pressure in the tanks. This has been tried experimentally, but never used in practice. Similarly, the gases produced in a separate gas generator can be used directly to pressurize the propellants.

3. A novel version of the turbopump system is the so-called jet blast type, in which the turbine is driven by the rocket exhaust jet. In the only fully developed unit of this type, the turbine is a single, high speed, axial flow wheel, only partially immersed in the hot blast. This partial immersion allows air cooling of the blades. Although it has the advantage of very light weight, a number of serious starting, regulating, and operating difficulties have put it in disfavor.
4. Another ingenious mechanism is the so-called rotojet with the rocket motors mounted on a rotatable shaft at a small angle, so as to create rotary shaft power to drive the pumps when thrust is being delivered. The development was not successful, due to mechanical difficulties, sealing problems, and erratic combustion. A similar device, running at even higher speed to create the required feed pressure by centrifugal action, was developed experimentally but ran into comparable difficulties.
5. It has been proposed to use the injector principle commonly used to inject feed-water into steam boilers, the propellants to be heated in a boiler surrounding the rocket motor. The starting problem and the difficulties of heating propellants to high temperature seem to make this scheme impractical.
6. Various types of mechanical power can be employed to drive the propellant pumps, depending on the conditions of the application. An auxiliary rocket engine for a turbojet airplane can draw pumping power from the main gas turbines by bleeding air from the compressor discharge to run an air turbine. Similarly, the pumps can be mounted as accessories on the gear box of the engine, if it is of the reciprocating type. A JATO rocket engine of this type was used in World War II. Finally, a separate auxiliary gas turbine or reciprocating engine can be used to power the pumps.

System selection. The optimum liquid rocket system for a particular application is determined on the basis of the following considerations:

1. Minimum weight: In the competition for minimum weight between the two main system types, compressed gas and turbopump, it is evident that the firing duration is the principal controlling factor. For short durations, the large fixed weight of the turbopump and gas generation system is a penalty; for long durations, the very heavy tank system of the compressed gas type is objectionable. The cross-over point is about 15 seconds for high thrust engines, 30 seconds for small engines. This comparison can be made only after each system is optimized

with respect to its operating parameters. The most important of these is the combustion pressure—the higher the pressure, the lower the propellant consumption rate for a given thrust; but at the same time, the greater the weight of the pressurization system. For the compressed gas type, the optimum pressure usually works out to lie between 250 and 400 lb/in.^{2}; for the turbopump system, the optimum is usually near 1000 lb/in.^{2}, but a lower pressure is always employed to ease the cooling problem. It is clear that the turbopump type is the most suitable one for long range, large size rockets with long firing durations.

2. Simplicity and ruggedness: The great simplicity of the compressed gas type makes it the most appropriate system for such applications as small anti-aircraft guided missiles, which have to be ready for instantaneous firing with high reliability with a minimum of prefiring inspection and adjustment. The same consideration applies to detachable JATO units.
3. Low cost: The consideration of cost is important mainly for civilian applications such as the take-off of heavy cargo planes, and perhaps rocket travel in the future. For military purposes, cost is significant mainly as an indication of manufacturing difficulty, and, therefore, for missiles or JATOs required in very large numbers, the gas-pressurized type is preferred.

G,8. Bibliography.

Article 1.

Dictionary of guided missile terms. Compiled by Ordnance Dept., U.S. Army. *Anti-Aircraft J.*, 1949.

Glossary of guided missile terms. Prepared by *Committee on Guided Missiles, Research and Develop. Board, U.S. Dept. of Defense, GM 51/3 and GM 51/8,* Sept. 1948.

Layton, J. P., and Youngquist, R. Proposed A.S.A. letter symbols for rocket propulsion. *Jet Propulsion 25*, 634 (1955).

Article 2.

Malina, F. J. Characteristics of the rocket motor unit based on the theory of perfect gases. *J. Franklin Inst. 230*, 433 (1940).

Seifert, H. S., and Crum, J. Thrust coefficient and expansion ratio tables. *Ramo-Wooldridge Corp.*, Feb. 1956.

Seifert, H. S., Mills, M. M., and Summerfield, M. The physics of rockets, Part I. *Am. J. Phys. 15*, 1 (1947).

Shapiro, A. H. *Dynamics and Thermodynamics of Compressible Fluid Flow*, Vol. I, Chap. 4. Ronald Press, 1953.

Sutton, G. P. *Rocket Propulsion Elements*, 2nd ed., Chap. 3. Wiley, 1956.

Article 3.

Banerian, G. Rocket performance with heat added to inlet propellants by regenerative and external means. *Jet Propul. 25*, 712 (1955).

Beckwith, I. E., and Moore, J. A. An accurate and rapid method for the design of supersonic nozzles. *NACA Tech. Note 3322*, Feb. 1955.

Dillon, P., and Line, L. E., Jr. Heat transfer between solid particles and gas in a rocket nozzle. *Jet Propul. 26*, 1091 (1956).

Durham, F. P. Thrust characteristics of underexpanded nozzles. *Jet Propul. 25*, 696 (1955).

Gilbert, M., Davis, L., and Altman, D. Velocity lag of particles in linearly accelerated combustion gases. *J. Am. Rocket Soc. 25*, 26–30 (1955).

Glassman, I. Impulse expressions for rocket systems containing a solid phase. *Jet Propul. 27*, 542 (1957).

Kogan, A. Boundary layer correction in supersonic nozzle scaling. *J. Aeronaut. Sci. 25*, 64 (1958).

Puckett, A. E. Supersonic nozzle design. *J. Applied Mech. 13*, A265-A270 (1946).

Seifert, H. S., and Altman, D. A comparison of adiabatic and isothermal expansion processes in rocket nozzles. *J. Am. Rocket Soc. 22*, 159–162 (1952).

Shapiro, A. H. *The Dynamics and Thermodynamics of Compressible Fluid Flow*, Vol. I, Chap. 7 and 15. Ronald Press, 1953.

Summerfield, M., Foster, C. R., and Swan, W. C. Flow separation in overexpanded supersonic exhaust nozzles. *Jet Propul. 24*, 319 (1954).

Sutton, G. P. *Rocket Propulsion Elements*, 2nd ed., Chap. 3. Wiley, 1956.

Article 4.

Battelle Memorial Inst. Physical properties and thermodynamic functions of fuels, oxidizers, and products of combustion. *Project Rand Repts. R-127, 129, 196*. Columbus, Ohio, 1949.

Charts of theoretical performance of several rocket propellant combinations. *Rocketdyne, Div. North Amer. Aviation, Inc.*, 1956.

Donegan, A. J., and Farber, M. Solution of thermochemical propellant calculations on a high speed digital computer. *Jet Propul. 26*, 164–172 (1956).

Fickett, W., and Cowan, R. D. Values of thermodynamic functions to 12,000°K for several substances. *Los Alamos Sci. Lab. Rept. 1727*, Sept. 1954.

Hottel, H. C., Williams, G. C., and Satterfield, C. N. *Thermodynamic Charts for Combustion Processes*. Wiley, 1949.

Huff, V. N., and Calvert, C. S. Charts for computation of equilibrium composition of chemical reactions in the C-H-O-N system from 2000°K to 5000°K. *NACA Tech. Note 1653*, July 1948.

Huff, V. N., and Gordon, S. Tables of thermodynamic functions for analysis of aircraft propulsion systems. *NACA Tech. Note 2161*, Aug. 1950. (Includes light metals and halogens, as well as C, H, O, and N compounds.)

Huff, V. N., and Morrell, V. E. General method for computation of equilibrium composition and temperature of chemical reactions. *NACA Tech. Note 2113*, June 1950.

Krieger, F. J. Chemical kinetics and rocket nozzle design. *J. Am. Rocket Soc. 21*, 179–185 (1951).

Lewis, B., and von Elbe, G. *Combustion, Flames and Explosions of Gases*, Chap. 13. Academic Press, 1951.

Natl. Bur. Standards. Selected values of properties of hydrocarbons. *Circular 461*, Gov't. Printing Office, 1947.

Natl. Bur. Standards. Tables of selected values of chemical thermodynamic properties, 3 vol. *Circular 500*, Gov't. Printing Office, 1952.

Penner, S. S. Thermodynamics and chemical kinetics of one-dimensional nonviscous flow through a Laval nozzle. *J. Chem. Phys. 19*, July 1951.

Pocket Data Book for Rocket Engines. Bell Aircraft Corp., Buffalo, N.Y., 1954. (Properties of fuels, oxidizers, and propellant combinations.)

Sutton, G. P., *Rocket Propulsion Elements*, 2nd ed., Chap. 4. Wiley, 1956.

Vichnievsky, R., Sale, B., and Marcadet, J. Combustion temperatures and gas composition. *J. Am. Rocket Soc. 25*, 105 (1955).

Article 5.

Baker, D. I. Mixture ratio and temperature survey of ammonia-oxygen rocket motor combustion chambers. *Jet Propul. 25*, 217 (1955).

Barrere, M., and Moutet, A. Instabilities of low frequency in a rocket motor. *Recherche aéronaut. Paris 44*, 29 (1955).
Baxter, A. D. Combustion in the rocket motor. *J. Brit. Interplanetary Soc. 10*, May 1951.
Berman, K., and Cheney, S. H. *Rocket Motor Instability Studies 25*, 513 (1955).
Crocco, L. Aspects of combustion instability in liquid propellant rocket motors. *J. Am. Rocket Soc. 21*, 163 (1951); *22*, 7 (1952).
Crocco, L. Considerations on the problem of scaling rocket motors. *Selected Combustion Problems*, Vol. 2. Butterworths, London, 1956.
Crocco, L., and Cheng, S. I. *Theory of Combustion Instability in Liquid Propellant Rocket Motors.* Butterworths, London, 1956.
Diplock, B. R., Lofts, D. L., and Grimston, R. A. Theory, design, and development of liquid propellant rocket motors. *J. Roy. Aeronaut. Soc. 57*, Jan. 1953.
Gunn, S. V. The effects of several variables upon the ignition lag of hypergolic fuels oxidized by nitric acid. *J. Am. Rocket Soc. 22*, 33–38 (1952).
Penner, S. S., and Fuhs, A. E. On generalized scaling procedures for liquid fuel rocket engines. *Combustion and Flame Quart. 1*, June 1957.
Ross, C. C. Scaling of liquid fuel rocket combustion chambers. *Selected Combustion Problems*, Vol. 2. Butterworths, London, 1956.
Ross, C. C., and Datner, P. P. Combustion instability in liquid propellant rocket motors—A survey. *Selected Combustion Problems*, Vol. 1. Butterworths, London, 1954.
Stehling, K. R. Injector spray and hydraulic factors in rocket motor analysis. *J. Am. Rocket Soc. 22*, 132–138 (1952).
Summerfield, M. A theory of unstable combustion in liquid propellant rocket motors. *J. Am. Rocket Soc. 21*, 108 (1951).
Sutton, G. P. *Rocket Propulsion Elements*, 2nd ed., Chap. 7. Wiley, 1956.
Tischler, A. O., and Bellman, D. R. Combustion instability in an acid-heptane rocket with a pressurized gas propellant pumping system. *NACA Tech. Note 2936*, 1953.
Tischler, A. O., and Dalgleish, J. E. Experimental investigation of a lightweight rocket chamber. *NACA Research Mem. E52L19A*, Dec. 1952.
Trent, C. H. Investigation of combustion in rocket combustion chambers. *Ind. Eng. Chem. 48*, 749 (1956).
Tsien, H. S. Servo-stabilization of combustion in rocket motors. *J. Am. Rocket Soc. 22*, 256 (1952).

Article 6.

Bartz, D. R. A simple equation for rapid estimation of rocket nozzle convective heat transfer coefficients. *Jet Propul. 27*, 49 (1957).
Boden, R. H. Heat transfer in rocket motors and the application of film and sweat cooling. *Trans. Am. Soc. Mech. Engrs. 73*, May 1951.
Cheng, C. M. Resistance to thermal shock. *J. Am. Rocket Soc. 21*, 147 (1951).
Dunn, L. G., Powell, W. B., and Seifert, H. S. Heat transfer studies relating to rocket power plant development. *Proc. Third Anglo-American Aeronaut. Conference*, Brighton, England, 1951. (Published by Roy. Aeronaut. Soc.)
Duwez, P. and Wheeler, H. L., Jr. Experimental study of cooling by injection of a fluid through a porous material. *J. Aeronaut. Sci. 15*, 509–521 (1948).
Eckert, E. R. G. Convective heat transfer at high velocities. *Heat Transfer Symposium*, Univ. Mich. Press, 1952.
Eckert, E. R. G. Transpiration and film cooling. *Heat Transfer Symposium*, Univ. Mich. Press, 1952.
Ellion, M. E. New technique for obtaining heat-transfer parameters of the wall and combustion gas in a rocket motor. *Trans. Am. Soc. Mech. Engrs. 73*, 109–114 (1951).
Elsasser, W. M. Mean absorption and equivalent absorption coefficient of a band spectrum. *Phys. Rev. 54*, 126 (1938).
Geckler, R. D. Transient, radial heat conduction in hollow circular cylinders. *Jet Propul. 25*, 31–35 (1955).

Greenfield, S. Determination of rocket-motor heat-transfer coefficients by the transient method. *J. Aeronaut. Sci. 18*, 512–518 (1951).
Jakob, M. *Heat Transfer*, Vol. I, Chap. 4. Wiley, 1949.
John, R. R., and Summerfield, M. Effect of turbulence on radiation intensity from propane-air flames. *Jet Propul. 27*, 169 (1957).
Kennard, E. H. *Kinetic Theory of Gases*. McGraw-Hill, 1938.
Kreith, F. and Summerfield, M. Heat transfer to water at high flux densities with and without surface boiling. *Trans. Am. Soc. Mech. Engrs., 71*, 805 (1949).
Manson, S. S. Behavior of materials under conditions of thermal stress. *Heat Transfer Symposium*, Univ. Mich. Press, 1952.
McAdams, W. H., *Heat Transmission*, 3rd ed., Chap. 4, 7, and 8. McGraw-Hill, 1954.
Natl. Bur. Standards. Tables of thermal properties of gases. *Circular 564*, U.S. Gov't. Printing Office, 1955.
Penner, S. S. Emissivity calculations for diatomic gases. *J. Appl. Mech. 18*, 53 (1951).
Rohsenow, W. M. Heat transfer with evaporation. *Heat Transfer Symposium*, Univ. Mich. Press, 1952.
Saunders, O. A. Heat transfer on a nozzle at supersonic speeds. *Engineering 173*, Aug. 1952.
Summerfield, M. Fundamental problems in rocket research. *J. Am. Rocket Soc. 81*, 79–98 (1950).
Summerfield, M. Recent developments in convective heat transfer with special reference to high temperature combustion chambers. *Heat Transfer Symposium*, Univ. Mich. Press, 1952.
Sutton, G. P. *Rocket Propulsion Elements*, 2nd ed., Chap. 7. Wiley, 1956.
Symposium on Transport Properties in Gases. Northwestern Univ. Press, 1958.
Von Bahr, E. Influence of pressure on the absorption of infra-red rays by gases. *Ann. Physik 33*, 585 (1910).
von Karman, Th. The analogy between fluid friction and heat transfer. *Trans. Am. Soc. Mech. Engrs. 61*, 705 (1939).

Article 7

Gore, M. R., and Carroll, J. J. Dynamics of a variable thrust, pump-fed, bipropellant liquid rocket engine system. *Jet Propul. 27*, 35 (1957).
Reichel, R. H. The importance of mixture ratio control for large rocket vehicles. *Jet Propul. 25*, 291–293 (1955).
Ross, C. C. Principles of rocket-turbopump design. *J. Am. Rocket Soc. 84*, 21–33 (1951).
Stehling, K. R., and Diamond, P. M. Flow controls. *J. Am. Rocket Soc. 23*, 178–183 (1953).
Sutton, G. P. *Rocket Propulsion Elements*, 2nd ed., Chap. 8. Wiley, 1956.
Tsien, H. S. *Engineering Cybernetics*, Chap. 5 and 8. McGraw-Hill, 1954.

SECTION H

SOLID PROPELLANT ROCKETS

C. E. BARTLEY
M. M. MILLS

CHAPTER 1. GENERAL FEATURES OF SOLID PROPELLANT ROCKETS

H,1. Introduction. Despite their fundamental simplicity, solid propellant rockets encompass a vast variety of detailed techniques. Just because of this simplicity, a realistic discussion of rockets requires a discussion of at least some of this technology, but the space available here is much too limited for a comprehensive treatment.

The choice has been made to emphasize the "power plant" aspects of the rocket. Propellant characteristics, theory of stability, internal ballistics, propellant grain, and combustion chamber design are discussed in some detail. On the other hand, vehicle problems, propulsive efficiency, missile accuracy, and related problems are mentioned only in a very qualitative way. Some discussion of these problems is required, because they often have a strong influence on the internal design of the rocket.

During World War II, an intensive rocket development effort was undertaken by most of the belligerents, chiefly aimed at the development of artillery rockets. Applications to long range missiles, guided missiles, and boosting, such as the assisted take-off of aircraft, have grown with the continuing improvements in the art of rocketry.

Artillery rockets, rocket-propelled missiles employed in much the same manner as gun artillery, have become important as an adjunct to guns in mobile warfare. Such missiles may range from 2 to 14 in. in diameter, from 20 to 120 in. in length, and from 3 to 1500 lb in weight. Since the rocket launcher weighs about the same as one round of rocket ammunition, artillery rocket weapons are much lighter and more mobile than corresponding guns. The rocket launcher does not have to confine the propelling gases, as the gun tube, nor does it have to absorb a large recoil; it serves merely as an aiming device. For example [*1*], the 75-mm gun fires a 13.5-lb missile and weighs nearly 2700 lb. An artillery rocket with a 14-lb payload would weigh about 24 lb total, and could probably be fired from a 26-lb launcher. Artillery rockets are usually inferior to guns in range, accuracy, and target penetration; but their superior mobility can compensate for this in many circumstances.

Rockets fired from aircraft, because the aircraft velocity may be added to the rocket velocity, are nearly as effective as a gun-fired shell of the corresponding caliber. A medium bomber can fire a salvo of 5-inch diameter rockets equivalent to a destroyer broadside. Rocket fire power can be made very large by utilizing multiple launchers or automatic launchers. Firing rates of 200 rounds of 5-inch missiles per minute are easily obtained.

Boosting applications are becoming increasingly important. The take-off of fast aircraft, otherwise requiring long runways, may be assisted by solid propellant rockets. Missiles frequently employ solid propellant booster rockets for launching, with sustaining thrust supplied by a variety of power plants such as the ramjet, the turbojet, and the liquid propellant rocket or solid propellant rocket.

Solid propellants obtain their name from the fact that they have mechanical properties resembling the mechanical properties of solid materials. Some propellant materials are actually very viscous fluids with peculiar thixotropic characteristics. A combination of properties is usually desired. A fairly rigid material is necessary so that the burning surface may be accurately defined and controlled, but a moderate degree of elasticity is required to permit the charge to accommodate itself to pressure and thermal stresses without cracking.

The chief advantage of solid propellant rockets, as compared to liquid propellant rockets, is their greater ease of utilization. After preparation, at the point of manufacture, they are easily transported and may usually be fired with simple electrical circuits. Elaborate field servicing equipment is usually not needed. For some applications, solid propellant rockets also can be superior to liquid propellant rockets in total impulse-to-total weight ratio.

The major advances in the development of solid propellant rockets during World War II were: the development of new propellants which were more suitable than gun propellants or black powder for rocket propulsion, the discovery and quantitative analysis of special characteristics of these propellants which strongly affect rocket design and performance, the invention of propellant materials with good mechanical properties capable of withstanding pressure and acceleration forces without failure, and the discovery of suitable techniques which could be employed to limit burning to certain surfaces of the propellant charge. Among the most significant advances was the development of methods to fabricate large propellant grains, of high quality, to precise geometrical shape.

Secondary developments of considerable importance were the development of suitable ignitor arrangements, methods of retaining propellant charges within the rocket motor, and techniques of exhaust nozzle and combustion chamber design. In addition, for artillery rockets, new fuse mechanisms for arming the explosive warhead had to be developed, since conventional shell fuses would not operate satisfactorily at low rocket accelerations.

The theory and semitheoretical correlation of factors important for both exterior and interior ballistics proceeded quite rapidly, and the fundamental understanding of rocket behavior and its dependence upon various parameters is now quite good. Rocketry grew from an art to a science, although much still remains to be done.

The central problem of interior ballistics is the combustion process of the solid propellant. Empirical relations, adequate for rocket design and analysis, were quickly established, but fundamental theoretical understanding is just now being reached [*2*,*3*,*4*,*5*] (see especially II,M). Additional problems, in the rocket motor, arise from the interaction of the gas flow resulting from combustion and the combustion process itself. The combustion mechanism is difficult to analyze because it is a two-phase process, the solid propellant transforms to a gas, and, because combustion is a rate process, the powerful tools of conventional thermodynamics are not applicable. Furthermore, the combination of chemical and gas dynamical features important in combustion make it a very complex process [*6*].

Exterior ballistic problems, very important for unguided missiles, are, in principle, rather more straightforward than problems of interior ballistics. These studies center around the problem of accuracy, and are concerned chiefly with the motion of a rigid body subject to aerodynamic forces and the rocket thrust force. As a matter of fact, the large number of factors that must be considered and the intricacy of the motion of a rocket missile has made these studies very difficult and at times has led to misunderstandings.

It is the purpose of the following discussion to sketch the main features of rocket theory and design. The large number of detailed considerations, quite important in any practical design, are mentioned.

H,2. Outline of Construction and Operation.

Shape. A rocket motor usually has the form of a circular cylinder, with length-to-diameter ratio considerably in excess of unity. In order to achieve a thrust exceeding its weight, and also good propellant utilization, the rocket must operate at internal pressures of several atmospheres. It follows that the combustion chamber must be a pressure vessel. Since it is desirable to have a large over-all impulse-to-weight ratio, the weight of the combustion chamber should be kept to a minimum. Stress considerations then indicate that spherical or circular cylindrical combustion chamber shape is desirable. A regular shape for the propellant charge is desirable, so that the burning surface area will not change widely during combustion. A large variation in burning surface area will produce extremely large variations in chamber pressure. The combustion chamber must then be designed to withstand a high peak combustion chamber pressure with a resultant increase in weight. Since it is much more feasible to design a regular propellant charge shape for the cylindrical combustion chamber than for a sphere, the cylindrical combustion chamber is usually preferred, despite a somewhat larger weight per contained volume than for a sphere. Rockets utilized to propel missiles usually have additional requirements of aerodynamic drag and stability, which make a cylindrical shape almost mandatory. Considerations of convenience in storage and transportation, and availability of seamless tubing, also make the cylindrical shape desirable.

Manner of operation. Before discussing solid propellant rockets further, it is desirable to describe their manner of operation and to outline the qualitative relationships which govern rocket performance.

The rocket unit consists of a charge of solid propellant within a combustion chamber and an exhaust nozzle through which the products of combustion of the solid propellant escape. The reaction due to the expulsion of the combustion products supplies the thrust of the rocket. The propellants used in rockets, just as those used in guns, do not explode but instead burn away at a definite rate on those surfaces which are exposed to the hot gases or flame within the combustion chamber. The burning rate depends upon the chamber pressure, increasing with higher pressures. Now, the thrust of a rocket motor may be considered as equal to the product of exhaust velocity and mass flow, so that, in order to get a large thrust, a large burning surface must be used to obtain a large mass flow. Similarly, to obtain a long duration of thrust only a small portion of the propellant charge must burn at a time. Since a combustion chamber of a given size contains only a certain weight of propellant, the thrust may be made large for a short time by providing a large burning surface, or small for a long time by providing a small burning surface.

The general pressure level within a rocket unit depends strongly on the ratio of the area of the burning surface to the cross-sectional area of the exhaust nozzle throat. For this reason, propellant-charge shapes are usually designed so that the area of the burning surface does not change extensively as the charge burns away. This requirement usually leads to regular geometric shapes for the propellant charge and to the arrangement of layers or strips of poorly combustible materials placed on the charge to prevent burning of certain areas (these layers are often called restrictors). In Fig. H,2a a variety of charge shapes which have been employed for rockets is shown.

The thrust of a rocket is proportional to the area of the burning surface, and depends upon the chamber pressure, the kind of propellant, and the internal geometry of the combustion chamber and propellant

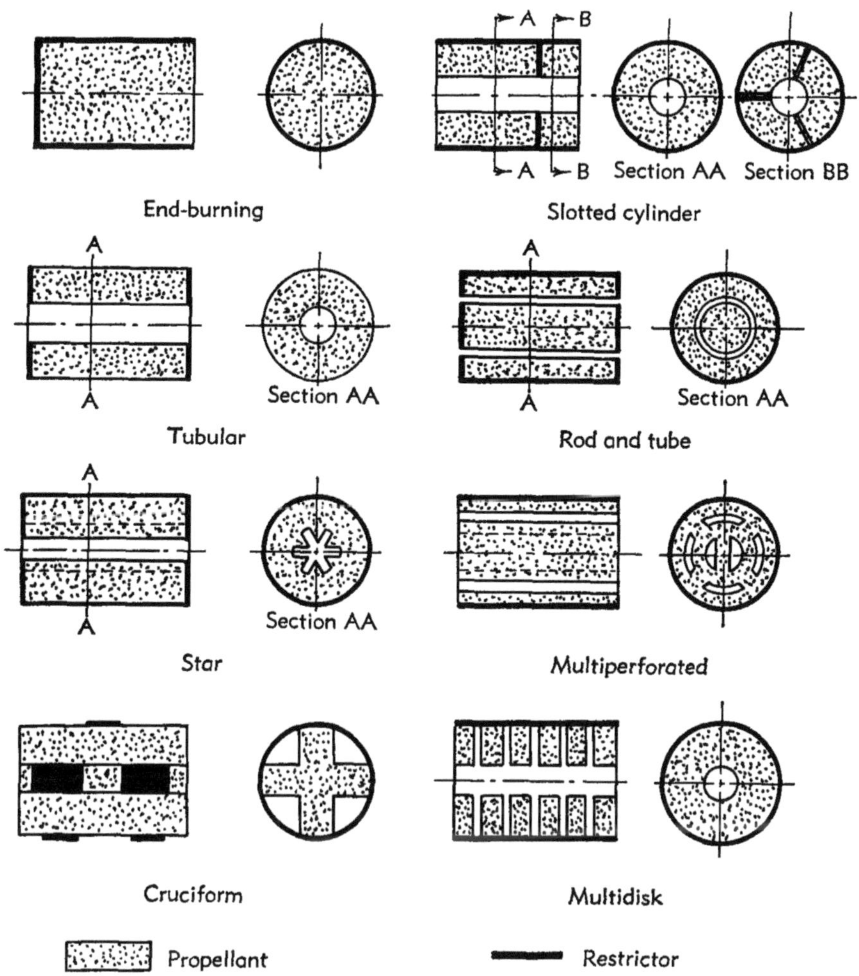

Fig. H,2a. The geometry of solid propellant charges.

grain. The duration of thrust burning time is proportional to the web thickness of the propellant grain (the distance that the burning surface must recede in order to consume the propellant).

As mentioned above, the chamber pressure depends strongly on the ratio of the area of the burning surface to the exhaust nozzle discharge area, increasing rapidly as this ratio increases. The pressure also depends upon the internal geometry of the combustion chamber and the propellant charge.

Referring again to Fig. H,2a the simplest arrangement is the end-burning, or cigarette-burning geometry. Since a long grain may be used, the end-burning grain is an excellent method to obtain long duration of thrust. Durations from 1 to 500 seconds have been obtained in this way.

Internal-burning grains, star center and multiperforated in the examples shown, are becoming the most widely used. Although this geometry may lead to combustion stability problems, it minimizes the problem of combustion chamber heating. The grain itself insulates the combustion chamber wall until the final bits of propellant are consumed. In general, arrangements other than end-burning lead to complex problems associated with gas flow past the burning grain.

Practical developments. The solid propellant rocket, in its essential features, is remarkably simple. However, the development of a practical

Table H,2a. General characteristics of two types of solid propellant rocket motors.

Type	Radial-burning (tubular, cruciform, internal)	End-burning
Thrust	50 to 100,000 lb	10 to 5000 lb
Duration	0.05 to 50 sec	1 to 500 sec
Impulse	30 to 1,000,000 lb-sec	1000 to 100,000 lb-sec
Propellants commonly employed	Ballistite (extruded); Cordite (British, Russian, German); NDRC moulded composite; thermosetting, cast, composite; British plastic composite	Cordite; NDRC moulded composite; asphalt thermoplastic composite; synthetic rubber thermosetting composite
Applications	Artillery rockets, armor-piercing bombs, antiaircraft rockets, air to ground rockets, rockets for assisted take-off of light aircraft, glide bombs, suicide aircraft, booster rockets, long range rockets	Aircraft assisted take-off and superperformance, underwater projectiles, glide bombs, large airborne missiles, booster rockets, gas generation, sounding rockets
Jet blast*	Proportional to thrust, usually severe exhaust flame from ballistite	Proportional to thrust, long duration units usually have mild exhaust flame to reduce heating of mechanical components

* Most of the composite propellants produce a smoky exhaust jet.

rocket motor that is reliable, convenient to use, and moderately easy to manufacture has been a lengthy and complex process. Because of the importance of these practical considerations, discussion of solid propellant rockets must necessarily refer to the state of the art.

In Table H,2a an outline is given of the general scope of performance and application attained by solid propellant rockets. Rapid development makes it difficult to separate experimental units or propellants from successful utilization, and it also makes a table of this type subject to rapid obsolescence. However, as a rough general survey, it should prove helpful.

The flexibility in design that is both useful and possible is very great. By utilization of suitable designs for various applications it is possible to obtain thrust variations and burning time variations ranging over a factor of 10^4. Total impulse, in different designs, can vary by factors greater than 10^6. Approximately 14 different applications and 10 different propellants are listed; both applications and propellants represent main lines of effort, and the number of each would increase considerably if all the various modifications and models were taken into account.

It will be noted that the ratio of total impulse to total loaded weight has not varied over such a wide range, due to the limitation on the energy per unit mass that can be obtained from solid chemical fuels. Because of this, features of rocket performance, other than specific impulse, receive considerable attention.

Table H,2b. Characteristics of some solid propellants.*

Property	*Ballistite extruded JPN*	*Asphalt base thermoplastic composite*	*NDRC moulded composite*
Specific impulse, sec	220	186	170
Exhaust velocity, ft/sec	7100	6000	5500
Density, lb/ft^3	101.5	110	100 to 115
Specific impulse/unit volume, per cent	100	96	80 to 90
Burning rate, in./sec	1.4	1.6	0.25 to 1.32
Burning rate exponent	0.73	0.76	0.40
Area ratio†	185	175	200 to 1000
Low pressure combustion limit, $lb/in.^2$	500	1000	no lower limit
Flame temperature, °F	5300	3000	3000
Exhaust jet	smokeless	smoky	smoky

* All pressure-dependent properties assume good exhaust nozzle design, 2000 $lb/in.^2$ abs combustion pressure, and 14.7 $lb/in.^2$ abs external pressure. The combustion process is temperature-sensitive, and 70°F has been assumed.

† Ratio of burning surface area to exhaust nozzle throat area.

Table H,2b lists the characteristics of some solid propellants, which serve to provide orientation concerning propellant performance. The variation in available specific impulse amounts to only 25 per cent. Burning rates, at 2000 $lb/in.^2$ abs, are all less than 2 in./sec. It is possible to achieve low burning rates, but very high burning rates, for example 20 in./sec, are not available. Such high burning rates could be very useful in certain applications.

Table H,2c describes an artillery rocket and an assisted take-off rocket. The examples chosen do not represent either the largest or the smallest of their kind, and very many other rockets could be cited.

The curves in Fig. H,2b show the variation of thrust and chamber pressure for a large, end-burning, asphaltic composite propellant rocket. The regularity and smooth operation of this long duration (23-sec) unit is evident.

In Fig. H,2c the curves show the variation of chamber pressure with time for a cruciform propellant charge. The difference in performance for different initial propellant charge temperatures is noteworthy. This illustrates one of the fundamental problems in rocket design,

Table H,2c. Characteristics of two rockets.

Rocket description	3.5 in. Fin-stabilized	12 AS-1000 D-1
Charge type	Cruciform	End-burning
Purpose	Artillery	Assisted take-off of aircraft
Thrust, lb	2340	1000 (rated)
Burning time, sec	0.8 sec at 70°F	12 (rated)
Impulse, lb-sec	1870	16,000 (actual)
Propellant	JPN	Asphaltic composite
Outside tube diameter, in.	3.5	9.6
Length, in.	54.5	36
Gross weight, lb	53.8	205
Propellant weight, lb	8.5	90
Payload weight, lb	20.0	—
Launcher weight, lb	15 (air)	—
Range, yd	9000*	—
Burnt velocity, ft/sec	1175	—
Dispersion, mils		
Air-fired	4	—
Ground-fired	20	—

* Estimated.

temperature sensitivity. Except at the extreme upper propellant temperature, it will be noted that these performance curves are very regular.

There are three characteristics, peculiar to solid propellant rockets, which limit their utilization. These limitations may be briefly designated by the following terms:

1. Temperature sensitivity—the variation of performance with propellant charge temperature.
2. Temperature limits. A rocket unit will malfunction if its temperature, before firing, does not lie between certain limiting temperatures.
3. The combustion limit. Many propellants do not burn satisfactorily at low pressures.

Propulsion by detonation. From time to time, suggestions are put forward that a "rocket" should be propelled by detonation. The advent of the shaped charge with particle ejection velocities of 30,000 ft/sec has given further impetus to these suggestions. There are, of course, some practical objections to such an arrangement. The violence of detonation

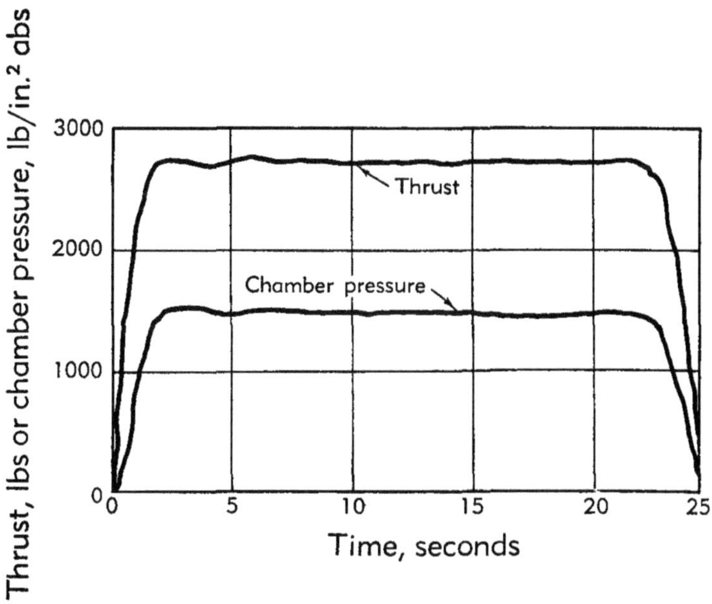

Fig. H,2b. Variation of thrust and chamber pressure with time for a large, end-burning, thermoplastic, asphaltic, solid propellant rocket. Performance data: p_c = 1465 lb/in.^{2} abs, F = 2660 lb, I = 62,912 lb sec, W_c = 349.5 lb, I_{sp} = 180 sec., A_{th} = 1.179 in.^{2}, C_F = 1.54.

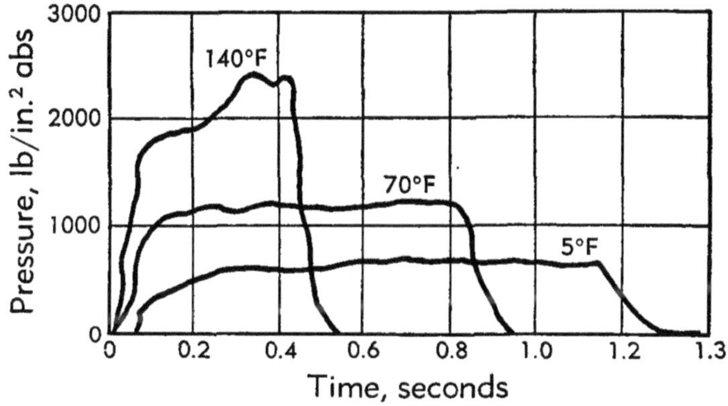

Fig. H,2c. The variation of chamber pressure with time for a 9-lb cruciform, semi-restricted-burning, ballistite grain for the 3.25-in. rocket motor. Note the effect of propellant charge temperature on performance.

would be very destructive to the launcher, and probably destructive to the rocket itself. In the case of the shaped charge, it has been shown [7] that the high speed penetrating jet is balanced by a much larger low speed jet emitted in the opposite direction. A feasible method to utilize the low speed jet to "push" a vehicle has yet to be demonstrated.

However, there is a more fundamental objection. Any jet propulsion device is limited by the conservation laws: conservation of energy, mass, and momentum. The detonating material will be superior to the common rocket propellants only if it can release more energy per unit mass than the latter. This is not generally the case. The improved explosives achieve their greater efficacy by having a greater rate of detonation, and the energy per unit mass is about the same as in the case of propellants. Indeed, ballistite has a very high specific energy and few solid materials exceed it in energy per unit mass.

It may be argued that the pressure level in a detonation is much higher than in the case of rocket motors and that, therefore, a thermodynamic gain in efficiency is to be looked for. However, the pressures commonly utilized in rocket motors usually are high enough so that a gain in exhaust velocity of 5 or 10 per cent is the most that can be expected on going to very high pressures.

The only remaining advantage would seem to be the possibility of achieving a different distribution of energy over the ejected mass rather than the uniform distribution obtained with rocket motors. Consider an abstract case: Suppose one has a mass M and a total amount of energy E. What is the optimum distribution of energy over this mass in order to achieve the maximum impulse from a rocket device? The corresponding variational problem seeks to maximize the total impulse $I = \int V_e(m)dm$, where m is the differential mass coordinate such that $\int dm = M$, with the constraint that the total energy is constant, $E = \frac{1}{2}\int V_e^2 dm =$ const. The solution of this problem states that the maximum impulse is obtained when every particle, dm, is ejected with the same exhaust velocity $V_e^2 = 2E/M$.

In conclusion, one may say that from a fundamental point of view there are at best very minor advantages in going to a detonating explosive as a means of propulsion, and that there are very great practical difficulties in the way of such a development.

H,3. Effect of Utilization on Rocket Design. The fundamental purpose of a rocket unit is to supply a required amount of impulse, force × time or mass × velocity, to a vehicle. Rockets are not usually employed if the fundamental purpose is to do work (move a force through a distance) but rather, from a general point of view, they are used to accelerate a mass to some velocity. The final velocity should be moderately high (an appreciable fraction of the exhaust velocity of the rocket) if the rocket is to be used efficiently. However, in many cases, convenience rather than efficiency is the feature of the rocket which recommends it for a particular application.

Rockets have been utilized to assist the take-off of an aircraft, to supply an initial boost to a missile, and to supply the main propulsive force for a missile. A modification of the rocket arrangement, in which the high pressure gas resulting from combustion is utilized to operate pneumatic devices, has also been employed. This last purpose, gas generation, is at complete variance with the missile applications mentioned above, but it has grown as a by-product of rocket development as the reliability of rocket devices has increased.

General requirements. Safety and reliability, with adequate performance, are nearly always demanded from rockets. Sensitive or unstable propellants must be avoided, even though their use might improve certain aspects of rocket performance. Safety also limits rocket ignitor design and the composition of ignition powder. The electric squib commonly used with ignitors usually requires a minimum of 0.5 amperes (resistance of about 1 ohm implies 0.5 volt) for ignition to minimize accidental firing. Squibs with smaller energy requirements can be obtained and would place a smaller load on firing circuits. However, use of such squibs would increase the danger of accidental firing.

The ignitor squib ignites the ignitor powder. Here, there are advantages, reduced ignition time and reduced ignitor blast, which may be gained by employing special powders.

Since rockets are self-propulsive, they can constitute a hazard to a wide area if accidentally set off. They are frequently shipped in a partially disassembled, nonpropulsive condition to minimize this danger. In handling rockets it is well to remember that both the front (missile) and nozzle (blast) ends are dangerous.

The basic simplicity of solid propellant rockets has made it possible to attain a high degree of reliability with them. There remain limits in the range of conditions under which they must be stored and used if reliable behavior is to be obtained. The most important requirement is usually that suitable temperatures, neither too high nor too low, be maintained. Rugged design and resistance to shock and vibration are required, and this requirement can usually be met.

The attainment of reproducible performance has been a slow and difficult part of rocket development. Even now, what has been achieved is essentially identical performance under given conditions (the most important condition is rocket temperature before firing), but not identical performance under all conditions. Reproducibility requires meticulous control of propellant composition and processing.

A rocket must meet certain performance requirements. For the rocket itself, the most natural specification is total impulse. Independent specification of thrust and duration of thrust is often necessary. Since a given weight of propellant will deliver, rather closely, the same impulse under a wide variety of conditions (although the burning time and thrust may vary oppositely over a wide range), specifications frequently take the form of a required impulse with an additional requirement of minimum and maximum thrust or else maximum burning time. Separate thrust and duration requirements then serve only to define the area of interest.

Rockets to assist the take-off of aircraft usually have large impulse but moderate thrust (to avoid large, sudden loads on the aircraft structure) requirements (12,000 lb-sec, 1000 lb), and the long burning time is most easily attained by end-burning grains. For missile boosters, a minimum thrust may be specified (to get the missile up to a speed at which aerodynamic control is effective). For reasons of accuracy, artillery rockets must attain a certain speed on the launcher, and a maximum burning time (which implies a minimum thrust) may be required. Rockets for boosters and missiles usually require such a large impulse and short burning time that internal-or radial-burning grains are used.

Prompt ignition is nearly always desirable. For assisted take-off of aircraft, ignition delays up to $\frac{1}{2}$ second are tolerable. For missiles, which may be fired at moving targets or in ripple salvos, delays should not exceed 10 to 50 milliseconds (depending on the application). The sustaining rocket in a boosted missile should start within a similar time interval, since its start-up must synchronize with the release of the booster.

Requiring short ignition delay usually implies very energetic ignitors, which, in turn, may damage the propellant charge, and even the structures at the launching site to the rear of the rocket. The ignitor blast may be more severe than the rocket blast (exhaust jet). Fragments of the ignitor (sometimes burning) may be ejected through the exhaust nozzle, thus constituting a hazard.

Ideally the exhaust jet of the rocket should not be visible (smoke, flash), and should not contain toxic or corrosive chemicals. Most rocket exhausts do contain carbon monoxide. Smoke is particularly undesirable for the assisted take-off of shipboard aircraft, since it may obscure the flight deck.

High velocity rocket missiles usually require a large ratio of total impulse to total weight together with a small frontal area, and a burning time of about one second. Internal-burning, traverse-burning, or radial-burning grains must be used, and the high internal gas flow velocity resulting from the large impulse and small frontal area requirements may give rise to combustion stability problems. The long grain is also difficult to support during the entire burning period, and may break up before delivering its full impulse.

Very short burning time rockets, designed to burn completely on the launcher, are subject to similar difficulties.

The proper balance between propellant performance and cost is always a difficult problem. When very large impulse is required (aircraft assisted take-off, large boosters), cost usually is given increasing consideration.

Artillery rockets. In addition to the usual requirements outlined above, the problem of accuracy introduces additional requirements and refinements for artillery rockets. The accuracy problem of exterior ballistics has been analyzed in great detail by Davis, Blitzer, and Follin [*8*], and by Rosser, Newton, and Gross [*9*]. Only the implications for rocket design are considered here.

Since range is closely correlated with impulse-to-weight ratio, both the propellant weight and total rocket weight must be controlled.

For most artillery rockets, only a small fraction of the rocket motor impulse is delivered while the rocket is on the launcher. For rockets fired from stationary launchers, more than 90 per cent of the dispersion (inaccuracy) may be attributed to motion during the burning period. This implies that proper alignment of the exhaust jet is important to a much larger degree than for booster rockets for a guided missile. As a rough order of magnitude, for a missile stabilized by external fins, if a line along the direction of thrust misses the center of mass of the rocket by 0.10 inch, a lateral deviation of 20 to 30 mils (1 mil is an angle corresponding to one yard at 1000 yards) is produced.

Thrust malalignment has been attributed to the following causes: manufacturing tolerance, gas malalignment, and thermal and pressure distortion of the rocket body.

Gas malalignment occurs when the exhaust jet does not coincide with the axis of a mechanically perfect exhaust nozzle. There is some indication that a high turbulence level in the gas flow reduces gas malalignment. Thermal and pressure distortion of the combustion chamber, if nonuniform, can make one side longer than the other, canting the exhaust nozzle to one side. For example, a 60°C mean temperature difference on two sides of a steel combustion chamber 3 feet long and 3 inches in diameter, can produce thrust malalignment of 0.06 inches (corresponds to 10- to 20-mil deviation). Insulation of the combustion chamber wall can reduce thermal distortion. To minimize pressure distortion, all parts must expand equally; closures, lugs, and fittings must be designed with this in mind, and the combustion chamber wall should have uniform thickness.

The effect of thrust malalignment may be reduced by spinning the rocket about its longitudinal axis. This is automatic with spin-stabilized rockets, and even fin-stabilized rockets can be given a slow spin.

Multiple exhaust nozzles may also be used to reduce thrust malalignment. The individual imperfections neutralize statistically.

High velocity fin-stabilized rockets may have a length-to-diameter ratio of 10 to 14 with attendant interior ballistics problems. Spin-stabilized rockets with a length-to-diameter ratio of 5 to 8 have easier interior ballistic problems.

The ignitor may contribute to inaccuracy if a portion of the ignitor, passing through the exhaust nozzle, disturbs the exhaust jet.

The support of the propellant charge throughout the burning period is also important. Break-up of the grain at the later stages of burning produces slivers of propellant which can disturb the exhaust jet on passing through the nozzle, and also represent a loss of impulse (range). Cruciform charge shapes, rather than tubular, were a first step to provide better charge support. Internal-burning grains now appear to be an even better solution (Art. 14).

CHAPTER 2. INTERIOR BALLISTICS THEORY

H,4. Scope of the Theory. The theory presented here is idealized in that perfect gas laws, simple burning rate behavior, ideal thermodynamic propellant characteristics, and steady state flows are assumed. With this simplified framework, analytic expressions may be derived which give more insight into the basic features of rocket behavior than the more precise numerical techniques. In many instances, the theory can be made quantitatively useful by semiempirical choice of constants; indeed, the burning rate expressions that are employed are essentially empirical in character.

The simplified theory is also extrapolated to very high pressures where it has only a schematic meaning. However, by doing this, one may obtain a broad survey of rocket processes. Most of the qualitatively different phenomena associated with steady state processes may be studied in this way.

Unsteady state processes, which can often be of great practical importance, are not treated here. Resonance burning, a combustion instability problem of rocket design, is not analyzed. The theory is complex and still controversial [*10; 11; 12*, pp. 893–906; *13*, pp. 29–40; *14; 15; 16*] (also II,M). Certain practical design arrangements to avoid this difficulty are listed below in Art. 10.

In addition to providing insight, the theory provides a helpful framework and useful analytical expressions for the semiempirical analysis of rocket data. A more comprehensive account of some of these problems is given by Crawford [*17*], Wimpress [*10*], and Price [*18,19,20,21*]. Very brief accounts are given by Sutton [*22*] and Seifert, Mills, and Summerfield [*23*].

H,5. Combustion of Solid Propellants. The general manner of operation of a rocket unit has been outlined. In most solid propellant rockets, no arrangement is made for the control of the combustion process during the burning of the propellant. If there are no mechanical valves or other devices installed in the unit (aside from certain safety devices) by which the burning of the propellant may be regulated, the control of the burning of the propellant is possible only to the extent to which a suitable grain and combustion chamber design, along with some arrangement of inhibitor coatings, is prepared before the unit is used. After ignition, the unit is "left to itself" and usually no further adjustment on the part of the operator is attempted. Provision for special control of the unit is a difficult design problem and could make the rocket unit undesirably complicated. One of the chief attractions of solid propellant rockets has been their simplicity and reliability.[1]

During the burning period, the solid propellant rocket regulates itself. The fundamental property of the solid propellant involved in this self-control of the burning process is its rate of burning. The rate of burning of a solid propellant is defined as the rate at which a burning surface recedes in a direction normal to itself. This is usually expressed as a velocity, e.g. in./sec.

Experimentally, burning rates can be determined in static firing tests of the rocket units or by strand burners (see Art. 12). If the geometry of the propellant charge is known, the burning time can usually be estimated from the pressure vs. time curve obtained in the test. From this information the rate of burning of the propellant may be calculated.

From special experimental studies it has been found that the burning rate r depends upon: the gas pressure on the propellant p, the temperature of the propellant charge before ignition T_p, the velocity of gas flow past the charge surface V, the time since the start of burning t, the position in the charge $\mathbf{x}$, and the direction in which the normal to the burning surface points $\mathbf{\Omega}$. In mathematical form this may be expressed as

$$r = r(p, T_p, V, t, \mathbf{x}, \mathbf{\Omega})$$

It is important to remember that all these parameters can affect the burning rate. However, not all of them are important for every propellant.

The variation in burning rate with time after start of combustion is believed to be due to gradual heating of the propellant and motor tube by radiant heat transfer from the hot propellant gas. Avery [*24,25*] has pointed this out, and a summary of studies of this problem has been prepared by Penner [*26*]. It is a more important effect for hot propellants such as ballistite and cordite than for cool propellants. The propellant grain must be slightly transparent to radiation to permit internal heating. (Very transparent propellants are subject to subsurface ignition by radiation, which must be avoided. See Art. 12.)

[1] Sometimes a spring-loaded valve discharging into an auxiliary exhaust nozzle is employed with rockets. The effect of this is to allow the mass outflow to increase with pressure more rapidly than the linear variation which is obtained with ordinary nozzles. This improves the stability of operation, but it does not constitute independent "external" regulation.

Both the variation with position and direction are presumably due to inhomogeneities and anisotropies introduced into the structure of the propellant during its manufacture. For solventless ballistite, extruded in the form of a tube, the burning rate may be 25 per cent higher at the center of the web than at the surface [10]. (However, this may be due to the radiant effects mentioned above.) The burning rate is also some 15 per cent higher in the direction of extrusion than it is at right angles. Cast, polymerized propellants utilized in end-burning charges show a higher burning rate near the mold wall (Art. 12). The asphaltic, thermoplastic, cast propellants do not seem to show the dependence of burning rate on time, position, or direction.

Although the detailed burning rate behavior of a propellant grain must be known in order to reliably design a rocket motor, it will simplify the general theory if only the variables of chamber pressure, propellant temperature, and gas flow velocity are retained. There are indications [10,27,28,29,30,31] (see also II,M) that the effect of temperature is separable from the pressure and velocity dependence:

$$r = f_1(T_p)f_2(p, V)$$

Experimental studies on the burning rate of propellants show that the temperature dependence can be expressed by

$$f_1(T_p) = \frac{A}{T_1 - T_p}$$

Here A and T_1 are empirical constants.

The pressure and velocity dependence seem to be separable in some cases [7], but not in others. Two forms are commonly employed [32,33]:

$$f_2(p_c, V) = a'\left(\frac{p}{1000}\right)^n (1 + k\rho V) \quad \text{or} \quad a'\left(\frac{p}{1000}\right)^n f_3(V)$$

Here the new parameter ρ is the density of propellant gas flowing past the grain, and a' and k are empirical constants. The density may be expressed in terms of pressure and velocity so that the first expression is consistent with f_2.

In the second form, $f_3(V)$ has a peculiar dependence:

$$f_3(V) = \begin{cases} 1 & \text{if } V < V_0 \\ 1 + k_V(V - V_0) & \text{if } V > V_0 \end{cases} \tag{5-1}$$

Here k_V and V_0 are empirical constants, and V_0 is a threshold velocity. Frequently an expression without a threshold velocity is employed:

$$f_3(V) = 1 + kV$$

The threshold situation is outlined in Fig. H,5a. The experimental curve is fitted rather well by Eq. 5-1 over most of the velocity region of interest.

The experimental results indicate that no acceleration of the burning rate takes place (referred to as erosive burning) until a velocity of 600 ft/sec is reached. This may be due to the buffer effect of propellant gas coming directly from the surface.

The factor $1 + k\rho V$, on the other hand, has more intuitive appeal, since transfer of mass, momentum, and heat across a boundary layer often show this dependence on ρV. The dimensional erosion constant k usually ranges from about 0.05 to 0.7 in.^{2} sec/lb. The theory of solid

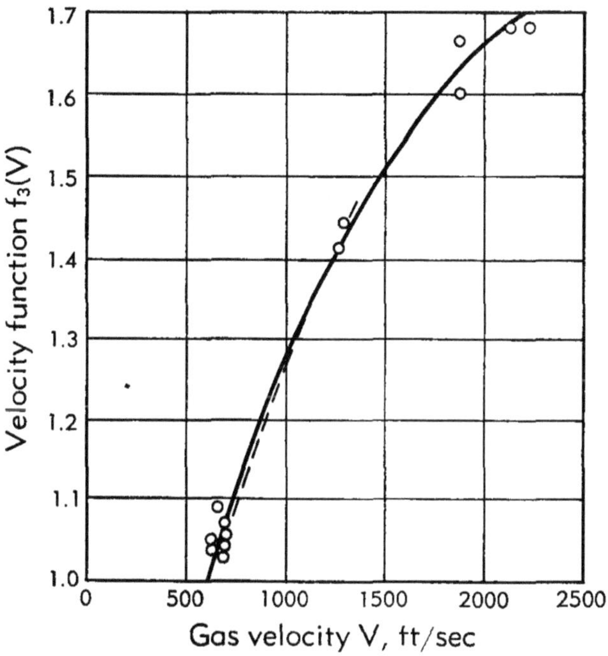

Fig. H,5a. Erosive burning effect, velocity-dependent only, of gas velocity on the burning rate of JPN propellant. After [*10*] by permission.

propellant burning without gas flow past the surface [*30,34,35,36,37,38*] is still being formulated. Both the theory [*6*] and facts [*32,33,39,40,41*] of erosive burning are still unsettled. Corner [*6*] indicates the theoretical possibility for either velocity or mass flow density erosive effects depending upon the propellant. The latter should usually be accompanied by an index n in the pressure dependence of 0.5 or less. The variation of k or k_V with propellant temperature is also not settled. Here they are assumed independent of propellant temperature.

Green [*38*] has carefully studied erosive burning effects (dependence of burning rate on gas flow velocity is frequently termed erosion), and he suggests that the peculiar threshold effect (Eq. 5-1) may be due to the shape of the gas flow velocity profile at the forward end of the propellant grain, and that the intrinsic velocity dependence is of the form $1 + kV$.

He was unable to decide conclusively whether or not the form should be $1 + kV$ or $1 + k\rho V$, although experiments performed by Bartley indicate conclusively that the latter expression is more suitable for certain composite propellants.

Green also suggests that the proper function should not be velocity but the dimensionless mass flow density (Art. 8). This would correspond to an erosive effect independent of density.

In the following discussion, both mass flow density $1 + k\rho V$ and velocity $1 + kV$ erosion effects will be included. The former has some experimental basis and also gives rise to an erosive instability which can be of practical importance for certain propellants. The "threshold" form (Eq. 5-1) will be chosen for the velocity-dependent erosion, since, according to Green, this is an empirical method to take into account the two-dimensional flow at the forward end of the grain. It also provides a qualitative comparison between the effect of gas flow on erosive and on nonerosive propellants.

The burning rate expressions adopted here may then be written

$$r = \frac{a'}{T_1 - T_p}\left(\frac{p}{1000}\right)^n f_3(V) \tag{5-2a}$$

$$r = \frac{a'}{T_1 - T_p}\left(\frac{p}{1000}\right)^n [1 + k\rho V] \tag{5-2b}$$

For negligible velocity these reduce to

$$r = a\left(\frac{p}{1000}\right)^n \tag{5-3}$$

with

$$a = \frac{a'}{(T_1 - T_p)}$$

The pressure is to be expressed in psia and the parameter a has values ranging from 0.1 to 1.0 in./sec. Values of n are less than unity, frequently ranging from 0.2 to 0.8 (Table H,2b and Art. 12).

In Eq. 5-2 the form chosen for temperature dependence is suggestive, indicating an infinite burning rate at $T_p = T_1$. Efforts to identify T_1 with an ignition temperature, T_i, have not always been quantitatively successful. For JP ballistite the ignition temperature was determined to be 330°F [*42*], while the constant T_1 determined from temperature sensitivity studied was found to be 440°F. However, smaller values of T_1 do indicate a larger degree of temperature sensitivity.

Special propellants have been devised which exhibit a constant burning rate, independent of pressure, over certain pressure ranges. Such propellants are particularly desirable since they yield extremely stable

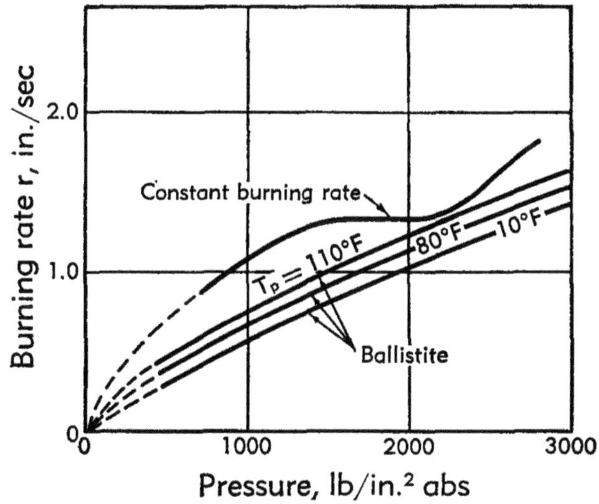

Fig. H,5b. Variation of burning rate with propellant gas pressure and propellant charge temperature for ballistite (JPN), together with an example of "constant burning rate" propellant.

rocket unit behavior. In Fig. H,5b burning rate curves for ballistite and a curve for such a constant burning rate propellant are shown.

H,6. Stability of the Burning Surface.

Combustion geometry. The solid propellant burning surface, in many cases, moves in a direction normal to itself at a uniform rate at every point. Propellants which are not macroscopically homogeneous and isotropic exhibit a different, although regular, behavior. It is important that the regular behavior of the burning surface should be stable, so that the burning surface area does not change in an unpredictable manner. If a disturbance is represented by an irregularity of the surface (this might be caused by a fault or large inhomogeneity in the propellant) simple geometrical considerations, assuming burning in a normal direction, indicate that the surface will smooth out and continue its regular recession as long as the desired surface is flat. Fig. H,6a shows this for an end-burning propellant charge. The burning surface behavior predicted for the cup-shaped burning surface (second from the top) has been verified experimentally. These same general considerations are valid for charges of other shapes as long as the disturbance is small compared to the radius of curvature of the charge surface. On the other hand, a large disturbance, on a cylindrical surface for example, will grow although the area of the burning surface will decrease slightly so that dangerously high pressures are not developed. A constant area for the burning surface is the important requirement, since a small change in the burning surface area will produce a large change in the chamber pressure. One may conclude that the burning surface is stable (see also Art. 12).

Charge deformation problems. The propellant charge is subject to pressure, and in many cases axial and radial acceleration forces. These forces produce deformation of the charge. In the case of end-burning propellants, sealed to the combustion chamber walls, the application of chamber

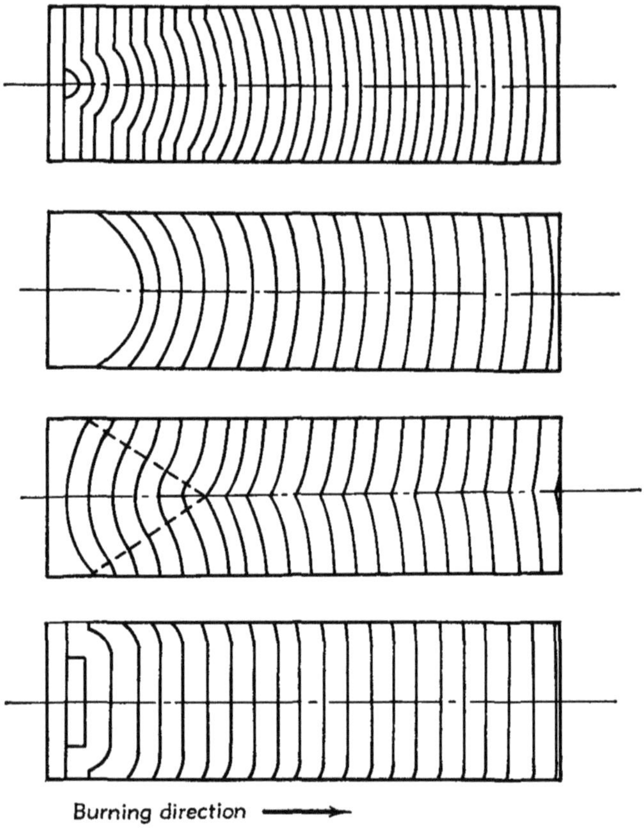

Fig. H,6a. Burning surface stability for end-burning grains.

pressure can lead to a reduction in charge length of 5 to 10 per cent. If the charge is contained within a flexible casing, not sealed to the combustion chamber wall, then the build-up of chamber pressure will apply a hydrostatic load to the propellant with little deformation and very little volume reduction since the Poisson ratio is near $\frac{1}{2}$ for many propellant materials.

In the case of tubular and internal-burning propellants, in which gas must flow through ducts along the grain, the situation is more serious. The general pressure level does not lead to a deformation of the charge, but the chamber pressure at the forward end of the rocket is larger than the chamber pressure at the rear or exhaust nozzle end of the charge.

The unbalance of pressure does lead to a nonisotropic, longitudinal compressive stress, bulging the charge into the gas duct area. For artillery rockets there are large acceleration forces which have the same effect. Skin drag and any impact forces on propellant irregularities from the rapidly flowing gas stream also add to the longitudinal compression, as illustrated in Fig. H,6b.

The bulging of the propellant charge into the gas channel leads to a restriction which further increases the pressure drop, etc. At high propellant temperatures the burning rate is higher, the propellant is softer,

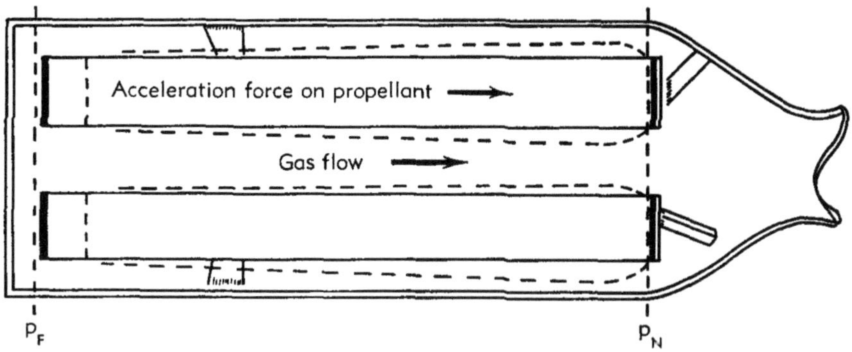

Fig. H,6b. Deformation of the propellant by pressure drop, gas drag, and rocket acceleration forces. The solid line represents the undistorted charge; the dashed line represents the distorted charge.

the normal duct area is reduced by thermal expansion of the propellant (which is much larger than the expansion of the combustion chamber). The higher pressure produces larger accelerations. All these effects result in increased pressure drop and bulging, and may lead to failure of the rocket.

This problem can be worked out in detail. For illustration, acceleration and skin friction forces are omitted and only the pressure drop due to the necessary gas flow velocity is considered. At the critical time, initiation of burning, frictional forces are small.

If the cross-sectional area of the propellant is normally A_{pn}, then the effect of pressure drop is to increase this to A_p, where

$$A_p = A_{pn}\left(1 + \frac{2\mu}{E}\sigma\right) \tag{6-1}$$

$$\sigma = p_F - P_N \tag{6-2}$$

The front pressure is p_F, the rear pressure p_N (Fig. H,6b), and the stress component leading to deformation is therefore σ. Poisson's ratio and the elastic modulus are indicated by μ and E. If the internal cross-sectional area of the combustion chamber tube is A_q, then the duct area

A_d is

$$A_d = A_q - A_p, \quad A_{dn} = A_q - A_{pn} \tag{6-3}$$

The subscript n corresponds to the normal (unbulged) state. For illustrative purposes we assume that $A_q = 2A_{pn}$, a value not far from those employed in rockets. This will simplify the algebra, but it will not affect the fundamental character of the results. If the propellant were perfectly rigid, the pressure drop would be less, σ_n, since no bulge would obstruct

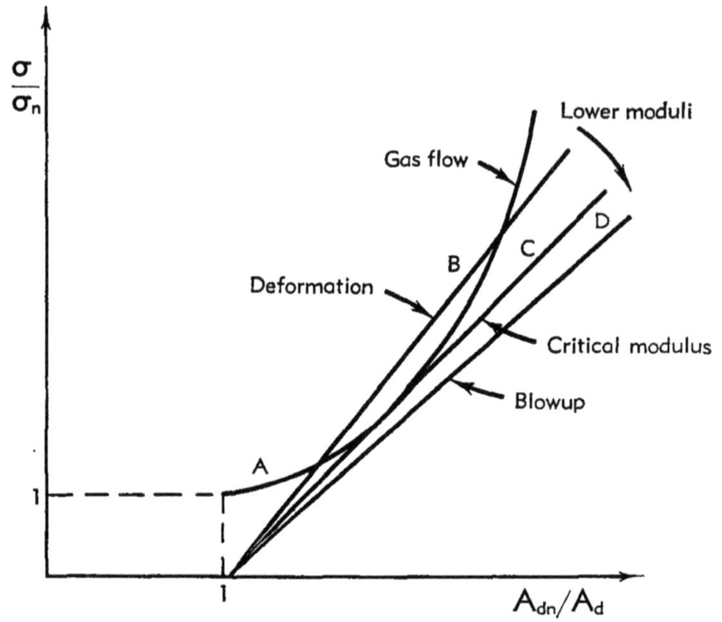

Fig. H,6c. Schematic curves showing the variation of stress in the propellant grain with gas duct area A_d. The curved line is the stress due to gas flow, the nearly straight lines represent the stress required to maintain the corresponding gas duct area. The latter curves rotate clockwise for lower values of the elastic modulus, and the curve C corresponds to the critical modulus.

the duct area. Eq. 6-1 and 6-2 may be combined to give

$$\frac{\sigma}{\sigma_n} = \frac{1}{2\mu}\frac{E}{\sigma_n}\left[1 - \frac{1}{(A_{dn}/A_d)}\right] \tag{6-4}$$

It is necessary to use a result, which is derived below (Art. 9), to complete the analysis. This is the relation between pressure drop σ and the gas duct area when the remaining parameters of the rocket (the burning surface area S_b and the exhaust nozzle throat area A_t) remain fixed. This result, together with the result of Eq. 6-4, is shown in Fig. H,6c. The curve labeled "gas flow" follows from the interior ballistics study (Art. 9), and the curve labeled "deformation" is the result of Eq. 6-4. An important deduction from this study is that there is a critical elastic modulus, and if the actual modulus of the propellant is less than this the rocket will blow up.

Consider the intersection point *A*. If the duct area were to decrease slightly by a disturbance (this corresponds to a motion to the right of *A* since the duct area is in the denominator of the horizontal coordinate), then the stress due to gas flow will be less than the elastic stress required to maintain the reduced duct area. Consequently the compression will relax and the conditions return to the intersection point. Similarly a disturbance causing a displacement to the left of *A* is followed by a decrease in duct area until *A* is reached. *A* is the stable intersection point and corresponds to the actual duct area and the actual compressive stress when charge deformation is taken into account. From the stress and Eq. 6-2 the actual chamber pressure distribution may be determined.

A disturbance, leading to a displacement to the right of the intersection point *B*, yields a higher compressive stress due to gas flow than is required to maintain the reduced duct area. The duct is reduced further, the pressure increases further, and the rocket blows up. Below *B*, pressures decrease until *A* is reached. *B* is an unstable equilibrium point.

Inspection of Eq. 6-4 shows that a reduction in the elastic modulus rotates the "deformation" curve clockwise, bringing the points of intersection of *A* and *B* closer together. If these points come too close together, a disturbance of the true equilibrium conditions at *A* might throw the system to the right of *B* and lead to a blowup. Finally, further reduction of the modulus leads to curve *C*, just tangent to the gas flow curve. An ideal rocket might operate here, but a slight disturbance could lead to a blowup. Further reduction in the elastic modulus of the propellant would remove all points of intersection of the "gas flow" and "deformation" curves (see curve *D*). The compressive stress from gas flow would always exceed the stress necessary to maintain the reduced duct area, leading to further reduction in duct area and further increase in pressure until the rocket blows up. The modulus, corresponding to curve *C*, is the critical modulus. A rocket should be proportioned so that the critical elastic modulus is substantially less than the actual modulus. The critical modulus depends only upon the pressure drop, duct area characteristics of the unit, and is determined by the geometry of the arrangement and the burning characteristics of the propellant.

It was mentioned above that flight acceleration and increased propellant temperature tend to increase the degree of propellant deformation. Wimpress [*10*] has computed curves taking all these effects into account, and his results are shown in Fig. H,6d. The intersection of the actual moduli curves with the critical modulus curve indicates the upper temperature limit (blowup) for the rocket. Agreement between theoretical upper temperature limits determined from these curves and firing tests is reasonably good. It should be noted that reduction in actual modulus

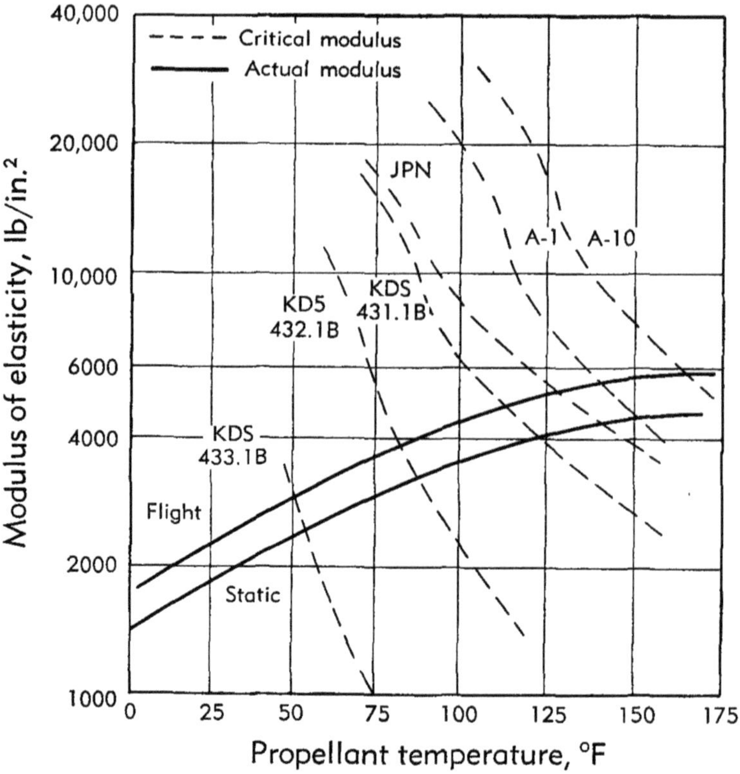

Fig. H,6d. Variation of the critical modulus for the Mk13 propellant grain in the 3.25-in. rocket motor. The variation of the actual elastic modulae for a number of propellants with propellant temperature is shown on the same scale. The intersection "critical-actual" corresponds to the upper temperature limit (after Wimpress [*10*] by permission).

with temperature is a more important effect than increase of critical elastic modulus with temperature.

H,7. Steady State Dynamics for End-Burning Grains. The theory [*43*] and stability criteria for the end-burning charge are relatively clear and simple, and they are discussed first. The notation that is used is indicated in Fig. H,7a.

It will be shown, provided that the exponential index n in the burning rate expression, Eq. 5-3, is equal to or less than unity and the burning surface area is constant, that a stable, steady chamber pressure is obtained during the combustion of the propellant. In a practical rocket, the index should be appreciably less than unity. It is assumed that the temperature of the combustion gas is T_0, a constant. An analysis of transient startup conditions indicates that in less than 0.04 sec after ignition, the cool gas, usually air, initially present in the combustion chamber has been swept out.

The equation for the conservation of mass is simply:

$$\begin{Bmatrix}\text{Mass}\\ \text{combustion}\\ \text{rate}\end{Bmatrix} = \begin{Bmatrix}\text{Rate of increase of the}\\ \text{mass of combustion products}\\ \text{inside combustion chamber}\end{Bmatrix} + \begin{Bmatrix}\text{Rate of mass}\\ \text{flow out}\\ \text{exhaust nozzle}\end{Bmatrix}$$

In symbolic form,

$$rS_b\rho_p = \frac{d}{dt}(\rho_c v_c) + \frac{p_c A_t}{c^*} \tag{7-1}$$

Here r is the burning rate, S_b the area of the burning surface assumed constant, ρ_c the density of combustion gas in the combustion chamber, ρ_p the density of the solid propellant before combustion, v_c the volume

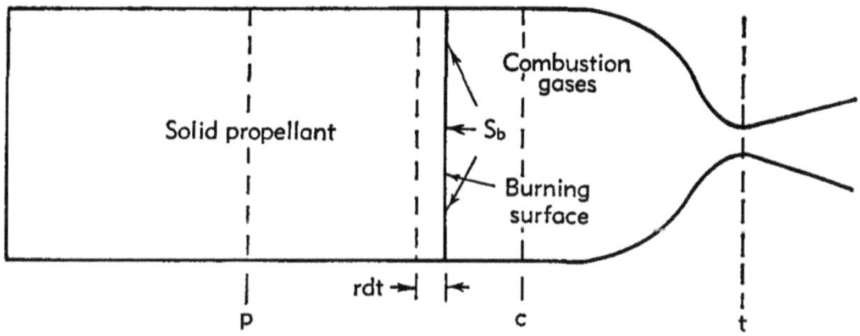

Fig. H,7a. Notation employed in the discussion of end-burning solid propellant grains.

of the combustion chamber from the burning surface to the exhaust nozzle throat section, p_c the pressure within the combustion chamber, A_t the exhaust nozzle throat area, and c^* the characteristic velocity for the products of combustion[2] (see Sec. G).

The ratio of nozzle throat area to burning surface area is so small, in end-burning grains, that the approach velocity of the combustion gas inside the combustion chamber is negligible, so that p_c may be employed for the stagnation pressure or reservoir pressure in computing nozzle mass flow.

The products of combustion are treated as ideal gases so that the

[2] The exhaust nozzle flow equations are not developed here (see Sec. G). Often a discharge coefficient C_D is employed so that the mass flow $\dot{m}$ is given by $\dot{m} = C_D p_c A_t$. The coefficient, as well as c^* have a theoretical basis, but may also be determined experimentally. For real propellants they depend weakly on chamber pressure; this is neglected here. For ideal gases the relation between the velocity of the gas at the exhaust nozzle throat V_t ($= a_t$), the speed of sound in the combustion chamber a_c, and c^* is given by $c^* = a_c/\sqrt{\gamma} = \dfrac{V_t}{\sqrt{\dfrac{2\gamma}{\gamma+1}}}\Gamma$. Γ is defined in Eq. 7-5c. Note that the form of Eq. 7-4 implies the use of absolute units. For engineering units, c^* should be replaced by c^*/g which has the physical dimensions of sec.

equation of state may be employed:

$$p_c = \rho_c \mathcal{R} T_0 \tag{7-2}$$

where $\mathcal{R}$ is the engineering gas content (containing the average molecular weight) and T_0 is the adiabatic, constant pressure, flame temperature for the propellant. It is also convenient to introduce the abbreviation,[3]

$$p_p = \rho_p \mathcal{R} T_0 \tag{7-3}$$

If Eq. 7-1 is multiplied through by $\mathcal{R}T_0$, and Eq. 7-2 and 7-3 employed,

$$rS_b p_p = \frac{d}{dt}(v_c p_c) + \frac{\mathcal{R} T_0 A_t}{c^*} p_c \tag{7-4}$$

The T_0 may be taken inside the differentiation sign since it is assumed constant.

It is convenient to introduce the ideal expression for c^*

$$c^{*2} = \frac{\mathcal{R} T_0}{\Gamma^2} \tag{7-5a}$$

$$\Gamma^2 = \gamma \left(\frac{2}{\gamma + 1} \right)^{\frac{\gamma+1}{\gamma-1}} \tag{7-5b}$$

$$a_c = \sqrt{\gamma}\, \Gamma c^* \tag{7-5c}$$

where γ is the ratio of specific heats—constant pressure to constant volume. For most rocket gases γ is near 1.2 and the corresponding $\Gamma^2 = 0.422$.

A convenient notation is

$$l_c = \frac{v_c}{S_b} \tag{7-6a}$$

Note that

$$\frac{dl_c}{dt} = r \tag{7-6b}$$

and with the further notation

$$K = \frac{S_b}{A_t} \tag{7-6c}$$

The combination of Eq. 7-4, 7-5, and 7-6 yields

$$\frac{dp_c}{dt} = \frac{p_p}{l_c} \left[\left(1 - \frac{p_c}{p_p} \right) r - \frac{\Gamma^2}{K} c^* \frac{p_c}{p_p} \right] \tag{7-7}$$

[3] For an ideal propellant of the type considered here, p_p is related to the maximum explosion pressure by: p (maximum explosion) $= \gamma p_p$, where γ is the ratio of the specific heats (constant pressure/constant volume) for the combustion products. The parameter p_p, in some degree, is an index of merit for the propellant. For asphalt potassium perchlorate propellants it is 125,000 lb/in.² abs and for ballistite it is 230,000 lb/in.² abs.

This equation, together with Eq. 7-6b, governs the internal ballistics of the end-burning rocket.

It is most convenient to analyze Eq. 7-7 by considering the two terms in the parentheses separately. Define $\dot{m}_b$, $\dot{m}_o$ as follows:

$$\dot{m}_b = \left(1 - \frac{p_c}{p_p}\right) r \tag{7-8a}$$

$$\dot{m}_o = \frac{\Gamma^2}{K} c^* \frac{p_c}{p_p} \tag{7-8b}$$

The function $\dot{m}_b$ is proportional to the net rate at which mass enters the combustion chamber due to combustion of the solid propellant. The term

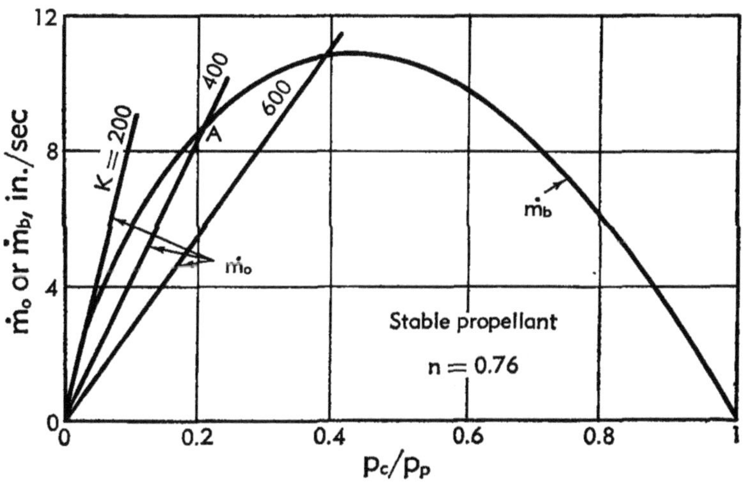

Fig. H,7b. The variation of net mass burning rate $\dot{m}_b$, and of the net mass outflow $\dot{m}_o$, with chamber pressure. (Assumed characteristics: a = 0.937 in./sec, n = 0.76, c^* = 3750 ft/sec, γ = 1.2, p_p = 125,000 lb/in.² abs.)

$[1 - (p_c/p_p)] = [1 - (\rho_c/\rho_p)]$ is a correction to the gross burning rate to take into account the volume previously occupied by the solid propellant, which, after combustion, must be filled with combustion gas. The function $\dot{m}_o$ is proportional to the rate at which mass leaves the combustion chamber through the exhaust nozzle. Clearly, if $\dot{m}_o = \dot{m}_b$, then $dp_c/dt = 0$, and the chamber pressure is constant. The question of the stability of this equilibrium will now be discussed.

In order to represent $\dot{m}_b$, the expression for the burning rate, Eq. 5-3, must be used:

$$r = a\left(\frac{p_c}{1000}\right)^n \tag{5-3}$$

Assume first that $0 < n < 1$. Then the functions $\dot{m}_b$ and $\dot{m}_o$ may be plotted as a function of p_c as shown in Fig. H,7b. The function $\dot{m}_o$ is plotted for three values of the area ratio S_b/A_t. Consider the intersection of the $\dot{m}_o$ and $\dot{m}_b$ curves at A. A disturbance, leading to a slight increase in chamber pressure, results in a situation in which $\dot{m}_o > \dot{m}_b$; the mass outflow exceeds the rate of mass combustion. Therefore (see Eq. 7-7 and 7-8) the chamber pressure decreases until the point A is reached. Similarly, if a disturbance leading to a decrease in pressure occurs, $\dot{m}_o < \dot{m}_b$, and the pressure rises until point A is reached. The steady state condition, $\dot{m}_o = \dot{m}_b$, is stable as long as the shape of the $\dot{m}_b$ curve is as indicated in Fig. H,7b. The shape of the curve is determined by the exponent n in the burning rate expression, Eq. 5-3.

The fundamental requirement for stability is that at the point of intersection of the $\dot{m}_o$ and $\dot{m}_b$ curves, the steady state point, the slope of the $\dot{m}_o$ curve shall be larger than the slope of the $\dot{m}_b$ curve. At the pressure p_c, defined by

$$\frac{\Gamma^2}{K} c^* \frac{p_c}{p_p} = \left(1 - \frac{p_c}{p_p}\right) a \left(\frac{p_c}{1000}\right)^n \tag{7-9a}$$

there must obtain:

$$\frac{d}{dp_c}\left(\frac{\Gamma^2}{K} c^* \frac{p_c}{p_p}\right) > \frac{d}{dp_c}\left[\left(1 - \frac{p_c}{p_p}\right) a \left(\frac{p_c}{1000}\right)^n\right] \tag{7-9b}$$

By carrying out the differentiation and utilizing Eq. 7-9, the inequality may be written as

$$n < \frac{1}{\left(1 - \frac{p_c}{p_p}\right)} \tag{7-10a}$$

The parameter p_p has numerical values ranging from 100,000 to 300,000 lb/in.^{2} abs, so that for all situations of practical interest, $p_c < 3000$ lb/in.^{2} abs, say, the ratio p_c/p_p is a small number. The essential stability requirement is then

$$n < 1 \tag{7-10b}$$

It is true that even if $n > 1$ there will always be one stable steady state pressure. However, this will lie at an unacceptably high value. Taking the equality sign in Eq. 7-10a and solving for the pressure yields

$$p_c \text{ (minimum, stable)} = \frac{n - 1}{n} p_p \tag{7-11}$$

If $n < 1$, Eq. 7-11 predicts negative stable pressures. This is merely mathematical continuation, but the physical meaning is that a rocket unit could operate in a stable manner at any required low pressure. If $n = 1$, stable pressures down to zero are predicted. But consider the case $n = 1.1$, $p_p = 125{,}000$ lb/in.^{2} abs (a moderate value), then the minimum stable pressure is at 11,300 lb/in.^{2} abs, a value too high for practical rocket design.

An example for hypothetical propellant with a burning rate exponent, $n = 2$, is indicated in Fig. H,7c. The intersections of the $\dot{m}_o$, $\dot{m}_b$ curves, marked U, are unstable. A slight increase in pressure results in more rapid mass production than can escape from the exhaust nozzle. The intersection points marked S are stable steady pressure points. The minimum steady stable pressure corresponds to tangency of the $\dot{m}_o$, $\dot{m}_b$ curves and for this example is at $\frac{1}{2}p_p \cong 62{,}500$ lb/in.^{2} abs. This also indicates that there is a minimum area ratio $S_b/A_t = K$ (= 80 for this example) below which $\dot{m}_o$ is always greater than $\dot{m}_b$. A rocket designed with $K < 80$

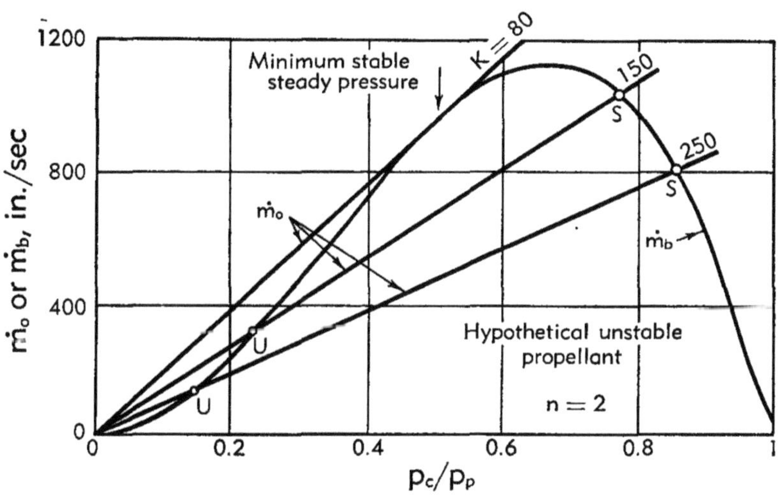

Fig. H,7c. The variation of net mass burning rate $\dot{m}_b$, and of net mass outflow $\dot{m}_o$, with chamber pressure for a hypothetical unstable propellant. (Assumed characteristics: $a = 0.0473$ in./sec, $n = 2$, $c^* = 3750$ ft/sec, $\gamma = 1.2$, $p_p = 125{,}000$ lb/in.^{2} abs.)

would always stop burning immediately after ignition (for this hypothetical propellant).

Experimental propellants have been prepared for which $n > 1$, and the prediction of very high pressure development has been verified. The curves (Fig. H,7c) also indicate difficulty in igniting the rocket since pressure, during the ignition phase, must rise at least to U since $\dot{m}_o > \dot{m}_b$ for lower pressures.

These curves (Fig. H,7c and H,7b) have been computed assuming that the ideal expressions for $\dot{m}_o$, $\dot{m}_b$ remain valid for the whole pressure range up to p_p. This assumption is known to be incorrect at low pressures (see Art. 12 on the combustion limit) and is probably wrong at high pressures, but it does illustrate the whole regime of rocket operation.

The criterion established for n ($n \leqq 1$) for the stability, is theoretically correct. However, values of n near unity lead to extreme sensitivity of the rocket to small disturbances. Low values of n are desired, and a propellant for which $n > 0.8$, say, is rather unsuitable.

Variation of chamber pressure with area ratio. Having established the stability criterion, one may now compute the stable steady pressure as a function of K, the ratio of the burning surface area to the exhaust nozzle throat area. The steady pressure obtains when $dp_c/dt = 0$ in Eq. 7-7. The resulting expression may be arranged to be

$$K = \frac{\Gamma^2 c^* \dfrac{p_c}{p_p}}{\left(1 - \dfrac{p_c}{p_p}\right) r} = \frac{p_c}{\rho_p c^* \left(1 - \dfrac{p_c}{p_p}\right) r} \tag{7-12}$$

Experimental values of r and c^* as a function of pressure may be used, or Eq. 5-3 and 7-5 may be employed (p_p may be eliminated by utilizing

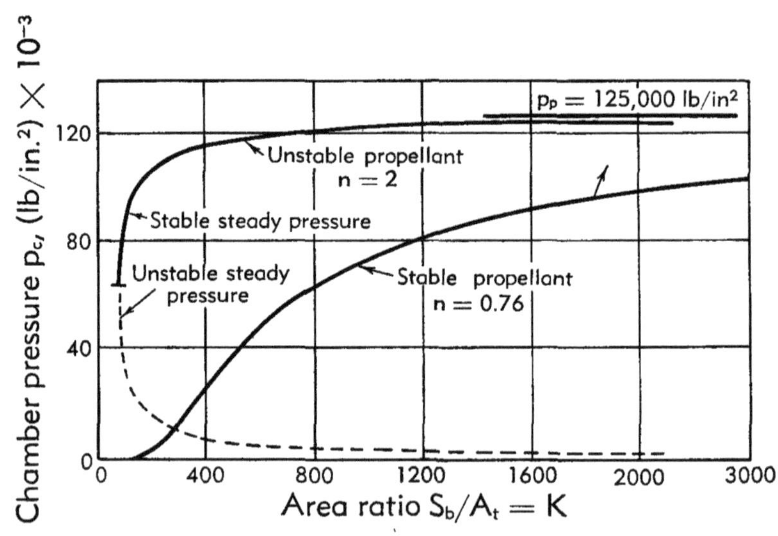

Fig. H,7d. Variation of steady state chamber pressure with area ratio K, for a stable and for an unstable propellant. The unstable propellant exhibits no stable steady state pressure below 62,500 lb/in.^{2} abs. (The assumed propellant characteristics are given in Fig. H,7b and H,7c.)

$p_p = \rho_p \Gamma^2 (c^*)^2$). Numerically, it is easier to assume p_c and compute K. Examples of the area ratio curves corresponding to Eq. 7-12 are shown for a stable and an unstable propellant in Fig. H,7d. The scale employed again covers the whole range of pressures. For practical rocket work, curves not extending above 4000 lb/in.^{2} abs are all that are needed. An illustrative set is shown in Fig. H,7e. The very rapid variation of chamber pressure with changes in K may be noted.

In the low pressure region, $p_c \ll p_p$, the expression for the burning rate (Eq. 5-3) may be inserted into Eq. 7-12 and the expression solved for the chamber pressure as a function of area ratio. As noted above,

p_p may be eliminated by utilizing Eq. 7-3 and 7-5. The result is

$$\frac{p_c}{1000} = \left(\frac{\rho_p a c^* K}{1000}\right)^{\frac{1}{1-n}} \tag{7-13}$$

where a and n are the parameters in the burning rate law. This indicates that for $n \cong 0.75$, the chamber pressure varies with the 4th power of

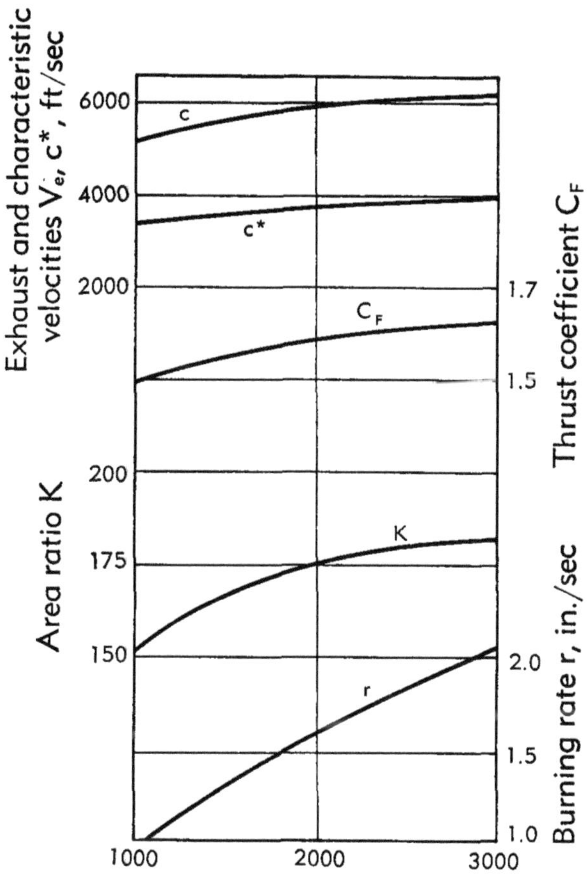

Fig. H,7e. Variation of propellant performance characteristics with combustion chamber pressure for an asphaltic, thermoplastic, composite propellant (end-burning).

the area ratio, giving a very rapid change of pressure with area ratio. This indicates the sensitive balance between gas evolution and outflow in a rocket motor, even for stable propellants. In experimental studies, data have been analyzed by log p_c vs. log K and log r vs. log p_c plots and fitted by least square methods. In a typical example n from K vs. p_c data was 0.80 and n from r vs. p_c data was 0.76. The higher value of n for the

K vs. p_c plot is probably due to the increase of c^* with chamber pressure indicated in Fig. H,7e.

H,8. Steady State Dynamics for Radial-Burning Grains. For internal-burning and radial-burning grains, the internal gas flow introduces complications into the theory. Erosive burning, if it is mass flow dependent, can also introduce a special kind of instability—erosive instability—into the combustion process.

A definite arrangement, of the type considered here, is shown in Fig. H,8a. The analysis extends in a straightforward manner to other, related, arrangements.

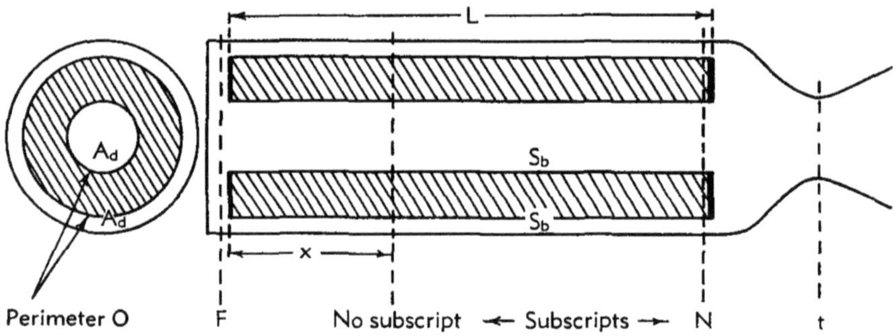

Fig. H,8a. Notation employed in the discussion of radial-burning solid propellant grains.

If the burning rate remains the same at every point on the surface, a hollow cylindrical charge, restricted from burning on the ends, maintains a constant burning surface throughout the burning period. Other arrangements can be devised to achieve something of the same result. It is assumed that the total area of the burning surface S_b is constant.

Even with this assumption, the gas flows do not correspond to steady state conditions. As the solid propellant burns away, the duct area A_d, for gas flow parallel to the axis of the rocket, increases. However, in the usual design regime for radial-burning rockets, the velocity of gas flow at the exhaust nozzle end of the charge is very much larger than the velocity of recession of the burning surface. During the time necessary for the combustion of 1 per cent of the solid propellant, the gas evolution corresponds to about twice the duct volume within the motor. That is, during the time of passage of a fluid particle from the burning surface to the exhaust nozzle throat, the duct area changes by only $\frac{1}{2}$ per cent. This would seem to indicate that assumption of steady state gas flow conditions for interior ballistics analysis is a satisfactory approximation, and steady state analysis is employed here.

In addition, it is assumed that the properties of the combustion gas are the same across any given cross section of the duct. This implies mixing and a high degree of turbulence. These conditions are usually fulfilled for rocket motor gas flows.

Finally, ideal gas behavior is assumed for the propellant gas. Actually, even if the gas composition changes as a function of pressure and temperature, its behavior can be closely approximated by selecting a suitable value for the ratio of specific heats [*10*].

The skin drag and frictional forces of the internal gas flow on the propellant grain and on the motor tube are neglected. The Reynolds number for internal gas flows in rockets is of the order of 10^6 and the corresponding friction factor is about 0.005. A calculation indicates that, during the critical period immediately after ignition, when the duct area is small, the frictional loss corresponds to 2 to 5 per cent of the pressure drop along the grain due to the dynamics of the gas flow. This omission of frictional effects, then, does introduce appreciable numerical error, but the qualitative features of rocket behavior are not affected.

Because of the internal flow, the analysis is complex. By defining an average value for the burning rate, explicit expressions are easily derived. The more intricate problem of finding suitable expressions for the average burning rate and related functions is then carried out as a separate problem and inserted into the preceding analysis.

Area ratio expressions. The area ratio curve is easily derived by the introduction of the concept of average burning rate and exhaust nozzle stagnation pressure. Consider the arrangement in Fig. H,8a. Between the front of the combustion chamber, marked F, and the exhaust nozzle end of the charge, marked N, the pressure of the gas stream decreases and the velocity of the gas flow increases. The actual distribution will be derived, shortly, but from the section N to the section t at the exhaust nozzle throat no further combustion mass is added to the flow. The flow between N and t corresponds to flow through a portion of an exhaust nozzle (of rather peculiar shape it is true) in which the entering gas velocity is not zero. The sudden expansion of gas as it leaves the end of the grain and enters the region just forward of the exhaust nozzle actually results in loss of stagnation pressure, but this is neglected in the analysis. Now the mass flow through an exhaust nozzle is given by

$$\dot{m}_o = \frac{p^0 A_t}{c^*} \tag{8-1}$$

The stagnation pressure p^0 is related to the pressure and velocity at section N, and through this to conditions at section F. The relation will be derived shortly.

The mass flow out of the exhaust nozzle must equal the rate of mass combustion, which in terms of the average of the burning rate over the propellant surface is given by

$$\dot{m}_b = \bar{r} S_b \rho_b \tag{8-2}$$

Here $\bar{r}$ is the average burning rate, S_b the area of the burning surface, and ρ_p the density of the solid propellant. Eq. 8-2 neglects the effect of increase of duct volume as the charge burns away. This would introduce a factor $[1 - (\rho/\rho_p)]$ on the right (ρ is the density of gas in the ducts), but at normal design pressures $\rho/\rho_p < 0.01$, and this factor is neglected. Equating mass outflow to mass burned gives

$$\frac{p^0 A_t}{c^*} = \bar{r} S_b \rho_p$$

This may be written, in terms of $K = S_b/A_t$,

$$K = \frac{p^0}{\rho_p \bar{r} c^*} \tag{8-3}$$

It is now convenient to introduce the parameters θ_r and θ^*, defined by

$$\theta^* = \frac{p_F}{p^0} \quad \text{or} \quad p^0 = \frac{p_F}{\theta^*} \tag{8-4a}$$

$$\theta_r = \frac{\bar{r}}{r_F} \quad \text{or} \quad \bar{r} = \theta_r r_F \tag{8-4b}$$

$$\theta = \theta^* \theta_r \tag{8-4c}$$

Here r_F and p_F represent the burning rate and gas pressure at the front of the grain, section F (see Fig. H,8a). In terms of these parameters,

$$K = \frac{p_F}{\rho_p c^* \theta^* r_F \theta_r} = \frac{p_F}{\rho_p c^* r_F \theta} \tag{8-5}$$

The parameters θ_r, θ^* may be determined in terms of the ratio to gas duct area to exhaust nozzle throat area A_d/A_t, and the front chamber pressure. Therefore Eq. 8-5 represents the area ratio curve for front pressure.

Internal gas flow. In general, the burning rate is not the same at every section of the grain, so that as burning progresses the gas duct area is different from section to section. This variation is neglected. However, the conditions at the nozzle end of the grain are correctly given although averages taken over the length of the grain are slightly in error. That is, the parameter θ^* is correct, but θ_r and θ are slightly in error due to this approximation. These errors are usually small, relative to the errors introduced by the neglect of friction effects.

The analytical technique used is to find the variation in conditions along the grain, relative to front conditions, in terms of the local Mach number. The Mach number at the nozzle end of the grain is determined by the ratio of exhaust nozzle throat area to the gas duct area. This then unifies the entire flow pattern.

The equations for conservation of mass, momentum, and energy at section x (Fig. H,8a), together with the equation of state, are, respectively

$$A_d \frac{d(\rho V)}{dx} = \rho_p r \frac{dS_b}{dx} = \rho_p r O \tag{8-6a}$$

$$OL = S_b \tag{8-6b}$$

$$p_F - p = \rho V^2 \tag{8-6c}$$

$$\tfrac{1}{2}V^2 = c_p(T_0 - T) \tag{8-6d}$$

$$p = \rho \mathcal{R} T \tag{8-6e}$$

Here ρ, V, p, and T are the gas density, velocity, pressure, and temperature at section x. A_d is the gas duct area, r the burning rate at position x, dS_b/dx is constant, and O the perimeter of the burning surface. $\mathcal{R}$ is the engineering gas constant (includes the effect of molecular weight), $T_0(= T_F)$ is the adiabatic constant pressure flame temperature for combustion of the solid propellant, the subscript F refers to front conditions, and the other symbols have been defined previously.

Eq. 8-6b is exact if the gas duct area is constant along the grain. In general the grain tapers slightly during combustion, producing an opposite taper in the gas duct.[4] This effect is neglected.

It is convenient to introduce the local Mach number M and the dimensionless mass flow density g, defined as follows:

$$M^2 = \frac{V^2}{a^2} = \frac{V^2}{\gamma R T} \tag{8-7a}$$

$$g = \frac{\rho V}{\rho_F a_F} \tag{8-7b}$$

Here a is the local velocity of sound in the gas (not to be confused with the burning rate parameter), γ is the ratio of specific heats (constant pressure/constant volume) for the propellant gas, a_F is the velocity of sound in the gas at section F, and ρ_F is the gas density at section F.

With these parameters Eq. 8-6 can all be solved for local conditions in terms of front conditions and local Mach number. The parameter g can also be expressed in terms of local Mach number, so that the variation of condition along the grain is now determined as a function of g.

[4] The burning rate is affected by compensating factors. At the front, the higher pressure tends to accelerate the rate; and at the rear, lower pressure region, the erosive effect of gas flow accelerates the rate. Usually the compensation is not exact.

The results are[5]:

$$\frac{p}{p_F} = \frac{1}{1 + \gamma M^2} \tag{8-8a}$$

$$\frac{T}{T_F} = \frac{T}{T_0} = \frac{1}{1 + \frac{\gamma - 1}{2} M^2} \tag{8-8b}$$

$$\frac{\rho}{\rho_F} = \frac{1 + \frac{\gamma - 1}{2} M^2}{1 + \gamma M^2} \tag{8-8c}$$

$$\frac{V}{a_F} = \frac{M}{\sqrt{1 + \frac{\gamma - 1}{2} M^2}} \tag{8-8d}$$

$$g = \frac{M}{1 + \gamma M^2} \sqrt{1 + \frac{\gamma - 1}{2} M^2} \tag{8-8e}$$

$$M^2 = \frac{\sqrt{1 - 2(\gamma + 1)g^2} - (1 - 2\gamma g^2)}{[(\gamma - 1) - 2\gamma^2 g^2]} \tag{8-8f}$$

The local stagnation pressure p^0 is defined by

$$\frac{p^0}{p} = \left(\frac{T_F}{T}\right)^{\frac{\gamma}{\gamma - 1}}$$

and may be expressed in terms of p_F to give

$$\frac{p^0}{p_F} = \frac{\left(1 + \frac{\gamma - 1}{2} M^2\right)^{\frac{\gamma}{\gamma - 1}}}{1 + \gamma M^2} \tag{8-8g}$$

The local stagnation pressure lies about halfway between the local pressure and the front pressure. If $M \ll 1$, one may expand (Eq. 8-8g) and dropping higher orders of M^2 find

$$p^0 = p_F - \tfrac{1}{2}(p_F - p)$$

In the limiting case, $M = 1$, the factor $\frac{1}{2}$ is replaced by 0.36. The approximate validity of the factor $\frac{1}{2}$ is reasonable since mass is being added along the gas streams at varying pressures above p, and a linear average would give just this value.

In Fig. H,8b the flow characteristics are plotted as a function of local Mach number. The local Mach number and other flow parameters are shown in Fig. H,8c plotted as functions of the dimensionless mass flow density g.

Internal sonic flow. The range of Mach numbers shown does not

[5] The algebra is simplied if expressions for p and T in terms of M are found first. See [*44*].

exceed unity. Physically, supersonic flow is not attained adjacent to a burning surface where mass is being added. Differentiation of Eq. 8-8e yields

$$\frac{dM}{M} = \frac{(1 + \gamma M^2)\left(1 + \frac{\gamma - 1}{2} M^2\right)}{1 - M^2} \frac{dg}{g} \tag{8-9}$$

It is clear that if $M < 1$, the Mach number increases with g; but if $M > 1$, the Mach number decreases as the mass flow increases. The flow cannot

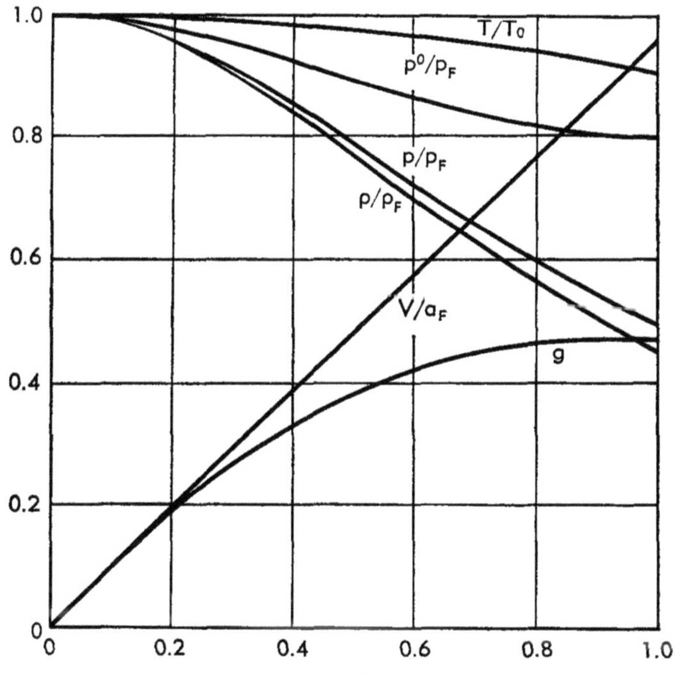

Fig. H,8b. Variation of flow characteristics with local Mach number.

become supersonic. At $M = 1$, the critical value of g is $\sqrt{1/2(\gamma + 1)}$. If the burning surface or the burning rate is large enough, or if the duct area is small enough, the conditions for sonic flow may be met for some point along the grain. The sonic position moves to the nozzle end of the grain and then jumps to the exhaust nozzle throat, provided that $A_t < A_d$. If $A_t > A_d$, then a sonic flow will be established near the nozzle end of the grain. Variation of the duct area, due to variation in burning rate, may cause the sonic position to jump from place to place along the grain. Uniform flow can be established, if $A_d > A_t$ and if the pressure level rises, since g is dimensionless and the actual mass flow can increase with increasing pressure. Since sonic flow corresponds to $M = J = 1$ (see below), the expression for the front pressure becomes

$$\frac{S_b}{A_d} = \frac{p_F}{\rho_p c^* r_F \theta}$$

where $\theta = \theta^*\theta_r = \theta^*(1)\theta_r(1, p_F)$. The value 1 corresponds to $J = 1$. This would enable one, in principle, to solve for the steady state front chamber pressure. The usual area ratio K is replaced by S_b/A_d.

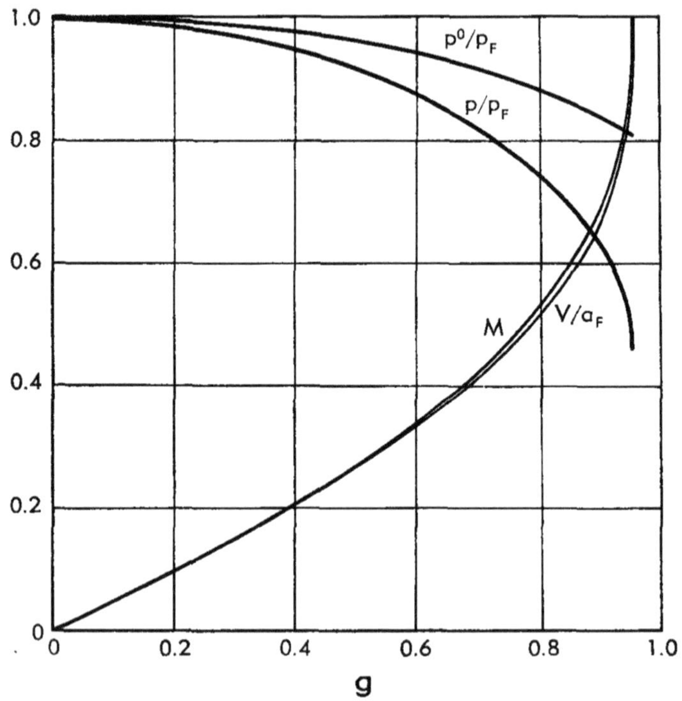

Fig. H,8c. Variation of flow characteristics with the dimensionless mass flow density parameter g.

It is clear that the internal geometry of the rocket should be arranged so that supersonic flow is not attained at any section of the grain. This is accomplished by insuring that $A_d > A_t$ at all times. Difficulties, such as unstable combustion and loss of propellant, may be encountered if this requirement is not met.

Throat-to-duct area ratio. Having established that sonic flow will occur only at the exhaust nozzle throat, it is easy to determine the Mach number M_N at the exhaust nozzle end of the propellant. The equality of mass flow at section N and at the nozzle throat, section t, yields (using Eq. 8-1)

$$\rho_N V_N A_d = \dot{m}_o = \frac{p_N^0 A_t}{c^*}$$

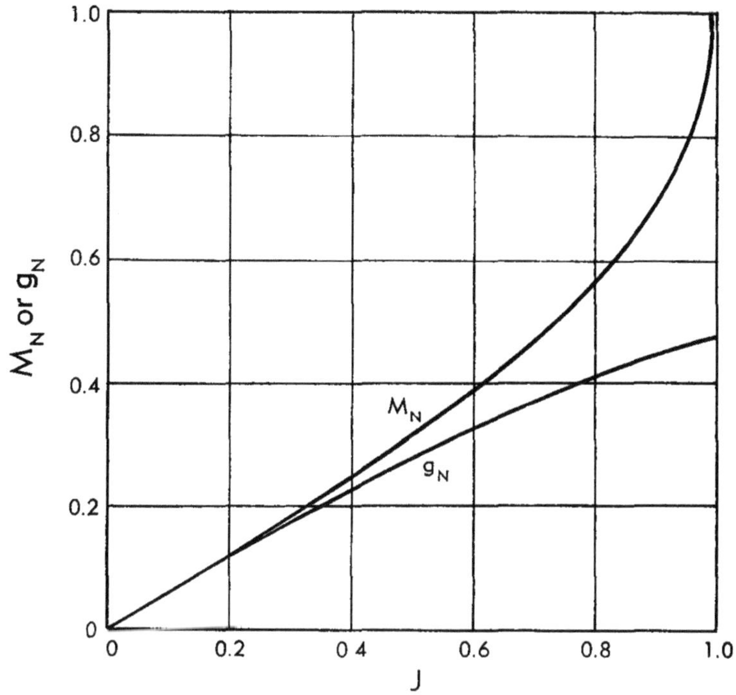

Fig. H,8d. Variation of Mach number at the exhaust nozzle end of the charge, and of the dimensionless mass flow parameter at the exhaust nozzle end of the charge, g_N with the ratio, J of exhaust nozzle throat area to gas duct area (for $\gamma = 1.2$).

Here p_N^0 is the stagnation pressure of the gas flow at the exhaust nozzle end of the grain. By utilizing the definition of c^* (Eq. 7-5) and Eq. 8-8c, 8-8d, and 8-8g the Mach number M_N may be determined from

$$J = \frac{A_t}{A_d} = \frac{M_N}{\left(\frac{2}{\gamma + 1} + \frac{\gamma - 1}{\gamma + 1} M_N^2\right)^{\frac{\gamma+1}{2(\gamma-1)}}} \tag{8-10}$$

From the definition (Eq. 8-4), θ^* may be written (Eq. 8-8g)

$$\theta^* = \frac{1 + \gamma M_N^2}{\left(1 + \frac{\gamma - 1}{2} M_N^2\right)^{\frac{\gamma}{\gamma-1}}} = \theta^*(J) \tag{8-11}$$

By knowing M_N from Eq. 8-10, θ^* may be expressed as a function of the throat-to-duct area ratio as indicated.

The variation of M_N with J is shown in Fig. H,8d, and the variation of θ^* with M_N and with J is shown in Fig. H,8e.

The average burning rate and $g(x)$. If the burning rate were the same at all sections along the grain, then the actual mass flow density ρV and the dimensionless mass flow density g would be proportional to x, the distance from the front of the grain (Fig. H,8a). The parameter g is the natural independent variable for description of the internal gas flow and local burning rate (g and M are connected and either may be used. See Eq. 8-8).

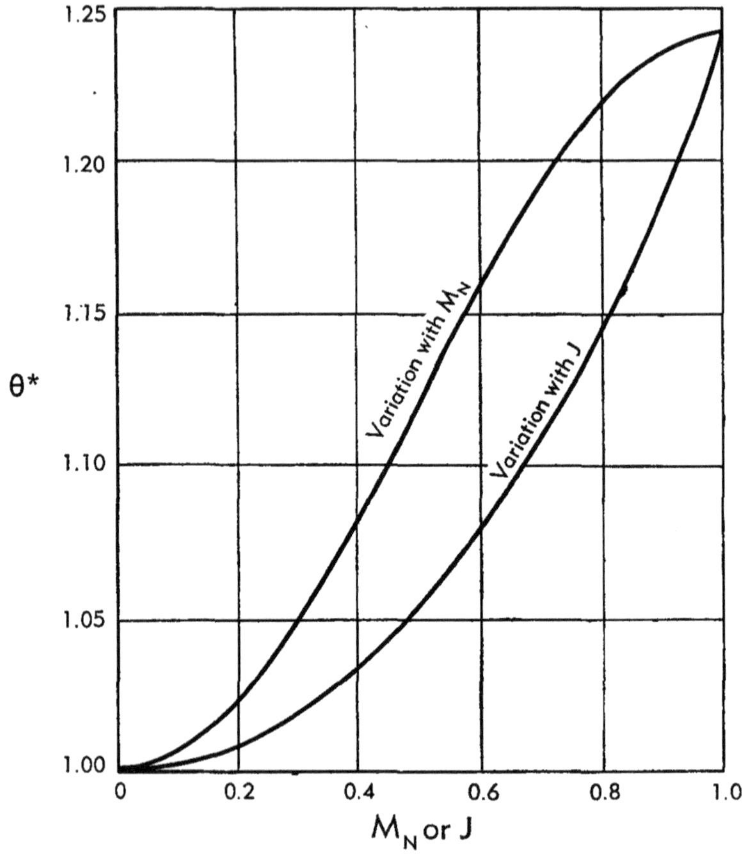

Fig. H,8e. Variation of the stagnation pressure flow parameter θ^* with Mach number at the exhaust nozzle end of the charge and also with the ratio J of exhaust nozzle throat area to gas duct area (for $\gamma = 1.2$).

The connection between g and x is derived first. The burning rate expressions (Eq. 5-1 and 5-2) may be written in terms of the front pressure and g. Let this be understood, then Eq. 8-6a may be written

$$dx = \frac{A_d}{\rho_p O} \frac{d(\rho V)}{r}$$

From the definition of g (Eq. 8-7b),

$$dx = \frac{A_d \rho_F a_F}{\rho_p O} \frac{dg}{r}$$

By Eq. 8-6b, and integration, this becomes

$$\frac{x}{L} = \frac{A_{\mathrm{d}}}{S_{\mathrm{b}}} \frac{\rho_F a_F}{\rho_{\mathrm{p}}} \int_0^g \frac{dg}{r} \tag{8-12}$$

This is the connection between the mass flow parameter and x. It may be simplified by calculation of the average burning rate $\bar{r}$.

The rate of mass flow past the nozzle end of the grain must equal the rate at which mass is consumed over the entire burning surface. Introducing the average burning rate $\bar{r}$ and the definition of g, this gives

$$\bar{r}\rho_{\mathrm{p}} S_{\mathrm{b}} = A_{\mathrm{d}}(\rho V)_N = A_{\mathrm{d}}\rho_F a_F g_N \tag{8-13}$$

If the integration in Eq. 8-12 is extended to g_N, then $x = L$. Combining Eq. 8-13 and 8-12 then gives

$$\frac{1}{\bar{r}} = \frac{1}{g_N} \int_0^{g_N} \frac{dg}{r} \tag{8-14a}$$

From the definition of θr (Eq. 8-4b),

$$\frac{1}{\theta_r} = \frac{1}{g_N} \int_0^{g_N} \frac{dg}{(r/r_F)} \tag{8-14b}$$

Now r/r_F depends only on M (or g), so that the integration may be carried out explicitly (although numerically). The burning rate expressions employed here may be combined with the flow functions (Eq. 5-1, 5-2 and 8-8) to give, for mass flow density erosion,

$$\left(\frac{r}{r_F}\right)_{\mathrm{me}} = \left(\frac{1}{1+\gamma M^2}\right)^n (1 + k\rho_F a_F g) \tag{8-15a}$$

and for velocity erosion,

$$\left(\frac{r}{r_F}\right)_{\mathrm{ve}} = \left(\frac{1}{1+\gamma M^2}\right)^n f_3\left(a_F \frac{V}{a_F}\right) \tag{8-15b}$$

By substituting the expressions for a_F, ρ_{p} in terms of c^*, p_{p} (Eq. 7-3 and 7-5), and by employing the expression for V/a_F (Eq. 8-8d), the burning rate ratios may be written, for mass flow density erosion,

$$\left(\frac{r}{r_F}\right)_{\mathrm{me}} = \left(\frac{1}{1+\gamma M^2}\right)^n \left(1 + \lambda_1 \frac{p_F}{p_{\mathrm{p}}} g\right) \tag{8-16a}$$

$$\lambda_1 = \sqrt{\gamma}\,\Gamma k \rho_{\mathrm{p}} c^* \tag{8-16b}$$

so that ultimately

$$(\theta_r)_{\mathrm{me}} = (\theta_r)_{\mathrm{me}}\left(g_N, \frac{p_F}{p_{\mathrm{p}}}\right) = (\theta_r)_{\mathrm{me}}\left(J, \frac{p_F}{p_{\mathrm{p}}}\right) \tag{8-16c}$$

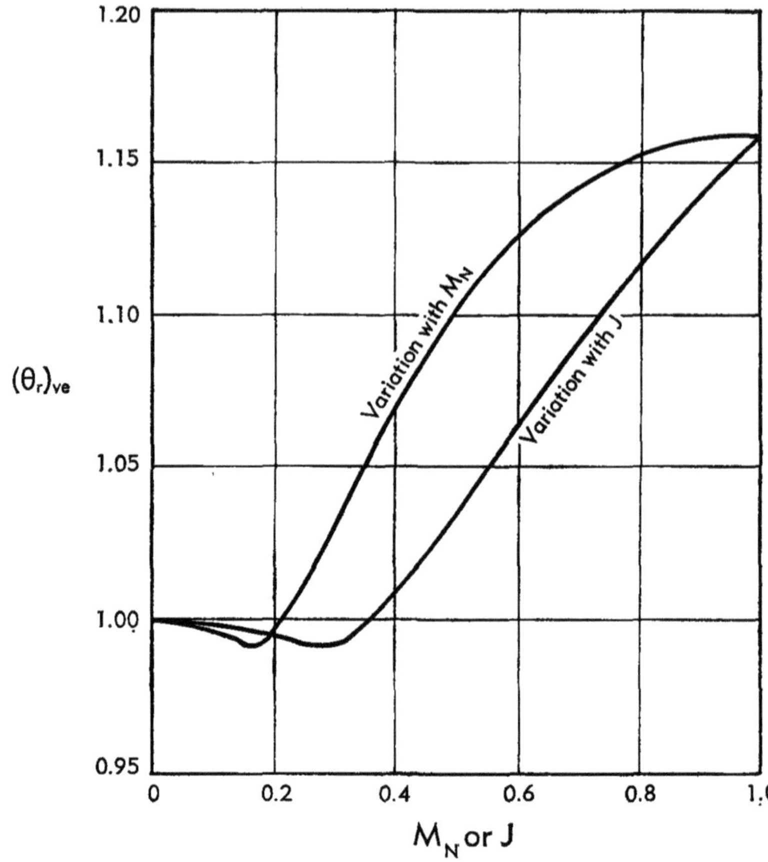

Fig. H,8f. Velocity erosion. Variation of the velocity erosive burning rate parameter θ_r with Mach number at the exhaust nozzle end of the charge and also with the ratio J of exhaust nozzle throat area to gas duct area. (For $\gamma = 1.2$, $a = 0.651$ in./sec, $n = 0.69$, $c^* = 5200$ ft/sec, $V_0 = 600$ ft/sec, $k_v = 5.72 \times 10^{-4}$ sec/ft, $\rho_p = 101.5$ lb/ft³, $p_p = 230{,}000$ lb/in.² abs.)

and for velocity dependent erosion,

$$\left(\frac{r}{r_F}\right)_{ve} = \left(\frac{1}{1 + \gamma M^2}\right)^n f_3(u) \tag{8-17a}$$

$$f_3(u) = \begin{cases} 1 & , u < u_0 \\ 1 + \lambda_2(u - u_0), & u > u_0 \end{cases} \tag{8-17b}$$

$$u = \frac{M}{\sqrt{1 + \frac{\gamma - 1}{2} M^2}} \tag{8-17c}$$

$$\lambda_2 = \sqrt{\gamma}\, \Gamma k_V c^* \tag{8-17d}$$

$$u_0 = \frac{V_0}{\sqrt{\gamma}\, \Gamma c^*} \tag{8-17e}$$

so that ultimately

$$(\theta_r)_{ve} = (\theta_r)_{ve}(g_N) = (\theta_r)_{ve}(J) \tag{8-17f}$$

The new erosion constants λ_1 and λ_2 are now dimensionless. The burning rate factor may be computed by combining Eq. 8-12, 8-16, and 8-17. The results of this computation are shown in Fig. H,8f for a velocity-dependent erosive propellant; and in Fig. H,8g for a mass flow density erosive propellant. The former does not depend upon the pressure, but the latter shows a pressure dependence (Eq. 8-16).

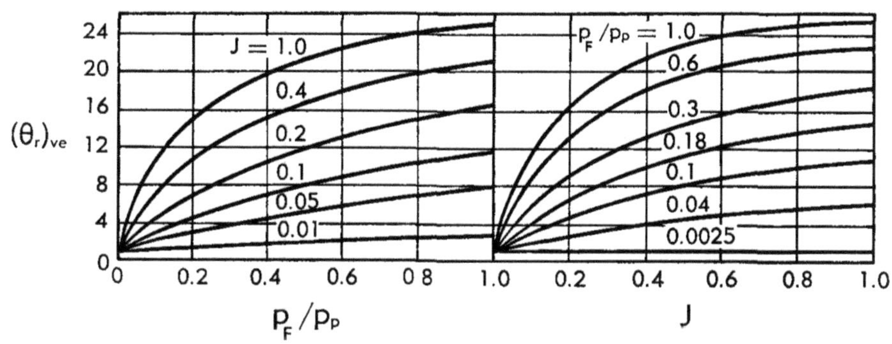

Fig. H,8g. Mass flow density erosion. Variation of the mass flow density erosive burning rate parameter θ_r with chamber pressure at the forward end of the charge p_F, and also with the ratio J of exhaust nozzle throat area to gas duct area. (For $\gamma = 1.2$, $a = 0.15$ in./sec, $n = 0.50$, $c^* = 3650$ ft/sec, $k = 0.61$ in.^{2} sec/lb, $\rho_p = 103$ lb/ft^3, $p_p = 112{,}000$ lb/in.^{2} abs.)

The expression (Eq. 8-12) connecting x and g may be simplified by dividing through by the same equation for $x = L$ and $g = g_N$. This gives

$$\frac{x}{L} = \frac{\int_0^g \frac{dg}{r}}{\int_0^{g_N} \frac{dg}{r}}$$

Multiplying numerator and denominator by r_F, and letting $\theta_r(g)$ correspond to the integral (Eq. 8-14b) for $g = g$ rather than $g = g_N$ gives

$$\frac{x}{L}\,\frac{\theta_r(g_N)}{\theta_r(g)}\,\frac{g}{g_N} \tag{8-18}$$

This gives x/L in terms of the integral, which would be calculated anyway as a function of g_N for various values.

It is interesting to note that x/L is nearly proportional to g/g_N, differing only by a factor equal to the mean burning rate divided by the mean of the burning rate from the front up to the section at which the mass flow is g. In Fig. H,8h the variation of g with x is shown for a mass flow density erosive propellant.

Comments on the average burning rate dependence. The flow pattern in the rocket is essentially kinematic and geometrical. The explosion enthalpy $c_pT_0 = c_pT_F$ of the propellant determines the speed of sound a_F at the forward end of the grain. This normalizes the velocity distribution and also the connection between density and pressure. With this established, and the pressure level set by the front pressure, all other parameters depend only on local Mach number M or the dimensionless mass flow density g. These are uniquely connected and are fixed by the geometrical ratio of throat-to-duct area ratio. For propellants subject to

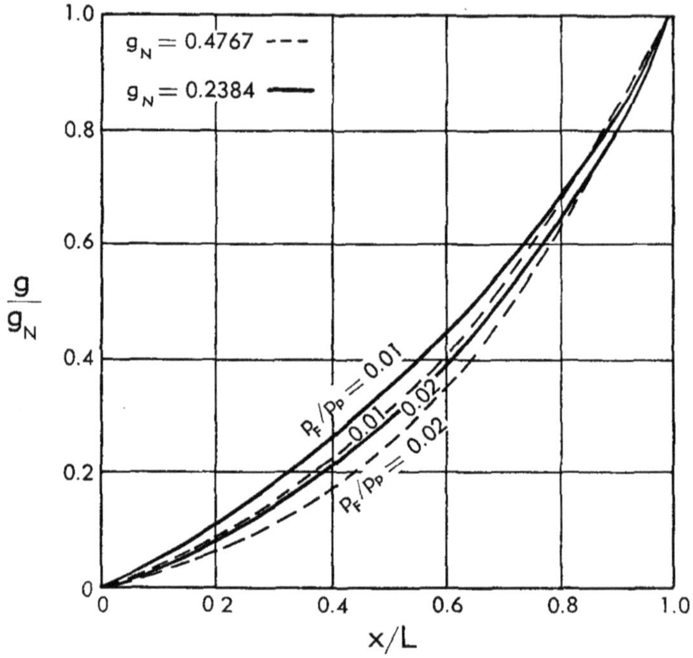

Fig. H,8h. Variation of the dimensionless mass flow density parameter g with the geometric position x, measured from the front of the grain for a mass flow density erosive propellant. (For propellant properties listed in Fig. H,8g.)

velocity erosion alone, the over-all effect is to simply modify the constants for end-burning grains. For propellants subject to mass flow density erosion, the density introduces an additional pressure dependence on the burning rate. The remaining effects are kinematic and geometric.

H,9. Area Ratio Dependence. Erosive Instability. In the case of unrestricted burning rockets, the erosive effect of internal gas flow is encountered. If the erosive effect depends only on gas velocity, the stability considerations are the same as for end-burning grains (θ_r depends only on J) and no new effects are encountered. However, if the erosive effect depends upon the mass flow density ρV (Eq. 3-2b), the over-all effect is to introduce additional pressure dependence for the average burning rate. The additional pressure dependence enters through the density factor of the erosive term. The erosive instability is a more fundamental limitation of rocket design than resonant burning since it is connected with a first order differential equation, and is based upon the conservation of mass.

In order to discuss the whole pressure regime, it is convenient to reintroduce the factor $1 - (p/p_p)$ which takes account of the increase of gas volume within the combustion chamber as the charge burns away. The area ratio curve may be expressed as

$$K = \frac{p_F}{\rho_p c^* \theta^* r_F \theta_r \left(1 - \frac{p_F}{p_p}\right)} \tag{9-1a}$$

$$\theta_r = \theta_r \left(J, \frac{p_F}{p_p}\right) \tag{9-1b}$$

Here p_F is the gas pressure at the forward end of the charge. J is the ratio A_t/A_d of nozzle throat area to gas duct area along the charge (see Fig. H,8a), and θ_r is determined numerically (Fig. H,8g).

This equation is of the same form as Eq. 7-12 for end-burning grains and has the same significance. The stability problem may be analyzed in the same way. The expressions for net mass production $\dot{m}_b$, and mass outflow $\dot{m}_o$, may be written by analogy:

$$\dot{m}_b = \left(1 - \frac{p}{p_p}\right) \theta_r r_F \tag{9-2a}$$

$$\dot{m}_o = \frac{\Gamma^2}{K} \theta^* c^* \frac{p}{p_p} \tag{9-2b}$$

The stability requirement is again that at the steady state point, $\dot{m}_o = \dot{m}_b$, that the slope of the $\dot{m}_o$ curve shall be greater than the slope of the $\dot{m}_b$ curve. The procedure is identical to Eq. 7-9. Although numerical results are available, it is helpful to obtain an approximate expression, analogous to Eq. 7-10. From Eq. 8-14 and 8-16 the expression for θ_r is

$$\frac{1}{\theta_r} = \frac{1}{g_N} \int_0^{g_N} \frac{(1 + \gamma M^2)^n dg}{\left(1 + \lambda_1 \frac{p_F}{p_p} g\right)}$$

If the factor depending explicitly on Mach number is assumed constant at a suitable average value, this may be evaluated to give

$$\frac{1}{\theta_r} \cong \overline{(1 + \gamma M^2)^{\bar{n}}} \frac{\ln (1 + x)}{x} \tag{9-3a}$$

$$x = \lambda_1 g_N \frac{p_F}{p_p} \tag{9-3b}$$

For the stability criteria, the logarithmic derivative of θ_r is required so that the Mach number factor approximately cancels out, as it does exactly to the assumed approximation. This gives

$$\frac{d \ln \theta_r}{d \ln p_F} = -\left[1 - \frac{x}{(1+x)\ln(1+x)}\right] \tag{9-4a}$$

$$x = \lambda_1 g_N \frac{p_F}{p_P} \tag{9-4b}$$

Taking the logarithmic derivatives of the two mass flow equations (Eq. 9-2), and employing the result (Eq. 9-4) gives the requirement for stability:

$$n - \frac{p_F/p_P}{(1 - p_F/p_P)} + \left[1 - \frac{x}{(1+x)\ln(1+x)}\right] < 1 \tag{9-5a}$$

$$x = \lambda_1 g_N \frac{p_F}{p_P} \tag{9-5b}$$

The term in the square brackets increases monotonically with x. It is shown in Fig. H,9a. The lowest pressure for which instability will

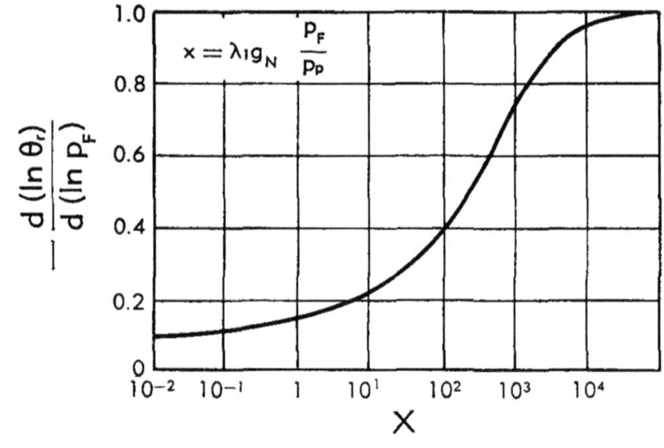

Fig. H,9a. Mass flow density erosive instability. Variation of the correction to the burning rate power index due to mass flow density erosion with dimensionless pressure parameter x.

occur is for the maximum value of g_N (= 0.4767). Setting the left side equal to unity and inserting the propellant characteristics given in Fig. H,8g, the minimum unstable pressure is found at $p_F/p_P = 7.8 \times 10^{-3}$, More accurate calculations employed for Fig. H,8g give 7.4×10^{-3}, which corresponds to a pressure of only 880 lb/in.^{2} abs.

According to Eq. 9-1 the pressure depends upon both K and J. Therefore certain regions of the K, J plane are unstable.

The situation is made clearer if the mass flow functions (Eq. 9-2) are plotted. The result is shown in Fig. H,9b. The erosive effect changes the

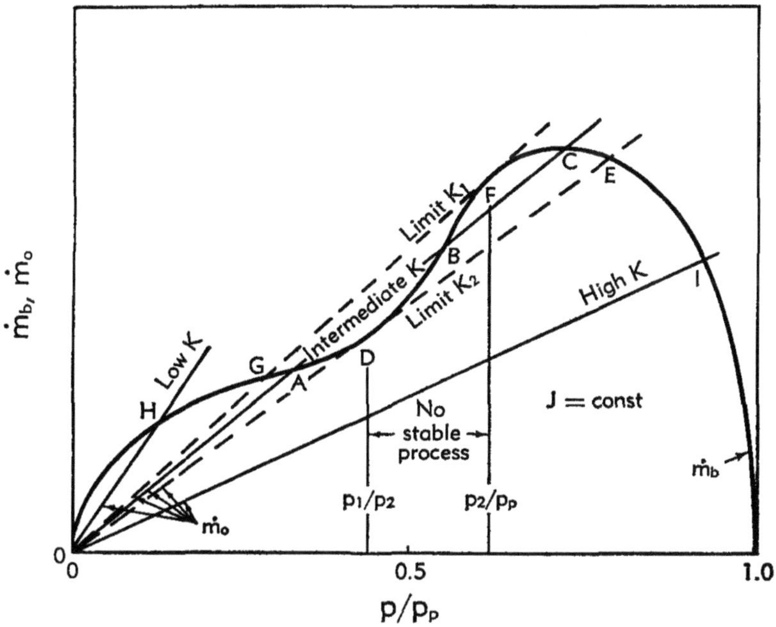

Fig. H,9b. Mass flow density erosive instability. Variation of net burning rate $\dot{m}_b$ and net outflow $\dot{m}_o$ with front chamber pressure p_F (schematic).

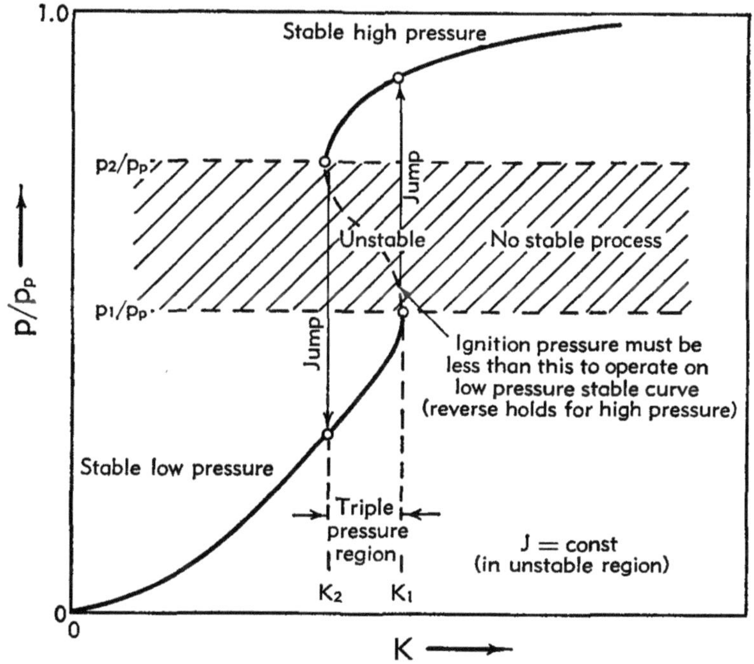

Fig. H,9c. Mass flow density erosive instability. Variation of steady state front pressure p_F with ratio K of burning surface area to exhaust nozzle throat area for a fixed J in the instability region (schematic). Note limit K_1, K_2 and limit p_1, p_2 reversal of magnitude.

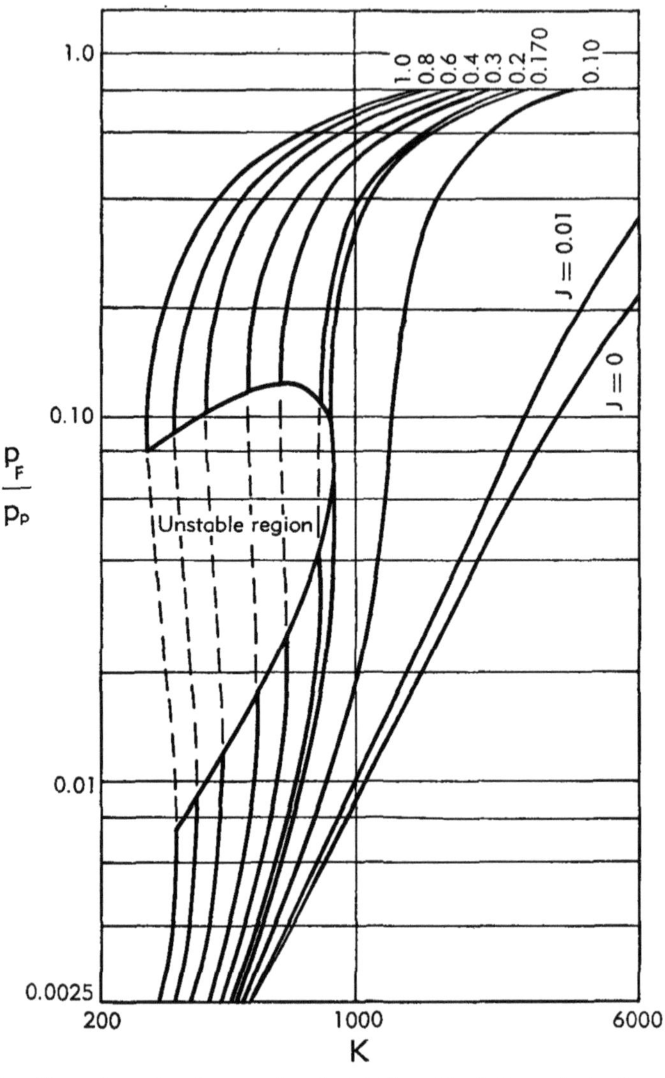

Fig. H,9d. Mass flow density erosive instability. Variation of steady state front pressure with the area ratios *K* and *J*. The critical *J* value is 0.170. Below this value, instability effects are absent. (Propellant characteristics listed in Fig. H,8g.)

$\dot{m}_b$ curve from a simple curve, concave to the pressure axis, to one of more complicated form. If the effect of erosion is changed by changing the ratio of nozzle throat area to gas duct area along the charge, then a *family* of $\dot{m}_b$ curves is developed. For large duct areas, the erosive effect decreases and, below a critical throat-duct area ratio J, the $\dot{m}_b$ curves become simple in character and the erosive instability is no longer encountered.

Now consider the curve shown (Fig. H,9b), which shows a marked erosive effect. The $\dot{m}_o$ and $\dot{m}_b$ intersections at very high and very low K value are simple and stable (intersections H and I). The $\dot{m}_o$ line at intermediate K values makes three intersections, A, B, and C with the $\dot{m}_b$ curve. There are two stable operating pressures A and C, separated by one

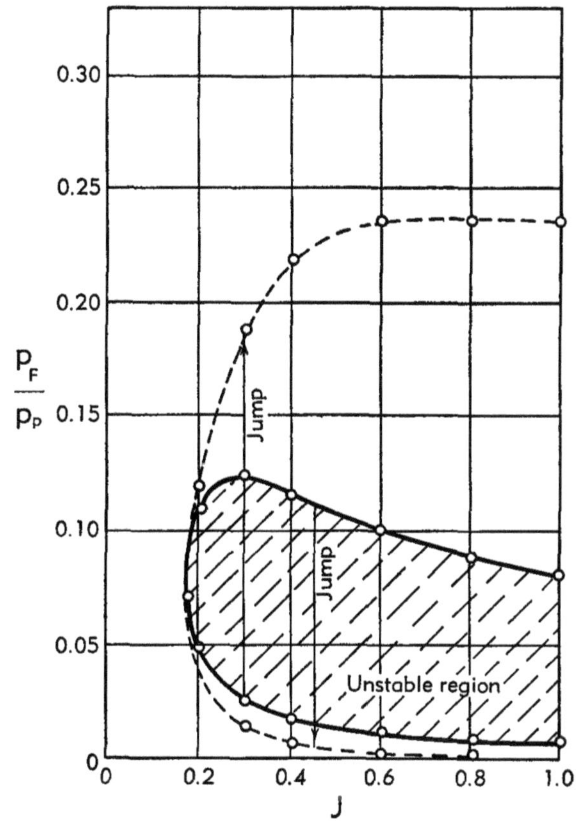

Fig. H,9e. Mass flow density erosive instability. The instability region in the front pressure J plane. The value of K is not fixed, but is adjusted to give the desired pressure. The pressure jump from the low pressure stable curve to the high pressure stable curve, for an infinitesimal change in K at the upper stability limit, is indicated by the arrows and dotted curves. The converse is indicated on going from high to low pressure stable burning. (Propellant properties listed in Fig. H,8g.)

unstable steady pressure B for a rocket of given geometry and given propellant.

If the ignition process is very severe, the ignition pressure may start the operation of the rocket at a pressure in excess of B, and the pressure rises to the stable pressure C, which may be disastrously high. If the ignition pressure lies below pressure B, the rocket drops in pressure and operates at pressure A. The stable pressures fall into two regions, high pressure and low pressure. There are two limiting $\dot{m}_o$ curves just tangent to the $\dot{m}_b$ curve, which are determined by taking the equality sign in Eq. 9-5. It is clear that the point D corresponds to the maximum stable pressure in the low pressure region. If an attempt is made to increase the pressure above D by increasing K (rotating the $\dot{m}_o$ line clockwise), the stable pressure jumps from D to E. (A similar argument follows for points F

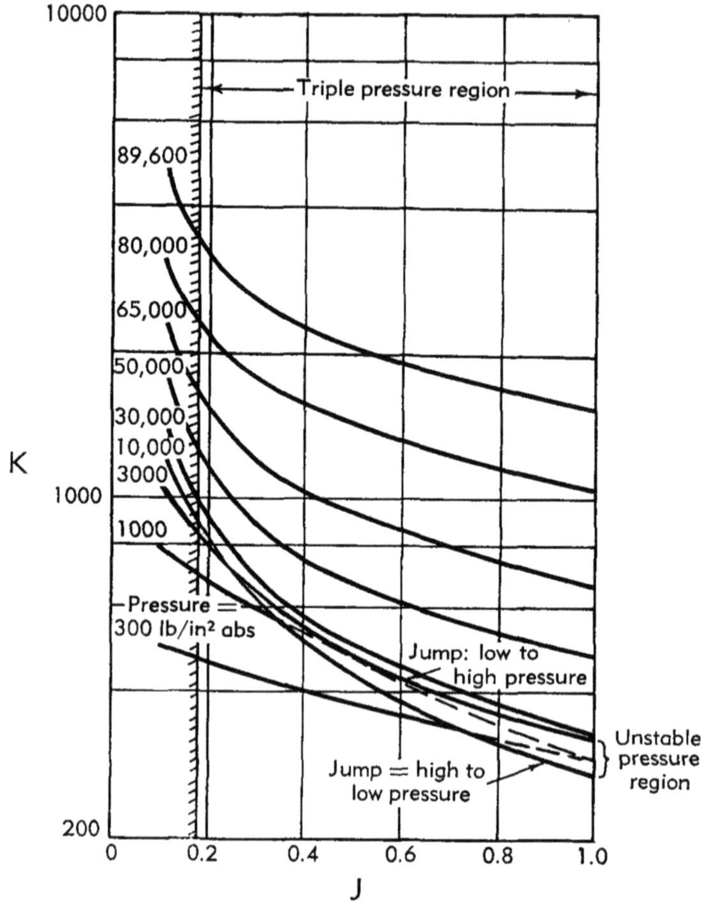

Fig. H,9f. Mass flow density erosive instability. The unstable pressure region in the K, J plane. At the edges of this region, an infinitesimal increase in K causes a discontinuous jump from the stable low pressure region to the stable high pressure region. The converse holds for attempts to continue the high pressure region to low pressures. (Propellant properties listed in Fig. H,8g.)

and G if one attempts to extend the stable high pressure region to low pressures by decreasing K.)

Consider Fig. H,9c, where the variation of steady pressure with K is shown for fixed throat-to-duct ratio. Starting in the low pressure region, the pressure increases with K. A limit K is reached, and the low pressure curve continues smoothly, but now a stable high pressure curve also

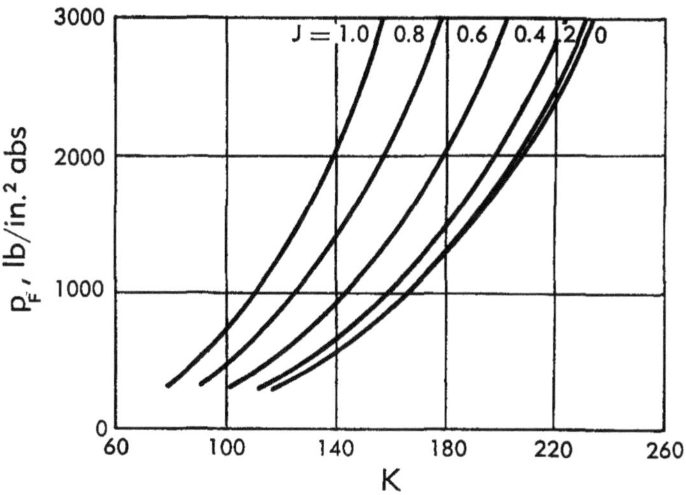

Fig. H,9g. Variation of front chamber pressure with area ratios K and J, for a propellant subject only to velocity erosion. (For propellant properties listed in Fig. H,8f.)

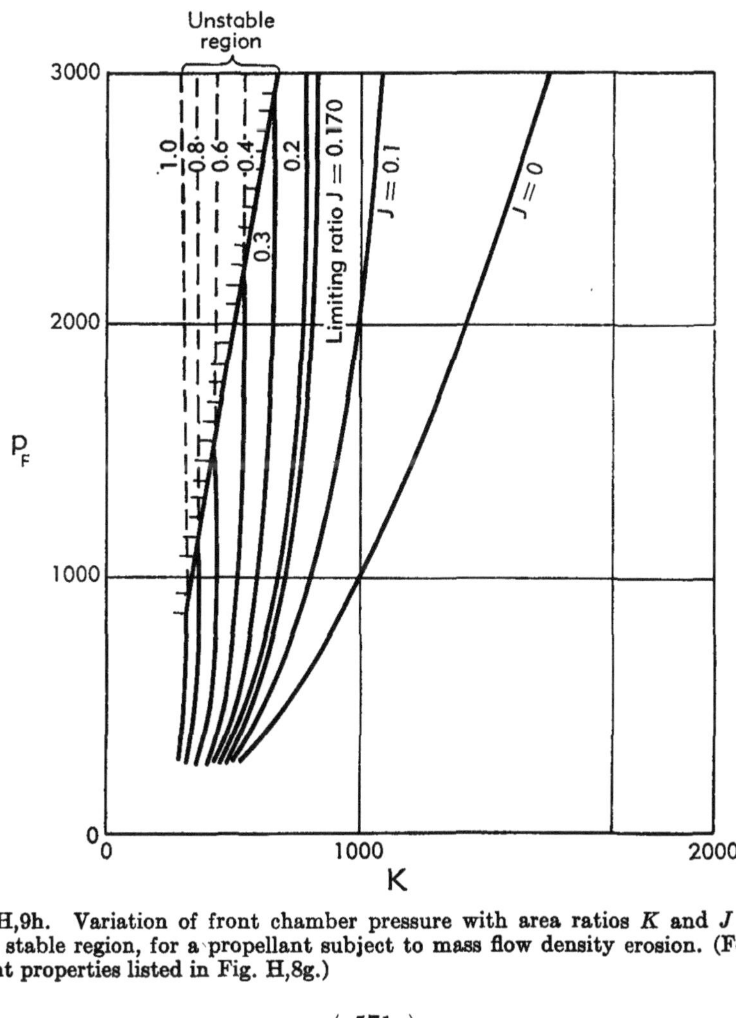

Fig. H,9h. Variation of front chamber pressure with area ratios K and J in the lower stable region, for a propellant subject to mass flow density erosion. (For propellant properties listed in Fig. H,8g.)

appears. As the second limit K is reached, the pressure must jump discontinuously to the high pressure curve. A pressure region is also indicated which cannot be reached by a stable combustion process. To operate on the low pressure curve in the triple (two stable, one unstable) pressure region, the ignition pressure must be less than the unstable pressure. The high pressure curve is of no practical interest because it lies at pressures

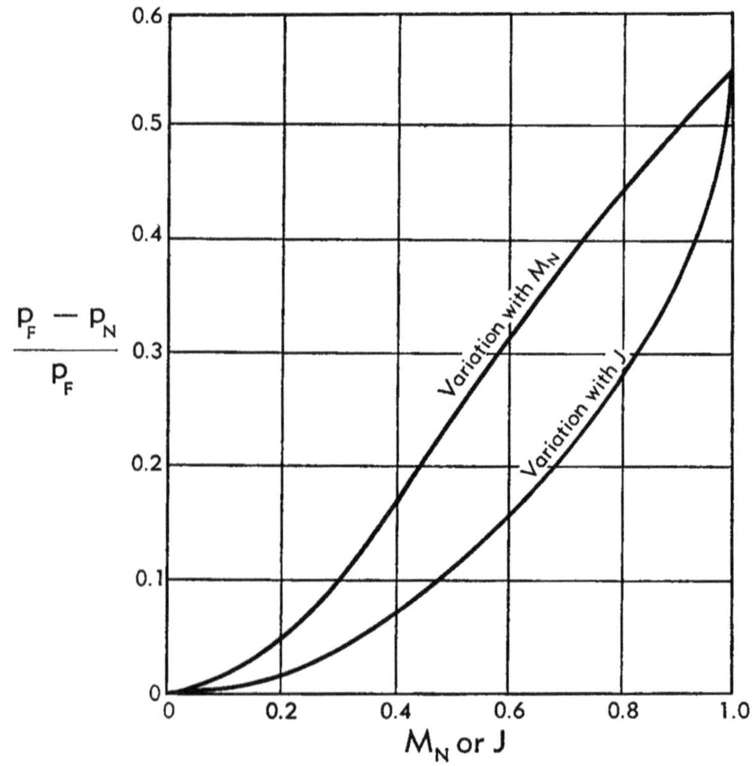

Fig. H,9i. Variation of the ratio: front minus nozzle pressure divided by front pressure with throat-to-duct area J or with Mach number at the nozzle end of the grain, M_N. (For $\gamma = 1.2$.)

comparable with p_p, too high for practical rocket design. The complete family of curves is shown to scale in Fig. H,9d.

In Fig. H,9e the limits of the stable low and high pressure regions are shown as a function of nozzle throat-to-gas duct area ratio. The dotted curves indicate the pressure jump from the stability curve to the high pressure (low pressure) region for a rocket of fixed J and infinitesimal change in K. The critical J value for erosive instability is indicated, for lower J values no erosive instability effects are encountered.

In Fig. H,9f the triple pressure region is outlined, as well as constant pressure contours, on the J, K plane. The critical J value is again indicated, and for J values less than this the $\dot{m}_b$ curve (Fig. H,9b) again has a simple shape.

It is clear that the effect of erosive instability will aggravate the effects of charge deformation and places severe limits on rocket motor design.

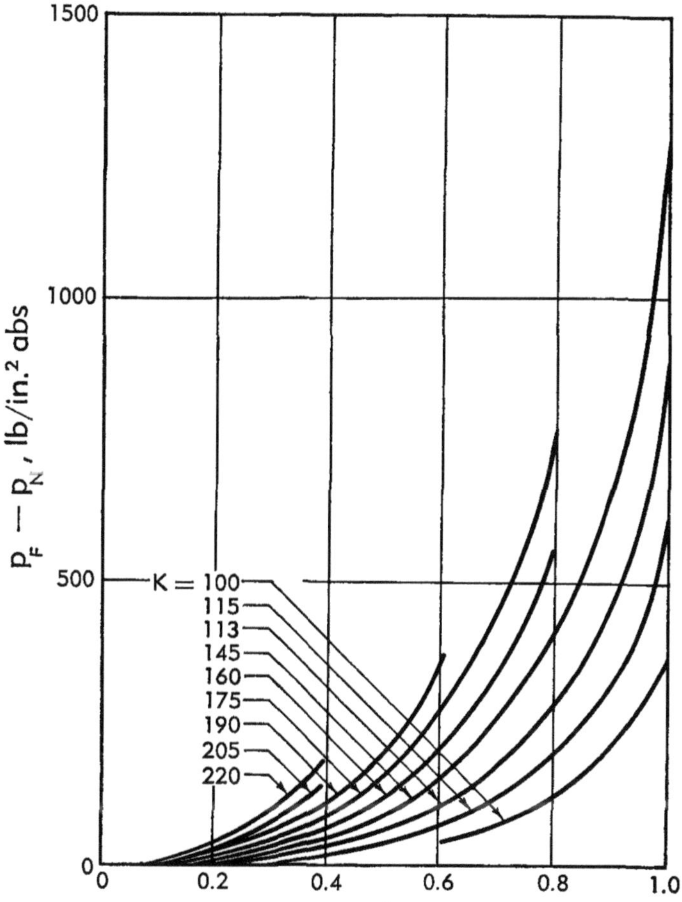

Fig. H,9j. Variation of the pressure difference, front minus nozzle pressure, with throat-to-duct ratio J for various K values, for a velocity-dependent erosive propellant. This is used in the discussion of charge deformation effects. (Propellant properties listed in Fig. H,8f.)

Not only is this a limit on design geometry, but in the triple pressure region it places a limit on ignition pressure.

The limitations implied by the erosive instability phenomenon are somewhat eased in practical rocket operation because the erosive burning enlarges the gas duct area rapidly, particularly at the exhaust nozzle end of the charge. Nevertheless, many blowups can be traced to this effect, charge deformation effects, or a combination of the two.

After the instability characteristics have been studied, the front chamber pressure in the practical low pressure region may be plotted as a function of K and J from Eq. 9-1. The results are shown in Fig. H,9g for a propellant affected only by velocity erosion and therefore not subject to erosive instability. A similar set of curves is shown in Fig. H,9h for a propellant subject to mass flow erosion and erosive instability. After the front pressure is determined, the general variation in pressure may be determined from Eq. 8-8. The front-nozzle pressure difference relative to the front pressure is shown in Fig. H,9i. The absolute front-nozzle pressure difference is shown in Fig. H,9j as a function of J for various values of K for a velocity-dependent erosive propellant. Curves of this type are employed in the discussion of charge deformation problems.

H,10. Temperature Sensitivity, Transients, Thin Web Grains, Resonant Burning, Chuffing, and Gas Generation.

Temperature sensitivity. A rocket of fixed geometry and given propellant shows a variation in performance with changing propellant temperature. This is due to the variation of burning rate with propellant temperature. Increasing the temperature increases the burning rate which increases the pressure, this in turn increases the burning rate further, etc. The essential result may be obtained as follows: At the pressures usual in rocket design, the expression for the area ratio curve may be simplified to yield a power law for the connection between pressure and area ratio. Eq. 7-13 provides this solution:

$$\frac{p_c}{1000} = \left(\frac{\rho_p a c^*}{1000} K\right)^{\frac{1}{1-n}} \tag{7-13}$$

The assumption has been made that the burning rate depends on the propellant temperature T_p only through a. (The indication is that the index n and the erosive constant k are not temperature-dependent for many propellants). Consequently the intrinsic temperature dependence of the burning rate is defined to be

$$\pi_r = \frac{1}{r}\left(\frac{dr}{dT_p}\right)_p = \frac{1}{a}\frac{da}{dT_p} = \frac{1}{T_1 - T_p} \tag{10-1}$$

The subscript p indicates the derivative at constant pressure. The equation also defines the burning rate temperature sensitivity coefficient π_r.

Logarithmic differentiation of Eq. 7-13 gives

$$\pi_p = \frac{1}{p_c}\left(\frac{dp_c}{dT_p}\right)_K = \frac{1}{1-n}\frac{1}{a}\frac{da}{dT_p} \tag{10-2}$$

The subscript K indicates the derivative at constant area ratio. The equation also defines the pressure temperature sensitivity coefficient π_p.

From Eq. 10-1 and 10-2 the result follows

$$\frac{1}{p_c}\left(\frac{dp_c}{dT_p}\right)_K = \frac{1}{1-n}\frac{1}{r}\left(\frac{dr}{dT_p}\right)_p \tag{10-3}$$

The factor $1/(1 - n)$ is an amplification factor for the effect of burning rate variation on rocket motor pressure. For real propellants the characteristic velocity c^* increases with combustion pressure; this effect makes the amplification larger. For radial-burning grains, and velocity erosion, the result of Eq. 10-3 is not changed. For mass flow density erosive propellants, however, there is an additional pressure dependence not included in Eq. 7-13. This may be taken into account by replacing n in Eq. 10-3 by the expression which follows from the stability study (Eq. 9-5).

$$n \to n' = n + \left[1 - \frac{x}{(1+x)\ln(1+x)}\right] \tag{10-4a}$$

$$x = \lambda_1 g_N \frac{p_F}{p_p} \tag{10-4b}$$

Geckler and Sprenger [*44*] have given a critical review of this and related problems.

Start-up and pressure recovery. In the treatment of steady state dynamics, it was assumed that the temperature of the gas in the combustion chamber or at the forward end of the grain did not vary with time. Actually, just before ignition, the combustion chamber is filled with a cold gas, usually air, and for a short time after ignition the gas temperature remains below the flame temperature. An analysis of this complex problem will not be attempted here. Von Kármán and Malina [*45*] have studied the problem in some detail and find that the average gas temperature rises quite rapidly to its normal value. This is illustrated in Fig. H,10. The time scale is longer the larger the ratio of initial volume to burning surface area.

The important conclusion that may be drawn from Fig. H,10 is that the temperature attains its steady value much more quickly than does the pressure. In further studies it will be assumed that the temperature is always at the adiabatic flame temperature value, which will only slightly distort results at the beginning of combustion.

With this assumption, analysis of pressure variation is simple. From Eq. 7-7, the theoretical value for $p_p = \Gamma^2\rho_p c^{*2}$, and neglecting the term p_e/p_p compared to unity, follows:

$$\frac{dp}{dt} = \frac{\Gamma^2\rho_p c^{*2}}{l_c}\left[r - \frac{c^*}{K}\frac{p}{\rho_p c^{*2}}\right] \tag{10-5}$$

The subscript has been omitted from p, which represents the chamber pressure for end-burning grains or the front pressure for radial-burning grains. The equilibrium or steady state condition is

$$K = \frac{p_e}{\rho_p c^* r_e} \tag{10-6}$$

where the subscript $_e$ designates the steady state value.

The effective burning rate may be written

$$r = a'\left(\frac{p}{1000}\right)^{n'} \tag{10-7}$$

If no erosive burning is present, set $a' = a$, $n' = n$; if velocity erosive

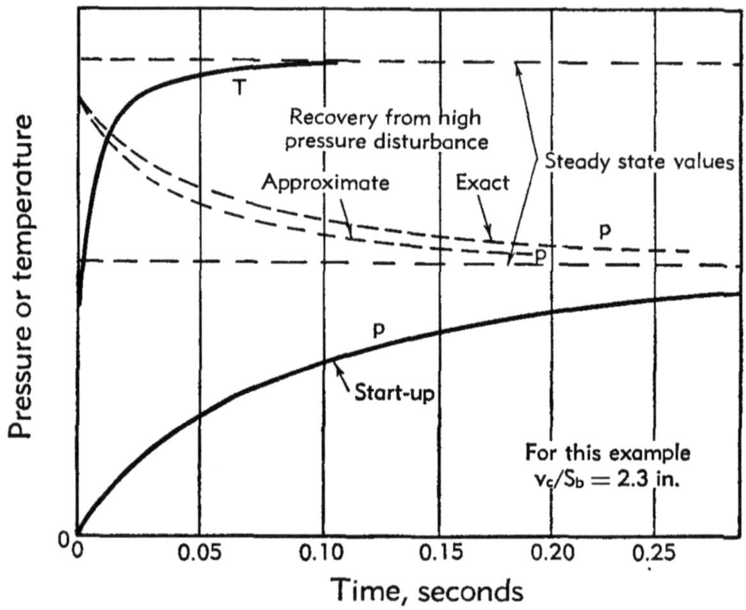

Fig. H,10. Variation of average gas temperature and chamber pressure with time subsequent to ignition (solid curves). Variation of pressure with time subsequent to a high pressure disturbance (constant gas temperature, dashed curves). The "exact" curve takes into account the variation of combustion chamber gas volume with time (dashed curves).

burning and pressure drop along the tube is present, set $a' = a\theta$; if mass flow erosive burning is present set $a' = a\theta$, and n' may be taken from Eq. 10-4.

It is most convenient to fit a tangent line to the burning rate curve at the equilibrium burning rate. This expression for the burning rate is

$$\frac{r}{r_e} = (1 - n') + n'\frac{p}{p_e} \tag{10-8}$$

Utilizing the equilibrium condition (Eq. 10-6), Eq. 10-5 may be written

$$\frac{dp}{dt} = \frac{\Gamma^2 c^*}{K} \frac{p_e}{l_c} \left(\frac{r}{r_e} - \frac{p}{p_e} \right)$$

From Eq. 10-8 this may be written

$$\frac{dp}{dt} = \frac{\Gamma^2 c^*}{K l_c} (1 - n')(p - p_e)$$

and from a differentiation theorem and Eq. 7-6

$$\frac{d}{dt}(p - p_e) = -\frac{p - p_e}{\tau} \tag{10-9a}$$

$$\tau = \frac{K l_c}{\Gamma^2 c^*} \frac{1}{1 - n'} \tag{10-9b}$$

$$\frac{dl_c}{dt} = r, \qquad l_c = \frac{V_c}{S_b} \tag{10-9c}$$

Although l_c varies with time, this variation is so slow that a constant value may be assumed. With this assumption, integration of Eq. 10-9a yields:

$$(p - p_e) = (p - p_e)_0 e^{-t/\tau} \tag{10-10}$$

where the subscript $_0$ denotes initial value.

The pressure, if disturbed, approaches the equilibrium value exponentially. Values of τ range from 0.01 sec to 0.50 sec. At startup, Eq. 10-10 may be applied and yields a pressure-time curve resembling that of Fig. H,10. Recovery from a high pressure disturbance is indicated by the dotted lines, and the effect of variation of l_c is shown on the "exact" curve.

Burnup. At the end of burning, the combustion chamber remains filled with high pressure gas. If the variation of gas pressure throughout the volume of the combustion chamber is neglected, then the pressure time variation, as the residual gas is exhausted, may be easily computed.

Assuming the adiabatic relation for the residual gas, and recalling that c^* is proportional to the square root of the temperature, the residual mass m in the chamber volume v obeys:

$$\frac{dm}{dt} = v_c \frac{d\rho}{dt} = -A_t \frac{p}{c^*} \tag{10-11a}$$

$$c^* = c^{*0} \left(\frac{\rho}{\rho^0} \right)^{\frac{\gamma - 1}{2}} \tag{10-11b}$$

$$p = p^0 \left(\frac{\rho}{\rho^0} \right)^{\gamma} \tag{10-11c}$$

The superscript ^{0} denotes initial values. Combination of these equations yields

$$\frac{dx}{dt} = -\frac{1}{\tau}(x)^{\frac{\gamma+1}{2}} \tag{10-12a}$$

$$\frac{1}{\tau} = -\left(\frac{1}{m}\frac{dm}{dt}\right)^0 = \frac{A_t p^0}{\rho^0 v_c c^{*0}} \tag{10-12b}$$

$$x = \frac{\rho}{\rho^0} \tag{10-12c}$$

Integration yields

$$x = \frac{1}{\left[1 + \left(\frac{\gamma - 1}{2}\right)\frac{t}{\tau}\right]^{\frac{2}{\gamma-1}}} \tag{10-13}$$

and the adiabatic relation

$$\frac{p}{p^0} = \frac{1}{\left(1 + \frac{\gamma - 1}{2}\frac{t}{\tau}\right)^{\frac{2\gamma}{\gamma-1}}} \tag{10-14}$$

Neglecting the degradation of thrust coefficient with decreasing pressure, the impulse is given by

$$I = \int_0^\infty C_F p A_t dt = C_F p^0 A_t \int_0^\infty \frac{dt}{\left(1 + \frac{\gamma - 1}{2}\frac{t}{\tau}\right)^{\frac{2\gamma}{\gamma-1}}}$$

$$I = \frac{2}{\gamma + 1} F^0 \tau \tag{10-15a}$$

$$F^0 = C_F p^0 A_t \tag{10-15b}$$

None of these expressions are valid after the pressure falls to the exhaust nozzle critical pressure, but the impulse estimate should not be very sensitive to this approximation.

Merits of thin-webbed grains. Very thin-webbed grains may be employed to reduce the effect of temperature sensitivity of a propellant [*17*]. Consider Eq. 10-10 applied to startup conditions, but with $t \ll \tau$. This gives

$$p = p_e \frac{t}{\tau} \tag{10-16}$$

It is assumed that the web of the grain is so thin that it is burned up before the equilibrium pressure is reached so that Eq. 10-16 is applicable to the entire burning period.

The burning stops at the time t_b, the web burns through, and at that time the peak combustion chamber pressure p_b is reached. Erosive effects are usually a minimum for thin-webbed motor arrangements, so that one may take Eq. 5-3 for the burning rate. Employing Eq. 10-16 for the pressure in the burning rate expression, the burning time t_b is determined from

$$\frac{w}{2} = \int_0^{t_b} r dt = \int_0^{t_b} a \left(\frac{p}{1000}\right)^n dt = \frac{r_e \tau}{n+1} \frac{t_b^{n+1}}{\tau} \tag{10-17}$$

where w is the web thickness, and r_e the burning rate corresponding to the equilibrium pressure. From Eq. 10-16 and 10-17 the peak pressure p_b at t_b is found:

$$p_b = p_e \left(\frac{n+1}{2} \frac{w}{r_e \tau}\right)^{\frac{1}{n+1}}$$

or

$$p_b = \left(\frac{n+1}{2} \frac{w}{r_e \tau} p_e\right)^{\frac{1}{n+1}} (p_e)^{\frac{n}{n+1}} \tag{10-18}$$

From the area ratio curve (Eq. 10-6) and the value of τ (Eq. 10-9), this may be written

$$p_b = \left(\frac{1-n^2}{2} \Gamma^2 \frac{2}{l_0} \rho_p c^{*2}\right)^{\frac{1}{n+1}} (p_e)^{\frac{n}{n+1}} \tag{10-19}$$

None of the terms in parentheses depend upon propellant temperature. Therefore

$$\frac{1}{p_b}\left(\frac{dp_b}{dT_p}\right)_K = \frac{n}{n+1} \frac{1}{p_e}\left(\frac{dp_e}{dT_p}\right)_K \tag{10-20}$$

For a propellant with n about 0.6, the thin web peak pressure will vary only 0.37 as much with temperature as the equilibrium pressure. Thin web arrangements are sometimes employed to reduce the temperature sensitivity of rockets.

Resonant burning. There is one peculiar phenomenon encountered with unrestricted burning grains, particularly on the interior of cylindrical perforations, which is sometimes referred to as resonant burning. The burning process becomes unstable. Charges, which have had the burning stopped part way through the combustion process, exhibit a wavy irregular burning surface (very crudely characterized by a wavelength and amplitude of $\frac{1}{8}$ inch in some cases), and sometimes the charge is split apart. The effect on rocket operation is to produce sudden high pressure peaks which may burst the motor, loss of propellant which affects range, and general erratic behavior which reduces accuracy.

The theoretical development of this problem is not attempted here. Various studies ([*10*; *11*; *12*, pp. 893–906; *13*, pp. 29–40; *14*; *15*; *16*] and II,M) have clarified the fundamental character of the phenomenon. Practical methods to overcome this difficulty [*10*] consist in placing a rod of nonburning material through the perforation (a "resonance rod," steel will do), making the perforation irregular in shape rather than circular, and by drilling radial holes through the web of the grain at frequent intervals. The diameter of the radial hole need not be larger than 0.4 times the diameter of the central hole, one hole is as good as several at any longitudinal section, the axes of the holes may be coplaner but are usually spiraled around the grain to promote smoother gas flow, and the longitudinal separation between holes may be uniform but the magnitude of the separation must usually be found by experiment. Cooler and slower burning propellants exhibit resonant burning effects to a smaller degree, and for many such propellants this difficulty does not arise. Solid particles in the exhaust gases act to suppress or eliminate resonance.

Chuffing. Sometimes when a rocket motor is fired with a very large exhaust nozzle throat opening, it burns in a series of sudden pressure rises (chuffs) separated by intervals of low pressure combustion. The time interval between chuffs may amount to a few seconds to half an hour depending upon the arrangement. This behavior is usually attributed to essential stopping of combustion by an ejection of the hot combustion gas during a chuff, followed by gradual reignition from hot fittings in the motor or by hot residue from the propellant (if it is not a clean burning material). The time interval between chuffs is sometimes very regular, and the explanation may be more complex.

In the case of artillery rockets, chuffing may be dangerous. The first pressure rise may serve merely to push the rocket off the launcher, and subsequent pressure pulses may then propel the rocket around in an unpredictable manner. Even with normal rocket arrangements, chuffing may occur at low propellant temperatures.

Gas generation. Solid propellants may be employed to operate pneumatic machines. Examples are: aircraft engine starters, pilot ejectors, and fluid ejection (fire-fighting equipment). The stability and temperature sensitivity requirements depend upon the flow characteristics of the system. For example, if a solid propellant is employed to force a liquid from a tank through an orifice, the flow rate will be proportional to the square root of the pressure, rather than directly proportional to the pressure as in a rocket motor. The stability requirement is then that the burning rate index n be $n < \frac{1}{2}$, and the amplification factor for temperature sensitivity $(\frac{1}{2} - n)^{-1}$.

Actually in arrangements of this type, propellant gases are emitted into a region with cold walls, very large in area compared to the burning surface area. The cooling and condensation of gas at these surfaces results in a pressure reduction which is roughly proportional to the pressure. The net effect is to make such an arrangement about as stable as a rocket.

CHAPTER 3. SOLID PROPELLANTS

H,11. Composition and Preparation. Solid propellants may be classified into three categories: homogeneous, heterogeneous, and molecular. General comments on these types and reference to selected compositions are made in this article.

HOMOGENEOUS PROPELLANTS.

Composition of double-base powders. The most common homogeneous propellant is popularly called the "double-base powder" which is a colloid formed by gelatinizing nitrocellulose with nitroglycerin, both of which are self-combustible. Nobel patented this process to make smokeless powder in 1888. Many modifications of double-base powders have been made which have varied to a large degree both their mechanical and ballistic properties. A high nitrogen content in the nitrocellulose and a high percentage of nitroglycerin each tend to increase the energy and burning rate of the propellant. A high nitrocellulose content provides high strength propellants. The source of nitrocellulose is important; for example, when

Table H,11a. Representative composition of extruded and cast double-base propellants.

Ingredients	*Extruded JPN ballistite*	*Experimental casting propellant*
Nitrocellulose (13.25 per cent N)	51.5	
Nitroglycerin	43.0	37.7
Diethyl phthalate	3.25	
Ethyl centralite	1.0	1.0
Potassium sulfate	1.25	
Carbon black*	0.2	0.3
Candelilla wax*	0.08	
Dimethyl phthalate		14.0
Nitrocellulose (12.6 per cent N)		47.0

* Added for extruded propellant.

it is made from cotton linter it provides better physical properties than when it is based on wood pulp. In addition to the nitrocellulose and explosive plasticizer, these propellants contain a stabilizer such as diphenylamine or centralite, auxiliary plasticizers such as phthalate esters, triacetin, and dinitrotoluene, and other compounds of this nature which contribute to lower the flame temperature of combustion. For suppressing flash and to promote smooth burning at low temperatures, one or two per cent of a potassium salt has been helpful. Opacifiers such as carbon black and nigrosine dye have been found necessary for the high energy powders to prevent radiant energy from being transmitted into the propellant, where a minute flaw would absorb sufficient radiation to cause the propellant to ignite internally and to break up under the resulting gas pressure.

The composition of two double-base propellants which have been prepared by different methods—as described in the subsequent paragraphs—is given in Table H,11a.

Methods of preparation. There are three different processes by which double-base propellants are made. These are *solvent extrusion, solventless extrusion,* and *casting.*

In the solvent extrusion process, a volatile solvent is added to the mixture of nitrocellulose and nitroglycerin and the whole mix is stirred. The solvent swells the nitrocellulose and, with the breakdown of the fibrous structure by the mechanical agitation, a soft colloidal paste results, which is extruded through dies to form the desired shape and size and then cut to the desired length. The resulting charges are dried to remove the solvent. The advantages of the solvent process are that the solvent desensitizes the explosive mixtures to reduce manufacturing hazards and the charges are physically stronger than those made by the other processes. The disadvantages are that the propellant cracks during drying if the maximum propellant thickness exceeds about $\frac{1}{2}$ inch and that control of shape and dimensions is difficult, because of the shrinkage which accompanies solvent removal.

In the solventless process the nitrocellulose-nitroglycerin colloid is formed by mixing these ingredients first in a water slurry and then transferring them to a heated roller mill. The propellant is removed in sheet form, either rolled like a carpet or flaked into small pieces and inserted into a press, the barrel of which is heated between 115 and 130°F, and then extruded at pressures ranging from 4000 to 6000 lb/in.^{2} The temperature of the extruded propellant is usually about 30°F less than the press barrel temperature. The space between the ram and the propellant is initially evacuated to prevent formation of bubbles and "pinholes" in the extruded powder. The extruded charge is finally annealed to relieve stresses. The advantages of the solventless process is that large charges can be made (the limitation being the size of the press) and that exact control of size and shape is possible. The disadvantages are physical properties inferior to solvent-processed propellant, heavy and costly equipment, and hazardous operations. However, most of the rocket propellant used in World War II was made by the solventless process.

Heterogeneous Propellants.

Composition. These propellants generally consist of an organic fuel binder and crystalline oxidizers which have been intimately mixed together. A large number of propellants of this type are possible because of the many fuels and oxidizers which are available. Thermoplastic mixes such as asphalt-oil and ethyl cellulose-castor oil, and thermosetting materials including several synthetic resins and elastomers have been feasible as fuel binders. Perchlorates and nitrates have been used most extensively as oxidizers, among the most common being ammonium nitrate, ammonium perchlorate, potassium perchlorate, and potassium and sodium nitrates. For example, a product of reaction of potassium perchlorate is potassium chloride, a white solid; hydrogen chloride from ammonium perchlorate condenses in a moist atmosphere to produce a white fog. Nevertheless, because of the simplicity of processing and the wide range of burning rates, these propellants can compete with smokeless double-base powders in many applications. The composition of two propellants which have been made by the casting and molding processes is given in Table H,11b.

Table H,11b. Composition of two heterogeneous propellants.

Ingredients	*Castable asphalt-perchlorate*	*Molded composite*
Asphalt	16.8	
Oil	7.2	
Potassium perchlorate	76.0	
Potassium nitrate		36.4
Ammonium picrate		54.6
Calcium stearate		3.6
Ethyl cellulose		5.4

Preparation. Heterogeneous propellants are prepared by *casting* and *molding* techniques.

In the cast process the propellant is usually fluid at some stage of the preparation so that it can be poured into molds of any desired size and conform to any shape. The fuel is usually the ingredient that provides fluidity. Thermoplastic fuels are usually originally solid, are melted to be mixed with the oxidizer, and are then cooled to a solid after casting [*46*]. Propellants using thermoplastic fuels are cooled gradually to avoid severe thermal stresses which cause cracks or pull-away from molds. Thermosetting fuels are usually liquid at normal temperatures and are solidified by chemical polymerization. The polymerization of thermosetting propellants is usually exothermic, so that the reaction has to be controlled to avoid high thermal stresses from this source and to prevent autoignition of propellant; therefore, control of the amount, chemical purity, particle size (if solid), and dispersion of the catalysts, which promote polymerization and control of the temperature at which the polymerization occurs, are essential.

Oxidizers are pulverized to very fine powders before being incorporated into the fuel. The distribution of oxidizer particle size often ranges from 5 to 40 microns. The classification of oxidizers on a large scale into narrow particle size ranges is an operation which has not been satisfactorily controlled and is a good subject for research. Both the mixing and burning properties of solid propellants depend largely upon the particle size distribution of the oxidizer. Many of the oxidizers are hygroscopic and the control of the humidity during processing is mandatory.

Considerable work has been done in an attempt to relate the properties of solid propellants with viscosity of the fuel, the volume concentration of the oxidizer, and the particle size distribution of the suspended oxidizer. The first work relating the viscosity of suspension with concentration of the suspended materials is due to Einstein. However, the Einstein equation was only applicable to extremely dilute suspensions. Later Vand [*47*] developed equations for the viscosity of suspensions as a function of concentration which held to about 30-volume-per cent solids. Since the Vand equations do not take into account non-Newtonian flow, they are not applicable to the high concentration commonly encountered in composite solid propellants. However, they are useful for first order approximation. Although considerable effort has been expended toward attempting to derive mathematical formulas to describe the behavior of suspensions as concentrated as those encountered with solid propellant, no suitable equations have been developed to date.

In addition to the concentration of the suspended particles, the particle size distribution of the particles is extremely important. Dalla Valle [*48*, pp. 100–122] has written a comprehensive book on the effect of varying particle size distribution on various properties (such as bulk density) which influence the rheology of suspensions. Furnas [*49*] and Mooney [*50*] have worked on the viscosity of suspension of spheres. However, since the oxidizers are not entirely spherical in shape, none of the equations for particle size distribution as a function of viscosity hold rigidly. The results obtained from spherical particles, however, can be used as a first approximation of obtaining optimum results with solid propellants.

Settling of solid propellants prior to solidification presents a problem since any significant sedimentation results in nonhomogeneous castings. The basic equation for the sedimentation velocity of particles in suspension is given by Stokes' law. Stokes' law does not, however, hold for concentrated suspensions, and alternate equations have been developed by Steinour [*51*] and Hawksley [*52*].

Compromises usually must be made between viscosity of the liquid fuel and particle size distribution of the oxidizer, in order to obtain the proper mixture ratio for ballistic performance and yet retain adequate mixability and castability without settling. In practice, a mixture ratio of 80 per cent by weight of oxidizer and 20 per cent by weight of fuel is about the maximum feasible for a system having: (1) an oxidizer having an approximate specific gravity of 2.0, (2) the oxidizer being blended in the ratio 70:30 per cent by weight with each fraction having a particle size of 50:5 microns, respectively, (3) a fuel with an approximate specific gravity of 1.0, and (4) the fuel having a viscosity of about 1000 centipoises.

For the mixing operation it is generally necessary to heat or cool the propellant mixture and often necessary to evacuate the air from it to provide a nonporous propellant. Mixers with paddle-type agitators and double counter-rotating differential speed blades have been used most extensively for these cast propellants.

The chief advantages of heterogeneous cast propellants are that the equipment cost is very low and the size of propellant charge appears to be unlimited. As a matter of interest, some of these propellants are the only ones existing which do not depend upon nitric acid. The big disadvantage is that most of these propellants produce smoke.

In the pressure-molding process of heterogeneous propellants, pressures of the order of 10,000 $lb/in.^2$ are used to consolidate the propellant. Consequently, heavy metal parts and large presses are often required which are cumbersome and make this process much more expensive than the cast process. Also, the charge size may be limited by the size of the press. However, sometimes molding is desirable for special propellants which could not be made by the cast process, an example being a material with nearly all of the ingredients crystalline and only a small amount of fuel binder permissible.

Two mixing methods have been popular for molded heterogeneous propellants. One method makes use of the intensive mixing of an edge-runner mill; the other, a differential roller mill. A propellant system consisting of ammonium picrate, sodium nitrate, and a small portion of resinous binder—popularly called composite propellants—has been successfully processed using the edge-runner mill. The resultant propellant mixture is a powdery product that is formed by compression, molding at about 4000 to 8000 $lb/in.^2$. The molded charges are cured for about 4 days at 175°F. A charge length about equal to the diameter has been the most satisfactory for composite propellants and charges with an l/d greater than 1 are made by cementing two or more charges together.

When using a differential roller mill the oxidizers and miscellaneous additives are blended into the fuel and the final mixture of propellant is removed from the roll as a sheet, similar to the solventless double-base powder process. If the fuel is thermoplastic, it may be molded under fairly low pressure directly into and bonded to rocket motors. These charges are of puttylike consistency and are often subject to plastic flow, but when the propellant is restrained and forced by thermal changes to expand or contract, dimensional changes in the direction of unrestraint usually do not set up stresses large enough to cause cracking. If the fuel is thermosetting, it can be forced into a mold which is subsequently locked and put into an oven where the temperature initiates polymerization. As the temperature expands the confined propellant, the propellant fuses to form one solid piece. After polymerization, the propellant is cooled and removed from the mold.

Molecular Propellants. Molecular propellant is the name given to a propellant consisting of one basic molecule. The single-base smokeless gun powder cellulose nitrate is an example of such a propellant, but it does not burn properly at pressures below 5000 lb/in.[2], the regime in which most rocket motors are designed. The molecular propellant would be ideal for processing if it were liquid at normal temperatures and could be polymerized, when desired, to a solid. Any size and shape of charge could be made with a minimum of effort. No published information is available on the status of the research with molecular propellants, but it is apparent that much work remains to be done before this type of propellant becomes practical.

Restrictors. After the propellant is cast to size and shape there are usually many cutting, drilling, and sawing operations before the charge is a finished product ready to be inserted into a rocket motor. One of the most important operations performed is that of applying a restrictor, which is a noncombustible layer applied to a portion of the propellant surface where burning is not desired. The process of applying a restrictor is called inhibiting. Many blowups of rocket motors have been due to failure of the restrictor.

Cellulose acetate has been satisfactory for restricting double-base propellants. For extruded double-base charges, the cellulose acetate—usually molded to a given size or cut to the desired shape from sheet stock—is often cemented to the propellant. For the cast double-base powders, the cellulose acetate in tube form often serves as the casting mold as well as for the restrictor.

The fuel with an inert filler substituted for the oxidizer has been found to be a very satisfactory restrictor for many of the heterogeneous propellants. For cast heterogeneous propellants the fluid restrictor may be applied in a thin layer ($\frac{1}{32}$ to $\frac{1}{16}$ inch) directly to the rocket motor wall by draining or spinning methods; the liner bonds to the motor wall when it is solidified by cooling or polymerization. The propellant then adheres to the liner after it is poured into the motor and solidified. In a few cases, the propellant itself may bond satisfactorily to the metal of the motor wall, thus eliminating the need for a restrictor. Heterogeneous propellants are often cast separately from the motor, however, and in this case the mold may serve as the restrictor, or the charge may be inhibited by dipping it into a pool of restriction material; often, additional wrappings of tape over the restrictor are found desirable. For molded heterogeneous propellants, restrictors may be made of special materials which are bonded in shaped pieces to the charges, for example cellulose acetate is bonded to double-base powders. For some elastomeric propellants, placing sheets of fuel at desired places on the propellant charge prior to molding has proved to be simple and satisfactory.

H,12. Propellant Properties. The efficiency and usefulness of a solid propellant rocket depend to a large extent upon the properties of the propellant. The combustion, mechanical, and physical properties of the propellant must be known in order to design and properly use such rockets. A discussion of these properties with reference to specific propellant systems is presented here.

COMBUSTION.

Basic propellant characteristics. By examination of the burning rate and equilibrium pressure relationships given by Eq. 5-1, 5-2a, 5-2b, 5-3, 7-13 and 9-1, it is evident that the basic characteristics which define the combustion of a given propellant are a, n, k or k_V and V_0, c^*, and ρ_p.

The burning rate coefficient a is the propellant characteristic which varies most significantly with propellant temperatures, so that it is an extremely important factor in rocket motor design. The sensitivity of burning rate to propellant temperature, T_p, change may be expressed as:

$$\pi_r = \frac{1}{r}\left(\frac{\partial r}{\partial T_p}\right)_p = \frac{1}{a}\frac{\partial a}{\partial T_p} = \frac{1}{T_1 - T_p}$$

where the last result follows from Eq. 5-2. Recall that T_1 is an experimental property of the propellant. π_r is defined as the *temperature sensitivity of burning rate at constant pressure.* The values of T_1 for some different propellant systems are approximately as follows: 400 to 600°F for double-base propellants; 500 to 600°F for propellants utilizing ammonium nitrate or ammonium perchlorate for oxidizers; greater than 700° F for propellants utilizing potassium perchlorate for oxidizer. It is evident that the constant T_1 should be high compared to the useful range of the propellant temperatures, T_p.

In addition to affecting the burning rate as the propellant temperature changes, the burning rate coefficient partly determines the magnitude of the burning rate. Since a choice of burning rate is convenient in rocket motor design, it is desirable to be able to vary the constant a over a wide range.

The burning rate exponent n is equally as important as the burning rate coefficient because of its influence on the variables which determine the equilibrium chamber pressure of a rocket motor. From Eq. 7-13 these variables are raised to be $1/(1 - n)$th power. As $n \to 1$, the pressure is extremely sensitive to small design changes. For example, if $n = 0.95$ and the nozzle throat area decreased by 1 per cent, the pressure would increase 73 per cent. *The temperature sensitivity of pressure at constant K value* is very important in rocket design. From Eq. 10-3,

$$\pi_p = \frac{1}{p}\left(\frac{\partial p}{\partial T_p}\right)_K = \frac{1}{1 - n'}\left(\frac{\partial r}{\partial T_p}\right)_p = \frac{1}{1 - n'}\pi_r$$

where n' is defined in Eq. 10-4. It is evident that for ideal operating characteristics the burning rate exponent should approach zero.

Constant burning rates ($n = 0$) have been obtained over limited pressure regions for both heterogeneous and homogeneous propellant regions. Small amounts of particular additives produce the constant burning rate effect for both the double-base and heterogeneous propellants. When the propellant gases flow parallel to the burning surface of a solid propellant, the burning rate of the propellant at a given pressure is increased (see Eq. 5-1 and 5-2). The proportionality factor k is called the erosive burning constant. Due to erosive burning the additional mass flow increases the rocket motor chamber pressure according to Eq. 9-1. The effect of erosive burning on the pressure-time characteristics of a rocket motor is indicated by Fig. H,12a.

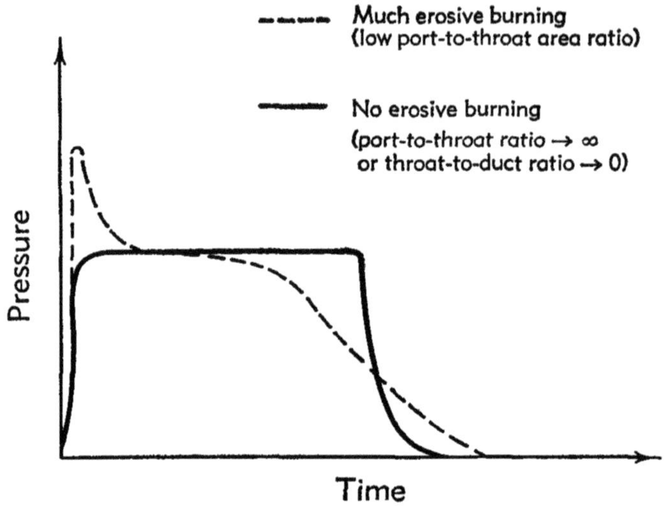

Fig. H,12a. Effect of erosive burning on the pressure-time curve of a rocket motor. Charge design conditions identical except for throat-to-duct area ratios.

The erosive burning constant can be determined experimentally by stopping the burning of a radial-burning solid propellant charge and measuring the distance burned versus the length of charge—which has been found to be nearly a linear relationship—then calculating from geometry the average mass flow density at any desired point along the charge for the time of burning, and finally making the calculation for the erosive constant using the basic relation of Eq. 5-1 and 5-2. The propellant at the front end of the charge is assumed to burn under nonerosive conditions and any significant pressure drop along the charge has to be taken into account.

In a given propellant system using different ratios of ingredients, the erosive constant may vary by a factor of 5. In general, it has been found that low energy propellants have high erosive constants, and high energy propellants have low erosive constants. The resistance of the propellant to the shear forces exerted on it by the gases may also be an important factor. Values of k usually range from nearly zero to 0.7 $in.^2$ sec/lb.

The characteristic velocity c^* is a measure of the intrinsic energy of the propellant, and is determined by the thermodynamic properties. For an idealized propellant model, these are T_0, γ, and $\mathfrak{M}$ (see Sec. G). It is essentially independent of propellant temperature and varies little in the pressure range normally encountered with rockets. This insensitivity of energy to initial propellant temperature is in contrast to the possible large dependency of the rate processes upon temperature. It follows that the total impulse of a rocket unit is nearly always the same, but that the rate of expenditure of energy varies with propellant temperature.

It is apparent that a high c^* is desirable. However, since $c^* \sim \sqrt{T_0/\mathfrak{M}}$, a high value of T_0 may cause excessive heating of parts of the rocket motor that are exposed directly to the hot gases. Nozzle erosion, in particular, is severe with conventional materials when T_0 approaches or exceeds 4000°F and the duration of rocket motor operation is more than a few seconds. For this reason, it may be more desirable to increase c^* by reducing $\mathfrak{M}$ than to increase T_0. In general, c^* values greater than 4000 ft/sec are easily attained for solid propellants, but values significantly exceeding 5000 ft/sec are not likely to be common in the foreseeable future.

The product of c^* and ρ_p should be high to provide a high impulse per unit volume. A high energy propellant of low density may not provide an efficient rocket motor because of the additional weight of the rocket motor chamber necessary to enclose the volume occupied by the propellant. The density should be high. The density of most solid propellants ranges between 0.05 lb/in.^{3} and 0.07 lb/in.^{3}

The products of combustion are important because they determine smoke, toxicity, and corrosiveness of the exhaust. For example, potassium chloride from potassium perchlorate gives a dense white exhaust, carbon monoxide is very toxic, and hydrogen chloride from ammonium perchlorate is very corrosive to materials such as iron. Carbon monoxide reacts vigorously with steel nozzles to form iron carbonyl when small amounts of sulfur are present in the exhaust.

Special combustion characteristics. There are some special combustion properties which sometimes impose major limitations on rocket design. Such characteristics as the combustion limit, pressure limit, temperature limits, and nonuniform burning properties are discussed in this article.

If a number of rockets, identical except for the exhaust nozzle throat diameter, are fired, a graph such as Fig. H,12b can be plotted showing the chamber pressure as a function of the throat diameter [*42*]. It is found, as might be expected, that the pressure decreases as the throat diameter increases. However, when a certain throat diameter is exceeded, the chamber pressure is found to be far below the pressure predicted from an extrapolation of the high pressure portion of the curve. The pressure corresponding to this critical thrust diameter is called the *combustion limit.*

If a number of units are fired with this same large throat diameter the chamber pressure varies erratically from unit to unit. Finally, if the exhaust nozzle throat is made large enough, individual units do not burn continuously, but in an irregular manner with a chugging noise. This last phenomenon is called "chuffing."

Thus it appears that a given propellant cannot be used at an arbitrarily low pressure, and the designer of a rocket unit must assume chamber pressures above the combustion limit if reproducible performance from unit to unit is to be obtained. When low over-all rocket weight is

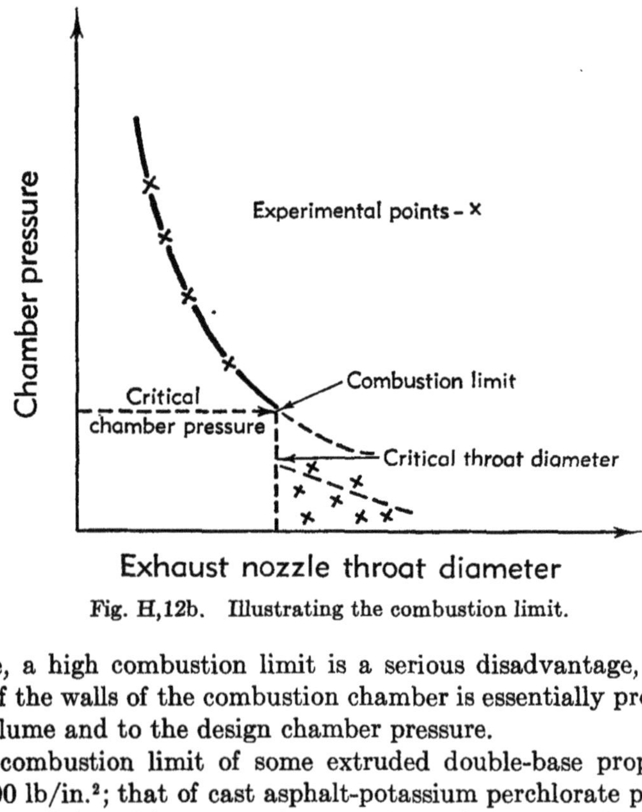

Fig. H,12b. Illustrating the combustion limit.

desirable, a high combustion limit is a serious disadvantage, since the weight of the walls of the combustion chamber is essentially proportional to its volume and to the design chamber pressure.

The combustion limit of some extruded double-base propellants is about 500 lb/in.^{2}; that of cast asphalt-potassium perchlorate propellants about 1000 lb/in.^{2} The composite propellants have a combustion limit not exceeding 100/in.^{2} The combustion limit of the straight nitrocellulose propellants frequently used in guns exceeds 5000 lb/in.^{2}; therefore such propellants are not suitable for rockets.

Some propellants may safely be used only below a critical chamber pressure, this pressure being called the *pressure limit.* If the pressure limit is exceeded, the propellant burns in a violent and unpredictable manner. Brittle propellants with a granular structure are particularly subject to this effect. For most of the commonly used rocket propellants the pressure limit exceeds 5000 lb/in.^{2}, and therefore it is not a problem in rocket unit design.

Special limitations on the temperature range within which a rocket may be used are sometimes introduced by a change of mechanical properties of the propellant with temperature. Certain types of propellants, those having thermoplastic fuels, may soften and deform sufficiently when stored at high temperature so that an abnormally large burning area is exposed. Upon ignition, the resulting high chamber pressure may cause failure of the unit. Some double-base propellants may soften sufficiently at high temperatures so that the large pressure gradient along the charge, which is present in radial-burning units, may cause the charge to break up, exposing a large burning area and leading to failure of the unit. At

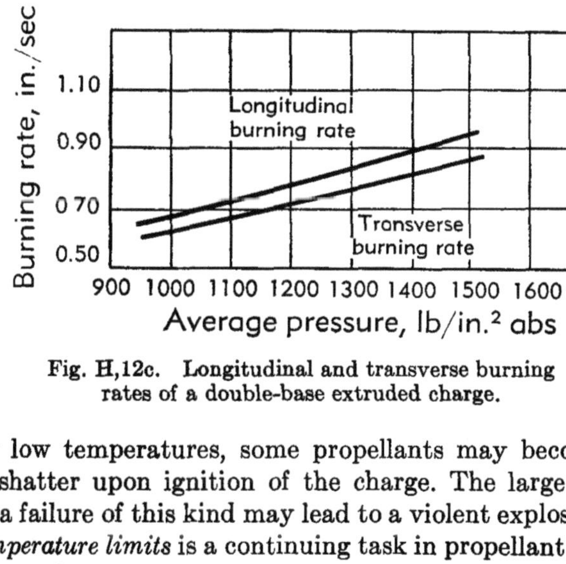

Fig. H,12c. Longitudinal and transverse burning rates of a double-base extruded charge.

sufficiently low temperatures, some propellants may become so brittle that they shatter upon ignition of the charge. The large burning area exposed in a failure of this kind may lead to a violent explosion. Improvement of *temperature limits* is a continuing task in propellant development.

Propellant charges may not always have uniform burning characteristics. For example, the burning rate in the longitudinal direction of extruded double-base charges differs from the rate in the transverse direction, an example of the difference being shown in Fig. H,12c [*53*]. As a consequence, these propellants, burning in a radial direction, burn differently than those burning from one end. Also, the burning rates depend upon whether the propellant is extruded in a direction parallel or perpendicular to the propellant sheet obtained from the mixing rolls. Presumably the orientation of the colloidal structure affects the burning rate.

METHODS FOR DETERMINING COMBUSTION CHARACTERISTICS. After it has been determined by calculation that a given propellant system has acceptable energy (c^*), the burning properties have to be evaluated. The burning rates of laboratory quantities of propellant are often determined by preparing pencil-size specimens and then measuring the time required for given lengths to burn from one end to the other at different pressures and temperatures; this method is referred to as the *strand burner method* [*31*]. The strand burner has proved to be very useful but has limitations, because the pressurizing gas (usually nitrogen) never duplicates the combustion gases upon which the burning characteristics are dependent. Sometimes, propellants which will not burn within a given pressure range in the strand burner will burn satisfactorily at the same pressure in rocket motors.

A simple method for determining the rocket propellant characteristics over a wide pressure range is to burn an internal-burning tubular charge which is restricted on the ends in a rocket motor using high port-to-throat ratio[6] (negligible erosive burning). The weight of propellant burned at any time is proportional to $\int_0^t p\,dt$; therefore the distance burned at any time can be calculated from the weight burned and geometry. As a consequence, the burning area corresponding to the pressure at any time can be calculated; then the burning rate exponent can be determined by the relation $p \sim A_b^{1/(1-n)}$. Since the calculation of ρ_p and c^* is straightforward, the burning rate coefficient can be determined from the relation

$$p = \left(\frac{c^* a \rho_p K}{g}\right)^{\frac{1}{1-n}}$$

This method of determining the burning characteristics is very useful for finding the exact limits of "constant burning rate" regions, because the change in burning rate exponent is shown immediately by the pressure-time curve (see Fig. H,12d). The method depends upon a constant throat area during burning, which is difficult to obtain in practice when the throat area is small (say less than 0.2 in.); therefore it is applicable to charges which are usually larger than those made in the chemical laboratory.

Storage Stability. The mechanical and internal ballistic properties of a propellant may be dependent upon storage time, storage temperature, and humidity.

The storage time permissible for double-base propellants is strongly dependent upon temperature because these propellants decompose chemically. The chemical decomposition is autocatalytic, and stabilizers such as diphenylamine have to be added to neutralize the catalytic effect of the decomposition products; otherwise these propellants may autoignite. This decomposition leads to the formation of gas which tends to fissure the charges, particularly those having thick sections of propellant. Another problem has been the exudation of the nitroglycerin into the

[6] The term port is synonomous with duct in rocket terminology. The port-to-throat ratio, $A_d/A_t = H$, is the reciprocal of the throat-to-duct (or throat-to-port) ratio: $A_t/A_d = J = 1/H$. In Art. 8 on theory, J was used since it ranges from 0 to 1.

restrictor to such a degree that the restrictor burns readily. The stability of the double-base propellants is being constantly improved, but storage at temperatures exceeding 140°F is inadvisable.

The composite propellants do not decompose chemically on prolonged storage but, in an atmosphere of high humidity, the sodium nitrate absorbs moisture and the charges containing it become soft and weak. Propellants containing ammonium nitrate are very hygroscopic, and the phase changes of this material make its use impractical for propellants

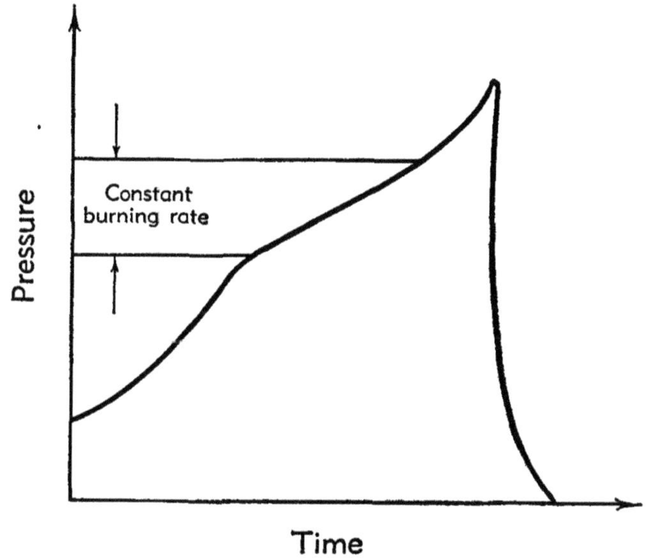

Fig. H,12d. Example of the constant burning rate pressure region of the burning rate for a propellant, as determined from a rocket motor test employing an internal-burning tubular charge.

stored and used over a wide temperature range unless the fuel binder can accommodate the volume changes.

Propellants containing volatile ingredients lose weight on prolonged storage at high temperatures, resulting in changes for both mechanical and burning properties.

The heterogeneous propellants with nonhygroscopic stable oxidizers and nonvolatile fuels usually have excellent storage stability. Satisfactory storage for longer than a year at temperatures exceeding 160°F is feasible.

Mechanical Properties. Constancy of mechanical properties over a wide temperature range is desirable so that brittleness at low temperature and plastic flow at high temperatures do not limit the use of the rocket propellant to a narrow temperature range.

In withstanding flight accelerations, the mechanical properties required of a rocket propellant depend upon the particular charge design being used. For illustration, three charges: end-burning, internal-burning, and tubular-burning (burning on both outside and inside) are considered, as shown in Fig. H,12e. If rocket motors containing these charges are accelerated at y times gravity, the following situations exist:

For the end-burning charges. The opposing forces are the acceleration "pressure" $y\rho_p l_p$ acting to throw the charge rearward and the chamber pressure p holding the charge forward. If the thrust which accelerates the motor is from the end-burning charge only, the pressure force is

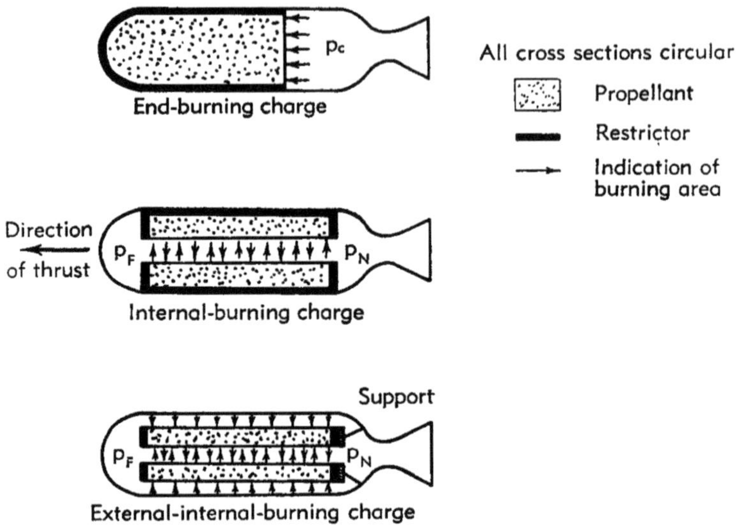

Fig. H,12e. Various propellant arrangements discussed in connection with the effect of forces on the charge.

always greater than the acceleration because in the expression

$$\frac{p}{yl_p\rho_p} = \frac{p}{\dfrac{F}{W_T}\dfrac{W_p}{S_b}} = \frac{pS_bW_T}{C_FpA_tW_p} = \frac{S_bW_T}{C_FA_tW_p}$$

the numerator is greater than the denominator, since

$$\frac{1}{C_F} > \frac{1}{2}, \quad \frac{S_b}{A_t} > 100, \quad \frac{W_T}{W_p} > 1$$

In the above formulas S_b is the burning surface of the propellant, W_T the total weight of the rocket motor, W_p the weight of the propellant, and C_F the nozzle thrust coefficient. The main requirement for this end-burning charge is that the propellant be "plastic" enough or have enough elongation to change dimensions without cracking as required by the rocket motor expansion under pressure. High elongation of propellant is desirable to "accommodate" the differential thermal expansion properties between the motor chamber and the propellant.

For the internal-burning charge. The forces tending to throw this charge rearward are due to the pressure drop along the charge and acceleration. These are:

Pressure drop force (see Art. 8 and Fig. H,9j)

Acceleration force $= yW_p$

Opposing these forces is the adhesion of the propellant to the motor wall or the cohesion to itself, whichever is weaker. It is evident that the adhesion strength in shear and the shear strength of the propellant are very important. For identical conditions, the stress increases as the charge becomes larger in diameter. Since this propellant is bonded to the combustion chamber, it should have good elongation to follow, without cracking, the expansion of the rocket motor under pressure and to withstand the dimensional change with temperature change. It is noted that J becomes less as the chamber wall expands, thus tending to decrease the pressure drop force.

For the tubular charge. The forces tending to throw the charge rearward are the same as for the internal-burning charge. Because the charge is supported by the cross-sectional area at the nozzle end, it must take the load in compression. It follows that the compressive strength should be high. However, the compressive strength for such a charge is not always the mechanical property, which is of prime significance [*54*]. Contrary to the internal-burning charge bonded to the motor wall, the gas passage becomes smaller as chamber pressure becomes higher, thus making J larger (see Art. 6). This leads to increased pressure, increased pressure drop force, and increased acceleration force. It follows that the modulus of elasticity should be high to prevent "bulging" of the charge under high compressive force. Also, at a time near the end of burning of tubular charges, a very thin propellant section is left which may buckle as a column; since the strength of a column is directly proportional to the modulus of elasticity and inversely proportional to its slenderness ratio, the significance of modulus of elasticity and design is again emphasized.

From the preceding discussion, it is evident that the most important mechanical properties may range from high elongation, high adhesive and cohesive shear strength to high modulus, and high compressive strength. It should be added that, since the mechanical properties of propellants are never constant with temperature, their variation with temperature must be considered in design studies.

Unfortunately, conventional test methods for evaluating the strength properties of materials do not give quantitative design information when applied to propellants. This is true because most propellants have complex molecular structures which do not obey conventional laws of elasticity. Properties such as tensile strength and elongation are significantly dependent upon the rate at which the test specimen is loaded and its complete past history with respect to temperature, humidity, and time. The variation of elastic modulus with propellant temperature is shown in Fig. H,6d. Regardless of the fact that the measured mechanical properties of propellants have not been usable for absolute design, knowledge of the behavior of the properties in a qualitative sense is extremely valuable for rating one propellant against another. Development of test methods from which quantitative designs can be made is a subject needing much research.

Miscellaneous Propellant Properties. Some additional, important properties of propellants are discussed briefly in this article.

Shrinkage. Some of the polymeric fuels become more dense as the polymer chains grow, thus resulting in shrinkage. If volatile materials evaporate from propellants, shrinkage occurs. The amount of shrinkage is useful for calculating final dimensions of charges and for estimating stresses which may result. When a propellant which may shrink is restrained from shrinkage—such as when a solid core is used for shaping the internal perforation of a charge—stresses develop in the propellant, sometimes high enough to produce cracks. Core withdrawal may be increasingly difficult as shrinkage continues.

Thermal properties. The thermal coefficient of expansion is required in order to allow proper tolerances between the propellant charge and the metal parts, and to estimate stresses when the propellant is bonded to the motor wall or when the propellant is free to contract around the cores used for shaping the charge.

Thermal diffusivity is needed to determine temperature distributions [*55*] according to the equation (circular symmetry, no axial heat flow)

$$\frac{\partial T}{\partial t} = \kappa\left(\frac{\partial^2 T}{\partial R^2} + \frac{1}{R}\frac{\partial T}{\partial R}\right)$$

where $\kappa = k/c_p\rho_p$. The thermal conductivity and specific heat are the basic thermal properties that determine the value of thermal diffusivity. Typical values of thermal coefficient of expansion, thermal diffusivity, and specific heat for propellants are 10^{-4} in./in./°F, 25×10^{-5} in.^{2}/sec, and 0.3 cal/g°C, respectively.

The specific heat of reaction c_r. It is necessary to know this in order to find the maximum internal temperature rise possible during the polymerization of propellants.

$$\Delta T = \frac{c_r}{c_p}$$

The maximum temperature should be well below the maximum safe storage temperature for the propellant.

Ignition properties. The ignition temperature of a propellant varies with the time of exposure. The temperature below which ignition is unlikely to occur should be found for every propellant. Unfortunately no standard ignition method has been established, so that ignition values quoted by different laboratories should not usually be compared. In the absence of a standard method, the value of T_1 (Eq. 5-2) gives an order of magnitude of the autoignition temperature, because from this equation (which is really determined from temperature sensitivity data rather than autoignition data),

$$r \to \infty \quad \text{as} \quad T_p \to T_1$$

However, it is dangerous to assume that T_1 has any physical significance with regard to the temperature below which a propellant will ignite. Some propellants may ignite when stored at temperatures several hundred degrees below T_1. Adiabatic heating brought about by chemical decomposition may cause ignition. Great care should be utilized in interpreting ignition data: size, shape, and storage conditions of the propellant are very significant.

CHAPTER 4. DESIGN OF ROCKET MOTORS

H,13. Discussion of Requirements. The primary function of the rocket motor usually is to supply a given propulsive impulse, and the

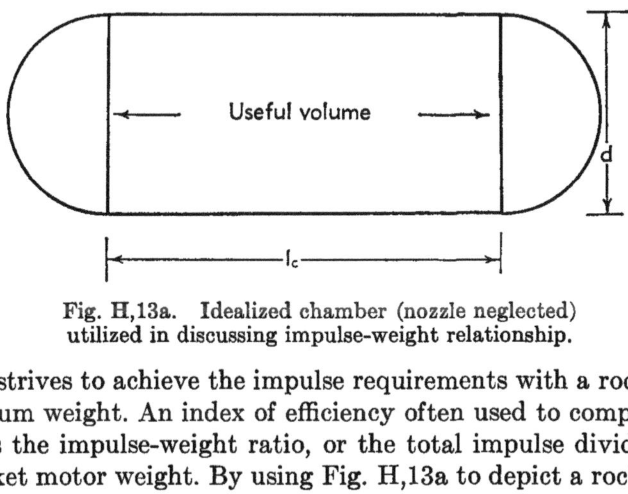

Fig. H,13a. Idealized chamber (nozzle neglected) utilized in discussing impulse-weight relationship.

designer strives to achieve the impulse requirements with a rocket motor of minimum weight. An index of efficiency often used to compare rocket motors is the impulse-weight ratio, or the total impulse divided by the total rocket motor weight. By using Fig. H,13a to depict a rocket motor, the impulse-weight ratio of a rocket motor may be shown to be [*42*]

$$\frac{I}{W_T} = \frac{I_{sp}}{\left(2 + \frac{d}{l}\right)\frac{\rho_w}{\rho_p}\frac{p_c}{\sigma_w}\frac{1}{v_\epsilon} + 1}$$

Here d is the useful volume diameter, l is the useful volume length, ρ_w is the density of the combustion chamber wall material, σ_w is the design stress for the wall material, p_c is the design chamber pressure, ρ_p is the propellant density, and v_ϵ is the fraction of *total* volume occupied by the propellant. An impulse-weight ratio greater than 100 seconds usually represents good design. It is evident that the propellant should have a high specific impulse and a high density, that the combustion chamber should be of materials having a high strength-to-weight ratio, that the volume available for propellant be used efficiently, and that the diameter-to-length ratio be low. It appears that the pressure should be as low as possible, but I_{sp}, d/l, and v_e each vary with pressure in such a way that an optimum pressure does exist. For the majority of applications rocket motor operation is most efficient between 500 and 1500 lb/in.^{2} abs. In addition to total impulse, the time rate at which the impulse is expended is usually specified, and it is this requirement that presents the most difficult design problem because of the limitations of available burning rates. Examples of some of the conflicting requirements follow.

Tubular charges have often been used for artillery rockets for which a reasonably short burning time is desired to obtain accuracy. Also, a high velocity is necessary for penetration and range. However, these requirements are not "compatible," because a short time can be achieved only by a thin propellant web (low propellant weight) and a high pressure (heavy metal parts), both of which result in reduced velocity and range.

For a high altitude rocket, a small frontal area is desirable to minimize the aerodynamic drag; this often leads to a small diameter requirement which restricts the web thickness for radial-burning charges. A thin web limits the burning time and propellant weight in such a way as to increase the significance of aerodynamic drag, so that, again, the requirements conflict.

Since the rate of expenditure of a fixed amount of a given propellant depends primarily upon the product of burning surface and burning rate, and the burning rate is limited by basic propellant properties, the designer must use ingenuity to obtain a practical charge geometry which has the necessary burning surface. In many radial-burning designs it is difficult to maintain a constant burning surface, and, as a consequence, the pressure is not constant; in order to keep the pressure below the maximum design pressure and above the combustion limit, the allowable pressure variation (or burning surface variation) is limited.

An optimum motor design. From the preceding discussion on requirements for rocket motor design, it is evident that an optimum rocket motor exists. The optimum motor is by no means *one* motor, however, and depends entirely upon the use for which it is intended. A free flight rocket is designed differently from a rocket used for auxiliary aircraft power; or, in fact, optimum design for rockets intended for a given purpose varies according to, for example, payload. Many rocket motors have been inefficiently used in applications for which they were not primarily designed.

The question often arises, "What rocket motor will give the longest range for a given payload?" For the purpose of illustrating a solution to this problem it is assumed that the rocket is fired horizontally in a vacuum a fixed distance above the earth. The rocket having the highest burnt velocity will achieve the longest range. The burnt velocity may be expressed by

$$V_b = -gI_{sp} \ln (1 - \nu)$$

For a given propellant the specific impulse depends only slightly upon chamber pressure and therefore the propellant-to-gross weight ratio ν is the main variable.

$$\nu = \frac{W_p(1 - \sigma)}{W_g} = \frac{W_p(1 - \sigma)}{W_p + W_n + W_c + W_l + W_m}$$

or

$$\nu = \frac{1 - \sigma}{1 + \frac{W_n}{W_p} + \frac{W_c + W_l + W_m}{W_p}}$$

By assuming that the weight of the nozzle is proportional to the total impulse

$$\nu = \frac{1 - \sigma}{1 + \phi I_{sp}(1 - \sigma) + \frac{W_c + W_l + W_m}{W_p}}$$

The velocity is greatest when ν is maximum or when $W_c + W_l + W_m/W_p$ is minimum. Now for a given payload, pressure, chamber materials, propellant and restriction materials, duct-to-throat area ratio, and initial-to-final burning surface ratio, the value of $(W_c + W_l + W_m)/W_p$ is solely a function of length-to-diameter ratio which can be optimized to give maximum velocity. For the purpose of illustrating the solution, a study [*56*] has been made by selecting two propellants of extremely different burning characteristics and by using the configuration and conditions shown in Fig. H,13b. After optimum values of l/d were obtained at various pressures, fictitious maximum ranges[7] R' were calculated by

$$R' = g[I_{sp} \ln (1 - \nu)]^2$$

and plotted vs. optimum l/d ratios and corresponding chamber pressures. The results are shown in Fig. H,13c and H,13d for the two propellants with zero payload and with a payload corresponding to one caliber of a material weighing 0.5 lb/in.[3] It is seen that the best value of l/d becomes lower as the chamber pressure and payload decrease and as the burning rate exponent n becomes larger. Of course, these curves indicate only general trends for other specified cases where different values for duct-to-throat area ratio and initial-to-final burning surface ratio are considered.

[7] The range R' corresponds approximately to the maximum range possible if the rocket had zero burning time and were fired upward at optimum quadrant angle.

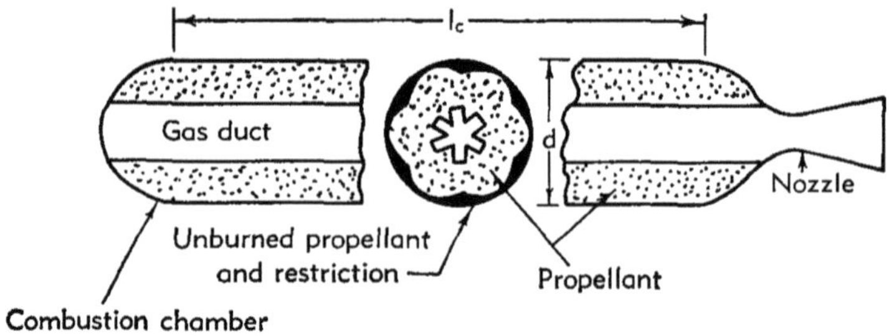

Fig. H,13b. Conditions for optimum rocket motor design study.

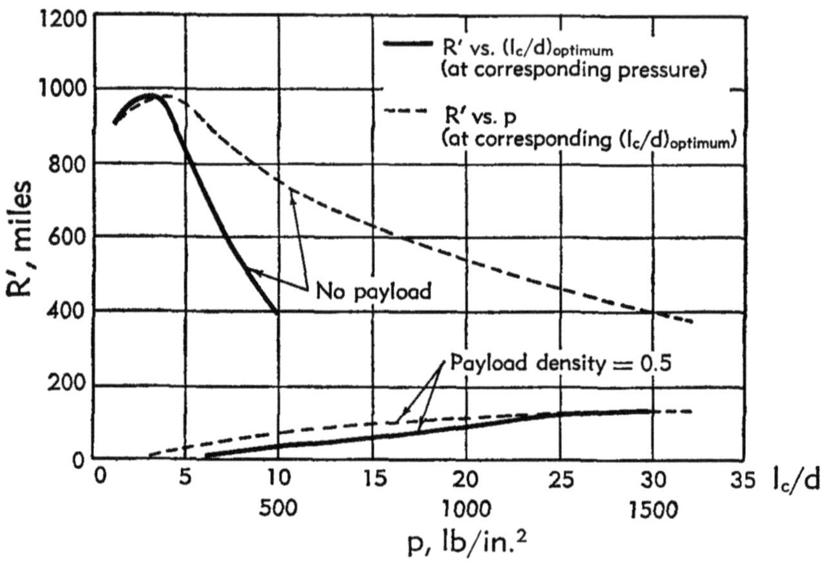

Fig. H,13c. Optimum drag-free range for rockets utilizing "slow burning" propellant with $r_{1000\text{psi}} = 0.35$ in./sec (see Fig. H,13b).

For vertical flight in atmosphere, the burning time partly controls the *drag deceleration. The weight of loading per unit cross-sectional area* μ also partly determines the deceleration due to drag and may be expressed by

$$\mu = \frac{\text{total weight of rocket}}{\text{cross-sectional area}} = \text{const} \times \left[\frac{W_p}{d^3} 1 + \phi I_{sp}(1 - \sigma) + \frac{W_c + W_l + W_m}{W_p}\right]$$

Therefore, in a study where drag and gravity are considered, the burning time, propellant weight, and diameter are important rocket motor characteristics which may entirely change the optimum l/d for maximum range. The variation of t_b/d and W_p/d^3 vs. l/d at various pressures is

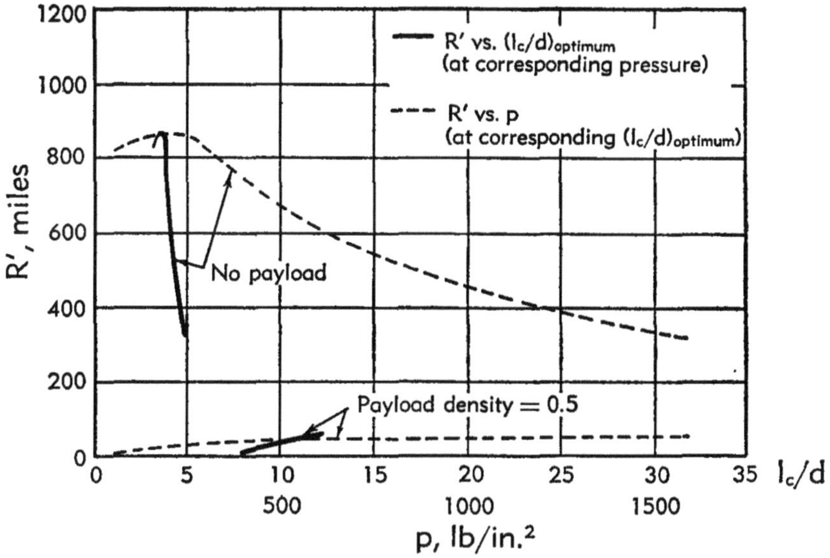

Fig. H, 13d. Optimum drag-free range for rocket utilizing "fast burning" propellant with $r_{1000psi}$ = 0.75 in./sec (see Fig. H,13b).

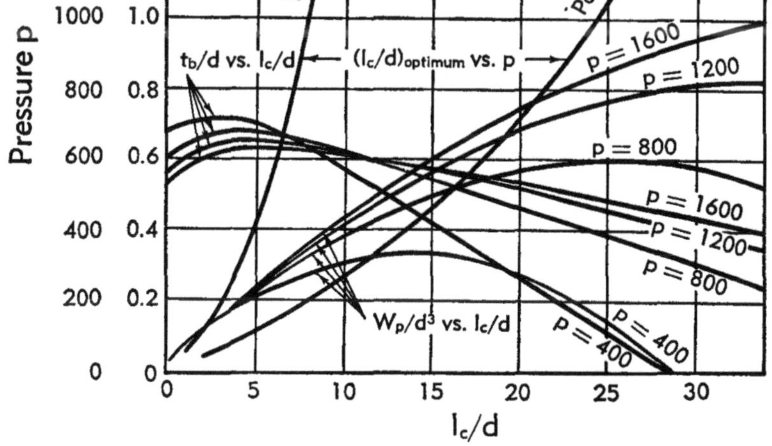

Fig. H,13e. Rocket motor characteristics for "slow burning" propellant with $r_{1000psi}$ = 0.35 in./sec (see Fig. H,13b).

given in Fig. H,13e and H,13f for the study which has just been presented for horizontal flight, vacuum conditions. Within the limitations imposed, it is seen that low chamber pressure, small l/d, and large diameter each increase the burning time; the propellant weight decreases as the l/d ratio becomes less than optimum and as the chamber pressure is reduced.

With regard to cost, it may be desirable to design rocket motors with l/d's less than optimum, since it has been shown that the propellant weight and gross weight then decrease. For rockets having high payloads

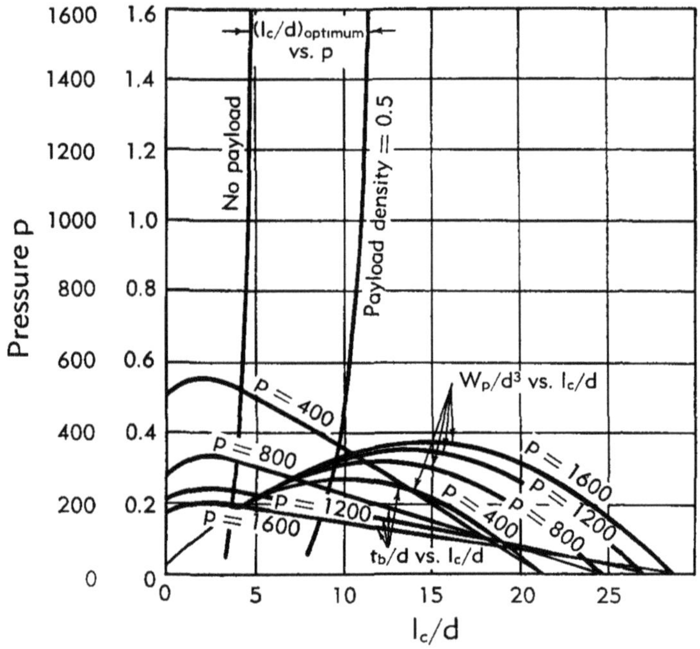

Fig. H,13f. Rocket motor characteristics for "fast burning" propellant with $r_{1000psi}$ = 0.75 in./sec (see Fig. H,13b).

(for example, a rocket used for boosting a missile), a considerable decrease in l/d is possible without sacrificing much performance (see Fig. H,13c). In general, the smallest diameter rocket possible should be used—consistent with payload and burning time required—to obtain the most economical rocket.

H,14. Design of Propellant Grains.

Design procedure. After the specifications for thrust, burning time, and chamber pressure have been determined, the details of rocket motor design can be worked out. For the design of the propellant charge the basic propellant characteristics as described in Art. 12 must be known. Useful design curves which may be prepared from these basic characteristics are shown in Fig. H,14a, H,14b, and H,14c. These curves are prepared from data obtained from actual rocket motor tests. Theory, alone, does not give sufficiently precise practical design data.

The critical propellant characteristics which must be known are propellant weight, web thickness,[8] and burning surface. The propellant

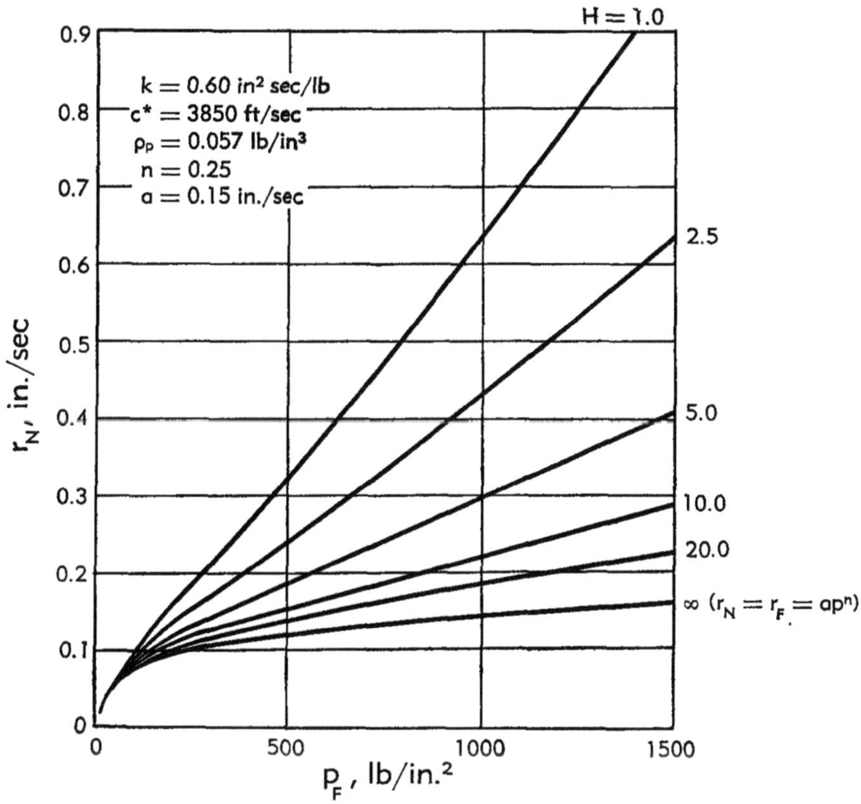

Fig. H,14a. Variation of burning rate, at the exhaust nozzle end of the charge, with front pressure and area ratio H, for a heterogeneous propellant (70°F). Note: $H = 1/J$.

weight can be calculated from the basic thrust equation $F = \dot{m}Vg$ by

$$W_p = \frac{gFt_b}{C_F c^*}$$

where C_F may be estimated to be about 98 per cent of the theoretical value. The web thickness is determined simply by

$$w = rt_b$$

[8] The web thickness w is the thickness of propellant measured normal to the burning surface.

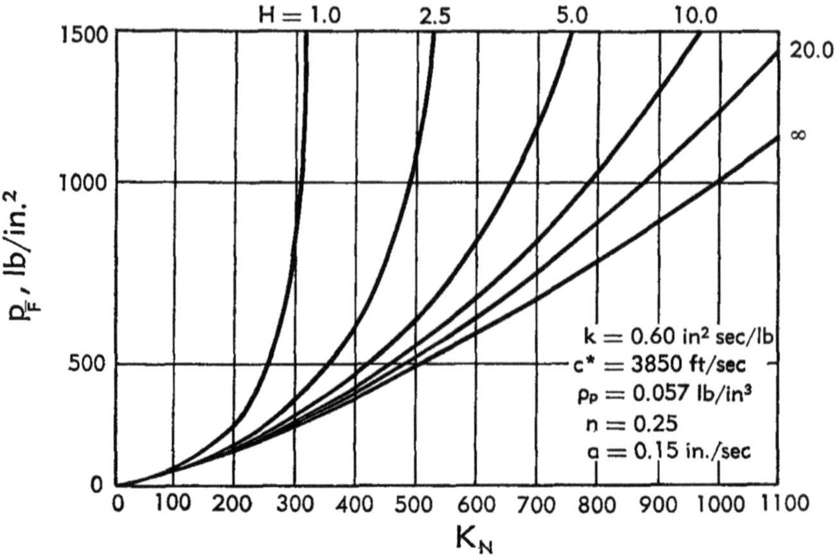

Fig. H,14b. Front-chamber pressure as a function of the area ratios *K* and *H* for a heterogeneous propellant (70°F). Note: $H = 1/J$.

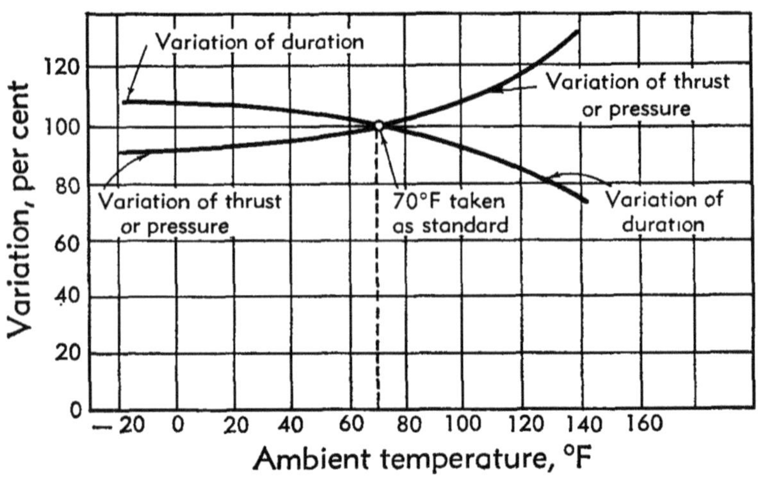

Fig. H,14c. Variation of performance with ambient temperature. (Erosive amplification is neglected.)

where the burning rate is selected from Fig. H,14a. Knowing the design pressure, the ratio $K = S_b/A_t$ can be found from Fig. H,14b; then the burning area is calculated by

$$S_b = KA_t = K\frac{F}{C_F p_c}$$

The variation of thrust and burning time vs. temperature is estimated from Fig. H,14c. Upon referring to Fig. H,14a and H,14b, the question of what gas duct-to-nozzle throat area ratio to use no doubt occurs to the viewer. For many propellants, erosive burning is negligible when $H > 5$, so that the curve for $H = \infty$ is often the only curve necessary. When the effect of erosive burning is significant, the pressure and burning rate variation with gas duct area change make it necessary to use curves such as Fig. H,14a and H,14b for a trial and error solution of w and S_b. Care must be exercised not to enter the region of erosive instability (see Art. 9).

Having the weight, burning area, and web thickness, the designer must determine the diameter and length dimensions which are dependent upon the particular design chosen. A discussion of charge geometry follows.

Propellant geometry. Because of the limited range of burning rates that exist for solid propellants, much work has been done with geometry to obtain design versatility. The propellant charge may burn on one end only (end-burning or "cigarette" burning), from the inside out (internal-burning), from the outside in (external-burning), or in any combination of the three (such as a completely unrestricted tubular charge). Charge shapes which have been found useful are shown in Fig. H,2a; all of these shapes can be designed to give nearly constant burning surface as the web burns away.

The end-burning charge has been found useful for long duration, low thrust rocket motors. It probably would be used universally if any burning rate desired were available, because it utilizes approximately 100 per cent of chamber volume and because it is simple to design: it exhibits no erosive burning and has a constant burning surface. However, with high thrust units, too large a diameter is required. The diameter required is given by

$$d_p = d_t\sqrt{K} = \frac{F\sqrt{K}}{C_F p_c}$$

The length of the end-burning charge is the web thickness. Another disadvantage of the end-burning charge is that the combustion gases flow past the chamber wall as the propellant burns away, and this requires heavy walls to absorb heat.

The double-base propellants were extruded into tubular shapes for use in artillery rockets during World War II. Because, for this design, the gases flow along the charge both internally and externally, it is necessary to so design that the pressure outside the charge is approximately equal to that inside.[9] This condition requires that the ratio of burning area

[9] In some designs it has been found necessary to drill radial holes through these charges to help equalize the inside and outside pressures, as well as to eliminate resonant burning effects.

to gas duct area be the same for the inside and outside channels, or expressed as a condition on the mean diameter z of the tubular propellant charge [57].

$$\frac{z}{d_c} = \frac{1}{2}\left[\left(\frac{w^2}{d_c^2} + 2\right)^{\frac{1}{2}} - \frac{w}{d_c}\right]$$

The corresponding limit on the length of the charge can be expressed by

$$\frac{l_p}{d_c} = \frac{12.5}{(z/d_c)} - \frac{50w}{d_c}$$

These limitations on charge diameter and length are plotted vs. web thickness in Fig. H,14d.

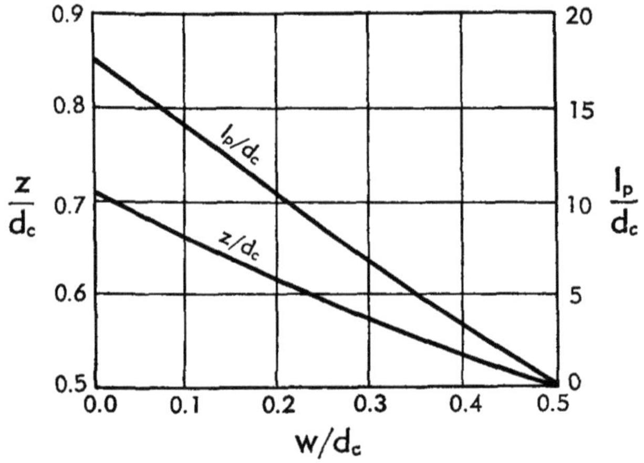

Fig. H,14d. Optimum mean diameter and length of single tubular grains of a given web thickness.

The simplicity of design and manufacture are the principal advantages of tubular charges. Some of the disadvantages are:

1. The gas flow between the charge and chamber wall produces high heat input to the chamber. As a consequence, the chamber wall gets hot quickly, making long duration rockets impractical because of the heavy wall required.
2. High volumetric efficiency is impossible, values greater than 65 per cent usually being rare. Thus low impulse weight ratios are characteristic of all motors using tubular charges.
3. Complicated arrangements are often necessary to hold these charges in place in the rocket motor.
4. Impulse is often lost by charge breakup near the end of burning when the web is very thin.

The cruciform charge has been found slightly superior to the tubular charge with respect to volumetric loading and the ability to withstand deformation, but it has the disadvantage of burning externally. Both the cruciform and tubular charges have been discussed in detail in [10].

The internal-burning charges such as the star, rod-and-tube, slotted-cylinder, multiperforated, and multidisk have the very attractive feature that the propellant itself insulates the combustion chamber wall from the hot combustion gases. With available burning rates, high thrust, short duration units can be designed with the multiperforated or multidisk

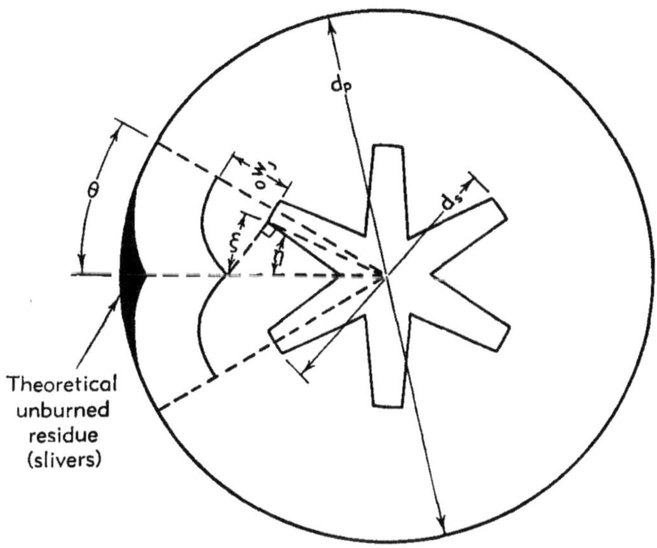

Fig. H,14e. Notations employed in the design of internal-burning, star-center propellant charges.

charges or long duration and low thrust can be achieved by using the star, slotted-cylinder, and rod-and-tube shapes. Also, high volumetric efficiencies (80 per cent and higher) can be achieved. As a consequence, the trend is to use internal-burning designs in most applications.

A geometric study of cross-sectional loading and the burning surface as a function of web thickness is necessary before preparing to design these charges. For illustration consider a study that has been made [57] of the internal star shape. If the star parameters are defined as given in Fig. H,14e, the variation in gas duct area and burning surface per unit length of charge (perimeter) vs. the amount of web burned may be expressed by

$$_0(A_d)_J = A_{d0} + \left(\frac{S_b}{d_p}\right)_0 w + \frac{M}{2} w^2$$

Note here that w refers to linear displacement of burning surface from its original position (web distance burned).

$$_J(A_d)_f = N\left[\left(\frac{d_s}{2} + w^2\right)(\theta - \xi) + w^2\left(\xi + \sin^{-1}\frac{d_s \sin \xi}{2w}\right)\right.$$
$$\left. + \frac{1}{2} d_s \sin \xi \left(R \cos \xi + w \cos \sin^{-1}\frac{d_s \sin \xi}{2w}\right)\right]$$

$$_0\left(\frac{S_b}{l_p}\right)_J = \left(\frac{S_b}{l_p}\right)_0 + Mw$$

$$_J\left(\frac{S_b}{l_p}\right)_f = 2N\left[\frac{1}{2}(\theta - \xi)d_s + w\left(\theta + \sin^{-1}\frac{d_s \sin \xi}{2w}\right)\right]$$

where

$$\left(\frac{S_b}{l_p}\right)_0 = Nd_s\left(\theta - \xi\frac{\sin \xi}{\sin \eta}\right) = \text{initial perimeter}$$

$$M = 2N\left(\frac{\pi}{2} + \theta - \eta - \cot \eta\right)$$

$$= \text{initial slope of perimeter versus web distance burned}$$

The subscript $_0$ refers to the initial condition and $_0(\;\;)_J$ indicates the interval over which the "flat" sides of the inner star points exist. After the "flat" sides disappear, the propellant burns outward until the motor wall is contracted; this interval is designated by $_J(\;\;)_f$. The junction point where the "flat" sides disappear occurs when the web distance is

$$w_J = \frac{d_s}{2}\frac{\sin \xi}{\cos \eta}$$

When the motor wall is contacted, the amount of propellant left is called the "theoretical unburned residue." It is desirable to keep the theoretical unburned residue to a minimum because the actual unburned residue is often roughly proportional to the theoretical amount.

For quick estimation of design, the long expressions for A_d and S_b are not convenient, but graphical methods are most useful. For a given number of star points N and initial-to-final perimeter ratio λ, the volumetric loading $1 - \alpha$, unburned residue σ, minimum perimeter ratio as the web burns away μ, and values for critical angles ξ and η can be shown as a function of web fraction $1 - \delta$; an example is given in Fig. H,14f [*58*]. The use of these geometrical design curves is demonstrated in the following paragraphs which show how the pressure-time curve is calculated.

Calculation of the pressure-time curve. The minimum chamber wall thickness is obtained for a given average pressure if the pressure-time curve is constant. One may therefore conclude that the most efficient rocket motor should operate at a constant pressure; however, for an internal-burning star charge, for example, constant pressure (or constant burning surface under nonerosive burning conditions) can be obtained only at a sacrifice of high volumetric loading and low unburned propellant residue. The impulse-weight ratio of a rocket motor can often be improved by adding propellant weight and allowing the pressure to

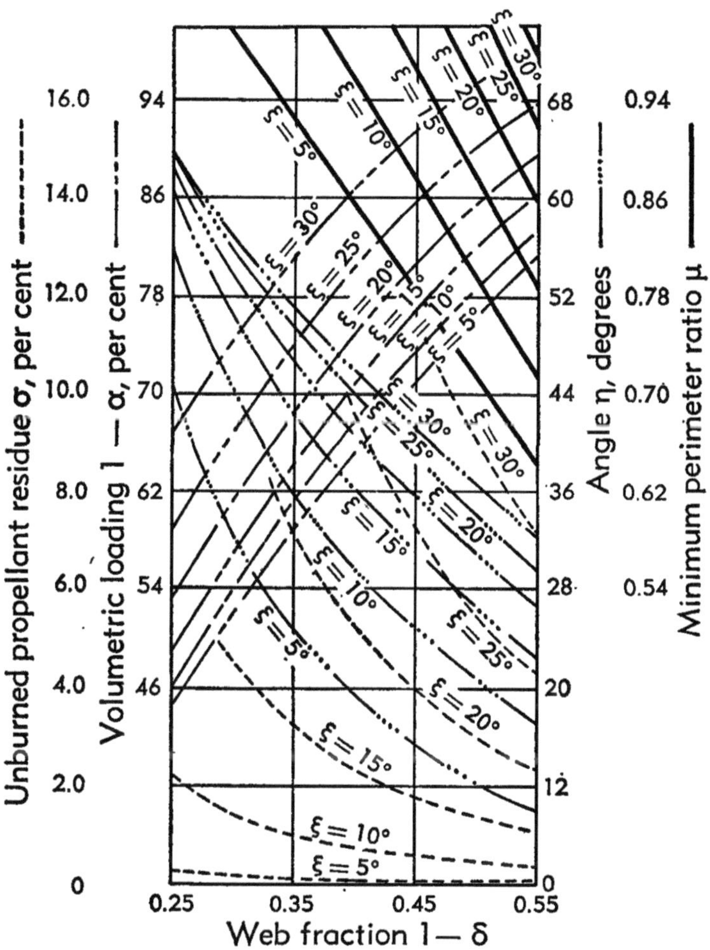

Fig. H,14f. *Relationship of geometrical variables for an internal-burning, star-center propellant charge (see Fig. H,14e). ($N = 6$, $\lambda = 0.80$.)*

increase with time. Fortunately, the results for some applications, such as assisted take-off of aircraft and high altitude rockets, are improved with an increasing pressure-time curve. The port area is small when high propellant loading is obtained with internal-burning charges; as a consequence, the port-to-throat area ratio is often small, thus leading to high erosive burning.

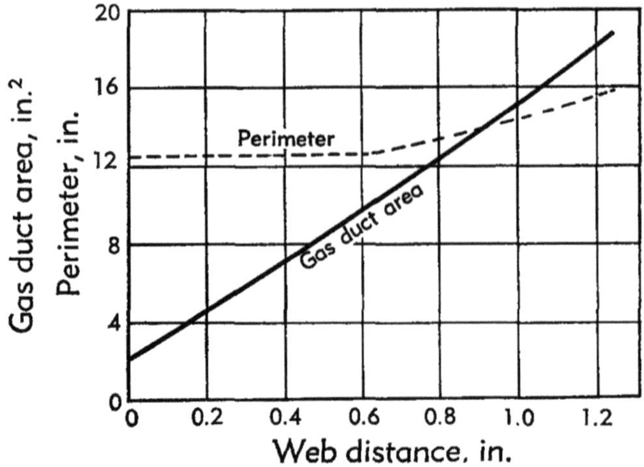

Fig. H,14g. Variation of perimeter and gas duct area with web distance for the star-center charge (see Fig. H,14e).

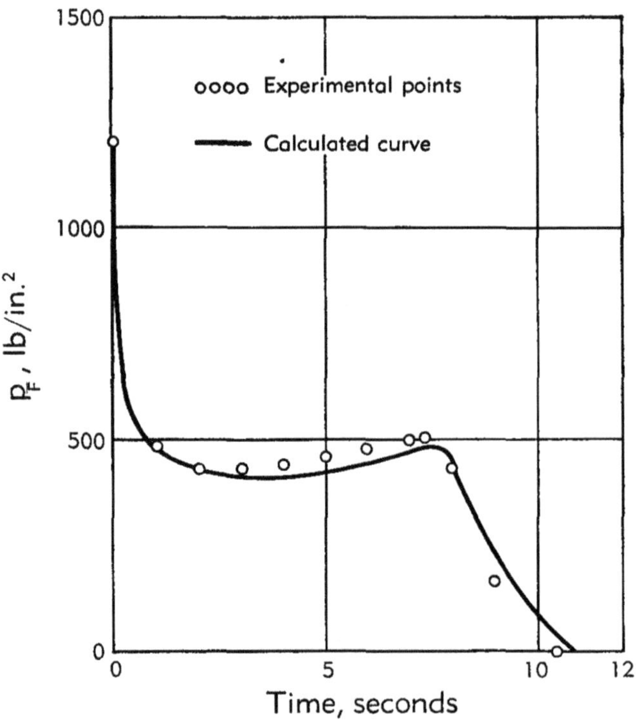

Fig. H,14h. Comparison between calculated and experimental pressure-time curve for an internal-burning, star charge (see Fig. H,14e and text).

An example of the calculation of a pressure-time curve for an internal-burning star charge will now be made. From the design chart of Fig. H,14f, a design is chosen which has a shape identical to the one of Fig. H,14e for this design $1 - \delta = 0.5$, $1 - \alpha = 89$ per cent, $\sigma = 5$ per cent, $\mu = 1$, $\lambda = 0.8$, $N = 6$, $\theta = 30°$, $\eta = 33.5°$, and $\xi = 24°$. The variation of port area and burning surface per unit length vs. web distance is plotted in Fig. H,14g. If the propellant chosen has the characteristics given by Fig. H,14a and H,14b, and then a charge diameter of 5 in., a length of 40 in., and a port-to-throat ratio of 2 are selected, all necessary information for the calculation is available. By the use of Eq. 9-1 the initial front chamber pressure and front burning rate are determined. The burning rate and pressure at the nozzle end of the charge are determined by Eq. 8-13, 8-15 and 9-1. Because the port area changes continually and the pressure and burning rate depend upon port-to-throat area ratio, incremental time intervals must be selected over which the calculated conditions are assumed to prevail. By successive recalculations at the end of each time interval, the pressure-time curve shown in Fig. H,14h is obtained. The pressure-time curve resulting from an actual test is also shown and the comparison between theory and experiment is evidently very good.

It is interesting to note that the shape of the pressure-time curve does not look much like that of the burning surface in Fig. H,14g; this non-similarity is due to the strongly erosive burning character of the propellant. It is also of interest that about 15 per cent of the impulse is obtained after the propellant burns through to the chamber wall at the nozzle end of the charge; thus the calculated propellant residue has little meaning when the propellant exhibits significant erosive burning. In order to achieve an abrupt cessation of burning, it may be necessary to taper the gas duct so that the propellant burns out to the chamber wall all along its length at the same time.

H,15. Mechanical Design. The mechanical components of a rocket motor are comprised of the combustion chamber, exhaust nozzle, and miscellaneous items such as safety release devices and charge supports. Views of the mechanical components of three typical motors are shown in Fig. H,15a, H,15b, and H,15c.

The combustion chamber. The combustion chamber is essentially a pressure vessel containing the propellant, which must be able to withstand the high pressure and temperature of the gases. The wall thickness τ of the tube which constitutes the main portion of the combustion chamber can be estimated by the hoop tension formula

$$\tau = \frac{p_{max} d_w}{2\sigma}$$

when the wall thickness computed is less than 5 per cent of the diameter, which is the usual case for well-designed motors. The chamber pressure upon which to design should be the maximum expected, plus a safety allowance. Normally, $p_{max} = p_c(1 + f_p)(1 + f_{bv})(1 + \pi_p \Delta T)f_s$, where f_p is a factor

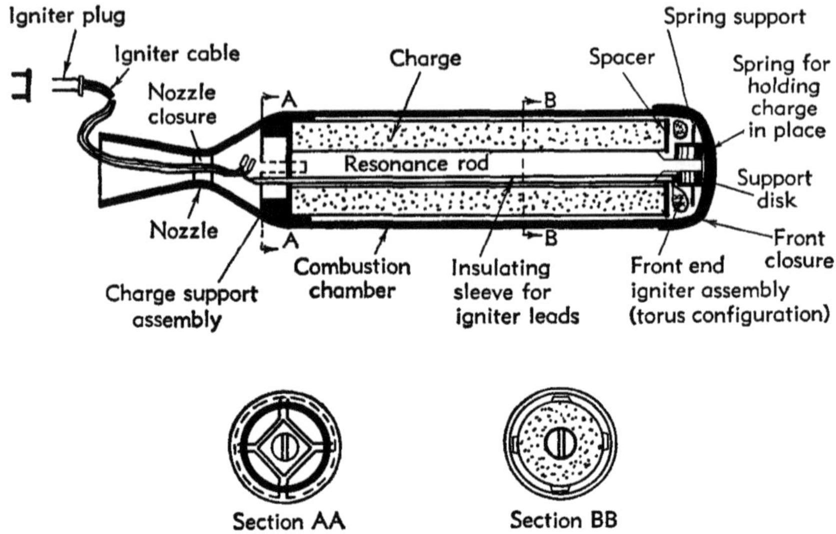

Fig. H,15a. Mechanical design characteristics of a large rocket employing a tubular charge. (After Jato Manual M1.)

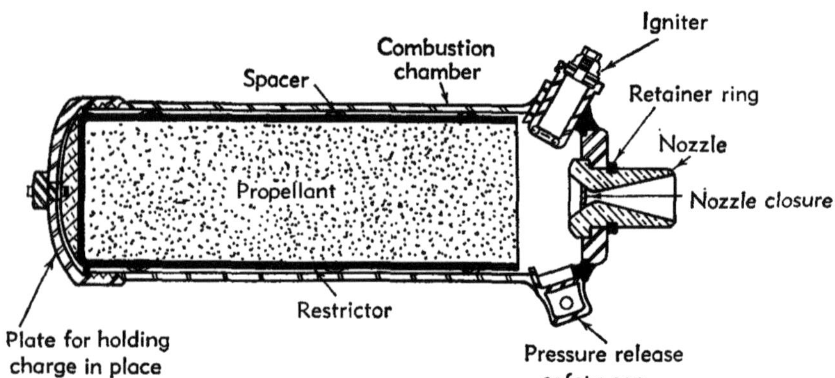

Fig. H,15b. Mechanical design characteristics of a rocket employing an end-burning charge. Note that propellant is not sealed to combustion chamber walls. (After Aerojet Jato design.)

indicating increase in chamber pressure above the average, f_{bv} is a batch variation factor for propellants, and f_s is a safety factor allowance. However, if the motor is spin-stabilized the effective increase in internal pressure caused by centrifugal force must also be considered; this may be estimated to be $r\omega^2 d_w \rho_w / 2g$. The inner chamber diameter d_w is determined

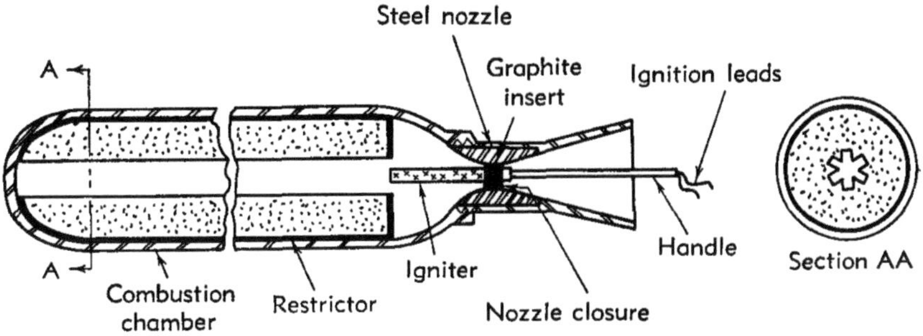

Fig. H,15c. Mechanical design characteristics of a rocket motor employing a star-center, internal-burning charge sealed to combustion chamber wall.

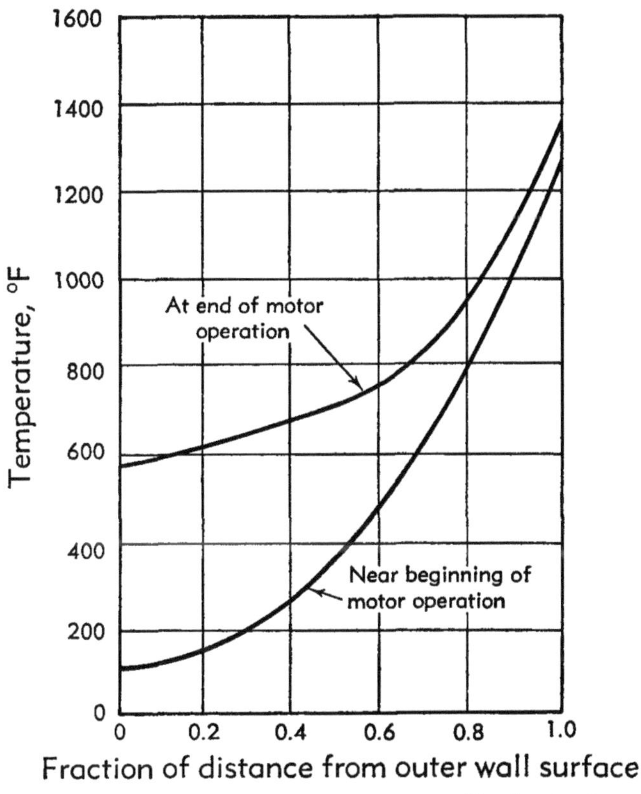

Fig. H,15d. Illustration of temperature distribution in the combustion chamber wall of a rocket motor.

by the sum of the charge diameter, the restriction thickness, and the space required to allow for the thermal expansion and flow of gases (external-burning charges). The value of the tensile strength depends upon the effective temperature of the combustion chamber wall at the end of burning and the rate at which the motor wall is heated. Since a temperature distribution, such as that shown in Fig. H,15d, may be present in some rocket motor walls, it is obvious that a complicated stress distribution may exist due to both the chamber pressure and the thermal gradient. It is usually adequate, however, to use the mean wall temperature for estimating the strength of the material being used. It remains, then, to determine accurately the value of the mean wall temperature for any particular design study.

The transfer of heat from the hot gases to the combustion chamber can be expressed by

$$q = h(T_r - T_w)$$

where T_r is the recovery temperature of the gases in the motor, T_w is the temperature of the motor wall at the inner wall surface, and h is the heat transfer coefficient which may be expressed [*10*] as:

$$h = \frac{\text{const} \times c_p G^n}{D^{1-n}}$$

where the constant depends upon the viscosity and thermal conductivity of the gases, G is the mass flow density, and c_p is the specific heat of the gases at constant pressure, and D is the effective duct diameter through which the gases pass. For double-base and "perchlorate" propellants, the heat transfer coefficient along the cylindrical portion of the combustion chamber can be approximated [*10*] by[10]

$$h \cong 0.046 c_p G^{0.8}$$

Since the diameter variation is not usually large, $D^{0.2}$, it can be neglected without much error. The value of G can be determined easily at the nozzle end (where the wall temperature is highest) of a cylindrical combustion chamber from the pressure-time curve and charge design.

$$G = \frac{W}{A_d} = \frac{g}{c^*\theta^*} p_F J$$

As the chamber wall heats up, however, the value of T_w in the equation for q changes continually, thus causing the heat input to vary. A practical solution to this transient heat flow problem has been obtained by Hall [*59*] by applying the numerical method of Dusinberre to hollow cylinders. Analytic solution of the Poisson-Fourier equation, applicable to a cylindrical system having circular symmetry and no axial heat flow,

$$\frac{\partial T}{\partial t} = \kappa \left[\frac{\tilde{\partial} T}{\partial R^2} + \frac{1}{R} \frac{\partial T}{\partial R} \right]$$

[10] The h's are greater than indicated by the equation when $G < 1200p^{\frac{1}{2}}$, where p is the pressure in atmospheres. In the relationship shown here, h is expressed in BTU/ft²/hr/°F and G in lb/ft²/hr. In rocket work the use of in.² and sec may be more convenient than ft² and hr.

is discussed in [*60*]. The numerical method does not require mathematical skill, gives acceptable accuracy, and can easily take care of changing thermal properties; in contrast, most analytic solutions require laborious numerical integration. Tsien and Cheng have developed a useful similarity law for this problem [*61*]. If the chamber wall is very thin, of thickness τ_c, an approximate conservative solution can be obtained simply by considering the heat flow over very small time intervals, assuming the

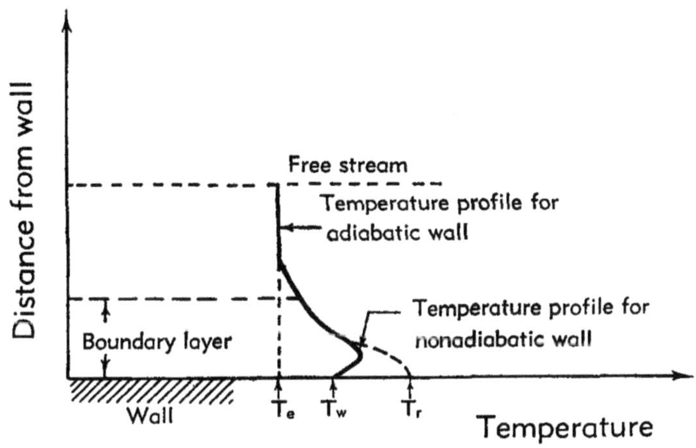

Fig. H,15e. Illustration of a temperature profile for gas flow along a solid boundary.

wall of heat capacity c_c to heat uniformally across its thickness, calculating a temperature rise, and then repeating the operation for the next time interval with the new wall temperature. By equating the heat input to the heat absorbed,

$$q\Delta t = \rho_c\tau_c c_c \Delta T$$

or

$$\Delta T = \frac{q\Delta t}{\rho_c\tau_c c_c}$$

An example of how the wall temperature distribution curve may vary with time was given in Fig. H,15e.

The joints, attaching the front and rear closures to the chamber body, may provide a difficult design problem requiring detailed stress analysis. In general, the principal load on the joint is longitudinal. In addition to the pressure, the effects of acceleration, internal drag forces, and external drag forces must be considered.

The exhaust nozzle. The thrust of a rocket motor is significantly dependent upon nozzle design, as evident from the equation (see G,1)

$$F = C_F p_c A_t$$

where

$$C_F \cong 0.98 \left[\frac{1}{2}(1 - \cos\psi)\right]\left[\Gamma' \sqrt{\frac{2}{\gamma - 1}\,\eta_{is}} + \frac{p_e - p_\infty}{p_c}\frac{A_e}{A_t}\right]$$

The highest possible value of C_F is desired, consistent with low nozzle weight. The half angle of the exit section ψ should be between 8 and 20° for good nozzle performance. Very small angles make the downstream portion of the nozzle excessively long from the weight and heat transfer standpoint; in general $\gamma \cong 15°$ is satisfactory. The half angle of the inlet section of the nozzle β does not appear in the above equation. Although β is not critical from a thrust standpoint, a smooth contour for the inlet is desirable; in general $20° \leqq \beta \leqq 45°$ is satisfactory. The value of the

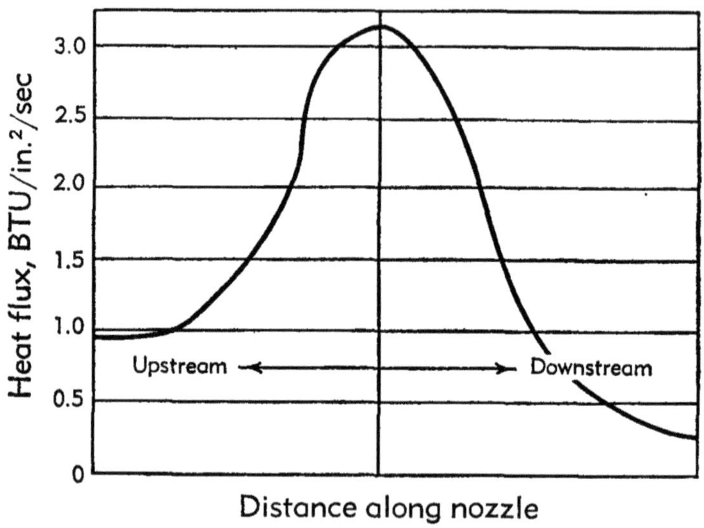

Fig. H,15f. The variation of heat flux with position along an exhaust nozzle.

expansion ratio A_e/A_t to give optimum thrust coefficient depends upon the particular application. If the exit pressure p_∞ is constant during the motor operation, the optimum value of A_e/A_t is [*42*]:

$$\frac{A_e}{A_t} = \frac{\left(\frac{2}{\gamma + 1}\right)^{\frac{\gamma+1}{2(\gamma-1)}}}{\left(\frac{p_\infty}{p_c}\right)^{\frac{1}{\gamma}} \sqrt{\frac{2}{\gamma - 1}\left[1 - \left(\frac{p_\infty}{p_c}\right)^{\frac{\gamma-1}{\gamma}}\right]}}$$

For $\gamma = 1.2$, $p_c = 2000$ lb/in.², and $p_\infty = 14.7$ lb/in.², the value of $A_e/A_t = 15$ for optimum C_F.

The heat transfer problem in the exhaust nozzle is severe, since the mass flow density G is greatest at the nozzle throat. An example of the heat flux vs. position along nozzle length is illustrated by Fig. H,15f.

As with the combustion chamber, the heat input to the nozzle can be given by the equation $q = h(T_r - T_w)$. However, the heat transfer coefficient, which applies for the tubular combustion chambers, does not apply to exhaust nozzles. The results of experimental tests at 500 psi [*59*] have indicated that the following relationships may apply to nozzles:[11]

Entrance portion of nozzle (subsonic flow): $h \cong 2.16 c_p G^{0.5}$

Exit section of nozzle (supersonic flow): $h \cong 0.00106 c_p G$

By the use of these equations, the temperature rise in a given nozzle can be estimated in a manner similar to that described for the combustion chamber.

An example of the effect of h upon the variation of temperature with time at the throat section of a copper nozzle is shown in Fig. H,15g[12]. The temperature of the metal at the surface next to the gas stream is of particular interest because, as it approaches the melting point, the nozzle erodes. In order to keep the inner wall temperature to a minimum, the thermal conductivity should be high. To keep the nozzle volume and weight small, the density and specific heat should be high.

Molybdenum, graphite, copper, and iron have been found to be the most useful materials for the heat-absorbing nozzle. Some of their properties are listed in Table H,15. Molybdenum and graphite have very high

Table H,15. Approximate properties of some nozzle materials.

Material	*Thermal* conductivity, BTU/ft hr °F*	*Specific* weight, lb/ft³*	*Specific* heat, BTU/lb °F*	*Melting point, °F*
Copper	223	560	0.09	1981
Graphite	80	97	0.16	6600 (sublimes)
SAE 1020 steel	30	491	0.11	2760
Iron	46	487	0.11	2795
Molybdenum	84	636	0.06	4748

* At 70°F.

melting (or subliming) temperatures and generally make the lightest nozzles. Unfortunately, molybdenum is not plentiful or easy to fabricate in large sizes. Graphite has very low strength and, consequently, must be used as an insert (Fig. H,15c). For combustion periods less than 5 sec, iron may be most practical. Copper is satisfactory for longer periods (30–60 sec), but its weight is usually prohibitive.

Heat-resistant materials offer the possibility of lighter weight than

[11] In these equations h is expressed as BTU/ft² hr °F and G as lb/ft² hr.

[12] Fig. H,15g suggests a convenient method for determining the value of h: by merely recording the temperature of the outside wall for a given section, which is insulated (such as by air space) against axial heat flow, h can be estimated by fitting the experimental temperature-time curve on a set of curves calculated for several values of h.

that obtained with heat-absorbing nozzles, but, in general, they are not of adequate strength to withstand thermal stresses. Silicon carbide has been used for some uncooled nozzles. A comprehensive treatment of exhaust nozzle materials is given in [*62*].

Charge support. When the charge is not sealed to the combustion chamber it usually must be held in place by some mechanical means.

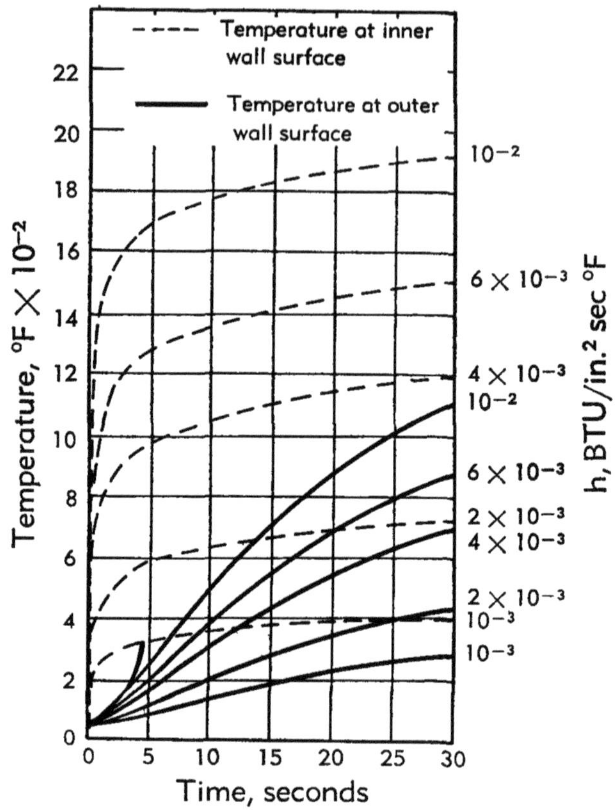

Fig. H,15g. Computed variation in temperature at the throat section of an exhaust nozzle, with time for several values of the heat transfer coefficient *H*. (Gas temperature 3200°F, copper nozzle, throat diameters: inside 1.28 in., outside 5.50 in.)

External-burning charges, in particular, sometimes require elaborate methods of support which may add considerable weight to the rocket motor. The supports must be designed to withstand the inertia forces and drag forces of gases such as those discussed in Art. 13, and also the high temperature combustion gases. Methods of supports for a few charge designs are illustrated in Fig. H,15a, H,15b, and H,15c.

Pressure release device. When a rocket motor is to be used near personnel, it usually has a safety device to release the chamber pressure if it should approach the maximum allowable design value. Copper diaphragms have been found useful as burst diaphragms, because copper can be obtained easily in high purity. When used, copper must be insulated from the combustion gases. For a rupture disk of high purity (99.5 per cent) copper, the burst pressure is given by [*42*]

$$\text{burst pressure, lb/in.}^2 = \frac{98{,}400(t - 0.0035)}{d}$$

where t = disk thickness, inches, and d = shear area diameter of disk, inches. This empirical formula is satisfactory over the pressure range 1500–4000 lb/in.^{2} for disks from 0.6 to 2.0 in. in diameter.

In general, the gases which escape through the pressure release device should be directed so that they produce no net thrust on the rocket motor.

Ignitors. The purpose of the ignitor is to bring the exposed propellant surface to the ignition temperature as quickly as possible without excess pressure peaks, which might rupture the motor or charge. The following remarks refer to practical methods, although the theory of ignition [*63*] is developing rapidly.

The principal components of the typical ignitor consist of a squib, ignition granules, and the ignitor case. The squib usually consists of an electrical wire heater imbedded in a small amount of heat-sensitive primary ignition. When ignited, the squib flashes to ignite the main ignition charge. The gases from the main ignition charge burst the container and throw the ignited particles against the propellant surface. For minimum delay time, the ignition particles should be at high temperature and should cover as large a portion of propellant surface as possible. A typical ignition history for a rocket motor is shown in Fig. H,15h.

The principal factors which determine the ignition characteristics are the type, granulation, amount, moisture content, and packing of the ignition powder; the location and construction of the ignitor case; the charge geometry; the ignition temperature of the propellant; the surface condition of the propellant (the presence of inert films such as wax from mold release); and the relationship between ignition pressure peak, nozzle closure blowout pressure, and the operating pressure for the motor. Since many of these variables are interdependent, ignitor design is at the mercy of the experience and judgment of the designer. Some qualitative guides for ignitor design may be obtained from [*10*] and the remaining portion of this section.

Not much attention has been given to the preparation of new ignition powder. The most commonly used material is black powder, which is readily available in several granulations. For some special applications, mixtures of metals with oxidizing agents (such as magnesium and aluminum with potassium perchlorate) have been useful; the combustion of such mixtures produces high temperatures, relatively small amounts of gas, and ignition times less than those obtained from black powder. The metal perchlorate mixtures are often extremely hazardous, and are used only for special experimental studies.

The ignition time can usually be kept to a minimum by using a nozzle closure to prevent the gases from exhausting during the ignition period. An approximate rule is to have the pressure from ignition be about 30 to 40 per cent of the expected motor pressure, and the nozzle closure blow-out pressure be about 90 to 100 per cent of the expected motor pressure. For black powder, the number of grams of ignition (W_i) required to raise

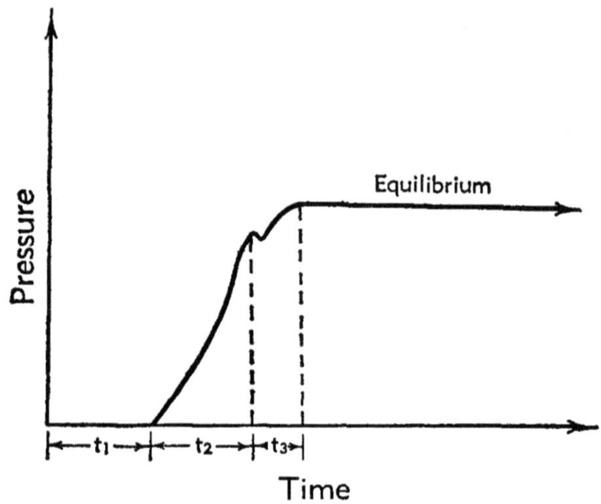

Fig. H,15h. Typical ignition characteristics for a solid propellant rocket motor. t_1 = interval between application of electric current and pressure rise = 20 milliseconds, t_2 = interval between rise of pressure and nozzle closure blowout = 50 milliseconds, t_3 = interval between nozzle closure blowout and establishment of equilibrium = 20 milliseconds.

the pressure (lb/in.^{2}) to a given value in a given volume v (in.^{3}) may be found from

$$p = 2000\left(\frac{W_i}{v}\right)^{0.86}$$

If the amount of ignition is too high, the nozzle closure blows out before enough of the propellant can burn to sustain combustion, resulting in a misfire.

The ignitor is usually located at the nozzle end of the motor, the head end, or distributed along the length of the charge; two of these types are illustrated in Fig. H,15a, H,15b, and H,15c. The ignitor which is located in the nozzle of the motor is usually economical, because it can be made simply and because no special machining of the combustion chamber is necessary to hold it. One objection to this type is that the flying fragments from the ignitor may be destructive. For internal-burning charges, a "shotgun" blast from the nozzle-end ignitor into the central perforation provides effective ignition. For external-burning charges, the forward-end ignitor has been popular because the ignitor flame must pass over the exposed propellant surface to escape through the nozzle. The ignitor which is distributed along the length of the charge is desirable when no nozzle closure is allowable; a "flash" type of ignition (such as the metal-perchlorate mixture) is often useful in this type of ignitor.

CHAPTER 5. DEVELOPMENT TRENDS

H,16. Trends in Solid Propellant Rocket Development.

Practical trends. Present efforts in rocket development are directed toward achieving flexibility in design, high performance, reliability, and low cost. In summary, the major factors which are furthering the attainment of these ideal goals are as follows:

1. Internal-burning charge design. (Combustion chamber case insulated from hot gases.)
2. Propellants with burning rates insensitive to pressure charges. (Operating conditions insensitive to temperature change and design factors.)
3. Casting processes. (Large and small charges made with nearly equal ease. Cost possibly much lower than high pressure molding and extrusion.)
4. High energy, high density propellants. (Energy per unit volume and per unit weight high, results in decreased over-all motor weight and, usually, decreased cost.)
5. Motor construction materials of high strength and low weight. Improved fabrication techniques.
6. High strength, low conductivity, heat-resistant materials. (Allows low weight for parts exposed to combustion gases, such as the exhaust nozzle.)
7. Improved quality control to enhance reliability and useful applications.

Design flexibility depends a great deal upon the ability to obtain a wide range of burning rates and still retain smokeless, noncorrosive exhaust gases; however, much work must still be done in order to combine these desirable characteristics. The trend at present is to use complex charge geometry, when high mass flows are required, to compensate for the low burning rate propellants which are available.

The cost of solid propellants has been high, at the very best, due to the safety requirements necessary in their manufacturing processes and the high basic material cost. The trend is to attempt to use less expensive materials in propellants for some applications, even though relatively low performance results. Because of the infancy of some processes—the cast process, a notable example—streamlining of manufacture has not been perfected. Regardless of the high propellant charge cost ($1 to $5 per pound being a reasonable range in 1948), the motor construction is so simple that the over-all cost of the complete solid propellant rocket unit may be less than a comparable liquid propellant rocket, which uses much less expensive propellant materials. Furthermore the simplicity of operation makes the solid propellant rocket motor particularly desirable for field use. In addition to rocket propulsion, the gas generation capabilities of solid propellants indicate that they may be used for many novel applications in the future.

Fundamental problems. The fundamental problem of solid propellant rockets is the detailed, quantitative understanding of the combustion process under a wide variety of conditions. Ultimately, one may hope to control the burning rate over a wide range of speeds by means of propellant composition and processing.

Understanding of the combustion process will require clarification of many diverse scientific problems, such as chemical kinetics, diffusion of heat and molecular species, radiation, the behavior of small particles in a gas stream and at a solid surface, and the hydrodynamic problem of gas flow past a burning surface.

A second class of fundamental problems concerns the control of propellant quality during manufacture. Relatively small changes may have large effects upon rocket motor operation. The technology of mixing small particle, high polymers and the vast arts of chemical engineering will require continued application and development.

H,17. Cited References.

1. Hayes, T. J. *Elements of Ordnance.* Wiley, 1938.
2. Symposium on kinetics of propellants. *J. Phys. and Colloid Chem. 54,* 847 (1950).
3. Boys, S. F., and Corner, J. *Proc. Roy. Soc. London A197,* 90 (1949).
4. Corner, J. *Proc. Roy. Soc. London A198,* 388 (1949).
5. Corner, J. *Theory of Interior Ballistics of Guns.* Wiley, 1950.
6. Corner, J. The effect of turbulence on heterogeneous reaction rates. *Trans. Faraday Soc. 43,* 635 (1947).
7. Birkoff, G., MacDougall, D. P., Pugh, E. M., and Taylor, G. *J. Appl. Phys. 19,* 363 (1948).
8. Davis, L., Jr., Blitzer, L., and Follin, J. W., Jr. *Exterior Ballistics of Rockets.* McGraw-Hill. To be published.
9. Rosser, J. B., Newton, R. R., and Gross, G. L. *Mathematical Theory of Rocket Flight.* McGraw-Hill, 1947.
10. Wimpress, R. N. *Internal Ballistics of Solid-Fuel Rockets.* McGraw-Hill, 1950.
11. Grad, H. *Commun. on Pure and Appl. Math. 2,* 79 (1949).
12. *Fourth Symposium on Combustion.* Williams & Wilkins, 1953.
13. *Fifth Symposium on Combustion.* Reinhold, 1955.
14. Cheng, S. I. High-frequency combustion instability in solid propellant rockets. *J. Am. Rocket Soc. 24,* 27–32, 102–109 (1954).
15. Cheng, S. I. An unstable burning of solid propellants. *J. Am. Rocket Soc. 25,* 79 (1955).
16. Green, L., Jr. Unstable burning of solid propellants. *J. Am. Rocket Soc. 24,* 252 (1954).

17. Crawford, B. L., Jr. Rocket fundamentals. *George Washington Univ. N. O. R. C. Div. 3, Sec. H, Office Sci. Research and Develop. Rept. 3711*, June 1944.
18. Price, E. W. Steady-state one-dimensional flow in rocket motors. *J. Appl. Phys. 23*, 1942 (1952).
19. Price, E. W. Theory of steady state flow with mass addition applied to solid propellant rocket motors. *J. Am. Rocket Soc. 23*, 237 (1953).
20. Price, E. W. Charge geometry and ballistic parameters for solid propellant rocket motors. *J. Am. Rocket Soc. 24*, 16 (1954).
21. Price, E. W. One-dimensional, steady flow with mass addition and the effect of combustion chamber flow on rocket thrust. *J. Am. Rocket Soc. 25*, 61 (1955).
22. Sutton, G. P. *Rocket Propulsion Elements*. Wiley, 1949.
23. Seifert, H. S., Mills, M. M., and Summerfield, M. *Am. J. Phys. 15*, 1 (1947).
24. Avery, W. H. Radiation effects in propellant burning. *J. Phys. and Colloid Chem. 54*, 917 (1950).
25. Beek, J., Jr., Avery, W. H., Dresher, M. J., McClure, F. T., and Penner, S. S. Studies of radiation phenomena in rockets. *Office Sci. Research and Develop. Rept. 5817*, 1946.
26. Penner, S. S. *J. Appl. Phys. 19*, 278, 392, 511 (1948).
27. Avery, W. H., and Hunt, R. E. Effect of pressure and temperature on the rate of burning of double-base powders of different compositions. *Office Sci. Research and Develop. Rept. 1993*, Oct. 1943.
28. Crawford, B. L., Jr., Huggett, C., and McBrady, J. J. Observations on the burning of double-base powders. *Office Sci. Research and Develop. Rept. 3544*, Apr. 1944.
29. Gibson, R. E. Introductory remarks, Symposium on kinetics of propellants. *J. Phys. and Colloid Chem. 54*, 847 (1950).
30. Crawford, B. L., Jr., Huggett, C., and McBrady, J. J. The mechanism of burning of double base propellants. *J. Phys. and Colloid Chem. 54*, 854 (1950).
31. Crawford, B. L., Jr., Huggett, C., Daniels, F., and Wilfong, R. E. Direct measurement of burning rates by an electric timing method. *Anal. Chem. 19*, 630 (1947).
32. Thompson, R. J., and McClure, F. T. Erosive burning of double base powders. *Office Sci. Research and Develop. Rept. 5831*, Dec. 1945.
33. Wilfong, R. E., Penner, S. S., and Daniels, F. An hypothesis for propellant burning. *J. Phys. and Colloid Chem. 54*, 863 (1950).
34. Klein, R., Mentser, M., von Elbe, G., and Lewis, B. Determination of the thermal structure of a combustion wave by fine thermocouples. *J. Phys. and Colloid Chem. 54*, 877 (1950).
35. Rice, O. K., and Ginell, R. The theory of burning of double-base rocket powders. *J. Phys. and Colloid Chem. 54*, 885 (1950).
36. Parr, R. G., and Crawford, B. L., Jr. A physical theory of burning of double-base rocket propellants. Part 1. *J. Phys. and Colloid Chem. 54*, 929 (1950).
37. Boys, S. F., and Corner, J. Unpublished work.
38. Green, L., Jr. Erosive burning of some solid composite propellants. *J. Am. Rocket Soc. 24*, 9 (1954).
39. Muracour, H. Sur les lois de combustion des poudres colloidales. *Bull. soc. chim. France 41*, 19–51 (1927).
40. Touchard, M. L. Sur les anomalies de vivacité des poudres tubulaires. *Mém. artillerie franç. 26*, 297 (1952).
41. Longwell, P. A., Sage, B. H., and Lacy, W. N. The temperature of spontaneous ignition of several samples of American ballistite. *Natl. Defense Research Comm., Div. A, Sec. H, Calif. Inst. Technol. JDC 8*, Apr. 1942.
42. *Jet Propulsion*. Prepared by Staffs of Jet Propul. Lab. and Guggenheim Lab., Calif. Inst. Technol., for Air Tech. Service Command, 1946.
43. Shapiro, A. H., and Hawthorne, W. R. The mechanics and thermodynamics of steady one-dimensional gas flow. *J. Appl. Mech. 14*, A317–A336 (1947).
44. Geckler, R. D., and Sprenger, D. F. The correlation of interior ballistic data for solid propellants. *J. Am. Rocket Soc. 24*, 22 (1954).

45. von Kármán, Th., and Malina, F. J. Characteristics of the ideal solid propellant rocket motor. *Calif. Inst. Technol. Jet Propul. Lab. Rept. 1–4*, 1940.
46. Mills, M. M. The preparation and some properties of an asphalt base solid propellant Galcit 61-C. *Calif. Inst. Technol. Galcit Project 1, Rept. 22*, May 1944.
47. Vand, V. *J. Phys. and Colloid Chem. 52*, 277, 300 (1948).
48. Dalla Valle, J. M. *Micromeritics*. Pitman, 1943.
49. Furnas, C. C. *Ind. Eng. Chem. 23*, 1502 (1931).
50. Mooney, M. *J. Colloid Sci. 6*, 162 (1951).
51. Steinour, H. H. *Ind. Eng. Chem. 36*, 618, 840, 901 (1944).
52. Hawksley, P. G. W. *Nature 164*, 220 (1949).
53. Processing of rocket propellants. *Calif. Inst. Technol. Office Sci. Research and Develop. Rept. 2552*, 1946.
54. McClure, F. T., Rosser, J. B., and Kincaid, J. F. The drag of the propellant gases on the power charge in rockets. *Office Sci. Research and Develop. Rept. 5872*, Feb. 1946.
55. Altman, D. The thermal diffusivities of some solid propellants. *Calif. Inst Technol. Propul. Lab. Progress Rept. 9-32*, Apr. 1949.
56. Thackwell, H. L. Unpublished data. *Calif. Inst. Technol. Jet Propul. Lab.*
57. Avery, W. H., and Beek, J., Jr. Propellant charge design of solid fuel rockets. *Office Sci. Research and Develop. Rept. 5890*, June 1946.
58. Shafer, J. I. Unpublished data. *Calif. Inst. Technol. Jet Propul. Lab.*
59. Hall, K. P. Unpublished data. *Calif. Inst. Technol. Jet Propul. Lab.*
60. Perry, R. L., and Berggren, W. P. Transient heat conduction in hollow cylinders after sudden change of inner-surface temperatures. *Univ. Calif. Publs. in Eng.*, 1944.
61. Tsien, H. S., and Cheng, C. M. A similarity law for stressing rapidly heated thin-walled cylinders. *J. Am. Rocket Soc. 22*, 144 (1952).
62. Mills, M. M. A study of materials for jet motor exhaust nozzles. *Air Corp. Calif. Inst. Technol. Jet Propul. Research, Galcit. Project 1, Rept. 18*, Aug. 1943.
63. Frazer, J. H., and Hicks, B. L. Thermal theory of ignition of solid propellants. *J. Phys. and Colloid Chem. 54*, 872 (1950).

SECTION I

THE RAMROCKET

IRVIN GLASSMAN
JOSEPH V. CHARYK

I,1. Introduction.[1] Conceptually, the principle underlying the ramrocket or ducted rocket scheme of propulsion is rather simple. In its basic form the exhaust from a rocket motor is used as the source of fuel for air entering the combustion chamber due to the ram effect and the inducting influence of the high velocity rocket exhaust jet. Comparing such a device with a pure ramjet and assuming identical conditions for the air at the inlet it is clear that, if the energy release associated with the combustion in the burner is the same in the two cases, the ramrocket is superior in that the mixing process results in a higher pressure at the end of the burner than in the case of the ramjet. Since the total temperature is the same, the exit momentum and hence the thrust are therefore greater for

[1] Editor's note: In these days of specialization, items are designed and tailored for a particular purpose. If the demand for such items is large and improvements in performance are great, such specializations are well worth the additional cost and development trouble. Power plants are no exceptions. Not only have power plants of different sizes and basic types been designed and built, but many hybrids of these types have been studied to show areas of application in which they yield an improved performance. There are many possible hybrid engines. By a hybrid engine, we mean here an engine which is part rocket and part air flow engine, or part jet and part rotary machine, propelled either by reaction or impulse.

Three of the most attractive engines are ramrocket, jet rotor, and air turborocket. Sections of this volume are devoted to the first two of these listed. A brief description of the latter is given here.

A turborocket, or more precisely, an air turborocket, is a modified turbojet in which the turbine which powers the compressor is driven by the exhaust products of a rocket motor. The turbine may be driven by the jet produced from the decomposition of a monopropellent, or by partial burning of a bipropellant fuel in which the temperature is maintained within an acceptable limit by fuel-rich burning. In either case the residual parts of the rocket fuels are then burned by combining them with the oxygen in the air which passes through the compressor. Additional fuel may also be burned in the air stream, where the whole combustion system aft of the turbine is similar to an afterburner. This engine is superior to a turbojet in that the combustion efficiency can be maintained to very high altitudes, the turbine power is practically independent of altitude and flight speed, and the operator is no longer subjected to the temperature limitation introduced by high flight velocities. As the altitude increases, the decrease in the thrust from the air flow is partially balanced by an increase in rocket thrust. Thus, the air turborocket is a lightweight power plant, with characteristics intermediate between the turbojet and the rocket, suitable for propulsion at higher altitudes and greater Mach numbers than a pure turbojet.

the ramrocket with the same consumption of fuel and air than for the ramjet. However, the difference is dependent on the per cent effect on the exit velocity, due to the mixing process occurring in the case of the ramrocket. The difference is obviously much more significant at relatively low speeds, where the ram effect is smaller, than at supersonic speeds. As a matter of fact, the performance of the two power plants becomes essentially identical on an ideal basis at sufficiently high forward speeds and air-fuel ratios. The situation will of course be altered if the energy release in the burner is different in the two cases and if losses and realistic combustion efficiencies are considered. Offhand it would be suspected that the ramrocket system would require longer burners and hence result in more significant frictional losses than in the case of the ramjet. This might be countered by lower burner drag because of elimination of flame holders and baffles and better combustion efficiencies over a wider range of fuel-air ratios, since the fuel is introduced in a very simple form and at elevated temperatures. The choice of propellant for the ramrocket is obviously rather limited, since one asks for a respectable specific impulse as a rocket coupled with an energy release in the afterburning process equivalent to that of conventional ramjet fuels. If the latter condition does not prevail, the ramrocket performance, of course, becomes inferior to that of the ramjet at some flight Mach number. On the other hand, perhaps one of the more attractive features of the ramrocket lies in its inherent flexibility. Since the power plant straddles the propulsion spectrum between a rocket and a ramjet it can develop a very high thrust-per-unit area as compared to a pure ramjet, particularly at low speeds, and, on the other hand, can operate at low specific fuel consumption if this is the more important criteria, however with a reduced thrust-per-unit area.

Many more complex versions of the general rocket decomposition and subsequent afterburning scheme have been proposed and studied. In all these devices, however, certain basic problems are present and the answers to these bear directly on the potential scope of usage of this class of power plants. These problems include special fuel selection, the mechanism of mixing of the rocket exhaust products with secondary air; the attainment of good combustion efficiencies over a wide range of fuel-air ratios and burner inlet velocities; the mechanism of initiation of afterburning with the determination of limits of spontaneous ignition and the combustion initiation requirements at other burner operating conditions; burner lengths required for mixing and combustion; the effect of additives to fuel on the above characteristics including the case where operation is desired utilizing various fuel-oxidizer ratios in the rocket motor; factors determining the size of possible solid particles in the rocket exhaust, the control of such size and the effect of such size on the subsequent processes; and the theoretical determination of rocket performance in cases where a sizable portion of the mass of the rocket exhaust products may be in the solid phase. These are among some of the major questions involved in a detailed evaluation of such propulsion devices, not all of which have been answered as yet. However, it appears reasonable to first make a general, more or less idealized theoretical performance study of the characteristics of such a power plant and to compare it with similar idealized calculations for related power plants.

Before proceeding to a general discussion of the characteristics of the ramrocket and comparisons with other related power plants, it is essential, in order to appropriately evaluate the end result, to discuss the theoretical questions involved in the analyses, and to present limitations and possible extensions. Much of the work to be discussed here was carried out at Princeton University and Experiment, Inc. The work at Princeton was entirely under the sponsorship of Project Squid.

I,2. Theoretical Analysis of Combustor Processes. The combustion chamber of the ramrocket (Fig. I,2) involves the problem of mixing

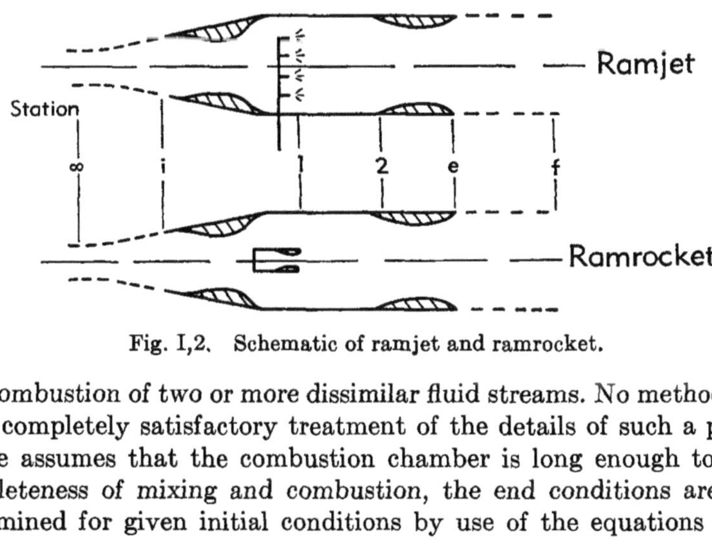

Fig. I,2. Schematic of ramjet and ramrocket.

and combustion of two or more dissimilar fluid streams. No method exists for a completely satisfactory treatment of the details of such a process. If one assumes that the combustion chamber is long enough to insure completeness of mixing and combustion, the end conditions are easily determined for given initial conditions by use of the equations for the conservation of mass, momentum, and energy. The normal one-dimensional scheme, however, is limited to the cases of constant area mixing and constant pressure integral over the internal combustion chamber surface known. However, a slight modification to the simplest form of the conservation relations following the procedure of Crocco (see III,B) permits the effect of the combustion chamber geometry to be assessed. It is also an improvement over the simpler method in that variations in the various quantities across the duct can be included. Thus, the effect on the thrust of the nonuniformity of the velocity and temperature profiles across the duct can be estimated.

The mass flow m, momentum I, and energy E of a gas stream of cross-sectional area A, passing a given section in unit time may be written (in the remainder of this article the convention with regard to the use of the specific gas constant is the same as in III,B):

$$\begin{aligned} m &= \int_A \rho V dA = \int_A dm \\ I &= \int_m V dm \\ E &= \int_m \left(c_p T + \frac{V^2}{2} \right) dm \end{aligned} \tag{2-1}$$

If mean quantities are defined as follows:

$$\begin{aligned} \bar{p}A &= \int_A p dA \qquad m\bar{V} = \int_m V dm \\ \bar{\rho} &= \frac{m}{\bar{V}A} \qquad \overline{\mathcal{R}T} = \frac{\bar{p}}{\bar{\rho}} = \frac{\bar{p}\bar{V}A}{m} \\ m\bar{\mathcal{R}} &= \int_m \mathcal{R} dm \qquad m\bar{c}_p = \int_m c_p dm \\ \frac{\bar{\gamma}}{\bar{\gamma}-1} &= \frac{\bar{c}_p}{\bar{\mathcal{R}}} \qquad \bar{M} = \frac{\bar{V}}{\sqrt{\gamma\overline{\mathcal{R}T}}} \end{aligned} \tag{2-2}$$

the integral forms are eliminated to yield

$$\begin{aligned} m &= \bar{\rho}\bar{V}A = \bar{p}A\,\frac{\sqrt{\bar{\gamma}}\,\bar{M}}{\sqrt{\overline{\mathcal{R}T}}} \\ I &= m\bar{V} = m\sqrt{\overline{\mathcal{R}T}}\,\sqrt{\bar{\gamma}}\,\bar{M} \\ E &= m\,\frac{\bar{\gamma}}{\bar{\gamma}-1}\,\overline{\mathcal{R}T}\left[\eta + \xi\,\frac{\bar{\gamma}-1}{2}\,\bar{M}^2\right] \end{aligned} \tag{2-3}$$

where

$$\begin{aligned} \xi &= \frac{m\overline{V^2}}{(m\bar{V})^2} = \frac{m\int_m V^2 dm}{\left(\int_m V dm\right)^2} \\ \eta &= \frac{\int c_p T dm}{m\bar{c}_p T} = \frac{\bar{\gamma}-1}{\bar{\gamma}}\,\frac{m\int_m c_p T dm}{\int_m V dm \int_m (\mathcal{R}T/V) dm} \end{aligned} \tag{2-4}$$

It is noted that if the velocity is assumed constant across the section, $\xi = 1$, and if $\mathcal{R}$ and c_p are also constant, $\eta = 1$.

The equations expressing the conservation of mass, momentum, and energy between the initial and final stations may now be written:

$$\begin{aligned} m_i &= m_f \\ I_f - I_i &= \bar{p}_i A_i - \bar{p}_f A_f + \int_i^f p_w dA_w \\ E_f - E_i &= Q \end{aligned} \tag{2-5}$$

where Q is the energy released and p_w is the normal pressure on duct walls.

The last term in the second of Eq. 2-5 presents any general treatment of the combustor flow problem. However, the influence of combustor geometry can be assessed if it is assumed that

$$\frac{p_w}{p_i} = \left(\frac{A_w}{A_i}\right)^{\frac{\epsilon}{1-\epsilon}} \tag{2-6}$$

and hence

$$\int_i^f p_w dA = (1 - \epsilon)[\bar{p}_f A_f - \bar{p}_i A_i] \tag{2-7}$$

It is readily seen that if $\epsilon = 0$, the case of constant pressure mixing results and if $\epsilon = 1$, the case of constant area mixing is obtained.

Utilizing the concept of stream thrust,

$$J = I + \epsilon \bar{p} A$$

or

$$J = \bar{p} A[\epsilon + \overline{\gamma M^2}] \tag{2-8}$$

the conservation relations simplify to:

$$\begin{aligned} m_i &= m_f \\ J_i &= J_f \\ E_i + Q &= E_f \end{aligned} \tag{2-9}$$

The effects of wall friction can also be included in the above formulation very simply if a special form for the friction law is used. Specifically, if we take

$$\int_i^f f dA_s = \frac{k}{2} m_f V_f \tag{2-10}$$

i.e. a friction loss proportioned to the exit momentum, the momentum equation can still be written as

$$J_i = J_f$$

where the stream thrust J, however, is redefined as

$$\begin{aligned} J_i &= m_i \sqrt{\overline{\mathcal{R}_i T_i}} \left[\sqrt{\bar{\gamma}_i}\, \overline{M}_i + \frac{\epsilon}{\sqrt{\bar{\gamma}_i}\, \overline{M}_i} \right] \\ J_f &= m_f \sqrt{\mathcal{R}_f T_f} \left[\sqrt{\gamma_f}\, M_f \left(1 + \frac{k}{2}\right) + \frac{\epsilon}{\sqrt{\gamma_f}\, M_f} \right] \end{aligned} \tag{2-11}$$

Eq. 2-9 permits a determination of M_f and p_f for given initial conditions and a value for ϵ. Solutions, however, do not always exist. The conditions for existence can best be expressed in terms of a function H, defined as

$$H = \frac{J^2}{mE}$$

$$= \frac{\gamma - 1}{\gamma} \frac{(\epsilon + \gamma M^2)^2}{\gamma M^2 \left[\eta + \xi \frac{\gamma - 1}{\gamma} M^2\right]} \tag{2-12}$$

In order to have real solutions,

$$H_f \geqq [H_{\min}]_f$$

where the minimum value of H corresponds to a certain critical Mach number, M_{cr}

$$M_{cr} = \frac{1}{\left[\frac{\gamma}{\epsilon} - \frac{\xi}{\eta}(\gamma - 1)\right]^{\frac{1}{2}}} \tag{2-13}$$

If a real solution exists, it may either be single-valued or double-valued. The former situation prevails if

$$\overline{M}_i < M_{cr}$$

Otherwise, two solutions exist, one supersonic and one subsonic. Whether a supersonic value for M_f can be realized in practice or not has not been established as yet. It is obvious, however, that if this should be possible, the potentialities of such a power plant on a thrust-per-unit frontal area basis would be tremendously enhanced over the normal cases treated in all present calculations, where the subsonic solution is invariably selected. There is some limited evidence available to suggest that the supersonic case is by no means impossible, and this is definitely an area in which research might fruitfully be directed.

With the relations available, conditions at the end of the ramrocket burner can be expressed simply in terms of the initial conditions. The peculiar characteristics of this type of power plant can be assessed nicely by reference to such plotted results.

It is obvious, however, that the above techniques cannot predict burner lengths required for achievement of a given percentage of the total energy release possible. The analytical treatment in detail of the combined mixing and combustion process is beyond the scope of our present knowledge. Experimental studies may give important guidance to analytical efforts to treat this complicated problem.

I,3. Combustor Performance Calculations. As already pointed out, if the energy release associated with the burning process in the combustion chamber is the same in the case of the ramrocket as in the ramjet, the margin of difference between the two power plants is directly related to the difference in total pressure at burner exit. This defines the exit momentum and hence the specific impulse. It is convenient, therefore, to make comparisons based on the total pressure change across the burner.

This total pressure change is obtained readily from the one-dimensional continuity and momentum equations. The one-dimensional momentum equation is given by

$$(pA + \rho V^2 A)_1 + (pA + \rho V^2 A)_r = (pA + \rho V^2 A)_2 \tag{3-1}$$

or

$$[pA(1 + \gamma M^2)]_1 + [pA(1 + \gamma M^2)]_r = [pA(1 + \gamma M^2)]_2 \tag{3-2}$$

and the continuity equation by

$$(\rho VA)_1 + (\rho VA)_r = (\rho VA)_2 \tag{3-3}$$

or

$$\left(\frac{pAM\sqrt{g\gamma \mathcal{R}T}}{\mathcal{R}T}\right)_1 + \left(\frac{pAM\sqrt{g\gamma \mathcal{R}T}}{\mathcal{R}T}\right)_r = \left(\frac{pAM\sqrt{g\gamma \mathcal{R}T}}{\mathcal{R}T}\right)_2 \tag{3-4}$$

where the subscript $_r$ refers to conditions at the rocket.

Continuity at any individual station is expressed by

$$m_x = \left(\frac{pAM\sqrt{g\gamma}}{\sqrt{\mathcal{R}T}}\right)_x \tag{3-5}$$

Eliminating the area terms in Eq. 3-2 by the expression gives

$$\left[\frac{(1 + \gamma M^2)m\sqrt{\mathcal{R}T}}{M\sqrt{g\gamma}}\right]_1 + \left[\frac{(1 + \gamma M^2)m\sqrt{\mathcal{R}T}}{M\sqrt{g\gamma}}\right]_r = \left[\frac{(1 + \gamma M^2)m\sqrt{\mathcal{R}T}}{M\sqrt{g\gamma}}\right]_2 \tag{3-6}$$

Now, introducing the isentropic relation between the static and stagnation temperatures, the parameter R^* and the Mach number function $N(M)$, i.e.

$$\frac{T^0}{T} = 1 + \frac{\gamma - 1}{2}M^2 \quad R^* = \frac{m_1}{m_r}$$

$$N(M) = M\left(1 + \frac{\gamma - 1}{2}M^2\right)^{\frac{1}{2}}(1 + \gamma M^2)^{-1}$$

one obtains

$$\sqrt{\gamma_2}\,N(M_2) = \frac{\sqrt{\mathcal{R}_2 T_2^0/\mathcal{R}_1 T_1^0}\,\sqrt{\gamma_1}N(M_1)\,\sqrt{\gamma_r}N(M_r)(R^* + 1)}{\sqrt{\gamma_r}N(M_r)R^* + \sqrt{\gamma_1}N(M_1)\,\sqrt{\mathcal{R}_r T_r^0/\mathcal{R}_1 T_1^0}} \tag{3-7}$$

Only the subsonic solution for $N(M_2)$ is considered in this section although the supersonic solution may be possible in certain cases, as explained in the previous section.

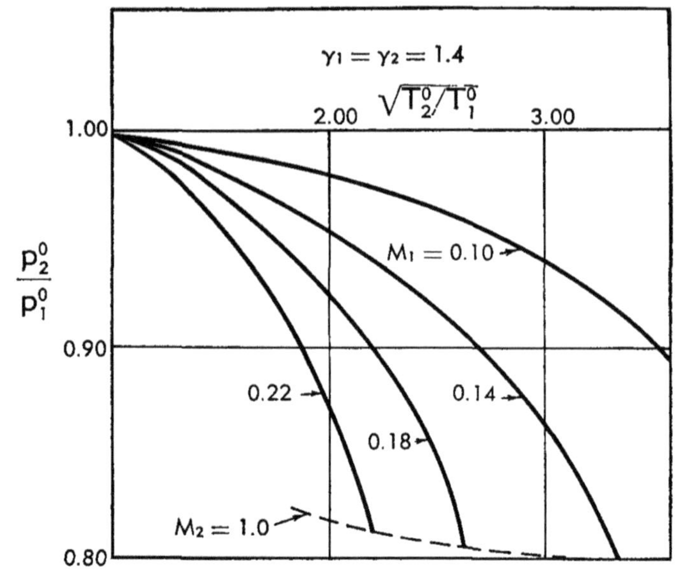

Fig. I,3a. Ramjet burner characteristics.

Thus, having determined M_2, it is now necessary to determine the burner characteristic of interest, p_2^0/p_1^0. Substituting M_2 into the expression for continuity given earlier,

$$\frac{A_1}{A_r} = \frac{\left(\frac{m\sqrt{\mathcal{R}T}}{Mp\sqrt{g\gamma}}\right)_1}{\left(\frac{m\sqrt{\mathcal{R}T}}{Mp\sqrt{g\gamma}}\right)_2} \quad \frac{A_2}{A_r} = \frac{\left(\frac{m\sqrt{\mathcal{R}T}}{Mp\sqrt{g\gamma}}\right)_2}{\left(\frac{m\sqrt{\mathcal{R}T}}{Mp\sqrt{g\gamma}}\right)_r} \tag{3-8}$$

Introducing the constant area condition

$$\frac{A_1}{A_r} + 1 = \frac{A_2}{A_r} \tag{3-9}$$

Thus, from Eq. 3-8 and 3-9 and further assuming that the rocket is perfectly expanded, $p_1 = p_r$, one obtains

$$\frac{p_2}{p_r} = \frac{p_2}{p_1} = \frac{(R^* + 1)\sqrt{\gamma_1/\gamma_2}\sqrt{\mathcal{R}_2T_2/\mathcal{R}_rT_r}(M_r/M_2)}{R^+\sqrt{\gamma_r/\gamma_1}\sqrt{\mathcal{R}_1T_1/\mathcal{R}_rT_r}(M_r/M_1) + 1} \tag{3-10}$$

Using the Mach number functions

$$A(M) = M\left(1 + \frac{\gamma - 1}{2}M^2\right)^{\frac{1}{2}}$$

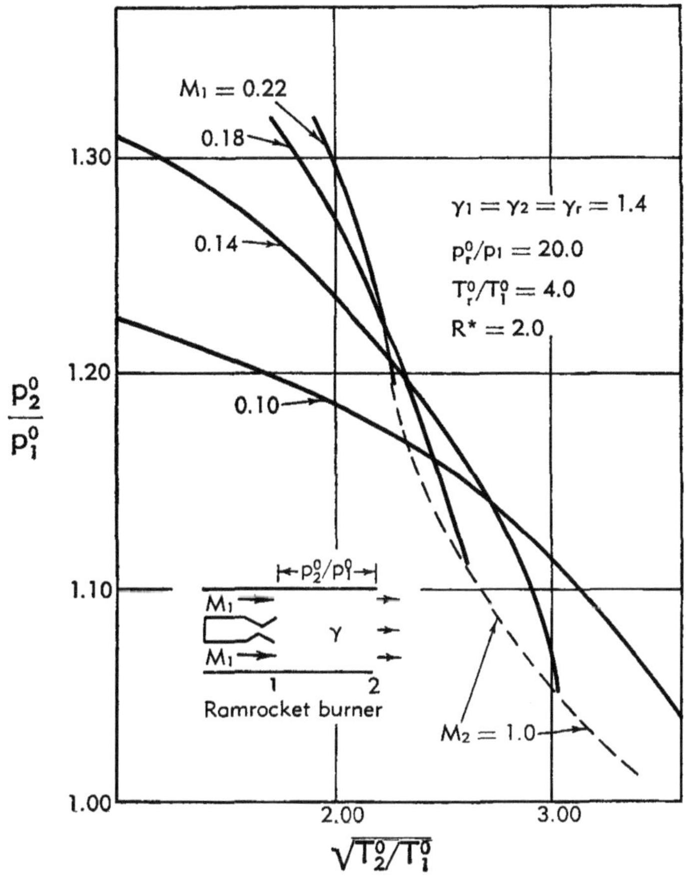

Fig. I,3b. Ramrocket burner characteristics.

and

$$D(M) = \frac{M\left(1 + \frac{\gamma - 1}{2} M^2\right)^{\frac{1}{2}}}{\left(1 + \frac{\gamma - 1}{2} M^2\right)^{\gamma/\gamma-1}}$$

and the isentropic relation

$$\frac{p^0}{p} = \left(1 + \frac{\gamma - 1}{2} M^2\right)^{\gamma/\gamma-1}$$

the burner characteristic becomes

$$\frac{p_2^0}{p_1^0} = \left\{\frac{[A(M_r)/D(M_r)]\sqrt{\mathcal{R}_2 T_2^0/\mathcal{R}_r T_r^0}\sqrt{\gamma_r/\gamma_2}(R^* + 1)}{[A(M_2)/D(M_1)]\sqrt{\mathcal{R}_1 T_1^0/\mathcal{R}_r T_r^0}\sqrt{\gamma_r/\gamma_1}(R^*) + 1}\right\} \tag{3-11}$$

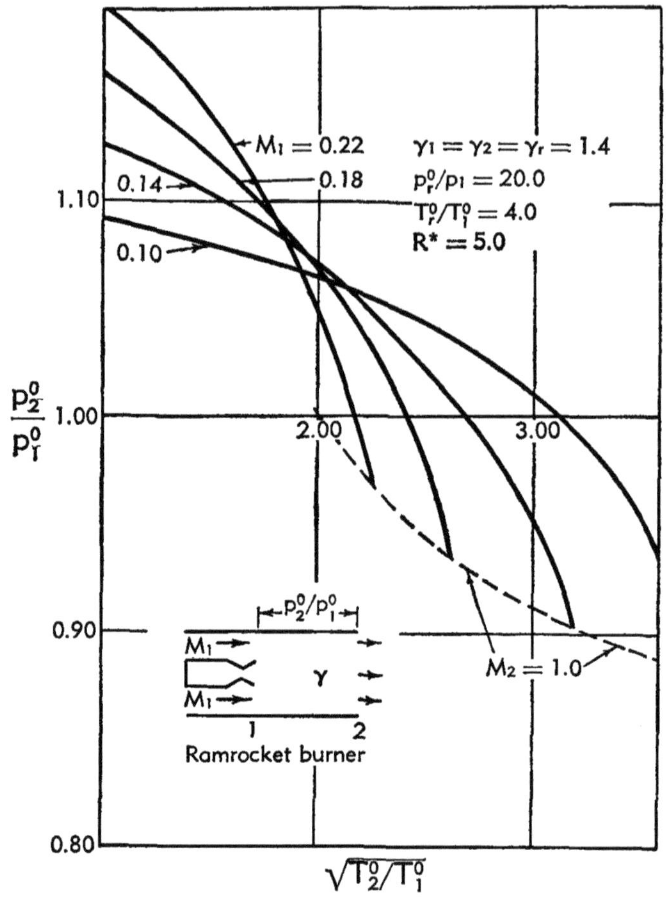

Fig. I,3c. Ramrocket burner characteristics.

In general then,

$$\frac{p_2^0}{p_1^0} = f\left(M_1, M_r, \frac{\mathcal{R}_2 T_2^0}{\mathcal{R}_1 T_1^0}, \frac{\mathcal{R}_r T_r^0}{R_1 T_1^0}, \gamma_1, \gamma_2, \gamma_r, \frac{m_a}{m_f}\right) \tag{3-12}$$

Actually, these quantities are not all independent since the choice of a propellant immediately establishes certain relationships. However, it is desirable to treat the problem completely generally at first. For illustrative purposes, the following values were selected arbitrarily.

$$\gamma_1 = \gamma_2 = \gamma_r = 1.4$$

$$M_r = 2.60; \quad \mathcal{R}_1 = \mathcal{R}_2 = \mathcal{R}_r$$

Hence

$$\frac{p_2^0}{p_1^0} = f\left(M_1, \frac{T_2^0}{T_1^0}, \frac{T_r^0}{T_1^0}, \frac{m_a}{m_f}\right) \tag{3-13}$$

In the limiting case of a ramjet, the air-fuel ratio determines T_2^0/T_1^0 and T_r^0/T_1^0 is not a parameter. Hence the total pressure change across the burner can be very simply represented as shown in Fig. I,3a. These curves represent the case of combustion in a constant area duct without mixing. It can be considered as the limit reached when the effect of the primary or rocket stream becomes negligible compared to the secondary stream.

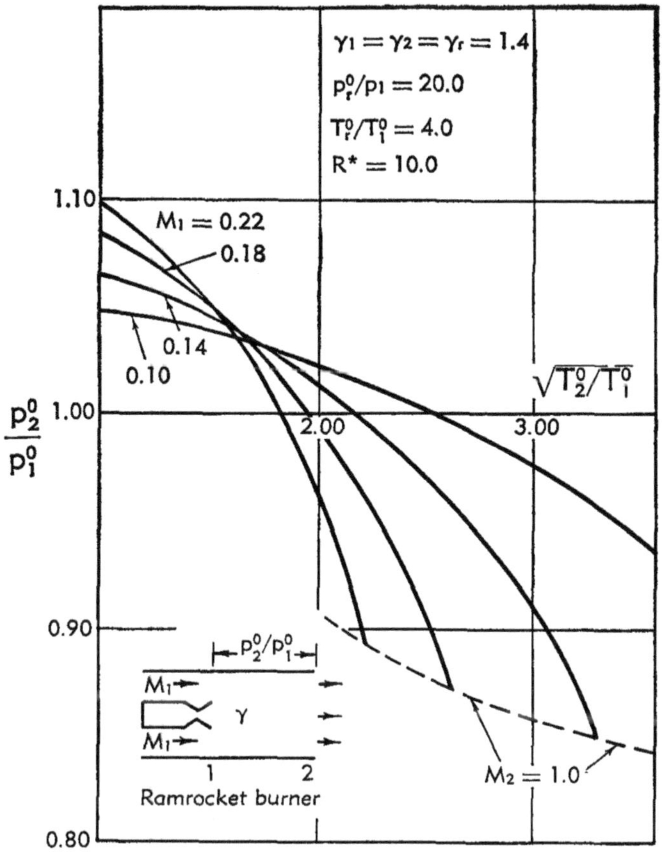

Fig. I,3d. Ramrocket burner characteristics.

It is noted, as discussed previously, that an increase in the ratio T_2^0/T_1^0 results in a loss of pressure or decrease in pressure ratio across the burner.

Fig. I,3b, I,3c, I,3d, I,3e, I,3f, and I,3g present the pressure ratio across the ramrocket burner again as a function of the combustion chamber inlet Mach number and the over-all temperature ratio; where the rocket temperature ratio T_r^0/T_1^0 and the air-fuel ratio, $m_a/m_f(\equiv R^*)$ are also included. The rocket temperature ratio is a measure of the velocity or momentum of the primary stream while the air-fuel ratio

m_a/m_f can be considered as a measure of the mixing effect of the rocket stream.

In Fig. I,3b, I,3c, I,3d, and I,3e T_r^0/T_1^0 is held constant at 4.0. For an inlet air temperature T_1^0 of 540°F this corresponds to a rocket flame temperature of about 2160°F or a rocket specific impulse of about 150 sec. Comparing Fig. I,3b with the plot of p_2^0/p_1^0 for the ramjet burner it is

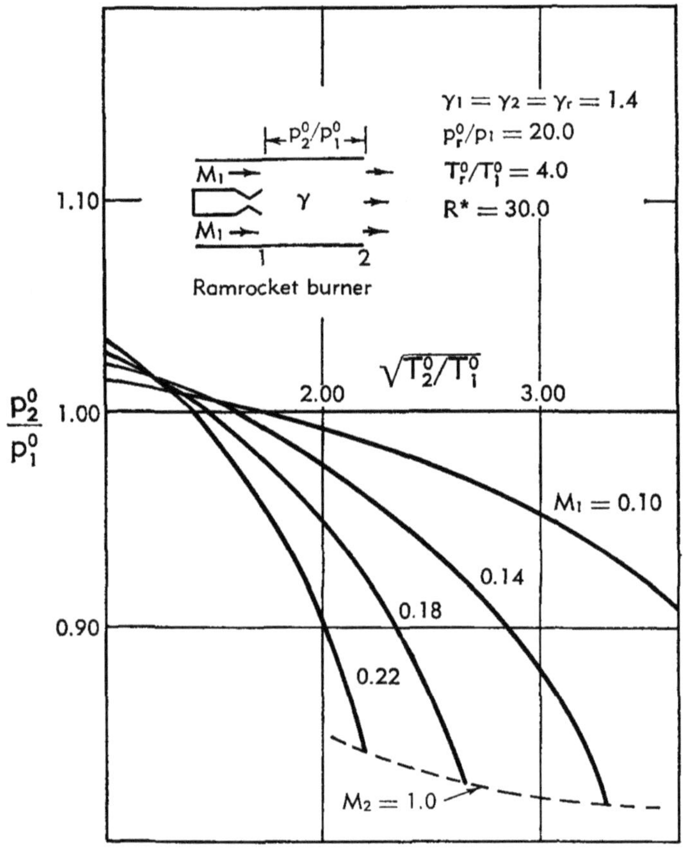

Fig. I,3e. Ramrocket burner characteristics.

noted that for a fixed value of the temperature ratio T_2^0/T_1^0, other factors remaining the same, the value of p_2^0/p_1^0 for the ramrocket is greater than that for the ramjet. Further, as one increases the air-fuel ratio, other factors remaining the same, this pressure rise due to mixing decreases. As m_a/m_f approaches large values the mixing effect tends to disappear and the data becomes identical to that for the case of the ramjet. For example, at an air-fuel ratio of 30 (see Fig. I,3e) the plots of the ramjet and ramrocket combustion chambers are nearly identical.

In Fig. I,3b and I,3g the rocket temperature ratio has been increased to 9.0 or a rocket specific impulse of approximately 225 seconds. The same trends are indicated as in the previous case, namely the decrease in mixing effect with increasing air-fuel ratio. However, since the specific impulse or momentum of the primary stream has increased, the over-all mixing effect has increased.

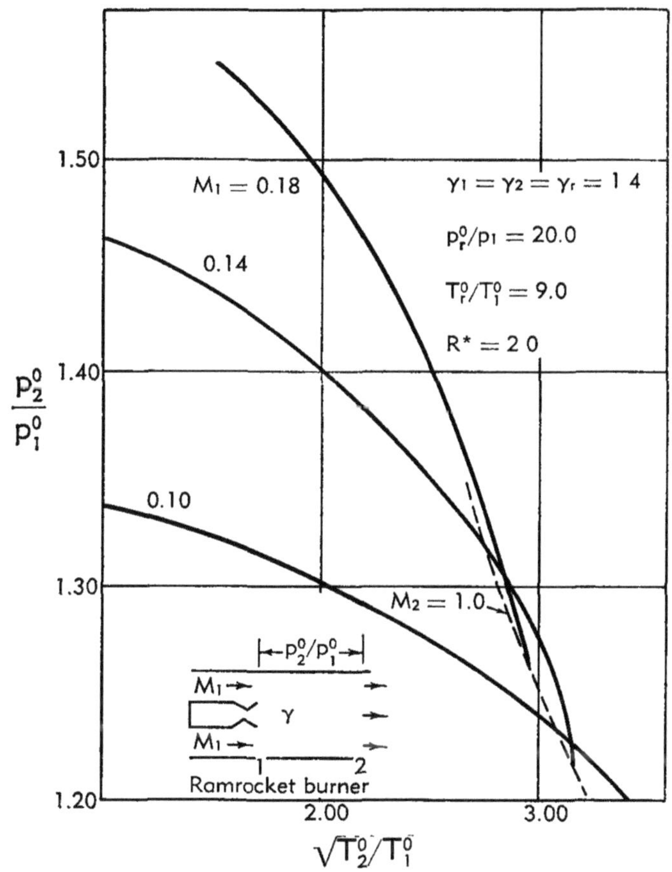

Fig. I,3f. Ramrocket burner characteristics.

The curves represent a few of a large series and readily illustrate some of the conditions under which ramrocket power plants are most attractive. It is clear that at high air-fuel ratios there is little or no difference between the ramrocket and the ramjet for the same energy release as far as specific fuel consumption is concerned. On the other hand, much smaller combustion chamber diameters could be utilized to accelerate up to such a design condition. In other words, if acceleration demands are placed on a ramjet, the required engine diameter could exceed substantially that demanded by the ramrocket. In the former case, therefore, a continuing

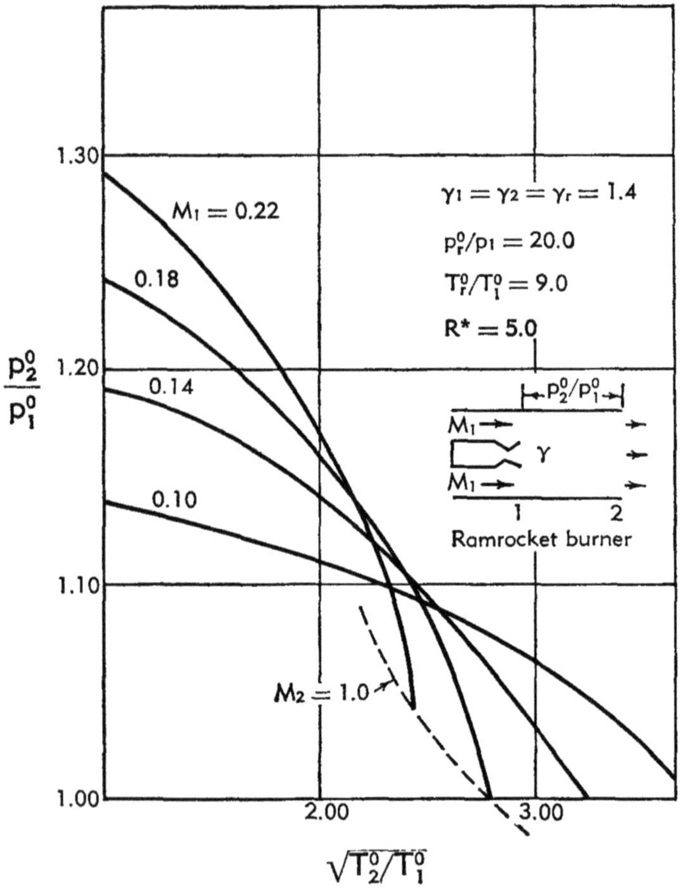

Fig. I,3g. Ramrocket burner characteristics.

and serious drag penalty at the cruise condition would have to be paid as a premium in the case of the ramjet, but not in the case of the ramrocket.

I,4. Fixed and Variable Configuration. In the determination of the thrust/cross-section area and specific impulse, two methods of attack are frequently used dependent upon the purpose of the particular calculation. One case may be considered as that of a variable configuration and the other that of a fixed configuration.

The former approach is more useful when one seeks to evaluate the optimum performance potentialities of a power plant. This appears to be the most desirable course if one is seeking to assess the merits or demerits of a particular scheme of propulsion.

In the case of the variable configuration the independent variable is a fixed quantity of working fluid. The fluid is passed through the power plant cycle, the individual processes taking place with some predetermined efficiency, and the performance is determined. The configuration is then designed to admit the given quantity of working fluid and to meet the required efficiencies. This system has its intrinsic advantages and disadvantages, its primary merit resting in the fact that it presents a method of comparison of the optimum performance of two distinct systems. The main disadvantage in preselecting this air flow and efficiencies is that the performance results can only be applied to a configuration which continually changes its shape. Therefore, performance determined assuming a variable configuration can best be used on a relative rather than an absolute basis.

In the case of the fixed configuration we have just the opposite situation. Here the independent variable is the configuration and the dependent variable is the amount of air passing through the system. In determining the performance for a preselected configuration the situation is then just the opposite from that above. In other words, this method of analysis leads to values of absolute performance over a wide range of operating conditions; however, it does not lend itself to making comparisons in that there is no way, except by trial and error, to determine how a particular configuration compares with the optimum.

Variable Configuration. For purposes of determining the performance of a "variable" configuration it is convenient to use the burner performance curves presented in a previous article and exemplified by Fig. I,3b, I,3c, I,3d, I,3e, I,3f, and I,3g. Prior to showing this analysis the two performance parameters, specific fuel consumption and thrust per cross-sectional, generally used to describe a power plant, are defined. The thrust per unit mass flow of fuel is given by

$$F = \frac{(m_a + m_f)V_e - m_a V_\infty}{m_f g} \tag{4-1}$$

or

$$F = \frac{(R^* + 1)V_e - R^* V_\infty}{g} \tag{4-2}$$

In terms of the stagnation conditions and Mach number, Eq. 4-2 may be written as

$$F = \frac{1}{g}\left\{\left[\frac{(R^* + 1)(M\sqrt{g\gamma\mathcal{R}T^0})}{\left(1 + \frac{\gamma - 1}{2}M^2\right)^{\frac{1}{2}}}\right]_e - \left[\frac{R^* M\sqrt{g\gamma\mathcal{R}T^0}}{\left(1 + \frac{\gamma - 1}{2}M^2\right)^{\frac{1}{2}}}\right]_\infty\right\} \tag{4-3}$$

The specific fuel consumption is then given by

$$\text{sfc} = \frac{3600}{\left[\frac{(R^* + 1)M\sqrt{g\gamma\mathcal{R}T^0}}{\left(1 + \frac{\gamma - 1}{2}M^2\right)^{\frac{1}{2}}}\right]_e - \left[\frac{R^* M\sqrt{g\gamma\mathcal{R}T^0}}{\left(1 + \frac{\gamma - 1}{2}M^2\right)^{\frac{1}{2}}}\right]_\infty} \tag{4-4}$$

The area required for a pound of fuel per second can be shown readily to be

$$A_e = \frac{(R^* + 1)(\sqrt{\mathcal{R}T})_e}{(pM\sqrt{g\gamma})_e} \tag{4-5}$$

Therefore, the area for each pound of thrust is

$$\frac{A_e}{F} = \frac{A_e \text{sfc}}{3600} \tag{4-6}$$

or

$$\frac{F}{A_e} = \frac{\left[pM\left(1 + \frac{\gamma - 1}{2}M^2\right)^{\frac{1}{2}}\sqrt{g\gamma}\right]_e 3600}{(\sqrt{\mathcal{R}T^0})_e \text{sfc}(R^* + 1)} \tag{4-7}$$

Now the previously mentioned analysis proceeds in the following fashion. The designer chooses values of the free flight Mach number M_∞, the altitude p_∞, combustion chamber inlet Mach number M_1, and the air-fuel ratio m_a/m_f ($\equiv R^*$). From the value of the free flight Mach number (M_∞) and an assumed value of the polytropic diffuser efficiency, it is possible to determine the stagnation pressure at the entrance to the combustion chamber (p_1^0). From the selected value of the air-fuel ratio, fuel type, and temperature of the air at the inlet to the combustion chamber (T_1^0) it is possible from purely energy consideration to determine the temperature at the end of the combustion process (T_2^0). Introducing this value into the burner curves it is possible to determine p_2^0/p_1^0 corresponding to the selected values of combustion chamber inlet Mach number and air-fuel ratio. Finally, assuming an isentropic or polytropic and correct expansion from the stagnation pressure p_2^0 to atmosphere it is possible to find the Mach number at station e, i.e.

$$\begin{aligned} \frac{p_e^0}{p_e} = \frac{p_e^0}{p_\infty} &= \frac{p_\infty^0}{p_\infty}\frac{p_1^0}{p_\infty^0}\frac{p_2^0}{p_1^0}\frac{p_e^0}{p_2^0} \\ &= \frac{p_\infty^0}{p_\infty}\eta_d\frac{p_2^0}{p_1^0}\eta_n \end{aligned} \tag{4-8}$$

and p_e^0/p_e gives M_e. Since $p_e = p_\infty$, then all parameters in Eq. 4-4 and 4-7 are known and the sfc and F/A can be calculated. For the case where M_∞ is supersonic and the shock is either frontal or swallowed, Eq. 4-8 is modified suitably to account for the stagnation pressure loss. Correction terms about the standard values stated in Fig. I,3b and I,3c have been calculated and can be used for more exact determinations.

The specific impulse and thrust per unit cross-sectional area are thus directly determined. These equations then give the optimum performance which can be expected from a ramrocket or ramjet power plant; as such the equations are of interest as the upper limit of the performance of an engine under the particular set of preselected operating conditions.

The method of analysis indicates quite clearly the pumping or mixing effect of the rocket. The velocity at the end of the exit nozzle, V_3, is a function of the stagnation pressure p_2^0, and the stagnation temperature T_2^0. Therefore, assuming both units are burning identical fuels under identical operating conditions, the ramrocket produces an improved performance over that of the ramjet in that the stagnation pressure at

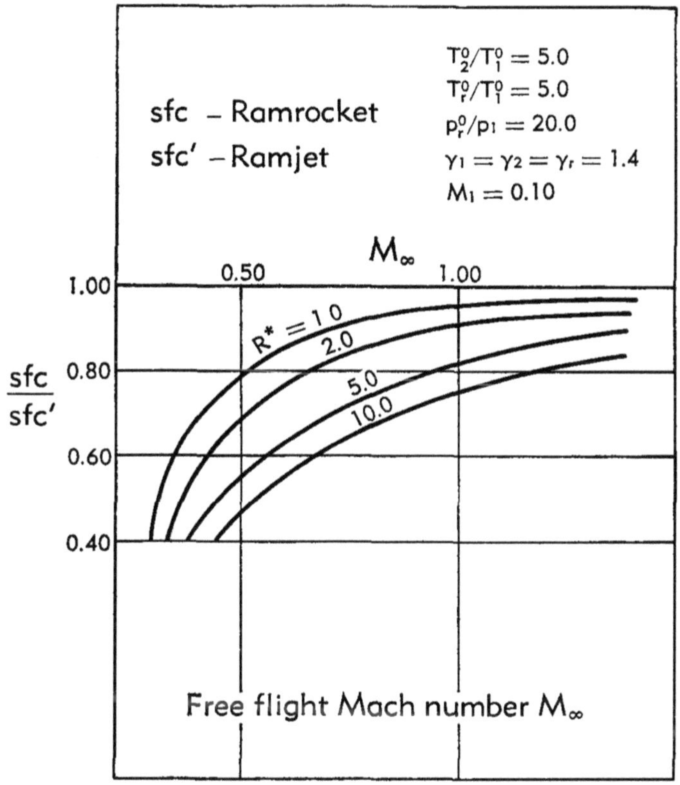

Fig. I,4a. Comparison of ideal specific fuel consumption.

the end of the combustion chamber is higher than in the case of the ramrocket.

From practical considerations it is of interest to have an estimate of this improvement in performance. Calculations which illustrate at least semi-quantitative trends are indicated in Fig. I,4a. In Fig. I,4a are curves of the ideal ramrocket specific fuel consumption compared to those of the ramjet. For purposes of comparison it was assumed that the free flight Mach number and air-fuel ratio could be varied while the other quantities were held constant. It is noted that at constant m_a/m_f, as the free flight Mach number is increased, the specific fuel consumption of the ramjet and ramrocket approach each other. This corresponds to the fact that at the higher free flight Mach numbers the pressure increase through the combustion chamber due to the mixing action of the rocket becomes insignificant compared with the pressure change through the diffuser. Again referring to Fig. I,4a, as the air-fuel ratio is increased, at constant free flight Mach number, the specific fuel consumption of the ramrocket approaches that of the ramjet. Physically this is expected since, as discussed previously, with increasing air-fuel ratio the mixing effect of the rocket stream tends to decrease.

Fixed Configuration. In the determination of the performance of a ramrocket or ramjet with a fixed configuration, it is found convenient to group the parameters in terms of the stream thrust ($F = pA + \rho V^2 A$) rather than treat them individually at different stations through the duct. Whereas in the case of a flexible configuration the designer can choose the air-fuel ratio and combustion chamber inlet Mach number independently of each other, in the present situation they are both dependent variables. Here it is found most satisfactory to choose as the independent variable, A_2/m_f, where A_2 is any characteristic combustion chamber cross section and m_f is the fuel flow into the system.

This form of analysis to determine the performance parameters F/A and sfc, introduced by Rudnick [1], develops as follows. The stream thrust expressed in terms of the Mach number is

$$F = pA(1 + \gamma M^2) \tag{4-9}$$

Applying the continuity equation at station 2, one obtains

$$m_a + m_f = p_2 A_2 M_2 \left(1 + \frac{\gamma - 1}{2} M_2^2\right)^{\frac{1}{2}} \sqrt{\frac{g\gamma_2}{\mathcal{R}_2 T_2^0}} \tag{4-10}$$

or

$$p_2 A_2 = (m_a + m_f) \sqrt{\frac{\mathcal{R}_2 T_2^0}{g\gamma_2}} \frac{1}{M_2 \left(1 + \frac{\gamma - 1}{2} M_2^2\right)^{\frac{1}{2}}} \tag{4-11}$$

Substituting Eq. 4-11 into Eq. 4-9 one has

$$F_2 = \frac{m_a \left(1 + \frac{m_f}{m_a}\right)}{M_2 \left(1 + \frac{\gamma_2 - 1}{2} M_2^2\right)^{\frac{1}{2}}} \sqrt{\frac{\mathcal{R}_2 T_2^0}{g\gamma_2}} (1 + \gamma_2 M_2^2) \tag{4-12}$$

If $M_2 = 1$, then

$$F_2(M_2 = 1) = m_a \left(1 + \frac{m_f}{m_a}\right) \sqrt{\frac{2(\gamma_2 + 1)}{g\gamma_2} \mathcal{R}_2 T_2^0} \tag{4-13}$$

The air specific impulse is defined as

$$S_a = \left(1 + \frac{m_f}{m_a}\right) \sqrt{\frac{2(\gamma_2 + 1)}{g\gamma_2} \mathcal{R}_2 T_2^0} \tag{4-14}$$

and is seen to be dependent upon m_a/m_f alone, since for a given fuel-air combination m_a/m_f determines T^0, γ, and $\mathcal{R}$.

Comparing the expression for stream thrust at station 2, corresponding to choked flow and the definition of air specific impulse, it is noted that the product of the air specific impulse and the air flow m_a corresponds to the value of the stream thrust at station 2 for choked flow, i.e.

$$F_2(M_2 = 1) = m_a S_a \tag{4-15}$$

Now if one represents the condition of choked flow as F_2^*, the

$$\frac{F_2}{F_2^*} = \frac{1 + \gamma_2 M_2^2}{\left(1 + \frac{\gamma - 1}{2} M_2^2\right) M_2 \sqrt{2(\gamma_2 + 1)}} = \phi(M_2) \tag{4-16}$$

F_2/F_2^* or $\phi(M_2)$ are functions which are conveniently tabulated in [2,3]. Thus,

$$F_2 = F_2^* \phi(M_2) = m_a S_a \phi(M_2) \tag{4-17}$$

Therefore, if the Mach number at the burner exit, M_2, and the air-fuel ratio are known it is immediately possible to determine the stream thrust at station 2. Similarly, if M_2 and F_2 have been determined it is possible to find the air-fuel ratio or amount of air passing through the system.

It can be readily shown for a constant area duct that for the ram-rocket configuration

$$F_2 = F_1 + F_r \tag{4-18}$$

or, if there is a burner drag, more properly,

$$F_2 = F_1 + F_r - F_D \tag{4-19}$$

The stream thrust of the air stream can be determined from the expression

$$F_1 = \eta_d \frac{p_\infty^0}{p_\infty} p_\infty (1 - b) B_s(M_1) A_2 \tag{4-20}$$

where b is the blockage and $B_s(M) = (p/p^0)(1 + \gamma M^2)$. For convenience, any pressure loss across a shock can be incorporated in η_d. In general, it is found convenient to group any internal drag diffuser efficiency and the blockage term, etc., into a blockage efficiency η_b. Over most ranges of operation, $B_s(M)$ is always very close to 1, so that

$$F_1 = \eta_b p_\infty^0 A_2 \tag{4-21}$$

The rocket stream thrust is defined as

$$F_r p_r A_r (1 + \gamma M_r^2) = m_f I_{r_{abs}} \tag{4-22}$$

Assuming that the flow does not separate in the rocket nozzle, the absolute specific impulse is only a function of the rocket nozzle area ratio, chamber pressure, and stagnation temperature. Unlike the conventional impulse the absolute is not a function of external pressure. This distinction is indicated in the following equation

$$I_{std} = \frac{(M\sqrt{g\gamma \mathcal{R} T})_r + A_r(p_r - p_1)}{g} \tag{4-23}$$

$$I_{abs} = \frac{(M\sqrt{g\gamma \mathcal{R} T})_r + A_r p_r}{g} \tag{4-24}$$

Thus one sees that the free flight Mach number, blockage efficiency, and the rocket stream thrust determine F_2 also.

With this introduction of the stream thrust concept, one can proceed now to determine the performance parameters.

Introducing the rates of fuel flow to combustion chamber area m_f/A_2, it is possible to find the value of

$$\frac{F_2}{A_2} = p_2(1 + \gamma_2 M_2^2) = \eta_b p_\infty^0 + \frac{m_f}{A_2} I_r \tag{4-25}$$

Next, introducing related boundary conditions it is possible to determine M_2 and thus $\phi(M_2)$.

1. *Unchoked flow* ($M_2 < 1.0$). If the flow is unchoked p_2 is determined from the known exit condition p_e. Therefore, it is possible to find p_e/p_2 and thus M_2:

$$1 + \gamma_2 M_2^2 = \eta_b \frac{p_\infty^0}{p_\infty} \frac{p_\infty}{p_2} + \frac{m_f}{p_2 A_2} I_r \tag{4-26}$$

For a straight tailpipe, i.e. no nozzle, $p_2 = p_0 = p_e$ and

$$1 + \gamma_2 M_2^2 = \eta_b \frac{p_\infty^0}{p_\infty} + \frac{m_f}{p_\infty A_2} I_r \tag{4-27}$$

2. *Choked flow*. For choked flow $M_{nozzle} = 1$, the values of M_2 can be determined directly from the exhaust nozzle area ratio. In a straight tailpipe the choking condition occurs when

$$\eta_b \frac{p_\infty^0}{p_\infty} + \frac{m_f}{p_\infty A_2} I_r > 1 + \gamma_2 \tag{4-28}$$

With the value of M_2 determined above, F_2/A_2 from Eq. 4-25, and a selected value of the design parameter A_2/m_f, it is possible to find a value of $(m_a/m_f)S_a$ from Eq. 4-17

$$\frac{m_a}{m_f} S_a = \frac{F_2}{A_2}\left(\frac{A_2}{m_f}\right)\frac{1}{\phi(M_2)} \tag{4-29}$$

From a plot of $(m_a/m_f)S_a$ vs. m_a/m_f, a graph of an independent energy calculation, it is possible to find m_a/m_f.

Having determined the value of m_a/m_f (since A_2/m_f is given, the air flowing through the vehicle is essentially determined) as a function of m_f/A_2, the blockage efficiency η_b, and M_∞, the problem is one of determining the Mach number at the inlet M_i, and the area of the free stream tube at ambient conditions A_∞. M_i and A_∞ can be determined in the following manner.

The value of the mass flow in a stream tube is given by

$$m = \rho VA = p^0 MA \sqrt{\frac{g\gamma}{\mathcal{R}T^0}} \frac{\left(1 + \frac{\gamma - 1}{2} M^2\right)^{\frac{1}{2}}}{\left(1 + \frac{\gamma - 1}{2} M^2\right)^{\frac{\gamma}{\gamma - 1}}} \tag{4-30}$$

or

$$m = p^0 \sqrt{\frac{g\gamma}{\mathcal{R}T^0}} D(M)A \tag{4-31}$$

where

$$D(M) = \frac{\left(1 + \frac{\gamma - 1}{2} M^2\right)^{\frac{1}{2}} M}{\left(1 + \frac{\gamma - 1}{2} M^2\right)^{\frac{\gamma - 1}{\gamma}}} \tag{4-32}$$

It is found convenient to use as a parameter the quantity of air in a free stream tube whose area is equal to the physical inlet area of the unit under consideration. This parameter has been termed the "cookie cutter" mass flow, m_{acc}. By definition then

$$m_{acc} = p_\infty^0 \left(\sqrt{\frac{g\gamma}{\mathcal{R}T_0}}\right)_\infty D(M_\infty)A_i \tag{4-33}$$

or

$$\frac{m_{acc}}{m_f} = \frac{M_\infty p_\infty g\gamma_\infty}{a_\infty} \frac{A_i}{A_2} \frac{A_2}{m_f} \tag{4-34}$$

The mass flow passing through the system is given by

$$m_a = \frac{\sigma_\infty p_\infty^0 \sqrt{g\gamma_\infty}\, D(M_i)A_i}{\sqrt{\mathcal{R}T_\infty^0}} \tag{4-35}$$

or

$$m_a = \frac{M_\infty p_\infty \gamma_\infty A_\infty}{a_\infty} \tag{4-36}$$

where σ is the normal shock recovery factor. Dividing Eq. 4-33 by Eq. 4-35

$$\frac{m_{acc}}{m_a} = \frac{D(M_\infty)}{D(M_i)\sigma_\infty} = \frac{A_i}{A_\infty} \tag{4-37}$$

Since m_a/m_f has been determined from Eq. 4-29 and m_{acc}/m_f from Eq. 4-37, and since the free flight Mach number and configuration are fixed, it is possible to determine both M_i and A_∞/A_i.

In the application of Eq. 4-39 it is necessary to check for the following limiting conditions which, in turn, are a function of the region of operations.

Subsonic free flight Mach numbers ($M_0 < 1.0$). In the case of subsonic flight the maximum value for the air-fuel ratio m_a/m_f is obtained when the physical inlet is therefore given by

$$\frac{m_a}{m_f} = \frac{m_{acc}}{m_f} = \frac{D(1.0)}{D(M_\infty)} \tag{4-38}$$

Supersonic free flight Mach number ($M_\infty \geqq 1.0$). Unlike the subsonic case, it is impossible for supersonic streamlines to converge. Therefore, in the case of supersonic flight $(A_\infty/A_i) \leqq 0$. If $(A_\infty/A_i) > 1$, it is assumed that the shock is swallowed and the entire free stream tube enters the vehicle. In this case

$$\frac{m_a}{m_f} = \frac{m_{acc}}{m_f}$$

$$M_i = M_\infty; \qquad p_i = p_\infty$$

Now, in terms of the stream thrust the thrust of a vehicle is given by

$$F = F_3 - F_\infty - p_\infty(A_3 - A_\infty) \tag{4-39}$$

then

$$\frac{F}{A_3} = \frac{F_2}{A_3} - \frac{F_\infty}{A_3} - p_\infty\left(1 - \frac{A_\infty}{A_3}\right) \tag{4-40}$$

and

$$\text{sfc} = \frac{3600}{\left(\dfrac{F}{A_3}\right)\left(\dfrac{A_3}{A_2}\right)\left(\dfrac{A_2}{m_f}\right)} \tag{4-41}$$

The use of this method is shown by two specific illustrations. The problem is to calculate the specific fuel consumption and thrust per cross-sectional area of a ramrocket using a pure hydrocarbon monopropellant as the fuel. The configuration is simply a straight tailpipe with a three-to-one inlet subsonic diffuser. The two flight conditions are $M_\infty = 0.4$ and 1.4 at sea level. All internal losses are neglected, and the monopropellant impulse of the liquid hydrocarbon rocket is taken as 220 seconds. It is further assumed that $p^0A_2/m_f = 1000$.

Calculation of subsonic $M_\infty = 0.4$. Rewriting Eq. 4-39 for the case under consideration,

$$\frac{F}{A_2} = \frac{F_2}{A_2} - \frac{F_\infty}{A_2} - p_\infty\left(1 - \frac{A_\infty}{A_2}\right) \tag{4-42}$$

Dividing Eq. 4-25 by p_∞,

$$\frac{F_2}{p_\infty A_2} = \frac{\sigma p_\infty^0}{p_\infty} + I_r\left(\frac{m_f}{p_\infty A_2}\right) \tag{4-43}$$

p^0_∞/p_∞ for $M_\infty = 0.4$ is 1.116; $\sigma = 1$; I_r is given as 220 sec, and $p_\infty A_2/m_f$ as 1000; therefore $F_2/p_\infty A_2 = 1.336$. Since the exhaust is unchoked F_2 can be expressed as

$$F_2 = p_\infty A_2(1 + \gamma_2 M_2^2) \tag{4-44}$$

or

$$(1 + \gamma_2 M_2^2) = \frac{F_2}{p_\infty A_2} \tag{4-45}$$

from which M_2 is found to be 0.524 and $\phi(M_2) = 1.197$. (γ_2 is a slowly varying function of m_a/m_f and can be estimated readily.)

$$F_2 = m_a S_a \phi(M_2)$$

or

$$\frac{F_2}{m_f} = \frac{m_a}{m_f} S_a \phi(M_2) \tag{4-46}$$

Rearranging

$$\left(\frac{F_2}{p_\infty A_2}\right)\left(\frac{p_\infty A_2}{m_f}\right) = \frac{m_a}{m_f} S_a \phi(M_2) \tag{4-47}$$

or

$$\frac{m_a}{m_f} S_a = \left(\frac{F_2}{p_\infty A_2}\right)\left(\frac{p_\infty A_2}{m_f}\right)\frac{1}{\phi(M_2)} \tag{4-48}$$

$F_2/p_\infty A_2$ was determined from Eq. 4-43; $p_\infty A_2/m_f$ is given, and $\phi(M_2)$ obtained from M_2, thus $(m_a/m_f)S_a = 1116$ sec. A plot of $(m_a/m_f)S_a$ vs. m_a/m_f for a liquid hydrocarbon monopropellant is shown in Fig. I,14b. Consequently m_a/m_f can be found equal to 5.65.

Eq. 4-34 can be rearranged to

$$\frac{m_{a\infty}}{m_f} = M_\infty\left(\frac{g\gamma_\infty}{a_\infty}\right)\left(\frac{A_i}{A_2}\right)\left(\frac{p_\infty A_2}{m_f}\right) \tag{4-49}$$

(A_i/A_2) is $\frac{1}{3}$, $m_{a\infty}/m_f = 5.37$, and the other groups known. Thus

$$\frac{\left(\dfrac{m_a}{m_f}\right)}{\left(\dfrac{m_{a\infty}}{m_f}\right)} = \frac{m_a}{m_{a\infty}} = \frac{A_\infty}{A_i} \tag{4-50}$$

Thus, $A_\infty/A_i = 1.052$, and then

$$\frac{A_\infty}{A_2} = \frac{A_\infty}{A_i}\frac{A_i}{A_2} = 1.052 \times \frac{1}{3} = 0.351 \tag{4-51}$$

Dividing Eq. 4-42 by p_∞ and substituting for F_∞

$$\frac{F}{p_\infty A_2} = \frac{F_2}{p_\infty A_2} - \frac{A_\infty}{A_2}(1 + \gamma_\infty M_\infty^2) - \left(1 - \frac{A_\infty}{A_2}\right) \tag{4-52}$$

from which $F/p_\infty A_2 = 0.258$ or $F/A_2 = 3.79$ lb/in.² Since

$$\left(\frac{F}{p_\infty A_2}\right)\left(\frac{p_\infty A_2}{m_f}\right) = \frac{F}{m_f} \tag{4-53}$$

$$\text{sfc} = \frac{3600}{(F/m_f)} = \frac{3600}{258} = 13.95 \text{ hr}^{-1} \tag{4-54}$$

Calculation for $M_\infty = 1.4$. The values of $p^0_\infty/p_\infty = 3.18$ and $\sigma = 0.957$ are introduced into Eq. 4-43 and $F_2/p_\infty A_2$ determined. This value

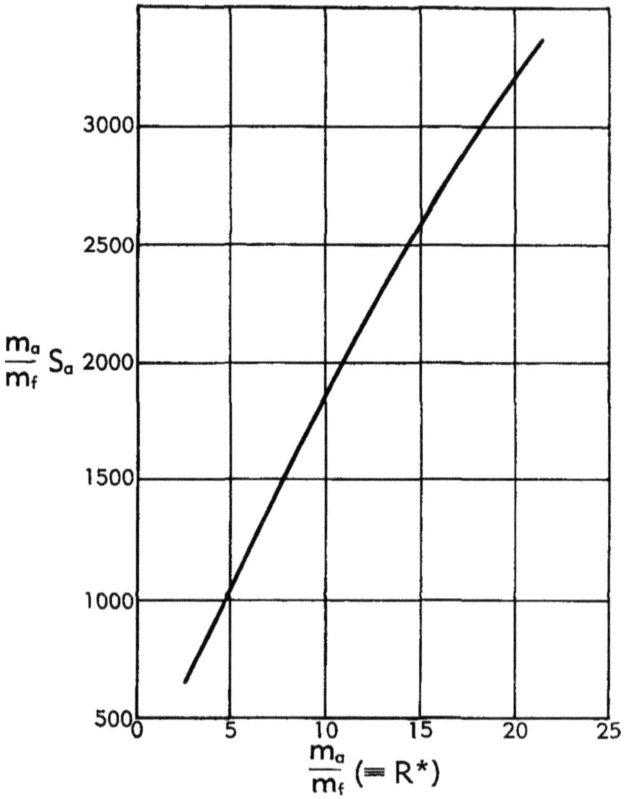

Fig. I,4b. Liquid hydrocarbon monopropellant-air system.

of 3463 inticates $M_2 > 1$; therefore the exhaust is choked and M_2 taken equal to 1, $\phi\ (M) = 1$. Thus $(m_a/m_f)S_a = 3263$ sec, and consequently $m_a/m_f = 21.25$. Calculating $m_{a\infty}/m_a$ as before, one finds this quantity equal to 18.91. This would require $A_\infty/A_i > 1$, a condition which cannot exist as explained before, consequently the frontal shock must be swallowed, and $A_\infty = A_i$. For this case the air entering the vehicle is given by $m_{a\infty}/m_f = m_a/m_f = 18.91$. Thus $(m_a/m_f)S_a = 3050$.

For $M_2 = 1$, $F_2 = m_a S_a$, or $F_2/m_f = (m_a/m_f)S_a$. Multiplying Eq. 4-42 by A_2/m_f

$$\frac{F}{m_f} = \frac{F_2}{m_f} - \left(\frac{p_\infty A_\infty}{m_f}\right)\left(\frac{A_\infty}{A_2}\right)(1 + \gamma_\infty M_\infty^2) - \left(\frac{p_\infty A_2}{m_f}\right)\left(1 - \frac{A_\infty}{A_2}\right) \quad (4\text{-}55)$$

All terms on the right-hand side of Eq. 4-56 are known and therefore $F_g/m_f = 1135$.

$$\text{sfc} = \frac{3600}{(F_g/m_f)} = 3.172 \text{ hr}^{-1} \quad (4\text{-}56)$$

$$\frac{F}{A_2} = \frac{(F_g/m_f)}{(p_\infty A_2/m_f)} p_\infty = 16.68 \text{ lb/in.}^2 \quad (4\text{-}57)$$

I,5. Fuel Selection. The selection of an ideal fuel for the ramrocket power plant or for one of its associated applications depends to a large degree upon the particular application or ultimate mission, and to some degree on the physical and handling characteristics, burning characteristics, cost, etc. Analysis of the problem readily reveals that the optimum fuel is one which can deliver the highest rocket impulse and the highest ramrocket specific impulse function $(m_a/m_f)S_a$. Since in steady flow air-

Table I,5a. Theoretical performance characteristics.

Propellants or propellant combinations	Rocket impulse I_{sp}, sec $p_c = 300$ lb/in.² abs	Total energy available per pound of propellants Q, BTU/lb	Stoichiometric air-propellant ratio R^*	Air-propellant ratio R^*	Air specific impulse S_a, sec	Fuel specific impulse function $\frac{m_a}{m_f} S_a$, sec
Liquid oxygen	200	8750	4.5	5	195	973
Octane, $R = 1.2$				14	130	1813
Liquid oxygen	224	8000	3.8	4	209	833
Octane, $R = 1.4$				10	145	1454
Ethylene oxide	160	11,700	7.9	8	181	1454
				13	155	2018
Ethylene oxide				4		
liquid oxygen		7800	3.8	10		
$R = 0.5$						
Liquid hydrocarbon	220	20,000	13.2	5	200	1000
monopropellant				13.2	182	2425

breathing engines there are always large quantities of air, the molecular weight of the combustion product may be considered relatively constant and the fuel specific impulse function may be considered proportional to the square root of the flame temperature. Since the flame temperature is a function of the heat of the combustion of the fuel, the evaluation of the rocket impulse and the heat of combustion may be sufficient for a ready comparison of various fuels. Table I,5a offers a comparison of various propellants on this particular basis. If for a particular application the amount of air is limited, it is possible to burn a fuel with less than the stoichiometric amount of air. Then the use of the heat of combustion as a parameter for comparison is not valid, since it is only suitable for conditions of stoichiometric and above. Thus it is quite possible to have a fuel with a low heat of combustion, burning at the stoichiometric condition, giving a higher flame temperature than a fuel with a higher heat of combustion burning at a value less than stoichiometric. Examination of the ethylene oxide and liquid hydrocarbon monopropellant calculations in Table I,5a substantiates this conclusion. Here, then, one must deal with the final flame temperature of the specific impulse function when comparing the various fuels. As discussed before, there is an alternative to the problem of insufficient air to burn the fuel necessary to produce the desired thrust, such as would be required in the take-off and acceleration phase of a missile. This alternative would be to add sufficient oxidizer in the rocket in order that all the fuel be burned and a high rocket thrust obtained. In general, the monopropellants exhibit the most suitable performance characteristics for the ramrocket because they may be operated as bipropellants over a wide range of mixture ratios and they meet the versatility demanded in certain applications described above. Bipropellant rocket systems, where the fuel is not a monopropellant, operated in the very rich region, may be a possibility for certain specific applications in the ramrocket spectrum, but these applications may be limited. The logistic consideration in fuels would undoubtedly favor the ordinary bipropellant fuel combinations since the monopropellants of interest are not usually the ordinary fuels or chemicals available at military establishments.

Ramrocket fuels may be divided into two types suggested by the discussion above, and these types may be characterized by their corresponding graphs of adiabatic flame temperature and energy release. Type 1 may be considered a pure monopropellant system of a hydrocarbon-type fuel, whereas type 2 covers the range of various bipropellant systems with the oxidizer-fuel ratios less than stoichiometric. A monopropellant may form a part of the bipropellant system and fall in the second category. If the ramrocket scheme is to be adapted for an ejector application, a third type of fuel may be considered—that of the overoxidized bipropellant in which there is no afterburning, but since this is not truely a propulsive device it is not considered here.

Fig. I,5a presents typical curves for the flame temperature and energy release of a monopropellant fuel as a function of the air-propellant ratio R^*. The temperature at $R^* = 0$ corresponds to the adiabatic decomposition temperature of the monopropellant. As air is added (R^* increased), the products of decomposition combine with the oxygen in the air, energy is released, and the final temperature is increased. A maximum temperature is obtained at R^* very close to stoichiometric; beyond this point air acts as a diluent and the flame temperature decreases. It is noted that the energy release increases to a maximum as R^* increases and then remains constant and independent of further increases in R^*. More generally the hydrocarbon monopropellant system tends to have a greater amount of energy available for afterburning than the bipropellant system. This follows from the fact that a given pound of monopropellants contains much less noncombustible species than a pound of a bipropellant combination. Also, for the monopropellants, since the oxidizer for the secondary combustion must come from the air, the point of maximum energy release and maximum adiabatic flame temperature tends to occur at

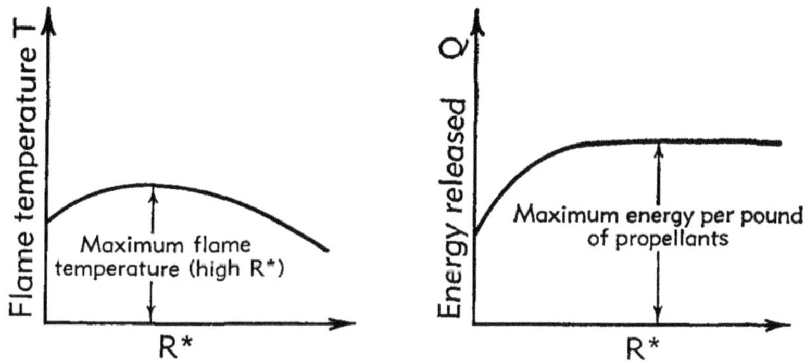

Fig. I,5a. Flame temperature and energy release of monopropellant ramrocket system.

higher values of R^* than in the case of other combinations. Since the air specific impulse function is directly proportional to the square root of the adiabatic flame temperature, it is evident that it follows the same trends as described above, i.e. it tends to reach the maximum value as higher values of R^* than other propellant combinations. However, monopropellants follow decomposition, and not combustion, processes in the rocket, and decomposition reactions are characteristically lower in temperature than combustion reactions; thus with no great variance in the molecular weight of exhaust products the rocket specific impulse is lower for monopropellants than bipropellants.

Fig. I,5b presents the adiabatic flame temperature and energy release as a function of R^* for a bipropellant system. In the monopropellant system after the fuel is selected, the shape of the temperature and energy release curves are fixed. However, in the bipropellant system it is possible by varying the oxidizer-fuel ratio R to change the shape of the flame temperature and energy release curves over a considerable range. As R is increased, the maximum flame temperature increases and tends to shift to lower values of R^*. Correspondingly, the values of the flame temperature at higher R^* tend to decrease with increasing values of R. Similarly with increasing values of R the maximum amount of energy available tends to decrease and to be released at lower values of R^*. These effects can be explained as follows.

For a unit weight of propellant combination there is a definite or stoichiometric amount of air needed to burn the propellant completely. As one increases the oxidizer-fuel ratio for a fixed mass of propellants, the value of a stoichiometric air-propellant ratio tends to decrease. Since the maximum flame temperature, in general, corresponds to the stoichiometric, the value of the stoichiometric R^* in turn decreases with increasing values of the oxidizer-fuel ratio. Therefore the maximum flame temperature can be expected to shift to lower values of R^* with increasing R, as shown in Fig. I,5b.

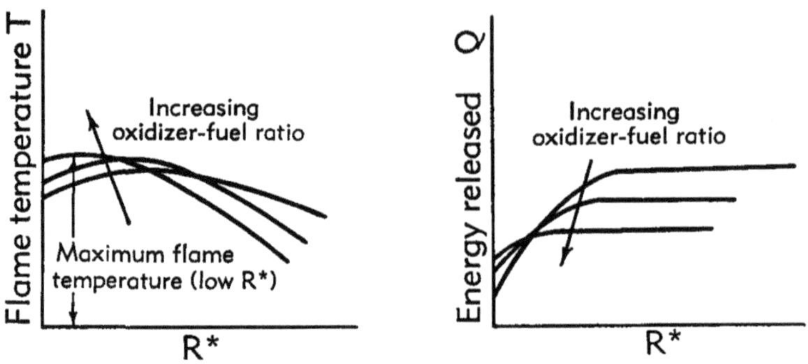

Fig. I,5b. Flame temperature and energy release of bipropellant ramrocket system.

Again, assuming unit weight of propellants, it is apparent that as the amount of oxidizer is increased and thus the amount of fuel reduced, the total amount of available chemical energy in a unit weight of propellant is reduced. Therefore the reduction with a maximum amount of energy release with increasing R is again expected. Similarly, since the point at which all the energy is released effectively corresponds to the stoichiometric point for the complete system, the point of complete energy release shifts to values of a decreasing R^* with increasing values of R as shown.

Finally it is observed that, as the oxidizer-fuel ratio is increased towards the stoichiometric condition in the rocket, the value of the maximum flame temperature is increased. This effect is not immediately apparent. As one increases the oxidizer-fuel ratio holding the total weight of the propellants constant, there are two opposing effects on the values of the maximum adiabatic flame temperature. The increase in oxidizer-fuel ratio tends to decrease the amount of available energy and thus to decrease the maximum final flame temperature; reduction in stoichiometric R^* means less diluent nitrogen which has the opposite effect. In general, for the case of the underoxidized bipropellant combination, the reduction in the amount of nitrogen outweighs the decrease in the amount of chemical energy per pound of propellants. Therefore the maximum flame temperature increases with increasing oxidizer-fuel ratio as shown.

In terms of the air and specific impulse functions and the rocket specific impulse the above effects can be interpreted as follows. As one increases the oxidizer-fuel ratio for a given weight of propellants, the maximum values of the air and fuel specific impulse functions tend to be higher and to occur at lower values of air-fuel ratio. Similarly, since the values of the temperature in the rocket combustion chamber increase substantially as the oxidizer-fuel ratio increases to stoichiometric, the rocket specific impulse increases.

From the preceding discussion it is apparent that if the fuel in a bipropellant system were a high energy monopropellant, it would be possible by varying the oxidizer-fuel ratio to make the ramrocket approach the high thrust per cross-sectional area range near that of a rocket or cruise at a high Mach number with the economical specific fuel consumption of a ramjet.

In dealing with ramrocket fuels, still another characteristic must be examined, and that is the state of the products of the decomposition or fuel-rich combustion process in the rocket. The reaction products are generally hot and gasified and flow at supersonic speeds. The first two conditions give the power plant a distinct advantage over devices which use liquid fuel sprays. As Lloyd and Mullins [*4*, p. 405] have pointed out, many of the problems of combustion and high altitudes are attributed to the liquid particle condition. It appears quite feasible to assume that high temperature gases burn more readily than liquids, even though the ignition temperatures of many of the gases in air are significantly higher than most liquid hydrocarbons.

The problem of the mixing of the supersonic gas stream with the surrounding air is discussed in a later section, but a combustion product of importance to consider here is any solid phase in the gaseous products and the effect of its presence on combustion. A very underoxidized bipropellant system, liquid oxygen-JP fuel, for example, would have an appreciable amount of carbon in its exhaust products. The burning of small solid particles is a field which has been investigated only sparsely, but certain characteristics of systems containing carbon appear evident. The time required to burn the solid particles is appreciably longer than the time to burn the gaseous products and thus long ramrocket burners may be necessary for this case. In fact, it is quite possible that when a solid is only one of several other products which are gaseous, the gaseous products react first, thus seriously depleting the air of oxygen and making the burning of the carbon even more difficult. There are, however, many ways of controlling the particle size in the products of a reaction and it is quite possible that one of these procedures could be used to control the carbon particle to a size suitable for rapid combustion.

The question of the direction in which investigations for new ramrocket fuel should proceed must necessarily arise. The most promising fuel would be an ordinary hydrocarbon which would have monopropellant characteristics and the highest possible heat of combustion. Another possibility would be the dispersion of metallic particles which are readily combustible in any of the ramrocket fuels.

Perhaps some mention should be made of the often-proposed scheme of vaporizing a hydrocarbon and ejecting it through a sonic nozzle to produce static thrust. This unit does not seem practical because the reheat scheme required for the fuel puts an unnecessary weight burden on the power plant and the attainment of static thrust is still virtually impossible.

I,6. Performance Evaluation. The question as to where in the power plant spectrum the ramrocket fits cannot be answered explicitly. From a

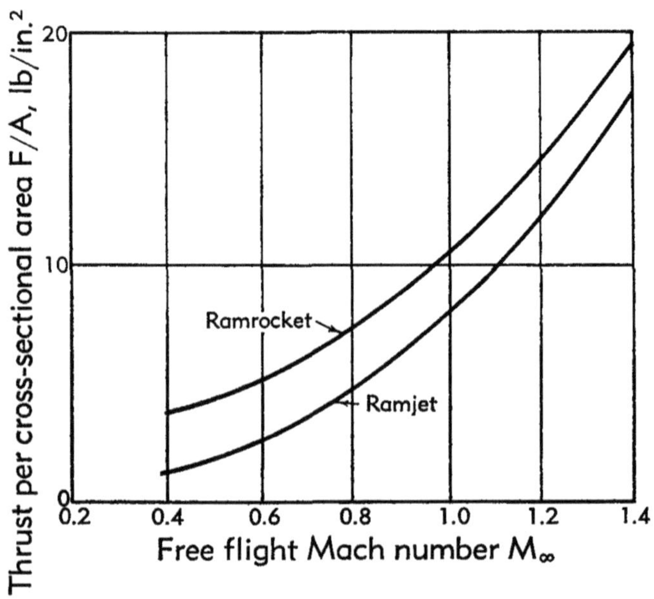

Fig. I,6a. Thrust per cross-sectional area comparison between liquid hydrocarbon monopropellant ramjet and ramrocket at sea level. 3:1 inlet diffuser, straight tailpipe.

military point of view, much depends on the application, but it does appear to show promise in the high subsonic range, particularly where lightweight, relatively inexpensive units are desired. For the high energy fuels, the performance of the ramrocket matches, and even slightly surpasses, that of a ramjet in the supersonic regime.

Perhaps the best manner in which to evaluate the performance of the ramrocket, then, is to compare it with a ramjet over a wide free flight

Mach number range. Rather than select a specific application, a comparison is made between a ramrocket and a ramjet flying at sea level and at 60,000 ft at Mach numbers from 0.4 to 1.5. The configuration of both is the same and is simply a straight tailpipe with a three-to-one inlet subsonic diffuser. This is, of course, not the optimum configuration for either power plant, but serves the purpose of enabling the reader to evaluate the capabilities of the ramrocket with respect to the ramjet. The fuel chosen for both cases is a liquid hydrocarbon which has a heat of combustion of approximately 20,000 BTU/lb and a monopropellant specific

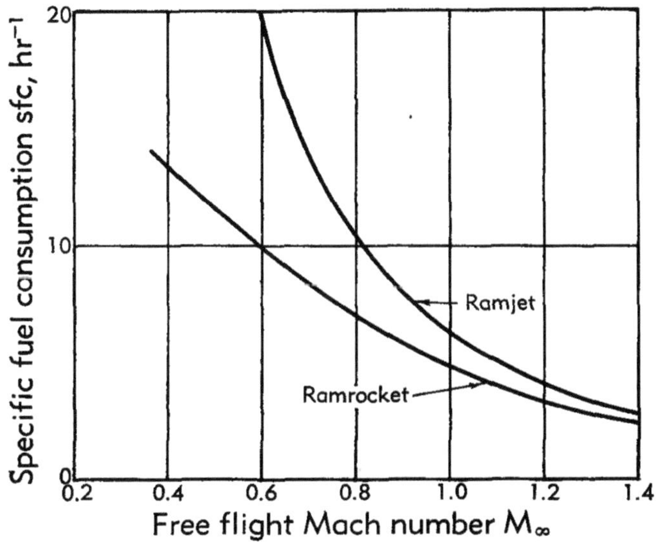

Fig. I,6b. Specific fuel consumption comparison between liquid hydrocarbon monopropellant ramjet and ramrocket at 60,000-ft altitude. 3:1 inlet diffuser, straight tailpipe.

impulse of 220 sec at 300 lb/in.^{2} abs chamber pressure. For the sake of further simplicity, the inlet diffuser in both cases was chosen to be 100 per cent efficient and both burners are operated at a 100 per cent efficiency. All internal drags, such as the flame holder drag for the ramjet, the rocket drag for the ramrocket, etc. are neglected.

Using the stream thrust method of calculation given at the beginning of this section and taking the parameter $p_\infty A_2/m_f = 1000$ seconds in all cases, one obtains the results represented in Fig. I,6a, I,6b, I,6c, and I,6d. The specific fuel consumption (sfc) results in Fig. I,6b and I,6c show that the ramrocket is substantially better than the ramjet in the high subsonic regime and the results rapidly approach each other in the supersonic range. The variations between the two power plants do not appear to be significantly different for the two altitudes chosen. Fig. I,6a and I,6d show that the improvement of the sfc for the ramrocket

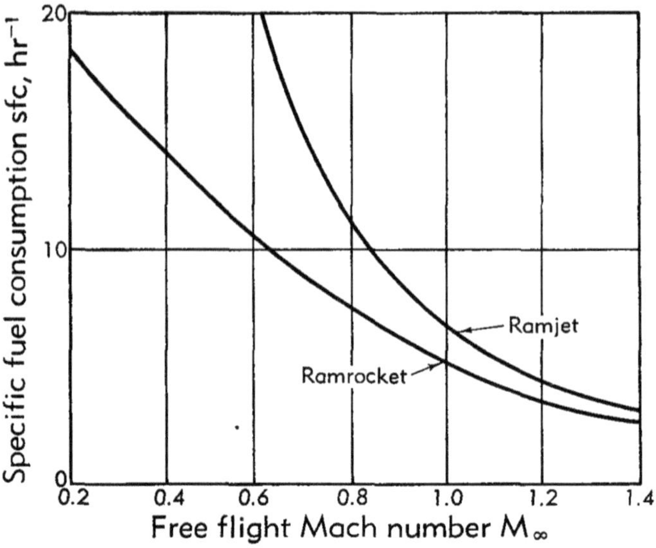

Fig. I,6c. Specific fuel consumption comparison between liquid hydrocarbon monopropellant ramjet and ramrocket at sea level. 3:1 diffuser, straight tailpipe.

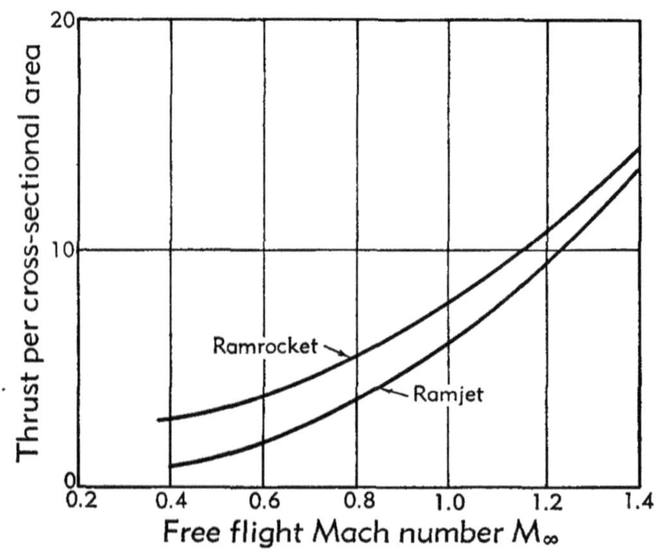

Fig. I,6d. Thrust per cross-sectional area comparison between liquid hydrocarbon monopropellant ramjet and ramrocket at 60,000-ft altitude, 3:1 inlet diffuser, straight tailpipe.

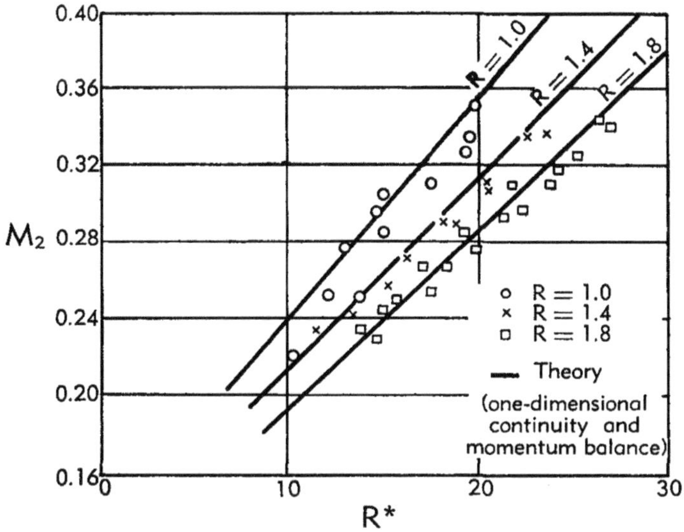

Fig. I,6e. Mach number M_2 at burner exit vs. R and R^*. Liquid oxygen-gasoline ramrocket system.

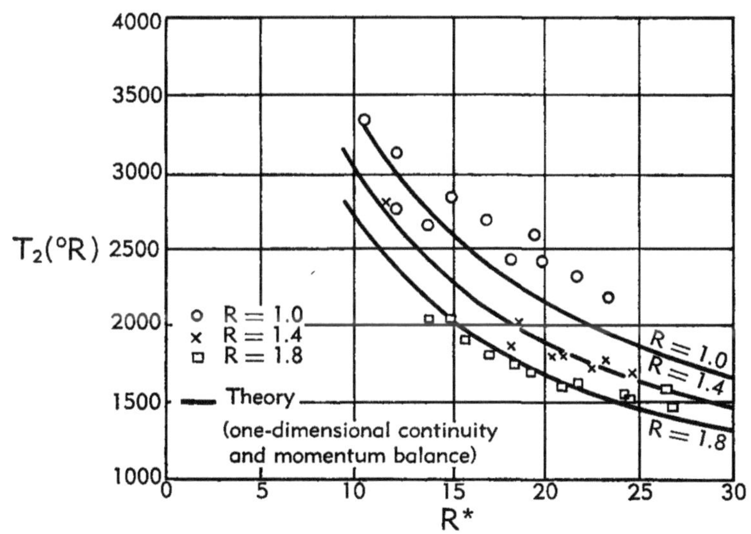

Fig. I,6f. Flame temperature T_2 at burner exit vs. R and R^*. Liquid oxygen-gasoline ramrocket system.

over the ramjet is not at the expense of the thrust per cross-sectional area (F/A) and that the same trends as above are observed.

Of course, the assumption of a 100-per cent burner efficiency is a rather crucial one and the efficiency of the ramrocket burner needs to be established experimentally. The only experimental ramrocket results appearing in the open literature are those due to Charyk, Glassman, and

John [*5*, p. 886] for the liquid oxygen-gasoline rocket-air system. These authors presented their results in the form of the Mach number and temperature at the end of the burner and compared them with the values predicted theoretically. The results which were corrected for losses to the water-cooled burner are given in Fig. I,6e and I,6f. Comparison with the theoretical values appears to indicate that the high burner efficiencies must have been obtained in this system. If these efficiencies are established experimentally for the high energy fuel, the ramrocket appears to have promise.

I,7. Experimental Burner Results. A problem that is unique with the ramrocket power plant is the mixing and burning of two fluid streams, one supersonic and one subsonic. The pure supersonic mixing case is in itself a most difficult one without the added complication of combustion. The recent investigations of Pai [*6*] and Warren [*7*] are just beginning to provide information on this problem. The only theoretical analysis of the mixing and combustion problem is attributed to Marble and Adamson [*4*, p. 111] for a very simplified, restrictive case.

The mixing problem is of considerable importance to the ramrocket, since it is the mixing and combustion of the streams which define the length of the burner required for the completion of the energy processes. Even though the ramrocket may show advantage over the ramjet in theoretical consideration, if the length of the burner required becomes excessively long this fact may rule out the ramrocket as a practical power plant. The large energy release associated with the mixing and burning process in the ramrocket power plants makes the systematic theoretical experimental approach to the problem quite difficult. Reference to Marble and Adamson [*4*, p. 111] indicates the theoretical difficulties. Experimentally, the pure mixing problem is usually attacked by taking profiles of velocity, temperature, and composition of the two streams as they mix. It would be well to extend this to the supersonic case with combustion, but the experimental techniques for measuring temperatures and sampling gases in high velocity streams have not developed to the point where they are considered precise and reliable experimental tools.

In an effort to circumvent these difficulties Charyk, Glassman, and John [*5*, p. 886] undertook to study the processes in the ramrocket burner by measuring the axial static pressure history through the burner. Their experiments have resulted in some interesting data. Recently the same techniques have been adapted by Barrere and Mestre [*4*, p. 405] in their study of ramjet combustion. Some typical results obtained by Charyk, et al. are shown in Fig. I,7a. They postulate that the mixing and burning process in the ramrocket burner proceeds in three integral parts—a short regime in which there is little or no mixing, a second regime where mixing predominates and where one obtains the characteristic pressure rise associated with mixing, and a third region where burning predominates and one obtains the characteristic pressure drop associated with combustion. From these static pressurecurves it was possible for these authors to estimate burner lengths required for 90-per cent combustion of the fuel. Their burner length curves for the liquid oxygen-gasoline rocket-air system are presented in Fig. I,7b. The requirement of longer burners with larger energy releases is interesting and is probably due to the increased

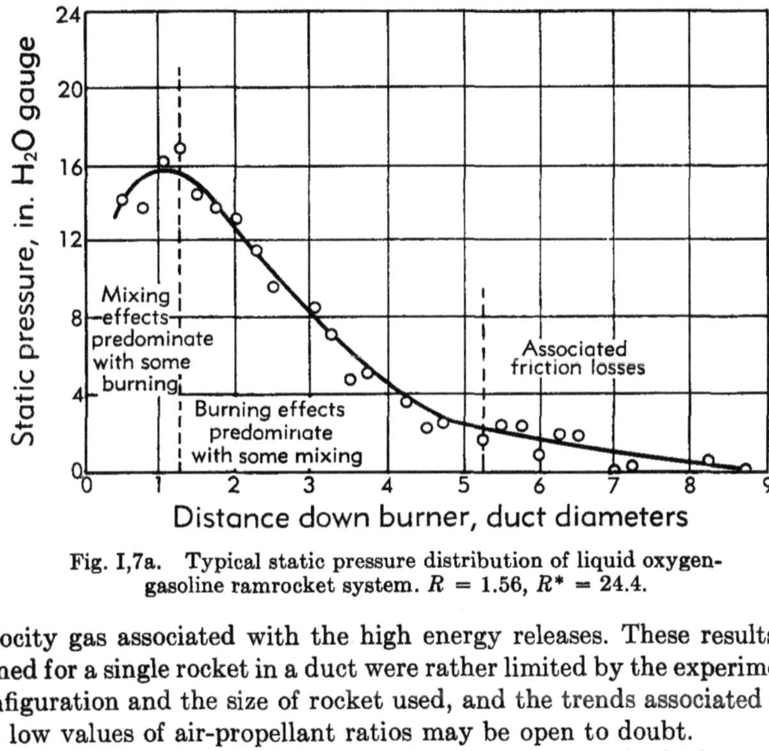

Fig. I,7a. Typical static pressure distribution of liquid oxygen-gasoline ramrocket system. $R = 1.56$, $R^* = 24.4$.

velocity gas associated with the high energy releases. These results obtained for a single rocket in a duct were rather limited by the experimental configuration and the size of rocket used, and the trends associated with the low values of air-propellant ratios may be open to doubt.

Scott [*8*] extended the work of Charyk, et al. to the higher energy fuels with a similar system. He was able to mix the rocket and air stream, both with and without combustion, and thus was able to gain further insight into the mixing problem. His resultant static pressure curves for combustion were the same as those shown in Fig. I,7a. The curves he obtained for pure mixing are typified by the curve presented in Fig. I,7c. Scott, whose data were taken on the rich side of stoichiometric, observed that the lengths required for the mixing of the rocket and air to come to completion were approximately the same as the lengths required for mixing and combustion to come to completion and that the lengths in either case did not vary appreciably with the equivalence of air-propellant ratio. He thus postulated that the rate of energy release or lengths required for completion of all processes is controlled by the rate of mixing of the rocket

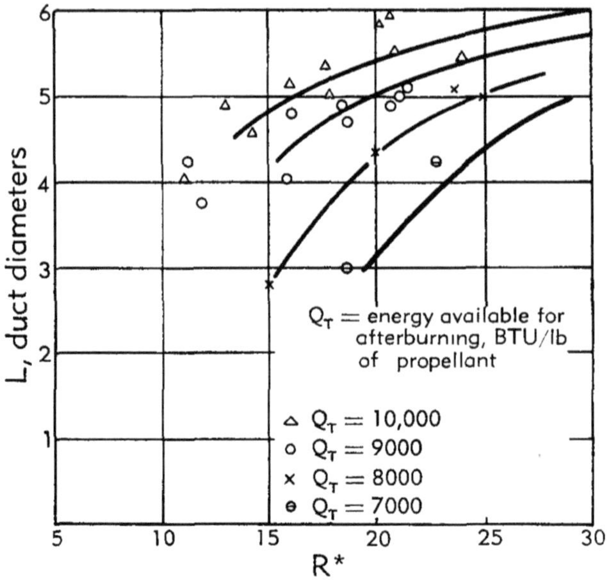

Fig. I,7b. Burner length based upon 90-per cent combustion.

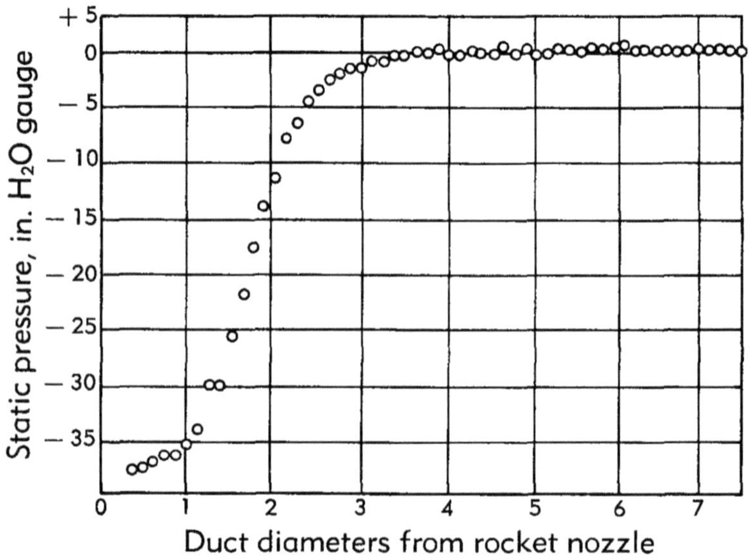

Fig. I,7c. Static pressure distribution; pure mixing.

and air streams. Thus it would appear that the chemical reaction time in the ramrocket burner using high energy fuel is negligible compared to the time required for sufficient mixing to take place so that chemical action can occur, the same conclusion that was reached for the simple turbulent diffusion flame.

Both Charyk, et al. and Scott have dealt with the case of a single rocket in a duct. Neither says what the effect of using multiple rockets would be, but if the postulate of Scott is correct in regard to mixing being the controlling factor, then it is probable that increasing the number of rockets would certainly increase the rate of mixing and thus shorten the length of burner required.

In the previous section the effect on burner length by the presence of solids in the rocket exhaust was postulated. This effect has been corroborated by some data obtained by Scott. In the very rich region where there is only sufficient air to burn the gaseous components of the solid-containing system, the burner lengths required are the same as the purely gaseous system. When sufficient air is present to burn all the gaseous components and some of the solid particles, longer burner lengths are obtained. In order to burn most of the particles, still longer burner lengths are required which indicates the slower rate of reaction of the solid phase. This problem can be answered only by further investigation of solid phase burning.

I,8. Cited References.

1. Rudnick, P. *J. Aeronaut. Sci. 14*, 540 (1947).
2. Keenan, J. H., and Kaye, J. *Gas Tables*. Wiley, 1948.
3. Handbook of supersonic aerodynamics, Vol. I. *Nav. Ord. Rept. 1488*, 1950.
4. *Selected Combustion Problems*. Butterworths, London, 1954.
5. *Fourth Symposium on Combustion*. Williams & Wilkins, 1953.
6. Pai, S. E. *The Fluid Dynamics of Jets*. Van Nostrand, 1953.
7. Warren, W. R., Jr. *Ph.D. Thesis*. Dept. of Aeronaut. Eng., Princeton Univ., 1957.
8. Scott, J. R., Jr. *Ph.D. Thesis*. To be submitted to Dept. of Aeronaut. Eng., Princeton Univ.

SECTION J

JET ROTORS

ALBERT GAIL

CHAPTER 1. INTRODUCTION

J,1. The Topic. A "jet rotor" is a system of jet-propelled blades which, like jet-propelled wings, fly around a common center of rotation to which one end of each blade is attached.

The propulsion of the blades of a jet rotor is in this sense a problem somewhat similar to the propulsion of a fixed wing aircraft. In principle, therefore, all of the known jet propulsion engines qualify as drivers for the blades of jet rotors.

The jet rotor as a whole is an engine which produces a resultant force which is the reaction to the momentum changes imparted upon the surrounding air by both the blades and their jet engines. Significantly, there exists no class designation or generic term which describes the role of this jet rotor engine.

Jet rotors may be pure propulsion engines (jet propellers) or pure lifting devices (the jet rotor of the McDonnell convertiplane), or both lifting and propelling devices (jet helicopter rotor), or alternating lifting or propelling devices (VTOL jet propeller). There is also the noteworthy case of a jet rotor which lifts the craft vertically during take-off and descent, accelerates it almost vertically in propeller fashion, and converts itself into a nonrotating jet-propelled wing in steady, level flight. This is the jet rotor VTOL according to Tank [*1*] (see Art. 16). Finally, there are auxiliary jet drives that operate in conjunction with other engines which normally drive the rotor by means of a torque acting on the rotor shaft. The rocket-boosted helicopter rotor [*2*] and the same configuration functioning as an emergency rocket rotor, in case of failure of the primary engine, are examples of its great potential usefulness in an auxiliary capacity.

An attempt will be made in Chap. 2 to treat the jet rotor engine in a general way and by selecting specific particular cases. These cases are usually chosen on the basis of historical precedent. Art. 2 shows why the usual choices are not necessarily the most reasonable ones.

J,2. History and Potential Future of Jet Rotors. According to a cartoon of 1855 depicting a helicopter with two side-by-side rotors driven by steam jets, a certain Mr. Phillips may have been the inventor of jet-driven rotors. However, the first jet-driven helicopter actually designed and built was that of Papin and Rouilly [*3*, pp. 169–172]. Their "Gyroptere" was a rather striking design: a single blade, counterbalanced by an engine-driven centrifugal blower. The reaction of the compressed cold air jet issuing from the blade tip propelled the blade. Flight tests were attempted on a lake in France in March, 1915. Unsatisfactory balancing was blamed for the failure of the flight test in which the helicopter was damaged.

Two decades later, Knight of the Georgia Institute of Technology designed and built a cold air jet-driven, single rotor helicopter. Most of the development effort over a period of six years was directed toward an efficient high speed axial compressor. The results of the ground tests in 1943 were disappointing because the engine did not deliver the specified power. Consequently, the blower did not reach peak efficiency and the rotor thrust did not reach even half of its expected value. Those inheriting the venture after Knight's untimely death lost heart and abandoned the project.

On September 23, 1943, a jet-driven helicopter, designed and built by von Doblhoff in Austria, made its first flights. Doblhoff built four different models in succession, all of the pressure-jet type [*1*]. An engine-driven compressor fed an air-fuel mixture through three rotor blades. This mixture was burned within the blade tips and expelled through nozzles. Since then, several ramjet- and pulse-jet-driven helicopters have successfully flown in the United States.

In 1913 Zselyi [*4*] built a jet propeller with combustion chambers in the hub, each feeding a nozzle intermittently through a hollow blade.

In the late 1920's a Mr. Bonney had a jet-driven propeller built by the Curtiss Corporation, then located on Long Island, New York. According to verbal accounts, Bonney's propeller was of aluminum and had a combustion chamber in the hub. The combustion gases were ducted through the hollow blades to orifices near the trailing edge of the blade tips. The propeller ran statically under its own power at rather high rates of revolution.

Diedrich [*2*] at the Argus Engine Company in Germany experimented with a pulse-jet-driven, two-bladed propeller in 1940. Difficulties with the high temperature strength of the blades, which were part of the pulse jet, led to the abandonment of that project.

Many jet-driven helicopter and convertible aircraft rotors have been built and flown in several countries in the past decade. To appraise the present status of jet rotor technology one must make the following distinction. The arbiter of theory in the physical sciences is observable fact, the success of a crucial experiment. In this sense, everyone of the flyable jet rotors developed to date is a successful confirmation of its scientific rationale. The arbiter of engineering rationale is the attainment of a widely accepted goal. At the time of this writing, none of the jet rotors that were successful physical experiments have attained the status of a successful design; at any rate none seem to have progressed beyond the prototype stage into production.

Since there have been many changes of accepted goals in the history of technology, it would seem rash to abandon jet rotors as foredoomed failures. Jet engines, particularly rockets, have a long history of dormant existence as mere curiosities. The need for them has only arisen in recent years after airplanes had approached transonic speeds. With this change of technological environment, the engine-propeller combination gradually failed to satisfy the new propulsion needs, and the formerly inferior jet engines became superior. Once on the scene, the jet engines created their own new opportunities in supersonic flight and missile technology.

There are many and potent reasons to expect that jet rotors may become similarly important major power plants with analogous revolutionary effects upon aeronautical technology. Until recently, jet rotors were considered primarily jet helicopter rotors. Although the usefulness of helicopters cannot be questioned, there is hardly a helicopter manufacturer today who does not attempt to escape from the intrinsic speed limitations of helicopters which are felt to hamper helicopter utility. All of these attempts depart more or less radically from the helicopter. The need for VTOL aircraft (i.e. for aircraft capable of vertical take-off and descent having adequate cruising speed, range, and load-carrying capacity) appears to exist. The survival of the VTOL may well hinge on the emergence of a design that imposes the least sacrifices of customary air transport capabilities in exchange for the vertical flight capability.

Judging by the intrinsic capabilities of jet rotors it appears almost compelling that the large VTOL offers the best opportunity for the development of the jet rotor and vice versa. Jet engines require at least transonic speeds for competitive fuel economy. Helicopters impose approximately 600 ft/sec as the speed limitation for the jet engines of their rotors. None of the advantages of jet rotors appear to be able to compensate for the waste of fuel which this helicopter operation dictates. Jet rotors for VTOL, operating like propellers in a flow with axial symmetry, permit transonic and supersonic jet engine speeds and high speed flight, which are the prerequisites for competitive fuel economy.

The design rationale includes a basic tenet: The failure to excel in competition with its rivals is as mercilessly damaging to an engineering design as the failure of crucial experiments is to a theory in the physical sciences.

In nascent technologies, such as jet rotors and VTOL, this awareness of competition is a source of concern. Indeed, until the competitively viable major design features of jet rotors have been identified, it would be premature to worry about details and the niceties of the physical and functional integrity of jet rotors. The emphasis on the competitive, goal-directed aspects in this section distinguishes it from those sections dealing with mature technologies in which problems of physical integrity predominate. As in other goal-directed activities (e.g. military strategy, social games, etc.), scientific method is frequently but not always applicable. However, impartiality can never be abandoned with impunity. The biased objective of establishing a valuable place for jet rotors is frankly stated as legitimate.

CHAPTER 2. INTRINSIC AND ELEMENTARY PROPERTIES OF JET ROTORS

J,3. Kinematics of the Blade. Represent the blade by a straight line, any point of which is located by the radius vector $\mathbf{r}$ having its origin at the rotor center. The blade tip is located by $\mathbf{R}$, the jet engine by $\mathbf{R}_j$. As a rule, $\mathbf{R}_j = \mathbf{R}$; the rare exception $\mathbf{R}_j \neq \mathbf{R}$ will be ignored. The blade turns about the rotor center with angular velocity ω. Constant ω is a good approximation for jet rotor blades in steady state operation. The velocity of the rotor center with respect to the earth is $\mathbf{V}$. Consequently, the velocity of any point of the rotor blade with respect to the earth is

$$\mathbf{U} = \mathbf{V} + \omega \times \mathbf{r} \tag{3-1}$$

of the blade tip and jet center in particular,

$$\mathbf{U}_j = \mathbf{V} + \omega \times \mathbf{R} \tag{3-1a}$$

For reasons discussed in Art. 12, there is a small "coning angle," β, described by

$$\omega \cdot \mathbf{r} = \omega r \sin \beta \cong \omega r \beta \neq 0$$

Adopt the convention that the rotor thrust and ω point in the same direction (right-handed screw); then $\beta > 0$.

In steady rotor operation, i.e. with $\mathbf{V}$ and ω constant,

$$\frac{d\mathbf{U}}{dt} = \omega \times (\omega \times \mathbf{r})$$

and (3-2)

$$\frac{d\mathbf{U}_j}{dt} = \omega \times (\omega \times \mathbf{R})$$

These relations represent the centripetal accelerations of any point of the blade and of the tip-mounted jet engine, respectively.

Since β is usually a small angle ($\cos \beta \cong 1$), the scalar magnitudes of these centripetal accelerations are given by $\omega^2 r$ and $(\omega R)^2/R$, respectively.

There are two classes of rotor operation: (1) "symmetrical" or propeller-type operation, defined by $\mathbf{V} \times \boldsymbol{\omega} = 0$; and (2) "unsymmetrical" or helicopter-type operation, defined by $\mathbf{V} \times \boldsymbol{\omega} \neq 0$. A helicopter rotor in vertical flight is an example of symmetrical operation. Unsymmetrical operation becomes symmetrical when $\mathbf{V} = 0$ which corresponds to the "static thrust" condition, such as the hovering state of a helicopter.

Symmetrical operation. The blade tip and jet engine trace a helix with the helical angle $\phi \equiv \tan^{-1} \lambda$, where $\lambda \equiv V/|\boldsymbol{\omega} \times \mathbf{R}| = V/\omega R$ for

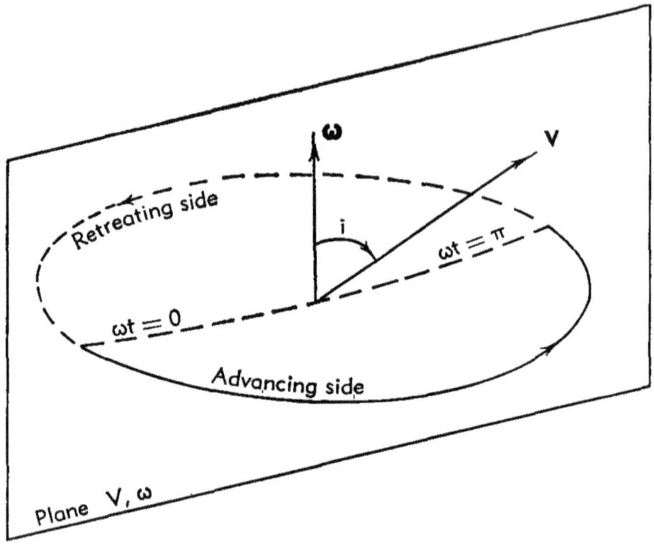

Fig. J,3a. Unsymmetrical blade motion.

small β. The resulting tip speed is given by

$$U_j = [V^2 + (\omega R)^2]^{\frac{1}{2}} = \omega R \sqrt{1 + \lambda^2} \tag{3-3}$$

Consider the case of constant tip speed, U_j = const. The possible values of V must then lie between zero and U_j. The conditions $V = 0$, and therefore $\lambda = \phi = 0$, correspond to static thrust. At the other extreme, $V = U_j$, and the conditions $\omega R = 0$, and therefore $\omega = 0$, $\phi = \frac{1}{2}\pi$ and $\lambda = \infty$ must be satisfied. Consequently, blade rotation has stopped, and the blade moves with constant translational speed $V = U_j$, like a fixed wing. The Tank jet rotor (Art. 16) operates in the whole domain from $V = 0$ to $V = U_j$.

Unsymmetrical operation. The blade traces a skew helical surface in the earth-fixed space, a shallow cone in the space which moves translationally with $\mathbf{V}$. The plane containing $\mathbf{V}$ and $\boldsymbol{\omega}$ slices the cone into an advancing side and a retreating side (Fig. J,3a), establishing the azimuth positions $\omega t = 0$ and $\omega t = \pi$. The blade positions $0 < \omega t < \pi$ represent the advancing side, and $\pi < \omega t < 2\pi$ the retreating side. The rotor incidence i is the angle between the vectors $\mathbf{V}$ and $\boldsymbol{\omega}$. By customary definition

$$V \sin i \equiv \mu\omega R$$

and (3-4)

$$V \cos i \equiv \lambda\omega R$$

Unsymmetrical operation is the consequence of i being unequal to either zero or π ($\mu \neq 0$).

Note that U and U_j are periodic in ωt:

$$\mathbf{U} = \mathbf{U}(\omega t, r) = \mathbf{U}(\omega t + n2\pi, r)$$

and (3-5)

$$\mathbf{U}_j = \mathbf{U}_j(\omega t) = \mathbf{U}_j(\omega t + n2\pi)$$

Imagine an observer located at the jet engine and rotating with the blade (see Fig. J,3b). The vector $\boldsymbol{\omega} \times \mathbf{R}$ will then appear to be constant,

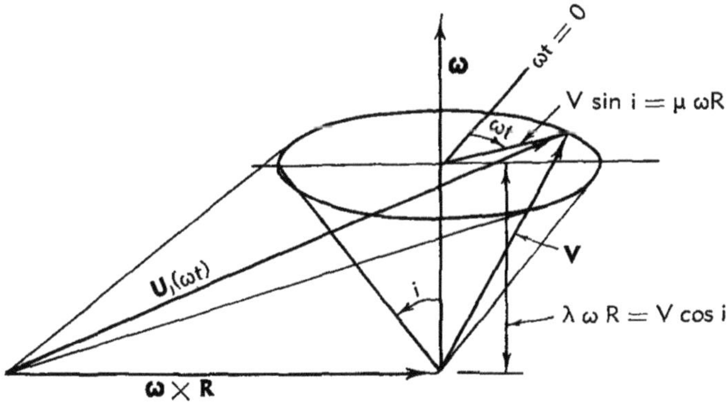

Fig. J,3b. Velocity vectors as "seen" from the jet engine mounted at the tip of the blade.

$\mathbf{V}$ appears to sweep a circular cone of angle i while it rotates with the angular velocity $-\boldsymbol{\omega}$. The hodograph of $\mathbf{V}$ is the base circle of this cone which has a radius $V \sin i = \mu\omega R$. The height of the cone is $V \cos i = \lambda\omega R$. The velocity vector of the observer is $\mathbf{U}_j = \mathbf{V} + \boldsymbol{\omega} \times \mathbf{R}$ which sweeps an elliptical cone during every revolution of the blade. The hodographs of $\mathbf{V}$ and $\mathbf{U}_j$ are the same circle.

It is important for all jet engines that they be well aligned with their direction of motion, but those engines which depend for their proper functioning upon air intake diffusers may be particularly harmed by bad alignment because the efficiency of present-day diffusers is very sensitive to misalignment. Perfect alignment of the jet engine in unsymmetrical operation would require a swivel-mounting of the engine, and alignment control, say, by vane stabilization, an uncomfortable design problem. Even with perfect alignment, the magnitude of the engine velocity relative to the surrounding air, $|\mathbf{U}_j|$, would periodically oscillate between

$\omega R\sqrt{\lambda^2 + (1 - \mu)^2}$ and $\omega R\sqrt{\lambda^2 + (1 + \mu)^2}$ which may degrade the operation of air-breathing engines.

Suppose that the jet engine is rigidly attached to the blade, and imagine $\mathbf{U}_j$ to be projected into the plane formed by $\boldsymbol{\omega}$ and $\boldsymbol{\omega} \times \mathbf{R}$ (see Fig. J,3b). The angle between this projection and $\boldsymbol{\omega} \times \mathbf{R}$ is the helical angle ϕ which is periodically oscillating if $\mu \neq 0$, about a mean value $\phi(0) = \phi(\pi) = \tan^{-1}\lambda$. The oscillation of ϕ need not cause serious misalignment of the rigidly attached engine since blade equilibrium requires periodic blade pitch oscillations in which the engine would participate. Such oscillations minimize the angular misalignment of the engine in this plane automatically.

The angle between U_j and the aforementioned projection of U_j is the yaw angle

$$\psi = \sin^{-1}[\mu\omega R \cos(\omega t)U_j^{-1}] = \psi(\omega t) \tag{3-6}$$

The extreme values of ψ occur at $\omega t = 0$ and $\omega t = \pi$, and are given by $\pm \sin^{-1}[\mu\omega R(\omega^2R^2 + V^2)^{-\frac{1}{2}}]$.

The mean value of ψ is zero (at $\psi(\pi/2)$ or $\psi(\frac{3}{2}\pi)$). One would, therefore, align the jet engine normal to $\mathbf{R}$ to minimize yaw misalignment of a rigid engine attachment. This yaw misalignment is a grave obstacle for the supersonic helicopter rotor, driven by ramjets or ramrockets. Take, for example, $\omega R = 2000$ ft/sec. The intake diffusers of ramjets at the corresponding Mach number of approximately 2 would tolerate a yaw misalignment of, at most, 4 degrees, so that $\psi(0)$ would have to be smaller than about 0.07. Thus, μ would also have to be smaller than 0.07, and V_{max} could not exceed $0.07\omega R$ which corresponds to 140 ft/sec or 95 mph. Subsonic helicopters may reach almost twice as large a maximum speed (Art. 10). This means that supersonic ramjet or ramrocket rotors with rigidly mounted engines are not suitable—as sometimes claimed—for removing the high speed limitations of subsonic helicopters. For this purpose, the misalignment tolerance of supersonic intake diffusers would have to be more than doubled.

It is seen from Fig. J,3b that

$$\begin{aligned} U_j^2 &= (\omega R + V\sin i \sin \omega t)^2 + (V\cos i)^2 + (V \sin i \cos \omega t)^2 \\ &= U_{j_{symm}}\left(1 + \frac{2\omega RV \sin i \sin \omega t}{U_{j_{symm}}^2}\right) \end{aligned} \tag{3-7}$$

where
$$U_{j_{symm}} = \sqrt{(\omega R)^2 + V^2}$$

This periodic function in ωt may be expanded in powers of μ and a Fourier series in ωt. For $\mu \ll 1$ such that $\mu^3, \mu^4, \ldots$ can be ignored, one obtains

$$\frac{U_j}{U_{j_{symm}}} = 1 - \frac{1}{4}\frac{(\omega R)^4\mu^2}{U_{j_{symm}}^4} + \frac{(\omega R)^2\mu}{U_{j_{symm}}^2}\sin \omega t + \frac{1}{4}\left(\frac{\omega R}{U_{j_{symm}}}\right)^4 \mu^2 \cos 2\omega t \tag{3-8}$$

Note that the mean value of the tip speed of a helicopter rotor, U_j, is less than the tip speed of the propeller flying with the same values of V and same ωR.

J,4. Blade Propulsion. The jet thrust $\mathbf{T}_j$ produced by the jet engine delivers to the blade the power

$$\mathbf{T}_j \cdot \mathbf{U}_j = \mathbf{T}_j \cdot \mathbf{V} + \mathbf{T}_j \cdot \boldsymbol{\omega} \times \mathbf{R} \tag{4-1}$$

This relation contains some noteworthy information:

1. $\mathbf{T}_j \cdot \boldsymbol{\omega} \times \mathbf{R} = Q_j\omega$ is the "turning power" of the jet thrust which could also have been supplied to the blade by a mechanical torque $Q_s = Q_j$ through the rotor shaft of a mechanically driven rotor. Since $\mathbf{T}_j \cdot \boldsymbol{\omega} \times \mathbf{R} = \boldsymbol{\omega} \cdot \mathbf{R} \times \mathbf{T}_j$, the torque of the jet thrust is $Q_j = \boldsymbol{\omega}/\omega \cdot \mathbf{R} \times \mathbf{T}_j$.
2. The term $\mathbf{T}_j \ \mathbf{V}$ is an intrinsic peculiarity of jet rotors. It is the *direct contribution of the jet thrust power to the power delivered by the jet rotor.*
3. The term $\mathbf{T}_j \cdot \mathbf{U}_j = T_jU_j\cos(\mathbf{T}_j, \mathbf{U}_j)$ has the many implications familiar to the student of propulsion. For maximum propulsive power at a given jet thrust T_j and speed U_j, the angle between these vectors ought to vanish. Jet engines are most economical for transonic and supersonic velocities U_j, different engine types being characterized by different domains of speed in which they excel. Engine types differ in their tolerance toward departures from design speed, i.e. in their tolerance of fluctuations of U_j. *The important opportunity given to jet engines by the jet rotor is the large U_j when V is small or even zero.* Unless, however, U_j is at least transonic, this opportunity is of no avail. The low tip speed ($U_j \cong 600$ ft/sec) of the developed and tried jet rotors is responsible for their notorious waste of fuel. The need for large U_j is the reason for locating jet engines at the tips of the blades.

Symmetrical operation. The blade sections, particularly those near the blade tip, require a small misalignment, viz. the section angle of attack α. Modern rotors have pitch control for the achievement of a desired small α under conditions of widely varying values of λ. Suitable rigid attachment of the jet engine on the blade tip at right angles to $\mathbf{R}$ assures almost perfect alignment within an error of less than α. Hence $\mathbf{T}_j \times \mathbf{U}_j = 0$ is both a realistic and a practical approximation. Consequently the power delivered to the blade is given by

$$\left.\begin{aligned} \mathbf{T}_j \cdot \mathbf{U}_j &= T_jU_j = T_jV\sin\phi + T_j\omega R\cos\phi \\ &= T_jU_j\sin^2\phi + T_jU_j\cos^2\phi \end{aligned}\right\} \tag{4-2}$$

the turning power by $Q_j\omega = T_j\omega R\cos\phi$, and the direct propulsive power by $VT_j\sin\phi$. The contribution to the *jet rotor thrust* by the *jet engine thrust* of this blade is $T_j\sin\phi$.

It should be noted that

1. At the static thrust condition, when $V = \phi = 0$, all the power delivered to the blade is turning power.
2. When $V = \omega R$; i.e. when $\phi = \frac{1}{4}\pi$; the turning power is equal to the direct power.
3. When $V = U_j$ then $\omega = 0$ and $\phi = \frac{1}{2}\pi$, all of the jet engine power is direct propulsive power. This is the fixed wing flight condition of the Tank VTOL (see Art. 3).
4. In general, the relation

$$\frac{\text{direct power}}{\text{turning power}} = \tan^2 \phi = \lambda^2 = \left(\frac{V}{\omega R}\right)^2 \tag{4-3}$$

applies.

Unsymmetrical operation. In Art. 3, the tip speed U_j was shown to be a function of μ and ωt, periodic in ωt. Consequently, $\mathbf{T}_j \cdot \mathbf{U}_j(\mu, \omega t)$ is also periodic in ωt. The mean value of the power delivered by the jet thrust during one revolution is

$$\frac{1}{2\pi}\int_{-\pi}^{\pi} \mathbf{T}_j \cdot \mathbf{U}_j d(\omega t) = \frac{1}{2\pi}\int_{-\pi}^{\pi} T_j U_j \cos \delta d(\omega t)$$

where δ represents the misalignment of the jet engine.

Suppose that the jet thrust is insensitive to U_j (as for a rocket) and that the alignment is uniformly perfect, then $T_j = \text{const}$, $\delta = 0$, and from Eq. 3-8,

$$\frac{1}{2\pi}\int_{-\pi}^{\pi} \mathbf{T}_j \cdot \mathbf{U}_j d\omega t = T_j U_{j_{\text{symm}}} \left[1 - \frac{1}{4}\left(\frac{\omega R}{U_{j_{\text{symm}}}}\right)^2 \mu^2\right] \tag{4-4}$$

The jet rotor in unsymmetrical operation therefore receives less power from its jet engine than in symmetrical operation if T_j, V, and ωR are the same in both cases.

This degradation in power is small and unimportant. Periodic variations of δ, caused by rigid blade attachment and periodic changes of T_j of air-breathing jet engines, modify the evaluation of the mean jet thrust power somewhat. The largest degradation of power in unsymmetrical operation is apt to arise from the misalignment of intake diffusers as discussed in Art. 3.

The gravest limitation to the jet thrust power in unsymmetrical operation stems from a limitation of ωR which will be treated in Art. 10.

J,5. Conversion of Blade Propulsion into Jet Rotor Forces and Powers. The rotor is formed by a number of b, identical jet-propelled blades, regularly spaced on the cone of rotation.

The resultant force of the jet rotor is

$$\mathbf{F}_{jr} = \mathbf{F}_r + \mathbf{F}_j \tag{5-1}$$

where $\mathbf{F}_r$ is the sum of the forces with which the air acts on the rotor blades, and $\mathbf{F}_j$ is the vector sum of the individual jet engine thrusts. Moreover, $\mathbf{F}_r$, $\mathbf{F}_j$, and $\mathbf{F}_{jr}$ are understood to be the mean values $(1/2\pi)\int_{-\pi}^{\pi}\mathbf{F}d\omega t$ of these forces. $\mathbf{F}_{jr}$ is acting on the rotor center only if the blades are articulated at the rotor center. There is no good reason for avoiding the design of jet rotors with articulated blades. $\mathbf{F}_{jr}$ is here considered to be the force exerted by the jet rotor center, or hub, upon the craft which it powers.

The power delivered by the jet rotor to the craft is given by

$$\mathbf{F}_{jr} \cdot \mathbf{V} = \mathbf{F}_j \cdot \mathbf{V} + \mathbf{F}_r \cdot \mathbf{V} \tag{5-2}$$

where $\mathbf{F}_j \cdot \mathbf{V} = b\mathbf{T}_j \cdot \mathbf{V}$ is the sum of the direct propulsion powers of all the jet engines. The blades of the rotor convert the turning powers of their jet thrusts into energy losses, $E + P_i$, and rotor propulsive power $\mathbf{F}_r \cdot \mathbf{V}$, so that

$$b\boldsymbol{\omega} \cdot \mathbf{R} \times \mathbf{T}_j = bQ_j\omega = E + P_i + \mathbf{F}_r \cdot \mathbf{V} \tag{5-3}$$

where E is the power dissipated by the aerodynamic drag, and P_i is the kinetic energy imparted to the air in unit time (induced power). So far as this power balance is concerned, bQ_j could as well be supplied by a shaft torque. Jet rotors have nothing essentially new to contribute to the aerodynamic mechanism of converting a driving torque power into rotor forces and powers. Propeller and helicopter rotor literature furnishes a great deal of aerodynamic information that is directly applicable to jet rotors.

J,6. Aerodynamics of the Rotor Disk. Consider a rotor to be represented by a circular surface (diameter $2R$) that maintains a pressure discontinuity Δp so that $F_r = \Delta p \pi R^2$ is normal to this surface, the "actuator disk."

The extension of the symmetrical case (Rankine-Froude theory) to the general, unsymmetrical flight of the actuator disk affords an informative perspective for the whole class of devices which both lift and propel by changing the fluid momentum.

The actuator disk in skew flow includes the circular lifting surface, characterized by zero net flow through the disk, as a special case. The kinetic energy waste left in the wake is then the induced power of an elliptically loaded wing of span $2R$,

$$P_i = \frac{F_r^2}{2\rho\pi R^2 V} \quad \text{for} \quad V \gg v_i \tag{6-1}$$

where v_i is the induced velocity, that is, the flow disturbance at the actuator disk, and ρ is the air density.

In the symmetrical case, the wake consists of a jet stream, the typical mark of a propulsive system. In the wing case there is no jet wake, but Prandtl has shown that one may imagine an equivalent curved stream tube of cross section πR^2 as a one-dimensional substitute for the flow of a wing having the span $2R$. Glauert and Küssner have suggested that the area of the affected stream tube is πR^2 regardless of the direction in which the actuator disk is flying. The mass of the affected fluid is then $\rho\pi R^2(|\mathbf{V} + \mathbf{v}_i|)$, and the resulting change of momentum is $F_r = \rho\pi R^2|\mathbf{V} + \mathbf{v}_i|2v_i$. The wasted kinetic energy is given by

$$P_i = 2\rho\pi R^2 V v_i^2 = F_r v_i$$
$$= \frac{F_r^2}{2\rho\pi R^2 V} \quad \text{for} \quad V \gg v_i \tag{6-2}$$

The Glauert-Küssner hypothesis covers the range between the two extremes of symmetrical flight and of winglike flight of the actuator disk.

In symmetrical rotor motion, when the vectors $\mathbf{V}$ and $\mathbf{F}_r$ are parallel, the restriction $V \gg v_i$ need not be imposed. The relations

$$\begin{aligned} P_i &= \frac{F_r^2}{2\rho\pi R^2(V + v_i)} \\ v_i &= -\tfrac{1}{2}V + \left[(\tfrac{1}{2}V)^2 + \frac{F_r}{2\rho\pi R^2}\right]^{\frac{1}{2}} \end{aligned} \tag{6-3}$$

then hold. In the particular case of static thrust ($V = 0$), one obtains

$$\begin{aligned} P_i &= \frac{F_r^2}{2\rho\pi R^2 v_i} = F_r^{\frac{3}{2}}(2\rho\pi R^2)^{-\frac{1}{2}} \\ v_i &= \left(\frac{F_r}{2\rho\pi R^2}\right)^{\frac{1}{2}} \end{aligned} \tag{6-4}$$

In general skew flight of the actuator disk, its wake consists of both an elliptic jet stream and an equivalent wing wake. The momentum change entrained in the jet wake corresponds to a propulsive force F_p, and the momentum change in the flow which does not touch the rotor corresponds to a lift force F_l so that

$$\mathbf{F}_r = \mathbf{F}_p + \mathbf{F}_l$$

Similarly, the sum of the kinetic energies inside and outside of the jet stream is given by

$$P_i = P_{ip} + P_{il}$$

With respect to the losses in the wake, with which we were concerned here, the details of the wake-producing devices are insignificant. The equivalent wing-propulsion system could be a conventional wing and propellers, a conventional wing and jet engines, or a wing with a jet flap [5,6,7], or a wing serving as corner vane for the deflection of a propeller slipstream or jet stream.

J,7. Blade Element Theory of Rotors. An infinitely long straight wing of chord c, moving at the velocity $\mathbf{U}_a$ in air of density ρ is subject to a force per unit length $\frac{1}{2}\rho U_a^2 c\mathbf{c}_R$, which is the vector sum of section lift and drag, i.e.

$$\mathbf{c}_R = \mathbf{c}_L + \mathbf{c}_D \tag{7-1}$$

where $\mathbf{c}_L$, the section lift coefficient and $\mathbf{c}_D$, the section drag coefficient, are interpreted as vectors. In wing theory, including rotor theory, it is assumed that any blade section is subject to the same force per unit length

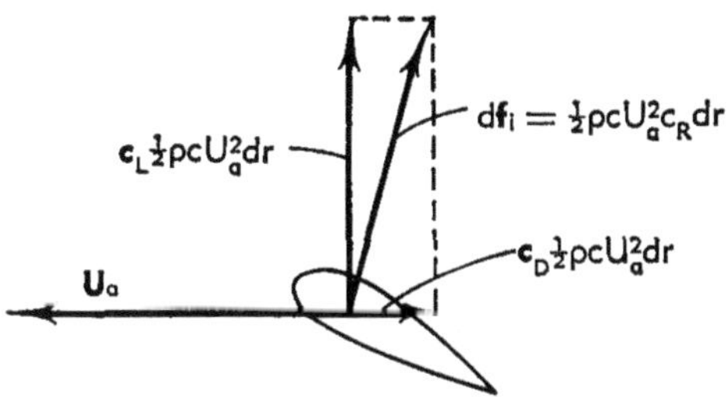

Fig. J,7. Vector diagram of the forces acting on a blade element.

as a two-dimensional section. Under this assumption, the instantaneous force acting on the ith blade is obtained as

$$\mathbf{f}_i = \int_0^R \tfrac{1}{2}\rho c U_a^2 \mathbf{c}_R dr \tag{7-2}$$

so that the mean rotor force becomes

$$\mathbf{F}_r = \frac{b}{2\pi} \int_{-\pi}^{\pi} \mathbf{f}_i d\omega t \tag{7-3}$$

The blade element moves in space with the velocity U (Eq. 3-1), and relative to the air near the rotor with

$$\mathbf{U}_a = \mathbf{U} - \mathbf{v}_i = \mathbf{V} + (\boldsymbol{\omega} \times \mathbf{r}) - \mathbf{v}_i \tag{7-4}$$

where $\mathbf{v}_i$ $(\omega t, r)$ is the induced velocity disturbance of the rotor.

Consider now the work done in unit time by the forces acting on a blade element during its motion through the air (see Fig. J,7)

$$\mathbf{U}_a \cdot d\mathbf{f}_i = -\frac{\rho}{2} U_a^3 c_D c dr$$

The power expended by the blades of the rotor

$$\frac{b}{2\pi} \int_0^R \int_{-\pi}^{\pi} \frac{d\mathbf{f}_i}{dr} \cdot \mathbf{U}_a dr d(\omega t) = -\frac{b}{4\pi} \rho \int_0^R \int_{-\pi}^{\pi} U_a^3 c_D c dr d(\omega t) \tag{7-5}$$

may be expanded into three terms by virtue of Eq. 7-4. The first,

$$\frac{b}{2\pi}\int_0^R\int_{-\pi}^{\pi}\mathbf{V}\cdot\frac{d\mathbf{f}_i}{dr}\,drd(\omega t) = \mathbf{V}\cdot\mathbf{F}_r \tag{7-6}$$

is the propulsive power of the rotor force $\mathbf{F}_r$. The second term containing $d\mathbf{f}_i \cdot \boldsymbol{\omega} \times \mathbf{r} = \boldsymbol{\omega} \cdot \mathbf{r} \times d\mathbf{f}_i$

$$\boldsymbol{\omega}\cdot\frac{b}{2\pi}\int_0^R\int_{-\pi}^{\pi}\mathbf{r}\times\frac{d\mathbf{f}_i}{dr}\,drd\omega t = \omega Q_r \tag{7-7}$$

is the turning power of the rotor blades. If $\omega Q_r > 0$, the air is accelerating the rotor, $\omega Q_r = 0$ signifies a rotor in autorotation, and $\omega Q_r < 0$ requires either a shaft-torque power or turning power of jet engines to maintain constant rotation. The torque balance for this case is given by

$$\omega Q_r + \omega b Q_j = 0 \tag{7-8}$$

The last term,

$$-\frac{b}{2\pi}\int_0^R\int_{-\pi}^{\pi}\mathbf{v}_i\cdot\frac{d\mathbf{f}_i}{dr}\,drd\omega t = P_i \tag{7-9}$$

represents that part of the rotor power which produces the flow disturbance, i.e. the kinetic energy waste P_i in the rotor wake.

The expansion of Eq. 7-5 yields, therefore,

$$\frac{b}{2\pi}\int_0^R\int_{-\pi}^{\pi}\frac{d\mathbf{f}_i}{dr}\cdot\mathbf{U}_a drd\omega t = \mathbf{V}\cdot\mathbf{F}_r - bQ_j\omega + P_i = -E \tag{7-10}$$

which demonstrates the previous assertion in Eq. 5-3.

The energy dissipated by the blade section drag is precisely defined by

$$E = \frac{b}{2\pi}\frac{\rho}{2}\int_0^R\int_{-\pi}^{\pi}U_a^3 c_D c\,drd\omega t$$

The propeller and helicopter rotor literature contains a large variety of assessments of E.

Let the following estimates be adopted as both adequate in this context and suitable for consistent comparison between propeller and helicopter rotor states:

For *subsonic rotors*, Bennett's formula [*8*] for the horizontal flight of lifting rotors ($V = \mu\omega R$) is well approximated by

$$E = \frac{1}{2}\left(\frac{1}{4} + \frac{4.6}{4}\mu^2\right)c_D\rho\sigma\pi R^2(\omega R)^3 \tag{7-11}$$

for $0 \leqq \mu < 0.6$. The rotor solidity is denoted by σ so that the quantity $\sigma\pi R^2 = bcR$ represents the total blade area. For propellers ($V = \lambda\omega R$, $\mu = 0$) Glauert [*9*, p. 204] estimates that

$$E = c_D\rho\sigma\pi R^2(\omega R)^3\tfrac{1}{2}f(\lambda)$$

where some values of the function $f(\lambda)$ are listed in Table J,7.

Table J,7

λ	0	0.1	0.2	0.3	0.4	0.5
$f(\lambda)$	0.25	0.258	0.282	0.325	0.393	0.488

In Eq. 7-11 and Table J,7, c_D has been assumed to be constant over the rotor disk. For subsonic airfoil sections for which the thickness ratio t/c lies between 0.04 and 0.15, at Mach numbers U/a_∞ between zero and M_D (the Mach number of drag divergence), and for $c_L < c_{L_{\text{stall}}}$, the approximate relation

$$c_D \cong 0.007(1 + c_L^2) \tag{7-12}$$

may be applied. Thus an appropriate mean drag coefficient may be chosen, based on Eq. 7-12 and a mean section lift coefficient of the rotor blades.

Both the rapid rise of the drag when U_t exceeds $a_\infty M_D$ [*10*], and the other inconveniences of transonic operation make it desirable to either avoid transonic rotor tip speeds or to go to sufficiently high supersonic velocities in order to move the transonic blade section well away from the blade tip domain which carries the burden of thrust and energy dissipation.

For *supersonic rotors*, $U_j > a_\infty$, one will choose supersonic blade sections at least at the blade tip domain, say biconvex sections, for which

$$c_D = c_f + \frac{16}{3\beta}\left(\frac{t}{c}\right)^2 + \beta\frac{c_L^2}{4} \tag{7-13}$$

where $c_f = 0.004$ is the approximate friction coefficient, $\beta \equiv \sqrt{M^2 - 1}$, and t/c is the section thickness ratio.

Since the losses E increase monotonically with c_D, they are obviously quite sensitive to increases of t/c. Thickness ratios greater than 0.03 are conducive to large losses and violate the validity of Eq. 7-13 which presupposes linearized flow.

Gail [*11*] approximated Eq. 7-13 by

$$c_D = \frac{K_1(t/c)^2}{M} + K_2c_L^2M$$

$$1 < M = \frac{U}{a_\infty} < 2.5$$

For $t/c = 0.03$ as a likely representative of supersonic rotors, this equation yields

$$c_D = 0.01296\frac{a_\infty}{U} + 0.222c_L^2\frac{U}{a_\infty} \tag{7-14}$$

Using Eq. 7-14, he obtained

$$E = E_0 + E_L$$

$$E_0 = 0.01296\left(\frac{1}{6} + \frac{\lambda^2}{2} + \frac{\mu^2}{2}\right)\rho\sigma\pi R^2(\omega R)^3\frac{a_\infty}{\omega R} \tag{7-15}$$

$$E_L = 0.222\left(\frac{1}{10} + \frac{\lambda^2\mu^2}{2} + \frac{\mu^2}{2}\right)\rho\sigma\pi R^2(\omega R)^3\frac{\omega R}{a_\infty}c_L^2$$

with the restrictions $\mu \leqq 0.25$, $t/c = 0.03$, and $c_L(r, \omega t) = \text{const}$. Eq. 7-15 overestimates the dissipation of the subsonic sections near the rotor hub (*11*, Fig. 8 and 9).

J,8. The Static Thrust of Jet Rotors. The static thrust ($\mathbf{F}_{jr}$ at $V = 0$) for perfectly aligned jet engines, for which $\mathbf{T}_j \times (\boldsymbol{\omega} \times \mathbf{R}) = 0$, is equal to $\mathbf{F}_r$; in addition, $\mathbf{F}_r \times \boldsymbol{\omega} = 0$ because of the symmetry of the system. Assuming a rectangular blade planform (c = const) and a constant value of c_L, one obtains for $v_i \ll U$

$$F_r = \rho\sigma\pi R^2(\omega R)^2\frac{c_L}{6} \tag{8-1}$$

which is valid for subsonic and supersonic rotors.

The energy balance yields

$$bT_j = \frac{E}{\omega R} + \frac{P_i}{\omega R} \tag{8-2}$$

the aggregate jet thrust required to maintain constant rotation. It follows from Eq. 6-4 and 8-1 that for subsonic and supersonic rotors

$$\frac{P_i}{\omega R} = F_r\frac{v_i}{\omega R} = F_r\sqrt{\frac{\sigma c_L}{12}} = \frac{F_r^{3/2}}{\omega R\sqrt{2\rho\pi R^2}} \tag{8-3}$$

Combining Eq. 7-11 or Table J,7 with Eq. 7-12 and 8-1, one obtains for subsonic rotors ($U_j < a_\infty M_D$)

$$\frac{E}{\omega R} = \frac{3}{4}F_r\frac{c_D}{c_L} = \frac{3}{4} \times 0.007F_r\left(\frac{1}{c_L} + c_L\right) \tag{8-4}$$

The required aggregate jet thrust as a fraction of the static jet rotor thrust is therefore,

$$\frac{bT_j}{F_r} = \frac{3}{4} \times 0.007\left(\frac{1}{c_L} + c_L\right) + \sqrt{\frac{\sigma c_L}{12}} \tag{8-5}$$

Fig. J,8, top, shows the reciprocal of this ratio as a function of the section lift coefficient c_L for different values of $\sigma \equiv bc/\pi R$.

For supersonic rotors ($\omega R > a_\infty$), and for $t/c = 0.03$, Eq. 7-15 and

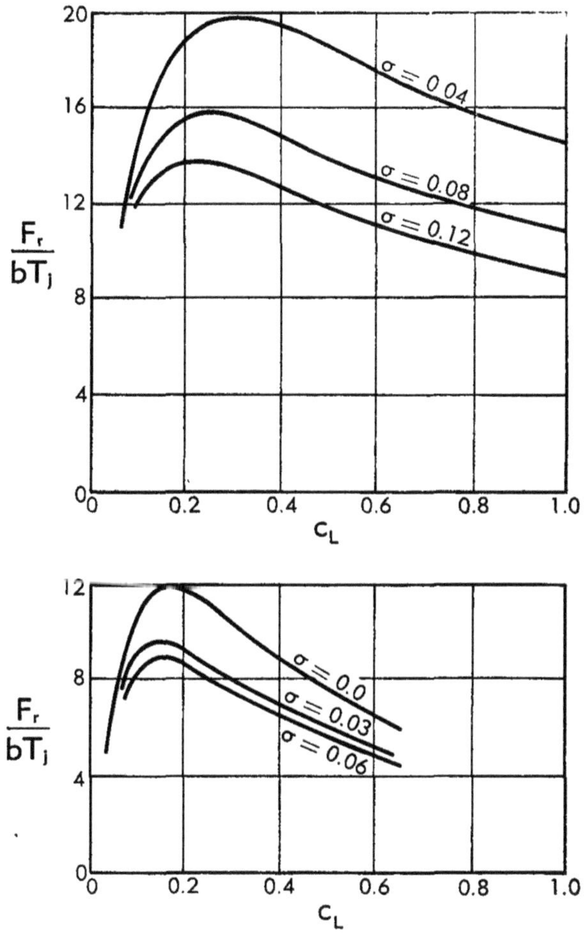

Fig. J,8. Static thrust of jet rotors. Top, subsonic rotors (Eq. 8-5); bottom, supersonic rotors (Eq. 8-8).

8-1 may be combined to give

$$\frac{E}{\omega R} = F_r\left(\frac{0.01296a_\infty}{c_L\omega R} + \frac{3}{5} \times 0.222c_L\frac{\omega R}{a_\infty}\right) \tag{8-6}$$

which yields

$$\frac{bT_j}{F_r} = \frac{0.01296}{c_L}\frac{a_\infty}{\omega R} + \frac{3}{5} \times 0.222c_L\frac{\omega R}{a_\infty} + \sqrt{\frac{\sigma c_L}{12}} \tag{8-7}$$

Specifically, one obtains for a supersonic jet rotor at sea level ($a_\infty = 1117$ ft/sec), and for $\omega R = 2000$ ft/sec

$$\frac{bT_j}{F_r} = \frac{0.00724}{c_L} + 0.2385c_L + \sqrt{\frac{\sigma c_L}{12}} \tag{8-8}$$

The reciprocal F_r/bT_j (c_L, σ) is plotted in Fig. J,8, bottom.

Note in both subsonic and supersonic cases that in order to produce the largest static thrust F_r at a given jet engine capacity bT_j, one should try to satisfy the following conditions:

1. The rotor solidity σ ought to be as small as practicable and, since $\sigma = bc/\pi R$,
   a. Two blades ought to be used unless a single blade with counterweight can be tolerated.
   b. The blades ought to be as slender as possible, that is, R/c as large as possible.
2. The blade lift coefficient C_L ought to lie between 0.2 and 0.3 for subsonic, and between 0.15 and 0.175 for supersonic rotors. The blade loading is given by (see Eq. 8-1)

$$\frac{F_r}{bcR} = \rho \frac{(\omega R)^2}{6} C_L$$

   so that desirable values for sea level operation are

$$\frac{F_r}{bcR} \cong 30 \text{ lb/ft}^2 \text{ for subsonic rotors with } \omega R \cong 600 \text{ ft/sec, and}$$

$$\frac{F_r}{bcR} \cong 240 \text{ lb/ft}^2 \text{ for supersonic rotors with } \omega R = 2000 \text{ ft/sec.}$$

3. The largest static thrust is, approximately, 20 bT_j for subsonic jet rotors and $10bT_j$ for supersonic rotors.

J,9. Jet Propeller Performance. Propeller operation means symmetrical operation (compare Art. 3 and 4) so that, with perfect alignment of the jet engines, the propeller thrust contribution of the jet engines and the jet rotor thrust is given by

$$F_j = bT_j \sin \phi \tag{9-1}$$

and

$$F_{jr} = F_r + bT_j \sin \phi, \tag{9-2}$$

respectively.

The equivalent shaft torque power supplied to the jet propeller

$$bQ_j\omega = bT_j\omega R \cos \phi = bT_jU_j \cos^2 \phi$$

is converted by the blades into propulsive powers and losses,

$$bT_jU_j \cos^2 \phi = F_rV + E + P_i \tag{9-3}$$

The ratio of the propulsive power to the power input

$$\frac{F_rV}{bT_jU_j \cos^2 \phi} = \eta_p \tag{9-4}$$

is called the propulsive efficiency of the jet propeller insofar as it acts as a pure propeller.

The total power output of the jet propeller is given by

$$F_{jr}V = F_rV + bT_jU \sin^2 \phi$$

and the total power input by bT_jU_j so that the efficiency of the jet propeller becomes

$$\begin{aligned} \eta_{jp} \equiv \frac{F_{jr}V}{bT_jU_j} &= \eta_p \cos^2 \phi + \sin^2 \phi \\ &= \eta_p + (1 - \eta_p) \sin^2 \phi \end{aligned} \tag{9-5}$$

One may estimate E and P_t to obtain

$$\eta_p = 1 - \frac{E + P_t}{bT_jU_j \cos^2 \phi} \tag{9-6}$$

or η_p may be known or estimated from propeller tests.

An idealized broken-line representation of a typical efficiency curve of a variable pitch propeller is shown in the lower part of Fig. J,9. Taking this η_p as given, η_{jp}, F_r/bT_j; F_{jr}/bT_j, and F_j/bT_j follow from Eq. 9-5, 9-4, 9-5, and 9-1, respectively. Also plotted in Fig. J,9 is the thrust ratio, T_P/bT_j, of a propeller that receives the same power bT_jU_j as the jet propeller, but as shaft power rather than by a number of jet engines mounted at the blade tips. It is seen that

1. $\eta_{jp} > \eta_p$ for $0 < \phi \leqq \pi/2$ since, of course, $\eta_p < 1$; $\eta_{jp} = 1$ for $\phi = \pi/2$.
2. $F_{jr} > T_P$ for $0 < \phi \leqq \pi/2$, particularly as ϕ approaches $(\pi/2)(V \to U_j)$. A shaft-driven propeller then loses all usefulness ($\eta_p \to 0$, $T_P \to 0$), whereas the jet propeller reaches peak efficiency ($\eta_{jp} \to 1$ and $F_{jr} \to bT_j$).

The reason for the superior high speed performance of jet rotors is the previously noted (Art. 4) increase with flight velocity of the direct contribution of the jet thrust to the propulsive power.

The flight condition $\phi = \pi/2$, $\omega = 0$, $U_j = V$ may be gradually reached (Art. 3), but it requires that rotor blades of special construction be used; for example, one might consider blades of limited slenderness with attendant lowered static thrust potential (Art. 8) to withstand air loads and jet engine loads without the benefit of the centrifugal stiffening (Art. 11 and 16) of a rotating blade.

Although an idealized function $\eta_p(\phi)$ has been deliberately chosen in Fig. J,9, the quasi-hyperbolic curve for $(F_r/bT_j)(V/U_j)$ must be expected as characteristic of all jet propellers, both subsonic and supersonic.

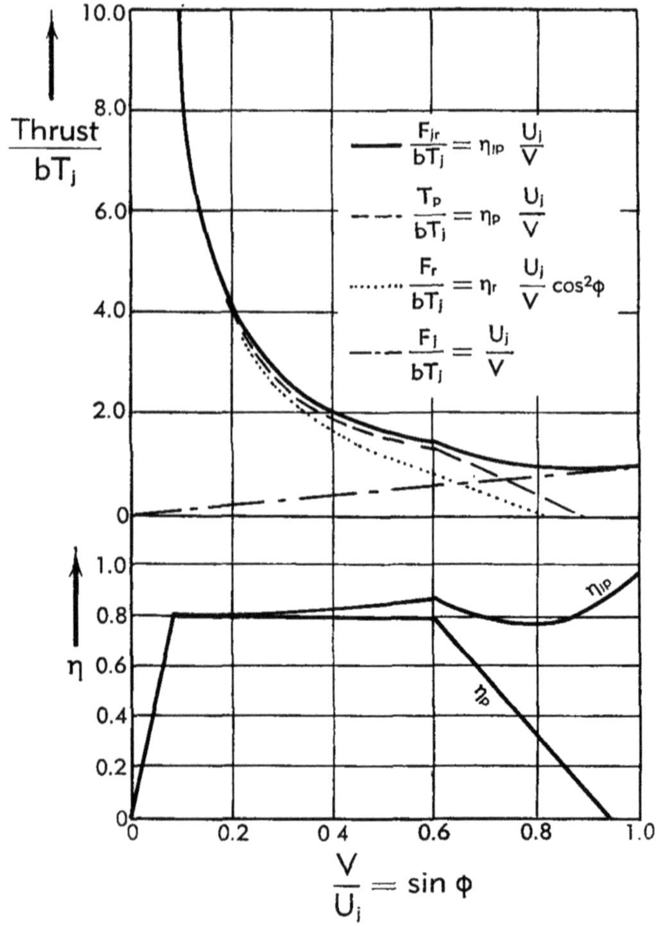

Fig. J,9. Performance of jet propellers.

Estimates for the performance of a supersonic jet propeller are given in [*11*].

J,10. Helicopter Jet Rotor Performance.

Steady level flight of a jet helicopter requires (Fig. J,10)[1]

$$\mathbf{F}_{jr} + \mathbf{W} + \mathbf{D} = 0 \tag{10-1}$$

The gross weight W of a helicopter is very much larger than its drag D, because the flight velocity is limited to less than 200 mph (see below). Consequently, the following approximations may be made:

[1] The jet helicopter is shown with a tail rotor which appears to be needed for yaw control. The absence of a reaction torque, which distinguishes jet rotors from shaft-driven ones, changes the role of the tail rotor from a device required for equilibrium and yaw control to one required for yaw control only.

$$
\begin{aligned}
i &= \frac{\pi}{2} - \frac{D}{W} \\
V &= \mu\omega R \\
\lambda &= \frac{D}{W}\frac{V}{\omega R} = \frac{D}{W}\mu \ll \mu < 1 \\
\mathbf{T}_j \cdot \boldsymbol{\omega} &= 0, \\
F_j &= 0, \text{ and} \\
F_{jr} &= F_r = W
\end{aligned} \tag{10-2}
$$

Note that $\mathbf{F}_r$ is parallel to $\boldsymbol{\omega}$ in hovering only when $V = 0$, $F_r = W$, and $D = 0$; it is almost, but not precisely, parallel to $\boldsymbol{\omega}$ when $\mu > 0$.

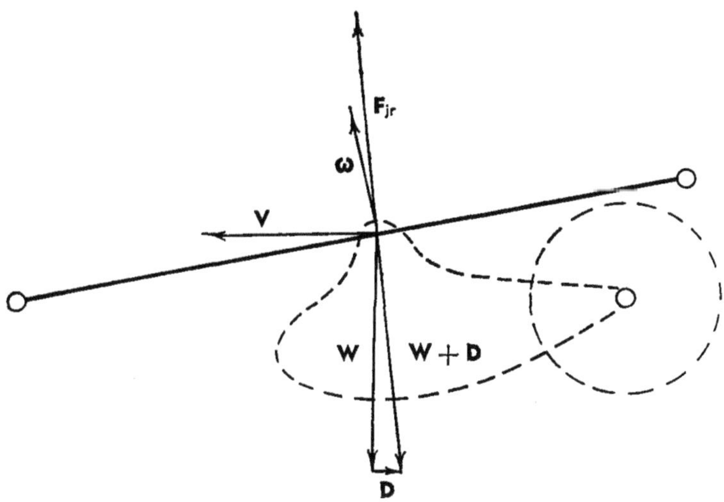

Fig. J,10. Jet helicopter in steady, level flight.

From Eq. 6-2 and 10-2

$$
P_i = \frac{W^2}{2\rho\pi R^2 \omega R \mu} \quad \text{for} \quad \mu > 0 \tag{10-3}
$$

$$
\begin{aligned}
\mathbf{F}_r \cdot \mathbf{V} &= -\mathbf{D} \cdot \mathbf{V} = DV \\
\frac{b}{2\pi}\int_{-\pi}^{\pi} \mathbf{T}_j \cdot \mathbf{V} d\omega t &= \mathbf{F}_r \cdot \mathbf{V} + P_i + E
\end{aligned} \tag{10-4}
$$

or, approximately, from Eq. 10-3, 10-4, and 7-11,

$$
bT_j\omega R = DV + \frac{W^2}{2\rho\pi R^2\omega R\mu} + \frac{\rho}{2}\sigma\pi R^2(\omega R)^2 c_D\left(\frac{1}{4} + \frac{4.6}{4}\mu^2\right)
$$

$\mu > 0$; subsonic rotor

Since $F_\tau \cong W$ for all values of μ, including $\mu = 0$, the mean lift coefficients of the blade section may be approximated by Eq. 8-1

$$c_L \cong \frac{GW}{\sigma\pi R^2(\omega R)^2\rho} \tag{10-5}$$

Thus the approximate required aggregate jet thrust of the jet helicopter rotor is, in dimensionless form,

$$\frac{bT_j}{W} = 3\,\frac{S_\pi}{\sigma\pi R^2}\,\frac{c_{D_\pi}}{c_L}\,\mu^3 + \frac{c_L\sigma}{12\mu} + \frac{3}{4}\,\frac{c_D}{c_L}\,(1 + 4.6\mu^2), \quad \mu > 0; \qquad \text{subsonic rotor} \tag{10-6}$$

where

$$D \equiv \frac{\rho}{2}\,S_\pi c_{D_\pi}(\omega R\mu)^2$$

Flight speed limitations of jet helicopters. The following phenomena may limit the flight speed of jet helicopters:

1. The advancing blade tip of a subsonic rotor may reach transonic speed at a flight velocity V_1,
2. The retreating rotor blade may stall at a flight velocity V_2,
3. The jet engine performance may deteriorate if at a flight velocity V_3 the yaw angle exceeds the yaw tolerance of the intake diffusor of the engine,
4. The required jet thrust power may exceed the available power at flight speeds in excess of V_4.

The usual situation for helicopters is that V_4 is larger than either V_1 or V_2. An example of performance degradation caused by yaw oscillations was encountered in Art. 3, where V_3, approximately equal to 95 mph, was estimated to limit a supersonic jet helicopter rotor. The transonic and stalling speed limitations are familiar obstacles in the technology of subsonic shaft-driven helicopters [*12*]. Flight speeds in excess of either V_1 or V_2 endanger the blade motion and the control of the helicopter.

The speed limit V_1 may be estimated to be that at which $U_j(\omega t = \frac{1}{2}\pi) = \omega R + V_1 = M_D a_\infty$, and, therefore,

$$V_1 = M_D a_\infty - \omega R \tag{10-7}$$

The stalling limit V_2 may be estimated by assuming that the blade lift on the advancing and retreating sides must be equal to make the rolling moment of the rotor vanish. This assumption permits one to generalize the tests reported by Gustafson and Myers [*13*] and to obtain the following semiempirical formula

$$c_{L_{\text{mean}}} = 1.2\left(1 - 1.67\,\frac{V_2}{\omega R}\right) \tag{10-8}$$

where $c_{L_{mean}}$ is the lift coefficient c_L of Eq. 10-5. Hence,

$$V_2 = \frac{\omega R}{1.67}\left(1 - \frac{c_{L_{mean}}}{1.2}\right) \tag{10-9}$$

Jet rotors favor $c_{L_{mean}} \cong 0.25$ (Fig. J,8, top), so that $V_2 \cong 0.475\omega R$. The largest flight speed which just avoids stall and transonic difficulties is

$$V_{max} = V_1 = V_2 = 0.475\omega R = M_D a_\infty - \omega R$$

or

$$(\omega R)_{max} = \frac{M_D a_\infty}{1.475}$$

and

$$V_{max} = \frac{0.475}{1.475} M_D a_\infty$$

For thin blade tip sections and low altitude, $M_D a_\infty \cong 900$ ft/sec [*10*], hence

$$\left.\begin{aligned} V_{max} &= 290 \text{ ft/sec} = 198 \text{ mph} \\ (\omega R)_{max} &= 610 \text{ ft/sec} \end{aligned}\right\} \tag{10-10}$$

This estimate of V_{max} is quite generous. No helicopter to date has reached this flight velocity.

The maximum yaw angle $\tan^{-1} [V_{max}/(\omega R)_{max}] \cong 25°$ may well be tolerable for subsonic intake diffusers.

The most damaging of the intrinsic helicopter limitations is, however, the restriction $\omega R < 610$ ft/sec. This condemns the jet engines of jet helicopters to a low speed domain in which they are notoriously uneconomical and inferior to other sources of power. Note that, in contrast to the subsonic jet helicopter, the same rotor used as a jet propeller allows $U_j \cong 900$ ft/sec and $V_{max} = U_j \cong 900$ ft/sec.

J,11. The Blade of Constant Tensile Stress. In shaft-driven as well as jet rotor blades, the dominant structural load is the tension caused by the centrifugal forces. Hence, blades designed for constant maximum allowable tensile stress σ_b are the lightest for their given strength.

Large rotors and most jet rotors carry a concentrated tip weight, W_t, as either a blade tip ballast or as weight of the jet engine, W_j (Art. 12 and [*12,14*]).

Small, shaft-driven blades and jet rotor blades with light tip weights (e.g. pressure jet nozzles or rocket nozzles) cannot practically be constructed with the paper-thin blade tip cross sections required by σ_b = const. Such rotor blades will have a blade tip domain of constant minimum practical mass [*12,14*]. Even for the small rotors with partly idle tip domain the constant stress blade with concentrated tip weight may be considered a reasonable approximation.

Consider the constant stress blade with tip weight W_t. Let

$S(x)$ = cross section of tensile load-carrying blade structure at $r = xR$

w = specific weight of blade material (load carrying or not), considered as a constant

$x = r/R$

$L = \sigma_b/w$ (see footnote 2)

$A = (\omega R)^2/2gL$

The equilibrium of a blade element requires

$$-(\omega R)^2 w S(x) x dx = g\sigma_b dS \tag{11-1}$$

whence

$$S(x) = S(1)e^{A(1-x^2)} \tag{11-2}$$

The cross section at the blade tip, $S(1)$, must accelerate the mass of the tip weight, so that

$$\sigma_b S(1) = \frac{W_t(\omega R)^2}{gR} \tag{11-3}$$

The weight of the blade including W_t is

$$W_b = wS(1)Re^A \int_0^1 e^{-Ax^2}dx + W_t$$
$$= W_t(\sqrt{\pi A}\, e^A + 1) \tag{11-4}$$

where the error function $2/\sqrt{\pi}\int_0^{\sqrt{A}} e^{-Ax^2}d(x\sqrt{A})$ has been approximated to be 1.0 which is both conservative and reasonable for the values of A greater than 0.8 that will be encountered.

The moment of inertia of the blade and its tip weight about the rotor center is

$$I = \frac{w}{g} S(1)R^3 \int_0^1 x^2 e^{A(1-x^2)}dx + \frac{W_t R^2}{g}$$
$$= W_t e^A \sqrt{\pi A}\, \frac{L}{\omega^2} \tag{11-5}$$

The blade subjects the rotor hub to the blade retention load,

$$\sigma_b S(0) = \frac{e^A W_t(\omega R)^2}{Rg} \tag{11-6}$$

The rotor hub may be estimated to weigh as much as the sum of its blades, hence the approximate rotor weight is

$$W_r = 2bW_b = 2bW_t(\sqrt{\pi A}\, e^A + 1) \tag{11-7}$$

[2] A cylindrical bar of length L and specific weight w hanging vertically from one end has a stress σ_b at the point of suspension.

Typical cases. Hafner [*12*] quotes as typical for subsonic blade construction,

$$L = \frac{\sigma_b}{w} = 7000 \text{ ft} \tag{11-8}$$

For supersonic slender jet rotor blades of solid steel construction (except for a radial fuel duct), Gail [*11*] estimates $\sigma_b = 60{,}000 \text{ lb/in.}^2$ and

$$L = 17{,}600 \text{ ft} \tag{11-9}$$

The large difference in L is not only caused by the larger allowable stresses in the limber supersonic blades, but also by the fact that w in the subsonic blade contains much blade material that carries no centrifugal loads, such as the chordwise stiffening structure and the chordwise ballast for the chordwise mass center control [*12*].

Typical configurations and their modes of operation are:

Case *a*. Subsonic helicopter rotor

$$\omega R = 600 \text{ ft/sec}$$
$$A = 0.8$$
$$L = 7000 \text{ ft}$$

Case *b*. Subsonic propeller-type rotor for VTOL application at $V = 0$

$$\omega R = 900 \text{ ft/sec}$$
$$A = 1.8$$
$$L = 7000 \text{ ft}$$

Case *c*. Supersonic propeller-type rotor for VTOL at $V = 0$

$$\omega R = 2000 \text{ ft/sec}$$
$$A = 3.53$$
$$L = 17{,}600 \text{ ft}$$

The corresponding weights of the individual blades and of the complete rotors in terms of the tip weight are listed in Table J,11.

Table J,11

	Case *a*	Case *b*	Case *c*
$\frac{W_b}{W_t} = e^A \sqrt{\pi A} + 1$	4.525	15.43	114.9
$\frac{W_r}{bW_t} = 2e^A \sqrt{\pi A} + 1$	9.05	30.86	229.8

J,12. Least Rotor Weights Required by Coning Restrictions. Rotor blades of adequate tensile strength (Art. 11) may be too light. The blade lift may either bend the blades too much (rigidly attached blades or seesaw-articulated blade pairs) or fold them up like rabbit ears (individually articulated blades). Rotor blades must be designed heavy enough to keep the coning angle β at or below a safe coning angle. At the equilibrium blade position, the moment of the centrifugal forces of the blade, $\omega^2 I \cos\beta \sin\beta$, equals the moment of the blade lift. The static thrust will be the crucial design condition corresponding to β^* and the vertical acceleration will be limited by a coning angle β_0 at which there is danger of rotor collapse.

The assumption is made that jet rotors as propellers or helicopter rotors will be used for vertical take-off and landing, i.e. including the hovering condition.

At the condition of static thrust, $V = 0$, $U = x\omega R \cos\beta$, and

$$F_r(\beta) = F_r(0) \cos^3 \beta \tag{12-1}$$

where $F_r(0)$ is the static thrust for β equal to or almost equal to zero (Eq. 8-1). The moment of the blade lift is

$$\tfrac{1}{8}\rho c C_L R^2(\omega R)^2 \cos^2\beta = \frac{F_r(\beta)}{b}\,\frac{3R}{4\cos\beta}$$

Thus, for blade equilibrium (strictly for individually articulated blades but because of the small contribution of the hub constraint, approximately valid also for seesaw and rigidly attached large blades)

$$\frac{F_r(\beta)}{b}\,\frac{3}{4\cos\beta} = \omega^2 I \cos\beta \sin\beta$$

or

$$F_r(\beta) = \frac{4}{3}\, b\,\frac{\omega^2 I}{R} \cos^2\beta \sin\beta \tag{12-2}$$

$$= K \cos^2\beta \sin\beta$$

For a given rotor and rotor speed, K is a constant so that $F_r(\beta)$ has a maximum

$$F_r(\beta_0) = K\tfrac{2}{3}\sqrt{\tfrac{1}{3}} \cong 0.385K$$

at

$$\beta_0 = \sin^{-1}\sqrt{\tfrac{1}{3}} \cong 35.5°$$

Thus if $F_r > F_r(\beta_0)$, no equilibrium blade motion is possible and the blades fold up ($\beta \to \pi/2$). If $F_r < F_r(\beta_0)$, the blade motion is stable at $\beta < \beta_0$. The equilibrium at $F_r(\beta_0)$ is neutrally stable.

The safe coning angle for hovering, β^*, should be substantially smaller than β_0 to avoid an inadvertent collapse of the rotor.

The accepted practice for helicopter rotors is to design for $\beta^* = 0.1$ radian or 6° in hovering and unaccelerated flight. For this small angle, $\beta^* = 0.1 = \sin \beta^* \cos^2 \beta^*$ and

$$F_r(\beta^*) = W = \frac{4}{30} b \frac{\omega^2 I}{R} = \frac{1}{3.85} F_r(\beta_0) \tag{12-3}$$

Such rotors would be in danger of folding up during a sustained vertical acceleration exceeding $2.85g$ if ω remained constant and at a smaller acceleration if ω decreases during this maneuver. This danger alone is sufficient justification for the stated limit of β^*. The assumed individual articulation of the rotor blades is no restriction. For seesaw or rigidly attached blades, the danger of rotor folding is replaced by the equally serious bending deformation of the blade roots.

The usually cited reason [*12*] for requiring $\beta^* \leqq 0.1$ is not the one quoted above, but the desire to limit the alternating blade loads in helicopter flight $\mu > 0$ which are indeed serious [*15*]. Our justification for the limitation of β^* has the merit of applying to all rotors used for vertical take-off and descent regardless of whether these operations are followed by unsymmetrical (helicopter) or symmetrical (propeller) flight.

If one substitutes Eq. 11-5 into Eq. 12-3, one obtains the least ballast weight, W_t^*,

$$\frac{bW_t^*}{W} = 7.5 \frac{R}{Le^A \sqrt{\pi A}} \tag{12-4}$$

The least blade weight W_b^*, and the approximate least rotor weight, W_r^*, then follow from Eq. 11-7 as

$$\frac{bW_b}{W} = 7.5 \frac{R(1 + e^A \sqrt{\pi A})}{Le^A \sqrt{\pi A}} \tag{12-5}$$

$$\frac{W_r^*}{W} = 2 \frac{bW_b^*}{W} \tag{12-6}$$

These are the least weights required to keep $\beta = \beta^* = 0.1$ in the crucial hovering condition.

The choice of ωR, made on other grounds of rotor performance, determines L and A and hence

$$\frac{bW_t^*}{W} = \frac{R}{K_1} \quad \frac{bW_b^*}{W} = \frac{R}{K_2} \quad \text{and} \quad \frac{W_r^*}{W} = \frac{R}{\frac{1}{2}K_2}$$

The parameters $K_1(\omega R)$ and $K_2(\omega R)$ for the typical cases a, b, and c (Art. 11) are listed in Table J,12.

Table J,12

	Case *a*	Case *b*	Case *c*
$K_1 = \dfrac{Le^A\sqrt{\pi A}}{7.5}$	3290 ft	13,468 ft	267,300 ft
$K_2 = \dfrac{K_1}{1 + e^A\sqrt{\pi A}}$	727 ft	872.8 ft	2326.6 ft
$\frac{1}{2}K_2$	363.5 ft	436.4 ft	1163.3 ft

Note that the least rotor weight for the same static thrust and rotor size decreases with increasing ωR (compare with [*12*, Fig. 24 and Eq. 28]).

For a given ωR the least ballast weights, blade weights, and rotor weights are proportional to the product WR with the following important consequences:

Consequence I.

There will be a rotor radius R_0 for which the tip ballast will equal the weight of the tip-borne jet engine, W_j, which could power the blade, i.e.

$$\frac{bW_t^*}{W} = \frac{R_0}{K_1} = \frac{bW_j}{W} = \frac{bT_j}{W}\frac{W_j}{T_j}$$

or

$$R_0 = \frac{bT_j}{W}\frac{W_j}{T_j}K_1 \tag{12-7}$$

R_0 is the optimum size of a jet rotor for the following reasons:

For $R = R_0$ the parasitic blade tip ballast W_t^* required to insure $\beta = \beta^* = 0.1$ is exactly and usefully replaced by the tip-borne jet engine weight W_j; the weight of the shaft-driven blade W_b^* equals the weight of the jet-driven blade W_{bj}; and the weight of the shaft-driven rotor W_r^* equals the weight of the jet rotor W_{jr}. The jet rotor as a powered rotor has a dry weight advantage over the shaft-powered rotor system which equals the weight of the shaft-driving engine and transmission.

Since $W_j < W_t^*$ for $R > R_0$, jet rotors of larger than optimum size require a parasitic ballast on the jet engine to keep $\beta = \beta^* = 0.1$.

If $R < R_0$, then $W_j > W_t^*$, $\beta < \beta^*$, $W_{bj} > W_b^*$, and $W_{jr} > W_r^*$, i.e., a jet rotor of smaller than optimum size is heavier than the equivalent shaft-driven rotor, though it is not likely to be heavier than the shaft-driven rotor engine transmission system.

Coning angles that are substantially smaller than 0.1 indicate a decreased ratio of the aerodynamic over the inertial forces acting on the blade and may lead to deficient rotor tilt control upon which jet rotors are prone to depend.

The weight/thrust ratio of the jet drive of the jet rotor is a characteristic of the type of engine in the particular environment at the blade tip. It may be computed from Eq. 12-7 for $F_r = W$ and for appropriate values of K_1 (Table J,12) by using the data from Fig. J,8a and Eq. 8-5 for a subsonic rotor or from Fig. J,8b and Eq. 8-8 for a supersonic rotor.

For the three typical cases (Art. 11), one obtains

$$\left.\begin{array}{lll} \text{Case } a & R_0 \geqq 164.5 \dfrac{W_j}{T_j}\text{ ft}; & \dfrac{bT_j}{W} \geqq \dfrac{1}{20} \\ \text{Case } b & R_0 \geqq 673.4 \dfrac{W_j}{T_j}\text{ ft}; & \dfrac{bT_j}{W} \geqq \dfrac{1}{20} \\ \text{Case } c & R_0 \geqq 26{,}730 \dfrac{W_j}{T_j}\text{ ft}; & \dfrac{bT_j}{W} \geqq \dfrac{1}{10} \end{array}\right\} \qquad (12\text{-}8)$$

Consequence II.

If the rotor is too big, the ratio of rotor weight over static thrust will be inferior to the weight/thrust ratio of rival power plants for VTOL. At the preposterous rotor size $R_1 = \frac{1}{2}K_2$, the rotor can just barely lift itself, $W_r = W$.

Suppose a turbojet or a cluster of turbojet engines is to be used as a rival power plant instead of a jet rotor. For a fair comparison of dry weights, one has to add the weight of hovering controls to the turbojet engine weights since jet rotors will obviously have the tilt control of helicopter rotors.

Suppose the weight/thrust ratio of this rival turbojet for VTOL is $\frac{1}{5}$. Jet rotors with tip-borne jet engines must satisfy the inequality $W_{jr}/W < \frac{1}{5}$ to possess a competitive dry weight. Consequently, the rotor radius R must be smaller than $R_2 = R_1/5 = K_2/10$, or

$$\left.\begin{array}{ll} \text{Case } a. & R < 72.7\text{ ft} \\ \text{Case } b. & R < 87.3\text{ ft} \\ \text{Case } c. & R < 232.7\text{ ft} \end{array}\right\} \qquad (12\text{-}9)$$

This constraint on jet rotor weight and size can be satisfied by jet rotors with optimum radius R_0 only if

$$\left.\begin{array}{lll} & \dfrac{W_j}{T_j} < \dfrac{K_2}{10K_1}\left(\dfrac{W}{bT_j}\right), \text{ or} & \\ \text{Case } a. & \dfrac{W_j}{T_j} < 0.442 \quad \text{and} & \dfrac{W}{bT_j} = 20 \\ \text{Case } b. & \dfrac{W_j}{T_j} < 0.13 \quad \text{and} & \dfrac{W}{bT_j} = 20 \\ \text{Case } c. & \dfrac{W_j}{T_j} < 0.0087 \quad \text{and} & \dfrac{W}{bT_j} = 10 \end{array}\right\} \qquad (12\text{-}10)$$

If W_j/T_j of chosen or available jet engines does not satisfy the inequality (Eq. 12-10), the jet rotor designer is left with these choices:

1. Attempt to modify the jet engine so as to decrease its W_j/T_j.
2. Make the jet rotor smaller than optimum dimension, i.e. $R < R_0$.
3. Compensate in other ways for the large dry weight of the jet rotor of optimum size R_0.

Or, a combination of these choices can be used.

J,13. Mutual Interaction between the Jet Rotor and Its Jet-Engines. The environment of jet engines mounted at the blade tip of jet rotors is dominated by the blade tip acceleration, G, measured in g's, so that

$$\left.\begin{aligned} G &= \frac{(\omega R)^2}{gR} \\ \text{and} \qquad \omega &= \frac{(\omega R)}{R} \end{aligned}\right\} \tag{13-1}$$

is the turning speed of the blade, the rotor, and the jet engine.

In Art. 12, the radius of a jet rotor was shown to be bounded by considerations of either optimum jet rotor weight

$$R \leqq R_0\left(\frac{W_j}{T_j}, \omega R\right) \tag{13-2}$$

or competitive rotor weight,

$$R < R_2(\omega R) \tag{13-3}$$

The jet engine speed U_j and hence ωR (for any flight velocity $\mathbf{V}$) are chosen with the aim of avoiding near sonic velocity U_j (Art. 7) and of operating with acceptable fuel economy.

The "specific fuel consumption" of all jet engines (i.e. the fuel flow divided by the engine power $\mathbf{T}_j \cdot \mathbf{U}_j$) decreases with increasing U_j at least within the bounds of practical U_j. Suppose we designate one "specific fuel consumption" as acceptable. Then there will be a corresponding least U_{acc} or $(\omega R)_{acc}$ at which a particular type of jet engine has this acceptable consumption. For reasons of fuel economy, the tip speed must then satisfy the condition

$$\omega R \geqq (\omega R)_{acc} \tag{13-4}$$

These considerations establish two lower bounds for G and ω

$$\left.\begin{aligned} G &\geqq \frac{(\omega R)^2_{acc}}{R_0[W_j/T_j, (\omega R)_{acc}]} \\ \omega &\geqq \frac{(\omega R)_{acc}}{R_0[W_j/T_j, (\omega R)_{acc}]} \end{aligned}\right\} \tag{13-5}$$

and

$$\left.\begin{aligned} G &> \frac{(\omega R)^2_{acc}}{R_2[(\omega R)_{acc}]} \\ \omega &> \frac{(\omega R)_{acc}}{R_2} \end{aligned}\right\} \qquad (13\text{-}6)$$

Note that these two pairs of inequalities (1) express the designer's desire for a jet rotor of acceptable fuel economy as well as competitive dry weight (Eq. 13-6) or optimum dimension (Eq. 13-5), (2) must both be satisfied, and (3) relate the least severity of the environment of the tip-borne jet engines to the properties W_j/T_j and $(\omega R)_{acc}$ which characterize any particular jet engine type.

Let G_{j_a} denote the "acceleration tolerance," that is, the maximum allowable acceleration continuously tolerated by a jet engine (as it is or after suitable modifications). Let ω_{j_a} denote the analogous "turning tolerance" of the jet engine. To perform reliably as tip-borne jet engine of a rotor, it is necessary that

$$\left.\begin{aligned} G_{j_a} &> G = \frac{(\omega R)^2}{R} \\ &\text{and} \\ \omega_{j_a} &> \frac{\omega R}{R} \end{aligned}\right\} \qquad (13\text{-}7)$$

It follows from the inequalities (Eq. 13-5, 13-6, and 13-7) that the jet engine characteristics of the environmental tolerance G_{j_a} and ω_{j_a}, the weight/thrust ratio W_j/T_j, and the acceptable economy $(\omega R)_{acc}$ are interrelated and those inequalities determine a basic requirement for the jet engine suitability as a power plant for a jet rotor.

Jet engine types differ widely in their environmental tolerance G_{ja} and ω_{j_a}. All of them are sensitive to blade tip acceleration and consequently favor the largest tolerable rotor radius $R = R_0$ or $R = R_1$. Jet rotors are thus intrinsically prone to be large [*11,14,16*]. The large size of jet rotors (compared to, say, shaft-driven propellers) makes it practically impossible to mount them in conventional propeller fashion. Jet rotors will have to be carried overhead like helicopter rotors during take-off and landing maneuvers; i.e. jet rotors are by necessity VTOL power plants. By fortunate coincidence, their large static thrust (Art. 8) and their control capabilities in the fashion of helicopter rotors make jet rotors uniquely qualified for this role (Art. 6).

CHAPTER 3. JET ROTOR DESIGN

J,14. An Accounting System. The purpose of any craft to be powered by a jet rotor or a rival engine is the transportation of a useful load from one place to another.

Let the take-off gross weight of the craft be

$$W = W_{lp} + W_f + W_u + W_o \tag{14-1}$$

where W_{lp} is the dry weight of the lifting-propelling system, W_f the fuel weight at take-off, W_u the useful load, and W_o the necessary other weight. The sum $W_{lp} + W_f = W_{lpf}$ may be called "gross weight of the lifting-propelling system," so that

$$\frac{W_u}{W} = 1 - \frac{W_o}{W} - \frac{W_{lpf}}{W} \tag{14-2}$$

The take-off fuel weight is consumed during the flight of trip length S_t or trip duration t_t, i.e.

$$\begin{aligned} \frac{W_f}{W} &= \int_0^{t_t} c \frac{P}{W} dt \\ &= \int_0^{S_t} c \frac{P}{W} \frac{dS}{V} \end{aligned} \tag{14-3}$$

where the specific fuel consumption c is defined by

$$c = \frac{dW_f}{dt} P^{-1}$$

and P is the power required at any instant during flight. Since P is also the power delivered to the lifting-propelling system, it is given by

$$P = \frac{b}{2\pi} \int_{-\pi}^{\pi} \mathbf{T}_j \cdot \mathbf{U}_j d\omega t$$

for jet rotors and by $P = Q_s\omega$ for shaft-driven rotors.

Transportation of the useful load over a reasonably long trip length, say more than 50 miles, is the usual transportation mission of widely acknowledged usefulness. For such trips, the level cruising flight condition is the major consumer of fuel and thence, approximately,

$$\begin{aligned} \frac{W_f}{W} &= c \frac{P}{W} \frac{S_t}{V} \\ \frac{W_{lpf}}{W} &= \frac{W_{lp} + cS_t(P/V)}{W} \end{aligned} \tag{14-4}$$

Eq. 14-4 establishes an accounting system by means of which a great variety of aircraft configurations employing jet rotors or rival engines may be compared. Either the transportation distance for a given useful load, or the useful load for a given transportation distance, may be used as an indication of the transportation merit of the system.

Such an accounting is incapable of weighing fuel costs, noise penalties, equipment cost, maintenance cost, development uncertainties, VTOL premium, operational inconveniences (e.g. tail-sitting VTOL) and other features which affect the competitive standing of jet rotors. However, the transportation merit of jet rotor aircraft must be sufficiently competitive with its rivals to satisfy the basic tenet of engineering design (Art. 2).

J,15. Jet Helicopters. The contemporary rival of the jet helicopter is the shaft-driven helicopter. (Potential future rivals are discussed in Art. 16.) The transport merits of these two types of helicopters will be compared in terms of their respective transport distance S_t. Let the quantities pertaining to shaft-driven and jet-driven rotors be distinguished by the subscripts $_s$ and $_j$, respectively. The comparison is based on the following assumptions:

$$\left(\frac{W_u}{W}\right)_s = \left(\frac{W_u}{W}\right)_j; \quad \left(\frac{W_o}{W}\right)_s = \left(\frac{W_o}{W}\right)_j$$

$$V_s = V_j = V = 120 \text{ mph (cruising speed)}$$

$$(\omega R)_s = (\omega R)_j = 600 \text{ ft/sec (see Art. 10)}$$

$$R_s = R_j \geqq R_o$$

$$P = P_s = \frac{b}{2\pi} \int_{-\pi}^{\pi} \mathbf{T}_j \cdot \mathbf{U}_j d(\omega t) \cong bT_j \omega R$$

As a result of these assumptions,

$$\left(\frac{W_{lpf}}{W}\right)_s = \left(\frac{W_{lpf}}{W}\right)_j \tag{15-1}$$

The left side of this equation may be expressed as

$$\left(\frac{W_{lpf}}{W}\right)_s = \frac{W_r^*}{W} + \frac{W_{et}}{W} + \frac{c_s P S_{ts}}{WV} \tag{15-2}$$

where W_{et} is the weight of the engine and transmission, and

$$\frac{W_r^*}{W} = \frac{W_{jr}}{W} \quad \text{for} \quad R \geqq R_0 \tag{15-3}$$

(see Art. 12). Similarly, one obtains for jet helicopters with tip-borne jet engines:

$$\left(\frac{W_{lpf}}{W}\right)_j = \frac{W_{jr}}{W} + \frac{c_j P S_{tj}}{W} \tag{15-4}$$

If Eq. 15-1, 15-2, 15-3, and 15-4 are combined, one obtains

$$S_{tj} = \frac{c_s}{c_j} V \frac{W_{et}}{W} \frac{W}{Pc_s} + \frac{c_s}{c_j} S_{ts} \tag{15-5}$$

One may make the following crude estimates:

$$\frac{W_{et}}{W} = 0.22, \text{ according to Hafner } [14]$$

$$\frac{W}{P} = 10 \text{ lb/hp, approximately representative for helicopters}$$

$$c_s = \tfrac{1}{2} \text{ lb/hp-hr}$$

$$\left.\begin{aligned} \frac{c_j}{c_s} &= 48 \text{ for ramjets, ramrockets, and rockets} \\ &= 6 \text{ for pulse jets} \\ &= 1.7 \text{ for turbojets} \end{aligned}\right\} \text{at } \omega R = 600 \text{ ft/sec}$$

Substitution of these estimates into Eq. 15-5 yields the following relations between the transport distances (in miles) for the two types of helicopters with different power plants

$$S_{tj} = 11 + \frac{S_{ts}}{48} \text{ for ramjets, ramrockets, and rockets}$$

$$S_{tj} = 88 + \frac{S_{ts}}{6} \text{ for pulse jets, and}$$

$$S_{tj} = 310 + \frac{S_{ts}}{1.7} \text{ for turbojets}$$

Accordingly, ramjet, ramrocket, and rocket helicopter rotors start with an 11-mile advantage over the equivalent shaft-driven rotor; in these 11 miles, the jet rotors consume a fuel weight that is equal to the weight of the engine and transmission of the shaft-driven rotor. The corresponding initial advantages of pulse jet and turbojet helicopters are 88 and 310 miles, respectively.

At some distance S_{t_0}, the transport distances of shaft-driven and jet-driven helicopters become equal. It follows from the foregoing that $S_{t_0} = 11.2$ miles for helicopters driven by ramjets, ramrockets, and rockets, 105 miles for pulse jet helicopters, and 754 miles for turbojet helicopters.

Jet helicopters are competitive (in transport merit) with shaft-driven helicopters for short-haul transportation, $S_t \lesseqgtr S_{t_0}$. Tip-borne turbojets, despite their almost double specific fuel consumption, are competitive up to distances that are near the maximum range of shaft-driven helicopters.

The fuel costs of all jet helicopters are substantially larger than those of the equivalent shaft-driven rotors, namely, in the proportion c_j/c_s.

Pressure jet rotors. The pressure jet system is a distinct class of jet rotors characterized by a powered gas generator located in the nonrotating airframe, gas ducts from the gas generator to the rotor hub, a sealed gas duct coupling at the rotor hub, ducts through the rotor blades, and gas discharge nozzles at the blade tips. The pressure jet system utilizes a turbojet engine that is modified to locate the moving parts of the engine in the nonrotating parts of the craft. This modification certainly circumvents many of the obvious difficulties of tip-mounted turbojet engines, but it incurs these obvious consequences:

1. The considerably decreased engine weight at the blade tips decreases the optimum rotor size R_0 drastically; large pressure jet helicopters with $R > R_0$ will need blade tip ballast and do not utilize the parasitic blade ballast weight in the manner of jet rotors with tip-borne engines.
2. The cross section and heat tolerance of the blade do not permit the attainment by pressure jet systems of thermal and propulsive efficiency levels characteristic of turbojets.

Therefore, pressure jet helicopters, compared to turbojet helicopters, are intrinsically smaller and have less transport merit. Moreover, for pressure jet rotors, the relation

$$\left(\frac{W_{\text{lpf}}}{W}\right)_{\text{pj}} = \frac{W_{\text{gg}}}{W} + \frac{c_{\text{pj}} P S_{\text{tj}}}{WV} + \frac{W_{\text{jr}}}{W} \tag{15-6}$$

applies where W_{gg} is the weight of the gas generator and nonrotating duct system, and $c_{\text{pj}}P$ is the fuel flow required to produce the power $P = bT_j\omega R$ at the discharge nozzles. For comparison between pressure jet and shaft-driven helicopters, combine Eq. 15-1, 15-2, 15-3, and 15-6, thus obtaining

$$S_{\text{tj}} = \frac{c_s}{c_{\text{pj}}} V \frac{W_{\text{et}} - W_{\text{gg}}}{W} \frac{W}{Pc_s} + \frac{c_s}{c_{\text{pj}}} S_{\text{ts}} \tag{15-7}$$

The dry weight advantage $W_{\text{et}} - W_{\text{gg}}$ is less than the weight of the transmission of the shaft-driven helicopter because the power and weight of the gas generator are somewhat larger than the power and weight of the shaft-driving engine. It appears that the ratio of empty to gross weights of the several prototype pressure jet helicopters is substantially the same as that of shaft-driven helicopters, suggesting $W_{\text{et}} - W_{\text{gg}} \cong 0$. Even if one were to concede a substantial dry weight advantage to the pressure jet rotor, its fuel consumption is, according to Miller [*17*],

$$2 < \frac{c_{\text{pj}}}{c_s} < 5$$

i.e. the transport merit of pressure jet rotors is somewhere between that of the pulse jet rotor and the turbojet rotor.

Much enthusiasm, sophisticated physical analyses, and developmental effort have been lavished on pressure jet systems. This author can find neither flaws nor hostile bias in Miller's reasoning [*17*] which led him to the prognosis that transport helicopters will not likely be powered by pressure jet systems.

Remarks about the turbojet helicopter. Turbojet engines are prone to be sensitive to the environment at the tips of rotor blades. Moreover, they cannot be built efficiently in the very small sizes required for small rotors. Fitzwilliams originated the proposal of turbojet rotors for helicopters [*14*] and suggested that giant rotors would be required. Hafner [*14*, Discussion] pointed out that the coning restrictions $\beta = \beta^* = 0.1$ require blade ballast in addition to the turbojet engines selected by Fitzwilliams for $R > 50$ ft.

Suppose one sets $W_{jr}/W = 0.2$ as the tolerable dry weight of the lifting-propelling system, then, with $bT_j/W = 1/20$ and $\omega R = 600$ ft/sec, one obtains

$$R_0 = R_2 = 72.7 \text{ ft} \qquad \text{(Eq. 12-11, Case } a.\text{)},$$

$$\frac{W_j}{T_j} = 0.442 \qquad \text{(Eq. 12-12, Case } a.\text{)},$$

$$G = \frac{600^2}{g72.7} = 154$$

$$\omega = \frac{600}{72.2} = 8.26 \text{ rad/sec}$$

The weight/thrust ratio of modern turbojet engines for conventional aircraft is substantially smaller than 0.442. Rather than attach ballast to the tip-borne turbojets, one would be tempted to modify the turbojet engine so as to increase its G_{j_a} and ω_{j_a} with the allowable weight margin. One would not, however, be tempted to modify the turbojet into a pressure jet.

Despite the great promise of the turbojet rotor for helicopter application, its promise as a jet propeller is so much more attractive that it would seem almost wasteful to consider the turbojet helicopter.

J,16. Two Jet-Propeller Types for VTOL Craft. Early studies [*18,19*] of jet propellers understandably failed to discover or to foresee the modern jet engines and their applicability, the VTOL role, the convertibility of the jet propeller into fixed wings, and the possibility of supersonic tip speeds for a jet propeller, i.e. precisely those features which hold the greatest promise for jet rotors in general and jet propellers in particular.

Throughout Chap. 2 the superiority of the propeller or symmetrical jet rotor operation over the unsymmetrical, or helicopterlike, operation became apparent. Substantially higher values of U_j and hence higher fuel economy, higher flight velocities, better jet engine alignment, substantially decreased alternating blade loads [*12*], almost all major aspects of design seem to favor the jet propeller.

The Tank propeller. A jet propeller lifting a VTOL craft vertically, accelerating it to a cruising speed $U_j = V$ at which the propeller converts automatically to a fixed wing in level flight (Art. 3, 4, and 9) is called a Tank propeller. Tank originally proposed [*1*] a three-blade ramjet propeller (Fig. J,16) for his VTOL craft, yet today one would probably prefer a two-blade propeller powered by turbojets.

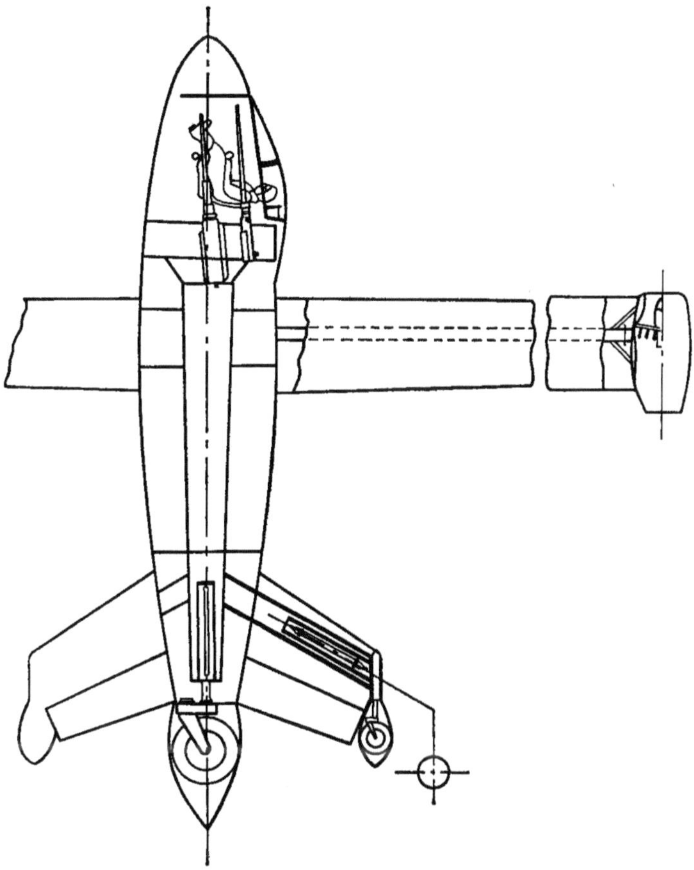

Fig. J,16. The VTOL craft and jet propeller proposed by Tank. After [*1*] by permission H. M. Stationery Office.

The cruising flight, during which the Tank propeller blades act as jet-propelled wings in translatory motion, is that of a conventional turbojet aircraft, i.e. just as economical and as fast. In the cruising condition, the blades are not stiffened by centrifugal tension and therefore their slenderness is limited by wing practice rather than rotor blade practice. An aspect ratio of $2R/c = 10$ would correspond to a solidity of $\sigma = 0.1274$. Thus, for vertical take-off acceleration, one reads from Fig. J,8a:

$$W < F_r \cong 2T_j 13$$

The jet thrust $2T_j$ of both turbojet engines of the Tank propeller which furnishes adequate vertical acceleration at low altitude is just adequate to power the fixed wing cruising flight at moderately high altitudes maintaining

$$U_j = V \cong 900 \text{ ft/sec} = 614 \text{ mph}$$

Because of its intrinsic thrust characteristics (Fig. J,9), the Tank propeller solves one of the thorniest problems of VTOL power plants, the matching of the power demands in vertical flight with those in cruising. The tentative promise of the Tank propeller is attractive enough to warrant more attention than can be paid here. Its turbojet problems are somewhat demanding but, probably, not prohibitive.

The major drawback of Tank VTOL craft is the apparently unavoidable tail-sitting feature which entails rather inconvenient loading and unloading problems—or requires the employment of the Tank VTOL as a tug or tractor aircraft towing a packaged useful load.

The ramprop. The ramprop is a supersonic ($\omega R = 2000$ ft/sec) jet propeller for VTOL aircraft driven by ramjets or ramrockets. This author was led to the discovery of the ramprop[3] by the almost compelling reasoning that ensues when one wants to use the simple propulsive ducts for jet rotors of competitive fuel economy. For lack of available space here, the analysis of the ramprop has been published elsewhere [*11*,*15*]. Like the Tank propeller it allows vertical take-off and descent, and transonic cruising flight with as much fuel economy and useful load capacity as contemporary transport aircraft possess.

The ramprop requires a separate wing for cruising flight since its slender blades do not permit their utilization as wings. Instead, the ramprop permits horizontal fuselage attitude by its tilting pylon suspension [*11*, Fig. 15].

The major drawback of the ramprop is its awful noise. If either the tail-sitting inconvenience of the Tank propeller or the noise of the ramprop can be tolerated, either or both of these jet propellers may well be successfully developed, assuring the success of VTOL aviation in turn.

J,17. Cited References.

1. Liptrot, R. N. Rotating wing activities in Germany during the period 1939–1945. *Brit. Int. Obj. Sub-comm. Rept. 8*, 1948.
2. Rocket engines on helicopter rotor tips. *Automotive Inds. 111*, 56–57 (1954).
3. Lamé, M. L. V. *Le Vol Vertical.* Blondel la Rougery, Paris, 1934.
4. Diedrich, G. The aero-resonator power plant of the V-1 flying bomb. *Project Squid Tech. Mem. PR-4*, 1948.
5. Davidson, I. M. The jet flap. *J. Roy. Aeronaut. Soc. 60*, 25–50 (1956).
6. Dimmock, N. A. An experimental introduction to the jet flap. *Natl. Gas. Turbine Establishment, England, Rept. R175*, 1955.

[3] The concept of the ramprop was disclosed in the first draft of this manuscript which was submitted to the Editors in March, 1951.

7. Helmbold, H. B. Über Den Auftrieb Eines Blasflügels. *Ingenieur Arch. 23*, 209 (1955).
8. Bennett, J. A. J. Rotary-Wing Aircraft. *Aircraft Eng. 12*, 7–9, 40–42, 65–67, 109–112 (1940).
9. Durand, W. F., ed. *Aerodynamic Theory*, Vol. 4. Springer, Berlin, 1934–1936.
10. Berggren, R. E., and Graham, D. J. Effects of leading edge radius and maximum thickness-chord ratio on the variation with Mach number of the aerodynamic characteristics of several thin NACA airfoil sections. *NACA Research Mem. A50D04*, 1950.
11. Gail, A. The ramprop, a supersonic jet driven propeller. *Z. Flugwiss. 4*, 341–355 (1956).
12. Hafner, R. The domain of the helicopter. *7th Louis Bleriot Lecture, J. Roy. Aeronaut. Soc.*, Oct. 1954.
13. Gustafson, F. B., and Myers, G. C., Jr. Stalling of helicopter blades. *NACA Tech. Note 1083*, 1946.
14. Fitzwilliams, O. L. L. The giant helicopter. *J. Helicopter Assoc. Great Brit. 5*, 391–424 (1952).
15. Daughaday, H., and Kline, J. An approach to the determination of higher harmonic rotor blade stresses. *Proc. 9th Ann. Helicopter Forum*, Washington, 1953.
16. Foa, J. V., Gail, A., and Goodman, T. R. Propulsive ducts at the tips of rotor blades. *J. Aeronaut. Sci. 23*, 381–383 (1956).
17. Miller, R. H. Some factors affecting helicopter design and future operations. *Anglo-American Aeronaut. Conference*, London, 1953.
18. Roy, M. Recherches théoretiques sur le rendement et les conditions de réalization des systèmes moto propulseurs à réaction. *Publs. sci. et tech. Ministère air, France, Serv. Recherches aéronaut.*, 1930. Transl. *NACA Tech. Mem. 1249*, 1950.
19. Sanders, T. C., and Sanders, N. D. A preliminary study of a propeller powered by gas jets issuing from the blade tips. *NACA Tech. Note 1155*, 1946.

SECTION K

ATOMIC ENERGY IN JET PROPULSION

RALPH ZIRKIND

CHAPTER 1. INTRODUCTION

K,1. Historical Remarks. The history of nuclear propulsion is still in the making and within the next decade the culmination of the first era in the atomic revolution should be reached. By that time every mode of transportation will probably have been subjected to an experimental nuclear system as a prime mover. To date, however, only the submarine has been propelled by nuclear power.

Although the peacetime potential of nuclear propulsion, including its application to aircraft, was realized during the existence of the Manhattan Project, it was not until about 1946 that serious consideration was given to this problem. At that time the ground work was being laid for two avenues of approach: (1) ship propulsion and (2) aircraft propulsion. The former progressed more rapidly so that, by 1955, two different systems were available to propel submarines. Experimentation in aircraft propulsion is steadily progressing, but at a slower rate because the problems encountered in this field are more difficult to solve. To fully appreciate this fact, consider that in 1946 the maximum temperature of operating reactors was about 200°F, whereas to adequately propel an aircraft about 1500°F is required, and about 4000°F for a rocket. The attainment of these high reactor temperatures, 1500°F for the sustained operation of aircraft and 4000°F for limited times in rockets, is one of the problems met in the application of atomic energy to jet propulsion. The standard high temperature materials, e.g. steels, are undesirable from a nuclear standpoint for reactor use and therefore are used sparingly. An associated problem is corrosion; that is, the corrosive rate of working fluids like liquid metals increases with increasing temperature and in the temperature range of interest here it becomes a serious problem. In addition to the problem of materials the weight of an airborne nuclear propulsion system, greatly influenced by radiation shielding, must be within reasonable limits so that the gross weight of aircraft or rockets propelled by nuclear power plants will be in the realm of practicability. In time the technological barriers will probably be eliminated, thus encouraging many applications of nuclear energy to airborne prime movers.

K,2. Scope. The basic concepts peculiar to an airborne nuclear system will be considered in this section, including a rather detailed discussion of reactor physics and design, heat transfer, and nuclear radiation shielding. In addition the influence of utilizing a nuclear system for several different arrangements will be examined. In view of this broad scope and the confinement to hypothetical situations the reader should refer to the bibliography for additional information.

K,3. Limitations. Government security restrictions on development systems preclude a full discussion of the problems and state of art associated with airborne systems. However, the unclassified literature is sufficiently extensive so that the basic principles can be discussed and the potential problems recognized.

CHAPTER 2. REACTOR PRINCIPLES

K,4. General Remarks. The reactor is the novel feature of a nuclear propulsion system and serves as its heat-generation mechanism similar to the combustion chamber in a jet engine. The similarity, however, ends there. In essence, a reactor is an assembly of suitable materials arranged in a manner efficient for fissioning the fuel, a fissionable isotope of uranium or plutonium. In the case of power generation an added condition is the generation of power in an efficient manner. At times, the resultant is a compromise between the two requirements.

For discussion purposes we shall consider only thermal or near-thermal reactors; that is, reactors where the fuel is fissioned primarily by neutrons having thermal or near-thermal energies. Since the neutrons emitted by the fissioned fuel have considerable kinetic energy, a *moderating material* is provided to slow down these neutrons. To extract the heat energy, a coolant is passed through the reactor. Normally, to protect the fuel and moderator from the corrosive properties of the coolant, they are placed in a suitable container. The last major constituent is a structural material, to provide structural integrity to the various reactor components. The coolant, container material, and structural material are neutron absorbers and are therefore deleterious to an efficient reactor. Our aim here is to present the principles and relationships which define the materials suitable for a reactor and their proper arrangement.

K,5. Physical Concepts. The nuclear particle of prime importance in a reactor is the neutron; its life history within the reactor and its interactions with the reactor materials determine the reactor characteristics.

The possible interactions which a neutron may have with the nuclei of the reactor constituents are (1) elastic scattering, (2) inelastic scattering, and (3) absorption. In the first instance, kinetic energy is conserved, whereas, in the second type, kinetic energy is not conserved because the energy of the neutron after collision is only a fraction of the precollision value and is usually accompanied by the emission of a gamma ray. In the last instance a true reaction occurs in the sense that the end product is not just a neutron and the original atom. Within the category of absorption several processes are possible: (1) charged particle emission, (2) radiative capture, i.e. gamma emission, and (3) fission. The probability of occurrence of a nuclear interaction per atom per incident particle per square centimeter is called the cross section.

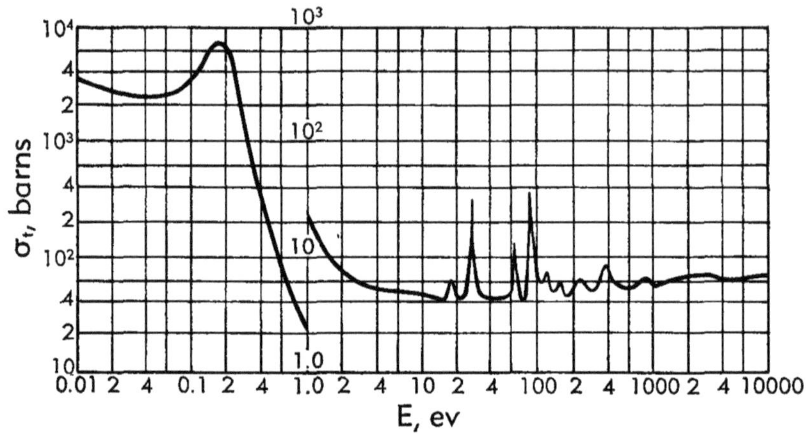

Fig. K,5a. Total cross section for cadmium [1].

The cross sections for the individual processes are denoted by the type of interaction, e.g. scattering cross section, σ_s, etc. The total probability is called the total cross section, σ_t. Since the nuclear radius is of the order of $1.5 \times 10^{-13}A^{\frac{1}{3}}$ cm where A is the isotopic mass number, the cross section is of the order of 10^{-24} cm², a unit often referred to as "the barn." The cross section is an energy-dependent function and, for each nucleus, depends on the structure of the interacting nucleus. The remainder of the discussion is particularized to the neutron as the incident particle.

To illustrate the behavior of the cross section, the total and scattering cross sections of cadmium are shown in Fig. K,5a and K,5b respectively. The ordinates are neutron cross section in barns and the abscissas, neutron energy in electron volts (ev).[1] In the thermal region, the scatter-

[1] It is standard practice to denote particle energies in electron volts. If one considers the distribution of particles at a given temperature as indicated by the Maxwell-Boltzmann law, then the most probable energy, $E = kT$, where k is the Boltzmann constant and T is in °K; that is, E (ev) $= 8.61 \times 10^{-5}T$. Hence, at 25°C the neutron energy would be 0.025 ev and is called a thermal neutron. In practice, if the medium is at a temperature of 1089°K, then the neutrons in thermodynamic equilibrium with the medium will be thermal at an energy of 0.094 ev. The energy region between 1 ev and 100 ev is denoted as the resonance region and the neutrons are called resonance

ing cross section is essentially constant whereas the total cross section decreases with increasing energy. The decrease is produced by the absorption process which is usually inversely proportional to the velocity or, $\sqrt{E}$, the so-called "$1/V$" law. At about 1 ev the magnitude of the cross-section value begins to increase and exhibit either capture or, at times, scattering resonances. A simple explanation for the resonance phenomenon is that the nucleus possesses energy levels similar to an atom but

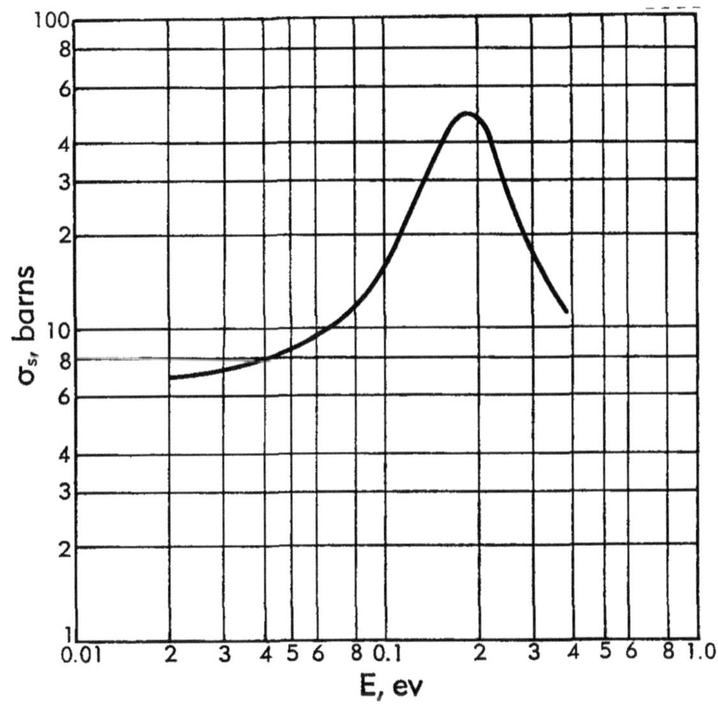

Fig. K,5b. Scattering cross section for cadmium [*1*].

of finite width. Consequently, when a neutron with an energy corresponding to a particular level above the ground state impinges upon the nucleus the probability of an interaction, either absorption or scattering, is increased significantly. The number of resonances and their spacing are indicative of the level within a particular nucleus. In the case of resonance capture, this is usually accompanied by the emission of a gamma ray. As the energy of the bombarding neutron increases, say greater than 0.5 mev, the elastic scattering remains constant, $\sigma_s \cong 2\pi R^2$; however, the inelastic scattering process can now enter the picture. No reference has yet been made to nuclear mass, isotopic variation, or other features of

neutrons. From 0.1–0.5 kev is the epithermal region, and above this is the fast region.

the nucleus. Consideration of these features with the detailed quantum mechanical theory of the nucleus is necessary to describe the variations in cross-section behavior for each isotopic mass. Suffice it to say that recent efforts by theorists [2] and [3] have shown that the general behavior can be obtained theoretically with the aid of nuclear spectroscopic data. To summarize the above discusson a brief table is given (Table K,5),

Table K,5. Qualitative summary of nuclear processes.

Neutron energy	*Light nuclei*	*Intermediate nuclei*	*Heavy nuclei*
0–1.0 kev	elastic	elastic gamma resonance	gamma elastic resonance
0.1–0.5 mev	scattering resonance	elastic gamma resonance	elastic gamma resonances
0.5–10 mev	scattering charge particle emission	elastic inelastic resonances for lower energy	elastic inelastic gamma

indicating the relative importance of the different processes for various energy regions. All the available unclassified cross-section data are summarized in [1].

We have described the cross section per atom, the microscopic cross section. If the target material contains N atoms per cm^3 then the probability of a collision per cm is $N\sigma = \Sigma$, the macroscopic cross section. The reciprocal of this quantity represents the mean free path, $\bar{l}$, or the mean distance a neutron travels between collisions, e.g. $\bar{l}_s$ is the scattering mean free path.

A thermal neutron undergoing scattering collisions within a target material will diffuse at a constant velocity until captured, whereas the fast neutron can lose energy on collision. The latter process will continue until the neutron is slowed down to thermal energies, if not captured by resonances in the epithermal region. If the target material is finite then escape or leakage must be added as another possible means to lose neutrons.

K,6. Collision Results. Assume that elastic scattering is isotropic in the center of the mass system which is valid to approximately 1 mev. Let E_0 be the initial energy of the incident neutron, then the energy after a single collision with a nucleus of mass A lies between E_0 and E_{min} where the residual minimum energy is given by

$$E_{min} = \left(\frac{A-1}{A+1}\right)^2 E_0 = \alpha E_0 \tag{6-1}$$

Hence the maximum energy loss is $E_0 - E_{min} = (1 - \alpha)E_0$, which corresponds to a head-on collision. The maximum fractional energy loss a neutron may suffer in the collision is

$$1 - \alpha = \frac{4A}{(A + 1)^2} \tag{6-2}$$

It should be noted that the assumption of isotropic scattering has the consequence that, after the collision, the neutron may take on all energy values between E_0 and αE_0 with equal probability.

A quantity employed in characterizing the degradation of neutrons is the average logarithmic energy decrement per collision, that is, the average for all collisions of $\ln E_1 - \ln E_2$ where 1 and 2 denote the neutron's energy before and after the collision respectively. If this quantity is denoted by ξ then

$$\xi = \overline{\ln \frac{E_0}{E}} = \frac{\int_{\alpha E_0}^{E_0} \ln \frac{E_0}{E} p(E)dE}{\int_{\alpha E_0}^{E_0} p(E)dE} \tag{6-3}$$

where $\int p(E)dE$ is the probability that the neutron will possess an energy value in the interval E_0, αE_0, which of course is unity. Further the probability that after one collision a neutron with initial energy E_0 will have an energy in the interval $(E, E + dE)$ is

$$p(E)dE = \frac{1}{1 - \alpha} \frac{dE}{E_0} \tag{6-4}$$

On substitution of Eq. 6-4 into Eq. 6-3 and carrying out the integration, we obtain for ξ the value

$$\begin{aligned} \xi &= 1 + \frac{\alpha}{1 - \alpha} \ln \alpha \\ &= 1 - \frac{(A - 1)^2}{2A} \ln \frac{A + 1}{A - 1} \\ &\cong \frac{2}{A + \frac{2}{3}} \quad (A > 2) \end{aligned} \tag{6-5}$$

It should be noted that the parameter ξ is independent of energy.

From this it immediately follows that the number of collisions required to reduce a neutron from energy E_0 to E is given by[2]

$$N = \frac{1}{\xi} \ln \frac{E_0}{E} = \frac{u}{\xi} \tag{6-6}$$

and by virtue of Eq. 6-5 increases as the atomic mass A increases. Thus an effective slowing-down material for a reactor, the moderator, should be of low atomic mass. This is true provided that the scattering cross

[2] The term u is called the lethargy of a neutron whose kinetic energy is E.

Table K,6. Moderators (from [4]).

Element or compound	*A, g/mole*	*Density, g/cm³**	*N** × 10⁻²⁴ atoms/cm³*	*σ_s, epithermal, barns*	*σ_a, 0.025 ev, barns*	*Σ_a, cm⁻¹*	*Σ_s, cm⁻¹*	ξ§	$N\xi\sigma_s$	$(\sigma_s/\sigma_a)\xi$	*Comments*
Hydrogen H_2	1.008	0.0090	0.0054	20.3	0.33	0.0018	0.11	1.000	0.11	61	H_2 gas at high pressure presents fire hazard.
Deuterium D_2	2.02	0.0180	0.0054	3.3	0.00046	2.5×10^{-6}	0.018	0.725	0.013	5200	D_2 gas at high pressure presents fire hazard.
Helium He	4.00	0.0180	0.0027	0.8	—	—	0.0022	0.425	0.0009	∞	Notice low slowing-down power.
Beryllium Be	9.01	1.85	0.124	6.1	0.009	0.0011	0.76	0.206	0.16	145	Good moderator.
Carbon (graphite) C	12.0	1.60	0.080	4.7	0.0045	0.00036	0.38	0.158	0.060	165	Graphite density varies.
Oxygen O_2	16.0	0.014	0.0054	3.8	0.0002	1.1×10^{-6}	0.021	0.12	0.0025	230	Low slowing-down power.
Sodium Na	23.0	0.97	0.0254	3.0	0.49	0.012	0.076	0.083	0.0063	0.53	Low slowing-down power.
Magnesium Mg	24.3	1.74	0.043	3.4	0.059	0.0025	0.15	0.073	0.011	4.4	Low slowing-down power.
Aluminum Al	27.0	2.70	0.060	1.35	0.215	0.013	0.081	0.071	0.0058	0.45	Low slowing-down power.
Beryllia BeO	25.0	2.8	0.067	9.9	0.009	0.00060	0.66	0.173	0.11	183	Good moderator.
Beryllium carbide Be_2C	30.0	2.4	0.048	16.9	0.023	0.0011	0.81	0.193	0.16	145	Good moderator.
Light water H_2O	18.0	1.00	0.033	44.4	0.66	0.022	1.47	0.925	1.36	62	Good moderator.
Heavy water D_2O	20.0	1.10	0.033	10.5	0.0011	36×10^{-6}	0.35	0.504	0.18	5000	Good moderator.
Sodium hydroxide NaOH	40.0	2.1	0.032	27.1	0.82	0.026	0.87	0.77	0.67	26	Fair moderator; neutron loss high
Polystyrene $(CH)_n$	17.0	1.07	0.038¶	25.0†	0.335†	0.013	0.95	0.842	0.80	62	Radiation damage a problem.
Paraffin $(CH_2)_n$	18.0	0.9	0.030‖	45.3‡	0.665‡	0.020	1.36	0.913	1.24	62	Radiation damage a problem.

* For the gases H_2, D_2, He, and O_2 an arbitrary density 100 times the density at NTP is assumed; i.e. a pressure on the order of 1500 lb/in.²

† Per (CH) unit.

‡ Per (CH_2) unit.

§ $\xi = 1 - [(A-1)^2/2A] \ln [(A+1)/(A-1)]$.

¶ Number of (CH) units/cm³.

‖ Number of (CH_2) units/cm³.

** Number of atoms or molecules per cm³.

section is not small and the absorption is low; that is, a good moderator is one whose moderating ratio $\xi\Sigma_s/\Sigma_a$ is large. The properties of some moderators are given in Table K,6.

Another result obtained from collision mechanics is that

$$\overline{\cos\theta} = \frac{2}{3A} \tag{6-7}$$

where θ is the scattering angle in the laboratory system. Thus the neutron after a collision persists in the direction of incidence; however, for heavy nuclei the scattering approaches isotropic conditions. To permit correction to the scattering cross section when anisotropic scattering occurs, the transport cross section has been introduced.

The transport mean free path is the average distance that a neutron would travel in an infinite number of collisions in the initial direction. Now, $\bar{l}_s\,\overline{\cos\theta}$ is the average projection of the free flight of the neutron after collision in the direction of travel before the collision, etc., so that

$$\begin{aligned} \bar{l}_{tr} &= \bar{l}_s + \bar{l}_s\,\overline{\cos\theta} + \bar{l}_s\,\overline{\cos\theta}^2 + \cdots \\ &= \bar{l}_s/1 - \overline{\cos\theta} \end{aligned} \tag{6-8}$$

where $\cos\theta$ is defined by Eq. 6-7. Hence the microscopic cross section is

$$\sigma_{tr} = \sigma_s(1 - \overline{\cos\theta}) \tag{6-9}$$

K,7. Slowing-Down. In the fission process the neutrons are emitted with energies in the epithermal and fast regions, whereas to be useful to a (thermal) reactor it is desirable to reduce their energies to thermal, or near thermal energies. The theory describing this degradation process is denoted by the "slowing-down" theory and we shall here present an approximate theory which yields surprisingly good results.

Let us denote by $n(\mathbf{r}, v)$ the number of neutrons per cm³ with velocity v, then at the point $\mathbf{r}$ *the neutron flux* $\phi(\mathbf{r}, v)$ is

$$\begin{aligned} \phi(\mathbf{r}, v) &= n(\mathbf{r}, v)v \\ &= \text{number of neutrons crossing a unit area in unit time} \end{aligned} \tag{7-1}$$

It should be noted that ϕ is a scalar. Physically, the flux represents the average distance traveled by the neutrons per unit volume per unit time. Hence the number of collisions expected to occur per cm³ per sec can be expressed as

$$\frac{\phi(\mathbf{r}, v)}{\lambda_i} = \frac{n(\mathbf{r}, v)v}{\lambda_i} = \Sigma_i\phi(\mathbf{r}, v) \tag{7-2}$$

where the subscript i may be any one of the nuclear interactions, scattering, absorption, etc. The average number of collisions a neutron makes in the lethargy interval du is du/ξ, and if $q(u)$ is the slowing-down density of neutrons of lethargy u about the interval du, then the total number of scattering collisions in du per cm^3 per sec is

$$\frac{q(u)du}{\xi} \tag{7-3}$$

This is by definition, however, equal to $\Sigma_s(u)\phi(u)du$, and therefore

$$q(u) = \xi\Sigma_s(u)\phi(u) \tag{7-4}$$

or, in energy terms,

$$q(E) = E\xi\Sigma_s(E)\phi(E)$$

Consider an infinite scattering medium with Q neutrons being introduced into the medium per cm^3 per sec, then the number of neutrons slowed down past each energy is $q(E) = Q$. Thus Eq. 7-4 becomes

$$\phi(E) = \frac{Q}{E\xi\Sigma_s(E)} \tag{7-5}$$

for energies less than E_0, the primary energy. If now some absorption exists at all energies, then the neutron balance in the interval dE is

$$\frac{d}{dE}\,q(E) = \Sigma_a(E)\phi(E) \tag{7-6}$$

For weak absorption, $\Sigma_a \ll \Sigma_s$, we assume that the relationship between $q(E)$ and $\phi(E)$ is given by Eq. 7-4. Substitution into Eq. 7-6 and integration yield the result

$$q(E) = Q\exp\left[-\int_E^{E_0}\frac{\Sigma_a(E)}{\Sigma_s(E)}\frac{dE}{\xi E}\right] \tag{7-7}$$

where the boundary condition $q(E_0) = Q$ has been used. Physically, the exponential term is the probability that a neutron will not be absorbed in slowing down from E_0 to E. For strong absorption, as in the region of resonance capture, a more suitable expression is

$$Q\exp\left[-\int_E^{E_0}\frac{\Sigma_a}{\xi\Sigma_t}\frac{dE}{E}\right] \tag{7-8}$$

If we now extend the energy interval to thermal energies E_{th}, then Eq. 7-8 becomes

$$q(E_{th}) = Q\exp\left[-\int_{E_{th}}^{E_0}\frac{\Sigma_a}{\Sigma_t}\frac{dE}{\xi E}\right] \tag{7-9}$$

where the exponential term in standard terminology is called the resonance-escape probability and is denoted by the symbol p.

We have presented a simplified asymptotic theory of slowing-down which provides good results in practical situations. For a more detailed, rigorous treatment see [*5*].

K,8. Spatial Variation. The Fermi age theory is an approximation to the spatial variation of the "slowing-down" neutrons. The inherent assumptions are: (1) the slowing-down process which is discontinuous can be replaced by a continuous one which approaches the actual process asymptotically and (2) scattering is isotropic.

Let us consider an infinite medium with weak capture and assume that the distribution can be described by a diffusion-type equation. Then the neutron balance is given by

$$-D\nabla^2\phi(\mathbf{r}, E) + \Sigma_a\phi(\mathbf{r}, E) = -\frac{\partial q(\mathbf{r}, E)}{\partial E} \tag{8-1}$$

where the first term on the left represents the neutron leakage per cm³ per sec. On rewriting Eq. 8-1 we obtain

$$\nabla^2\phi(\mathbf{r}, E) - 3\Sigma_a\Sigma_{tr}\phi(\mathbf{r}, E) = \frac{1}{D}\frac{\partial}{\partial E}q(\mathbf{r}, E) \tag{8-2}$$

where D, the diffusion coefficient, is defined by $l_{tr}/3$. If we now replace ϕ by the function $q(\mathbf{r}, E)$, as defined by Eq. 7-4, Eq. 8-2 becomes

$$\nabla^2 q(\mathbf{r}, E) - 3\Sigma_a\Sigma_{tr}q(\mathbf{r}, E) = \frac{\xi\Sigma_{tr}E}{D}\frac{\partial q}{\partial E}(\mathbf{r}, E) \tag{8-3}$$

or

$$\nabla^2 q_1(\mathbf{r}, \tau) = \frac{\partial}{\partial \tau}q(\mathbf{r}, \tau) \tag{8-4}$$

where

$$q_1 = q \exp\left[-\int_E^{E_0}\frac{\Sigma_a}{\xi\Sigma_{tr}}\frac{dE}{E}\right] \tag{8-5}$$

and, τ, the neutron age, is given by

$$\tau = \int_E^{E_0}\frac{D}{\xi\Sigma_s}\frac{dE}{E} = \int_E^{E_0}\frac{l_{tr}(E)l_s(E)}{3\xi}\frac{dE}{E} \tag{8-6}$$

From Eq. 8-5 we readily recognize that $q_1 = qp$. Since at each point in space the slowing-down density is identical $q_1(\mathbf{r}, \tau)$, it represents the asymptotic slowing-down density in a noncapturing medium. The variable τ is proportional to either the mean square distance a neutron will travel in being degraded from energy E_0 to E, or the actual time for the neutron to slow down. In fact, one can readily show that the age, τ, is defined by

$$\tau = \int_0^t Dv\,dt \tag{8-7}$$

The latter is a direct consequence of the first assumption. Mathematically, the assumption is

$$-du = v\xi\Sigma_s dt$$

and Eq. 8-6 tells us

$$d\tau = -\frac{Ddu}{\xi\Sigma_s}$$

from which Eq. 8-7 follows by direct substitution.

K,9. Point Source Solution. We consider now the solution of Eq. 8-4 for a monoenergetic point source of neutrons in an infinite medium. The boundary conditions are: (1) $q \to 0$ as $r \to \infty$ and (2) q remains finite at $r = 0$ and is given by $S\delta(\mathbf{r})$ where $\delta(\mathbf{r})$ is a delta function.

A particular solution of the age equation,

$$\nabla^2 q(\mathbf{r}, \tau) = \frac{\partial}{\partial\tau} q(\mathbf{r}, \tau) \tag{8-4}$$

is

$$q(\mathbf{r}, \tau) = R(\mathbf{r})T(\tau) \tag{9-1}$$

which, on substitution into Eq. 8-5 gives

$$\frac{\nabla^2 R(\mathbf{r})}{R(\mathbf{r})} = \frac{dT(\tau)}{T(\tau)} = -k^2 \tag{9-2}$$

where k^2, the separation constant, is positive. Thus we have immediately from Eq. 9-2 that

$$\begin{aligned} R(\mathbf{r}) &= e^{i\mathbf{k}\cdot\mathbf{r}} \\ T(\tau) &= e^{-k^2\tau} \end{aligned} \tag{9-3}$$

and

$$q(\mathbf{r}, \tau) = e^{i\mathbf{k}\cdot\mathbf{r} - k^2\tau} \tag{9-4}$$

The general solution $q^*(\mathbf{r}, \tau)$ may be expressed by a Fourier integral

$$q^*(\mathbf{r}, \tau) = \int_k A(\mathbf{k}) e^{i\mathbf{k}\cdot\mathbf{r} - k^2\tau} d\mathbf{k} \tag{9-5}$$

where $A(\mathbf{k})$, the Fourier transform of $S\delta(r)$, is given by

$$A(\mathbf{k}) = \frac{1}{8\pi^3}\int_{r'} S\delta(r) e^{-i\mathbf{k}\cdot\mathbf{r}} dr' = \frac{S}{8\pi^3} \tag{9-6}$$

It follows directly that

$$q^*(\mathbf{r}, \tau) = \frac{S}{8\pi^3}\int_k e^{-k^2\tau + i\mathbf{k}\cdot\mathbf{r}} dk \tag{9-7}$$

and

$$q^*(\mathbf{r}, \tau) = \frac{S}{(4\pi\tau)^{\frac{3}{2}}} e^{-r^2/4\tau} \tag{9-8}$$

The general solution (Eq. 9-8) states that the spatial variation is a Gaussian distribution. We can obtain directly from Eq. 9-8 that the mean square distance $\overline{R^2}$ is 6τ.

A point to consider is the adequacy of the Gaussian distribution in describing the neutron distribution, particularly at large distances from the source. The uncollided flux will have by definition an exponential distribution,

$$\phi(r, E_0) = e^{-r/l_s} \tag{9-9}$$

and the use of this kernel in place of the delta function would give the asymptotic behavior of q^* an exponential character. Physically, one may expect that the slowed-down neutrons will be in equilibrium with the nascent neutrons, which would require a larger number than that expected from the age theory. This has been verified experimentally [*6*].

From physical considerations, a possible correction to the age theory is to employ the first collided flux as the beginning of the aging process and to estimate the mean square distance traveled by a neutron during this first collision. If this is done, the over-all mean square distance is found to be

$$\overline{R^2} = 2l_s^2 + 6\tau \tag{9-10}$$

It should be noted that an analytical solution is not obtainable for the slowing-down density with the distribution in Eq. 9-9 (see [*7*]).

K,10. Diffusion. To describe the spatial variation of thermal neutrons or any univelocity group of neutrons with constant cross sections, one applies the Boltzmann transport equation. By application of this equation one may obtain details as to the angular distribution of particles, anisotropic effects, and the behavior of the neutron density in the vicinity of sources, sinks, and boundaries. However, these details can only be obtained at the expense of tedious numerical analysis. (For detailed references see [*8*].)

An alternate procedure is to derive an approximate equation, namely the diffusion equation of kinetic theory, and to superimpose the corrections obtained from the precise transport equation. The basic assumption of the diffusion theory is that

$$\mathbf{j} = -D_0 \operatorname{grad} n(\mathbf{r}) \tag{10-1}$$

or more generally,

$$\mathbf{j} = -D \operatorname{grad} \phi(\mathbf{r}) \tag{10-2}$$

where $\mathbf{j}$ is the current density and

$$D_0 = Dv$$

Transport theory shows that this assumption is not valid unless (1) the fractional change in density over one mean free path is negligible, (2) $\sigma_a \ll \sigma_s$, and (3) one mean free path away from sources or boundaries.

With the assumption in Eq. 10-2 let us consider the steady state condition in a medium and write the neutron balance for the neutron density,

$$\underset{\substack{\text{leakage per cm}^3 \\ \text{per sec}}}{\nabla \cdot \mathbf{j}} + \underset{\substack{\text{number of absorptions} \\ \text{per cm}^3 \text{ per sec}}}{\Sigma_a n(\mathbf{r}) v} = \underset{\text{sources}}{S(\mathbf{r})} \tag{10-3}$$

or

$$-D\nabla^2\phi(\mathbf{r}) + \Sigma_a\phi(\mathbf{r}) = S(\mathbf{r}) \tag{10-4}$$

Thus the basic diffusion equation is

$$D\nabla^2\phi(\mathbf{r}) - \Sigma_a\phi(\mathbf{r}) + S(\mathbf{r}) = 0$$

where the sources may be either introduced as thermal neutrons or result from the slowing-down process, i.e. $q(\mathbf{r}, \tau_{\text{th}})$.

For a source-free infinite medium Eq. 10-4 gives for the thermal flux distribution,

$$\phi(\mathbf{r}) = \frac{A}{r} e^{-r/L} \tag{10-5}$$

where L, the thermal diffusion length, is defined by

$$L = \sqrt{\frac{D}{\Sigma_a}} = (3\Sigma_a\Sigma_s)^{-\frac{1}{2}} \tag{10-6}$$

The parameter L is the characteristic length of diffusion theory. From Eq. 10-5 we see that the thermal flux is exponential and is determined by the diffusion length.

In Eq. 10-2 we have assumed a result which can readily be obtained, namely that the net current in the direction of the normal is

$$j(\mathbf{r}) = -D\nabla\phi(\mathbf{r})$$

where for a plane surface parallel to the x axis

$$\mathbf{j}_x(\mathbf{r}) = \mathbf{j}_{+x}(\mathbf{r}) - \mathbf{j}_{-x}(\mathbf{r}) \tag{10-7}$$

and

$$\begin{aligned} \mathbf{j}_{+x} &= \frac{1}{4}\,\phi(\mathbf{r}) - \frac{1}{6\Sigma_s}\left(\frac{\partial\phi}{\partial x}\right)_{\mathbf{r}} \\ \mathbf{j}_{-x} &= \frac{1}{4}\,\phi(\mathbf{r}) + \frac{1}{6\Sigma_s}\left(\frac{\partial\phi}{\partial x}\right)_{\mathbf{r}} \end{aligned} \tag{10-8}$$

The diffusion coefficient defined by Eq. 10-8 is $\frac{1}{3}\Sigma_s$ whereas the transport theory gives

$$\begin{aligned} D &\cong [3\Sigma_s(1 - \overline{\cos\theta})]^{-1} \\ &\cong \tfrac{1}{3}\Sigma_{\text{tr}} \end{aligned} \tag{10-9}$$

Eq. 10-2, 10-4, and 10-8 along with the appropriate boundary conditions represent the diffusion approximation. The boundary conditions to be met in a physical situation are:

1. The neutron flux is finite and non-negative in all regions in which the equations apply.
2. At a plane interface between two media with different diffusion properties the net current densities in the direction of the normal to the interface are equal and the neutron fluxes are also equal, i.e. $\phi_A = \phi_B$, $\mathbf{j}_A = \mathbf{j}_B$ at the interface.
3. If one of the media in (2) is a vacuum the flux varies in such a manner that linear extrapolation would require the flux to vanish at a definite distance beyond the physical boundary.

The first two boundary conditions require no explanation whereas the third is basic to the diffusion approximation. Let us consider diffusion

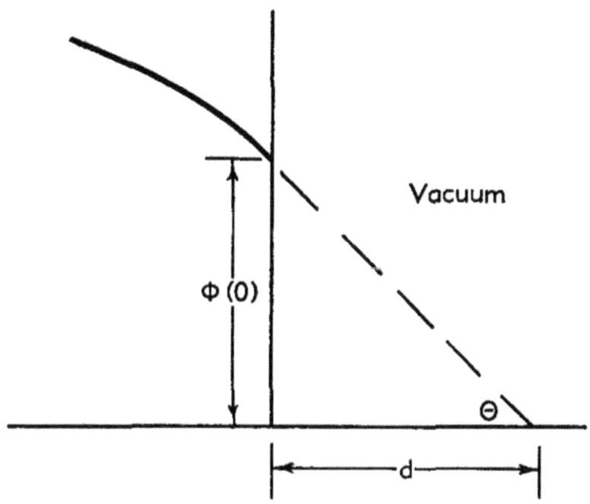

Fig. K,10. Extrapolation of flux.

from an interface located at $X = 0$ into a vacuum, then since there is no back-scattering from a vacuum we have from Eq. 10-1

$$\mathbf{j}_- = \frac{\phi(0)}{4} + \frac{\bar{l}_{tr}}{6}\frac{d\phi(0)}{dx} = 0$$

or (10-10)

$$\frac{d\phi(0)}{dx} = -\frac{6}{4\bar{l}_{tr}}\phi(0)$$

Now the flux is positive at the boundary and it follows from Eq. 10-10 that the slope is negative. If the flux is extrapolated into the vacuum using a straight line with the slope $d\phi/dx$, the flux vanishes at the distance d. From Fig. K,10, we have

$$\tan\theta = -\frac{\phi(0)}{d} = \frac{d\phi}{dx} = -\frac{6}{4\bar{l}_{tr}}\phi(0) \tag{10-11}$$

or

$$d = \tfrac{2}{3}\bar{l}_{tr} \tag{10-12}$$

where d is the linear extrapolation distance. A suitable correction to this boundary condition is to fit the solution of Eq. 10-4 to the asymptotic solution obtained from the transport equation. This leads to the boundary condition that the flux vanishes at the extrapolated distance equal to $0.7104\bar{l}_{tr}$ for a plane boundary. For a curved surface the distance increases, e.g. for infinite curvature.

$$d = \tfrac{4}{3}\bar{l}_{tr} \tag{10-13}$$

We shall apply these developments when reactor theory is considered, after a brief review of pertinent facts concerning fission.

K,11. Fission. The fission process is a significant nuclear reaction peculiar to several heavy isotopes at moderate neutron energies, say less than 10 mev. At thermal energies it is the predominant reaction for the isotopes U^{233}, U^{235}, and Pu^{239}. The significance is the emission of several fast neutrons whose spectral distribution as determined by experiment [*9*] is given by

$$N(E) \sim \sinh\sqrt{2.29E}\; e^{-E/0.965} \tag{11-1}$$

or the simpler expression

$$N(E) \sim \sqrt{E}\; e^{-E/1.29}$$

The average number of neutrons emitted per fission is denoted by ν. Of greater practical importance, however, is the number emitted per neutron absorbed by the nucleus. The relation between the two quantities is evidently

$$\eta = \nu\frac{\sigma_f}{\sigma_\gamma + \sigma_f} = \frac{\nu}{1+\alpha} \tag{11-2}$$

where σ_γ is the absorption cross section leading to gamma radiation. Representative values of the parameters in Eq. 11-2 are given in Table K,11a. The values of η and ν remain essentially constant with energy as

Table K,11a. Fission parameters at thermal energies.

Isotope	ν	$1+\alpha$	η	σ_f, barns
U^{233}	2.67	1.125	2.37	533
U^{235}	2.49	1.184	2.08	580
Pu^{239}	2.91	1.42	2.05	720

α is almost independent of energy. At energies greater than 0.5 mev, other isotopes of uranium, plutonium, and neptunium begin to fission; however, these play a minor role in highly enriched reactors.

The number of neutrons emitted, ν, includes all neutrons both prompt and delayed. The delayed neutrons comprise about 1 per cent of the instantaneous neutrons. They are related to particular fission fragments and, as to be expected, are emitted with definite mean lives. Experimental evidence indicates the existence of five major periods, which are given in Table K,11b along with the abundance per fission and energy.

Table K,11b. Delayed neutron periods [*10*].

Period, sec	U^{235}	Pu^{239}	Energy, mev
0.431	8.5×10^{-4}	4×10^{-4}	0.4
1.52	24.1×10^{-4}	10.5×10^{-4}	0.67
4.51	21.3×10^{-4}	11.2×10^{-4}	0.41
22	16.6×10^{-4}	9.4×10^{-4}	0.57
55.6	2.5×10^{-4}	1.2×10^{-4}	0.25

It is to be denoted that the energies of the delayed neutrons are less than that for the prompt neutrons. For the latter the most probable is about 2.0 mev.

The importance of the delayed neutrons lies in the practical role they play in controlling the reactor, provided that they are necessary to maintain the chain reaction. Since the lifetime of a neutron within a reactor is about a millisecond, the fluctuations of the neutron density will be of the same order of magnitude, which is difficult to control. With the added effect of the delayed neutrons, the allowable time now becomes several seconds.

K,12. Multiplication Constant. Under suitable conditions and with the proper materials it is possible to maintain a self-sustaining neutron-multiplying system. For an infinite homogeneous medium where fuel capture is the only process by which neutrons are lost, the multiplication constant k_∞ is

$$k_\infty = \nu \frac{\Sigma_f}{\Sigma_a} = \eta \tag{12-1}$$

where we have assumed a particular energy. For a multi-energy beam of neutrons, Eq. 12-1 takes the form

$$k_\infty = \frac{\int_0^\infty \nu\, \Sigma_f(E)\phi(E)dE}{\int_0^\infty \Sigma_a(E)\phi(E)dE} \tag{12-2}$$

Since a practical reactor contains other materials which absorb neutrons, Eq. 12-1 should be rewritten to reflect this added absorption. Let Σ_a^t be the total macroscopic absorption cross section per unit volume, that is,

$$\Sigma_a^t = N\sigma_a(\text{moderator}) + N\sigma_a(\text{fuel}) + N\sigma_a(\text{structure})$$

Table K, 12a. Properties of reactor

Element	Symbol	Atomic no.	Atomic wt.	Density, gm/cm³	Cost, $/lb*	Melting point, °C	Specific heat, cal/gm/°C	Thermal expansion coefficient, per °C × 10⁶	Thermal conductivity, cal/cm²/sec/°C/cm	Resistivity, μohm-cm
Aluminum	Al	13	26.98	2.70	0.30–0.60	660.2	0.22	24	0.54	2.65
Beryllium	Be	4	9.013	1.85	(175)	1300	0.5	12	0.3	4.0
Magnesium	Mg	12	24.32	1.74	0.25	651	0.25		0.35	
Molybdenum	Mo	42	95.95	10.2	3.00	2620	0.065	5.5	0.346	4.8–5.7
Nickel	Ni	28	58.69	8.9	0.60	1455	0.11	13	0.04	6.1–10
Tantalum	Ta	73	180.88	16.6	65	2996	0.036	6.5	0.130	12.5
Titanium	Ti	22	47.90	4.5	(15)	1725	0.129	8.5	(low)	48–90
Vanadium	V	23	50.95	6.02	?	1735	0.12			26
Tungsten	W	74	183.92	19.3	6.00	3410	0.034	4.0	0.476	5.5
Zirconium	Zr	40	91.22	6.5	20–125	1830		4.96–5.8	0.05	41
18-8 Stainless steel (Fe, Cr, Ni, Mn, Nb)				7.92	(0.60)	1399–1421	0.12	16.7	0.039	71
Inconel "X" (Ni, Cr, Fe, Ti, Nb)				8.3	(3.00)	1395–1425	0.11	13.9	0.035	122

* Cost figures in parentheses are estimated.

where N is the number of nuclei per cm³ of the respective materials:

$$k_\infty = \nu \frac{\Sigma_f}{\Sigma_a^f} \frac{\Sigma_a^f}{\Sigma_a^t} = \eta f \tag{12-3}$$

The ratio Σ_a^f/Σ_a^t denoted by f is the thermal utilization and represents a figure of merit in the competition of neutrons within a reactor. Evidently it is desired to make the value of this fraction approach one and the less absorption produced by extraneous materials the greater this possibility. The mechanical and nuclear properties of some reactor materials are given in Tables K,12a and K,12b.

From Eq. 12-3 it follows immediately that the critical equation for an infinite homogeneous reactor is

$$k_\infty = \eta f = 1 \tag{12-4}$$

provided that (1) no nonthermal fissions occur and (2) the resonance escape probability p is one. Normally, the nonthermal fissions may be neglected; however, the p factor should be included. Therefore Eq. 12-4 materials. (Reprinted from [4] by permission.)

Modulus of elasticity, 10^6 lb/in.^{2}	Poisson's ratio	Hardness, BHN	Tensile strength, 10^3 lb/in.^{2}		Elongation in 2 in., %		Crystal structure†	α-β phase transition	β phase structure	σ_a, neutron, barns
			Annealed	Cold worked	Annealed	Hardened				
10	0.3	20–25	13	24	35	5	FCC			0. 22
42		110	45		1–2		HCP			0.010
6.5	0.35	50	32–46	32–50	16–12	8–3	HCP			0.059
40–50	0.31	147	100	250	10–20	2–5	BCC			2.4
30		75	47	125	15	3	FCC			4.5
27		75–125	50	125	40	1	BCC			21
16.8		200	80	122	24	7.5	HCP	880	BCC	5.8
20–22	0.36	260‡					BCC			4.8
50–60		260	50	300			BCC			19
12	0.33		35	85	31	18	HCP	863	BCC	0.18
29	0.38	160	90		45		FCC			2.9
31	0.29	200–400	160–180		24		FCC			4.1

† The following abbreviations are used: FCC, face-centered cubic; BCC, body-centered cubic; HCP, hexagonal close packed.
‡ Cold worked.

becomes

$$k_\infty = \eta f p = 1 \tag{12-5}$$

Whenever $k < 1$, the system is said to be subcritical or convergent and for $k > 1$, the system is supercritical or divergent.

For a finite system some neutrons may leak out, thereby reducing the multiplication rate given in Eq. 12-4. Hence an effective multiplication constant k_{eff} is introduced which naturally is less than k_∞ in numerical value; that is,

$$k_{\text{eff}} = k_\infty \cdot (\text{total nonleakage probability}) \tag{12-6}$$

Analytical expressions for the bracketed expression are derived later.

K,13. Bare Reactors. The treatment to follow considers a homogeneous medium of fuel and moderator, and the sources are determined from the Fermi age theory. The steady state diffusion equation is from Eq. 10-4

$$D\nabla^2\phi(\mathbf{r}) - \Sigma_a\phi(\mathbf{r}) + pq(\mathbf{r}, \tau_{\text{th}}) = 0 \tag{13-1}$$

Table K,12b. Nuclear properties of structural materials. (Reprinted from [4] by permission.)

Element	Symbol	Atomic number	Natural element σ_a, barns	Isotope	Abundance in natural element, per cent	Activation cross section, barns	Hardest gamma rays, mev	Half life
Beryllium†	Be	4	0.01					
Boron‡	B	5	750					
Carbon†	C	6	0.0045					
Magnesium†	Mg	12	0.059	26	11.1	0.05	1.0 . . .	9.6 m
Aluminum†	Al	13	0.22	27	100	0.22	1.78	2.3 m
Titanium†	Ti	22	5.8	50	5.3	0.14	0.32	6 m
Vanadium*†	V	23	4.7	51	99.8		1.46	2.6 m
Chromium†	Cr	24	2.9	50	4.4	16	0.32	27 d
Manganese*†	Mn	25	12.6	55	100	12.6	3.0	2.6 h
Iron†	Fe	26	2.4	58	0.33	0.8	1.3 . . .	45 d
Cobalt*	Co	27	34	59	100	14	1.33	11 m
				59	100	20	1.33	5.2 y
Nickel†	Ni	28	4.5	64	1	3.0	1.5 . . .	2.56 h
Copper†	Cu	29	3.6	63	69	4.3	1.34 . . .	12.8 h
				65	31	2.1	1.04	5.17 m
Zinc*†	Zn	30	1.06	64	48.9	0.5	1.12	250 d
Zirconium†	Zr	40	0.18	94	17.4	0.07	0.92	65 d
Niobium†	Nb	41	1.1	93	100	1	0.9	6.6 m
Molybdenum†	Mo	42	2.4	98	23.8	0.13	0.73	67 h
				100	9.5	0.2	0.96	15 m
Silver*	Ag	47	60	107	51.4	30	0.61	2.3 m
				109	48.6	82	0.9	24 sec
				109	48.6	2	0.66	270 d
Cadmium‡	Cd	48	3100	106	1.2	1	0.85	6.7 h
				114	28.8	0.14	1.28	43 d
				114	28.8	1.1	0.52	54 h
				116	7.6	1.4	1.2	2.9 h
Indium*‡	In	49	190	115	95.8	145	2.1 . . .	54.3 m
Tin†	Sn	50	0.6	124	6.0	0.2	1.4 . . .	9.5 m
Antimony*	Sb	51	5.3	121	57.2	5.7	0.69 . . .	2.8 d
Lanthanum*	La	57	8.8	139	99.9	8.8	2.5 . . .	40 h
Praseodymium*	Pr	59	11	141	100	11	1.8 . . .	19.3 h
Samarium	Sm	62	10,000					
Europium*‡	Eu	63	4300	151	47.8	1400	0.94 . . .	9.3 h
				151	47.8	7000	1 . . .	13 y
				153	52.2	400	1.3 . . .	16 y
Gadolinium‡	Gd	64	36,000					
Hafnium‡	Hf	72	120	174	0.18	~500	0.34 . . .	70 d
				180	35.3	13	0.61	45 d
Tantalum*†	Ta	73	21	181	100	21	1.2 . . .	111 d
Tungsten*†	W	74	19	186	28.7	34	0.7 . . .	24 h
Rhenium*	Re	75	84	185	37.1	100	0.76 . . .	91 h
				187	62.9	63	2.4 . . .	17 h
Iridium*	Ir	77	440	191	38.5	700	0.6 . . .	75 d
				193	61.5	120	>2.2 . . .	19 h
Gold*	Au	79	95	197	100	95	1.1	2.7 d
Mercury‡	Hg	80	360					

* Major contributors to persistent induced activity even if present in relatively small quantities. Elements possessing isotopes which combine high activation cross section, high isotopic abundance, intermediate half life and relatively hard gamma rays.

† Major or common minor components of structural metals and alloys associated with reactor technology.

‡ Elements whose high neutron absorption cross section makes them particularly undesirable as an impurity (especially in moderators).

where the sources are defined by Eq. 8-5 and τ_{th} is the age from E_0 to thermal energies. Now, for an infinite medium all thermal neutrons are absorbed and the number per cm³ per sec is by definition

$$\Sigma_a \phi(\mathbf{r}) \tag{13-2}$$

and by virtue of Eq. 12-5 the number of fast neutrons (monoenergetic fission neutrons) per cm³ per sec is

$$\frac{k}{p} \Sigma_a \phi(\mathbf{r}) \tag{13-3}$$

The expression (Eq. 13-3) must represent the slowing-down density for the fission neutrons, i.e. for $\tau = 0$. Therefore

$$q(\mathbf{r}, 0) = \frac{k}{p} \Sigma_a \phi(\mathbf{r}) \tag{13-4}$$

will provide the coupling to the age equation. From Art. 9 the solution of the slowing-down density can be written as

$$q(\mathbf{r}, \tau) = AR(\mathbf{r})e^{-B^2\tau} \tag{13-5}$$

where B^2 is now the separation constant in keeping with classical terminology. If now $\tau = 0$, then we have from Eq. 13-4 that

$$\frac{k}{p} \Sigma_a \phi(\mathbf{r}) = AR(\mathbf{r}) \tag{13-6}$$

which, on substitution into Eq. 13-5, gives the significant result

$$q(\mathbf{r}, \tau) = \frac{k}{p} \Sigma_a \phi(\mathbf{r})e^{-B^2\tau} \tag{13-7}$$

If we rewrite Eq. 13-1 with the use of Eq. 13-7 then the reactor diffusion equation becomes on collecting terms

$$\nabla^2\phi(\mathbf{r}) + \frac{k_\infty e^{-B^2\tau_{\text{th}}} - 1}{L^2} \phi(\mathbf{r}) = 0 \tag{13-8}$$

If we now assume that $\phi(\mathbf{r})$ and $q(\mathbf{r}, \tau)$ satisfy the same boundary condition, namely vanish at the same extrapolated distance, then from the proportionality between the spatial variations given by Eq. 13-6 and 13-7 they satisfy the same wave equation,

$$\nabla^2\phi(\mathbf{r}) + B^2\phi(\mathbf{r}) = 0 \tag{13-9}$$

On examination of Eq. 13-8 and 13-9 one finds that $\phi(\mathbf{r})$ will satisfy both equations provided that

$$\frac{k_\infty e^{-B^2\tau_{\text{th}}} - 1}{L^2} = B^2 \tag{13-10}$$

or

$$\frac{k_\infty e^{-B^2\tau_{\text{th}}}}{1 + L^2B^2} = 1 \tag{13-11}$$

Eq. 13-11 is the critical equation for a bare homogeneous reactor as determined by the age-diffusion theory and is valid to within one mean free path of the boundary.

K,14. Critical Size. The value of B^2, the buckling, in Eq. 13-9 determines the "wavelength" of the neutron distribution, i.e. the neutron density has nodes on some surface for a given value of B^2. Only if the extrapolated reactor surface coincides with the first nodal surface of the neutron density will the solution of Eq. 13-9 satisfy Eq. 13-8 and the boundary conditions. The reactor must therefore have particular physical dimensions for a pre-assigned set of parameters, k, L^2, p, $e^{-B^2\tau_{\text{th}}}$, in order for the buckling of the solution which satisfies the boundary conditions to simultaneously satisfy Eq. 13-11. The buckling which satisfies the boundary conditions,

$$B_{\text{g}}^2 = -\frac{\nabla^2\phi}{\phi} \tag{14-1}$$

is called the geometrical buckling. The value of B^2 which satisfies Eq. 13-11 is called the material buckling B_{m}^2. Thus a reactor will be critical if only $B_{\text{g}}^2 = B_{\text{m}}^2$ and is positive.

For a finite reactor it can be shown that there will exist a finite critical size provided that $k_\infty > 1$. In Eq. 13-11 the term $(1 + L^2B_{\text{g}}^2)^{-1}$ is the nonleakage probability of the thermal neutrons. To demonstrate this, the ratio of thermal leakage to thermal absorption from Eq. 13-9 is

$$\frac{\text{thermal leakage}}{\text{thermal absorption}} = \frac{-D\nabla^2\phi}{\Sigma_{\text{a}}\phi} = \frac{DB_{\text{g}}^2\phi}{\Sigma_{\text{a}}\phi} = L^2B_{\text{g}}^2 \tag{14-2}$$

and therefore,

$$\frac{\text{thermal absorption}}{\text{thermal (absorption and leakage)}} = \frac{1}{1 + L^2B_{\text{g}}^2} \tag{14-3}$$

From Eq. 13-7 one can readily see that $e^{-B^2\tau}$ represents the nonleakage probability of slowing-down neutrons. Hence with the aid of Eq. 12-6 we obtain the result that, for a finite critical reactor,

$$k_{\text{eff}} = \frac{k_\infty e^{-B^2\tau_{\text{th}}}}{1 + L^2B^2} = 1 \tag{14-4}$$

and $k_\infty > 1$.

For a small reactor, B^2 is large and the leakage terms in Eq. 14-4 increase as would be expected, thereby depressing the multiplication constant. Similarly, if the age τ_{th} or L^2 is large, which naturally depends on the materials used, the leakage would increase from a finite reactor. This is so because $\sqrt{\tau + L^2}$ is the distance a neutron will migrate from creation as a fission neutron to thermal absorption, and therefore the probability of escape for neutrons created near a boundary will increase as the migration length increases.

K,15. Spherical Reactor. We shall treat a spherical reactor to illustrate the theoretical development in Art. 13. From symmetry considerations, the Laplacian in Eq. 13-9 is a function of the radial coordinate. Hence the wave equation becomes

$$\frac{d^2\phi(r)}{dr^2} + \frac{2}{r}\frac{d\phi(r)}{dr} + B_g^2\phi(r) = 0 \tag{15-1}$$

and a solution for the flux is

$$\phi(r) = \frac{A}{r}\sin B_g r + \frac{C}{r}\cos B_g r \tag{15-2}$$

From the boundary condition that $\phi(0) = 0$ and remain finite, $C = 0$, therefore

$$\phi(r) = \frac{A}{r}\sin B_g r \tag{15-3}$$

Now, applying the boundary condition that $\phi(\tilde{R}) = 0$, where $\tilde{R}$ is the extrapolated boundary of the reactor, then

$$\phi(R) = 0 = \frac{A}{\tilde{R}}\sin B_g\tilde{R} \tag{15-4}$$

Since A and $\tilde{R}$ are nonzero this means that

$$\sin B_g\tilde{R} = 0 \tag{15-5}$$

and

$$B_g = \frac{n\pi}{\tilde{R}} \tag{15-6}$$

where $n = 0, 1, 2$, etc. The lowest eigenvalue which leads to a nontrivial solution is $n = 1$, yielding the result:

$$B_g^2 = \left(\frac{\pi}{\tilde{R}}\right)^2 \tag{15-7}$$

The flux distribution is, by means of Eq. 15-7,

$$\phi(r) = \frac{A}{r}\sin\frac{\pi r}{\tilde{R}} \tag{15-8}$$

It should be noted that the lowest eigenvalue leads to the smallest reactor.

Reflectors. It is evident from the discussion in Art. 14 that neutrons will escape from a finite system which will reduce the multiplication constant. To maintain the multiplication constant two possibilities exist: (1) to increase the fuel concentration or (2) to reduce leakage. The first alternative is not always attractive. The second possibility can be accomplished by surrounding the reactor with a moderator type of material. By this mechanism, the neutrons scattered within the reflector have a good chance to be returned to the reactor. Qualitatively, since the effective thickness of a reflector is of the order of 1.5 migration lengths beyond this, no further benefits are derived. Further, by keeping the fuel concentration constant for the base reactor and the reflected system, the reflector savings on the critical dimension is equivalent to the reflector thickness.

The theoretical development for a reflected reactor is given as part of the two-group method which is now presented.

K,16. Two-Group Method. Another approach to reactor analysis is the "group" method; that is, the behavior of neutrons within the reactor is described by a number of diffusion equations. The number of equations correspond to the number of neutron energy groups in which one divides the energy interval from fission energy to thermal energy. Here the two-group model will be explored; that is, the neutron groups will correspond to (1) thermal energies and (2) nonthermal or fast energies. The merit of this method is that rapid estimates of a reactor system can be made to a reasonably good degree of accuracy where the accuracy will depend to a large extent upon the constants employed.

For a bare reactor the diffusion equations are

$$-D_{1\mathrm{F}}\nabla^2\phi_{1\mathrm{F}}(\mathbf{r}) + \Sigma_{1\mathrm{F}}(\mathbf{r}) = \nu\Sigma_f\phi_{1\mathrm{S}}(\mathbf{r}) \tag{16-1}$$

$$-D_{1\mathrm{S}}\nabla^2\phi_{1\mathrm{S}}(\mathbf{r}) + \Sigma_{1\mathrm{S}}\phi_{1\mathrm{S}}(\mathbf{r}) = p\Sigma_{1\mathrm{F}}\phi_{1\mathrm{F}}(\mathbf{r}) \tag{16-2}$$

where the subscripts F and S refer to the fast and thermal groups respectively, and $\Sigma_{1\mathrm{S}}$ is unambiguously the thermal absorption cross section. The cross section $\Sigma_{1\mathrm{F}}$ corresponds to the removal of fast neutrons from the fast group and can be chosen so that $\Sigma_{1\mathrm{F}}\phi_{1\mathrm{F}}$ can be represented by the slowing-down density. The source term for fast neutrons $\nu\Sigma_f\phi_{1\mathrm{S}}$ indicates only thermal fissioning and, for thermal neutrons, $p\Sigma_{1\mathrm{F}}\phi_{1\mathrm{F}}$ represents the actual number of neutrons slowing down to thermal energies. The factor p is the resonance escape probability.

We shall now assert that $\phi_{1\mathrm{F}}$ and $\phi_{1\mathrm{S}}$ are proportional to the same spatial function, that is,

$$\phi_{1\mathrm{F}}(\mathbf{r}) = \phi_{1\mathrm{F}}X(\mathbf{r}) \tag{16-3}$$

$$\phi_{1\mathrm{S}}(\mathbf{r}) = \phi_{1\mathrm{S}}X(\mathbf{r}) \tag{16-4}$$

and the function $X(\mathbf{r})$ satisfies the wave equation

$$\nabla^2 X(\mathbf{r}) + B^2 X(\mathbf{r}) = 0 \tag{16-5}$$

The parameter B^2 is the previously described geometric buckling. Hence both fluxes vanish at the extrapolated distance which is assumed, with slight error, to be identical.

Eq. 16-1 and 16-2 can now, by applying Eq. 16-3 and 16-4, be expressed as

$$D_{1F}B^2\phi_{1F} + \Sigma_{1F}\phi_{1F} = \nu\Sigma_f\phi_{1S} \tag{16-6}$$

$$D_{1S}B^2\phi_{1S} + \Sigma_{1S}\phi_{1S} = p\Sigma_{1F}\phi_{1F} \tag{16-7}$$

which on elimination of ϕ_{1F}, ϕ_{1S} gives

$$\frac{\nu\Sigma_f}{D_{1S}B^2 + \Sigma_{1S}} = \frac{D_{1F}B^2 + \Sigma_{1F}}{p\Sigma_{1F}} \tag{16-8}$$

If we now introduce a "fast" diffusion length in analogy to the thermal one, namely $L^2 = D_{1F}/\Sigma_{1F}$, then the critical equation for the bare reactor is

$$\nu\frac{\Sigma_f}{\Sigma_{1S}}\,p\,\frac{1}{(1 + L_{1F}^2B^2)}\,\frac{1}{(1 + L_{1S}^2B^2)} = 1 \tag{16-9}$$

In comparison with the form of Eq. 14-4, we have

$$k_{\text{eff}} = \frac{k_\infty}{(1 + L_{1F}^2B^2)(1 + L_{1S}^2B^2)} = 1 \tag{16-10}$$

where $(1 + L_{1F}^2B^2)^{-1}$ is the nonleakage probability of fast neutrons in place of $e^{-B^2\tau}$. It should be noted that for a large reactor, $B^2 \ll 1$, the two expressions are approximately the same, provided that $\tau \cong L_{1F}^2$.

To compute the parameters D_{1F} and Σ_{1F} we must determine their effective value over the entire nonthermal energy range. Thus, in terms of lethargy units, we have

$$D_{1F} = \int_0^{u_{\text{th}}} \frac{D(u')du}{u_{\text{th}}} \tag{16-11}$$

and

$$\Sigma_{1F} = \frac{D_{1F}}{\tau_{\text{th}}} \tag{16-12}$$

Some values of the parameters, τ_{th}, L_S and $\bar{l}_{\text{tr}}(E_{\text{th}})$ are given in Table K,16.

Table K,16. Nuclear parameters for some moderators.

Moderator	τ_{th}, cm²	L_S, cm	$\bar{l}_{\text{tr}}$, cm
H_2O	31	2.85	0.48
Be (ρ = 1.85)	97.2	20.8	1.43
BeO (ρ = 3)	105 ± 10	29	0.9
Graphite	364	54.4	—

Returning to the critical equation (Eq. 16-9) it is recognized as a second degree algebraic equation in B^2:

$$L_{1F}^2 L_{1S}^2 B^4 + (L_{1F}^2 + L_{1S}^2)B^2 + (1 - k) = 0 \tag{16-9}$$

the roots of which are

$$B_1^2 = \frac{1}{2}\left[-\left(\frac{1}{L_{1F}^2} + \frac{1}{L_{1S}^2}\right)^2 + \sqrt{\left(\frac{1}{L_{1F}^2} + \frac{1}{L_{1S}^2}\right)^2 + \frac{4(k-1)}{L_{1F}^2 L_{1S}^2}}\right] \tag{16-13}$$

$$B_2^2 = \frac{1}{2}\left[-\left(\frac{1}{L_{1F}^2} + \frac{1}{L_{1S}^2}\right)^2 - \sqrt{\left(\frac{1}{L_{1F}^2} + \frac{1}{L_{1S}^2}\right)^2 + \frac{4(k-1)}{L_{1F}^2 L_{1S}^2}}\right]$$

where B_1^2, B_2^2 are positive quantities.

The solution for the fluxes therefore, is

$$\phi_{1F}(\mathbf{r}) = \phi_0 X_1(B_1\mathbf{r}) \tag{16-14}$$

$$\phi_{1S}(\mathbf{r}) = \phi_0 \frac{D_{1F}B_1^2 + \Sigma_{1F}}{\frac{k}{p}\Sigma_{1S}} X_1(B_1\mathbf{r}) \tag{16-15}$$

where X depends upon core geometry.

We shall now consider a reflected reactor by this method. The diffusion equations to be satisfied in the reflector are

$$-D_{2F}\nabla^2\phi_{2F}(\mathbf{r}) + \Sigma_{2F}\phi_{2F}(\mathbf{r}) = 0 \tag{16-16}$$

$$-D_{2S}\nabla^2\phi_{2S}(\mathbf{r}) + \Sigma_{2S}\phi_{2S}\phi = p\Sigma_{2F}\phi_{2F}(\mathbf{r}) \tag{16-17}$$

with the boundary conditions

(a) $$\left.\begin{aligned} \phi_{1F} &= \phi_{2F} \\ \phi_{1S} &= \phi_{2S} \\ \mathbf{j}_{1F} &= \mathbf{j}_{2F} \\ \mathbf{j}_{1S} &= \mathbf{j}_{2S} \end{aligned}\right\} \text{at core-reflector interface}$$

(b) ϕ_{2F}, ϕ_{2S} vanish at the extrapolated distance

where 1 and 2 refer to the core and the reflector respectively. In view of Eq. 16-5 let the solution for the flux within the reactor core corresponding to B^2 be given by

$$\begin{aligned} \nabla^2 X &= -B_1^2 X \\ \nabla^2 Y &= B_2^2 Y \end{aligned} \tag{16-18}$$

where B_1^2, B_2^2 are positive quantities. The general solution for the flux can now be expressed as

$$\phi_{1F}(\mathbf{r}) = aX(B_1\mathbf{r}) + bY(B_2'\mathbf{r})$$

$$\phi_{1S}(\mathbf{r}) = a\frac{D_{1F}B_1^2 + \Sigma_{1F}}{\frac{k}{p}\Sigma_{1S}} X(B_1\mathbf{r}) + b\frac{D_{1F}B_2^2 + \Sigma_{1F}}{\frac{k}{p}\Sigma_{1S}} Y(B_2'\mathbf{r}) \tag{16-19}$$

where B_2' is the positive square root of B_2^2.

On application of boundary condition (*b*) we obtain directly the solutions for Eq. 16-16 and 16-17:

$$\begin{aligned}\phi_{2F} &= \alpha Z(\kappa_1 \bar{R})\\ \phi_{2S} &= \alpha \frac{\Sigma_{2F}}{D_{2S}\kappa_1^2 - \Sigma_{2S}} Z[\kappa_1(\bar{R} - r)] + \beta Z[\kappa_2(\bar{R} - r)]\end{aligned} \tag{16-20}$$

where $\bar{R}$ is the extrapolated distance and κ_1, κ_2 are the reciprocal diffusion lengths for fast and thermal neutrons respectively.

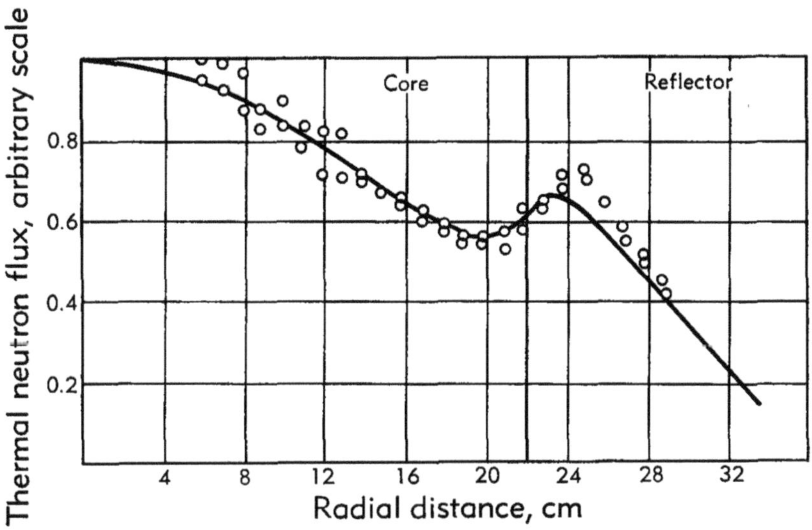

Fig. K,16. Measured and 2-group calculated radial distribution of thermal neutron flux in an H_2O reflected Zr-H_2O-U^{235} critical assembly. From [*11*].

We now have four equations and four unknowns, *a*, *b*, α, and β. From the four boundary conditions (*a*) at the core-reflector interface, we can obtain four homogeneous equations in these unknowns which after algebraic manipulation can only be satisfied by the determinant,

$$\begin{vmatrix} X & Y & Z & 0 \\ \dfrac{D_{1F}B_1^2 + \Sigma_{1F}}{\dfrac{k}{p}\Sigma_{1S}} X & \dfrac{D_{1F}B_2^2 + \Sigma_{1F}}{\dfrac{k}{p}\Sigma_{1S}} Y & -\dfrac{\Sigma_{2F}}{D_{2S}\kappa_1^2 - \Sigma_{2S}} & -Z_2 \\ D_{1F}X' & D_{1F}Y' & -D_{2F}Z_1' & 0 \\ D_{1S}\dfrac{D_{1F}B_1^2 + \Sigma_{1F}}{\dfrac{k}{p}\Sigma_{1S}} X' & D_{1S}\dfrac{D_{1F}B_2^2 + \Sigma_{1F}}{\dfrac{k}{p}\Sigma_{1S}} Y' & D_{2S}\dfrac{\Sigma_{2F}}{D_{2S}\kappa_1^2 - \Sigma_{2S}} Z_1' & -D_{2S}Z_2' \end{vmatrix} = 0 \tag{16-21}$$

where *X*, *Y*, *Z* and their first derivatives are evaluated at the interface. The primed functions are the first derivatives with respect to the spatial variables and Z_1 and Z_2 refer to the functions whose variables are κ_1 $(\tilde{R} - r)$ and κ_2 $(\tilde{R} - r)$ respectively. This determinant is the critical equation for a homogeneous reflected reactor as given by the two-group method.

To solve the determinant, one specifies the reactor constituents and then proceeds by a reiterative process to determine the critical radius, for a given reflector, that satisfies the determinant. To illustrate the reliability of the two-group method, the thermal flux distribution for a H_2O-reflected H_2O-Zr-U^{235} system as determined experimentally is compared with the theoretical prediction in Fig. K,16. It is interesting to note that the thermal flux has a point of inflection a short distance within the reflector. This results from the fast neutrons slowed down within the reflector; however, they are less strongly absorbed than in the reactor.

K,17. Physical Aspects. The physical aspects of the solutions in Art. 16 as presented in [*12*] will be discussed briefly. If we write Eq. 16-18 in a general form,

$$\nabla^2 X_i(B_i\mathbf{r}) = -B_i^2 X(B_i\mathbf{r}) \tag{17-1}$$

and multiply through by $X_i(B_i\mathbf{r})$, then upon integrating over the volume enclosed by the surface S and applying Green's theorem we obtain

$$B_i^2 = \frac{\int_V (\nabla X_i)^2 dV}{\int_V X_i^2 dV} = \frac{\int_S X_i \frac{\partial X_i}{\partial n} dS}{\int_V X_i dV} \tag{17-2}$$

Now, if either X_i or $\partial X_i/\partial n$ vanishes on S, B_i^2 being the quotient of two positive quantities must also be positive. If B_i^2 is negative neither X_i or $\partial X_i/\partial n$ can vanish over any closed surface. Thus the two solutions of Eq. 17-1 can be classified according to whether B_i^2 is positive or negative. Since $B_1^2 > 0$ and $B_2^2 > 0$, $X_1(B_1r)$ is a function which may pass through a zero on certain surfaces, whereas $X_2(B_2r)$ can never vanish. Consequently, for a bare reactor, $b = 0$, and we obtain Eq. 16-14 and 16-15 which are purely of an oscillatory nature. For the reflected system, $b \neq 0$ and the solutions X_2 or Y occur which are not oscillatory in character. Now the coefficients of X and Y in Eq. 16-19 are different and, if both solutions are admissible, the fast and slow neutron densities are not proportional throughout the reactor.

It is by this manner that the variation in neutron spectrum close to a boundary is taken into account by the group theory. Clearly, by increasing the number of groups, the more nearly can one describe the actual variation.

K,18. Other Aspects.

Kinetics. The static theory of reactors has already been presented. The kinetic behavior and the theory of control is beyond the scope of this article and therefore the reader is referred to [*5*]. It should be mentioned that the control of a reactor is effected by foreign bodies, e.g. movable rods, which alter the fission rate by perturbing the spatial variation of the neutron density. Normally the rods are made of absorbing materials like cadmium or boron steels; however, the multiplication rate can be altered by using rods of fissionable material.

Poisons. Of the fission products produced by a reactor some nuclides have enormous capture cross sections, e.g. for xenon, $\sigma_a = 3 \times 10^6$ barns for energies less than 1 ev, which act in the capacity of "poisons" with respect to the fission reaction. To override this build-up of poisons within the reactor, sufficient reactivity must be built into the system.

Temperature. If one contemplates a reactor to operate at elevated temperatures, then the microscopic properties of the reactor, k, Σ, B^2, etc. will vary. The temperature dependence results from the fact that these properties are proportional to some power of the material density. Since the reactor will be critical at ambient conditions, namely at start-up, and will remain critical at the operating temperature, it is desirable that the reactor possess a decreasing reactivity with increasing temperature. This is called a negative temperature coefficient.

Heterogenous systems. A heterogeneous system is one in which the fissionable material is separated from the moderator. The over-all theory presented for the homogeneous system is applicable; however, the detailed structure of the flux distribution will vary depending on the particular arrangement of the fuel within the moderator. As a general statement, the errors introduced by considering a high performance heterogeneous reactor with enriched fuel as a homogeneous one are to a first approximation small. Physically, the error arises from the fact that the fuel is not being used to the greatest advantage, and consequently if a "disadvantage" factor is introduced into the critical equation the error can be reduced.

CHAPTER 3. SHIELDING PRINCIPLES

K,19. Radiological Aspects. On completion of an efficient reactor design one must provide a protective barrier, i.e. a shield, if the reactor is to be put into operation. This is readily realized from the considerations in reactor theory namely, the neutron leakage, and also gamma ray leakage. Since these two radiations are deleterious to matter in general and humans in particular, suitable materials must be placed about the reactor to reduce the radiation at its exterior to a tolerable level. What precisely is the tolerable dose is still the subject of investigations by radiobiologists, however, the safe limit established by the Atomic Energy Commission is 300 milliroentgen per 40-hour week.

Before proceeding much further let us define a number of terms. A *roentgen* is a quantity of X or gamma radiation which will produce 1 esu of charge in 1 cm^3 of dry air at standard atmospheric conditions or deposit 83.7 ergs per gram of air. To generalize this unit to particle radiations the rep (Roentgen equivalent physical) was introduced. The rep is defined as the quantity of radiation which on absorption by body tissue will deposit 83.7 ergs per gram of tissue. The process of deposition is by direct ionization of the atoms within the tissue. Converting this into a physical flux term, we have

$$1 \text{ (roentgen)} = \frac{7.2 \times 10^8}{CE}, \text{ photons/cm}^2$$

$$C = 0.32\ (\mu - \sigma_s)10^5$$

$$\mu = \text{total coefficient of absorption, cm}^{-1}$$

$$\sigma_s = \text{Compton scattering coefficient of absorption, cm}^{-1}$$

$$E = \text{gamma ray energy, mev}$$

Now, neutron radiation performs its damaging functions in an indirect manner; that is, it must first either eject or scatter a proton from tissue molecules and in turn the protons ionize the tissue. This leads us to a dual definition; (1) physical and (2) biological. The physical dose for fast neutrons is directly related to the physical flux. To demonstrate this point, the energy flux of neutrons at energy E is $E \cdot N(E)$ where $N(E)$ is the flux of neutrons of energy E; however, the cross section $\sigma \sim E^{-1}$ and hence, the physical dose $\sim N(E)$. In the course of time, radiobiologists have found that the specification of physical dose is inadequate since various organs behaved differently or were more susceptible to damage by neutron radiation. As an artifice the concept of relative biological equivalent (RBE) has been introduced to place neutron radiation on an equal basis with gamma radiation; that is, gamma radiation has an RBE = 1. Several values of RBE for fast neutron radiation are given in Table K,19a.

Table K,19a. RBE values for fast neutrons.

Type of damage	*RBE*
Cataracts	5–10
Skin	3.5
Bone marrow	1.6

The ratio of roentgen equivalent physical (rep) to roentgen equivalent man (rem) defines the relative biological equivalent. The RBE for several different types of radiation is given in Table K,19b.

In view of the last chapter, the reactor emits a given dose of radiation and the function of the shield is to reduce it to a specified tolerance level.

Hence the initial conditions, namely sources of reactor radiation, must be specified along with the analytical tools so that the reduction effected by particular combinations of materials can be determined.

Table K,19b. The RBE for nuclear radiation.

Type of radiation	*RBE*
X or gamma	1
Beta	1
Fast neutrons	10
Thermal neutrons	5
Alpha (internal)	20

K,20. Sources of Radiation. The sources of radiation emanating from a reactor are: (1) prompt fission radiation, (2) fission fragment decay, and (3) reactor material activation.

The escaping, prompt fission radiation consists of neutrons and gamma rays. The former is a consequence of the finite reactor geometry, that is, the high probability that fission neutrons born within one mean free path of the boundary will escape. Further, we see from Eq. 11-1 that about 6 per cent of all fission neutrons are highly energetic, i.e. greater than 6 mev, and hence there is small probability of their being slowed down within the reactor proper. The energies of the escaping gammas are: (1) fission, about 2.0 mev, (2) decay gammas are about 2–3 mev, and (3) activation gammas are in excess of 3 mev (see Table K,20a and K,20b). Moteff [*14*] has analyzed the energy spectrum of the gammas emitted by fission products as a function of reactor operation time and times after shutdown.

Table K,20a. Gamma sources per fission.

Type	*Total energy*	*Average energy per photon*	*Reference*
Prompt fission	7.59 mev/fission*	~1.0	[*9*]
Fission products $t < 140$ msec	10^{-11} curies	3	
10 sec $< t <$ 24 hr	$0.9t^{-1.2}$ mev/sec	2	[*13*]

* Gamble gives an empirical formula for the energy distribution of prompt fission gamma rays; $N(E) = 8.2e^{-E}$ photons/fission/mev.

Since reactor materials are primarily of low atomic mass these gammas on the whole escape unaffected from the reactor. Thus we see that in the shielding problem it is the unusual gamma or neutron that is of interest in contradistinction to the reactor in which average events are significant.

K,21. Theory. The theoretical treatment of shielding is no different than any radiation transfer problem. The basic equation is the Boltzmann transport equation.

$$\nabla \cdot \mathbf{\Omega} N + \mu N = \iint_{4\pi} N(\mathbf{r}, E', \mathbf{\Omega}')\sigma(\mathbf{\Omega}', E'; \mathbf{\Omega}, E) dE d\mathbf{\Omega}' + S(\mathbf{r}, E, \mathbf{\Omega}) \tag{21-1}$$

Table K,20b. Capture gamma data [14].

Target	Thermal (n, γ) cross section, barns	Spectral type	Photons in designated, mev energy intervals per 100 captures					Highest energy gamma ray, mev	Average number of photons/capture	Fraction of energy emitted $S(E_m)$†		Decay gammas‡¶
			0–1	1–3	3–5	5–7	>7			E_m, mev	$S(E_m)$	
Aluminum	0.215	1	?	>13	77	21	35	7.724	~2	2.5	0.78	2.3-min Al^{28}: 100(1.80)
Antimony	6.4	3*	?	~80	36	12		6.80				Complicated
Arsenic	4.1	3*	?	~80	47	22	1	7.30	2.7			Complicated, uncertain
Barium	1.17	3*	?	~80	75	14	1	9.23				
Beryllium††	0.009	1	0	0	50	75	0	6.814				
Bismuth††	0.015	1	0	0	100	0	0	4.17				
Boron-10§	3990	4	0	0	0	0	0	0.478				Fast B^{11}§: 94(0.478)
Cadmium	3500	3*	>120	20	73	17	1	9.046	4.1			
Calcium	0.406	2	?	50	60	101	2.4	7.83		2.6	0.80	
Carbon-12	0.0045	1	?	<30	100	0	0	4.95	1.3	3.68	0.93	
Chlorine	32	2	?	20	13	18	21	8.56	3.1	3.3	0.26	
Chromium	2.9	1*	>37	16	12	18	69	9.716	>2	2.8	0.73	
Cobalt	34.8	2*	?	?	36	49	8	7.486		2.5	0.67	5.3-y Co^{60}: 100(1.17), 100(1.33)
Copper	3.59	1*	?	?	>23	22	42	7.914	2.6	3.4	0.67	
Fluorine	0.009	1	?	?	?	35	0	6.63				12-sec F^{20}: 100(1.63)
Gadolinium	36,300	3*	?	80	23	4	2	7.78	3.9			
Gold	94	3*	?	?	66	38	0	6.494	3.5			
Hydrogen-1††	0.330	1	0	100	0	0	0	2.230				
Indium	190	3*	?	?	36	4	0	5.86	3.3			54.3-min In^{116}: 18(2.09), 15(1.49), 54(1.27), 39(1.09)
Iron	2.43	1*	?	<10	24	22	50	10.16	1.7	3.0	0.78	
Lead††	0.17	1	0	0	0	7	93	7.38				
Lithium-6	910	4	0	0	0	0	0	0				
Magnesium	0.059	2	?	>59	110	25	11	9.216		2.6	0.99	9.58-min Mg^{27}: 10(0.84), 2(1.01)
Manganese	12.6	2*	?	?	>27	30	27	7.261	2.6	3.5	0.64	2.59-hr Mn^{56}: 100(0.845), 25(1.81), 15(2.13)
Mercury	380	3*	?	?	86	41	0	6.446	3.3			
Molybdenum	2.4	3*	?	?	84	26	3	9.15				

Nickel	4.8	1	?	?	>14	30	72	8.997		3.4	0.95	
Niobium	1.1	3*	?	?	54	14	0	7.19	2.6			
Nitrogen-14	0.1	2	?	<5	<35	90	39	10.8				
Phosphorus	0.193	2	?	?	115	43	11	7.94		2.5	0.66	
Platinum	8.1	3*	?	~120	45	15	1	7.920				Complicated
Potassium	1.89	2	?	36	36	32	12	9.23		3.00	0.48	
Praseodymium	11.2	3*	?	~80	34	8	0	5.83				19-hr Pr^{142}: 4(1.58)
Rhodium	150	3*	?	~70	38	10	0	6.792				Exist
Samarium	10,600	3*	?	~150	45	5	1	7.89	5.6			
Scandium	22	3*	?	?	63	29	14	8.85				85-d Sc^{46}: 100(1.12), 100(0.89)
Selenium	11.8	3*	?	?	65	27	11	10.483				
Silicon	0.160	2	?	>100	229	41	16	10.55		2.5	1.44	
Silver	60	3*	?	~90	70	17	0.5	7.27	2.9			Complicated 270-day Ag^{110}
Sodium	0.47	2	≅100	>50	61	29	0	6.41	<2			14.9-hr Na^{24}: 100(2.758), 100(1.380)
Strontium	1.16	(2–3)*	?	~140	62	49	13	9.22				
Sulfur	0.49	2	?	>19	80	91	8	8.64		2.7	0.76	
Tantalum	21.3	3*		~50	26	2	0	6.07				Complicated
Thallium	3.3	3*	?	~100	76	62	0	6.54				
Tin	0.65	3*	?	?	139	33	4	9.35				
Titanium	5.8	2*	>50	100	33	99	10	9.39		2.4	1.02	
Tungsten	19.2	3*	?	?	53	14.5	0.5	7.42				
Vanadium-51	4.7	3*	?	?	24	54	18	7.305	2.5	2.6	0.82	3.74-min V^{52}: 100(1.46)
Zinc	1.06	(2–3)*	?	?	48	29	17	9.51		2.8	0.60	250-d Zn^{65}: 25(1.12)
Zirconium**	0.18	3*	?	?	113	35	4	8.66				65-d Zr^{95}: ~98(0.750), 35-d Nb^{95}: 100(0.764)

* For elements thus marked, number of unresolved photons has been determined; for all others only the intensity of resolved lines is given.

† Fraction, S, of available energy emitted by gamma rays with energy greater than E_m. E_m is the minimum gamma-ray energy observed in the measurement, it has no deep significance.

‡ "Complicated" means decay gammas above 750 kev are claimed to exist but the multiplicity of isotopes and/or the complexity of the decay schemes render estimations of intensities difficult.

¶ Notation for decay gamma rays is as follows: half life of decay-gamma emitter is given first, then number of decay gamma rays of each energy >750 kev per 100 neutron captures. Energy is in parentheses. Thus "9.58-min Mg^{27}: 10 (0.84), 2 (1.01)" means isotope emits 10 0.84-mev and 2 1.01-mev decay gamma rays per 100 captures.

§ Boron-10 decay gamma is from Li^7 following alpha decay of B^{11}.

** Capture-gamma-ray intensities for zirconium may be off by 50%.

†† For these elements, the highest-energy gamma ray is the only gamma ray present in the highest occupied energy interval. For lead, *in addition*, the seven 5–7-mev gamma rays are all 6.73 mev; for beryllium, in addition, the fifty 3–5-mev gammas are all 3.37 mev.

where $N(\mathbf{r}, E, \mathbf{\Omega})$ = spatial flux of particles or photons per unit one of energy E, and moving in direction $\mathbf{\Omega}$ per unit solid angle.

$\sigma(\mathbf{\Omega}', E'; \mathbf{\Omega}, E)$ = differential energy and angle scattering cross section

$S(\mathbf{r}, E, \mathbf{\Omega})$ = source term, number of particles or photons emitted at E per unit volume in the direction $\mathbf{\Omega}$ per unit solid angle with energy per unit energy range of E.

To solve this equation, it is evident that one must be capable of defining the differential cross section and the energy dependence of the cross section. We shall now discuss the situation for gamma and neutron radiations.

The microscopic processes of gamma ray interaction with matter are well understood [*15,16*]. The three significant processes are (1) photoelectric effect, (2) Compton effect, and (3) pair production. The first is most significant for low energy photons and high atomic number. For intermediate energies the Compton process, which is primarily a scattering one, is most important. Pair production is the dominant effect for high energy photons and atomic number. In similar order, the cross section per atom is proportional to Z^2, Z, and $Z(Z + 1)$; therefore, for the first and last process the cross sections increase with atomic number or with materials of increasing weight. On the other hand, the Compton cross section is about the same for all materials as $Z/A \cong 0.5$, with the exception of hydrogen. Lastly, theory provides the angular dependence of the scattering cross section. We shall return later to gamma rays.

K,22. Neutron Attenuation. Our situation for neutrons is not as favorable: the detailed cross-section data, particularly for fast neutrons, are not available; and differential cross-section information is extremely meager. Recent theoretical developments in nuclear theory will aid materially to fill these gaps and, needless to say, so will an expanded experimental program. When sufficient data are available for neutrons, such as hydrogen and H_2O, theoretical solutions to Eq. 21-1 have been obtained. These calculations have shown that the neutrons with an initial energy of about 8 mev account for the deep penetrations.

To handle other materials, one must resort to experimentation. In this field the experiments under the direction of Blizard of the Oak Ridge National Laboratory [*17*] and their interpretation by Welton [*17*] have been a significant contribution. The experiments conducted had the following configuration: a fission source, in the form of a plane disk, a thin slab of material, and a thickness of water of about 140 cm. The detector located at the latter position in water normally measures the thermal flux which can be shown to correspond to the hard component.

The interactions which occur in the slab material, large angle scattering and inelastic scattering, are equivalent to absorption as viewed by the detector as the great depths of water filters all degraded neutrons; that is, the attenuation affected by the slab is exponential in character. This effective attenuation cross section is the removal cross section Σ_R. In this manner, the removal cross section has been obtained for a number of shielding materials (see Table K,22).

Table K,22. Removal cross sections [*18*].

Substance	Σ_R, *barns/atom*
Al	1.2
C	0.84
Cu	2.0
Fe	2.0
Pb	3.4

In the above discussion several apparent limitations appear: (1) slab thickness, (2) water depth, (3) multielement layers, and (4) distribution of nonhydrogenous material. The slab thickness may be as much as five reciprocal macroscopic removal cross sections with little variation. The water depth is not critical; however, sufficient thickness should exist to filter out the degraded neutrons. If a slab consisting of several layers of different elements is employed, the cross sections are additive. In a practical shield the neutron shielding may be located at some distance from the reactor. To establish that the removal cross section is unaltered, experiments were performed under extreme conditions, namely uniformly distributing the material. The results indicated that for the uniform distribution the cross section was somewhat less, about nine tenths of the slab value. Finally, what is the limitation upon a pure nonhydrogenous shield?

Blizard points out that for the model discussed the behavior of fast neutrons must be the controlling factor. When a nonhydrogenous shield is considered the relaxation length for fast neutrons is of the order of 7 cm. However, for intermediate fast neutrons the slowing-down length L_{SD} is the characteristic length, which is about 25–50 cm for high atomic mass materials. Under these circumstances at the extreme edge of the shield, neutrons of intermediate energies will be the controlling ones. Thus the requirement for a satisfactory shield is

$$L_{SD} < (\text{macroscopic removal cross section})^{-1} \tag{22-1}$$

To meet this requirement, it is necessary to add hydrogen, as one would expect from neutron diffusion theory. In fact, one can demonstrate that a considerable fraction should be hydrogenous. From a practical consideration, then, the addition of hydrogen is a necessity for a good shield.

Thus, to treat a neutron shield, the attenuation effected by hydrogen can be treated independently, either analytically or by the use of a measured value. This procedure is imposed because of the rapidly varying cross section, and therefore the attenuation varies rapidly with position. To handle the remainder, one can assume that the attenuation is defined by a single exponential with the appropriate divergence, i.e. take exp $(-\Sigma_R R)/4\pi R^2$ for a point source in an infinite medium, such as the kernel. This infers that the dose is a function of distance alone. Clearly, this is violated by a real shield as (1) a shield is finite and (2) homogeneity is lacking. These factors introduce errors which are shown by experiment not to be serious.

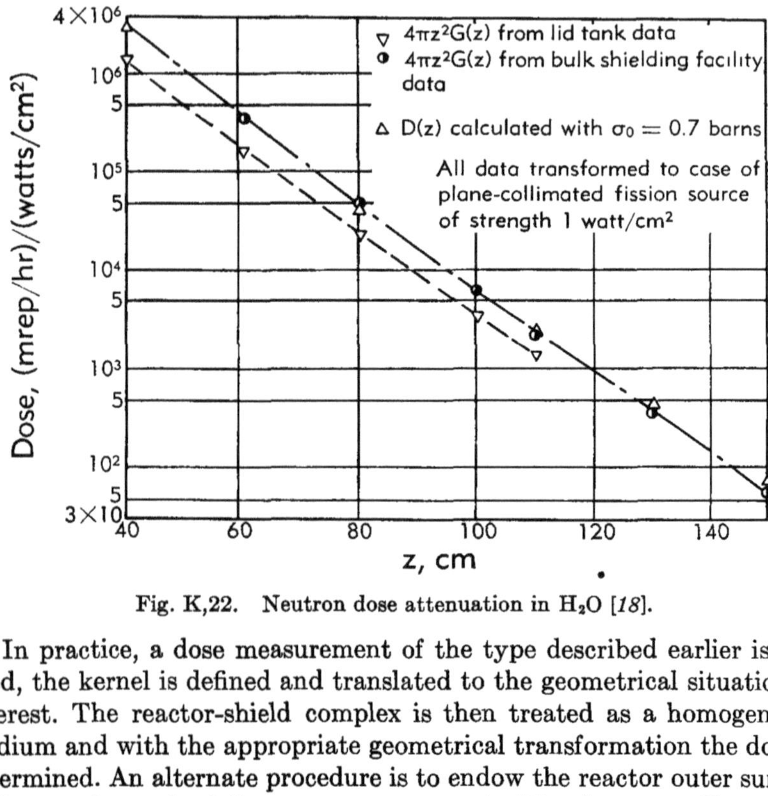

Fig. K,22. Neutron dose attenuation in H_2O [*18*].

In practice, a dose measurement of the type described earlier is utilized, the kernel is defined and translated to the geometrical situation of interest. The reactor-shield complex is then treated as a homogeneous medium and with the appropriate geometrical transformation the dose is determined. An alternate procedure is to endow the reactor outer surface with a predetermined flux or source strength. The dose at a distance r_2 from the center of the reactor, of radius r_1, is given by

$$D(r_2, r_1) = \frac{r_1}{r_2[D_{pl}(r_2 - r_1, \infty) - D_{pl}(r_2 + r_1, \infty)]} \tag{22-2}$$

where $D_{pl}(a, b)$ is the dose produced by a plane disk source of radii a, b.

Complete data on the plane disk is still in the classified literature. However, the dose as a function of distance from a fission plate for water is given in Fig. K,22.

The removal cross sections for several materials are given in Table

K,22. Now, to calculate the attenuation for a mixture of a heavy material and water, one applies the expression

$$I = I_0 e^{-\Sigma_R t'} D(t_w) \tag{22-2}$$

where

I_0 = effective source strength
Σ_R = removal cross section (cm^{-1})
t' = thickness of heavy material, cm
$D(t_w)$ = attenuation in water

In Fig. K,22 the attenuation of the neutron dose $D(t_w)$ is presented.

K,23. Gamma Rays. To return to the attenuation of gamma rays and the Boltzmann equation, the attenuation of the unscattered flux can readily be specified, that is, for a point source in an infinite homogeneous medium

$$N(E_0) = \frac{S}{4\pi R^2} e^{-\mu(E_0)R} \tag{23-1}$$

where

S = source
$\mu(E_0)$ = linear attenuation coefficient at energy E_0
R = distance from source

Now the attenuation coefficient includes both absorption and scattering collisions. To determine the total number of photons one must account for the photons which have undergone scattering collisions. A first approach is to define a build-up factor, B, which represents the totality of scattered photons, so that

$$B = \frac{\sum_{i=0}^{n} N(E_i)}{N(E_0)} \tag{23-2}$$

where $\sum_{i=0}^{n} N(E_i)$ is the actual flux over all energies. In similar fashion one can define the actual energy flux at a point R, namely

$$I(R) = B(R) I_0(R) \tag{23-3}$$

where

$$I_0(R) = N(R, E_0)E_0$$

$$I(R) = \int_0^E EN(R, E)dE + N(R, E_0)E_0 \qquad (E < E_0)$$

and E_0 is energy of source photons.

Of practical importance is the dose or energy deposition per unit volume. The latter is usually defined as energy flux times absorption coefficient and, for the dose build-up factor, one integrates the latter product over all energies. Now the photoelectric and pair production processes are true absorptions whereas in Compton scattering a relatively small amount of energy is lost depending upon the angle of scattering; that is, the absorption coefficient is given by

$$\mu(E) = \mu_{\text{p.e.}}(E) + \mu_{\text{p.p}}(E) + \int_0^{E_0} \left(\frac{d\mu}{d\Omega}\right) \frac{E_0 - E}{E} \, d\Omega \tag{23-4}$$

The term $(d\mu/d\Omega)$ in the integral is the differential Compton cross section per unit solid angle for scattering a photon of energy E_0 into a direction such that the scattered photon has an energy E. Thus it is evident that the energy flux is the desirable parameter to ascertain. The early attempts usually made approximations which overestimated the flux and thereby the dose. Recently, Spencer and Fano [*19*] undertook to develop a suitable procedure to solve the transport equation by the moments method. Goldstein and Wilkins [*20*] have now fully developed this technique and have carried out a monumental calculation program in conjunction with the National Bureau of Standards. Prior to presenting some of the results, the method will be described.

Since the interest is centered about the energy flux, Eq. 21-1 upon multiplying through with the energy E becomes

$$\nabla \cdot \mathbf{\Omega} I + \mu I = \int_{4\pi} I(\mathbf{\Omega}', E'; \mathbf{\Omega}, E) \sigma E \frac{dE'}{E} \, d\mathbf{\Omega}' + S(\mathbf{r}, E, \mathbf{\Omega}) \tag{23-5}$$

as $I = NE$.

To facilitate numerical calculations a change of variable is introduced, i.e. $dE'/E' = d\lambda'/\lambda'$. Also, the relationship of the energy and change of angle on scattering is given by

$$\lambda_{\text{final}} - \lambda_{\text{initial}} = 1 - \cos\psi \tag{23-6}$$

where ψ is the angle of scattering. Making use of the Klein-Nishina formula and the latter relationship, the cross section σ can be written as

$$\sigma(\mathbf{\Omega}', E'; \mathbf{\Omega}, E) = \frac{d\sigma}{d\Omega} \frac{\delta(1 + \lambda' - \lambda - \cos\psi)}{E^2} \tag{23-7}$$

Finally, a new function is introduced, namely

$$K(\lambda', \lambda) = 2\pi \frac{\lambda}{\lambda'} \frac{d\sigma}{d\Omega} \tag{23-8}$$

Upon substituting Eq. 23-7 and 23-8 into Eq. 23-5 and considering a uniform plane source, the transport equation becomes

$$\omega \frac{\partial I}{\partial x} + \mu(\lambda) I = \int_0 \int_{4\pi} I(x, \lambda', \omega') K(\lambda', \lambda) \frac{\delta(1 + \lambda' - \lambda - \cos\psi)}{2\pi} \, d\Omega' d\omega' + S(\lambda, \omega)\delta(x) \tag{23-9}$$

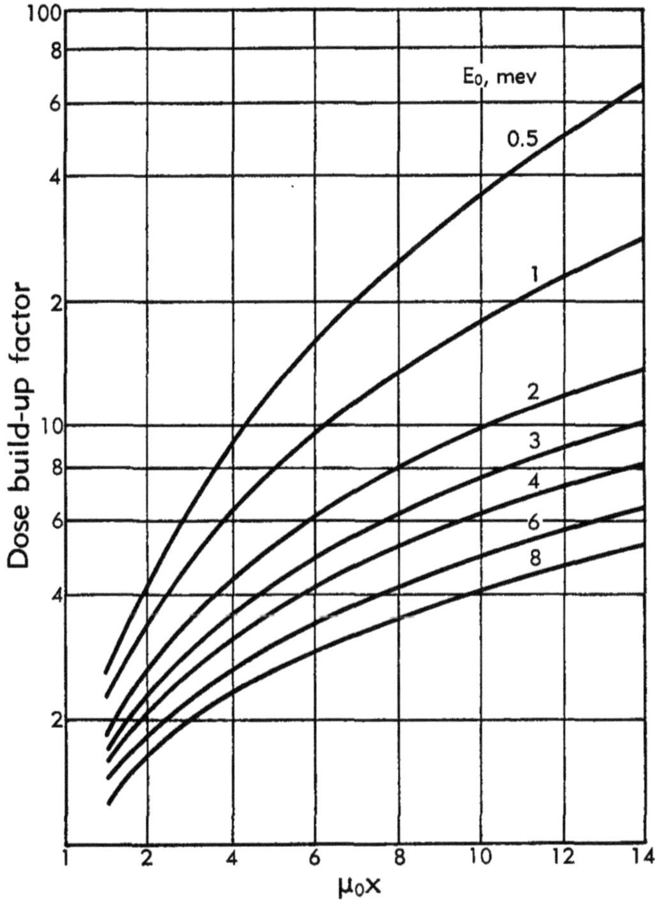

Fig. K,23a. Dose build-up factor, water. Plane monodirectional source [*20*].

It should be noted that the energy flux I is a function only of x, and S is a delta function in x.

The energy flux $I(x, \lambda, \omega)$ is now expanded as a Legendre polynomial series,

$$I(x, \lambda, \omega) = \sum_{l=0}^{\infty} \frac{2l + 1}{4\pi} I_l(x, \lambda) P_l(\omega) \qquad (23\text{-}10)$$

in which

$$I_l(x, \lambda) = 2\pi \int^{\infty} I(x, \lambda, \omega) P_l(\omega) d\omega \qquad (23\text{-}11)$$

It should be noted that I_0 (x, λ) is the energy density and I_1 (x, λ) is the energy current density. The substitution of Eq. 23-10 into Eq. 23-9, multiplying through with $P_l(\omega)$, and integrating over all solid angles gives a sequence of integro-differential equations in I_l. The only spatial

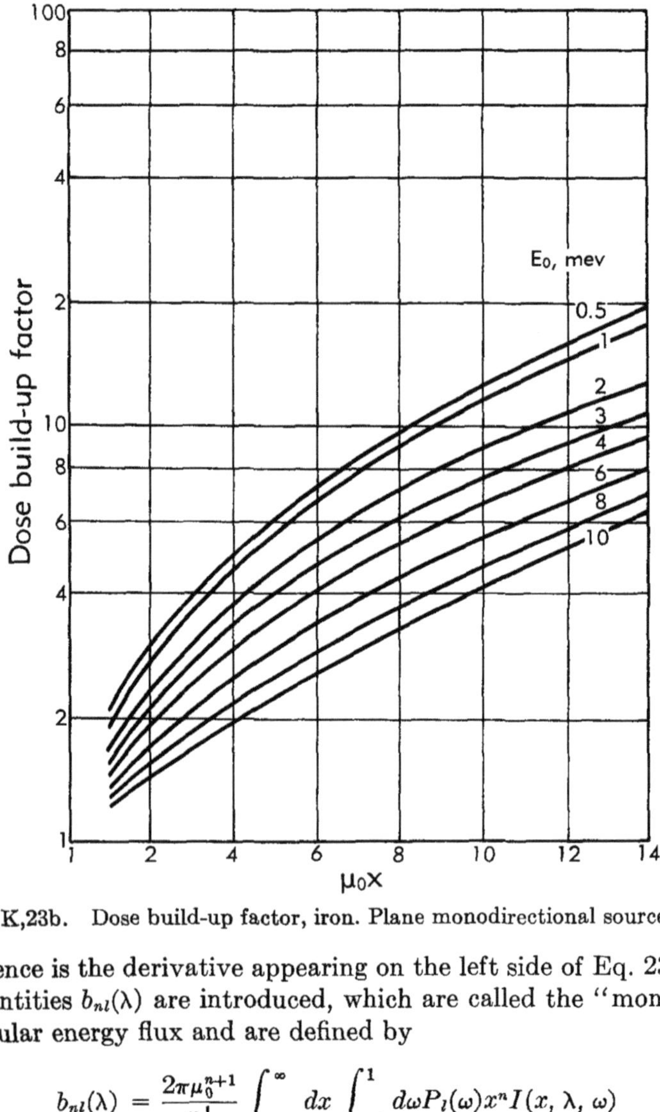

Fig. K,23b. Dose build-up factor, iron. Plane monodirectional source [20].

dependence is the derivative appearing on the left side of Eq. 23-9. Next the quantities $b_{nl}(\lambda)$ are introduced, which are called the "moments" of the angular energy flux and are defined by

$$b_{nl}(\lambda) = \frac{2\pi\mu_0^{n+1}}{n!}\int_{-\infty}^{\infty} dx \int_{-1}^{1} d\omega P_l(\omega) x^n I(x, \lambda, \omega) \qquad (23\text{-}12)$$

From the definition of b_{nl} and I_l the spatial dependence of Eq. 23-9 is completely eliminated and the original Boltzmann equation is reduced to a doubly indexed sequence of linear integral equations of the Volterra type

$$\mu(\lambda)b_{nl}(\lambda) = \int_0^{\lambda} K(\lambda', \lambda)P_l\,(1 + \lambda' - \lambda)b_{nl}(\lambda')d\lambda' + \frac{\mu_0}{2l+1}[(l+1)b_{n-1,l+1} + lb_{n-1,l-1}] + \mu_0 S(\lambda)\delta_{n,0} \qquad (23\text{-}13)$$

The equation is solved for a monoenergetic source and the scattered flux alone, thereby eliminating the delta function in the source. The energy flux $I_0(X, \lambda)$ is then reconstructed from the calculated moments, which is no simple job.

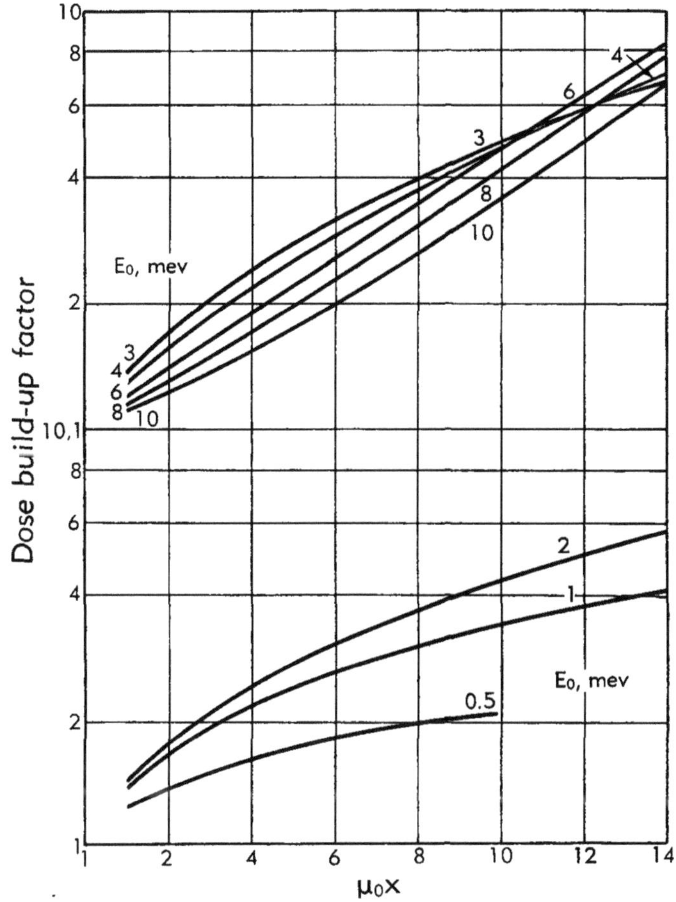

Fig. K,23c. Dose build-up factor, lead. Plane monodirectional source [20].

By this technique Goldstein and Wilkins have computed the build-up factor for photon energies of 0.5–10 mev and a large number of elements. Some representative values are given in Fig. K,23a, K,23b, and K,23c. The comparative results between theory and experiment have been gratifying.

For completeness, the experimental results for pure water and different arrangements of 2.22-cm iron slabs are given in Fig. K,23d. From the results it is evident that the iron slabs are more effective when displaced from source.

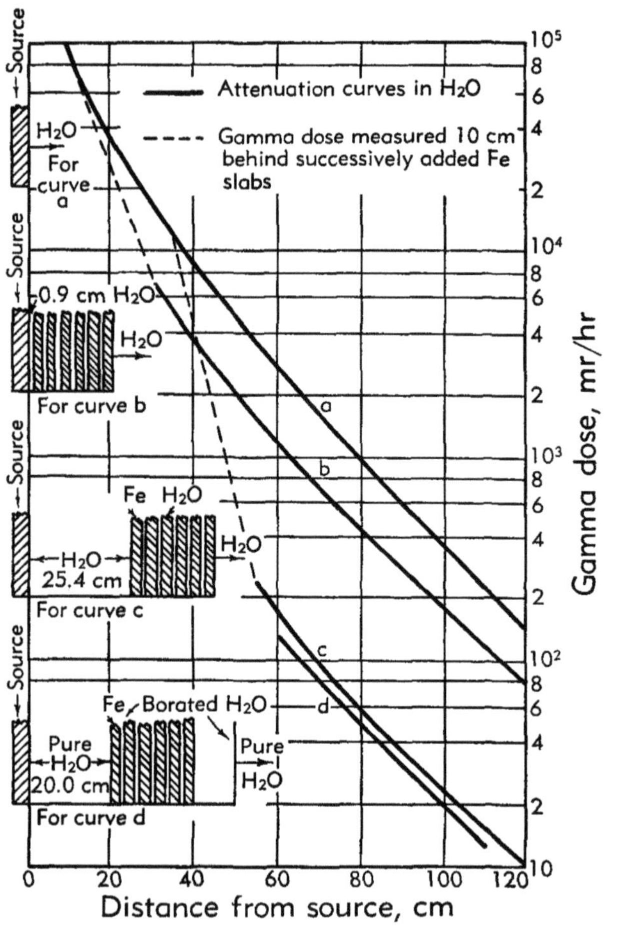

Fig. K,23d. Gamma attenuation in Fe-H_2O shield [11].

The results in Fig. K,23d and K,22 give the reader an opportunity, once the source strength of the reactor is specified, to estimate a shield composed of iron and water.

K,24. Shielding Materials. As indicated earlier, the theoretical treatment specifies the type of materials suitable for shielding against neutrons and gamma rays: hydrogenous materials for neutrons, and high *Z* substances for gammas. In addition, the thermal neutrons escaping from a reactor must be filtered at an early stage in the shield.

For the latter task a boron compound, e.g. boron carbide, B_4C, is most satisfactory. The gamut of materials that can be classified as hydrogenous is too numerous to list here; however, several categories can be named; hydrocarbons, plastics, cellulose, concretes, etc. (see Table K,24a). In

Table K,24a. Properties of some hydrogenous materials.

Material	*Density, g/cm^3*	*Molecular weight*	*Atomic density of hydrogen in compound, atoms/$cm^3 \times 10^{22}$*
Octane	0.703	114.22	6.68
Neoprene	0.84	88.54	4.2
Paraffin	0.89	—	7.82–8.01
Water	1.0	18.02	6.9
Lucite	1.2	100.1	5.78
Ammonia	0.771	17.03	8.18

comparison, the extent of desirable gamma materials is limited to iron, lead, uranium, tungsten, etc. (see Table K,24b).

Table K,24b. Calculated mass absorption coefficients of gamma rays for some materials.

Material	*Density, g/cm^3*	*Mass absorption coefficient per cm* 1 *mev*	3 *mev*	6 *mev*
Air	1.293×10^{-3}	3.36^{-5}	2.54^{-5}	2.09^{-5}
Al	2.7	0.166	0.0953	0.0718
Ammonia	0.771	0.0612	0.0322	0.0221
Be	1.85	0.104	0.0579	0.0392
B_4C	2.5	0.15	0.0825	0.0675
Gasoline	0.739	0.0537	0.0299	0.0203
	7.86	0.47	0.282	0.24
Pb	11.34	0.797	0.468	0.505
Lucite	1.19	0.0816	0.0457	0.0317
Neoprene	1.23	0.0813	0.0462	0.0333

The types of materials chosen depend on the reactor function. For a low power reactor, e.g. the Los Alamos water boiler, blocks of paraffin will suffice. The shields for the large reactors must possess long-life structural integrity in addition to its functional purpose. To satisfy these requirements the shield is normally constructed of hydrated concrete with a metallic ingredient. The concrete acts as a neutron shield and the metallic ingredient reduces the gamma flux. For this type of shield, the cost is important and, therefore, good design is not to be taken for granted.

In the case of mobile reactors, particularly aircraft reactors, weight and packaging are the basic factors. Needless to say, the shield problem for aircraft is the most severe one facing the shield designer; he must protect the flight crew from a high temperature, high power reactor with the least weight. To accomplish this task the most exotic materials are not excluded although standard materials can suffice.

K,25. Design Consideration. In dealing with the problems of design engineering of aircraft, the use of an integral shield is hypothetical because the ducting for reactor coolants may perforate the shield. Since the shield serves a dual purpose, the coolant passage either removes neutron-shielding or gamma-shielding depending on its nature, or possibly removes both. It is evident that a solution to a first approximation is to place patches of removed material so that the radiation escaping from the reactor has the same mass distribution. Further, the use of a devious path to some extent reduces the streaming of radiation through the duct.

From a design point of view, the arrangement of the shield materials is of interest; that is, is there an optimum placement? In this respect the production of secondary gammas plays an important role. To reiterate, capture gammas are formed within the shield (secondaries) by thermal neutrons or neutrons thermalized by the hydrogenous materials. Thus the first layer of shielding should be a thermal neutron absorber, e.g. boron compound. Secondly, the gamma absorber must be sufficiently thin to prevent the production rate of secondary gammas from equaling or exceeding the attenuation rate of primary gammas. Hence it would be best to space the gamma absorber between layers of neutron-shielding as shown in Fig. K,23d.

The radiation attenuated by the shield represents the kinetic energy transferred to the absorber atoms and molecules in the form of heat energy. The quantity of heat deposited per unit volume can readily be given as

$$Q(\mathbf{r}) = K \int_{E_0}^{E} \phi(\mathbf{r}, E)\Sigma(E)dE \tag{25-1}$$

where K is simply a conversion factor, 3.8×10^{-14} cal/mev, and $\Sigma(E)$ is the appropriate cross-section absorption, inelastic scattering, or both. A similar expression provides the same data for gamma rays. Nuclear heating of the shield may be quite a serious problem for high power reactors because 6–10 per cent of the total energy generated by the reactor may be stopped in the shield. It is obvious that the temperature rise will depend on the total shield mass and its specific heat.

Another consideration is that of post-shutdown radiations. Upon shutdown of a reactor the fission products and activated reactor materials provide a strong source of gamma rays. In a properly designed shield these radiations should present no greater hazard than when the reactor is at full power.

Lastly, the atmospheric molecules can act as scatters and therefore the flux received in free air will consist of direct plus scattered radiation.

CHAPTER 4. COOLANTS

K,26. Coolants. The variety of coolants considered for reactors is too numerous to list but the more significant ones are air, water, and liquid metals. Each of these coolants has particular advantages and disadvantages, with the final selection being determined by practical considerations.

For comparative purposes, let us employ the two fundamental equations,

$$Q = Wc_p\Delta T \tag{26-1}$$

$$Q = hA\Delta t \tag{26-2}$$

where

Q = heat removal, BTU/hr
W = coolant flow, lb/hr
c_p = specific heat, BTU/lb/°F
ΔT = coolant temperature rise, °F
h = heat transfer coefficient, BTU/ft^2 hr °F
A = heat transfer area, ft^2
Δt = average temperature difference between wall and bulk temperature, °F

If the flow area and the coolant are fixed, Eq. 26-1 states that $Q/V\Delta T$ for each coolant is in the ratio of

$$H_2O/\text{air}/\text{Na}::40/0.03/15 \tag{26-3}$$

Although water, on this basis, appears superior to the liquid metal, the high vapor pressure would impose limitations on the practicable temperature. Hence for a high performance reactor where maximum attainable temperature is a criterion, the liquid metal coolant is superior. From Eq. 26-3 one may deduce that sodium is superior to air; however, other considerations enter, such as the coolant velocity and the increased flow area. If one assumes that the velocity ratio is about 10, the required area ratio will be

$$\frac{A_{\text{air}}}{A_{\text{Na}}} \cong \frac{5}{1} \tag{26-4}$$

The disadvantage of a larger flow area requirement is an increased reactor diameter, but this is offset by the corrosive problems imposed by liquid metals.

From Eq. 26-2 one arrives at the same general conclusion with one additional consideration. Since the heat transfer coefficient for liquid metals is 40–50 times larger than for air, and if the transfer area ratio is taken to be about 10, the ratio of Δt is about 4–5. This implies that if the coolant film temperature drop for the liquid metal is 50°F then for air the drop is 250°F; that is, for an average bulk temperature of 1300°F the reactor wall temperature will be 1350°F for sodium and 1550°F for air. Obviously, this imposes an additional requirement on the design of satisfactory fuel elements.

Of the liquid metals, sodium, Na-K, and bismuth have received the

Table K,26a. Thermal, fluid properties of coolants [21].

	Water		Sodium	56 % Na 44 % K	Lithium	Pb-Bi Eutectic
Pressure, lb/in.^{2} abs	14.7	1500	14.7	14.7	14.7	14.7
Boiling point, °F	212	596.23	1621	1518	2403	3038
Freezing point, °F	32	—	208	66.2	354	257
Density, lb/ft^3 at						
100°F	62.0	—	—	56.1	—	—
500°F	—	48.6	55.3	52.9	31.3	646
1000°F	—	—	51.1	48.7	29.8	624
Specific heat, BTU/lb/°F at						
100°F	0.9976	—	—	0.2733	—	—
500°F	—	1.165	0.3150	0.2583	1.000	0.035
1000°F	—	—	0.3005	0.2486	1.000	0.035
Heat transfer coefficient,* BTU/hr/ft^2/°F at						
100°F	3970	—	—	7870	—	—
500°F	—	6030	15,200	7630	14,100	7150
1000°F	—	—	13,200	7400	13,900	8120
Pumping power for equivalent heat removal† per °F temp. rise at						
100°F	1.000	—	—	57.8	—	—
500°F	—	0.674	26.5	48.7	3.16	134
1000°F	—	—	34.0	63.4	3.41	133
Pumping power for equivalent volumetric flow‡ at						
100°F	1.00	—	—	0 758	—	—
500°F	—	0.544	0.717	0.662	0.457	7.39
1000°F	—	—	0.625	0.582	0.418	6.63
Pressure, lb/in.^{2} abs	14.7	100	500	14.7	100	500

	Air			Helium		
Density, lb/ft^3 at						
100°F	0.071	—	—	0.00978	—	—
500°F	0.0413	0.281	1.385	0.00570	0.0388	0.1940
1000°F	0.0272	0.185	0.920	0.00375	0.0255	0.1275
Specific heat, BTU/lb/°F at						
100°F	0.2405	—	—	1.241	—	—
500°F	0.2475	0.2495	0.2520	1.240	1.240	1.242
1000°F	0.2660	0.2680	0.2710	1.240	1.240	1.240
Heat transfer coefficient§, BTU/hr/ft^2/°F at						
100°F	23.4	—	—	25.9	—	—
500°F	20.6	95.8	342	16.5	79.3	289
1000°F	18.6	87.5	228	12.5	57.2	205
Pumping power for equivalent volumetric flow¶ at						
100°F	1.00	—	—	0.1002	—	—
500°F	0.72	3.62	13.90	0.1652	0.78	2.41
1000°F	0.625	2.81	10.26	0.226	0.614	1.82

* Water and metal coolant coefficients determined for a velocity of 20 ft/sec in a ½-in. inside diameter tube.

† Normalized to a value for water at 100°F and 14.7 lb/in.^{2} abs with a velocity of 20 ft/sec in a ½-in. inside diameter × 10-ft long tube.

‡ Water and metal coolants normalized to water at 100°F, 14.7 lb/in.^{2} abs with a velocity of 20 ft/sec in a ½-in. I. D. × 10-ft long tube.

§ Gas coolants coefficients determined for a velocity of 100 ft/sec in a ½-in. inside diameter tube.

¶ Gas coolants normalized to air at 100°F, 14.7 lb/in.^{2} abs with a velocity of 100 ft/sec in a ½-in. inside diameter × 10-ft long tube.

Table K,26b. Nuclear properties of coolants [21].

	Cross section		Fractional energy loss per collision	Moderating ratio
	Capture σ_c, barns	Scattering σ_s, barns	ξ	$\frac{\sigma_s \xi}{\sigma_c}$
H_2O	0.6016	164 (44)*	0.918	250 (67)*
D_2O	0.000920	15.3 (11)*	0.495	8250 (5930)*
Na	0.45	4	0.083	0.89
K	2.5	1.5	0.03	0.018
NaK (44%K)	1.1	3.2	0.0774	0.225
Li	65	1.5	0.26	0.006
Pb	0.2	11.0	0.00962	0.529
Bi	0.015	9.0	0.00958	5.75
Pb-Bi (44.5%Pb)	0.17	9.9	0.00960	0.56
He	0.008	1.5	0.43	83
Air	1.4	8.82	0.13	0.82

* At 100 ev.

greatest attention as to heat transfer properties and containment. This is thoroughly discussed in [11]. Some representative thermal and fluid properties are given in Table K,26a and the nuclear properties in Table K,26b.

CHAPTER 5. HEAT TRANSFER

K,27. Heat Production. From a pragmatic standpoint one may consider the heat transfer problems associated with a power reactor to be the prime concern of the entire system. As stated earlier, the reactor is only a heat source from which the energy must be extracted; however, since the energy density is extremely high very efficient coolants and sufficient heat transfer area are necessary. Because of these complicating factors, an actual high performance reactor is a compromise between the ideal design and the heat removal characteristics.

To briefly review the heat generation process, the fission process releases about 197 mev and, of this, approximately 167 mev is carried off by the fission fragments in the form of kinetic energy. The range of these fragments is about 0.025 cm. Similarly, the range of the other charged particles released in association with the fission process is within a fraction of a centimeter. Therefore, for a reactor with fuel elements, the heat which results from stopping these particles is almost entirely liberated within the element. The energy liberated in the form of neutrons and gamma rays is either lost to the system or produces heat in the nonfuel-bearing materials of the reactor. For example, in a thermal reactor, if all neutrons are stopped in the reactor proper then this energy or about 3 per cent of the total energy is deposited in the moderator. This is evident because the function of the moderator is to degrade the fast neutrons to thermal energy. Similarly, a portion of the gamma ray energy is deposited in the coolant and in the moderator. For the thermal or near-thermal high performance reactor with fixed fuel elements, approximately 90–95 per cent of the total energy remains in the core with about 5 per cent in the moderator and the remainder in the fuel element. For a liquid fuel, two situations may exist: (1) homogeneous and (2) heterogeneous. In the case of the homogeneous reactor the heat is produced entirely within the fuel-moderator solution except, of course, for the escaping radiations. In the heterogeneous case, in addition to the moderation energy deposited in the moderator, considerable heat will be transferred from the fuel mixture to the moderator by conduction.

K,28. Power Distribution. As discussed in Chap. 2, the power distribution is determined by the distributions of fuel and neutron fluxes, and perturbations of either will alter the power distribution; i.e. the fission rate per cm^3 is equivalent to the heat generation per cm^3.

In the case of a spherical reactor the thermal neutron flux is given by Eq. 15-8,

$$\phi(\mathbf{r}) = \frac{\phi_0}{r} \sin \frac{\pi r}{\tilde{R}} \tag{15-8}$$

and vanishes at $r = \tilde{R}$. When a reflector is added, this distribution is altered as shown in Fig. K,16. Consequently, the ratio of maximum to average flux within the reactor is significantly decreased. The latter has the effect of flattening the power distribution in the reactor proper. An additional procedure used to flatten the flux is nonuniform fuel-loading within the reactor. These steps reduce the difficulties of the heat transfer problem and the induced thermal stresses.

K,29. Temperature Distribution. In a reactor of practical geometry, say a cylinder, the power distribution along the longitudinal axis which is the direction of coolant flow is a cosine function with a maximum at the central plane. The temperature distribution of the coolant, fuel surface, and fuel along the axis of the reactor core is shown in Fig. K,29. From this figure it can be observed that: (1) the maximum fuel surface temperature does not match the coolant temperature and (2) a large temperature gradient across the cladding of the fuel element is possible. The maximum surface temperature actually depends on the heat transfer coefficient; that is, the better the heat transfer coefficient, the closer to the downstream end of the core will be the maximum surface temperature.

K,30. Design Factors. The introduction of a heat transfer system into a reactor interjects new problems concerning the effects of coolant volume and heat transfer surface on the reactor design. Since the fuel and fission products must be contained and separated from the coolant, the fuel is encased by a high temperature corrosion-resistant, low neutron-absorbing metal or refractory. Further, sufficient cladding surface must be

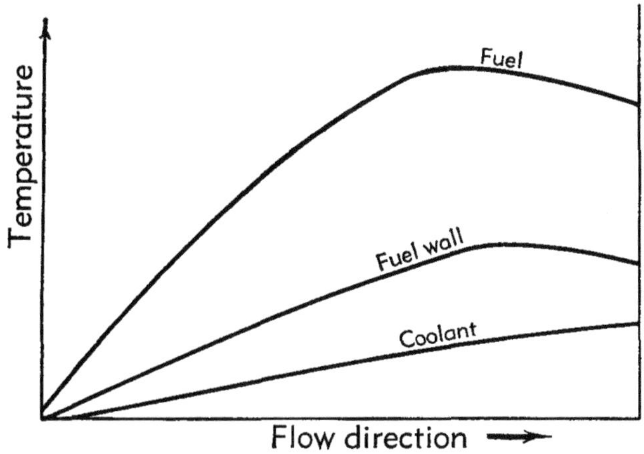

Fig. K,29. Typical temperature distribution along core length.

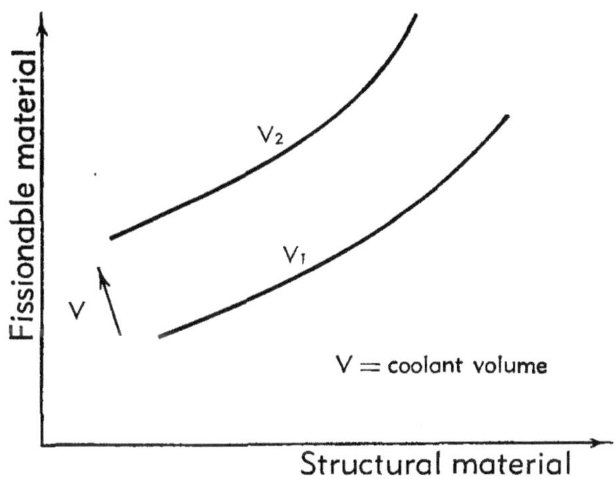

Fig. K,30. Typical effect of heat transfer system on fuel mass.

provided to satisfy the heat removal requirements. Thus the practical reactor contains a considerable fraction of foreign materials which tend to decrease its reactivity or to increase the mass of fissionable material (see Fig. K,30).

CHAPTER 6. SYSTEMS

K,31. Introduction. The propulsion systems which can be integrated with a nuclear reactor are gas turbines, ramjets, and rockets. The type of reactor employed, however, has a marked effect upon the over-all configuration. The applicable reactors may be cooled either by gases or by liquid metals and the selection of a particular coolant depends on the power requirements, the adaptability, and other engineering features of the system.

K,32. Gas-Cooled System. Of the many available gases, atmospheric air results in the simplest system, but the use of atmospheric air can only be considered for gas turbines and ramjets. Briefly, the air is compressed in either a compressor or diffuser, then enters the reactor and is heated to an elevated temperature. It is then exhausted through a turbine or nozzle which performs work in the process (see Fig. K,32).

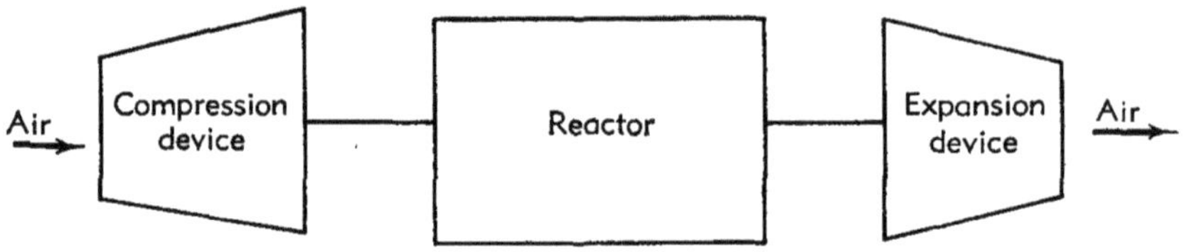

Fig. K,32. Schematic diagram for air-cooled system.

For this system the thermal energy normally supplied by the hydrocarbon fuel is replaced by nuclear fuel, i.e. one can interpose a reactor between compressor and turbine or diffuser and nozzle. From an engine standpoint, therefore, this system operates on a standard Brayton cycle and all concepts are directly applicable. In detail some differences exist, for instance the combustion process is replaced by a heat transfer. The latter substitution results in (1) increased pressure losses and (2) allowable range of air temperatures. It is for this reason that a one-to-one correspondence does not exist between the reactor power and the fuel flow; that is, one megawatt corresponds to approximately 183 pounds of hydrocarbon fuel per hour. However, until the appropriate losses are incorporated the reactor power is an indeterminate.

K,33. Liquid Metals. The liquid metal system is usually a ternary cycle; that is, an intermediate stage is introduced to reduce nuclear activation of the coolants exterior to the shield. A schematic arrangement is shown in Fig. K,33. Briefly, the mode of operation is: (1) the heated coolant leaves the reactor and passes through the intermediate heat exchanger, (2) the secondary coolant is heated by the primary one and enters the radiator, and (3) the compressed air is heated by the secondary coolant. Here the radiator has replaced the combustion chamber and therefore this portion of the system is susceptible to standard jet engine analysis. It is evident from practical considerations that this type of system would not be suitable for ramjet propulsion.

The primary difference between the above systems is only in the use of liquid metal circuits. In view of the low film resistance for liquid metals the temperature differential between the reactor coolant outlet and the radiator air outlet can be made comparable with the air-cooled system (Art. 26). In practice, however, the acceleration of the corrosion rate with elevated temperatures imposes temperature limitations on liquid metal systems. In [*11*] it is determined that the allowable temperature for liquid metal containers is in the vicinity of 1500–1600°F, whereas present gas turbine experience indicates that air exhibits little corrosive effects at temperatures in the neighborhood of 3000°F. Thus the air-cooled reactor

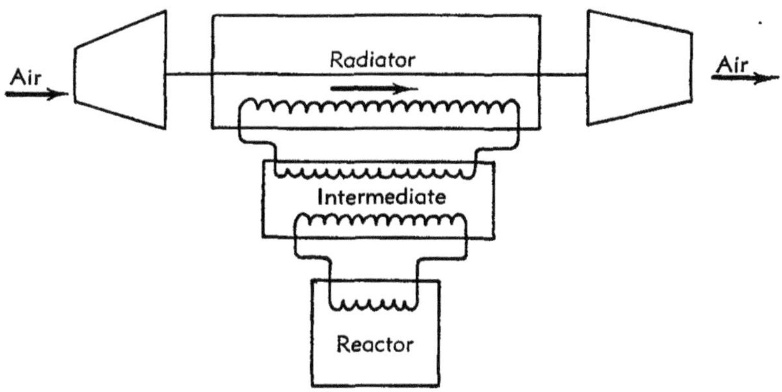

Fig. K,33. A schematic arrangement of a liquid metal system.

may be capable of providing air at temperatures consistent with turbine technology.

The last point to be considered is the pumping power required for the two systems. To achieve a high heat transfer coefficient for air a considerable pressure loss is sustained, say about 20 per cent. This loss can be considered, like burner losses, as a decrease in compressor efficiency and, consequently, an over-all system loss. The pumping power for liquid metals is nominal and may be added to the requirements for other accessories. The pressure losses sustained by the air in the radiator may be made comparable to the losses in a conventional combustion chamber. On the basis of these estimates, a calculation using generalized performance data [*22*] reveals that, at a compression ratio of 10, the pressure losses in the air-cooled system result in a cycle efficiency somewhat lower than for the liquid metal system, 22 per cent vs. 26 per cent. What is more important, the specific output (bhp/lb of air/sec) of the two systems is comparable. It should be noted that this comparison is only valid for the estimated values used here.

K,34. Closed Systems. A closed system is defined here as one with a working fluid other than atmospheric air. Under this system two possibilities exist: (1) turbopropeller aircraft and (2) rockets. In the first instance, a gas-cooled reactor, e.g. helium, may drive the turbines of the engine or a vapor, e.g. water or mercury, may be used to drive the turbines. The performance can readily be ascertained once the working fluid is selected.

The other possible use of a closed system is the rocket. Here again, nuclear energy replaces the chemical energy of combustion; that is, thermal energy is extracted from the reactor by a coolant and then exhausted through a nozzle. According to rocket principles it is well known that the attainment of high combustion chamber temperatures is vital to an efficient motor. The temperatures attained in chemical propellant rockets are in the vicinity of 5000–6000°R. Although a practical rocket may require only a few minutes of operation, the possibility of maintaining a conventional reactor fuel element at this temperature appears beyond the reach of present metallurgy. Graphite or ceramals impregnated with fissionable material, however, could provide temperatures of about 5000°R.

The reactor could take one of several forms: (1) a solid matrix with coolant passages, (2) a plate system, straight-through or folded, and (3) a bed of uranium-graphite pebbles with the coolant forced through the interstices. Since graphite is the moderator and the reactor must be small the intermediate region will dominate the fission spectrum and a reflector would be of great value. The selection of a coolant should satisfy several criteria, such as (1) low specific heat ratio γ, (2) low molecular weight preferably hydrogenous, and (3) low pumping losses. Among those available, methane and ammonia could be considered potentially suitable. Let us consider methane and compare it to an oxygen-hydrogen system where, as a basis of comparison, we shall utilize the concept of characteristic velocity c^*. The latter is defined by

$$c^* = \frac{\sqrt{g\gamma \mathcal{R} T_c}}{\sqrt{2\left(\frac{2}{\gamma+1}\right)^{\frac{2}{\gamma-1}}\left(\frac{\gamma^2}{\gamma+1}\right)}} \tag{34-1}$$

If an allowable combustion chamber temperature of 6300°R for the oxygen-hydrogen is used, then the ratio of the characteristic velocities is,

$$\frac{c^*_{\text{methane}}}{c^*_{\text{O-H}}} \cong 0.9 \tag{34-2}$$

Thus it is evident that the nuclear fuel is comparable to the best chemical propellant combination. Other coolants, high temperature materials, and practical considerations may make the nuclear-propelled rocket superior to that propelled by chemical means.

CHAPTER 7. PRELIMINARY DESIGN

K,35. Reactor Analysis. To illustrate and extend the principles discussed earlier, an approximate reactor calculation is now presented. Consider a small 100 per cent-enriched, beryllium carbide moderated reactor, which has a cylindrical geometry and is cooled by liquid lead. A hypothetical fuel element can be U^{235}, clad with beryllium in the shape of a hollow cylinder, which also serves as the coolant passage (Fig. K,35). The use of beryllium cladding is suggested by tests conducted at the National Physical Laboratory [*23*].

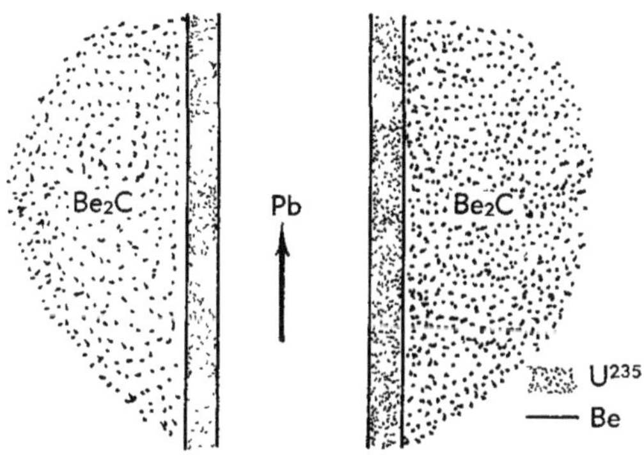

Fig. K,35. Section of hypothetical reactor.

The geometric buckling for a cylinder, upon solving Eq. 13-8, is found to be

$$B_g^2 = \left(\frac{\pi}{H}\right)^2 + \left(\frac{2.405}{R}\right)^2 \tag{35-1}$$

If we now consider that the effect of a reflector is only to reduce the core size, keeping the fuel concentration constant, then a 2-foot-square cylinder with a 2.5-inch reflector will have an effective bare radius equal to 14.5 inches plus 0.71 λ_{tr}. Thus the "extrapolated" dimensions of the reactor are

$$H = D \cong 73.66 + 1.32 = 74.98 \text{ cm} \tag{35-2}$$

which gives, on substitution of Eq. 35-2 into Eq. 35-1, a value of $B^2 \cong 5.14 \times 10^{-3}/\text{cm}^2$. We can now proceed to calculate the critical mass and, in this instance, employ the critical equation as defined by the age-diffusion theory, namely

$$k_{\text{eff}} = \frac{k_\infty e^{-B_g^2\tau}}{1 + L^2 B_g^2} = 1 \tag{14-4}$$

Our first task is to determine the parameters τ, k_∞, and L^2.

The value of τ_{th} for Be_2C, the moderator, is 89 cm² at 0.025 ev. As the average reactor temperature will be about 1300°F, the value of the age will be decreased, i.e.

$$\tau = 89 - \int_{0.025}^{0.08} \frac{1}{3\Sigma_s^2 \xi} \frac{dE}{E} \tag{35-3}$$

However, the coolant being a poor slowing-down material and occupying about 17.4 per cent of the core volume (Art. 36) will tend to increase this value. A calculation shows that this effect amounts to an increase of about 20 cm². Thus we shall use here a value of $\tau = 100$ cm².

From Eq. 12-3 we have the expression,

$$\begin{aligned} k_\infty &= \eta f \\ &\cong 2.1 \frac{\Sigma_{a,f}}{\Sigma_{a,f} + \Sigma_{a,m} + \Sigma_{a,c}} \end{aligned} \tag{12-3}$$

where the subscripts indicate absorption cross section for fuel f, moderator m, and coolant c. The parameter L^2 is defined by Eq. 10-6:

$$L^2 = (3\Sigma_a \Sigma_{tr})^{-1} \tag{10-6}$$

which for a three-component system becomes

$$L^2 = [3(\Sigma_{a,f} + \Sigma_{a,m} + \Sigma_{a,c})(\Sigma_{tr,f} + \Sigma_{tr,m} + \Sigma_{tr,c})]^{-1} \tag{35-4}$$

or

$$L^2 \cong \frac{L_m^2}{\left(1 + \frac{\Sigma_{a,f}}{\Sigma_{a,m}} + \frac{\Sigma_{a,c}}{\Sigma_{a,m}}\right)} \tag{35-5}$$

where the scattering is usually considered insignificant.

Let us denote the fraction $\Sigma_{a,f}/\Sigma_{a,m}$ by x; then for a critical system with $k_{eff} = 1$, Eq. 14-4 can with the aid of Eq. 12-3 and 35-5 be written as

$$\frac{2.1x}{1 + 1.1x + \left(\frac{\Sigma_{a,c}}{\Sigma_{a,m}}\right)} = 1.672 \left[1 + \frac{L_m'^2 5.141 \times 10^{-3}}{0.68\left(1 + x + \frac{\Sigma_{a,c}}{\Sigma_{a,m}}\right)}\right] \tag{35-6}$$

The following should be noted: (1) the factor 1.1, multiplying the x term on the left side of Eq. 35-6, is the so-called disadvantage factor described in Art. 18; and (2) $L_m'^2$ refers to its value at the theoretical density whereas the moderator density is at about 83 per cent of the theoretical value, or

$$L_m^2 = \frac{L_m'^2}{\rho} = \frac{L_m'^2}{0.68} \tag{35-7}$$

On substitution of the numerical values for $\Sigma_{a,c}/\Sigma_{a,m}$ and $L_m'^2$ into Eq. 35-6 we obtain the result:

$$x = 28.7 \tag{35-8}$$

Let α define the volume percentage of fuel in the reactor so that

$$\alpha = x \frac{\Sigma_m'}{\Sigma_t'} 100 \qquad (35\text{-}9)$$

$$= 7.36 \times 10^{-2} \text{ per cent}$$

where the primed quantities again refer to values at theoretical densities. Since the volume has been stipulated, one finds that the critical mass corresponding to Eq. 35-9 is 5.4 pounds. Under operation this reactor will produce poisons that tend to reduce the reactivity. To offset this, additional fuel is required. On inclusion of the poisoning effect into Eq. 35-6 the required critical mass is found to be 21.7 pounds. A further allowance must be made for fuel "burn-up" so that, if the reactor operates for 120 hours producing 30 megawatts of power, then approximately 0.33 pounds of fuel will be consumed. The latter is estimated on the basis that one megawatt-day is equivalent to one gram of fuel. The total fuel investment is now 22.03 pounds.

The multiplication constant, k_∞, for the "fresh" reactor differs from the "poisoned" one by about 5 per cent, where the latter is 1.92. Although the multiplication rate is quite high and may be difficult to control, we will briefly consider a possible control system.

A control rod is most effective at the core center since its effectiveness is proportional to the square of the neutron flux. In our case, however, let us provide control at the core-reflector interface. One finds, using standard control rod theory, that a 0.75-inch solid boron steel rod will effectively control about 2.5 per cent reactivity, or two rods can control this reactor during its excursion from start-up to a poisoned condition. A smaller rod at the center can serve as an additional regulating mechanism.

K,36. Heat Transfer Analysis. The possibility of extracting 1.04×10^8 BTU/hr from the reactor, limited to the determination of the size and number of coolant flow passages, will now be considered. Assume that at the outset the inlet and outlet temperatures are 1000°F and 1500°F respectively. It follows directly from Eq. 26-1 that the free flow area is

$$A = \frac{2.73}{V} \text{ ft}^2 \qquad (36\text{-}1)$$

and for a reasonable fluid velocity V of 5 ft/sec the flow area is 0.546 ft.^{2} Since the reactor frontal area is 3.14 ft^2 the free flow ratio is 17.38 per cent, which from geometrical considerations is also the coolant volume percentage. Similarly it follows from Eq. 26-2 that the required heat transfer surface area A^* is

$$A^* = \frac{1.04 \times 10^8}{h\Delta t} \text{ ft}^2 \qquad (36\text{-}2)$$

By definition the expression for Δt is

$$\Delta t = \frac{t_s - t_b}{2} \tag{36-3}$$

The value of t_s is estimated as 1450°F and t_b is found to be 1250°F. Consequently the value of Δt is

$$\Delta t = \frac{1450 - 1250}{2} = 100°\text{F} \tag{36-4}$$

and

$$A^* = \frac{1.04 \times 10^6}{h} \tag{36-5}$$

Now Martinelli [*24*] gives the correlation formula for liquid metal flow in pipes as

$$h = \frac{k}{d}\left[7 + 0.025\left(\frac{\rho V c_p d}{k}\right)^{0.8}\right] \tag{36-6}$$

i.e.

$$\text{Nusselt number} = [7 + 0.025\ (\text{Peclet number})^{0.8}]$$

where k is the thermal conductivity, V the fluid velocity ft/hr, and d is the pipe diameter. If we denote l as the length of the flow passage, on combination of Eq. 36-5 and Eq. 36-6 we obtain the result that

$$n\pi dl = \frac{1.04 \times 10^6}{\frac{k}{d}\left[7 + 0.025\left(\frac{\rho V c_p d}{k}\right)^{0.8}\right]} \tag{36-7}$$

where n, the number of passages required, is defined by

$$n = \frac{4}{\pi}\frac{0.546}{d^2} \tag{36-8}$$

The elimination of n from Eq. 36-7 gives an algebraic equation in d, which after substitution of the fluid constants evaluated at the bulk temperature t_b gives the result

$$d = 0.27 \text{ inches} \tag{36-9}$$

and consequently,

$$n = 1373$$

The heat transfer coefficient h can now be found and is 3290 BTU/hr/ft²/°F.

At this point for an aircraft application one may either have the air heated by the reactor coolant in a radiator, or introduce an intermediate heat exchanger. Practical considerations suggest the use of an intermediate heat exchanger with a liquid metal as the secondary coolant. In order to provide short liquid metal circuits for the primary coolant the intermediate heat exchanger is placed adjacent to and concentric with the reactor. Heat exchanger design practice would suggest the use of a counterflow, shell, and tube configuration with the secondary coolant flowing in the tubes. The specific dimensions would depend upon the requirements to be met by the over-all system.

A design calculation was performed using lithium as the secondary coolant with the conditions that the inlet and outlet temperatures are to be 950°F and 1450°F respectively and 1.04×10^8 BTU/hr be extracted from the primary coolant. The results of the calculation indicate that approximately 5000 tubes 2.5 feet long contained within a cylindrical shell 5.75 inches thick are required. The last stage in the heat transfer system is to pass the lithium through a radiator, as shown in Fig. K,33, and heat the compressed air. To complete the design a shield is required for personnel protection. An illustrative calculation of a shield design would certainly be desirable. However, in view of the fact that only a practical one would be informative and that this is precluded for security reasons, no design is presented.

K,37. Cited References and Bibliography.

Cited References

1. Hughes, D. J., and Harvey, J. A. Neutron cross sections. *Brookhaven Natl. Lab. Rept. BNL 325*, July 1955.
2. Feshbach, H., Peaslee, D. C., and Weisskopf, V. F. *Phys. Rev. 71*, 45 (1947).
3. Feshbach, H., and Weisskopf, V. F. *Phys. Rev. 76*, 1550 (1949).
4. Evans, G. E. *Nucleonics 11*, (*6*), 19, 20, 23 (1953).
5. Glasstone, S., and Edlund, M. C. *Elements of Nuclear Reactor Theory.* Van Nostrand, 1952.
6. Roberts, L. D., Hill, J. E., and McCammon, G. *Phys. Rev. 80*, 6 (1950).
7. Christy, R. F. Lecture series on nuclear physics. *Los Alamos Rept. 24*, 127 (1943).
8. Kourganoff, V. *Basic Methods in Transfer Problems.* Oxford Univ. Press, 1952.
9. Leachman, R. B., et al. Determination of fission quantities of importance to reactors. *International Conference on the Peaceful Uses of Atomic Energy, Paper P/592*, Geneva, 1955.
10. Kinsey, B. B., Hanna, G. C., and Van Patter, D. M. *Can. J. Research 26A*, 79 (1948).
11. The Reactor Handbook. *Atomic Energy Commission Rept. AECD 3646*, 1955.
12. Weinberg, A. M., and Noderer, L. C. Theory of neutron chain reactions. *Oak Ridge Natl. Lab. Rept. CF51-5-98, Vol. II*, 1 (1951).
13. Way, K., and Wigner, E. P. *Phys. Rev. 73*, 1318 (1948).
14. Moteff, J. *Nucleonics 13*, (*5*), 51 (1955).
15. Fano, U. *Nucleonics 11*, (*8*), 8 (1953).
16. Heitler, W. *The Quantum Theory of Radiation*, 2nd ed. Oxford Univ. Press, 1944.
17. Blizard, E. P. Nuclear radiation shielding. *Ann. Rev. of Nuclear Science*, Stanford Univ., 1955.
18. The Reactor Handbook. *Atomic Energy Commission Rept. AECD 3645*, 1955.
19. Spencer, L. V., and Fano, U. *J. Research Natl. Bur. Standards 46*, 446 (1951).
20. Goldstein, H., and Wilkins, J. E. Calculations of the penetration of gamma rays. *Nuclear Develop. Corp. Rept. NDA 15C-41*, 1954.
21. Woodruff, O. J., et al. *Nucleonics 11*, (*6*), 1953.
22. Mallinson, D. H. General performance calculations for gas turbine engine. *Ministry of Supply, Brit. Aeronaut. Research Council Rept. RM 2684*, 1954.
23. Udy, M. C., et al. *Nucleonics 11*, (*5*), 52 (1953).

24. Martinelli, R. C. Heat transfer to molten metals. *Trans. Am. Soc. Mech. Engrs. 69*, 947 (1947).

Bibliography

Blatt, J. M., and Weisskopf, V. F. *Theoretical Nuclear Physics*. Wiley, 1952.
Glasstone, S. *Principles of Nuclear Reactor Engineering*. Van Nostrand, 1955.
Holmes, D. K., and Meghreblian, R. V. Notes on reactor theory. *Oak Ridge Natl. Lab. Rept. CF54-7-88*, 1954.
Lyon, R. N. *Liquid Metals Handbook*, 2nd ed. Govt. Printing Office, 1952.
Rockwell, T. *Reactor Shielding Design Manual*. Van Nostrand, 1956.
Schultz, M. A. *Nuclear Reactor and Power Plant Control*. McGraw-Hill, 1955.
The Reactor Handbook. *Atomic Energy Commission Rept. AECD 3647*, 1955.
Wallace, P. R., and LeCaine, J. Elementary approximations in the theory of neutron diffusion. *Natl. Research Council of Canada, N.R.C. 1480*, (undated).

SECTION L

FUTURE PROSPECTS OF JET PROPULSION

F. ZWICKY

L,1. Introduction.[1] Research and development in the field of jet propulsion during the past decade have brought forth two types of significant results.

In the first place, many new engines for the generation of propulsive power have been invented, analyzed, and constructed. This includes both prime movers and auxiliary devices.

In the second place, a method of research and invention, the morphological method, has been developed which greatly enhances man's inventive capabilities and which puts progress on a far more systematic basis than was hitherto available. The new method or mode of thought also opens up many vistas into both the near and the far future; in fact, it elevates the art of prophecy from the level of mere speculation to a status very much more rational in nature.

Prophecy in the past consisted chiefly of intuitive guesses based on past experience. Some of the sounder types of prophecy limited themselves to the more or less far-reaching extrapolation of knowledge already available. Except for cautious predictions based on sound extrapolation, long views into the future were indeed speculative in nature and had little basis in scientific fact and method.

The morphological procedure introduces a new element of thought which makes possible the reasonable long range forecasting of scientific, technical, and even human developments. It is this procedure which must be followed if one wishes to make realistic statements as to the future prospects in jet propulsion, the task which is before us.

L,2. The Morphological Mode of Thought. This mode of thought has its starting point in a number of elementary observations which we shall state briefly as follows.

[1] The manuscript for this article was submitted to the Editors in January, 1951. Some parts of the original manuscript have been deleted, but nothing has been added except for the two footnotes describing the test in 1956 for monatomic oxygen in the upper atmosphere and the successful launching of artificial meteors into interplanetary space on October 16, 1957.

1. The number of detailed and communicable aspects of nature or of life is inexhaustible. This fundamental fact implies the possibility of continued discovery.

2. As a consequence of fact 1, it follows that the problem of absolute truth, if we attempt to state such truth in finite terms, must be handled with great caution. There always remains something to be done which was neglected when a so-called absolute truth was derived.

From this observation the morphological mode of thought derives one of its prime incentives, which must be met by striking hard where nothing is known.

3. Morphological thought in pure application has as its main aim the drive for rounded-out knowledge. Rounded-out knowledge in specific instances may be equivalent to total knowledge. Because of the inexhaustibility of the aspects of life, such total knowledge is of course possible only if the universe of discourse which one considers is strictly limited. The morphological method, when striving for total knowledge, thus starts from the strictest of possible definitions or statements of the problem which is to be solved.

The importance of a very precise definition of the problem to be solved cannot be stressed enough if one wishes to apply the morphological method successfully. Even in the simplest cases, the formulation of such definitions is far more difficult than the uninitiated is apt to think. As a mental exercise the reader might try to define in a generic way some classes of devices such as pumps, injectors, telescopes, and so on. In attempting, for instance, to formulate a definition of pumps or injectors, one will find out very quickly that sound knowledge of very many physicochemical phenomena is necessary to master the problem.

Only if the definition of the problem at hand is precise will the morphological method give unique results. These then can be systematically ordered in a multidimensional box, called the "morphological box" whose various axes represent the fundamental parameters of the problem. The same morphological box is arrived at, no matter how one starts its construction. If the statement of the problem is ambiguous, no definite morphological box can be constructed.

4. The morphological method initially investigates all of the solutions of a given problem without regard to the absolute or relative values of these solutions. Any prejudice defeats the possibility of a successful application of the morphological method from the outset. For practical applications, such an evaluation is of course necessary. Every evaluation must be carried out with respect to some clearly stated standard of values. Independent of the morphology of the solution of any given problem, a morphology of possible values must be executed once and for all.

5. Morphological thought postulates that most of the solutions which are derived for a given large scale problem are useful when seen in the proper perspective. Also, all of the solutions can actually be realized and

constructed unless the morphological analysis is vitiated by internal contradictions or if some fundamental principle such as the second law of thermodynamics has been contradicted.

6. The usefulness of all of the solutions of a given large scale problem must be judged after a general performance evaluation has been carried out. This evaluation is often one of the most difficult and laborious tasks of a morphological analysis. There are, however, certain cases where one succeeds with approximate performance calculations based on general principles.

The main purpose of this discussion is to inspire those who wish to become active in new fields of scientific research, technical development, and invention, and who wish to study the morphological method as a new and powerful tool of progress.

L,3. Definition of Propulsion. Strictly speaking, propulsion means the change of relative motions within a given set of bodies. To achieve this change of motion, thrust must be generated on some of the bodies.

The simplest case, with which we shall be concerned almost exclusively, involves the generation of thrust on a single body or vehicle. This thrust may accelerate or decelerate the vehicle in its motion relative to some permanently adopted system of reference Σ_0. It is advisable to choose for Σ_0 a so-called inertia system. In such a system all forces are real in the sense that they are directly traceable to matter. The thrust may also be just adequate to maintain the vehicle in a state of uniform motion against resistive forces exerted by some external medium. In the latter case it is this external medium whose state of motion is changed, while the state of motion of the vehicle on which the thrust is exerted remains unchanged.

The thrust generation may be fundamentally twofold in character. We consider (1) intrinsic thrust generation and (2) extrinsic thrust generation.

The thrust generation is intrinsic if all of the active mass which generates the propulsive power is carried on the vehicle.

The thrust generation is extrinsic if none of the active mass which furnishes the energy for the propulsion of a vehicle is carried on this vehicle. This active mass must be furnished freely by the medium through which the vehicle moves.

In many practical cases we have of course a mixed type of thrust generation. Furthermore, either one of the pure types may be *internal* or *external* in character. We therefore have four pure types of thrust generation

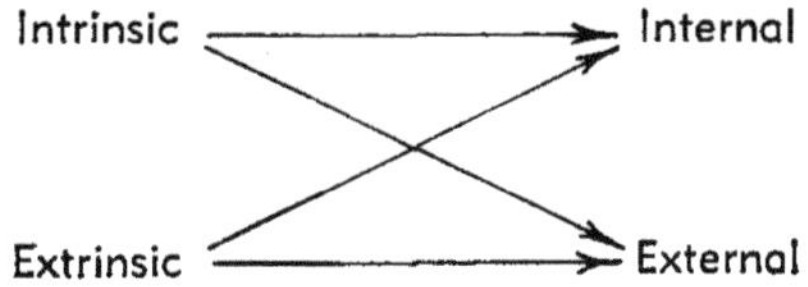

Some of the possible modes can be illustrated by the following examples.

1. The thrust generation is, for instance, *intrinsic and internal* in a common rocket which carries its own self-contained propellant (fuel plus oxidizer). All of the energy which ultimately appears as energy of propulsion is generated internally within the rocket.

2. The thrust generation is *intrinsic* and *external* in a self-contained aeroturbopropeller. In this device a gas turbine is activated through the chemical reaction or combustion of a propellant combination which is carried on the vehicle, liquid oxygen and oil, for instance. The turbine drives a propeller which by its motion through an external medium, air, water (or earth) generates the thrust.

3. The thrust generation is *extrinsic* and *internal* in an aeroduct (ramjet, aerothermodynamic duct) which flies through the very tenuous high strata of the atmosphere in which many atoms and molecules exist in excited states. Deexcitation of these states in the reaction chamber of the duct could be used to provide the energy for expelling the air passing through the duct with increased speed.

4. The thrust generation is *extrinsic* and *external* in the operation of a sailplane, because neither is the active mass carried by the plane nor is any reaction chamber being used, so that no internal but only external flow is used for the thrust generation.

As mentioned, most jet propulsion engines are of a combined type. In a turbojet the thrust generation is intrinsic as far as the fuel is concerned, but extrinsic in view of the fact that the oxidizer oxygen is derived from the atmosphere. The thrust generation is, of course, internal. In the aeroturbopropeller the thrust generation is combined intrinsic-extrinsic and also combined internal-external.

A secondary "fine structure" may immediately be introduced into the diagram on p. 759 by the following consideration. The active mass m_a may be subdivided into

$$m_a = m_c + \Delta m_a \tag{3-1}$$

where m_c is the mass of the actual propellants, that is, of those substances which participate in the chemical reaction that furnishes the energy of propulsion. On the other hand, Δm_a is chemically inert, but mechanically active in the sense that it is introduced to change (in general to augment) the thrust produced by m_c. Depending on whether Δm_a is carried on the vehicle or taken in from the surrounding medium, we speak of intrinsic or extrinsic thrust change, which again may be internal or external in character. Since, in addition, we must consider separately the case of no thrust "augmentation," $\Delta m_a = 0$, we arrive at 4×5 modes of thrust generation by means of energy produced in chemical reactions. (The case $\Delta m_a = 0$ does not necessarily follow as a limit of $\Delta m_a \neq 0$.)

L,4. The Morphological Box. After having clearly formulated the problem which is to be solved, the morphological method establishes the significant parameters which determine the qualitative character of all of the solutions of the given problem. Each parameter p in general has a discrete number of qualitative values $p_1, p_2, \ldots, p_n$. With these parameters a morphological box can be constructed which is a box occupying part of a multidimensional space, the various axes of which correspond to the various parameters. The morphological box contains a number of drawers which is equal to the product of all of the numbers of the different parameters. The morphological analysis of the possible solutions of a given problem is complete if each drawer contains either one solution or none. If there are several solutions in one drawer, one must introduce new parameters in order to make the box uniquely determined.

Since a multidimensional box cannot be geometrically drawn as such, the following procedure may be adopted. The matrices which represent the various parameters may be written down one beneath the other. Circling one element in each matrix and interconnecting these circles will result in a "chain of circles" which represents one basic solution of the problem at hand.

For the present we shall limit our discussion to propulsive power plants which are built up of single elements and which travel through pure media. One may also consider combination engines and in particular those devices which travel along two media such as water surface craft or automobiles. These we shall designate as interfacial engines.

The total scheme of possibilities for pure medium engines is somewhat as follows. The determining parameters are:

A with the elements
- A_1 intrinsic chemically active mass
- A_2 extrinsic chemically active mass

B with the elements
- B_1 internal thrust generation
- B_2 external thrust generation

C with the elements
- C_1 intrinsic thrust augmentation
- C_2 extrinsic thrust augmentation
- C_3 no thrust augmentation

D with
- D_1 internal thrust augmentation
- D_2 external thrust augmentation

Further, we have a parameter E whose elements are

$$E_1 = \text{positive jets}$$
$$E_2 = \text{negative jets}$$

We repeat that we are concerned here only with the morphology of pure medium jet engines of the simplest possible design. These engines move in one medium only and their elements of construction and operation are of the simplest kind. To make the case quite specific, we have in previous publications used only the simplest set of parameters to supplement those already stated. This simplest set is as follows.

$F(F_1, F_2, F_3, F_4)$	designates the medium through which the engine travels. We consider the vacuum F_1, the air F_2, bodies of water F_3, and the earth F_4.
$G(G_1, G_2, G_3, G_4)$	designates the mode of motion of the principal engine part relative to the working fluid, that is, G_1 no motion; G_2, G_3, G_4 translatory, rotatory, and oscillatory motion, respectively.
$H(H_1, H_2, H_3)$	tells us whether the chief propellants are gaseous, liquid, or solid.
$I(I_1, I_2)$	is a matrix of two elements I_1 = continuous operation and I_2 = intermittent operation.
$K(K_1, K_2)$	is a matrix of two elements which characterize self-igniting and non-self-igniting propellants.

All possible simple pure medium jet engines activated by chemical energy are now schematically obtained as all possible chains of circles containing one element of each matrix. One such chain of circles is shown in the diagram on page 763.

The propulsive power plant shown is an aeroduct operating continuously with a positive jet and internal thrust augmentation and using for its energy activated gaseous components of the earth's atmosphere, the energy from which must be extracted by artificial ignition.

If no internal contradictions existed in the scheme shown and if no physical obstacle appeared, the number of possible chains of circles would be

$$\mathfrak{N}_1 = 2 \times 2 \times 3 \times 2 \times 2 \times 4 \times 4 \times 3 \times 2 \times 2 = 9216 \qquad (4\text{-}1)$$

This then would be the total number of possible simple chemical jet engines, if we disregard for the present the modes of conversion of chemical into mechanical energy. However, we must introduce some exclusions; for instance, the starred elements may not be connected since if there is no thrust augmentation the distinction between D_1 and D_2 becomes meaningless. We thus have

$$\mathfrak{N}_2 = \mathfrak{N}_1(1 - \tfrac{1}{6}) = 7680 \qquad (4\text{-}2)$$

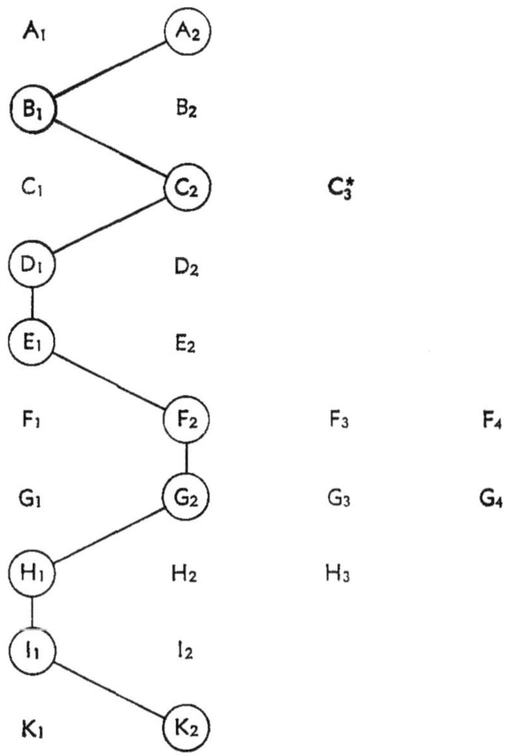

Furthermore, if the element F_1 designates strictly the vacuum, any chain of circles which contain F_1 may not at the same time include either A_1 or C_1. This cuts the possible number of chains of circles containing F_1 by $7\mathfrak{N}_1/48 = 1344$. The total number of engines thus becomes

$$\mathfrak{N}_3 = 7680 - 1344 = 6336 \tag{4-3}$$

There is one more exceedingly important distinction which we may make. This distinction refers to the mode of transformation of the available chemical energy into propulsive energy. Ideally two major modes exist, namely irreversible transformation and reversible transformation.

In practice a reversible process can of course never be fully realized, but it can be closely approximated. It is here that we introduce a value for the first time by disregarding in our survey all intrinsically irreversible conversions of the chemical energy into propulsive power. The question then remains of how many reversible conversions exist. The answer which thermodynamics gives is that there is an infinite number of such conversions, provided that we have at our disposal an infinite number of large "constant temperature reservoirs." In reality we can make use of only a very small number of such reservoirs such as the ocean, the atmosphere, and so on. As a consequence, the total number of ideally possible reversible processes is reduced to four. These are:

Thermomechanical conversion	*Thermoelectrical conversion*
(MI) isothermal	(EI) isothermal
(MA) adiabatic	(EA) adiabatic

We must leave it to the discerning reader to analyze this scheme in greater detail. We wish to call attention to the fact, however, that the tabulation of all possible processes of reversible conversions is a typical result of morphological thinking and that it suggests immediately possibilities which helter-skelter research has not uncovered in the past and presumably would take a long time to discover in the future. This statement may be illustrated by the following observations.

Most of the conscious efforts in the field of jet propulsion have been concentrated on devices which make use of the adiabatic thermomechanical (abbreviated MA) conversion. Although all of the above-mentioned types of conversions were pointed out some years ago [*1*], it is doubtful whether at the present the specific merits of the (MI) and (EI) conversions are recognized, not to speak of the (EA) process of which no mention of any kind seems to exist.

The (MA) conversion has the following typical characteristics and shortcomings. The major part of the chemical reaction is completed in a rocket chamber C that is as small as possible, while the adiabatic expansion and conversion of heat into mechanical energy take place through

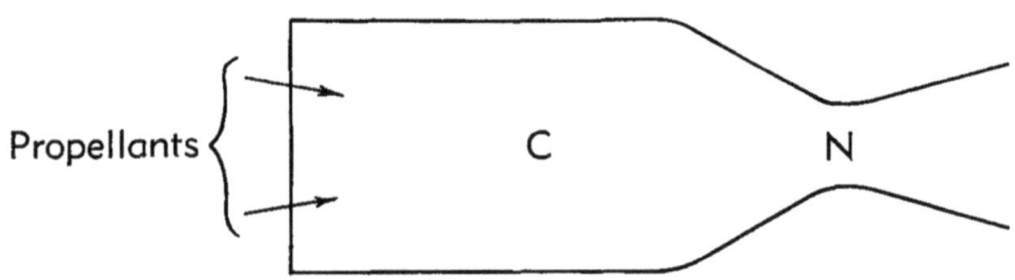

Fig. L,4a. Conventional rocket motor. C = combustion; N = nozzle (expansion).

the nozzle N. Although some chemical reactions take place during this expansion the reaction and the expansion are more or less neatly separated in time and space, which generally makes possible the use of simple motors and simple injection and control. But the (MA) process also suffers from the following shortcomings. At best only the enthalpy or the heat of reaction H can be used, never the free enthalpy of reaction $\Delta\psi = \Delta(h - Ts)$ where T is the absolute temperature and s the entropy, although $\Delta\psi$ may be considerably larger than Δh. Also the initial step of the reaction in the (MA) conversion is always irreversible. And finally, very high temperatures are generated in C and must be so generated if one wishes to attain high efficiencies. For the best propellants known at the present time these temperatures are too high to be withstood by any rocket chamber walls except if most elaborate systems of cooling are used. For future still better propellants, especially those of the metachemical type to be discussed later, the practical difficulties attending the (*MA*) conversion become greater yet.

The (*MI*) conversion makes use of what is known in thermochemistry as a Van't Hoff furnace in which the propellants are introduced and reacted reversibly while the extraction of the reaction products is conducted in an equally reversible manner. It is typical for this process as well as for the (*EI*) conversion that the maximum useful work to be gained is given by the free enthalpy of the reaction $\Delta\psi$. Even if the enthalpy (or heat) of reaction is zero or negative, power may still be

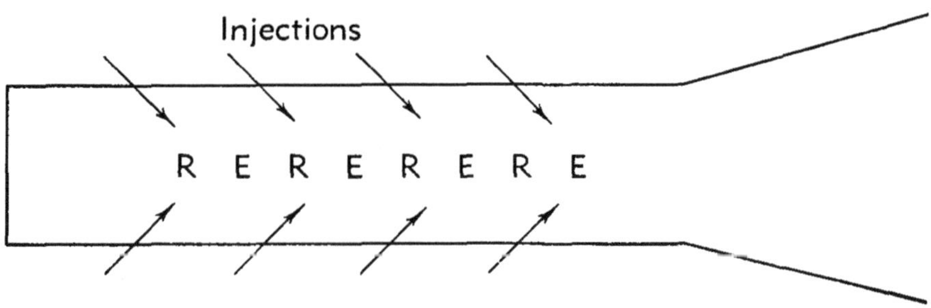

Fig. L,4b. Tubular motor with reaction on the flow. R = reaction; E = expansion.

extracted from the reaction which of course has to "borrow" the lacking heat or energy from the surrounding medium. One of the best examples for a propulsive power plant operating with the (*MI*) process is the so-called isothermal hydroduct, which we shall discuss in more detail later on.

If the (*MI*) conversion is to be used in an ordinary rocket motor activated by self-contained propellants, these propellants must be injected and reacted in a continued succession along a tubular motor, for instance, so that the reactions R and the expansion E follow one another in continued succession. This we call "reaction on the flow." The transformation of heat into mechanical energy can thus be accomplished in principle without giving rise to any excessive temperatures. It is obvious, of course, that much must be learned about the kinetics of chemical reactions before the (*MI*) conversion can be successfully mastered. It is of interest that, in certain special cases, quite favorable practical results have already been achieved.

Those interested in improving the conventional stationary power plants will also do well to familiarize themselves with the various characteristics of "reaction on the flow." Here lie new possibilities for efficient and smooth running engines which are free of knock, resonance, and detonation.

Of the conversion processes which generate electrical rather than mechanical energy, the (EI) process is of course well known. Its thermodynamic efficiency, for small currents, is as nearly equal to unity as that of any process known. In fact, free enthalpies ψ are in practice determined with great accuracy by measuring electrode potentials. In the practical application the (EI) process has been of negligible importance as compared with the use of the thermomechanical conversion in thermal combustion engines. Only as auxiliaries in many devices or as main emergency power plants (in submarines, etc.) have storage batteries been used extensively. Most of these batteries are of the self-contained propellant type and therefore excessively heavy, per unit of useful work delivered. Very little development has been done on continuous (external) propellant flow batteries, although some laboratories have achieved rather remarkable results with such devices [*2*]. It is the author's conviction that this much neglected field will prove in the future to be one of the gold mines in the field of production of power. Of course, what has enormously retarded the development of propellant flow batteries is the peculiar fact that, in spite of very prolonged efforts, the so-called problem of the carbon cell has not yet found any practical solution [*3*, p. 1306]. As long as neither carbon nor one of its cheaply available compounds (hydrocarbons) can efficiently be reacted with oxygen in a continuous propellant flow battery of light weight, this type of battery will be used only for the production of power in those very high grade applications of both a commercial and a military nature, in which low cost of operation is not a primary objective. Even with the best choices of propellants available today much needs to be learned about the chemical kinetics of reactions in electrolytes and liquid-solid interfacial processes before continuous propellant flow batteries can come into greater use.

With regard to the (EA) conversion the situation is somewhat of a puzzle. The present author has been unable to find any reference to systematic investigations of any such process anywhere. There seems to be simply a complete void into which the morphological method may help to introduce some new and lively concepts.

The basic process in the (EA) conversion is a *cooling by extraction of current* rather than cooling by adiabatic expansion or any other type of suitable polytropic expansion. While for the efficient operation of the latter a good proportion of the products of a chemical reaction must be gaseous, *the* (EA) *conversion is not subject to any such restrictions.* This is philosophically a very satisfactory result since it seems odd indeed that the heat from reactions, which produce solid and liquid products but no gas, should be convertible into mechanical work only after transfer of the heat of reaction to an auxiliary gaseous "working fluid."

Just like the (MA) conversion, the (EA) process can no doubt be used either as a continuous or as an intermittent operation. It seems, however, that a long series of theoretical and experimental investigations will be necessary to learn more about the intrinsic nature of these operations. It is possible that in this direction may lie one of the solutions of the old and unsolved problem of constructing efficient carbon cells.

One of the most remarkable facts of nature is the essentially isothermal transformation of chemical energy into mechanical energy and specifically into propulsive energy in living organisms. Just what the details of this transformation are is still a secret, the unraveling of which presents a scientific challenge of the highest order. The successful analysis of the thermomechanical phenomena in living organisms is presumably dependent on a real understanding of the conversion processes which we have described in the preceding.

As a result of the availability of four basic types of reversible conversions of heat into mechanical or electrical work the total number of possible elementary pure medium propulsive power plants is

$$\mathfrak{N} = 4\mathfrak{N}_3 = 25{,}344 \qquad (4\text{-}4)$$

L,5. The Combination Engines. The elementary pure medium jet engines represent only the basic skeleton of all possible propulsive power plants which are activated by chemical reactions. By combining simple elements of the basic engines one arrives at the combination engines, the possible number of which is naturally tremendous. For instance, an ordinary reciprocating motor turning a propeller is a combination engine since, of the simple modes of motion, not only one but two (rotation and oscillation) are being made use of. In this brief survey we must refrain from any further discussion of combination engines except for pointing out one special subclass of these engines which in many respects must be considered as the most important group of propulsive power plants. The devices which we have in mind operate on the boundary, or in the "interface" of two media. For this reason we call them *interfacial propulsive power plants*. For instance, the driving mechanism of a locomotive is an interfacial power plant. Indeed its "stationary" power plant which may be a steam engine, a Diesel engine, or a steam turbine (electric motors do not fall into the category of chemical reaction engines here considered) uses the oxygen and the nitrogen of the *air* for purposes of combustion and cooling, while the thrust is actually generated by the action of the wheels on the solid earth. The same is true for the driving mechanism of automobiles. Most marine propulsive power plants are likewise of the interfacial types which again use the various components of air while the thrust is generated through propellers or wheels acting on the water. We can only hint here that, when hydrofuels (water-reactive chemicals) are used in addition to conventional fuels, some most interesting vistas in the field of interfacial hydrojet-engines are opened.

L,6. Future Prospects of Jet Propulsion as Visualized by Morphological Analysis. We have stated previously that the construction of the morphological box of all propulsive power plants and the subsequent thorough exploitation of its contents represents the only rational way of prophesying the future of jet propulsion. The box, as we have shown, contains 25,344 basic chemical reaction power plants composed of single simple elements operating in single simple action.

It is not possible in this short study to describe in detail all of the engines contained in the morphological box. We shall only be able to illustrate with a few examples how the knowledge of the potential contents of the box allows us to put prophecy on a rational basis. In doing this we shall also automatically become aware of the additional problems which must be solved in order to really construct and operate the power plants which were "schematically invented" by the morphological method.

Before discussing some individual propulsive power plants of unusual interest we must say a few words about the question of a general nomenclature [*4*] and the problem of performance calculations.

NOMENCLATURE. It is of course not feasible in everyday usage to designate engines by strict morphological characterization, such as $(A_2, B_1, C_3, D_1, E_1, F_2, G_2, H_1, I_1, K_2, MA)$, which follows from the matrix representation of all possible propulsive power plants. Instead one will use adequate words to bring into focus at least the main features. For instance, propulsive power plants which operate in a vacuum will be called rockets, while the engines which work in air, water, and earth, respectively, will be distinguished as aero-, hydro-, and terra-engines. Likewise one will speak of self-contained propellants or simply propellants in the case of rockets while in the other engines only the fuel part must be mentioned, that is, simply "fuels" for the aero-engines and hydrofuels or terrafuels for the hydro- or terra-engines.

Types of motion may be designated by using the words duct, turbine, piston, and chamber to distinguish between translation, rotation, oscillation, and no motion at all.

This type of nomenclature results in the following designations which will be used throughout this section:

aeroduct (ramjet)	hydroduct
aeroresonator (pulsejet)	hydroresonator
aeropulse	hydropulse
aeroturbojet	hydroturbojet
aeropistonjet	hydropistonjet
aeroturbopropeller	hydroturbopropeller

SCHEMATIC DESIGNS AND OPERATIONS. To proceed from the abstract to the concrete, the visualization of the design and operation of the various propulsive power plants contained in the general morphological box is necessary. The less one knows about a particular engine the more abstract these designs will be. Under all circumstances it is useful to sketch whatever one may know as a first approximation and stepping stone for the crystallization of further ideas. As a simple example we here introduce some most elementary representations of hydroresonators activated by solid hydrofuels.

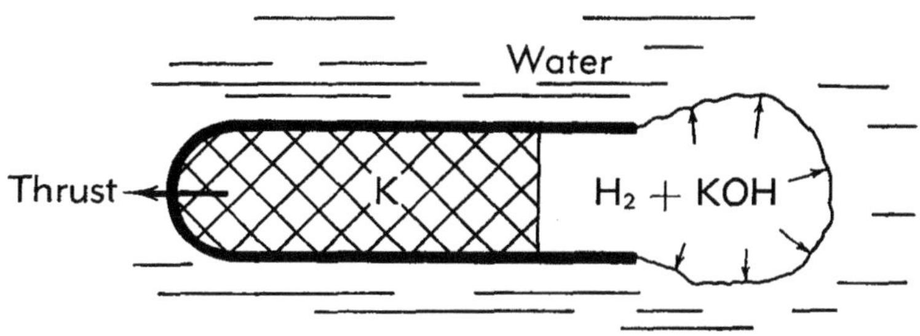

Fig. L,6a. Simple cartridge hydroresonator. K = potassium.

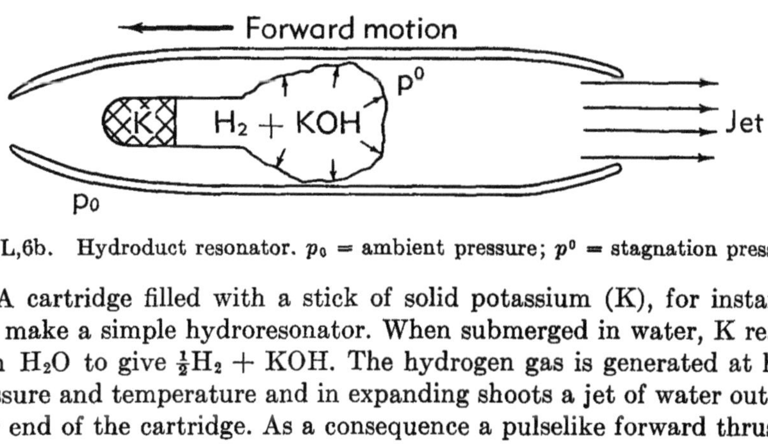

Fig. L,6b. Hydroduct resonator. p_0 = ambient pressure; p^0 = stagnation pressure.

A cartridge filled with a stick of solid potassium (K), for instance, will make a simple hydroresonator. When submerged in water, K reacts with H_2O to give $\frac{1}{2}H_2$ + KOH. The hydrogen gas is generated at high pressure and temperature and in expanding shoots a jet of water out the rear end of the cartridge. As a consequence a pulselike forward thrust is generated. Overexpansion of the hydrogen generates an underpressure near the surface of the potassium stick, the partial vacuum sucks in the water, and then the cycle repeats itself. Actually the first hydroresonator operated by the author in 1943 was nothing but a section of a straight pipe 3 cm in diameter and 20 cm long, which was closed at the front end and filled with Hg-amalgamated Al-foil.

Better streamlined and thus more efficient hydroresonators are shown in Fig. L,6b and L,6c.

The advantage of this design is that the hydrogen is generated at the stagnation pressure p^0, which increases rapidly with forward velocity, rather than at the ambient surrounding pressure p_0. Also the jet action is now limited to one direction and thus produces an increased thrust.

The valves in this model are actuated by the chemical reaction, closing under the pressure of the H_2 generated. The valves are not actuated artificially from the outside and the device is therefore a true resonator. This propulsive power plant has a higher efficiency at low forward speeds than the devices shown in Fig. L,6a and L,6b.

To those bent towards future inventions it is recommended that they build up for themselves a collection of schematic designs of possible jet engines as shown in the above primitive illustrations.

General Performance Characteristics of Jet Engines. Subsequent to the construction of the morphological box and the qualitative visualization of the operation of all individual jet engines, general performance calculations are indispensable if one wishes to choose wisely any specific engine most suited to perform a prescribed task.

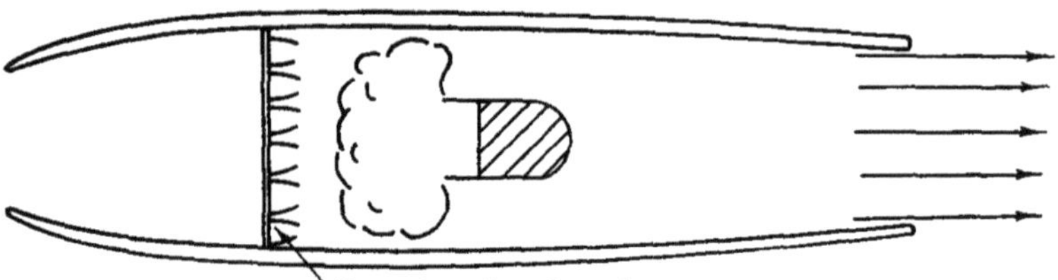

Fig. L,6c. Valved hydroresonator-duct.

Performance of jet engines must generally be known as a function of forward speed V_∞, location above or below sea level, duration of the operation, and character of the propellants used.

Propulsive power plants are designed to produce thrust. It is this thrust F in which we are primarily interested in our performance calculations. In particular, we wish to know F as a function of the chemical characteristics of the propellants used, in its relationship to the medium through which we move and as a function of forward speed, propellant consumption per unit time, and so forth. Often it is not the total thrust F which is most important but the specific cross-section thrust F/A where A is the area subtended by the vehicle in the line of motion.

The author has discussed in detail in other places the general methods of calculating thrust and propulsive performance [4]. Here we just mention briefly that through the derivation of a universal thrust formula it was possible to quickly obtain rather accurate estimates of the propulsive performance of various power plants. The universal thrust formula was developed on the basis of the following considerations.

Obviously it would be a very difficult and practically impossible task to analyze exactly the flow of the propellants and of a part of the surrounding fluid through a propulsive power plant, especially in view of the fact that the kinetics of the chemical reactions involved introduces enormous complications into the already complex conditions of a "nonreacting" flow. We must therefore endeavor to steer clear of the precise differential equations of the phenomena involved and limit ourselves to approximations which make use only of the integral first principles of mechanics and of thermodynamics. This is possible if we assume in a first approximation that all flows considered have uniform velocity profiles. Under these circumstances the following universal thrust formula is arrived at,

$$F = mV_\infty\left[\beta - 1 + \sqrt{1 - \beta + \frac{2\beta\Delta\epsilon}{V_\infty^2}}\right] \tag{6-1}$$

Here V_∞ is the forward velocity and m is the total mass of propellant and of external fluid passing through the engine per second. This mass m is composed of two parts

$$m = m_{\mathrm{fl}} + m_{\mathrm{p}} \tag{6-2}$$

where m_{fl} refers to the outside fluid and m_{p} is the mass of the propellant used per second. Also, the "propellant parameter" is defined as

$$\beta = \frac{m_{\mathrm{p}}}{m} \tag{6-3}$$

The quantity $\Delta\epsilon$ is the most sophisticated part of the universal thrust formula. Strictly speaking, it represents that part of the available "free enthalpy" ψ per gram of propellant which is ultimately transformed into mechanical energy. This thermodynamic potential ψ is defined as

$$\psi = h - Ts = e + pv - Ts \tag{6-4}$$

where h, e, and s are the enthalpy, internal energy, and entropy, while p, v, and T are the pressure, specific volume, and temperature. In practice, one thinks of $\Delta\epsilon$ mostly in terms of available enthalpies or ordinary heats of reaction. In order to show how erroneous this procedure may be, we shall apply the universal thrust formula to three cases for which

$$\begin{array}{lll} (\alpha) & \Delta h > 0, & \text{the conventional case} \\ (\beta) & \Delta h = 0 & \\ (\gamma) & \Delta h < 0 & \end{array} \tag{6-5}$$

These three cases may suffice to demonstrate the very considerable usefulness of the universal thrust formula. In order to refer the performance of a propulsive power plant to one gram of the propellant, the effective exhaust velocity

$$V_{\mathrm{eff}} = \frac{F}{m_{\mathrm{p}}} = \frac{V_\infty}{\beta}\left[\beta - 1 + \sqrt{1 - \beta + \frac{2\beta\Delta\epsilon}{V_\infty^2}}\right] \tag{6-6}$$

is often used. This is the velocity with which the propellant itself would have to be ejected in order to produce the same thrust as the mass m of propellant plus external fluid being ejected with the actual exhaust velocity. Often the specific impulse

$$I_{sp} = \frac{V_{eff}}{g} \tag{6-7}$$

is introduced, where g is the acceleration of gravity. The quantity I_{sp} is equal to the time during which any mass M_p of the propellant could be propelled with the acceleration g if the energy inherent in this mass were made available with the efficiency expressed by the universal thrust formula.

The adiabatic hydroduct. A duct with entrance cross section A_1, stagnation section $A^0 \gg A_1$, and appropriate exit area A_{exit} is moving with velocity V_∞ through water of density ρ_w. In the first approximation it is

$$m_{fl} = A_1 \rho_w V_\infty \tag{6-8}$$

We inject into the duct at A^0 a hydrofuel at the rate of m_p g/sec. Boron for instance is a potential hydrofuel. Let us assume that we succeed in reacting at sufficient speed boron (B) at ambient temperature with water. The reaction is

$$2B + 3H_2O = B_2O_3 + 3H_2 + 95 \text{ kcal} \tag{6-9}$$

if the equation is understood in terms of gram moles. Of the heat of reaction we assume 30 kcal to reside in the three moles of hydrogen. $\mu_p = 3.6$ grams of boron produce one gram of hydrogen. We assume conservatively, and for simplicity, that all of the heat residing in the solid reaction products of Eq. 6-9 is lost by transfer to the water. Only the heat residing in the hydrogen gas, which by the reaction is brought to a temperature of $T_1 = 2650°K$, is available for conversion into kinetic energy. This conversion takes place between the stagnation pressure p^0 and the ambient pressure p_∞. Inserting the resulting values into the universal thrust formula we obtain

$$V_{eff} = \frac{V_\infty}{\beta}\left\{\beta - 1 + \sqrt{1 - \beta + \frac{2\beta p_\infty T_1}{V_\infty^2 \rho_{g\infty} \mu_p T_\infty}\left[1 - \frac{1}{\left(1 + \frac{\rho_w V_\infty^2}{2p_\infty}\right)^{\frac{\gamma-1}{\gamma}}}\right]}\right\} \tag{6-10}$$

or numerically in our case

$$V_{eff} = \frac{V_\infty}{\beta}\left\{\beta - 1 + \sqrt{1 - \beta + 23.55 \times 10^6 \frac{\beta}{V_\infty^2}\left[1 - \frac{1}{(1 + 4.935 \times 10^{-3} V_\infty^2)^{0.259}}\right]}\right\} \tag{6-11}$$

where $\gamma = c_p/c_v$ is the ratio of the specific heats and $\rho_{g\infty}$ the density of the gas, at p_∞ and T_∞.

The isothermal hydroduct. We again operate the hydroduct through the injection of boron which in some way has been rendered fast water-reactive. We assume, however, still more conservatively than in the adiabatic case, that not only the heat of reaction residing in the solid reaction products but also the heat in the hydrogen gas is instantaneously lost by transfer to the water of the ambient temperature T_∞. Since the transfer of heat is assumed to be "instantaneous" the hydrogen gas which expands from p^0 to p_∞ remains at the constant temperature T_∞. The terms Δe and $\Delta(pv)$ in $\Delta\psi$ will consequently both be zero. The isothermal hydroduct thus derives its whole propulsive energy from the term $T_\infty \Delta s$ while the change in enthalpy $\Delta h = 0$. In our special case, using boron as the hydrofuel, we arrive at the following relation derived from the universal thrust formula.

$$V_{\text{eff}} = \frac{V_\infty}{\beta}\left[\beta - 1 + \sqrt{1 - \beta + \frac{2\beta p_\infty}{V_\infty^2 \rho_{g\infty} \mu_p} \ln\left(1 + \frac{\rho_w V_\infty^2}{2p_\infty}\right)}\right] \tag{6-12}$$

or numerically

$$V_{\text{eff}} = \frac{V_\infty}{\beta}\left[\beta - 1 + \sqrt{1 - \beta + 6.910 \times 10^5 \frac{\beta}{V_\infty^2} \ln\left(1 + 4.935 \times 10^{-3} V_\infty^2\right)}\right] \tag{6-13}$$

Although the over-all operation of the boron hydroduct involves values of $\Delta h > 0$, the essential operation proceeds according to $\Delta h = 0$. By injecting H_2 at the temperature T_∞, the whole operation develops strictly with $\Delta h = 0$.

The negative enthalpy hydroduct. In this case we inject liquid hydrogen into the water flowing through the stagnation cross section of the hydroduct. Heat will be extracted from this water to evaporate the liquid hydrogen and bring it to the temperature T_∞. Assuming that $\beta \ll 1$, so that the water itself does not freeze, we can thus operate our hydroduct with an over-all change in enthalpy $\Delta h < 0$.

The universal thrust formula applied to this case reads [*3*, p. 1306]:

$$V_{\text{eff}} = \frac{V_\infty}{\beta}\left[\beta - 1 + \sqrt{1 - \beta + \frac{2\beta R T_\infty}{V_\infty^2 \mathfrak{M}} \ln \frac{p^0}{p_\infty}}\right] \tag{6-14}$$

where R is the universal gas constant and $\mathfrak{M}$ the molecular weight of H_2.

The propulsive efficiencies of all three hydroducts as a function of forward speed are all similar. We here reproduce some curves for case γ.

The examples discussed illustrate the great versatility of the universal thrust formula. They also point towards a new problem, that of efficient hydrofuels. All along the line it becomes obvious that, whenever a proper morphological analysis of a certain large scale problem is carried out, a great number of auxiliary problems automatically present themselves. In the case of the propulsive power plants the problem of adequate propellants is one of the most important. This problem is discussed briefly later on.

A most useful method of comparing the relative performance of various jet engines is that of the "topological performance charts." This was developed independently by von Kármán, Tsien, and the author

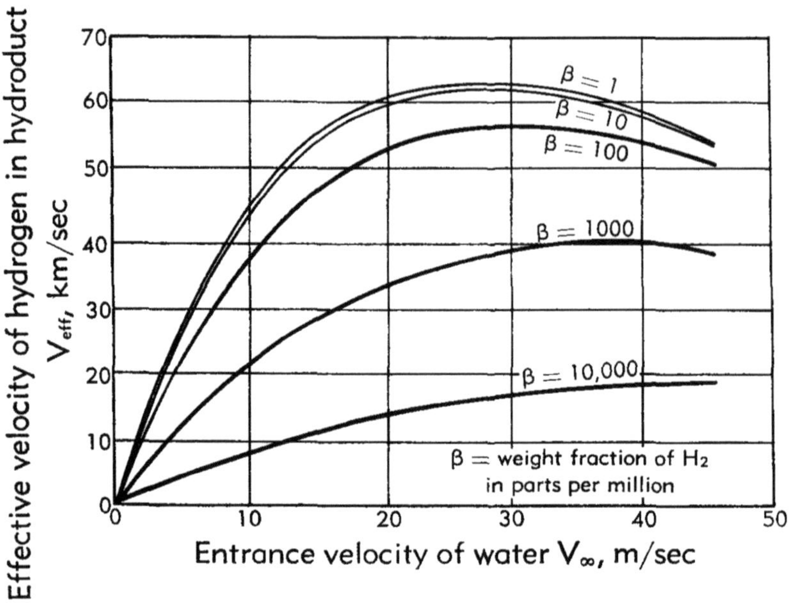

Fig. L,6d. Performance curves of negative enthalpy. Hydroduct operated with liquid hydrogen.

in the United States and by Neugebauer and Reindorf [5] in Germany. The essential quantity considered is the so-called invested mass

$$M = M_0 + m_p t \tag{6-15}$$

where M_0 is the mass of the empty vehicle and $m_p t$ is the total mass of propellant expended during a "journey" of duration t at the uniform forward speed V_∞. At given V_∞, t, and height h (or depth in the ocean) of the journey we are interested in those engines which operate with the smallest value of M. The space (h, t, V) is subdivided into many regions bounded by surfaces

$$M_i = M_k \tag{6-16}$$

where the subscripts i and k refer to two different power plants i and k. Consider, for instance, the boundary line in the h, t plane between the pure rocket and the propeller driven by an internal combustion engine. Near the earth's surface at $h = 0$ the invested mass of the rocket is smaller than the M for the engine propeller for very small durations $t < t_0$ only.

However, at great altitudes the propeller cannot produce thrust at all. The line $M_1 = M_2$ therefore is of the type shown in Fig. L,6e.

SOME SPECIAL PROPULSIVE POWER PLANTS. Within the limits set for this study, we can only sketch a few of the more unusual power plants which are brought to light by the morphological analysis.

The pure rockets. As mentioned already, in practice only the adiabatic thermomechanical conversion of the enthalpy has been considered seriously so far. Much work therefore remains to be done on the additional three types of reversible conversions which we have discussed in Art. 4.

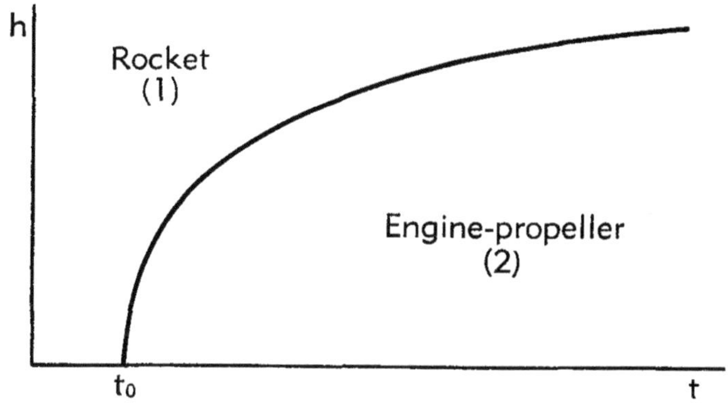

Fig. L,6e. Subdivision of the topological performance space.

Some additional, most important problems are the following. Propellant combinations must be made easily available, in which the impulse is greater than 300 seconds and the density impulse I_d specific impulse times density is likewise greater than 300 g-sec/cm³. The most hopeful approach within the realm of ordinary chemistry lies in the investigation of fuel compounds containing light elements of the first two rows of the periodic system. For more powerful oxidizers, the morphology of fluorine compounds promises to provide an enormous field which is little explored. It also will be most interesting to investigate monopropellants containing light elements combined with oxygen and fluorine.

To reach still higher performance, *metachemical* propellants may have to be produced. For this purpose special particles such as excited atoms and molecules must be stabilized in macroscopic bulk. In contradistinction to ordinary chemical propellants which pack one to ten kilocalories per gram, metachemical propellants can liberate as much as several hundred kilocalories. The stabilization of metachemical aggregates may presumably be achieved through the use of very low temperatures.

The rocket pulse. In our terminology this propulsive power plant strictly speaking is an interfacial engine since it operates partly in the vacuum and partly in air. While an ordinary rocket operates on $m_f + m_{ox}$

$= m_p$ g/sec where m_f is the mass flow of fuel and m_{ox} refers to the oxidizer, the rocket pulse operates with a propellant mass flow

$$m'_p = \xi_1 m_p + \xi_2 m_f \tag{6-17}$$

where $\xi_2 \neq 0$. The two limiting cases of $\xi_1 = 0$ and $\xi_2 = 0$ represent a pure aerial engine and a pure rocket, respectively.

The rocket pulse has the following advantages. As long as it operates in air it makes use of some free oxygen and its specific impulse may therefore lie in the range of that of a pure rocket and the much higher value for an aerial engine. The rocket pulse, because of its intrinsic potentiality of generating underpressures, may also automatically act as its own pump for the propellant injection.

All in all, it may be guessed that the rocket pulse will be most useful in the range of specific impulse 500 seconds $< I_{sp} <$ 1500 seconds. It obviously will be far superior to completely self-contained rocket motors for use in long range missiles and perhaps in interceptor fighters as well as for the boosting of missiles and the assisted take-off of planes.

The "interplanetary" aeroduct. This engine in our formalized nomenclature would bear the designation:

$$[A_2, B_1, C_2, D_1, E_1, F_2, G_2, H_1, I_1, K_1, (MA)] \tag{6-18}$$

The interplanetary aeroduct or ramjet is an engine of unusual interest. It operates actually on metachemical propellants, that is on ions, excited atoms, and molecules which are freely available in the upper atmosphere. By deexciting these atoms and mutually neutralizing the ions, energy is made freely available in the air trapped by the forward motion of the aeroduct. In the tenuous atmosphere the drag is at most equal to

$$\text{drag} = A\rho_\infty V_\infty^2 \tag{6-19}$$

where A is the cross section normal to the forward velocity V_∞, and ρ_∞ is the density of air. The thrust to be generated on the other hand is of the order of

$$F = A\rho_\infty V_\infty (V_{exit} - V_\infty) \tag{6-20}$$

where

$$V_{exit} \sim \sqrt{2E} \tag{6-21}$$

and E is the energy available per gram of air. If $E \geqq 25$ kilocalories per gram then $V_{exit} \geqq 10$ km/sec which is of the order of the escape velocity from the earth. As long as $V_\infty < V_{exit}$ the maximum possible drag will always be smaller than the optimum thrust. The interplanetary aeroduct literally is an engine of infinite specific impulse. It is in this sense superior to any other power plant, including those operated by nuclear energy.

The study of the excited particles available in the upper atmosphere, the methods of trapping, deexciting, and reemitting them promises to become one of the most exciting ventures in physical chemistry.

It will of course be noted that the aeroduct just described, in an indirect manner, makes use of solar energy. It is not usually thought that this energy is easily available. The reason for this is that a major part of solar radiation which reaches the lower strata of the atmosphere and the earth's surface is absorbed more or less uniformly. As a consequence of this, the general level of temperature is raised, but no very steep gradients are generated. Such gradients are necessary, however, for thermomechanical power plants to operate with good thermodynamic efficiency. On the other hand, in the upper tenuous atmosphere the ultraviolet and the corpuscular rays from the sun first raise the level of excitation of individual atoms and molecules.[2] These energies of excitation are available to their full amount for exploitation in properly designed power plants. The excited levels correspond, so to speak, to exceedingly high "local" temperatures T_1 while their immediate microscopic neighborhoods (unexcited particles and the radiation space density) are at a very moderate temperature T_∞. The ratio $(T_1 - T_\infty)/T_1$ which expresses the thermodynamic maximum efficiency can therefore approach unity.

Since the aeroduct described makes use of energy available at the fringes of the earth's atmosphere and perhaps also in interplanetary space, we call it the interplanetary aeroduct. This designation is perhaps also most appropriate because this type of power plant gives promise of providing propulsion for vehicles intended to leave the earth.

The aeropulse. The aeropulse and the aeroresonator (V-1 buzz-bomb engine) are historically of interest because their feasibility and efficiency were derived in 1943 by the present author from his morphological analysis of jet engines [*6*].

Although the V-1 engine demonstrates the feasibility of this type of power plant, its efficiency is only about one third of what the theory predicts. As pointed out in [*1,4*], the difficulty lies in replacing the actions of compressors and pumps by a proper phasing of the significant reaction and thrust augmentation processes in the aeropulse. Failure to achieve the predicted performance clearly must be ascribed to our ignorance of some of the most important phases of chemical kinetics [*7,8*], a field which needs much development.

Hydrojet engines. The principal idea here is to introduce free water both as a chemical reagent (oxidizer) and as a thrust augmentor. The possibility of using water as an "oxidizer" is of course contingent upon the existence of elements or hydrofuels, the oxides of which represent energy states low enough to render the reaction between the hydrofuel and water highly exothermic. It is well known that many hydrofuels of this type are available; boron as shown by Eq. 6-9 is one of them.

[2] In the spring of 1956, successful experiments were conducted by M. Zelikoff of the United States Air Force Cambridge Research Center in inducing the recombination of monatomic oxygen through the dispersal of NO from an Aerobee rocket into the atmosphere at a height of about 110 km. These experiments constitute a first direct confirmation of the views expressed in the present articles.

In principle, therefore, all of the hydrojet engines contained in our morphological box are feasible. Among them are the hydroduct, staggered hydroduct, hydroresonator, hydropulse, hydroturbojet (single-ducted and double-ducted), hydropistonjet, hydroturbopropeller, and so on. The possible variations of design of these engines are far greater than those of the analogous propulsive power plants in air. This is due to the fact that the working fluid consists generally of two phases, *gas* and *water*, the relative proportions of which can be varied.

The hydrojet engines are of importance for the entire field of underwater operations. The morphological outlook visualizes the "opening up of the oceans" both in regard to seeing through them as well as to moving and operating in them. The resources available within the ocean as well as the untapped deposits available on the very extended ocean floors will in the not-too-distant future provide considerable incentive for the development of hydrojet engines.

Terrajet engines. These power plants are to move through the earth by using earth-reactive fuels or terrafuels. Terra-engines, never seriously considered before, are typical brain children of morphological thinking. It is one of the essential characteristics of this type of thinking to insist that everything is possible and should be tried unless very definite evidence, such as the interference of the second law of thermodynamics, demonstrates that insurmountable obstacles lie in the way of the realization of one's ideas. It appears to the author that no such obstacles lie in the way of terrajet engines.

The main problem obviously is the production of powerful terrafuels which react with suitable components of the solid earth, liberating sufficient amounts of energy during the process. Such terrafuels exist, the element fluorine being one of the many possible. The problem at the present time is complicated by our lack of knowledge of the exact geochemical and mineralogical composition of the earth. Developments presumably will follow the line of construction of individual engines for purposes of drilling, excavation, and so on, which will be useful in themselves and which will establish the pegs of knowledge necessary for the systematic application of the morphological method to the problem of terra-engines.

Auxiliary Problems. It is inevitable that, in any large field, in addition to the principal problem a great number of auxiliary problems appear. In jet propulsion the crucial issue is obviously that of all possible propulsive power plants. The successful realization of these power plants depends on the following auxiliary issues:

1. The chemistry and general logistics of adequate propellants.
2. The thermochemistry of the propellants.
3. The chemical kinetics of the propellants.

4. Auxiliary devices such as pumps, injectors, valves, controls, heat exchangers, storage, and feed systems, etc. It so happens that from the morphological point of view all of these issues present problems which are more difficult and scientifically more basic than that of the mechanics of the jet engines themselves.

Morphological thought naturally does not stop with the analysis of the power plants as such. It is vitally concerned with the part played by these plants in the whole vehicle which is to be propelled. This interrelation is obviously of great complexity and presents a great number of problems as yet unsolved and even unchartered. Because of the limited space we must content ourselves with a few remarks on the morphology of propellants.

Morphology of Chemical and Metachemical Propellants. So far no systematic analysis of the field of chemical propellants has been carried out. The application of the morphological method to this problem promises to be most illuminating. The morphological approach will not only produce deeper insights into the nature of the substances and of the processes which are involved in the conversion of chemical energy into mechanical energy, it almost certainly will result in the discovery of many new and unsuspected propellant combinations.

The limited length of the present article allows us only to sketch briefly how we might proceed. The following lines should therefore be considered merely as advisory suggestions to those who intend to make use of the morphological approach and who in particular are interested in an over-all perspective of the field of chemical propellants.

A given amount of matter constitutes a potential chemical propellant if, from an initial state S_i it can reach or be made to reach a final state S_f, thereby generating a macroscopic quantity of mechanical or of electrical energy. S_f is to differ only chemically from S_i. The final state S_f therefore must contain the same number of various kinds of atoms as the initial state S_i. In L,4 we have already discussed the four practically possible reversible processes which we may use to transform the liberated chemical energy into propulsive energy. There it was assumed that S_i and S_f are stationary states in the thermodynamic sense. There is of course room for generalization, inasmuch as in a more general morphological approach we might investigate the following scheme:

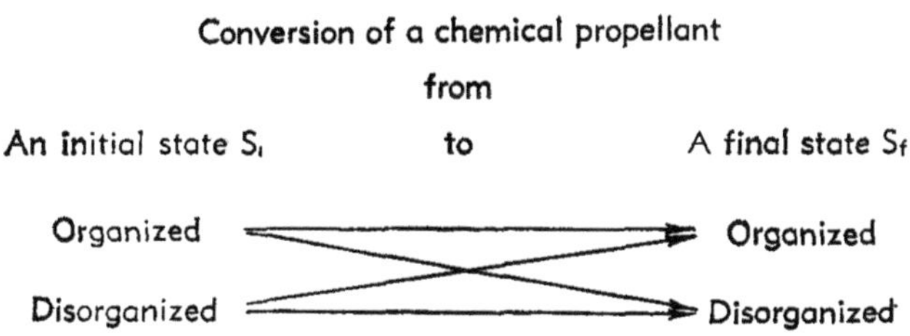

Thus four generic types of transformations exist, as the arrows indicate. The multiplicity of the possibilities grows as we consider various types of order and of disorder in the states S_i and S_f. We only mention in passing that order and disorder may be present either on a microscopic or on a macroscopic scale. They may also be physical or chemical in character. Disregarding refinements of this kind, our parameter matrix $[p_1]$ = "conversion" has the four elements

$$p_1^1 = (S_{io} \rightarrow S_{fo}), \quad p_1^2 = (S_{io} \rightarrow S_{fd}), \text{ etc.}$$

The indices $_o$ and $_d$ refer respectively to order and disorder.

The concept of initially disordered states S_i is particularly significant in the case of extrinsic propellants. It is of less importance for completely self-contained power plants or rockets.

In order to illustrate the character of some of the additional parameters which are vital in a morphological study of chemical propellants we shall now restrict ourselves to the consideration of highly organized initial states S_i. We assume these to be dynamically stable although thermodynamically they may be only pseudostable. The state S_i may contain one or more chemical components. Our second parameter thus has the following elements

$$p_2^1 = \text{one component}, \qquad p_2^2 = \text{two components and so on}$$

There is a third parameter p_3, the elements of which are simply whole numbers. These numbers designate whether we introduce separately into the reaction chamber of our propulsive power plant one or more chemically or physically different propellant combinations. It is thus

$$p_3^1 = 1, \quad p_3^2 = 2, \ . \ . \ . \text{ etc.}$$

The combination p_2^1, p_3^1 means a true monopropellant, such as nitromethane CH_3NO_2, which is a dynamically stable but thermodynamically only pseudo stable molecule and which therefore can decompose with a net liberation of energy. On the other hand, any coupling of p_2^k with p_3^1 stands for an apparent monopropellant which for $k \neq 1$ consists of one substance containing more than one chemical component. In the latter case a finer distinction would result from the question of whether or not these components can coexist in arbitrary proportions or whether they necessarily coexist in determined ratios as a consequence of thermodynamic equilibrium requirements.

The parameter matrix p_4 might perhaps be used to describe all of the possible basic power plants. We have discussed the character of this matrix previously. From here on at least two avenues are open to us for a further morphological classification.

The first approach would attempt to establish groups of basic chemicals which act alike such as

$$\alpha_1 = \text{noble gases}, \qquad \alpha_2 = \text{alkali atoms, and so on}$$

Or the α_i's might actually designate all of the individual atoms. A basic molecule therefore would be written as

$$(n_1\alpha_1, n_2\alpha_2, n_3\alpha_3, \ . \ . \ . \ ) = p_5$$

and the matrix p_s would contain the elements $p_s^{n_1,n_2,\cdots}$ with the n_i's assuming the values of all integers, and zero. The presence of $n_i\alpha_i$ in the molecular matrix means that the molecule in question contains n_i atoms, different or alike, which belong to the group α_i.

The first approach which we have just described is subject to the objection that it does not readily lend itself to the next step in the morphological analysis. This step, as we have discussed it previously, involved the generalized performance calculation for all propellants relative to one of the four possible reversible conversions of the types (MA), (MI), (EA), and (EI). Prospects for a general performance evaluation of propellants become brighter if instead of using purely constitutional chemical parameters of the propellants we introduce some of their basic physicochemical characteristics. Some of these characteristics are as follows:

1. The number of the chemical components may be greater than, equal to, or smaller than the number of the reaction products.

2. Among the "oxidation" products of a reaction used there may be either one single "oxide" (or fluoride, chloride, etc.) of each atom or there may be several competing oxides, fluorides, etc.

Hydrocarbons, for instance, because of the existence of CO and CO_2 and the free enthalpy characteristics of the water gas equilibrium are only medium good fuels if the (MA)-conversion is used. Propellants that contain light elements such as Li which have one dominating oxide such as Li_2O, on the other hand, promise to be among the best for the (MA)-conversion.

3. The reaction products may contain gaseous, liquid, and solid phases; or some of these phases may be missing.

4. The heat (or enthalpy) of reaction Δh may have the same or the opposite sign of the free enthalpy $\Delta\psi$ of reaction; or Δh may be zero. We have already illustrated the significance of these three possibilities in our discussion of the isothermal and of the adiabatic hydroduct.

5. The stability of the molecules used in a propellant combination is one of the most important characteristics. This stability may be viewed in the light of the following fundamental aspects.

(α) All of the atoms or molecules involved may be both dynamically and thermodynamically stable lithium or H_2 (parahydrogen), and O_2 or

F_2 in their ground states could be used as completely stable fuels and oxidizers.

(β) Some of the molecules in the original propellant might be dynamically stable but thermodynamically pseudostable. Acetylene C_2H_2 or nitromethane CH_3NO_2 are examples of this sort. The heats of formation from the elements of C_2H_2 and of CH_3NO_2 are respectively negative and positive, but both molecules under the proper circumstances can be induced to decompose with a net liberation of energy. Both are thus potential explosives.

The substances falling into the two categories α and β constitute the *conventional chemical propellants.*

(γ) There are groups of atoms such as CH_3, OH, etc., and all of the radicals which are not usually thought of as regular molecules. We propose to introduce for them the designation *molecular fragments* or simply *fragments* and also to talk about *fragment chemistry* or *chemistry of fragments* as distinct from ordinary chemistry.

Fragments as isolated particles may be dynamically stable, that is, a single CH_3 radical, for instance, can exist in empty space without ever falling apart. Fragments are, however, thermodynamically pseudostable. For instance a group of free CH_3 radicals can react and form more stable groups liberating a net amount of energy in the process.

To distinguish radicals and generally fragments from pseudostable molecules, it is perhaps best to use the existence or nonexistence of a paramagnetic moment or of unpaired and paired electrons as the distinguishing characteristics.

Fragment propellant chemistry, which is quite a new field, will be concerned with two tasks. In the first place the stabilization in bulk density of the fragments must be attempted. In the second place free fragments exist in nature, for instance in the upper atmosphere. These may be conceivably used in propulsive power plants of the type of the interplanetary aeroduct which we have discussed [*1* and Bibliography 3].

(δ) Some of the atoms or molecules which constitute a propellant may be quantum-mechanically stable or metastable. The metastability is the consequence of "forbidden transitions." Higher order transitions are nevertheless not excluded and result in a finite, more or less extended lifetime of the metastable states. The "delayed" quantum jumps of metastable atoms or molecules back to their normal states in the absence of any collisions with other molecules may involve the electrons or the atoms; or a reorientation of magnetic moments may take place, such as for instance in the spontaneous transition of orthohydrogen into parahydrogen.

Metastable states which involve the jump of an electron are, for instance, those of O_{III} which result in the delayed emission of the nebulium lines N_1 and N_2. Another case of a "macroscopic metastable state" is a piece of metal which is placed in a very strong electric field and which as a consequence is induced to "cold" electron emission. The H_3 and He_2 molecules are perhaps cases of the type in which atomic quantum jumps can occur.

The two main problems of metachemistry as related to the field of propellants are the stabilization of metastable states in macroscopic bulk and the exploitation of the freely available excited particles in the upper atmosphere for the generation of propulsive power.

In conclusion we emphasize again that in addition to conventional propellant chemistry the morphological outlook suggests perspectives and many interesting investigations in the new fields of fragment chemistry and metachemistry [*7,8*].

Applications of Jets. A study of the possible application of all types of jets will command considerable interest. These applications may be grouped roughly into those which concern themselves with the generation of propulsive power, and all others. Among the latter the uses of jets for cutting are of importance. For instance the metallic particle jets extruded at velocities of the order of ten kilometers per second from suitable metallic inserts in exploding "shaped charges" penetrate the heaviest armors. The author some time ago proposed to use the particles in these jets as artificial meteors [*9,10*] to explore the tenuous atmosphere at great heights, to probe the electromagnetic field around the earth and to send them as test particles into interplanetary space.[3] Also, the acceleration of these particles during extrusion is billions of times that of gravity on the earth's surface, a fact which opens up a novel field of experimentation.

Generally speaking all applications might be profitably discussed by first classifying jets in order of magnitude. At the lower end of the scale we have beams of electrons, mesons, nucleons, ions, molecules, and other elementary particles. These *microscopic jets* during the past decades have become of immense importance in the investigation of the properties of matter. Going up the scale we come to the *colloidal jets,* of which those propelled by shaped charges represent one typical class. Macroscopic jets carrying grams to tons per second are well known in the general field of rocketry, jet propulsion, and artillery. Finally there are the *cosmic jets* emitted by all types of stars and especially by the novae and the supernovae [*1,10*]. In the latter, which are the most stupendous stellar explosions known, masses of hot and ionized gas of the order of the sun's total mass are hurled into space with velocities of thousands of kilometers per second. The study of these cosmic jets has only just begun. It promises returns of significance for many fields of science.

[3] The present author and some of his colleagues were recently given the opportunity by Maurice Dubin of the Cambridge Field Station, U.S. Air Force, to install three special-shaped charges in an Aerobee rocket at the Holloman Air Force Base in New Mexico. This resulted, on the night of October 16, 1957, in the first successful launching of pellets into interplanetary space, away from the gravitational pull of the earth. For an account of this fundamental experiment, see [*11*].

L,7. Conclusions. As we stated in the beginning, the purpose of this study is twofold. In the first place, we endeavored to throw light on various fields of jet propulsion which are not generally known. In the second place, we attempted to accomplish our purpose through the use of the morphological method and to stimulate other investigators to use this method. Many elements of this method have of course been used for a long time in all branches of science, engineering, and in all human planning. In its totality the method, however, has been applied so far to only *limited* problems, such as the analysis with the group theory of all possible crystals or crystallographic systems and many other mathematical problems. In physics, Faraday, in particular, thought along morphological lines by investigating systematically and with great persistence the interrelations between various classes of physical phenomena. Goethe studied the morphological aspects of plant life, and the phenomena of heredity have been viewed in the same light. However, to the objection that the morphological method is "nothing new under the sun," it can only be said that in the sense here presented the method with all of its implications has never yet been applied to any general problem except to that of the propulsive power plants which are activated by chemical energy. The method in this field has produced a great number of unforeseen and unusual results and it promises similar success in many other fields.

L,8. Cited References and Bibliography.

Cited References

1. Address presented at the banquet held during the *1950 Annual Convention of the Am. Rocket Soc.*, Nov. 26–Dec. 1, 1950. Also [*3*, p. 1306].
2. Heise, G. W., et al. *J. Electrochem. 94*, 99 (1948).
3. Stodola, A. *Steam Turbines*, 6th ed., Vol. II. Peter Smith, New York, 1945.
4. Zwicky, F. *Intern. Congress for Appl. Mech.*, Paris, Sept. 1946; *Helv. Phys. Acta 21*, 299 (1948). See also *U.S. Patent 2,461,797.*
5. Reindorf, H. The evaluation of motors in a space diagram according to velocity, altitude and range. *Wright Field Air Document Center*, 1947.
6. Zwicky, F. *Aeronaut. Eng. Rev. 6*, (*6*), 1947. Also *Aviation*, June 1947 and *Army Ord. J.*, Nov.–Dec. 1947.
7. Zwicky, F. *Chem. Eng. News 28*, 156 (1950.) Also *Eng. and Science Monthly, Calif. Inst. Technol. 13*, 3 (1950).
8. The labile molecule. *Discussions Faraday Society 2*, (*2*), 1947.
9. Zwicky, F. *Publ. Astron. Society of Pacific 58*, 260 (1946); *59*, 32, 64 (1947). Also *Army Ord. J.*, July–Aug. 1947.
10. Rivkins, H. *Army Ord. J.*, Apr. 1947.
11. Zwicky, F. *Engineering and Science Monthly, Calif. Inst. Technol.*, Jan. 1958.

Bibliography

Zwicky, F. *Aviation Age*, Jan. 1952.
Zwicky, F. *Experientia 10*, 326 (1954).
Zwicky, F. *J. Am. Rocket Soc. 84*, 3–20 (1951).
Zwicky, F. *J. Am. Rocket Soc. 22*, 339–342 (1952).
Zwicky, F. *Science and Eng. Monthly, Calif. Inst. Technol.*, *16*, 20 (1953).

INDEX

GPSR Authorized Representative: Easy Access System Europe - Mustamäe tee 50, 10621 Tallinn, Estonia, gpsr.requests@easproject.com

www.ingramcontent.com/pod-product-compliance
Ingram Content Group UK Ltd.
Pitfield, Milton Keynes, MK11 3LW, UK
UKHW020911270925
463356UK00006B/137
9780691626314